UNIVERSITY LIBRARY
UW-STEVENS POINT

ADVANCES IN CHEMICAL PHYSICS

VOLUME 123

EDITORIAL BOARD

BRUCE J. BERNE, Department of Chemistry, Columbia University, New York, New York, U.S.A.

KURT BINDER, Institut für Physik, Johannes Gutenberg-Universität Mainz, Mainz, Germany

A. WELFORD CASTLEMAN, JR., Department of Chemistry, The Pennsylvania State University, University Park, Pennsylvania, U.S.A.

DAVID CHANDLER, Department of Chemistry, University of California, Berkeley, California, U.S.A.

M. S. CHILD, Department of Theoretical Chemistry, University of Oxford, Oxford, U.K.

WILLIAM T. COFFEY, Department of Microelectronics and Electrical Engineering, Trinity College, University of Dublin, Dublin, Ireland

F. FLEMING CRIM, Department of Chemistry, University of Wisconsin, Madison, Wisconsin, U.S.A.

ERNEST R. DAVIDSON, Department of Chemistry, Indiana University, Bloomington, Indiana, U.S.A.

GRAHAM R. FLEMING, Department of Chemistry, University of California, Berkeley, California, U.S.A.

KARL F. FREED, The James Franck Institute, The University of Chicago, Chicago, Illinois, U.S.A.

PIERRE GASPARD, Center for Nonlinear Phenomena and Complex Systems, Brussels, Belgium

ERIC J. HELLER, Institute for Theoretical Atomic and Molecular Physics, Harvard-Smithsonian Center for Astrophysics, Cambridge, Massachusetts, U.S.A.

ROBIN M. HOCHSTRASSER, Department of Chemistry, The University of Pennsylvania, Philadelphia, Pennsylvania, U.S.A.

R. KOSLOFF, The Fritz Haber Research Center for Molecular Dynamics and Department of Physical Chemistry, The Hebrew University of Jerusalem, Jerusalem, Israel

RUDOLPH A. MARCUS, Department of Chemistry, California Institute of Technology, Pasadena, California, U.S.A.

G. NICOLIS, Center for Nonlinear Phenomena and Complex Systems, Université Libre de Bruxelles, Brussels, Belgium

THOMAS P. RUSSELL, Department of Polymer Science, University of Massachusetts, Amherst, Massachusetts

DONALD G. TRUHLAR, Department of Chemistry, University of Minnesota, Minneapolis, Minnesota, U.S.A.

JOHN D. WEEKS, Institute for Physical Science and Technology and Department of Chemistry, University of Maryland, College Park, Maryland, U.S.A.

PETER G. WOLYNES, Department of Chemistry, University of California, San Diego, California, U.S.A.

ADVANCES IN CHEMICAL PHYSICS

Edited by

I. PRIGOGINE

Center for Studies in Statistical Mechanics and Complex Systems
The University of Texas
Austin, Texas
and
International Solvay Institutes
Université Libre de Bruxelles
Brussels, Belgium

and

STUART A. RICE

Department of Chemistry
and
The James Franck Institute
The University of Chicago
Chicago, Illinois

VOLUME 123

AN INTERSCIENCE PUBLICATION
JOHN WILEY & SONS, INC.

This book is printed on acid-free paper. ∞

An Interscience Publication

Copyright © 2002 by John Wiley & Sons, Inc. New York. All rights reserved.

Published simultaneously in Canada.

No part of this publication may be reproduced, stored in a retrieval system or transmitted in any form or by any means, electronic, mechanical, photocopying, recording, scanning or otherwise, except as permitted under Sections 107 or 108 of the 1976 United States Copyright Act, without either the prior written permission of the Publisher, or authorization through payment of the appropriate per-copy fee to the Copyright Clearance Center, 222 Rosewood Drive, Danvers, MA 01923, (978) 750-8400, fax (978) 750-4744. Requests to the Publisher for permission should be addressed to the Permissions Department, John Wiley & Sons, Inc., 605 Third Avenue, New York, NY 10158-0012, (212) 850-6011, fax (212) 850-6008, E-Mail: PERMREQ@WILEY.COM.

For Ordering and customer service, call 1-800-CALL WILEY

Library of Congress Catalog Number: 58-9935

ISBN 0-471-21453-1

Printed in the United States of America.

10 9 8 7 6 5 4 3 2 1

CONTRIBUTORS TO VOLUME 123

ANDRZEJ R. ALTENBERGER, Department of Chemical Engineering and Materials Science, University of Minnesota, Minneapolis, MN

ELI BARKAI, Department of Chemistry, Massachusetts Institute of Technology, Cambridge, MA

A V. BARZYKIN, National Institute of Advanced Industrial Science and Technology (AIST), Tsukuba, Ibaraki, Japan

R. STEPHEN BERRY, James Franck Institute, University of Chicago, Chicago, IL

PETER BOLHUIS, Department of Chemical Engineering, University of Amsterdam, Amsterdam, The Netherlands

JOHN S. DAHLER, Department of Chemical Engineering and Materials Science, University of Minnesota, Minneapolis, MN

CHRISTOPH DELLAGO, Department of Chemistry, University of Rochester, Rochester, NY

P. A. FRANTSUZOV, National Institute of Advanced Industrial Science and Technology (AIST), Tsukuba, Ibaraki, Japan

JÜRGEN GAUSS, Institüt fur Physikalische Chemie, Universität Mainz, Mainz, Germany

PHILLIP L. GEISSLER, Department of Chemistry and Chemical Biology, Harvard University, Cambridge, MA

YOUNJOON JUNG, Department of Chemistry, Massachusetts Institute of Technology, Cambridge, MA

TAMIKI KOMATSUZAKI, Nonlinear Science Laboratory, Department of Earth and Planetary Sciences, Faculty of Science, Kobe University, Nada, Kobe, Japan

MARCO ANTONIO CHAER NASCIMENTO, Departamento de Fisico-Quimica, Insituto de Quimica, Universidade Federal do Rio de Janeiro, Predio do CT-Bloco A, Sala 408, Cidade Universitaria, Rio de Janeiro, Brazil

K. SEKI, National Institute of Advanced Industrial Science and Technology (AIST), Tsukuba, Ibaraki, Japan

ROBERT J. SILBEY, Department of Chemistry, Massachusetts Institute of Technology, Cambridge, MA

CLARISSA OLIVEIRA SILVA, Departamento de Fisico-Quimica, Instituto de Quimica, Universidade Federal do Rio de Janeiro, Predio do CT-Bloco A, Sala 408, Cidade Universitaria, Rio de Janeiro, Brazil

JOHN F. STANTON, Institute for Theoretical Chemistry, Department of Chemistry and Biochemistry, University of Texas at Austin, Austin, TX

M. TACHIYA, National Institute of Advanced Industrial Science and Technology (AIST), Tsukuba, Japan

MIKITO TODA, Department of Physics, Nara Women's University, Nara, Japan

INTRODUCTION

Few of us can any longer keep up with the flood of scientific literature, even in specialized subfields. Any attempt to do more and be broadly educated with respect to a large domain of science has the appearance of tilting at windmills. Yet the synthesis of ideas drawn from different subjects into new, powerful, general concepts is as valuable as ever, and the desire to remain educated persists in all scientists. This series, *Advances in Chemical Physics*, is devoted to helping the reader obtain general information about a wide variety of topics in chemical physics, a field that we interpret very broadly. Our intent is to have experts present comprehensive analyses of subjects of interest and to encourage the expression of individual points of view. We hope that this approach to the presentation of an overview of a subject will both stimulate new research and serve as a personalized learning text for beginners in a field.

I. Prigogine
Stuart A. Rice

CONTENTS

TRANSITION PATH SAMPLING *By Christoph Dellago, Peter Bolhuis, and Phillip L. Geissler*	1
CHEMICAL REACTION DYNAMICS: MANY-BODY CHAOS AND REGULARITY *By Tamiki Komatsuzaki and R. Stephen Berry*	79
DYNAMICS OF CHEMICAL REACTIONS AND CHAOS *By Mikito Toda*	153
A STOCHASTIC THEORY OF SINGLE MOLECULE SPECTROSCOPY *By YounJoon Jung, Eli Barkai, and Robert J. Silbey*	199
THE ROLE OF SELF-SIMILARITY IN RENORMALIZATION GROUP THEORY *By Andrzej R. Altenberger and John S. Dahler*	267
ELECTRON-CORRELATED APPROACHES FOR THE CALCULATION OF NMR CHEMICAL SHIFTS *By Jürgen Gauss and John F. Stanton*	355
COMPUTATIONAL CHEMISTRY OF ACIDS *By Clarissa Oliveira Silva and Marco Antonio Chaer Nascimento*	423
COOPERATIVE EFFECTS IN HYDROGEN BONDING *By Alfred Karpfen*	469
SOLVENT EFFECTS IN NONADIABATIC ELECTRON-TRANSFER REACTIONS: THEORETICAL ASPECTS *By A. V. Barzykin, P. A. Frantsuzov, K. Seki, and M. Tachiya*	511
AUTHOR INDEX	617
SUBJECT INDEX	649

TRANSITION PATH SAMPLING

CHRISTOPH DELLAGO

Department of Chemistry, University of Rochester, Rochester, NY 14627

PETER G. BOLHUIS

Department of Chemical Engineering, University of Amsterdam, 1018 WV Amsterdam, The Netherlands

PHILLIP L. GEISSLER

Department of Chemistry, Massachusetts Institute of Technology, Cambridge, MA 02139

CONTENTS

- I. Introduction
- II. Defining the Transition Path Ensemble
  - A. Dynamical Path Probability
  - B. Reactive Path Probability
  - C. Deterministic Dynamics
  - D. Stochastic Dynamics
  - E. Monte Carlo Dynamics
  - F. Defining the Initial and Final Region
- III. Sampling the Transition Path Ensemble
  - A. Shooting Moves
    1. Shooting Moves for Deterministic Dynamics
    2. Selecting Phase-Space Displacements
    3. Momentum Rescaling
    4. Efficiency of Deterministic Shooting Moves
    5. Shooting Moves for Stochastic Dynamics
  - B. Shifting Moves
    1. Shifting Moves for Deterministic Dynamics
    2. Shifting Moves for Stochastic Dynamics

C. Memory Requirements
   D. Stochastic Trajectories as Sequences of Random Numbers
   E. Other Algorithms
      1. Local Algorithm for Stochastic Pathways
      2. Dynamical Algorithm
      3. Configurational Bias Monte Carlo
   F. Parallel Tempering
   G. Generating an Initial Path
   H. Transition Path Sampling with an Existing Molecular Dynamics Program
IV. Computing Rates and Reversible Work
   A. Population Fluctuations
   B. Reversible Work
   C. Umbrella Sampling
   D. A Convenient Factorization
   E. Correspondence with Reactive Flux Theory
   F. How Long Should Pathways Be?
V. Analyzing Transition Pathways
   A. Reaction Coordinates and Order Parameters
   B. The Separatrix and the Transition State Ensemble
   C. Computing Committors
   D. Committor Distributions
   E. Path Quenching
VI. Outlook
Acknowledgments
References

1. INTRODUCTION

In this chapter, we review in detail the theory and methodology of transition path sampling. This computational technique is an importance sampling of reactive trajectories, the rare but important dynamical pathways that bridge stable states. We discuss the statistical view of dynamics underlying the method. Within this perspective, ensembles of trajectories can be sampled and manipulated in close analogy to standard techniques of statistical mechanics. Because transition path sampling does not require foreknowledge of reaction mechanisms, it is a natural tool for studying complex dynamical structures of high-dimensional systems encountered at the frontiers of physics, chemistry, and biology.

The dynamics of many such systems involve rare but important transitions between long-lived stable states. These stable states could be, for example, distinct inherent structures of a supercooled liquid, reactants and products of a chemical reaction, or native and denatured states of a protein. In each case, the system spends the bulk of its time fluctuating within stable states, so that transitions occur only rarely. In order to understand such processes in detail, it is necessary to distinguish reaction coordinates, whose fluctuations drive transitions between stable states, from orthogonal variables, whose fluctuations may be viewed as random noise. In principle, computer simulations can provide such insight. Because the times separating successive transitions are long, however, conventional simulations most often fail to exhibit the important dynamics of interest.

A straightforward approach to such problems is to follow the time evolution of the system with molecular dynamics simulations until a reasonable number of events has been observed. The computational requirements of such a procedure are, however, impractically excessive for most interesting systems. For example, a specific water molecule in liquid water has a lifetime of about 10 h, before it dissociates to form hydronium and hydroxide ions. Thus, only a few ionization events occur every hour in a system of, say, 100 water molecules. Since the simulation of molecular motions proceeds in time steps of roughly 1 fs, approximately 10^{18} steps would be required to observe just one such event. Such a calculation is beyond the capabilities of the fastest computers available today and in the foreseeable future.

A different strategy, often used to study chemical reactions, is to search for the dynamical bottlenecks the system passes through during a transition. For a simple system, with an energy landscape as depicted in Figure 1.1(a), this can often be accomplished by enumerating stationary points on the potential energy surface [1,2]. Neglecting the effects of entropy, local minima exemplify stable (or metastable) states. Saddle points exemplify transition states, activated states from which the system may access different

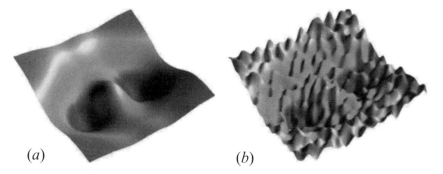

(a) (b)

Figure 1.1. Prototypical potential energy surface of a simple system (a) and of a complex system (b). In a simple, low-dimensional system, dynamical bottlenecks for transitions between long-lived stable states most often coincide with saddle points on the potential energy surface. Locating these stationary points reveals the reaction mechanism. In a typical complex system, the potential energy surface is rugged and has countless local minima and saddle points. Nevertheless, there can be well-defined long-lived stable states and rare transitions between them. Such transitions can occur via a multitude of different transition pathways.

stable states via small fluctuations. One can often infer the mechanism of a reaction by comparing stable and transition states. Transition rates can subsequently be calculated by computing the reversible work to reach the transition state from a stable state, and then initiating many fleeting trajectories from the transition state [3].

The situation is dramatically different for complex systems, classified by Kadanoff [4] as having "many chaotically varying degrees of freedom interacting with one another." Figure 1.1(b) shows how one might envision the energy landscape of such a system. As in the simple system, long-lived stable states are separated by an energetic barrier. But the stationary points exemplifying this barrier comprise only a small fraction of the total set of saddle points, as is generally the case for complex systems. An incomplete enumeration of stationary points is thus insufficient to locate transition states of interest. One might hope instead to guide the search for transition states using physical intuition, in effect postulating the nature of reaction coordinates. But these variables can be highly collective, and therefore difficult to anticipate. In the case of electron transfer, for example, the relevant coordinate is an energy gap that depends on many atomic coordinates. A specific value of the energy gap can be realized in many different ways. Similarly, reaction coordinates for protein folding are expected to depend on many protein and solvent degrees of freedom.

In order to overcome these problems inherent to the study of rare events in complex systems, we have developed a computer simulation technique based on a statistical mechanics of trajectories [5]. In formulating this method, we have recognized that transitions in complex systems may be poorly characterized by a single sequence of configurations, such as a minimum energy pathway. Indeed, a large set of markedly different pathways may be relevant. We term the properly weighted set of reactive trajectories the *transition path ensemble*. Defining this ensemble does not require prior knowledge of a reaction coordinate. Rather, it is sufficient to specify the reactants and products of a transition. This is a crucial feature of this method, since knowledge of a reaction coordinate is usually unavailable for complex systems.

To sample the transition path ensemble, we have developed efficient Monte Carlo procedures [6,7] that generate random walks in the space of trajectories. As a result of this "transition path sampling", one obtains a set of reactive trajectories, from which the reaction mechanism (or mechanisms) can be inferred. Since trajectories generated in the transition path sampling method are true dynamical trajectories, free of any bias, the ensemble of harvested paths can also be used to calculate reaction rates. The high efficiency of these algorithms has significantly widened the range of processes amenable to computer simulation. As was demonstrated in applications of the methodology [5–26] the spectrum of tractable problems now

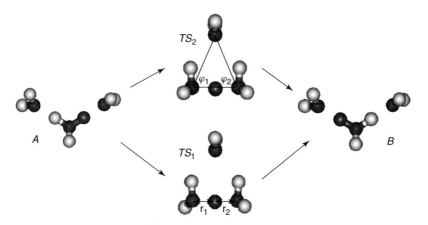

Figure 1.2. The protonated water timer consists of three water molecules and one excess proton. In the equilibrium configuration of this cluster, the excess proton (dark gray) is bound to a central water molecule forming a well-defined hydronium ion (configuration A) [14]. Proton transfer from this central water molecule to one of the other molecules can occur through two different transition state regions denoted by TS_1 and TS_2 in the figure [10]. In the final state (configuration B), a different water molecule holds the excess proton. The transfer of the proton, shown in dark gray, requires rearrangement of the cluster's hydrogen-bonding structure. The angles φ_1 and φ_2 indicated in the upper transition state configuration, TS_2, are used to define an order parameter characterizing the stable states A and B. The distance r_1 and r_2 indicated in the lower transition state configuration are the distances between the transferring proton and the donating and accepting water molecule, respectively.

includes chemical reactions in solution, conformational transitions in biopolymers, and transport phenomena in condensed matter systems.

In this chapter, we present the foundations and methodology of transition path sampling comprehensively, including details important for its implementation. Readers interested in a broad overview of the perspective exploited by the method, and several of its applications, are encouraged to consult a recent review [18]. In the following sections, we first discuss the theoretical basis of transition path sampling, namely, a statistical mechanics of trajectories. We then describe how reactive trajectories may be efficiently sampled, and subsequently analyzed. The practical simplicity of the method is emphasized by outlining essential algorithms in separate schemes. Computer code exemplifying the application of these algorithms can be downloaded from the website http://www.pathsampling.org.

Applications of the method are not discussed separately in this chapter, but are instead used to exemplify important aspects of the method. In particular, proton transfer in the protonated water trimer as shown in Figure 1.2 serves as an illustration for many of the techniques discussed in this chapter. Details of this process are discussed in the caption of Figure 1.2.

II. DEFINING THE TRANSITION PATH ENSEMBLE

In an ergodic system, every possible trajectory of a particular duration occurs with a unique probability. This fact may be used to define a distribution functional for dynamical paths, upon which the statistical mechanics of trajectories is based. For example, with this functional one can construct partition functions for ensembles of trajectories satisfying specific constraints, and compute the reversible work to convert between these ensembles. In later sections, we will show that such manipulations may be used to compute transition rate constants. In this section, we derive the appropriate path distribution functionals for several types of microscopic dynamics, focusing on the constraint that paths are reactive, that is, that they begin in a particular stable state, A, and end in a different stable state, B.

A. Dynamical Path Probability

Let us denote a trajectory of length $\mathcal{T}$ by $x(\mathcal{T})$. While in principle the time evolution of the system is continuous, it is convenient to discretize time and represent a trajectory by an ordered sequence of states, $x(\mathcal{T}) \equiv \{x_0, x_{\Delta t}, x_{2\Delta t}, \ldots, x_{\mathcal{T}}\}$. Consecutive states, or *time slices*, are separated by a small time increment, Δt. Accordingly, such a representation consists of $L = \mathcal{T}/\Delta t + 1$ states. Each of the states x along the trajectory contains a complete set of variables describing the system. For a system that evolves according to Newtonian dynamics, for example, the state $x \equiv \{r, p\}$ consists of the coordinates, r, and momenta, p, of all particles. For a system that evolves according to the rules of Brownian dynamics, x denotes only the configuration of the system. A trajectory connecting stable states A and B is schematically depicted in Figure 1.3.

The statistical weight, $\mathcal{P}[x(\mathcal{T})]$, of a particular trajectory $x(\mathcal{T})$, depends on the distribution of initial conditions and on the specific propagation rules describing the time evolution of the system. Consider, for example, a Markovian process in which state x_t evolves into state $x_{t+\Delta t}$ over a time Δt with probability $p(x_t \to x_{t+\Delta t})$. In this case, the dynamical path probability can be expressed as a product of short-time transition probabilities,

$$\mathcal{P}[x(\mathcal{T})] = \rho(x_0) \prod_{i=0}^{\mathcal{T}/\Delta t - 1} p(x_{i\Delta t} \to x_{(i+1)\Delta t}) \tag{1.1}$$

Here, $\rho(x_0)$ denotes the distribution of states x_0 serving as starting points for trajectories $x(\mathcal{T})$. For example, these initial conditions might be distributed according to the canonical ensemble, $\rho(x_0) \propto \exp\{-\beta H(x_0)\}$, where

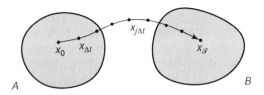

Figure 1.3. A transition pathway $x(\mathcal{T}) \equiv \{x_0, x_{\Delta t}, \ldots, x_{j\Delta t}, \ldots, x_{\mathcal{T}}\}$ connecting stable regions A and B.

$H(x)$ is the Hamiltonian of the system. Both $\rho(x_0)$ and $p(x_{i\Delta t} \to x_{(i+1)\Delta t})$ are assumed to be normalized.

B. Reactive Path Probability

Since we are interested only in reactive trajectories connecting A and B, we now restrict the path ensemble to trajectories beginning in region A at time zero and ending in region B at time $\mathcal{T}$

$$\mathcal{P}_{AB}[x(\mathcal{T})] \equiv h_A(x_0)\mathcal{P}[x(\mathcal{T})]h_B(x_{\mathcal{T}})/Z_{AB}(\mathcal{T}) \qquad (1.2)$$

Here, $h_A(x)$ and $h_B(x)$ are the population functions, or characteristic functions, of regions A and B, respectively. The function $h_A(x) = 1$ if its argument, x, lies in A, and vanishes otherwise. The characteristic function $h_B(x)$ is similarly defined. In most cases, $h_A(x)$ and $h_B(x)$ depend only on the configuration part r of state x, but situations may arise in which it is advantageous to introduce characteristic functions that depend on momenta as well. In practical applications, the stable regions A and B must be characterized carefully. We will return to this issue in Section II.F.

In Eq. (1.2), $Z_{AB}(\mathcal{T})$ is a factor normalizing the distribution of trajectories,

$$Z_{AB}(\mathcal{T}) \equiv \int \mathcal{D}x(\mathcal{T})h_A(x_0)\mathcal{P}[x(\mathcal{T})]h_B(x_{\mathcal{T}}) \qquad (1.3)$$

The fact that $Z_{AB}(\mathcal{T})$ has the form of a partition function is important for the calculation of rate constants, as we will discuss later. The notation $\int \mathcal{D}x(\mathcal{T})$, borrowed from the theory of path integrals, indicates a summation over all pathways $x(\mathcal{T})$. For a discretized path, this summation corresponds to a integration over states at each time slice of the path.

The probability functional from Eq. (1.2) is a statistical description of all pathways of length $\mathcal{T}$ connecting reactants with products. We call this set of appropriately weighted paths the *transition path ensemble*. Pathways $x(\mathcal{T})$, which do not begin in A or do not end in B, have zero probability in this ensemble. Reactive trajectories, on the other hand, may have a non-vanishing probability, depending on the dynamical path weight $\mathcal{P}[x(\mathcal{T})]$.

The perspective exploited by transition path sampling, namely, a statistical description of pathways with endpoints located in certain phase-space regions, was first introduced by Pratt [27], who described stochastic pathways as chains of states, linked by appropriate transition probabilities. Others have explored similar ideas and have constructed ensembles of pathways using ad hoc probability functionals [28–35]. Pathways found by these methods are reactive, but they are not consistent with the true dynamics of the system, so that their utility for studying transition dynamics is limited. Trajectories in the transition path ensemble from Eq. (1.2), on the other hand, are true dynamical trajectories, free of any bias by unphysical forces or constraints. Indeed, transition path sampling *selects* reactive trajectories from the set of all trajectories produced by the system's intrinsic dynamics, rather than *generating* them according to an artificial bias. This important feature of the method allows the calculation of dynamical properties such as rate constants.

C. Deterministic Dynamics

Consider a system whose time evolution is described by a set of ordinary differential equations,

$$\dot{x} = \Gamma(x) \tag{1.4}$$

where $\dot{x}$ indicates the time derivative of x and $\Gamma(x)$ is a function of x only. The time evolution of such a dynamical system is deterministic in the sense that initial conditions x_0 completely determine the trajectory for all times. Newton's equations of motion, for example, have this form.

$$\dot{r} = \frac{\partial H(r, p)}{\partial p}, \qquad \dot{p} = -\frac{\partial H(r, p)}{\partial r} \tag{1.5}$$

Other examples for this type of dynamics include the equations of motion based on the extended Lagrangian of Car and Parrinello [36] and equations of motion for various thermostatted systems [37]. The equations of motion for hydrodynamic flow can also be cast in this form [38].

Solving the equations of motion (1.4) yields the propagator ϕ_t, which maps the initial state of the system to that at time t:

$$x_t = \phi_t(x_0) \tag{1.6}$$

Because this mapping takes the state x_t into exactly one state $x_{t+\Delta t}$ at time Δt later, the short time transition probability is represented by a Dirac delta function:

$$p(x_t \to x_{t+\Delta t}) = \delta[x_{t+\Delta t} - \phi_{\Delta t}(x_t)] \tag{1.7}$$

Note that here the argument of the delta function is a high-dimensional vector. Accordingly, the delta function of Eq. (1.7) is actually a product of delta functions, one for each coordinate. The reactive path probability for a deterministic trajectory is therefore

$$\mathscr{P}_{AB}[x(\mathscr{T})] = \rho(x_0) h_A(x_0) \left\{ \prod_{i=0}^{\mathscr{T}/\Delta t - 1} \delta[x_{(i+1)\Delta t} - \phi_{\Delta t}(x_{i\Delta t})] \right\} h_B(x_\mathscr{T}) / Z_{AB}(\mathscr{T}) \tag{1.8}$$

The normalization factor $Z_{AB}(\mathscr{T})$ is given by

$$Z_{AB}(\mathscr{T}) = \int dx_0 \rho(x_0) h_A(x_0) h_B(x_\mathscr{T}) \tag{1.9}$$

where integrations over the states along the path have been carried out at all times except time zero.

D. Stochastic Dynamics

Often, the analysis of molecular systems can be simplified by replacing certain degrees of freedom by random noise. With this replacement, the remaining degrees of freedom x evolve according to a Langevin equation. In the simplest case, the random noise is uncorrelated in time, giving

$$\dot{r} = \frac{p}{m}$$
$$\dot{p} = \mathscr{F}(r) - \gamma p + \mathscr{R} \tag{1.10}$$

Here, $\mathscr{F}(r)$ is the force derived from the potential energy $V(r)$. The friction constant γ and the random force $\mathscr{R}$ are related through the fluctuation dissipation theorem: $\langle \mathscr{R}(t)\mathscr{R}(0) \rangle = 2m\gamma k_B T \delta(t)$, where T is the temperature and k_B is Boltzmann's constant. The random thermal noise $\mathscr{R}$ compensates for the energy dissipated by the frictional term $-\gamma p$. Because we focus on finite segments of trajectories, the treatment of noise that is correlated in time is awkward within the specific methodology presented in this chapter.

But even in this non-Markovian case, every finite trajectory of the primary variables has a well-defined probability, and transition path sampling can in principle be carried out with sufficient generalizations.

Various integration algorithms have been derived to solve the equation of motion (1.10) over small time increments Δt [39]. Applying these operations repeatedly yields stochastic trajectories of arbitrary length. Typically, these integration algorithms have the form

$$x_{t+\delta t} = x_t + \delta x_S + \delta x_R \tag{1.11}$$

While the systematic part δx_S is fully determined by x_t, the random part δx_R is drawn from a distribution $w(\delta x_R)$. For Langevin dynamics, $w(\delta x_R)$ is a multivariate Gaussian distribution as derived by Chandrasekhar [40]. The random component of the small time step propagator "smears out" the time evolution of the system such that many different states are accessible starting from the same initial state. The single-step transition probability is given by

$$p(x_t \to x_{t+\Delta t}) = w(\delta x_R) \left| \frac{\partial \delta x_R}{\partial x_{t+\Delta t}} \right| \tag{1.12}$$

where the Jacobian on the right-hand side arises from the variable transformation from random displacement δx_R to phase space point $x_{t+\Delta t}$. For the most widely used integration algorithm [39] this Jacobian is unity, simplifying the path weight considerably [6]. Furthermore, the transition probability derived from the Langevin equation is normalized, and it conserves the canonical distribution as required by the fluctuation–dissipation theorem [6]. By concatenating transition probabilities of Eq. (1.12) as in Eq. (1.1) and imposing boundary conditions as in Eq. (1.2), we obtain the reactive path probability

$$\mathscr{P}_{AB}[x(\mathscr{T})] = \rho(x_0)h_A(x_0) \left\{ \prod_{i=0}^{\mathscr{T}/\Delta t - 1} w[x_{(i+1)\Delta t} - x_{i\Delta t} - \delta x_S] \right\} h_B(x_{\mathscr{T}})/Z_{AB}(\mathscr{T}) \tag{1.13}$$

In contrast to deterministic pathways, stochastic trajectories are not completely determined by the initial state x_0. Accordingly, the probability functional $\mathscr{P}_{AB}[x(\mathscr{T})]$ explicitly depends on the entire path $x(\mathscr{T})$.

E. Monte Carlo Dynamics

The dynamics of some complex systems, such as spin systems, lattice gases, and certain models of proteins, are most naturally studied using Monte

Carlo simulations [8, 41–43]. Such simulations proceed stepwise in a biased random walk through the space of possible states. In this walk, a trial state is generated from the current state of the system, and is accepted with a probability that depends on the relative weights of the trial and original state in the ensemble of interest. If the trial state is not accepted, then the original state is retained. The acceptance probability is constructed so that every state x is visited with a frequency proportional to its equilibrium weight $\rho(x)$ [44]. This procedure is called importance sampling, because the most important states, those with the largest weight, are visited most often. States with negligible weight are rarely observed.

Various acceptance rules have been devised for importance sampling, the simplest of which is the so-called Metropolis algorithm [45]. In this case,

$$p(x_t \to x_{t+\Delta t}) = \omega(x_t \to x_{t+\Delta t}) + \delta(x_t - x_{t+\Delta t})Q(x_t) \qquad (1.14)$$

where

$$\omega(x_t \to x_{t+\Delta t}) = \eta(x_t \to x_{t+\Delta t})\min\left[1, \frac{\rho(x_{t+\Delta t})}{\rho(x_t)}\right] \qquad (1.15)$$

is the probability for an accepted trial move from x_t to $x_{t+\Delta t}$ and

$$Q(x) = 1 - \int dx'\omega(x \to x') \qquad (1.16)$$

is the total rejection probability for a move starting from x. Trial states are generated by selecting a random displacement from the distribution $\eta(x_t \to x_{t+\Delta t})$, which is assumed to be symmetric, that is, $\eta(x \to x') = \eta(x' \to x)$. In simple Monte Carlo simulations $\eta(x \to x')$ is a rapidly decaying function of the magnitude of the displacement, but more complicated choices are sometimes useful [44]. The min-function in Eq. (1.15) returns the smaller of its arguments. Thus, a trial state is always accepted if it has a larger weight than the original state, but is rejected with a finite probability if its weight is lower. The transition probability from Eq. (1.14) is normalized and conserves the equilibrium distribution $\rho(x)$.

By concatenating Monte Carlo transition probabilities according to Eq. (1.1), one obtains the probability of a particular stochastic path $x(\mathcal{T})$ generated in a Metropolis Monte Carlo simulation. The time variable t describing the progress of this stochastic process is artificial. This *Monte Carlo time* can be approximately mapped to a physical timescale by comparing known dynamical properties such as transport coefficients [8,41].

F. Defining the Initial and Final Region

Transition path sampling does not require knowledge of a *reaction coordinate* describing the progress of a transition through the dynamical bottleneck region. It is only necessary that the initial and final states of the transition are characterized carefully. While this requirement is considerably less stringent, its satisfaction is crucial. Typically, regions A and B are defined by distinct ranges of a low dimensional *order parameter* q. Identifying an order parameter that truly discriminates between A and B can be quite challenging.

A successful order parameter must satisfy several criteria. First, regions A and B must be large enough to accommodate typical equilibrium fluctuations in their corresponding basins of attraction. The basin of attraction of a specific stable state consists of all configuration from which trajectories nearly always relax into that stable state. If this first criterion is not met, important transition pathways will likely be overlooked. Second, and more importantly, the regions spanned by $h_A(x)$ and $h_B(x)$ should be located entirely within the corresponding basins of attraction. In other words, region A should not overlap with the basin of attraction of B, and vice versa. If this second criterion is not met, transition path sampling may harvest nonreactive trajectories. This situation is illustrated in Figure 1.4, which shows two pathways on a free energy surface $w(q, q')$. While both

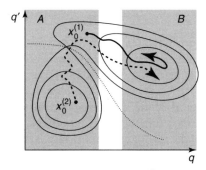

Figure 1.4. Contour lines of a free energy landscape $w(q, q')$ in which the coordinate q fails to unambiguously separate the basins of attraction of the stable states. Regions A and B (gray) accommodate most equilibrium fluctuations in the respective stable states, but the overlap of A with the basin of attraction of the final state leads to pathways that are not truly reactive. Although the trajectory depicted as a solid line with initial state $x_0^{(1)}$ starts in A and ends in B it does not cross the transition state surface (dotted line) separating the basins of attraction. The trajectory depicted as a dashed line with initial condition $x_0^{(2)}$, on the other hand, starts in the basin of attraction of A and ends in the basin of attraction of B. With the definition of the initial region A and the final region B depicted in the figure the transition path sampling algorithm is most likely to sample nonreactive trajectories rather than reactive ones.

pathways begin in A and end in B, only one of them actually crosses the transition state surface (dashed line). Although the other path (solid line) also begins with $h_A = 1$, its initial point $x_0^{(1)}$ is in fact located in the basin of attraction of state B. Thus, this pathway does not exhibit a true transition.

Suitable definitions of regions A and B may require considerable trial and error. Fortunately, it is straightforward to diagnose an unsuccessful order parameter. For example, most short trajectories initiated from the state $x_0^{(1)}$ will quickly visit states with values of q characteristic of state B. In other words, the probability to relax into B is close to one. (This relaxation probability plays an important role in the analysis of transition pathways, as will be discussed in detail in Section V.) In contrast, the probability to relax into B from $x_0^{(2)}$ is negligible. When relaxation probabilities indicate that definitions of A and B do not exclude nonreactive trajectories, the nature and/or ranges of the order parameter must be refined.

III. SAMPLING THE TRANSITION PATH ENSEMBLE

Transition path sampling is an importance sampling of trajectories, akin to the importance sampling of configurations described in Section II.E. Specifically, it is a biased random walk in the space of trajectories, in which each pathway is visited in proportion to its weight in the transition path ensemble. Because trajectories that do not exhibit the transition of interest have zero weight in this ensemble, they are never visited. In this way, attention is focused entirely on the rare but important trajectories, those that are reactive.

We accomplish the random walk through trajectory space as follows: Beginning with a trajectory $x^{(o)}(\mathcal{T})$ [here, the superscript (o) stands for old] whose weight $\mathcal{P}_{AB}[x^{(o)}(\mathcal{T})]$ in the transition path ensemble is nonzero, we generate a new trajectory $x^{(n)}(\mathcal{T})$. Using the terminology of conventional Monte Carlo techniques we call this procedure to create a new trajectory a "trial move". Efficient methods for such generation of new trajectories by modifying existing ones will be discussed in the following sections. Next, we accept the newly generated path with a certain probability. There are many ways to construct an appropriate acceptance probability. The simplest is based on *detailed balance* of moves in trajectory space. This criterion requires that the frequency of accepted moves from $x^{(o)}(\mathcal{T})$ to $x^{(n)}(\mathcal{T})$ is exactly balanced by the frequency of reverse moves:

$$\mathcal{P}_{AB}[x^{(o)}(\mathcal{T})]\pi[x^{(o)}(\mathcal{T}) \to x^{(n)}(\mathcal{T})] = \mathcal{P}_{AB}[x^{(n)}(\mathcal{T})]\pi[x^{(n)}(\mathcal{T}) \to x^{(o)}(\mathcal{T})]$$

(1.17)

where, $\pi[x(\mathcal{T}) \to x'(\mathcal{T})]$ is the conditional probability to make a move from $x(\mathcal{T})$ to $x'(\mathcal{T})$ given an initial path $x(\mathcal{T})$. In our case, $\pi[x(\mathcal{T}) \to x'(\mathcal{T})]$ is a product of the probability to generate $x'(\mathcal{T})$ from $x(\mathcal{T})$, $P_{\text{gen}}[x(\mathcal{T}) \to x'(\mathcal{T})]$, and the probability $P_{\text{acc}}[x(\mathcal{T}) \to x'(\mathcal{T})]$ to accept the trial path $x'(\mathcal{T})$:

$$\pi[x(\mathcal{T}) \to x'(\mathcal{T})] = P_{\text{gen}}[x(\mathcal{T}) \to x'(\mathcal{T})] \times P_{\text{acc}}[x(\mathcal{T}) \to x'(\mathcal{T})] \quad (1.18)$$

From the detailed balance condition one obtains a condition for the acceptance probability:

$$\frac{P_{\text{acc}}[x^{(o)}(\mathcal{T}) \to x^{(n)}(\mathcal{T})]}{P_{\text{acc}}[x^{(n)}(\mathcal{T}) \to x^{(o)}(\mathcal{T})]} = \frac{\mathcal{P}_{AB}[x^{(n)}(\mathcal{T})] P_{\text{gen}}[x^{(n)}(\mathcal{T}) \to x^{(o)}(\mathcal{T})]}{\mathcal{P}_{AB}[x^{(o)}(\mathcal{T})] P_{\text{gen}}[x^{(o)}(\mathcal{T}) \to x^{(n)}(\mathcal{T})]} \quad (1.19)$$

This condition can be satisfied conveniently using the Metropolis rule [45]:

$$P_{\text{acc}}[x^{(o)}(\mathcal{T}) \to x^{(n)}(\mathcal{T})] = \min\left\{1, \frac{\mathcal{P}_{AB}[x^{(n)}(\mathcal{T})] P_{\text{gen}}[x^{(n)}(\mathcal{T}) \to x^{(o)}(\mathcal{T})]}{\mathcal{P}_{AB}[x^{(o)}(\mathcal{T})] P_{\text{gen}}[x^{(o)}(\mathcal{T}) \to x^{(n)}(\mathcal{T})]}\right\}$$
(1.20)

Since the old trajectory $x^{(o)}(\mathcal{T})$ is reactive, that is, $h_A[x_0^{(o)}] = 1$ and $h_B[x_{\mathcal{T}}^{(o)}] = 1$, the acceptance probability can be written as

$$P_{\text{acc}}[x^{(o)}(\mathcal{T}) \to x^{(n)}(\mathcal{T})] = h_A[x_0^{(n)}] h_B[x_{\mathcal{T}}^{(n)}]$$
$$\times \min\left\{1, \frac{\mathcal{P}[x^{(n)}(\mathcal{T})]}{\mathcal{P}[x^{(o)}(\mathcal{T})]} \frac{P_{\text{gen}}[x^{(n)}(\mathcal{T}) \to x^{(o)}(\mathcal{T})]}{P_{\text{gen}}[x^{(o)}(\mathcal{T}) \to x^{(n)}(\mathcal{T})]}\right\} \quad (1.21)$$

where $x_0^{(n)}$ and $x_{\mathcal{T}}^{(n)}$ are the first and last state of the new trajectory, respectively. An acceptance probability of the form $\min[1, \alpha]$ may be realized by accepting a new pathway unconditionally when $\alpha > 1$, and with probability α when $\alpha < 1$. A simple way to implement this rule is to accept a trial move whenever a random number ζ drawn from a uniform distribution in the interval $[0, 1]$ is smaller than α. According to Eq. (1.21), only new pathways connecting regions A and B have a nonzero acceptance probability. If a new pathway does not begin in A or end in B it is rejected. A summary of the algorithm is given in Scheme 1.1.

An important feature of the acceptance probability in Eq. (1.21) is that a new pathway with lower statistical weight than the old one is accepted with finite probability. As a result, "barriers" in path space can be surmounted, facilitating "relaxation" toward the most important regions in path space. Thus, it is not essential that the first reactive trajectory have high statistical weight.

1. Generate a new pathway $x^{(n)}(\mathcal{T})$ from the current one, $x^{(o)}(\mathcal{T})$, with generation probability $P_{gen}[x^{(o)}(\mathcal{T}) \to x^{(n)}(\mathcal{T})]$.
2. Accept or reject the new pathway according to a Metropolis acceptance criterion obeying detailed balance with respect to the transition path ensemble $\mathcal{P}_{AB}[x(\mathcal{T})]$.
3. If the new trajectory is accepted, it becomes the current one. Otherwise the old trajectory is retained as the current trajectory.
4. Repeat starting from 1.

Scheme 1.1. Metropolis Monte Carlo sampling algorithm for transition pathways.

In order to implement the algorithm described above, one must be able to generate trial paths from existing ones. While in principle there are many ways to do this, the efficiency of the algorithm depends crucially on the nature of trial moves. Specifically, it is important that the average acceptance probability for generated paths not be too small. For transition path sampling, satisfying this condition requires that trial paths have a reasonable weight $\mathcal{P}[x(\mathcal{T})]$ in the dynamical path ensemble, as well as a reasonable chance of connecting states A and B.

Desirable trial trajectories for transition path sampling have significant dynamical weights, $\mathcal{P}[x(\mathcal{T})]$, and thus resemble natural trajectories. The simplest way to obtain such paths is to apply the very same propagation rules that define the natural dynamics of the system. We have employed this strategy to construct two basic types of trial moves, which we call "shooting" and "shifting". In both cases, new trajectories are obtained by applying dynamical propagation rules to a phase space point that is taken (and possibly modified) from an existing transition pathway. The resulting paths have significant dynamical weights by construction. In addition, trial paths generated by shooting and shifting have a good chance of exhibiting successful transitions from A to B, since they are grown from phase-space points on or near reactive trajectories. In Sections III.A. and III.B., we describe shooting and shifting moves in detail, and indicate how they may be successfully implemented for various classes of dynamics.

A. Shooting Moves

In a shooting move, a phase-space point $x_{t'}^{(o)}$ is selected at random from the chain of states comprising an old pathway, $x^{(o)}(\mathcal{T})$. This state may be modified in some way, for example by displacing the atomic momenta by a small amount. Trajectory segments are then "shot off" forward and backward in time from the modified state $x_{t'}^{(n)}$, applying the appropriate dynamical rules until the new path extends from time zero to time $\mathcal{T}$. This procedure is schematically depicted in Figure 1.5.

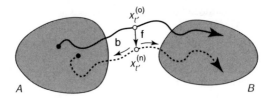

Figure 1.5. In a shooting move, a new pathway (dashed line) is generated from an old one (solid line) by first selecting a time slice $x_{t'}^{(o)}$ of the old path and modifying it to obtain $x_{t'}^{(n)}$. Then, a new pathway is constructed by generating forward (f) and backward (b) trajectory segments initiated at the modified time slice $x_{t'}^{(n)}$. If the new trajectory is reactive, that is, if it starts in A and ends in B, the path is accepted with a finite probability and rejected otherwise.

Because trajectories obtained by shooting reflect the system's underlying dynamics, the corresponding generation and acceptance probabilities are relatively simple. Specifically, the generation probability for the forward trajectory segment, beginning at time t' and ending at time $\mathcal{T}$, is

$$P_{\text{gen}}^{\text{f}}[o \to n] = \prod_{i=t'/\Delta t}^{\mathcal{T}/\Delta t - 1} p[x_{i\Delta t}^{(n)} \to x_{(i+1)\Delta t}^{(n)}] \quad (1.22)$$

This generation probability is identical to the dynamical path weight for the forward trajectory. Similarly, the generation probability for the backward trajectory segment, beginning at time t' and ending at time zero, is

$$P_{\text{gen}}^{\text{b}}[o \to n] = \prod_{i=1}^{t'/\Delta t} \bar{p}[x_{i\Delta t}^{(n)} \to x_{(i-1)\Delta t}^{(n)}] \quad (1.23)$$

Here, the appropriate small time step probability, $\bar{p}(x \to x')$, describes the evolution of the system backward in time.

By combining the generation probability for the modified time slice $x_{t'}^{(n)}$ with the generation probabilities of the forward and backward segments of the new trajectory, one obtains the complete generation probability for the new trajectory,

$$P_{\text{gen}}[x^{(o)}(\mathcal{T}) \to x^{(n)}(\mathcal{T})] = p_{\text{gen}}[x_{t'}^{(o)} \to x_{t'}^{(n)}] \prod_{i=t'/\Delta t}^{\mathcal{T}/\Delta t - 1} p[x_{i\Delta t}^{(n)} \to x_{(i+1)\Delta t}^{(n)}]$$

$$\times \prod_{i=1}^{t'/\Delta t} \bar{p}[x_{i\Delta t}^{(n)} \to x_{(i-1)\Delta t}^{(n)}] \quad (1.24)$$

Here, $p_{\text{gen}}[x_{t'}^{(o)} \to x_{t'}^{(n)}]$ denotes the probability to obtain state $x_{t'}^{(n)}$, which we also call the *shooting point*, by modification of state $x_{t'}^{(o)}$.

The generation probability from Eq. (1.24) can now be used to determine the acceptance probability of a shooting move with the help of Eq. (1.21). For this purpose, the ratio appearing as the second argument of the min-function in Eq. (1.21) must be determined. By using the dynamical path probability from Eq. (1.1) for the old and the new paths and the generation probability from Eq. (1.24), we obtain

$$\frac{\mathcal{P}[x^{(n)}(\mathcal{T})]P_{\text{gen}}[x^{(n)}(\mathcal{T}) \to x^{(o)}(\mathcal{T})]}{\mathcal{P}[x^{(o)}(\mathcal{T})]P_{\text{gen}}[x^{(o)}(\mathcal{T}) \to x^{(n)}(\mathcal{T})]} = \frac{\rho[x_0^{(n)}]p_{\text{gen}}[x_{t'}^{(n)} \to x_{t'}^{(o)}]}{\rho[x_0^{(o)}]p_{\text{gen}}[x_{t'}^{(o)} \to x_{t'}^{(n)}]}$$

$$\prod_{i=0}^{t'/\Delta t - 1} \frac{p[x_{i\Delta t}^{(n)} \to x_{(i+1)\Delta t}^{(n)}]}{\bar{p}[x_{(i+1)\Delta t}^{(n)} \to x_{i\Delta t}^{(n)}]} \times \frac{\bar{p}[x_{(i+1)\Delta t}^{(o)} \to x_{i\Delta t}^{(o)}]}{p[x_{i\Delta t}^{(o)} \to x_{(i+1)\Delta t}^{(o)}]} \quad (1.25)$$

where factors have canceled because the trial trajectory $x^{(n)}(\mathcal{T})$ was generated using the propagation rule of the underlying dynamics.

This general result simplifies considerably if the dynamics conserve a stationary distribution, $\rho_{\text{st}}(x)$. This condition is very general, applying to systems at equilibrium, nonequilibrium systems in a steady state, and nonequilibrium systems relaxing to equilibrium with time-translationally invariant dynamics. In this case, p and $\bar{p}$ are related in a simple way by microscopic reversibility,

$$\frac{p(x \to y)}{\bar{p}(y \to x)} = \frac{\rho_{\text{st}}(y)}{\rho_{\text{st}}(x)} \quad (1.26)$$

By substituting this relation into Eq. (1.25) and by using Eq. (1.21), the acceptance probability for a shooting move is written

$$P_{\text{acc}}[x^{(o)}(\mathcal{T}) \to x^{(n)}(\mathcal{T})] = h_A[x_0^{(n)}]h_B[x_{\mathcal{T}}^{(n)}]$$

$$\min\left\{1, \frac{\rho[x_0^{(n)}]}{\rho[x_0^{(o)}]} \frac{\rho_{\text{st}}[x_0^{(o)}]}{\rho_{\text{st}}[x_0^{(n)}]} \frac{\rho_{\text{st}}[x_{t'}^{(n)}]}{\rho_{\text{st}}[x_{t'}^{(o)}]} \frac{p_{\text{gen}}[x_{t'}^{(n)} \to x_{t'}^{(o)}]}{p_{\text{gen}}[x_{t'}^{(o)} \to x_{t'}^{(n)}]}\right\} \quad (1.27)$$

The above acceptance probability depends only on phase-space points at two times, at 0 and t'.

Often, the distribution of initial conditions is an equilibrium or steady-state distribution, and is therefore identical to the conserved distribution, $\rho(x) = \rho_{\text{st}}(x)$. In this case, the acceptance probability simplifies still further,

$$P_{\text{acc}}[x^{(o)}(\mathcal{T}) \to x^{(n)}(\mathcal{T})] = h_A[x_0^{(n)}]h_B[x_{\mathcal{T}}^{(n)}]$$

$$\times \min\left\{1, \frac{\rho[x_{t'}^{(n)}]}{\rho[x_{t'}^{(o)}]} \times \frac{p_{\text{gen}}[x_{t'}^{(n)} \to x_{t'}^{(o)}]}{p_{\text{gen}}[x_{t'}^{(o)} \to x_{t'}^{(n)}]}\right\}. \quad (1.28)$$

For symmetric generation probabilities of the shooting point, this equation becomes

$$P_{\text{acc}}[x^{(o)}(\mathcal{T}) \to x^{(n)}(\mathcal{T})] = h_A[x_0^{(n)}]h_B[x_{\mathcal{T}}^{(n)}] \min\left\{1, \frac{\rho[x_{t'}^{(n)}]}{\rho[x_{t'}^{(o)}]}\right\}. \quad (1.29)$$

Next, we will assume that generation probabilities for phase-space modifications are symmetric. If they are not, the asymmetry must be taken into account as prescribed by Eq. (1.28). The simplicity of the acceptance probability in Eq. (1.29) entails an algorithmic simplicity for shooting moves. In order to evaluate P_{acc}, one needs only to compute the relative weights of old and new phase-space points at time t' and determine whether the new path begins in region A and ends in region B.

Equation (1.29) also suggests an efficient implementation of shooting moves. A shooting move is initiated by selecting a time slice along the existing path. After generating the shooting point $x_{t'}^{(n)}$ from the old time slice $x_{t'}^{(o)}$ a first acceptance–rejection decision can be made. The shooting point is accepted if $\rho[x_{t'}^{(n)}]/\rho[x_{t'}^{(o)}] > 1$ or if a random number ζ drawn from a uniform distribution in the interval $[0, 1]$ is smaller than $\rho[x_{t'}^{(n)}]/\rho[x_{t'}^{(o)}]$. Otherwise the trial shooting move is aborted. If, on the other hand, the shooting point is accepted, one proceeds with growing either one of the forward or backward trajectory segments. If the appropriate boundary condition is satisfied (e.g., if the forward trajectory segment ends in region B), the other segment is grown. The path is finally accepted if the boundary condition for this latter trajectory segment is satisfied as well. A rejection at any stage of this procedure allows subsequent steps to be skipped. In this case, the trial move is rejected and the old path is retained. Using this sequential algorithm reduces the cost of rejected moves, saving considerable computational effort (typically 10–30%).

Naively, one might expect that a gain in computational efficiency might be obtained by growing the shorter of the two trajectory segments first. In case of rejection, the longer and more expensive of the two segments need not be grown. In most cases, however, the shorter trajectory segment has a much higher probability to reach the appropriate stable state than the longer one, which is due to the fact that the shorter trajectory segment has less time to diverge from the old path than the longer segment. Also, the short trajectory segment is less likely to have to pass the dynamical bottleneck separating the stable states. Consequently, early rejection is not likely to occur when growing the short segment first and the long trajectory segment must be determined in most cases. This effect can compensate the potential advantage of occasional early rejections. Indeed, for some systems it might even be advantageous to grow the longer trajectory segment first and proceed with growing the shorter one only in case of acceptance. Since in general the associated efficiency increase is only of the order of a few percent, it is good practice to carry out integration of, say, the backward segment first and after that the forward segment regardless of which one of the two is shorter.

Shooting moves derive their efficiency from the tendency of trajectories to diverge in phase space, so that subsequent paths may be quite different. But it is important that trial paths are not too different, so that they have reasonable chance of connecting A and B. In deterministic systems, the degree of divergence between old and new trajectories in shooting moves depends on the magnitude of modification of $x_{t'}^{(o)}$. Arbitrarily small modification gives new paths that are arbitrarily similar to old paths. Large displacements give very different paths. It is thus possible to tune the magnitude of modifications to give a desired acceptance probability on average, as will be discussed in Section III.A.4. For stochastic systems, small path displacements may not be possible, since even an unmodified state can yield a different path.

1. Shooting Moves for Deterministic Dynamics

For deterministic dynamics, applying dynamical propagation rules to a phase-space point $x_{t'}^o$ of an existing trajectory simply regenerates that trajectory. Shooting moves must therefore include modification of the shooting point, for example, by adding a random perturbation δx to $x_{t'}^{(o)}$,

$$x_{t'}^{(n)} = x_{t'}^{(o)} + \delta x \qquad (1.30)$$

This can be done in a symmetric way by drawing δx from a distribution $w(\delta x)$ for which $w(\delta x) = w(-\delta x)$. In this procedure, the generation probability of the new shooting point $x_{t'}^{(n)}$ from $x_{t'}^{(o)}$ is identical to that for the reverse move, that is, $p_{\text{gen}}[x_{t'}^{(o)} \to x_{t'}^{(n)}] = p_{\text{gen}}[x_{t'}^{(n)} \to x_{t'}^{(o)}]$. Selecting symmetric perturbations can be complicated when allowed values of x are restricted by internal constraints of the system. Methods for generating such displacements are discussed in Section III.A.2.

It is usually sufficient to modify only the momentum part of the selected time slice $x_{t'}^{(o)}$, leaving the configurational part unchanged. In this case, $p_{t'}^{(n)} = p_{t'}^{(o)} + \delta p$ and $r_{t'}^{(n)} = r_{t'}^{(o)}$ (see Fig. 1.6). For some applications, however, it is advantageous to change both the configuration and momentum parts of $x_{t'}^{(o)}$ [13].

For deterministic dynamics, a trial trajectory is obtained by integrating the equations of motion from the shooting point. The forward segment of

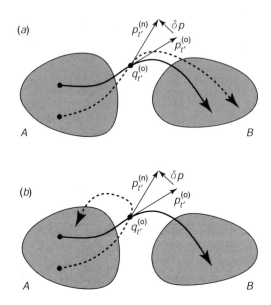

Figure 1.6. In a shooting move for deterministic trajectories, a time slice $x_{t'}^{(o)} = \{q_{t'}^{(o)}, p_{t'}^{(o)}\}$ on the old path (solid line) is selected at random and the corresponding momenta $p_{t'}^{(o)}$ are changed by a small random amount δp. Integration of the equations of motion backward to time 0 and forward to time t starting from the modified state $x_{t'}^{(n)} = \{q_{t'}^{(n)}, p_{t'}^{(n)}\}$ yields the new trajectory $x^{(n)}(\mathcal{T})$ (dashed line). If this trajectory is connecting A and B, it will be accepted with a nonvanishing probability (panel *a*). Otherwise, it will be rejected (panel *b*).

the trajectory is generated simply by integrating for the appropriate number of small time steps. The backward segment, on the other hand, must be generated with the direction of time inverted, that is, with a negative time step. In the case of time-reversible dynamics, this is accomplished by first inverting all momenta, and then integrating forward in time. Formally, $\bar{p}(x \to y) = p(\bar{x} \to \bar{y})$. Here, $\bar{x} = \{r, -p\}$ is obtained from $x = \{r, p\}$ by inverting all momenta. In the resulting chain of states, momenta are then inverted so that the backward trajectory segment evolves properly in time. The complete protocol for performing shooting moves with deterministic dynamics is summarized in Scheme 1.2.

We verify the soundness of this procedure by demonstrating that the reversibility condition in Eq. (1.26) is satisfied. A small step in the backward trajectory segment described above has probability

$$\bar{p}(y \to x) = \delta[x - \phi_{-\Delta t}(y)] = \delta[x - \phi_{\Delta t}^{-1}(y)] \quad (1.31)$$

1. Randomly select a time slice $x_{t'}^{(o)}$ on a existing trajectory $x^{(o)}(\mathcal{T})$.
2. Modify the selected time slice by adding a random displacement: $x_{t'}^{(n)} = x_{t'}^{(o)} + \delta x$. The random displacement must be consistent with the ensemble of initial conditions and should be symmetric with respect to the reverse move.
3. Accept the new shooting point with probability $\min[1, \rho(x_{t'}^{(n)})/\rho(x_{t'}^{(o)})]$. Abort the trial move if the shooting point is rejected.
4. If the shooting point is accepted, integrate the equations of motion forward to time $\mathcal{T}$ starting from $x_{t'}^{(n)}$.
5. Abort the trial move if the final point of the path segment, $x_{\mathcal{T}}^{(n)}$, is not in B and continue otherwise.
6. Integrate the equations of motion backward to time 0 starting from $x_{t'}^{(n)}$.
7. Accept the new trajectory if its initial point $x_0^{(n)}$ is in A and reject it otherwise.
8. In case of a rejection, the old trajectory is counted again in the calculation of path averages. Otherwise the new trajectory is used as the current one.

Scheme 1.2. Shooting algorithm for deterministic trajectories.

For the forward trajectory,

$$p(x \to y) = \delta[y - \phi_{\Delta t}(x)] = \delta[\phi_{\Delta t}^{-1}(y) - x]|\partial\phi_{\Delta t}(x)/\partial x|^{-1} \quad (1.32)$$

where $|\partial\phi_{\Delta t}(x)/\partial x|$ is the Jacobian associated with time evolution of duration Δt. This Jacobian describes the contraction or expansion of an infinitesimal comoving volume element in phase space. Combining these small time step probabilities,

$$\frac{p(x \to y)}{\bar{p}(y \to x)} = \left|\frac{\partial\phi_{\Delta t}(x)}{\partial x}\right|^{-1} = \frac{\rho_{st}(y)}{\rho_{st}(x)} \quad (1.33)$$

The final equality results from the fact that the phase-space flow generated by time evolution conserves probability. Because the reversibility criterion is satisfied, the simple acceptance probabilities from Eqs. (1.27) and (1.29) can be used, provided a stationary distribution ρ_{st} exists.

For some applications, the distribution of initial conditions is distinct from the stationary distribution preserved by a system's dynamics. For example, one might be interested in the relaxation of a system that has been driven away from equilibrium. One might even be interested in an unstable dynamics that does not preserve a stationary distribution. In both cases, it is possible to sample trajectories using the same type of shooting move described above, provided the distribution of initial conditions $\rho(x)$ is well defined. To derive an appropriate acceptance probability for this case, we return to Eq. (1.25). By using Eq. (1.32) and the chain rule for the Jacobian, we obtain

$$P_{acc}[x^{(o)}(\mathcal{T}) \to x^{(n)}(\mathcal{T})] = h_A[x_0^{(n)}]h_B[x_{\mathcal{T}}^{(n)}]\min\left\{1, \frac{\rho[x_0^{(n)}]}{\rho[x_0^{(o)}]}\frac{|\partial x_{t'}^{(o)}/\partial x_0^{(o)}|}{|\partial x_{t'}^{(n)}/\partial x_0^{(n)}|}\right\} \quad (1.34)$$

where we have again assumed a symmetric generation probability, that is, $p_{gen}[x_{t'}^{(o)} \to x_{t'}^{(n)}] = p_{gen}[x_{t'}^{(n)} \to x_{t'}^{(o)}]$.

The Jacobian $J(t') \equiv |\partial x_{t'}/\partial x_0|$ appearing in Eq. (1.34) can in principle be calculated by integrating

$$\frac{dJ(t)}{dt} = \Lambda(x)J(t) \quad (1.35)$$

along the trajectory of interest with $J(0) = 1$. In Eq. (1.35), $\Lambda(x) \equiv \text{Tr}(\partial \dot{x}/\partial x)$ is the so-called *phase-space compressibility* of the dynamical system at x. In general, $\Lambda(x)$ is nonzero, and the Jacobian consequently differs from unity. But for Newtonian dynamics, Liouville's theorem guarantees that the volume of comoving phase-space elements is conserved [46]. The Jacobian is thus unity, and P_{acc} has a particularly simple form:

$$P_{acc}[x^{(o)}(\mathcal{T}) \to x^{(n)}(\mathcal{T})] = h_A[x_0^{(n)}]h_B[x_{\mathcal{T}}^{(n)}]\min\left\{1, \frac{\rho[x_0^{(n)}]}{\rho[x_0^{(o)}]}\right\} \quad (1.36)$$

Remarkably, even when a stationary distribution does not exist, the acceptance probability for Newtonian shooting moves depends only on the relative weights of the initial conditions, and on the reactivity of the trial trajectory.

2. Selecting Phase-Space Displacements

For the shooting algorithm we have described, acceptance probabilities are particularly simple if phase-space modifications have a symmetric generation probability. If an asymmetry is present, that is, $p_{gen}[x_{t'}^{(o)} \to x_{t'}^{(n)}] \neq p_{gen}[x_{t'}^{(n)} \to x_{t'}^{(o)}]$, and is not accounted for in the acceptance probability, the trajectories harvested by path sampling will not be correctly weighted. As indicated above, modifying states in a symmetric way can be challenging when constraints are present. A general formal procedure for taking into account linear constraints on momenta was presented in the Appendix of [10]. Here, we describe straightforward procedures for displacing momenta symmetrically, while satisfying common constraints such as fixed total linear or angular momentum, or rigid intramolecular bonds. These simple procedures are equivalent to the more general methods presented in [10]. For clarity, we will discuss this issue using proton transfer in the protonated water trimer as discussed in Section I as an illustrative example.

Proton transfer in the protonated water trimer has been studied extensively with transition path sampling using empirical and ab initio models [10, 15]. In these studies, shooting moves were implemented by using momentum displacements δp only. Since in the classical limit an isolated cluster evolves at constant energy E according to Newton's equation of motion, the simulations were carried out in the microcanonical ensemble, that is, $\rho(x) \propto \delta[E - H(x)]$. Furthermore, the dynamics conserves the total linear momentum $\mathbf{P}$ and the total angular momentum $\mathbf{L}$. Thus the complete distribution of initial conditions is

$$\rho(x) \propto \delta[E - H(x)]\delta[\mathbf{P}(x)]\delta[\mathbf{L}(x)] \qquad (1.37)$$

where both the total linear momentum and the total angular momentum are assumed to vanish. The delta functions in expression (1.37) take into account the reduced dimensionality of the accessible phase space caused by conservation of these quantities.

To obtain a nonvanishing acceptance probability for shooting moves the perturbation δp used to construct a new shooting point $x_{t'}^{(n)}$ must be consistent with the ensemble of initial conditions. For the cluster at constant energy, total linear momentum, and total angular momentum, the displacement δx must be chosen to conserve these quantities. Furthermore, as mentioned above, it must produce a symmetric path generation probability, which can be accomplished as follows.

First, one selects a momentum displacement δp from a Gaussian distribution that is added to the old momentum vector $p_{t'}^{(o)}$. The total linear momentum of the new momentum p' is set to zero by subtracting $\Sigma_i \mathbf{p}'_i / N$ from all single particle momenta. Next, the total angular momentum $\mathbf{L}' = \Sigma \mathbf{r}_i \times \mathbf{p}'_i$ is set to zero. This can be accomplished with a procedure proposed by Laria et al. [25] and used in simulations of water clusters. For this purpose, one first calculates the angular velocity $\boldsymbol{\omega} = I^{-1}\mathbf{L}'$, where $I \equiv \Sigma_i m_i(\mathbf{r}_i^2 - \mathbf{r}_i\mathbf{r}_i)$ is the inertia tensor. Note that calculation of the angular velocity $\boldsymbol{\omega}$ requires inversion of the inertia tensor. Then, one calculates new momenta

$$\mathbf{p}''_i = \mathbf{p}'_i - m_i \boldsymbol{\omega} \times \mathbf{r}_i \qquad (1.38)$$

Finally, the new momenta are rescaled to obtain the appropriate total kinetic energy. It can be shown that provided the center of mass of the system is located in the origin, the new momenta obey all constraints dictated by the conserved total energy, linear, and angular momentum of the system.

This procedure guarantees that the generation probability for new trajectories is equal to the probability of the reverse move. It therefore explicitly satisfies detailed balance. Similar procedures can be used to remove components of δp violating the constraint of constant bond lengths. Molecules with fixed bond lengths are often used in molecular simulations to eliminate the fast oscillatory motion induced by stiff intramolecular bonds. If the distribution of initial conditions is microcanonical and the dynamics is Newtonian, the acceptance probability for a shooting move generated according to this procedure is

$$P_{\text{acc}}[x^{(o)}(\mathcal{T}) \to x^{(n)}(\mathcal{T})] = h_A[x_0^{(n)}]h_B[x_\mathcal{T}^{(n)}] \tag{1.39}$$

This acceptance probability implies that trajectories are always accepted if they connect regions A and B.

Phase-space displacements obeying the linear constraints of vanishing total linear momentum and total angular momentum, and of fixed bond lengths can be also generated using an iterative procedure such as RATTLE [47].

3. Momentum Rescaling

If the distribution of initial conditions $\rho(x_0)$, such as the canonical distribution, allows for variations of the energy, shooting points with different energies must be created. This can be accomplished by adding a momentum displacement δp chosen from an appropriate distribution to a given momentum $p_{i'}^{(o)}$ without rescaling the momenta to a fixed total energy. Large momentum changes, however, most likely produce large changes in the total energy of the system and therefore lead to a low acceptance probability. This problem can be solved by alternating constant energy shooting moves with moves in which the energy is changed by rescaling the momenta. This approach allows one to control the change in momentum and the change in energy independently.

Such an approach was necessary in recent work by Geissler and Chandler [16] in which they studied the nonlinear response of a polar solvent to electronic transitions of a solute with the transition path sampling method. In this study, nonequilibrium trajectories relaxing from states obtained by inversion of the solute dipole were harvested with a variant of the shooting algorithm. Since such solvation dynamics occurs rapidly, the trajectories of interest are only tens of femtoseconds in duration. Accordingly, large momentum displacements at the shooting point are required to generate new pathways sufficiently different from the old ones.

The energy changing move alternated with constant energy shooting moves described in Section III.A can be carried out as follows. First, a time slice $x_{t'}^{(o)}$ along the old path $x^{(o)}(\mathcal{T})$ is randomly chosen. At time slice $x_{t'}^{(o)}$, the system has a kinetic energy of $K_{t'}^{(o)}$. Then, a new kinetic energy $K_{t'}^{(n)}$ at time t' is selected from a distribution $\psi(K)$. Although any form of $\psi(K)$ can be selected, a smart choice of $\psi(K)$ can increase the efficiency of the simulation considerably. In general, the distribution $\psi(K)$ should be similar to the actual distribution of kinetic energies observed in the system under study. In the study carried out by Geissler and Chandler [16], for example, $\psi(K)$ was chosen to be the distribution of kinetic energies in an equilibrium system in contact with a heat bath. Next, a new time slice $x_{t'}^{(n)}$ is generated by rescaling the old momenta $p_{t'}^{(o)}$ with a factor of $\gamma = (K_{t'}^{(n)}/K_{t'}^{(o)})^{1/2}$ to give the kinetic energy $K_{t'}^{(n)}$. Note that rescaling of the momenta does not affect the total linear momentum and the total angular momentum provided that they vanish.

In the above procedure, shooting points are generated with relative probability,

$$\frac{p_{\text{gen}}[x_{t'}^{(n)} \to x_{t'}^{(o)}]}{p_{\text{gen}}[x_{t'}^{(o)} \to x_{t'}^{(n)}]} = \frac{\psi[K_{t'}^{(o)}]dK_{t'}^{(o)}}{\psi[K_{t'}^{(n)}]dK_{t'}^{(n)}} \times \frac{\delta\left[p_{t'}^{(o)} - \sqrt{\frac{K_{t'}^{(o)}}{K_{t'}^{(n)}}}\,p_{t'}^{(n)}\right]}{\delta\left[p_{t'}^{(n)} - \sqrt{\frac{K_{t'}^{(n)}}{K_{t'}^{(o)}}}\,p_{t'}^{(o)}\right]} \quad (1.40)$$

Due to the presence of the delta function the differential $dK_{t'}^{(o)}$ of the old kinetic energy can be expressed as function of the differential $dK_{t'}^{(n)}$ of the new kinetic energy,

$$dK_{t'}^{(o)} = \sum_\alpha \frac{1}{m_\alpha} p_{\alpha,t'}^{(o)}\,dp_{\alpha,t'}^{(o)} = \left[\frac{K_{t'}^{(o)}}{K_{t'}^{(n)}}\right] \sum_\alpha \frac{1}{m_\alpha} p_{\alpha,t'}^{(n)}\,dp_{\alpha,t'}^{(n)} = \left[\frac{K_{t'}^{(o)}}{K_{t'}^{(n)}}\right] dK_{t'}^{(n)} \quad (1.41)$$

where $p_{\alpha,t'}$ is component α of momentum $p_{t'}$ and m_α is the mass associated with degree of freedom α. By using this result and by applying Eq. (1.32) to simplify the ratio of delta functions in Eq. (1.40), we obtain

$$\frac{p_{\text{gen}}[x_{t'}^{(n)} \to x_{t'}^{(o)}]}{p_{\text{gen}}[x_{t'}^{(o)} \to x_{t'}^{(n)}]} = \frac{\psi[K_{t'}^{(o)}]}{\psi[K_{t'}^{(n)}]} \times \left[\frac{K_{t'}^{(o)}}{K_{t'}^{(n)}}\right]^{f/2-1}. \quad (1.42)$$

where f is the number of independent degrees of freedom. In determining f, the number of constraints acting on the system must be properly taken into account.

Assuming the dynamics conserves phase-space volume and correctly taking into account the asymmetric generation probability from the above equation yields an acceptance probability of

$$P_{acc}[x^{(o)}(\mathcal{T}) \to x^{(n)}(\mathcal{T})] = h_A[x_0^{(n)}]h_B[x_{\mathcal{T}}^{(n)}]$$
$$\times \min \left\{ 1, \frac{\rho[x_0^{(n)}]}{\rho[x_0^{(o)}]} \times \frac{\psi[K_{t'}^{(o)}]}{\psi[K_{t'}^{(n)}]} \times \left[\frac{K_{t'}^{(o)}}{K_{t'}^{(n)}}\right]^{f/2-1} \right\} \quad (1.43)$$

Thus, new pathways connecting A and B will be accepted depending on the change in kinetic energy at time t' and on the change of the probability of the initial conditions at time 0.

4. Efficiency of Deterministic Shooting Moves

The efficiency of shooting moves can be controlled by tuning the magnitude of phase-space displacements δx. An optimal magnitude minimizes the correlations between harvested paths. If the magnitude of δx is very small the new trajectory essentially retraces the old one. Although the average acceptance probability is near unity in this case, the random walk in trajectory space will progress slowly because of strong correlations between subsequent trajectories. If, on the other hand, the magnitude of δx is large, the new trajectory will drastically differ from the old one due to the chaotic nature of the underlying dynamics. In this case, the new trajectory most likely will not connect A with B and the resulting acceptance probability is near zero. Pathways are repeated many times before a new path is accepted. Again, correlations between subsequent pathways decay slowly. The optimum magnitude of the perturbation δx lies between these two extremes, which is completely analogous to the situation in conventional Monte Carlo simulations. In that case, the optimum acceptance probability is often near 50% [44].

The effect of the magnitude of δx on the efficiency of transition path sampling can be systematically analyzed by calculating correlation functions of various quantities as a function of the number of simulations cycles. Ideally, such correlation functions decay quickly, indicating that path space is sampled with high efficiency. In [11], we carried out such an efficiency analysis for transition path sampling of isomerizations of a model dimer immersed in a soft sphere liquid. In that study, we calculated correlation functions

$$c(n) \equiv \frac{\langle \delta G(0) \delta G(n) \rangle_{AB}}{\langle \delta G^2 \rangle_{AB}} \quad (1.44)$$

where n is the number of simulation cycles. The fluctuation δG is defined as $\delta G(n) \equiv G(n) - \langle G \rangle_{AB}$ and $G[x(\mathcal{T})]$ is a quantity depending on the path coordinates $x(\mathcal{T})$. The notation $\langle G \rangle_{AB}$ indicates a weighted average of the path functional $G[x(\mathcal{T})]$ over all pathways in the transition path ensemble,

$$\langle G \rangle_{AB} \equiv \int \mathcal{D}x(\mathcal{T}) \mathcal{P}_{AB}[x(\mathcal{T})] G[x(\mathcal{T})] \tag{1.45}$$

Since the sequence of pathways $\{x_i(\mathcal{T})\}$ generated in a transition path sampling procedure visits pathways according to their weight in the transition path ensemble, path averages $\langle G \rangle_{AB}$ can be calculated as averages over the sequence of pathways $\{x_i(\mathcal{T})\}$,

$$\langle G \rangle_{AB} = \lim_{N \to \infty} \frac{1}{N} \sum_i^N G[x_i(\mathcal{T})] \tag{1.46}$$

where N is the total number of simulations cycles. Accordingly, the correlation function $c(n)$ can be expressed as

$$c(n) = \frac{\lim_{N \to \infty} (1/N) \sum_i^N \delta G[x_i(\mathcal{T})] \delta G[x_{(i+n)}(\mathcal{T})]}{\lim_{N \to \infty} (1/N) \sum_i^N \delta G^2[x_i(\mathcal{T})]} \tag{1.47}$$

Correlation functions $c(n)$ are shown in Figure 1.7(a) for three different quantities G: the final region characteristic function $h_B(x_{\mathcal{T}/2})$ at the path midpoint $x_{\mathcal{T}/2}$, the potential energy V_{tr} of the system when the dimer surpasses the potential energy barrier separating the stable states, and the transition time τ. The transition time τ is the time required to reach final region B after the system has left the initial region A. These correlation functions were obtained in a simulation for which the average acceptance probability for shooting moves was $\sim 40\%$.

From the correlation functions $c(n)$ one can compute the number of correlated cycles, n_c by determining at which cycle n the correlation function falls below a certain threshold, say, 0.5. The number n_c of correlated cycles is a measure of the efficiency of the simulation. For the isomerization of the model dimer, we have studied how n_c depends on the magnitude of the momentum displacement. To obtain a more generally usable criterion for the choice of δp, we have plotted n_c as a function of the average acceptance probability P_{acc} of shooting moves. This acceptance probability is monotonically related to the magnitude of the momentum displacement δp. A small δp yields a high acceptance probability and a large δp leads to a small one.

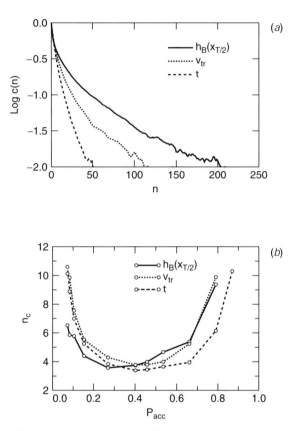

Figure 1.7. (a) Typical correlation functions for the three quantities $h_B(x_{\mathcal{T}/2})$ (solid line), V_{tr} (dotted line), and τ (dashed line) as a function of the number of simulation cycles calculated for isomerizations in a model dimer [11]. (b) Number of correlated cycles as a function of the acceptance probability P_{acc}. The different line types refer to results obtained from analyzing correlation functions of $h_B(x_{\mathcal{T}/2})$ (solid line), V_{tr} (dotted line), and τ (dashed line).

The results of this analysis are depicted in Figure 1.7(b), which shows the number of correlated cycles n_c as a function of the average acceptance probability P_{acc}. The three different curves represent results obtained by analyzing the correlation functions of $h_B(x_{\mathcal{T}/2})$, V_{tr}, and τ. In all three cases, the number of correlated cycles is high for low and high acceptance probabilities and has a minimum for intermediate acceptance probabilities. These results suggest that, as a rule of thumb, the magnitude of the random displacement δx should be chosen to obtain an acceptance probability of $\sim 40\%$. Since rejected moves are computationally less expensive on the average than accepted moves, acceptance probabilities even lower than 40% might be optimal in some cases.

The curves depicted in Figure 1.7 indicate that in transition path sampling a statistically independent trajectory is obtained after a small number (say three to five) of accepted trial moves. Thus, the computational cost of harvesting N independent trajectory of length $\mathcal{T}$ is comparable to the effort required to compute a single long trajectory of length $N\mathcal{T}$. This linear scaling in trajectory length $\mathcal{T}$ and number of trajectories N allows application of the transition path sampling method to the simulation of complex, high-dimensional systems, such as chemical reactions in solution or isomerizations of biomolecules.

In some cases, the applicability of current shooting algorithms can be limited by the Lyapunov instability of the underlying dynamics. This mechanical instability leading to chaotic behavior, is best quantified with the so called Lyapunov exponents λ describing the divergence in phase-space of neighboring trajectories. On the average, the separation δ between two initially close points in phase-space grows in time like $\delta \propto \exp(\lambda t)$. Due to this fast exponential divergence, trajectories generated by "shooting" from only slightly different phase-space points become completely different after a relatively short time. In typical simple liquids, Lyapunov exponents are of the order of $\lambda \sim 1\text{ps}^{-1}$ [48]. Accordingly, a perturbation of the order of the computer precision, $\sim 10^{-15}$ for double precision arithmetic (i.e., the smallest possible perturbation on a digital computer), needs of the order of 10 ps to grow to the scale of the liquid structure. Doubling the number of bits in the representation of numbers would increase this time only by a factor of 2. The average shooting acceptance probability of trajectories longer than this characteristic time is therefore low. Accordingly, processes in which the crucial fluctuations take much longer to occur than this critical time are difficult to study with current shooting algorithms [17].

5. *Shooting Moves for Stochastic Dynamics*

Shooting moves are efficient for stochastic dynamics as well. In most aspects, these moves are identical to the shooting moves employed for deterministic dynamics. Some important details, however, are different. These differences will be discussed in this section. The shooting procedure is schematically illustrated in Figure 1.8 and summarized in Scheme 1.3.

Imagine a stochastic path $x^{(o)}(\mathcal{T})$ of length $\mathcal{T}$ starting in region A and ending in region B. One can randomly select a point $x_{t'}^{(o)}$ along this path, initiate a stochastic trajectory starting from $x_{t'}^{(o)}$ and integrate it forward to time $\mathcal{T}$ using the propagation rule corresponding to the underlying dynamics, for example Eq. (1.11) or (1.14). Then the backward segment of the

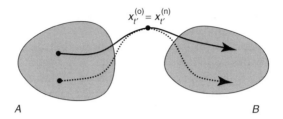

Figure 1.8. In a shooting move for stochastic dynamics forward and backward trajectory segments can be initiated from an unchanged shooting point $x_{t'}^{(n)} = x_{t'}^{(o)}$.

trajectory is generated using the time inverted dynamics satisfying Eq. (1.26). Alternatively, the backward segment can be grown before the forward segment. In contrast to the case of deterministic dynamics, it is not necessary to modify $x_{t'}^{(o)}$ before shooting, because the random nature of the dynamical propagation will alone cause the new path to diverge from the old one. In this case, the generation probability $p_{\text{gen}}[x_{t'}^{(o)} \to x_{t'}^{(n)}]$ is trivially symmetric and the acceptance probability of Eq. (1.28) simplifies to

$$P_{\text{acc}}[x^{(o)}(\mathcal{T}) \to x^{(n)}(\mathcal{T})] = h_A[x_0^{(n)}]h_B[x_{\mathcal{T}}^{(n)}] \qquad (1.48)$$

Thus, any new pathway connecting regions A and B can be accepted.

For Brownian dynamics described in Section II.D, the backward trajectory segment can be generated by integration of the equations of motion with an appropriate small time step algorithm [6]. Before initiating a trajectory, the momenta belonging to the shooting point are inverted. Then, a new trajectory segment of length t' is initiated at the

1. Select a time slice $x_{t'}^{(o)}$ at random from the time slices of an existing path $x^{(o)}(\mathcal{T})$.
2. Compute a new stochastic trajectory segment from t' to $\mathcal{T}$ starting from $x_{t'}^{(o)}$ using the progation rule of the underlying dynamics.
3. The new trajectory segment is rejected if its last state does not lie in region B, that is, if $h_B[x_{\mathcal{T}}^{(n)}] = 0$. Otherwise one proceeds with the next step.
4. If necessary, invert momenta at the selected time slice $x_{t'}^{(n)}$.
5. Compute a stochastic trajectory back to time 0 starting from the configuration with inverted momenta using the propagation rule of the underlying dynamics.
6. Invert the momenta along the newly generated trajectory segment so that the whole path evolves in forward direction.
7. The new trajectory is accepted if its initial state lies in region A, that is, if $h_A[x_0^{(n)}] = 1$.

Scheme 1.3. Shooting algorithm for stochastic trajectories.

shooting point with inverted momenta and is integrated back to time 0. After that, the momenta in this newly generated path segment are reversed such that the whole path evolves in forward direction. Accordingly, the small time step generation probability associated with this type of backward shot is

$$\bar{p}(x \to y) = p(\bar{x} \to \bar{y}) \qquad (1.49)$$

where, again, $\bar{x} = \{r, -p\}$ is obtained from $x = \{r, p\}$ by inverting all momenta. It can be easily shown that this generation probability satisfies condition (1.26) with the canonical distribution as the stationary distribution [6]. Hence, the acceptance probability from Eq. (1.48) is valid for stochastic trajectories generated by numerical solution of Langevin's equation of motion.

In the high friction limit, the inertial term appearing in the Langevin equation can be neglected leading to simplified equations of motion in which the state of the system is completely described by the particle positions alone. In this case, the above procedure can be carried out using the unmodified forward integration algorithm to generate the backward trajectory segment.

As for Langevin dynamics in the high friction limit time does not appear in the Metropolis Monte Carlo transition probability from Eq. (1.14). Forward and backward propagation are therefore indistinguishable and the backward trajectory is generated with the forward propagation rule. Accordingly, the backward generation probability is identical to the forward generation probability, that is, $\bar{p}(x \to y) = p(x \to y)$. Condition (1.26) therefore becomes identical to the detailed balance condition which the Monte Carlo algorithm satisfies by construction,

$$\frac{p(x \to y)}{p(y \to x)} = \exp\{-\beta[H(y) - H(x)]\} \qquad (1.50)$$

Hence, Eq. (1.48) is valid also for Monte Carlo dynamics and a new trajectory can be accepted if it connects region A with region B.

Another important characteristic of shooting moves for stochastic trajectories is that it is not necessary to shoot forward and backward simultaneously, that is, replacing only the forward of backward segment of an old path gives a trial path with significant dynamical weight. In contrast, growing only one of the two segments for deterministic dynamics yields trajectories with zero dynamical weight.

B. Shifting Moves

A shifting move translates an existing pathway forward or backward in time. As depicted in Figure 1.9, this move is graphically similar to the "reptation" motion of a polymer confined to a microscopic tube [49]. In a shifting move, a trial trajectory is obtained by first deleting a segment of length δt from the beginning (or end) of an existing trajectory. A new trajectory segment of length δt is then grown from the opposite end of the old path, by applying the dynamical propagation rules. These operations effectively shift the pathway forward (or backward) in time.

As in the case of shooting moves, the generation and acceptance rules for shifting moves are quite simple. The generation probability for a forward shifting move is just the dynamical weight of the newly generated trajectory segment appended to the end of the path:

$$P_{\text{gen}}^{\text{f}}[x^{(\text{o})}(\mathcal{T}) \to x^{(\text{n})}(\mathcal{T})] = \prod_{i=(\mathcal{T}-\delta t)/\Delta t}^{\mathcal{T}/\Delta t - 1} p[x_{i\Delta t}^{(\text{n})} \to x_{(i+1)\Delta t}^{(\text{n})}] \quad (1.51)$$

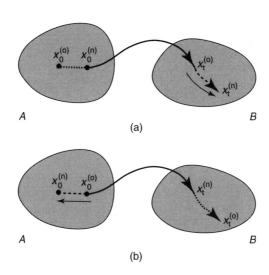

Figure 1.9. In a forward shifting move as shown in panel (*a*), a new trajectory is generated by removing $\delta t/\Delta t$ time slices (dotted line) from the beginning of the old path $x^{(\text{o})}(\mathcal{T})$ and regrowing the same number of time slices (dashed line) at the end of the old path. If the time interval δt is small, most part of the new trajectory coincides with the old one and only a small segment of the new trajectory must be integrated. Panel (*b*) shows a backward shifting move in which $\delta t/\Delta t$ time slices (dotted line) are removed from the end of the old path and a new path segments if grown backward starting from the beginning of the old pathway (dashed line).

Here, it is understood that for times $t' = 0$ through $t' = t - \delta t$, the new path is identical to a portion of the old path,

$$x_{i\Delta t}^{(n)} = x_{i\Delta t + \delta t}^{(o)} \quad \text{for} \quad i = 0, \ldots, (\mathcal{T} - \delta t)/\Delta t \tag{1.52}$$

In a backward shifting move, a new trajectory segment affixed to the beginning of the path is obtained by time-reversed propagation. Correspondingly, the dynamical weight of this segment is composed of time-reversed transition probabilities $\bar{p}(x \to y)$. The generation probability for backward shifting is thus

$$P_{\text{gen}}^{b}[x^{(o)}(\mathcal{T}) \to x^{(n)}(\mathcal{T})] = \prod_{i=1}^{\delta t/\Delta t} \bar{p}[x_{i\Delta t}^{(n)} \to x_{(i-1)\Delta t}^{(n)}] \tag{1.53}$$

where it is understood that

$$x_{i\Delta t + \delta t}^{(n)} = x_{i\Delta t}^{(o)} \quad \text{for} \quad i = 0, \ldots, (\mathcal{T} - \delta t)/\Delta t \tag{1.54}$$

An appropriate acceptance probability for shifting moves is most easily obtained by requiring that forward and backward moves are performed with equal frequency. We also require that the shifting length, δt, which is required to be a multiple of Δt, is drawn from the same distribution for forward and backward moves. With these restrictions, it is sufficient that detailed balance is satisfied for each pair of forward and backward moves

$$\mathcal{P}_{AB}[x^{(o)}(\mathcal{T})]P_{\text{acc}}^{f}[x^{(o)}(\mathcal{T}) \to x^{(n)}(\mathcal{T})]P_{\text{gen}}^{f}[x^{(o)}(\mathcal{T}) \to x^{(n)}(\mathcal{T})]$$
$$= \mathcal{P}_{AB}[x^{(n)}(\mathcal{T})]P_{\text{acc}}^{b}[x^{(n)}(\mathcal{T}) \to x^{(o)}(\mathcal{T})]P_{\text{gen}}^{b}[x^{(n)}(\mathcal{T}) \to x^{(o)}(\mathcal{T})] \tag{1.55}$$

By substituting Eqs. (1.51) and (1.53) into Eq. (1.55), we obtain

$$\frac{P_{\text{acc}}^{f}[x^{(o)}(\mathcal{T}) \to x^{(n)}(\mathcal{T})]}{P_{\text{acc}}^{b}[x^{(n)}(\mathcal{T}) \to x^{(o)}(\mathcal{T})]}$$
$$= h_A[x_0^{(n)}]h_B[x_{\mathcal{T}}^{(n)}] \times \frac{\rho[x_0^{(n)}]}{\rho[x_0^{(o)}]} \times \prod_{i=0}^{\delta t/\Delta t - 1} \frac{\bar{p}[x_{(i+1)\Delta t}^{(o)} \to x_{i\Delta t}^{(o)}]}{p[x_{i\Delta t}^{(o)} \to x_{(i+1)\Delta t}^{(o)}]} \tag{1.56}$$

where many terms have canceled since $(\mathcal{T} - \delta t)/\Delta t$ time slices of the new and the old path are identical (but have different indices). Further cancelation is due to the identity of path probability of the new path and the generation probability associated with the newly grown part of the new path.

As is the case of shooting moves, this ratio simplifies considerably if $p(x \to y)$ and $\bar{p}(y \to x)$ are related by the microscopic reversibility criterion from Eq. (1.26). In this case, satisfactory acceptance probabilities for forward and backward shifting moves are

$$P^{\text{f}}_{\text{acc}}[x^{(\text{o})}(\mathcal{T}) \to x^{(\text{n})}(\mathcal{T})] = h_A[x_0^{(\text{n})}]h_B[x_{\mathcal{T}}^{(\text{n})}]\min\left\{1, \frac{\rho[x_0^{(\text{n})}]}{\rho[x_0^{(\text{o})}]} \times \frac{\rho_{\text{st}}[x_0^{(\text{o})}]}{\rho_{\text{st}}[x_0^{(\text{n})}]}\right\}, \tag{1.57}$$

and

$$P^{\text{b}}_{\text{acc}}[x^{(\text{n})}(\mathcal{T}) \to x^{(\text{o})}(\mathcal{T})] = h_A[x_0^{(\text{o})}]h_B[x_{\mathcal{T}}^{(\text{o})}]\min\left\{1, \frac{\rho[x_0^{(\text{o})}]}{\rho[x_0^{(\text{n})}]} \times \frac{\rho_{\text{st}}[x_0^{(\text{n})}]}{\rho_{\text{st}}[x_0^{(\text{o})}]}\right\}, \tag{1.58}$$

where we have used the fact that $x_{\delta t}^{(\text{o})} = x_0^{(\text{n})}$. If, further, the distribution of initial conditions $\rho(x)$ is identical to the stationary distribution $\rho_{\text{st}}(x)$, Eqs. (1.57) and (1.58) become

$$P^{\text{f}}_{\text{acc}}[x^{(\text{o})}(\mathcal{T}) \to x^{(\text{n})}(\mathcal{T})] = h_A[x_0^{(\text{n})}]h_B[x_{\mathcal{T}}^{(\text{n})}], \tag{1.59}$$

and

$$P^{\text{b}}_{\text{acc}}[x^{(\text{n})}(\mathcal{T}) \to x^{(\text{o})}(\mathcal{T})] = h_A[x_0^{(\text{o})}]h_B[x_{\mathcal{T}}^{(\text{o})}] \tag{1.60}$$

From these results, it follows that shifting should be accepted provided the trial path $x^{(\text{n})}(\mathcal{T})$ connects states A and B. The average acceptance probability of a shifting move can be easily controlled by changing the magnitude of the time interval δt.

Short shifting moves are computationally inexpensive and facilitate convergence of quantities averaged over the transition path ensemble. However, shifting moves are essentially orthogonal to shooting moves; alone, small shifting moves do not change the portions of trajectories near the transition state region. Therefore shifting moves do not sample trajectory space ergodically and must be used in conjunction with other trial moves, such as shooting moves.

Shifting moves are slightly different for stochastic and deterministic dynamics. First, we discuss the application of the shifting algorithm to systems evolving deterministically and then discuss shifting moves for stochastic dynamics.

1. Shifting Moves for Deterministic Dynamics

In a forward shifting move of length δt for deterministic dynamics, the new trajectory segment is generated by integrating the equations of motion of the system for $\delta t/\Delta t$ time steps starting from the endpoint $x_{\mathcal{T}}^{(o)}$ of the old path. Similarly, the backward shifting move is carried out by integrating the equations of motion for $\delta t/\Delta t$ time steps backward in time starting from $x_0^{(o)}$. Accordingly, the single step generation probabilities for the forward and the backward move are

$$p(x \to y) = \delta[y - \phi_{\Delta t}(x)] \qquad (1.61)$$

and

$$\bar{p}(y \to x) = \delta[x - \phi_{-\Delta t}(y)] \qquad (1.62)$$

respectively. As shown in Section III.A.1, the generation probabilities $p(x \to y)$ and $\bar{p}(y \to x)$ associated with this type of procedure satisfy the reversibility condition expressed in Eq. (1.26). In equilibrium, the acceptance probability for a shifting move for deterministic trajectories is therefore simply given by Eqs. (1.59) and (1.60), that is, the new pathway obtained by shifting the old one in time can be accepted whenever it connects A with B. The shifting algorithm is summarized in Scheme 1.4.

Shifting moves for deterministic trajectories are particularly inexpensive when carried out in long uninterrupted sequences. In this case, the pathway is constrained to a one-dimensional (1D) manifold defined by the equations

Forward Shifting

1. Randomly select a time interval δt from a distribution $w(\delta t)$.
2. Copy the $(\mathcal{T} - \delta t)/\Delta t$ last time slices of the old path to the first $(\mathcal{T} - \delta t)/\Delta t$ time slices of the new path, that is, $x_{i\Delta t + \delta t}^{(n)} = x_{i\Delta t}^{(o)}$ for $i = 0, \ldots, (\mathcal{T} - \delta t)/\Delta t$.
3. Integrate the equations of motion forward for $\delta t/\Delta t$ time steps starting from $x_{\mathcal{T} - \delta t}^{(n)}$.
4. Accept the new path $x^{(n)}(\mathcal{T})$ if it is reactive and reject it otherwise.

Backward Shifting

1. Randomly select a time interval δt from a distribution $w(\delta t)$.
2. Copy the $(\mathcal{T} - \delta t)/\Delta t$ first time slices of the old path to the last $(\mathcal{T} - \delta t)/\Delta t$ time slices of the new path, that is, $x_{i\Delta t}^{(n)} = x_{i\Delta t + \delta t}^{(o)}$ for $i = 0, \ldots, (\mathcal{T} - \delta t)/\Delta t$.
3. Integrate the equations of motion backward for $\delta t/\Delta t$ time steps starting from $x_{\delta t}^{(n)}$.
4. Accept the new path $x^{(n)}(\mathcal{T})$ if it is reactive and reject it otherwise.

Scheme 1.4. Shifting algorithm for deterministic trajectories.

of motion. By storing states along this manifold as they are obtained, shifting moves can eventually be performed with no cost, since the time evolution of new trajectory segments is already known.

If no stationary distribution $\rho_{st}(x)$ exists or if it is unknown, a different approach must be taken. For this purpose, we return to Eq. (1.56) and insert the forward and backward generation probabilities from Eqs. (1.61) and (1.62). Using Eq. (1.32) and applying the chain rule for products of Jacobians, we finally obtain the following expression for the acceptance probability of a forward shifting move,

$$P_{acc}^{f}[x^{(o)}(\mathcal{T}) \to x^{(n)}(\mathcal{T})] = h_A[x_0^{(n)}]h_B[x_{\mathcal{T}}^{(n)}] \min\left\{1, \frac{\rho[x_0^{(n)}]}{\rho[x_0^{(o)}]} \times \left|\frac{\partial x_{\delta t}^{(o)}}{\partial x_0^{(o)}}\right|\right\} \quad (1.63)$$

For a new trajectory $x^{(n)}(\mathcal{T})$ generated with a backward shifting move from the old trajectory $x^{(o)}(\mathcal{T})$, the acceptance probability is slightly different. By using analogous manipulations one obtains

$$P_{acc}^{b}[x^{(o)}(\mathcal{T}) \to x^{(n)}(\mathcal{T})] = h_A[x_0^{(n)}]h_B[x_{\mathcal{T}}^{(n)}] \min\left\{1, \frac{\rho[x_0^{(n)}]}{\rho[x_0^{(o)}]} \times \left|\frac{\partial x_{\delta t}^{(n)}}{\partial x_0^{(n)}}\right|^{-1}\right\}$$

(1.64)

If the dynamics is phase-space volume conserving, that is, $|\partial x_t / \partial x_0| = 1$, such as for Newtonian dynamics, the acceptance probability for both forward and backward shifting moves reduces to

$$P_{acc}[x^{(o)}(\mathcal{T}) \to x^{(n)}(\mathcal{T})] = h_A[x_0^{(n)}]h_B[x_{\mathcal{T}}^{(n)}] \min\left\{1, \frac{\rho[x_0^{(n)}]}{\rho[x_0^{(o)}]}\right\}. \quad (1.65)$$

In this case, a new trajectory obtained with a shifting move is accepted with a probability of $\min\{1, \rho[x_0^{(n)}]/\rho[x_0^{(n)}]\}$ provided the new trajectory connects A with B.

2. Shifting Moves for Stochastic Dynamics

Analogous shifting moves can be carried out also for stochastic trajectories. In a forward reptation move for stochastic trajectories time slices removed from one end of the path are regrown on the other using the integration rules of the underlying dynamics. In a backward reptation move, on the other hand, a new path segment is grown backward using the same procedures used to generate the backward trajectory of a shooting move. As was shown in Section III.A, generation probabilities associated with these

Forward Shifting

1. Randomly select a time interval δt from a distribution $w(\delta t)$.
2. Copy the $(\mathcal{T} - \delta t)/\Delta t$ last time slices of the old path to the first $(\mathcal{T} - \delta t)/\Delta t$ time slices of the new path, that is, $x^{(n)}_{i\Delta t + \delta t} = x^{(o)}_{i\Delta t}$ for $i = 0, \ldots, (\mathcal{T} - \delta t)/\Delta t$.
3. Generate $\delta t/\Delta t$ new time steps starting from $x^{(n)}_{\mathcal{T} - \delta t}$ using the propagation rule of the underlying stochastic dynamics.
4. Accept the new path $x^{(n)}(\mathcal{T})$ if it is reactive and reject it otherwise.

Backward Shifting

1. Randomly select a time interval δt from a distribution $w(\delta t)$.
2. Copy the $(\mathcal{T} - \delta t)/\Delta t$ first time slices of the old path to the last $(\mathcal{T} - \delta t)/\Delta t$ time slices of the new path, that is, $x^{(n)}_{i\Delta t} = x^{(o)}_{i\Delta t + \delta t}$ for $i = 0, \ldots, (\mathcal{T} - \delta t)/\Delta t$.
3. Invert the momenta belonging to $x^{(n)}_{\delta t}$
4. Generate $\delta t/\Delta t$ time steps backward starting from $\bar{x}^{(n)}_{\delta t}$ by applying the propagation rule corresponding to the underlying stochastic dynamics.
5. Invert the momenta in the newly generated path segment.
6. Accept the new path $x^{(n)}(\mathcal{T})$ if it is reactive and reject it otherwise.

Scheme 1.5. Shifting algorithm for stochastic trajectories.

procedures satisfy the reversibility condition from Eq. (1.26). Therefore, the acceptance probability from Eqs. (1.59) and (1.60) can be used for stochastic reptation moves based on Langevin and Monte Carlo dynamics. According to this acceptance probability a shifting move for stochastic trajectories is always accepted if both the starting point and the end point of the new path lie in their respective stable states. The reptation algorithm for stochastic trajectories is summarized in Scheme 1.5.

C. Memory Requirements

Molecular dynamics simulations proceed in discrete time steps that are comparable to the shortest characteristic time of atomic motions, often ~ 1 fs. A 1-ps pathway is thus represented by $\sim 10^3$ microstates. Storing such pathways in computer memory quickly becomes costly for large systems. Fortunately, it is not necessary for the purposes of transition path sampling to store every microstate belonging to a given pathway. The algorithms we have described are exact even when a path is represented by just a few states. Clearly, though, a shooting move can only be initiated from a microstate that has been stored. For efficient sampling, it is thus advantageous to store states with a time resolution that captures typical fluctuations of the reaction coordinate. In our experience with molecular liquids, storage of states is necessary only about every 10 fs leading to considerable memory savings. Note that such a reduction of memory requirements is not possible for other algorithms such as the local algorithm or the dynamical algorithm discussed in Section III.F. In these cases, it is necessary to maintain full trajectories in memory at all times.

For some systems, storing even just tens of microstates can be burdensome. For example, in the density functional theory based method of Car and Parrinello [36] the occupied single-particle Kohn–Sham orbitals $\{\psi_i\}$ are propagated in time together with the nuclear positions using a set of fictitious equations of motion. Shooting moves to harvest Car–Parrinello trajectories can be carried out by selecting a time slice along an existing path and changing only the nuclear momenta before computing a new trajectory. Then, the new path is accepted according to the criteria described in the preceding paragraphs. Due to the large amount of memory necessary to store the Kohn–Sham orbitals and their time derivatives, it is, however, not possible to keep a complete pathway consisting of typically 100 copies of the system in memory with current computational resources. Therefore, a few time slices for shooting are selected randomly before calculating the trajectory and a complete set of data including the Kohn–Sham orbitals is stored only at these specific time slices [15]. Shooting moves can then be initiated from the preselected states. In the unlikely case that shooting moves are rejected from each of these states, one regenerates the current path, storing a complete description of the system at a new set of randomly selected times along the path. Such a technique has been used to harvest Car–Parrinello trajectories for proton transfer in the protonated water trimer [15] and for autoionization in liquid water [24].

D. Stochastic Trajectories as Sequences of Random Numbers

In order to use shooting moves with the simple acceptance probability in Eq. (1.29), one must be able to generate forward and backward trajectories that satisfy microscopic reversibility. So far, we have considered several dynamical systems for which this task is straightforward. But there are many systems for which we do not know how to construct time-reversed trajectory segments that satisfy Eq. (1.26). For example, a stochastic system driven by a nonconservative force may converge to a steady state, but the corresponding stationary distribution generally cannot be written down. This class of dynamical phenomena, which may exhibit rare but important fluctuations [50, 51], are in principle amenable to transition path sampling. But a different type of trial move is needed to implement the method. Crooks and Chandler [26] devised such a move.

The basic idea of the Crooks–Chandler algorithm described in [26] is to represent a pathway by the sequence of random numbers, the noise history, used to generate it. In this representation, local changes of the noise history can be employed to generate nonlocal changes of trajectories. To see how this can be achieved, consider a system evolving according to Langevin's equation of motion in the high-friction limit:

$$\dot{r} = \mathscr{F}(r, t) + \mathscr{R}(t) \tag{1.66}$$

where r represents the configuration of the system and $\mathscr{R}$ is a stochastic force with correlations $\langle \mathscr{R}(0)\mathscr{R}(t)\rangle = \varepsilon\delta(t)$. The system is driven out of equilibrium by a systematic force $\mathscr{F}(r,t)$ that is either explicitly time-dependent or nonconservative. The above equations of motion can be integrated over short times by representing $\mathscr{R}$ with a set of random numbers ξ [38]. The appropriate distribution for ξ is Gaussian, with zero mean and a variance ε. Because a random number must be drawn for each degree of freedom, ξ is a vector with the dimensionality of configuration space, f. The sequence of random numbers along a trajectory specifies the noise history $\mathscr{R}(t)$. Accordingly, the probability to observe a certain stochastic trajectory $x(\mathscr{T})$ is proportional to the probabilities $p(\xi)$ of all vectors of random numbers ξ used to generate that trajectory,

$$\mathscr{P}[x_0; \xi(\mathscr{T})] = \rho(x_0) \prod_{i=0}^{\mathscr{T}/\Delta t - 1} p(\xi_{i\Delta t}) = \rho(x_0) \prod_{i=0}^{\mathscr{T}/\Delta t - 1} \frac{1}{(2\pi\varepsilon)^{f/2}} \exp\{-|\xi_{i\Delta t}|^2/2\varepsilon\} \tag{1.67}$$

Here, $|\xi_t|$ is the magnitude of the vector ξ drawn for time t.

This representation of the dynamical path probability emphasizes specific realizations of the random force $\mathscr{R}$. It enables an interesting class of trial moves in trajectory space: A new path $x^{(n)}(\mathscr{T})$ is obtained from an old path $x^{(o)}(\mathscr{T})$ by replacing the noise $\xi_{t'}^{(o)}$ at one randomly chosen time t' by a new set of random numbers $\xi_{t'}^{(n)}$ drawn from the distribution $p(\xi)$. The noise histories of the new and the old path are identical except at time t', and the new path coincides with the old one from time 0 to time t'. From time $t' + \Delta t$ to time $\mathscr{T}$, the new path is obtained by integrating the equation of motion with the modified noise history. All new time slices after time t' will differ from those of the old path. A local modification in noise space thus generates a small but global move in trajectory space.

Due to the dissipative nature of these dynamics, trajectories initiated from two nearby points converge rapidly if they possess the same noise history. As a consequence, a new trajectory generated from a reactive old trajectory has a high probability to be reactive as well. Such moves in noise space therefore have a high acceptance probability. This algorithm is summarized in Scheme 1.6.

The generation probability for trial paths obtained from this type of move is

$$P_{\text{gen}}[x^{(o)}(\mathscr{T}) \to x^{(n)}(\mathscr{T})] = p(\xi_{t'}^{(n)}) \tag{1.68}$$

1. Randomly select a time slice $x_{t'}^{(o)}$ on the existing trajectory $x^{(o)}(\mathcal{T})$.
2. Replace the set of random numbers $\xi_{t'}^{(o)}$ belonging to time slice $x_{t'}^{(o)}$ by a new set of random numbers $\xi_{t'}^{(n)}$ drawn from a Gaussian distribution with the appropriate mean and variance.
3. Determine the new path $x^{(n)}(\mathcal{T})$ from $t' + \Delta t$ to $\mathcal{T}$ using the modified set of random numbers to integrate the equations of motion.
4. Accept the new path if the final point $x_{\mathcal{T}}^{(n)}$ is in region B.

Scheme 1.6. Crooks–Chandler noise history algorithm for stochastic trajectories.

where $\xi_{t'}^{(n)}$ is the new random number at time t'. By inserting this generation probability and the path probability from Eq. (1.67) into Eq. (1.21), we obtain a corresponding acceptance probability

$$P_{\text{acc}}[x^{(o)}(\mathcal{T}) \to x^{(n)}(\mathcal{T})] = h_A[x_0^{(n)}]h_B[x_{\mathcal{T}}^{(n)}]$$

$$\times \min\left\{1, \frac{\rho[x_0^{(n)}]\prod_{i=0}^{\mathcal{T}/\Delta t - 1} p[\xi_{i\Delta t}^{(n)}]}{\rho[x_0^{(o)}]\prod_{i=0}^{\mathcal{T}/\Delta t - 1} p[\xi_{i\Delta t}^{(o)}]} \times \frac{p[\xi_{t'}^{(o)}]}{p[\xi_{t'}^{(n)}]}\right\} \quad (1.69)$$

Most factors in the second argument of the min-function cancel because the noise histories of the new and the old trajectories differ only at time t'. Because such a move does not modify the initial time slice of the path $[x_0^{(n)} = x_0^{(o)}]$, Eq. (1.69) simplifies to

$$P_{\text{acc}}[x^{(o)}(\mathcal{T}) \to x^{(n)}(\mathcal{T})] = h_B[x_{\mathcal{T}}^{(n)}] \quad (1.70)$$

Thus, a new path obtained by applying local changes in noise space is accepted if it ends in region B.

As described, this algorithm harvests an ensemble of trajectories with identical initial states x_0, and is therefore incomplete. But sampling the distribution of initial conditions is in fact very simple for the system considered, since $\rho(x) = \rho_{\text{st}}(x)$. One need only apply dynamical propagation rules [with a randomly selected realization of $\mathcal{R}(t)$] to an existing initial state. States along the resulting trajectory are correctly weighted by $\rho(x)$. In order to make the algorithm complete, these new initial states are occasionally used to grow trial trajectories with an existing noise history. Again, the dissipative nature of these dynamics ensures that such a trial trajectory has a reasonable chance of exhibiting the rare fluctuation of interest. If the path does satisfy the desired boundary conditions, it is accepted with unit probability. With this complete algorithm, Crooks and Chandler [26] have successfully sampled large fluctuations in nonequilibrium systems that were previously inaccessible.

E. Other Algorithms

Preceding sections have described the most efficient algorithms we have devised to sample transition paths in complex systems. Other algorithms, while sound in principle, are far less efficient than shooting and shifting. For completeness, we discuss these less efficient algorithms briefly.

1. Local Algorithm for Stochastic Pathways

It is possible to sample stochastic transition pathways by making only local displacements of trajectories. For example, a randomly chosen time slice $x_{t'}^{(o)}$ of an existing pathway may be modified by adding a small displacement δx to positions and momenta, $x_{t'}^{(n)} = x_{t'}^{(o)} + \delta x$. All other time slices remain unchanged. This modification, which is local in time, gives a different but finite path probability $\mathscr{P}_{AB}[x(\mathscr{T})]$. If the displacement δx is chosen from a symmetric distribution, $w(\delta x) = w(-\delta x)$, then the generation probability for the new path is identical to that for the reverse move. According to Eq. (1.21), the new path should be accepted with probability

$$P_{\text{acc}}[x^{(o)}(\mathscr{T}) \to x^{(n)}(\mathscr{T})] = \min\left\{1, \frac{p[x_{t'-\Delta t}^{(o)} \to x_{t'}^{(n)}]p[x_{t'}^{(n)} \to x_{t'+\Delta t}^{(o)}]}{p[x_{t'-\Delta t}^{(o)} \to x_{t'}^{(o)}]p[x_{t'}^{(o)} \to x_{t'+\Delta t}^{(o)}]}\right\} \quad (1.71)$$

where we have assumed that an intermediate time slice is modified, $0 < t' < \mathscr{T}$. Acceptance probabilities for displacing initial and final time slices are somewhat different [5].

The local algorithm suffers from several shortcomings. The most serious of these is its inefficiency. In contrast to shooting and shifting moves, local moves sample phase space with a rate proportional to L^3, where L is the number of time slices in a path [52]. This unfavorable scaling makes the local algorithm impractical for all but the simplest problems. Furthermore, all time steps of a path must be stored in computer memory, a burdensome task for high-dimensional systems. Finally, the local algorithm cannot be applied to deterministic trajectories. To date, we have only been able to sample deterministic trajectories with shooting and shifting moves.

A local algorithm for Metropolis Monte Carlo trajectories must be constructed carefully. Due to the finite probability of exactly repeated states in these paths, the corresponding transition probability includes a singular term [see Eq. (1.14)]. The generation algorithm for local path moves must take this singularity into account properly. Appropriate acceptance probabilities are given in [5]. (H. C. Andersen has drawn our attention to an omission in [5]. In Metropolis Monte Carlo trajectories sequences of multiple rejections can occur. Attempts to modify time slices in the interior of such a sequence always lead to rejections. More specifically, Eqs. (17) and (18) of [5] must be modified to include sequences of the form $r_{\tau-1} = r_\tau = r_{\tau+1}$ (in the notation of [5]). In that case, the acceptance probability for a local move from r_τ to r'_τ vanishes, that is, $P_{\text{acc}}[r_\tau \to r'_\tau] = 0$.)

2. Dynamical Algorithm

For stochastic pathways, such as those generated according to Langevin's equation of motion, the path probability functional $\mathscr{P}_{AB}[x(\mathscr{T})]$ can be written as

$$\mathscr{P}_{AB}[x(\mathscr{T})] \equiv \exp(-S_{AB}[x(\mathscr{T})]), \tag{1.72}$$

where the functional $S_{AB}[x(\mathscr{T})]$ depending on all path coordinates $x(\mathscr{T})$ is called the path "action". The form of Eq. (1.72), similar to the canonical distribution function $\exp[-\beta V(r)]$, suggests that we sample the path distribution $\mathscr{P}_{AB}[x(\mathscr{T})]$ with dynamical methods based on generalizations of Newton's equation of motion. For this purpose, path space is extended to include also "momenta" $y(\mathscr{T}) \equiv M\dot{x}(\mathscr{T})$ of the path coordinates $x(\mathscr{T})$. Here, M is an artificial mass associated with the path coordinates and the dot indicates a derivative with respect to an artificial time θ. This new space $\{x(\mathscr{T}), y(\mathscr{T})\}$ has twice the dimensionality of the original path space. One then defines a path Hamiltonian functional $H_P[x(\mathscr{T}), y(\mathscr{T})]$ by adding a "kinetic energy" to the path action,

$$H_P[x(\mathscr{T}), y(\mathscr{T})] \equiv \sum_\alpha \frac{y_\alpha^2}{2M} + S[x(\mathscr{T})] \tag{1.73}$$

From this Hamiltonian functional one can derive a set of equations of motion capable of moving the path $x(\mathscr{T})$ through path space in artificial time θ,

$$\frac{dx_t}{d\theta} = \frac{\partial H_P}{\partial y_t} = \frac{y_t}{M}$$

$$\frac{dy_t}{d\theta} = -\frac{\partial H_P}{\partial x_t} = -\frac{\partial S_{AB}[x(\mathscr{T})]}{\partial x_t} \tag{1.74}$$

If these equations are equipped with an appropriate thermostat, for example, a stochastic Andersen thermostat [53] or a deterministic Nosé–Hoover thermostat [37], the resulting distribution in path space is consistent with $\exp(-S_{AB}[x(\mathscr{T})]) = \mathscr{P}_{AB}[x(\mathscr{T})]$. Alternatively, a stochastic Langevin equation may be employed to drive the path through path space.

The forces necessary to integrate the artificial equations of motion are obtained by differentiating $S_{AB}[x(\mathcal{T})] = -\ln \mathcal{P}_{AB}[x(\mathcal{T})]$ with respect to the path coordinates x_t. Since $\mathcal{P}_{AB}[x(\mathcal{T})]$ is a product of transition probabilities, the distribution of initial conditions and the boundary conditions constraining the path to start in A and end in B, the path action $S_{AB}[x(\mathcal{T})]$ is a sum of terms originating from the different contributions to the path probability. The boundary conditions $h_A(x_0)$ and $h_B(x_{\mathcal{T}})$ act as hard walls confining the endpoints of the path to regions A and B, respectively. These hard walls require special attention when integrating the equations of motion (1.74).

Although correct in principle, dynamical path sampling algorithms are far inferior to shifting and shooting algorithms in terms of efficiency. One reason for this deficiency is that due to the strong coupling between subsequent time slices especially for low friction very small time steps must be used in order to reproduce the associated high-frequency oscillations in the path dynamics correctly. Thus, path motion in trajectory space proceeds slowly. The sampling rate is further decreased by the necessity of computing second derivatives of the potential in order to determine forces acting on the path variables. Such calculations can be onerous especially in the case of first principles molecular dynamics simulations. For these reasons dynamical algorithms have not been applied to processes occurring in complex systems so far.

3. *Configurational Bias Monte Carlo*

In its original form, the configurational bias Monte Carlo algorithm is a method to sample polymers in the melt in an efficient way [54]. The basic idea of the algorithm is to regrow entire polymers in a biased fashion in order to avoid unfavorable overlaps. Due to the formal similarity between polymers and stochastic pathways a configurational bias Monte Carlo scheme can be constructed to sample transition pathways [5]. To date, this algorithm has been only used to study the kinetics of hydrogen bonds in liquid water [8].

Also, algorithms familiar from path integral simulations [52], such as the staging algorithm [55], can be used to sample the transition path ensemble for stochastic trajectories. These algorithms are, however, less efficient than the shooting and shifting algorithms, which are the most efficient transition path sampling algorithms found so far.

F. Parallel Tempering

The transition path ensemble includes all pathways that make successful transitions on the time scale of the path length $\mathcal{T}$. For complicated energy surfaces, however, it may be difficult to sample all statistically relevant pathways. For example, proton transfer occurs in the protonated water trimer via two different classes of pathways [15]. Switching from one class of transition pathways to the other is hindered by a high energetic barrier. One might expect similar problems in sampling protein folding pathways for physiologically relevant temperatures. Such ergodicity problems are akin to those encountered in Monte Carlo simulations of supercooled liquids and glasses, where locally stable states are separated by high energy barriers. Various techniques have been developed to overcome these problems in the context of conventional Monte Carlo simulations, including J-walking [56], multicanonical sampling [57] and parallel tempering [58–60]. In principle, all these methods can be generalized to improve sampling of the transition path ensemble. Vlugt and Smit [23] recently showed that parallel tempering is especially suitable for transition path sampling.

In a conventional parallel tempering calculation, several simulations are performed in parallel at different temperatures. One allows exchange of configurations between different temperature levels with a probability satisfying detailed balance. At low temperatures, the system is confined to the vicinity of a local energy minimum, but at high temperatures the system can surmount energetic barriers readily. With parallel tempering, the system is able to move efficiently through configuration space, while sampling the statistically most important, low energy regions with correct probability.

Parallel tempering can be used to sample transition pathways analogously. Pathways are harvested in parallel at N different temperatures $T_1, T_2, \ldots, T_N$. Shooting and shifting moves are complemented with exchange of pathways between different temperature levels. Appropriate acceptance rules satisfying detailed balance can be easily derived [23]. Parallel tempering not only improves sampling efficiency at each temperature level, but also provides temperature-dependent results at basically no extra cost. The structure of this algorithms is ideal for implementation on massively parallel computers.

To test parallel tempering in conjunction with transition path sampling, we have applied it to a simple toy model. In this two dimensional (2D) system (see Fig. 1.10), a "molecule" is immersed in a fluid of purely repulsive disks. The molecule consists of three atoms bonded with harmonic springs that repel each other at short distance. All atoms have the same mass and size and the system evolves according to Newton's equation of motion. Initial conditions are weighted by a canonical distribution. The triatomic molecule can reside in two stable states: state A, in which atoms 1, 2, and 3 are arranged in a clockwise manner (as in Fig. 1.4), and state B, in which the arrangement is counterclockwise. The two stable states are distinguished

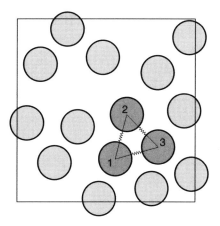

Figure 1.10. A molecule (dark gray) of three 2D atoms with short-range repulsion and connected with harmonic springs is immersed in a fluid of purely repulsive disks. The molecule can reside in two long lived enantiomers.

with the order parameter $q \equiv (y_{12}x_{13} - x_{12}y_{13})/r_{12}$, where $\mathbf{r}_{13} \equiv (x_2 - x_1, y_2 - y_1)$ and $\mathbf{r}_{13} \equiv (x_3 - x_1, y_3 - y_1)$ are the vectors from atom 1 to atom 2 and from atom 1 to atom 3, respectively. The order parameter q measures the position of particle 3 with respect to particles 1 and 2.

Fluid atoms act as a heath bath for the triatomic molecule, providing activation energy for transitions between states A and B and dissipation to stabilize the molecule following a transition. Three different pathways connect states A and B. In each of them, a different atom of the molecule squeezes its way through the gap formed by the other two atoms. In each of the transition states, the three atoms are collinear with a different atom at the center.

We have studied transitions between states A and B with transition path sampling in the canonical ensemble. For the parameters used in the simulation, the energetic barrier separating states A and B has a height of $\sim 22\varepsilon$ in Lennard–Jones units. We used a total of 16 particles (including the atoms of the molecule) with periodic boundary conditions. If the sampling is ergodic, that is, if all relevant regions in the space of all transitions paths are visited, one expects that the three classes of pathways occur with the same frequency. However, such ergodic sampling is observed only at temperatures larger than $k_B T \simeq 1.0\varepsilon$. For lower temperatures, the sampling is effectively confined to one of the three transition states, even for simulations of $> 10^6$ cycles. Results of these simulations are shown in Figure 1.11.

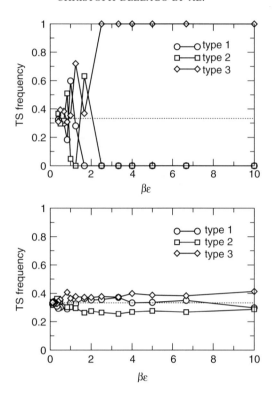

Figure 1.11. Fraction of transition pathways of type 1, 2, and 3 found in the simulations as a function of $\beta\varepsilon = \varepsilon/k_B T$. The panel on the top shows the results obtained without parallel tempering. The bottom panel depicts the results of parallel tempering simulations with 20 temperature levels.

These results demonstrate that parallel tempering can enhance sampling considerably. Without parallel tempering severe sampling problems occur for $\beta \equiv 1/k_B T > 1/\varepsilon$. With parallel tempering, all path space regions are visited with the correct probability.

G. Generating an Initial Path

The transition path sampling techniques we have described assume that an initial reactive pathway is available. Generating such a pathway is therefore an important step in applying the method. In the simplest cases, a trajectory connecting A and B can be obtained by running a long molecular dynamics (or stochastic dynamics) simulation. For most applications, however, the rarity of transitions rules out this straightforward approach, and an initial trajectory must be procured by other means.

The path used to initiate sampling need not be representative of the transition path ensemble. This fact is a particular strength of our methodology. In fact, the initial path need not even be a true dynamical pathway. Any chain of states connecting A and B is in principle sufficient to begin transition path sampling. Subsequent importance sampling will naturally guide pathways to the most important regions of trajectory space. This "equilibration" is directly analogous to the initial relaxation of conventional Monte Carlo simulations. Suppose, for example, one wants to harvest configurations of an atomic liquid. If a typical state of the liquid is unavailable, one may begin by placing atoms on the sites of a periodic lattice. Such a regular arrangement is clearly atypical of the liquid phase. But importance sampling will quickly carry the fluid from this unlikely configuration to more representative states. In a similar way, transition path sampling efficiently carries reactive pathways from unlikely regions of trajectory space to those representative of natural transitions.

An artificial chain of states connecting basins of attraction can be constructed in various ways. In fact, the many methods that focus on nondynamical transition paths may be useful for this task [28, 30, 31]. The more closely an initial pathway resembles natural reactive trajectories, the more quickly transition path sampling will reach the important regions of trajectory space. We describe here a few useful techniques for generating reactive chains of states.

One systematic way to generate an initial pathway is to gradually transform an ensemble of all trajectories starting in A to one including only those starting in A and ending in B. As discussed in Section IV, such a conversion of path ensembles is computationally demanding. But if the transformation is performed quasistatically, the resulting pathway will be truly representative of the transition path ensemble. In this case, no equilibration of the initial path is needed.

High-temperature pathways can also be used to initiate transition path sampling. Consider, for example, the folding of a protein. At physiological temperatures, folding and unfolding occur very rarely on a molecular timescale. But at higher temperature, unfolding is sufficiently facile that it can be simulated with straightforward molecular dynamics [61]. Such a high-temperature trajectory can be used to begin sampling protein folding pathways at more relevant temperatures. In some cases, high-temperature transition pathways are qualitatively different from those at lower temperatures. It may then be necessary to carry out a systematic cooling procedure, in which the ensemble of pathways is brought to lower temperatures in small steps.

Finally, a presumed notion of a reaction coordinate can sometimes be used to acquire an reactive pathway. By controlling such a variable, it may be possible to drive the system from one stable state to another. If this procedure is successful, shooting and shifting moves can be applied to the resulting chain of states.

In our experience, no general recipe for generating initial trajectories is ideally suited to all situations. It seems best to proceed by experimenting with various schemes. In this process, one often gains useful insight about the relevant basins of attraction and the coordinates that characterize them.

H. Transition Path Sampling with an Existing Molecular Dynamics Program

Transition path sampling based on the shooting and shifting algorithms can be easily implemented using existing molecular dynamics (MD) programs. One might, for example, desire to conduct transition path sampling studies using a commercial molecular dynamics package for which the source code is unavailable. Such a study is most conveniently done by developing a separate path sampling module and interfacing it with the existing MD program through system calls and input–output to and from disk.

In such combination of an existing MD program with a new path sampling module, a deterministic shooting move, for example, can be carried out as follows. In the path sampling module, a time slice is first selected from the existing path that is stored in memory or on disk. Next, the corresponding momentum is changed according to the rules described in Section III.A. and the modified phase-space point is written to a file from which it can later by read by the MD program. Then, a forward shot is carried out by first writing the desired number of integration steps into the input file of the MD program and then starting the MD program via a system call. The MD program integrates the equations of motion and stores the resulting trajectory segment on disk. Note that it is not necessary to store every time step of the new trajectory. Rather, it is sufficient to store the new trajectory only at a smaller number of regularly spaced points. When the trajectory is completed, control returns to the path sampling module that reads the forward trajectory from disk and verifies whether the last state of the new path, $x_{\mathcal{T}}^{(n)}$, lies within region B. If it does not, the move is rejected. Otherwise, one proceeds with the backward part of the shooting move. For this purpose, one inverts the momenta of the time slice from which the forward shot was initiated and the modified time slice is again written to a file. Then, the number of backward integration steps is specified in the appropriate input file and the MD program is again started with a system call. After that the newly generated backward trajectory segment is read by the path

1. Randomly select time slice $x_{t'}^{(o)}$ for shooting.
2. Modify momenta and write time slice $x_{t'}^{(n)}$ to disk.
3. Write number of integration time steps for forward shot into input file of MD-code.
4. Start MD-code with system call.
5. Read newly generated forward trajectory segment from disk.
6. Reject move if new endpoint $x_{\mathcal{T}}^{(n)}$ not in B and proceed with next point otherwise.
7. Invert momenta of time slice $x_{t'}^{(n)}$ and write modified time slice to disk.
8. Write number of integration time steps for backward shot into input file of MD-code.
9. Start MD-code with system call.
10. Read newly generated backward trajectory segment from disk and invert momenta.
11. If new initial point $x_0^{(n)}$ is in A, the path is accepted, and rejected otherwise.

Scheme 1.7. Shooting algorithm with existing molecular dynamics code.

sampling module and the momenta along this segment of the trajectory are inverted such that they point forward in the whole path. The shooting move is completed by accepting it if the new initial point, $x_0^{(n)}$, is in region A and rejecting it otherwise. A summary the algorithm for shooting moves in a path sampling interface is given in Scheme 1.7.

In a shifting move, the path sampling code is interfaced with the existing MD program in an analogous way. A Fortran code originally developed as a path sampling interface to the Car–Parrinello molecular dynamics code CPMD [62] can be downloaded from the web site http://www.pathsampling.org. Modifications necessary to use this interface in combination with other molecular dynamics packages are straightforward.

IV. COMPUTING RATES AND REVERSIBLE WORK

Because trajectories harvested by transition path sampling are true dynamical pathways, they can be used to compute dynamical observables. For transitions between two stable states A and B, the most important observables are rate constants k_{AB} and k_{BA} for forward and backward transitions, respectively. In this section, we show how rate constants can be calculated by, in effect, reversibly changing ensembles of trajectories. We accomplish this transformation by extending the importance sampling techniques discussed in previous sections. The calculation of rate constants with transition path sampling thus does not require an understanding of reaction mechanism, offering a significant advantage over conventional methods.

The standard prescription for computing a rate constant k_{AB} is a two-step procedure, based on the perspective of transition state theory. It requires the choice of a putative reaction coordinate, q, and its success depends on the validity of that choice. In the first step, one computes the reversible work $W(q)$ to bring the system from stable state A to a surface, $q = q^*$, dividing A and B. Here, q^* is the value of q for which the free energy $W(q)$ is locally maximal, so that $q = q^*$ approximates the transition state surface. From the computed reversible work, the transition state theory estimate of the rate constant, k_{TST}, can be determined for the particular choice of q,

$$k_{TST} = \frac{\langle|\dot{q}|\rangle}{2\langle\theta(q^*-q)\rangle} \langle\delta(q-q^*)\rangle = \frac{1}{2}\langle|\dot{q}|\rangle \frac{\exp\{-\beta W(q^*)\}}{\int dq' e^{-\beta W(q')}\theta(q^*-q')} \quad (1.75)$$

Here, $\dot{q}$ is the time derivative of the reaction coordinate q and $\theta(x)$ is the Heaviside step function. The estimate k_{TST} is accurate to the extent that reactive trajectories cross the surface $q = q^*$ only once during a transition. In the second step of the procedure, corrections to transition state theory are computed by initiating many fleeting trajectories from the $q = q^*$ surface. The fates of these trajectories determine the time-dependent transmission coefficient,

$$\kappa(t) = \frac{2}{\langle|\dot{q}|\rangle} \frac{\langle\dot{q}_0 \delta(q_0 - q^*)\theta(q_t - q^*)\rangle}{\langle\delta[q_0 - q^*]\rangle} \quad (1.76)$$

where q_t is the reaction coordinate at time t. Since after a short transient time τ_{mol} fleeting trajectories started from the dividing surface are committed to one of the stable states, $\kappa(t)$ becomes constant in the time range $\tau_{mol} < t \ll \tau_{rxn}$, where $\tau_{rxn} \equiv (k_{AB} + k_{BA})^{-1}$ is the reaction time. This plateau value of $\kappa(t)$ is the transmission coefficient κ. Note that in this formalism regions A and B are adjacent. The plateau value of

$$k(t) = \kappa(t) k_{TST} \quad (1.77)$$

is a formally exact expression for the (classical) rate constant k_{AB}. The practical utility of this "reactive flux" method depends on the size of κ. For the optimal choice of q, that is, the true reaction coordinate, κ is maximized and is typically near unity if the crossing of the dynamical bottleneck is not diffusive. In this case, hundreds of fleeting trajectories are sufficient to determine κ to within, say, 10%. For a poor choice of q, on the other hand, $\kappa \ll 1$, leading to severe numerical problems. In this case, even thousands of fleeting trajectories are insufficient to distinguish κ from zero. This failure reflects the fact that states with $q = q^*$ are poor approximations of true transition states. Most trajectories passing through these states are therefore not reactive.

Preceding sections described how reactive trajectories may be harvested without a priori knowledge of a reaction coordinate. In a similar way, it is possible to compute rate constants accurately without this knowledge. In Section IV.A, we begin by identifying the central time correlation function of kinetics as a ratio of partition functions for different ensembles of trajectories. As such, it may be considered as the exponential of a work, specifically the work to reversibly confine the final state of a trajectory to the product state B. The rarity of transitions ensures that this work is large, so it cannot be computed directly. Instead, we borrow and extend well-known techniques from the statistical mechanics of configurations, namely umbrella sampling and thermodynamic integration. With these techniques, and a convenient factorization reminiscent of that in the reactive flux method, rate constants can be calculated with reasonable computational cost even when the underlying mechanism is not well understood. We will conclude this section by discussing the question of how to find the path length t necessary to allow sampling of all important transition pathways. Note that all formulas and algorithms presented in this section are equally valid for stochastic and deterministic dynamics.

A. Population Fluctuations

In a system at equilibrium, the populations of stable states A and B fluctuate in time due to spontaneous transitions between them. The dynamics of transitions are therefore characterized by the correlation of state populations in time

$$C(t) \equiv \frac{\langle h_A(x_0) h_B(x_t) \rangle}{\langle h_A(x_0) \rangle} \qquad (1.78)$$

Here, $\langle \ldots \rangle$ denotes an average over the equilibrium ensemble of initial conditions. $C(t)$ is the conditional probability to find the system in state B at time t provided it was in state A at time 0. According to the fluctuation-dissipation theorem [63], dynamics of equilibrium fluctuations are equivalent to the relaxation from a nonequilibrium state in which only state A is populated. At long timescales, these nonequilibrium dynamics are described by the phenomenology of macroscopic kinetics. Thus, the asymptotic behavior of $C(t)$ is determined by rate constants k_{AB} and k_{BA}. At long times, and provided that a single dynamical bottleneck separating A from B causes simple two-state kinetics,

$$C(t) \approx \langle h_B \rangle (1 - \exp\{-t/\tau_{\text{rxn}}\}) \qquad (1.79)$$

At short times, $C(t)$ will reflect microscopic motions in the transition state region, which are correlated over a timescale τ_{mol}. τ_{mol} is essentially the time required to cross the dynamical bottleneck separating the stable regions and commit to one of the stable states. Equation (1.79) links the reaction time τ_{rxn} measured experimentally with the microscopic correlation function $C(t)$. At long times,

$$C(t) \approx k_{AB} t \qquad (1.80)$$

provided the kinetics of transitions from A to B are two state. The slope of $C(t)$ in this region is the forward reaction rate constant k_{AB}. If metastable states mediate these transitions, the corresponding kinetics will be more complicated. In such cases, sampling trajectories that linger in metastable states may be difficult. In general, sequences of transitions between stable and metastable states should be dissected. Considering component transitions separately facilitates both sampling and analysis of reactive trajectories.

In general, $C(t)$ contains all the information needed to determine kinetic coefficients measured experimentally. In the following sections, we describe the calculation of $C(t)$ using transition path sampling. Once $C(t)$ is determined for times greater than τ_{mol}, rate constants can be extracted by analysis of $C(t)$.

B. Reversible Work

In the context of transition path sampling, the time correlation function in Eq. (1.78) is naturally written in terms of sums over trajectories

$$C(t) = \frac{\int \mathscr{D}x(t) h_A(x_0) \mathscr{P}[x(t)] h_B(x_t)}{\int \mathscr{D}x(t) h_A(x_0) \mathscr{P}[x(t)]} = \frac{Z_{AB}(t)}{Z_A} \qquad (1.81)$$

The second equality in Eq. (1.81) follows from the definition of partition functions for ensembles of trajectories [see Eq. (1.3)]. Specifically, $Z_{AB}(t)$ is the partition function for the ensemble of trajectories that begin in region A and end in region B a time t later. Similarly, Z_A is the partition function for the ensemble of all trajectories that begin in region A without any restriction on the endpoint at time t. Z_A is written without time argument because the dynamical weight $\mathscr{P}[x(t)]$ is normalized, so that $Z_A = \int dx_0 \rho(x_0) h_A(x_0)$ is just the equilibrium probability to find the system in state A. While Z_A "counts" all trajectories with initial points in A, Z_{AB} counts only those trajectories which start in A *and* end in B.

The partition function Z_A transforms into the partition function Z_{AB} if the constraint $h_B(x_t)$ on the endpoint of the path is introduced. We therefore interpret the ratio of partition functions in Eq. (1.81) as the exponential of a reversible work $W_{AB}(t)$ to change between these two ensembles,

$$W_{AB} \equiv -\ln \frac{Z_{AB}(t)}{Z_A} \qquad (1.82)$$

$W_{AB}(t)$ is an effective change in free energy, describing the confinement of trajectory endpoints x_t to region B, while preserving the constraint that initial points x_0 lie in region A. Provided this confinement process is carried out reversibly, the corresponding work is independent of the specific procedure used. In Section IV.C, we describe an advantageous choice for this process that is closely related to the method of thermodynamic integration.

C. Umbrella Sampling

For ensembles of configurations, changes in free energy that are large compared to $k_B T$ are often computed by introducing an artificial bias potential, $U(x)$ [41, 44, 63]. In the simplest cases, $U(x)$ is chosen so that the appropriate importance sampling visits rare but interesting states with the same frequency as typical equilibrium states. Differences in free energy may then be computed directly, and correcting for the presence of $U(x)$ is straightforward. When the relevant phase space is very large, it is generally more efficient to implement umbrella sampling by dividing space into a series of overlapping regions, or "windows" [63]. In this case, the ith bias potential confines the system to the ith window, and ensures uniform sampling within the window. The distribution of states in the entire space is then constructed by requiring that distributions within windows are consistent in overlapping regions. These methods are readily generalized to the sampling of trajectories.

In order to make explicit the correspondence between umbrella sampling of configurations and umbrella sampling of trajectories, we consider a schematic example. Imagine that an order parameter $\lambda(x)$ successfully distinguishes between stable states A and B. For concreteness, suppose that these states are two different conformers of a single molecule. In the first conformational state, λ has a certain range of values, $\lambda_{\min}^A < \lambda < \lambda_{\max}^A$. A distinct range of values characterizes the second conformer, $\lambda_{\min}^B < \lambda < \lambda_{\max}^B$, with $\lambda_{\min}^B > \lambda_{\max}^A$. We could compute the reversible work, w_B, to confine the molecule to the latter conformational state using umbrella sampling in the conventional way. The total range of λ is first divided into narrow windows, $\lambda^{(i)} - \Delta/2 < \lambda < \lambda^{(i+1)} + \Delta/2$. Here, Δ is a small, positive quantity ensuring that adjacent windows overlap. The corresponding bias potential for the ith window, $U^{(i)}(x)$, is infinite for $\lambda < \lambda^{(i)} - \Delta/2$, as well as for $\lambda > \lambda^{(i+1)} + \Delta/2$ and vanishes inside the window. For a wise choice of the window boundaries $\lambda^{(i)}$, the distribution of λ, $P(\lambda)$, will not vary considerably within each window. Within a window, then, $P(\lambda)$ may be accurately computed up to a constant of proportionality with straightforward importance sampling. These proportionality constants are obtained by demanding continuity in the intervals of overlap between adjacent windows. Finally, w_B is calculated by appropriate integration over the order parameter distribution,

$$w_B = -k_B T \ln \int_{\lambda_{\min}^B}^{\lambda_{\max}^B} d\lambda\, P(\lambda) \tag{1.83}$$

Umbrella sampling for ensembles of trajectories can be carried out in close analogy to the procedure described above. For the purpose of computing rate constants (in this case the rate of conformational transitions), we focus on computing the reversible work to confine trajectories' endpoints to state B, given that these trajectories begin in state A. $P_A(\lambda, t)$ is defined to be the distribution of λ at the endpoints of trajectories initiated in A,

$$P_A(\tilde{\lambda}, t) = \frac{\int \mathscr{D}x(t) h_A(x_0) \mathscr{P}[x(t)] \delta[\tilde{\lambda} - \lambda(x_t)]}{Z_A} = \langle \delta[\tilde{\lambda} - \lambda(x_t)] \rangle_A \tag{1.84}$$

Here, $\langle \cdots \rangle_A$ denotes an average over all pathways beginning in state A. As before, we divide the range of λ into narrow, overlapping windows. Each window corresponds to a region $\mathscr{W}[i]$ in phase space. Within each window, $P_A(\lambda, t)$ can be computed accurately up to a constant of proportionality using the methods described in Section III. Specifically, importance sampling is used to harvest trajectories according to the path weight functional

$$\mathscr{P}_{A\mathscr{W}[i]}[x(t)] = \rho(x_0) \mathscr{P}[x(t)] h_A(x_0) h_{\mathscr{W}[i]}(x_t) \tag{1.85}$$

The path ensemble $\mathscr{P}_{A\mathscr{W}[i]}[x(t)]$, consists of pathways starting in A and ending in the window region $\mathscr{W}[i]$.

The distribution of the order parameter λ in region $\mathscr{W}[i]$ is,

$$\begin{aligned} P_{A\mathscr{W}[i]}(\tilde{\lambda}, t) &= \frac{\int \mathscr{D}x(t) h_A(x_0) \mathscr{P}[x(t)] h_{\mathscr{W}[i]}(x_t) \delta[\tilde{\lambda} - \lambda(x_t)]}{\int \mathscr{D}x(t) h_A(x_0) h_{\mathscr{W}[i]}(x_t)} \\ &= \langle \delta[\tilde{\lambda} - \lambda(x_t)] \rangle_{A\mathscr{W}[i]} \end{aligned} \tag{1.86}$$

Since $P_{A\mathscr{W}[i]}$ is the ensemble distribution function for pathways with endpoints in region $\mathscr{W}[i]$ the distribution $P_{A\mathscr{W}[i]}(\lambda, t)$ can have nonvanishing values only in the order parameter range corresponding to $\mathscr{W}[i]$ and must vanish for all other values of λ. As can be seen by comparing Eq. (1.84) with Eq. (1.86), inside window i the distribution $P_{A\mathscr{W}[i]}(\lambda, t)$ with path endpoints restricted to region $\mathscr{W}[i]$ is proportional to the complete order parameter distribution $P_A(\lambda, t)$ obtained as an average over pathways with unrestricted endpoints. By matching the distributions $P_{A\mathscr{W}[i]}(\lambda, t)$ in the overlapping regions, one can thus obtain the complete distribution $P_A(\lambda, t)$. To reduce the number of required windows it is sometimes convenient to introduce a bias $U(x_t)$ in the sampling. Of course, such a bias must be appropriately corrected for in the calculation of path averages.

By integrating, this distribution function over those values of the order parameter λ belonging to region B one can obtain the probability for a path starting in A to reside in B at a time t later

$$C(t) = \frac{\langle h_A(x_0) h_B(x_t) \rangle}{\langle h_A \rangle} = \int_{\lambda_{\min}^B}^{\lambda_{\max}^B} d\lambda \, P_A(\lambda, t) \tag{1.87}$$

Accordingly, the reversible work required to confine the endpoint of paths starting in A to region B is,

$$W_{AB}(t) = -\ln \int_{\lambda_{\min}^B}^{\lambda_{\max}^B} d\lambda \, P_A(\lambda, t) \tag{1.88}$$

where we have used the fact that $P_A(\lambda, t)$ is normalized.

Figure 1.12 shows the results of an umbrella sampling path sampling simulation carried out to determine $C(t)$ for proton transfer in the protonated water trimer described in Section I. The curves in (a) are the distributions $P_{A\mathscr{W}[i]}(\lambda, t)$ restricted to the window regions $\mathscr{W}[i]$. In this case, the order parameter $\lambda(x)$ was defined as the angular difference $\Delta\phi \equiv \varphi_2 - \varphi_1$, where the angles φ_1 and φ_2 were defined as indicated in Figure 1.2. Requiring continuity of $P_A(\lambda, t)$, where the regions $\mathscr{W}[i]$ overlap, one obtains the distribution $P_A(\lambda, t)$ shown on a logarithmic scale in (b). The correlation function $C(t)$ is then determined by integration of $P_A(\lambda, t)$ over values of λ corresponding to the final region B.

D. A Convenient Factorization

In principle, the time correlation function $C(t)$ may be calculated by repeating the procedure described above at each time t. But this umbrella sampling of trajectories is laborious, requiring independent sampling of pathways for each window of the order parameter. Indeed, it mimics a

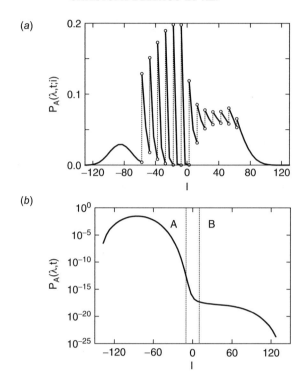

Figure 1.12. (a) Distributions of the order parameter $\lambda(x_t)$ in the window regions $\mathscr{W}[i]$ for proton transfer in the protonated water trimer. In this calculation, the energetics of the trimer were described by an empirical valence bond model [64]. The solid lines denote the individual distributions and the open circles mark points where adjacent order parameter windows overlap. Each distribution was obtained by sampling Newtonian transition pathways of length $t = 147\,\mathrm{fs}$ with shooting and shifting moves at an energy corresponding to a temperature of $T \approx 290\,\mathrm{K}$. The order parameter λ used in this illustrative calculation was defined as the angle difference $\Delta\varphi \equiv \varphi_2 - \varphi_1$, where φ_1 and φ_2 are angles describing the geometry of the OOO-ring of the cluster (see Fig. 1.2). Region A was defined by $\Delta\varphi < -10°$ and region B by $\Delta\varphi > 10°$. (b) The distribution $P_A(\lambda, t)$ shown on a logarithmic scale is obtained by requiring continuity at the overlap of adjacent windows. In (b) regions A and B are indicated by vertical dotted lines. Integration of $P_A(\lambda, t)$ over the values of λ corresponding to region B yields $C(t) = 9.8 \times 10^{-17}$.

quasistatic process in which the constraint on trajectory endpoints is introduced reversibly. Repeating this process many times, and especially for long trajectories, would incur significant computational cost. In this section, we describe a far less expensive scheme for computing $C(t)$ in a time interval $0 < t < \mathscr{T}$, requiring that umbrella sampling be performed only once and only for relatively short trajectories.

In Section IV.C, we took advantage of the relationship between $C(t)$ and an effective reversible work, $W_{AB}(t)$, to constrain the state of a system at time t. In effect, we gradually applied this constraint to the ensemble of trajectories of length t that begin in state A, obtaining a new ensemble of reactive trajectories. Here, we consider a different process connecting the same two ensembles of trajectories. Because this process is also reversible, it involves the same work $W_{AB}(t)$. In the new process, trajectories of length $\mathcal{T} > t$ that begin in state A are first constrained to visit state B at a time t'. The behavior of these trajectories at times later than t' is unconstrained. The work associated with this first step is thus $W_{AB}(t')$, and it may be computed using umbrella sampling. In the second step, the time at which trajectories are constrained to visit state B is shifted from t' to t. We denote the work associated with the second step as $\Delta W_{AB}(t, t')$. We will show that $\Delta W_{AB}(t, t')$ can be computed for all t at once with little cost, by sampling trajectories that visit state B at any time prior to $\mathcal{T}$. As a result, computing the total work,

$$W_{AB}(t) = W_{AB}(t') + \Delta W_{AB}(t, t') \tag{1.89}$$

for all times $t < \mathcal{T}$ requires umbrella sampling only for a single time t'. In addition, the time t' may be chosen to be small, further reducing the computational cost.

The quantity $\Delta W_{AB}(t, t')$ is the effective reversible work to shift the time at which trajectories are required to reach state B. It can be written in terms of correlation functions $C(t)$ and $C(t')$:

$$\exp[-\Delta W_{AB}(t, t')] = \frac{C(t)}{C(t')} = \frac{\langle h_A(x_0) h_B(x_t) \rangle}{\langle h_A(x_0) h_B(x_{t'}) \rangle} \tag{1.90}$$

For simplicity, we define a function

$$R(t, t') = \exp[-\Delta W_{AB}(t, t')] \tag{1.91}$$

such that $R(t, t') = C(t)/C(t')$.

In order to calculate $R(t, t')$ efficiently, we introduce a path functional $H_B[x(\mathcal{T})]$ that is unity if at least one state along the trajectory $x(\mathcal{T})$ lies in B and vanishes otherwise,

$$H_B[x(\mathcal{T})] \equiv \max_{0 \leq \bar{t} \leq \mathcal{T}} h_B(x_{\bar{t}}) \tag{1.92}$$

Because $h_B(x_t) = h_B(x_t) H_B[x(\mathcal{T})]$ for all $t < \mathcal{T}$, we may rewrite Eq. (1.91) as

$$R(t,t') = \frac{\langle h_A(x_0)h_B(x_t)H_B[x(\mathcal{T})]\rangle}{\langle h_A(x_0)H_B[x(\mathcal{T})]\rangle} \times \frac{\langle h_A(x_0)H_B[x(\mathcal{T})]\rangle}{\langle h_A(x_0)h_B(x_{t'})H_B[x(\mathcal{T})]\rangle} \quad (1.93)$$

Each of the factors in this equation can be written as an average over pathways beginning in region A and visiting region B at some time in the interval $[0,\mathcal{T}]$. These pathways are reactive, but are not required to reside in B at any specific time. Their weight functional is

$$\mathcal{P}^*_{AB}[x(\mathcal{T})] \equiv h_A(x_0)\mathcal{P}[x(\mathcal{T})]H_B[x(\mathcal{T})]/Z^*_{AB}(\mathcal{T}) \quad (1.94)$$

The partition function normalizing $\mathcal{P}^*_{AB}[x(\mathcal{T})]$ is given by

$$Z^*_{AB}(\mathcal{T}) = \int \mathcal{D}x(\mathcal{T})h_A(x_0)\mathcal{P}[x(\mathcal{T})]H_B[x(\mathcal{T})] \quad (1.95)$$

We denote averages in this ensemble by the subscript AB and an asterisk, for example,

$$\langle h_B(x_t)\rangle^*_{AB} \equiv \frac{\int \mathcal{D}x(\mathcal{T})\,h_A(x_0)\mathcal{P}[x(\mathcal{T})]h_B(x_t)H_B[x(\mathcal{T})]}{\int \mathcal{D}x(\mathcal{T})\,h_A(x_0)\mathcal{P}[x(\mathcal{T})]H_B[x(\mathcal{T})]} \quad (1.96)$$

The path distribution $\mathcal{P}^*_{AB}[x(\mathcal{T})]$ can be sampled with algorithms nearly identical to those described in Section III. We simply replace the characteristic function $h_B(x_t)$ with the functional $H_B[x(\mathcal{T})]$ in all expressions. In sampling this distribution, one may determine $R(t,t')$ for all times $t,t' < \mathcal{T}$ at once. Also, in this case efficiency can be increased by growing the backward segments of a shooting move first, because only in this case early rejection can be exploited to reduce the computational cost of shooting moves.

The population correlation function $C(t)$ can be efficiently determined over the entire interval $0 < t < \mathcal{T}$ as follows: First, $\langle h_B(x_t)\rangle^*_{AB}$ is computed by sampling pathways of a single length $\mathcal{T}$. In order to extract a rate constant from $C(t)$, $\mathcal{T}$ must be longer that the transient time τ_{mol}, as indicated by linear scaling of $\langle h_B(x_t)\rangle^*_{AB}$. In general, it is best to experiment with the value of $\mathcal{T}$, in order to access the asymptotic regime without wasting computational effort on unnecessarily long pathways. After computing $\langle h_B(x_t)\rangle^*_{AB}$, $C(t)$ is determined at a single time t' using the umbrella sampling methods described in Section IV.C. To reduce the cost of this calculation, t' may be chosen to be much shorter than $\mathcal{T}$. Combining results

1. Calculate the path average $\langle h_B(x_t) \rangle^*_{AB}$ using the transition path ensemble $\mathscr{P}^*_{AB}[x(\mathcal{T})]$.
2. Determine the time derivative $d\langle h_B(x_t) \rangle^*_{AB}/dt$.
3. If $d\langle h_B(x_t) \rangle^*_{AB}/dt$ displays a plateau, proceed with step 4, otherwise repeat procedure with a longer time $\mathcal{T}$ starting from step 1.
4. Calculate the time correlation function $C(t')$ at a specific time t' in the interval $[0, \mathcal{T}]$ with umbrella sampling.
5. Determine $C(t)$ in the entire interval $[0, \mathcal{T}]$ using $C(t) = C(t')\langle h_B(t) \rangle^*_{AB}/\langle h_B(t') \rangle^*_{AB}$.
6. Take the derivative of $C(t)$ and extract the rate constant k_{AB} from the plateau value of $\dot{C}(t)$.

Scheme 1.8. Algorithm for the calculation of transition rate constants.

of these two steps yields $C(t)$ in the whole interval of interest,

$$C(t) = \frac{\langle h_B(x_t) \rangle^*_{AB}}{\langle h_B(x_{t'}) \rangle^*_{AB}} \times C(t') \tag{1.97}$$

A rate constant can then be obtained from the asymptotic slope of $C(t)$. This algorithm for calculating rate constants is summarized in Scheme 1.8. We emphasize that $C(t)$ is calculated exactly in this prescription. Specifically, we do not assume an underlying separation of timescales. Rather, the behavior of $C(t)$ reveals whether such a separation exists.

The results of the first step of a rate constant calculation in the transition path sampling method are shown in Figure 1.13 for proton transfer in the protonated water trimer. Panel (a) depicts the path average $\langle h_B(t) \rangle^*_{AB}$ as a function of time. Initially $\langle h_B(t) \rangle^*_{AB}$ is zero because pathways must cross a gap separating region A from region B. At least 50 fs are necessary to cross this gap. After this time, $\langle h_B(t) \rangle^*_{AB}$ starts to grow and reaches a linear regime after ~ 200 fs. Figure 1.13(b) shows the time derivative of $\langle h_B(x_t) \rangle^*_{AB}$, which displays a plateau after 200 fs. A plateau value of $\langle \dot{h}_B(t) \rangle^*_{AB,\text{plateau}} = 0.34 \times 10^{-2}\,\text{fs}^{-1}$ can be inferred from the figure. To calculate the rate constant k_{AB}, we now use the time correlation function $C(t') = 9.83 \times 10^{-17}$ obtained by umbrella sampling for a particular time $t' = 147$ fs (see Fig. 1.12). Note that t' can be smaller than $\mathcal{T}$ and does not even need to be in the plateau regime. At t' the path average $\langle h_B(t') \rangle^*_{AB} = 0.16$. By combining these values, we obtain a rate constant of $k_{AB} = 2.1 \times 10^{-3}\,\text{s}^{-1}$.

E. Correspondence with Reactive Flux Theory

The factorization of $C(t)$ introduced in Section IV.D is reminiscent of the factorization used in conventional rate constant calculations [3]. In both cases, a reversible work calculation is needed to compute the correlation of

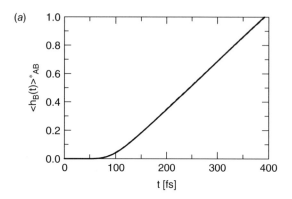

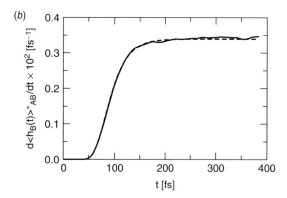

Figure 1.13. (a) Path average $\langle h_B(t)\rangle^*_{AB}$ for proton transfer in the protonated water trimer computed directly in a transition path simulation (solid line) and from the distribution of transition times (dashed line) as discussed in Section IV.F. The average was obtained by sampling Newtonian transition pathways of length $t \sim 390\,\text{fs}$ at a total energy equivalent to a temperature of $T \approx 290K$. Region A was defined by $\Delta\varphi < -10^0$ and region B by $\Delta\varphi > 10^0$. (b) Time derivative of the $\langle h_B(t)\rangle^*_{AB}$ curves shown in panel (a). The time derivative reaches a plateau only for times larger than the longest transition time τ, that is, for times larger than $\sim 200\,\text{fs}$.

population fluctuations at a specific time. The remaining time dependence of correlations can be computed independently of the absolute scale, and at much less cost. For the reactive flux method, a reversible work calculation at time 0^+ corresponds to transition state theory. (Trajectories of infinitesimal length do not recross the dividing surface $q = q^*$.) In the method based on transition path sampling, the specific time at which $C(t)$ is calculated remains arbitrary. In fact, this resemblance is more than superficial. For a particular choice of the boundaries defining states A and B, the two factorization schemes are intimately related.

If a single dividing surface, $q = q^*$, is chosen as the boundary of both states A and B, then the two states are adjoining: $h_A(x) + h_B(x) = \theta(q - q^*) + \theta(q^* - q) = 1$. For such a choice, the time derivative of the population correlation function in Eq. (1.78) exactly corresponds to the reactive flux through the $q = q^*$ surface, $k(t) = dC(t)/dt$. With the factorizations we have described,

$$k_{\text{TST}} \kappa(t) = C(t') \frac{dR(t, t')}{dt} \quad (1.98)$$

The factors $\kappa(t)$ and $dR(t, t')/dt$ both describe the full time dependence of this flux. They therefore differ only by a proportionality constant, $C(t')/k_{\text{TST}}$. For a choice of t' that is vanishingly small, this constant is just $\dot{C}(t')t'/k_{\text{TST}} = t'$, since $C(0^+) = 0$. For adjoining states and small t', our factorization of Eq. (1.78) is thus related to that of reactive flux theory by a simple multiplicative constant.

When states A and B are not adjoining, reactive trajectories must cross a region in which $h_A + h_B = 0$. In this case, the transient, short-time behavior of $\dot{C}(t)$ will differ from that of $k(t)$. In particular, $\dot{C}(0^+) = 0 \neq k_{\text{TST}}$, because infinitesimally short trajectories cannot bridge the gap between A and B. At long times, however, $\dot{C}(t)$ approaches $k(t)$ asymptotically, provided the region of phase space included in the gap has nearly vanishing weight. In this case, $\dot{C}(t)$ and $k(t)$ are described by the same phenomenological kinetics as $t \to \infty$. We have verified this asymptotic correspondence for several systems [6, 12].

F. How Long Should Pathways Be?

In harvesting transition pathways, it is important that trajectories are long enough to exhibit typical barrier crossing behavior. Trajectories that are constrained to connect stable states in a very short time may not be representative of the natural ensemble of reactive trajectories. In this section, we discuss quantitative criteria for the appropriate minimum length (in time) of trajectories. These criteria are based on the condition that correlations of population fluctuations have reached their asymptotic behavior. They are thus related to the corresponding condition for the reactive flux method, namely, that the flux is essentially constant on a molecular timescale. Depending on the chosen boundaries of stable regions, however, the appropriate duration of trajectories can be quantitatively different for transition path sampling.

In general, regions A and B defined in transition path sampling are not adjacent and the system must cross a gap between these regions during a transition. For example, the results depicted in Figure 1.13 were obtained from calculations in which region A was defined by $\Delta\varphi < -10°$ while region B was characterized by $\Delta\varphi > 10°$. Since in general the reaction coordinate describing the course of the transition is unknown, it is important that regions A and B do not overlap and do not contain states belonging to the basin of attraction of the other stable region. Due to the associated finite width of the gap the system needs a certain finite transition time τ to cross the gap. Therefore, the pathways in transition path sampling must be long enough to accommodate the transition times τ of the most important pathways. In the following paragraph, we will justify and analyze this requirement in greater detail.

Consider, again, a system in which transitions from A to B occur rarely. Along each possible pathway $x(\mathcal{T})$ of length $\mathcal{T}$ the system spends a certain time $\tau_A[x(\mathcal{T})]$ in A, then stays between A and B for a time $\tau[x(\mathcal{T})]$, and finally arrives in B spending the rest of the time, $\tau_B[x(\mathcal{T})]$, there. For simplicity, we assume that there are no multiply entries and exits to and from regions A and B. Hence, every pathway $x(\mathcal{T})$ has a single well defined transition time $\tau[x(\mathcal{T})]$. Since different pathways have different transition times it is convenient to introduce a distribution of transition times,

$$p(\tilde{\tau}; \mathcal{T}) \equiv \int \mathscr{D}x(\mathcal{T}) \mathscr{P}_{AB}[x(\mathcal{T})] \delta(\tilde{\tau} - \tau[x(\mathcal{T})]) \qquad (1.99)$$

where the argument $\mathcal{T}$ in $p(\tau; \mathcal{T})$ indicates that the distribution of transition times depends on the total path length $\mathcal{T}$. As an example for such a distribution Figure 1.14 shows $p(\tau; \mathcal{T})$ for proton-transfer pathways in the protonated water trimer described in Section I. As can be inferred from the figure, the system needs at least $\sim 50\,\text{fs}$ to cross the gap and some trajectories spend up to $200\,\text{fs}$ in the region between A and B.

Now, the path average $\langle h_B(x_t) \rangle^*_{AB}$ appearing in the expressions for the rate constant in Eq. (1.97) can be expressed as an integral over the distribution of transition times $p(\tau; \mathcal{T})$. Analysis of that expression then yields a criterion for the path length $\mathcal{T}$.

To demonstrate this we rewrite Eq. (1.96) as

$$\langle h_B(x_t) \rangle^*_{AB} = \int \mathscr{D}x(\mathcal{T}) \mathscr{P}^*_{AB}[x(\mathcal{T})] \theta(t - \tau[x(\mathcal{T})] - \tau_A[x(\mathcal{T})]) \qquad (1.100)$$

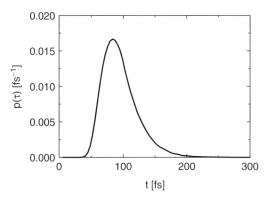

Figure 1.14. Distribution of transition times τ for proton transfer in the protonated water trimer for the same system as in Figure 1.13.

where $h_B(x_t)$ for the specific pathway $x(\mathcal{T})$ has been expressed using the Heaviside step function $\theta(t)$. In doing so we have assumed that trajectory $x(\mathcal{T})$ leaves region A and enters region B exactly once. We next insert two Dirac delta functions into the right-hand side of Eq. (1.100) and integrate over their arguments:

$$\langle h_B(x_t)\rangle^*_{AB} = \int d\tilde{\tau} \int d\tilde{\tau}_A \int \mathcal{D}x(\mathcal{T})\mathcal{P}^*_{AB}[x(\mathcal{T})]\theta(t - \tau[x(\mathcal{T})] - \tau_A[x(\mathcal{T})])$$
$$\times \delta(\tilde{\tau} - \tau[x(\mathcal{T})])\delta(\tilde{\tau}_A - \tau_A[x(\mathcal{T})]) \quad (1.101)$$

The delta functions allow us to replace the functionals $\tau[x(\mathcal{T})]$ and $\tau_A[x(\mathcal{T})]$ appearing in the argument of the Heaviside step function by the numbers $\tilde{\tau}$ and $\tilde{\tau}_A$, respectively. We therefore obtain

$$\langle h_B(x_t)\rangle^*_{AB} = \int d\tau \int d\tau_A \theta(t - \tau - \tau_A) p(\tau, \tau_A; \mathcal{T}) \quad (1.102)$$

where we have renamed the integration variables. The distribution function $p(\tau, \tau_A; \mathcal{T})$ is the probability to observe both τ and τ_A in a path with length $\mathcal{T}$,

$$p(\tau, \tau_A; \mathcal{T}) \equiv \int Dx(\mathcal{T})\mathcal{P}^*_{AB}[x(\mathcal{T})]\delta(\tilde{\tau} - \tau[x(\mathcal{T})])\delta(\tilde{\tau}_A - \tau_A[x(\mathcal{T})]) \quad (1.103)$$

For a given transition time τ, the time τ_A spent in region A can have values ranging from 0 to $T - \tau$. Assuming that τ_A is uniformly distributed in this interval and that it is furthermore statistically independent from τ, that is,

$$p(\tau, \tau_A; \mathcal{T}) = \frac{\theta(\mathcal{T} - \tau - \tau_A)\theta(\tau_A)}{\mathcal{T} - \tau} p(\tau; \mathcal{T}) \qquad (1.104)$$

One can carry out the integration over τ_A in Eq. (1.102) obtaining

$$\langle h_B(x_t) \rangle^*_{AB} = \int_0^{\mathcal{T}} d\tau \, p(\tau; \mathcal{T})\theta(t - \tau) \frac{t - \tau}{\mathcal{T} - \tau} \qquad (1.105)$$

Thus, the contribution to $\langle h_B(x_t) \rangle^*_{AB}$ of pathways with transition time τ is zero for $0 \leq t \leq \tau$ and grows linearly from 0 to 1 in the interval $\tau < t \leq \mathcal{T}$. The path average $\langle h_B(x_t) \rangle^*_{AB}$ computed from Eq. (1.105) for proton-transfer pathways in the protonated water trimer is shown along with its time derivative in Figure 1.13 as dashed lines. The excellent agreement of these results with the results obtained in a path simulation directly indicates that the assumptions made in Eq. (1.104) are correct for the example system.

Equation (1.105) can be used to obtain a criterion on the path length required to sample all important pathways. According to the reasoning of the preceding sections, pathways are sufficiently long if the time derivative $dC(t)/dt$ reaches a plateau. But $dC(t)/dt$ can reach a plateau only if $\langle h_B(x_t) \rangle^*_{AB}$ does so. In this plateau regime, the first derivative of $\langle h_B(x_t) \rangle^*_{AB}$ is constant and its second derivative must vanish. By calculating the second time derivative of $\langle h_B(x_t) \rangle^*_{AB}$ from Eq. (1.105), we obtain

$$\frac{d^2}{dt^2} \langle h_B(x_t) \rangle^*_{AB} = \frac{p(t; \mathcal{T})}{\mathcal{T} - t} \qquad (1.106)$$

Consequently, $dC(t)/dt$ can reach a plateau value only if the path length $\mathcal{T}$ is larger than the maximum transition time $\tau_{\max}$ after which the distribution of transition times $p(\tau; \mathcal{T})$ vanishes. More exactly, the existence of a plateau requires that $p(\tau)/(\mathcal{T} - \tau)$ vanishes in the whole plateau regime. The maximum transition time $\tau_{\max}$ can be considerably larger than the molecular transient τ_{mol} from reactive flux theory. For example, $\tau_{\max} \approx 250\,\text{fs}$ for proton transfer in the protonated water trimer as can be inferred from Figure 1.14 (and from Fig. 1.13). In contrast, the time-dependent transmission coefficient $\kappa(t)$ shown in Fig. 1.15 indicates that a trajectory initiated from the dividing surface defined by $\Delta \phi = 0$ commits to one of the stable states in less than 100 fs. In general, the path length should therefore be chosen to be larger then the maximum transition time or the time necessary to commit into one of the stable states, whichever is larger.

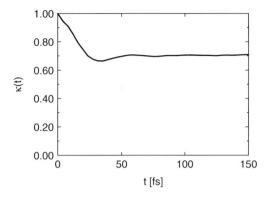

Figure 1.15. Transmission coefficient $\kappa(t)$ as a function of time for the same system as in Figure 1.14. After a molecular transient time of less than 100 fs $\kappa(t)$ reaches a plateau.

V. ANALYZING TRANSITION PATHWAYS

Thus far, we have described efficient methods for harvesting pathways that exhibit transitions of interest. We have also detailed how these trajectories may be used to characterize reactive dynamics at a macroscopic level by computing rate constants. In this section, we focus on gleaning mechanistic information from harvested pathways, in order to understand transitions at a microscopic level. For complex systems, this task is made challenging by the huge number of irrelevant degrees of freedom. It is generally not possible to recognize the coordinate that describes a transition's progress simply by visualizing trajectories (e.g., using computer graphics) or by following the behavior of preconceived order parameters. Rather, identifying such a reaction coordinate requires a thorough statistical analysis of pathways and transition states. Here, we describe several concepts and diagnostics that facilitate this analysis.

A. Reaction Coordinates and Order Parameters

A coordinate that successfully distinguishes between two basins of attraction does not necessarily characterize the dynamical bottleneck between them. We emphasize this fact by referring to discriminating coordinates as *order parameters*, and reserving the term *reaction coordinate* for the specific coordinate that describes a transition's dynamical mechanism. This distinction is illustrated in Figure 1.16. For the free energy landscapes in both panels of the figure, the coordinate q serves as a reasonable order parameter. Specifically, the range of values of q in the basin of attraction of state A does

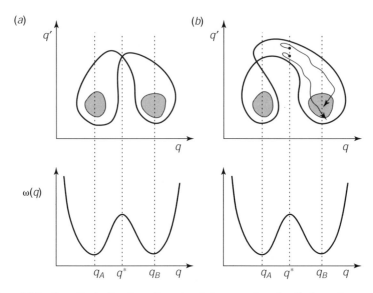

Figure 1.16. Two illustrative two dimensional free energies $w(q, q')$ depending on two variables q and q', and their corresponding reduced free energy functions $w(q) \equiv -k_B T \ln \int dq' \exp[-\beta w(q, q')]$. In both cases, $w(q)$ has the same bistable form, but in (a) the coordinate q is a reasonable reaction coordinate, as the transition state surface coincides with $q = q^*$. In (b), on the other hand, q is not at a reasonable reaction coordinate. The orthogonal variable q' is crucial for the mechanism of $A \to B$ transitions, and the maximum in $w(q)$ at $q = q^*$ does not coincide with the transition state surface. The trajectories initiated at configurations with $q(r) = q^*$ and all ending in B illustrate this.

not significantly overlap the range of values in the basin of state B. As a result, the reduced free energy $w(q)$, obtained by integrating out the orthogonal coordinate q', has a pronounced maximum, located at $q = q^*$. But the utility of q as a reaction coordinate differs greatly in the two scenarios. In (a), the ensemble of configurations with $q = q^*$ is a good approximation to the transition state surface dividing the basins of attraction. Here, q' is essentially irrelevant to the dynamical bottleneck, and q may be accurately called the reaction coordinate. In (b), however, barrier crossing occurs primarily in the direction of q'. Typical configurations with $q = q^*$ are poor approximations to true transition states, lying well within the basin of attraction of state B. In this case, the order parameter q does not coincide with the reaction coordinate. This latter situation is in fact common in physical systems. For example, symmetry and density successfully characterize fluid and solid phases of a simple material, but the bottleneck for crystallization of a supercooled liquid involves the size and structure of a crystal nucleus [65, 66]. One should thus not expect order parameters in general to function as reaction coordinates.

B. The Separatrix and the Transition State Ensemble

In a simple system, saddle points of the potential energy surface coincide with transition states. At these stationary points, forces vanish, and the system is not driven to either of the corresponding stable states. In other words, the stable states are equally accessible from the transition state. This concept of equal accessibility is readily extended for complex systems, in which individual saddle points are no longer relevant for transitions of interest. Specifically, transition states are defined to be configurations from which the system relaxes into one or the other stable state with equal probability [27, 67]. Transition states defined in this way are generally unrelated to the local topography of the potential energy surface, and may be strongly influenced by entropy. The set of all such transition states forms the *separatrix*, a high-dimensional surface dividing basins of attraction.

In order to formalize this definition of the separatrix, we introduce a function $p_B(r, t_s)$ called the *committor*:

$$p_B(r, t_s) \equiv \frac{\int \mathcal{D}x(t_s) \mathcal{P}[x(t_s)] \delta(r_0 - r) h_B(x_{t_s})}{\int \mathcal{D}x(t_s) \mathcal{P}[x(t_s)] \delta(r_0 - r)} \quad (1.107)$$

Since transition probabilities are normalized, the denominator on the right-hand side of this equation is just the equilibrium probability distribution for configuration r. The parameter $p_B(r, t_s)$ is the probability that a system with initial configuration r will reside in state B at time t_s. The analogous committor for state A, $p_A(r, t_s)$ is similarly defined. If the timescale of molecular fluctuations, τ_{mol}, is well separated from the typical time between spontaneous transitions, τ_{rxn}, then p_A and p_B will be nearly independent of t_s for $\tau_{mol} < t_s \ll \tau_{rxn}$. In this time regime, a trajectory will have committed to one or the other basin of attraction, even if it originated in the transition state region. Once such a trajectory has committed, subsequent spontaneous transitions are unlikely for $t_s \ll \tau_{rxn}$. In the discussions that follow, we assume that t_s has been chosen in this asymptotic regime, and omit the explicit dependence of committors on time. With this choice, p_A and p_B are essentially functions only of a configuration r, quantifying the propensity of that configuration to relax into a particular basin of attraction under the system's intrinsic dynamics. States A and B are thus equally accessible when $p_A(r) = p_B(r)$. This equation defines the separatrix.

The committor p_B is a direct statistical indicator for the progress of transitions from A to B. In this sense, it is an ideal reaction coordinate. But interpreting this highly nonlinear function of atomic coordinates in terms of molecular motions and intuitively meaningful fields is not straightforward. Understanding a transition's mechanism basically amounts to identifying the simple, low-dimensional coordinates that determine p_B. The following sections are concerned with this task. We first describe an efficient scheme for locating the separatrix, and then introduce unambiguous diagnostics for verifying the determinants of p_B. These tools greatly aid the interpretation of trajectories harvested by transition path sampling, but are not unique to that methodology. Indeed, any proposed reaction coordinate must satisfy the statistical criteria we describe in Section V.D.

C. Computing Committors

Locating the separatrix by screening typical configurations for $p_A \approx p_B$ would be an extraordinary computational challenge. Not only are such configurations rare at equilibrium, but evaluating committors requires the generation of many fleeting trajectories for each tested configuration. Fortunately, the pathways harvested by transition path sampling are guaranteed to include examples of the separatrix. In collecting a properly weighted ensemble of transition states, it is only necessary to screen the configurations along these relatively short, reactive pathways. Transition paths may even cross the separatrix more than once. In this case, each crossing provides a valid example of the transition state ensemble. Such multiple crossings are not uncommon in complex systems. For example, isomerization pathways of an alanine dipeptide cross the separatrix up to seven times [17].

The computational cost of identifying transition states is further reduced by the fact that configurations far from the separatrix may be excluded rather quickly. In practice, $p_B(r)$ is estimated for a particular configuration by initiating a finite number N of fleeting trajectories, $x^{(i)}(t_s)$, from that configuration

$$p_B^{(N)}(r) \approx \frac{1}{N} \sum_{i=1}^{N} h_B(x_{t_s}^{(i)}) \tag{1.108}$$

where we imagine that initial momenta are drawn at random from the appropriate distribution. This procedure is graphically illustrated in Figure 1.17. For large N, $p_B^{(N)}(r)$ is a Gaussian random variable according to the central limit theorem, with fluctuations of size

$$\sigma = \sqrt{\langle (p_B^{(N)} - p_B)^2 \rangle} = \sqrt{\frac{p_B(1 - p_B)}{N}} \tag{1.109}$$

Here, $\langle \ldots \rangle$ indicates an average over many independent calculations of $p_B^{(N)}$. These fluctuations describe the typical errors in our estimate of p_B using a

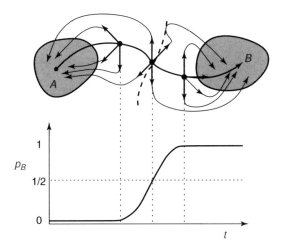

Figure 1.17. The committor p_B for a time slice at time t along a transition path (thick solid line, top panel) is computed by determining the fraction of fleeting trial trajectories (thin solid lines) initiated with randomized momenta that reach region B in a time t_s. Typically 10–100 of these fleeting trajectories are needed to obtain p_A and p_B with sufficient accuracy. While $p_B \approx 0$ for the left time slice in the top panel, because all trajectories started from that time slice end in A, $p_B \approx 1$ for the time slice on the right, because all the corresponding fleeting trajectories end in B. The configurations for which $p_B \approx \frac{1}{2} \approx p_A$ are part of the separatrix (thick dashed line) and are called transition states.

finite number of fleeting trajectories. The same fleeting trajectories $x^{(i)}(t_s)$ may be used to estimate p_A. For simplicity, we assume here that $p_A + p_B = 1$, that is, there are no additional basins of attraction. In this case, the error in our estimate of p_A is identical to that for p_B. Typical errors are only of size $\sigma \sim \sqrt{p_B/N} \ll 1$ when $p_A \approx 1$. Similarly, when $p_B \approx 1$, $\sigma \sim \sqrt{p_A/N} \ll 1$. We may thus determine that a configuration lies deep within the basin of attraction of A or B with a relatively small number of trajectories. (Of course, N must be large enough that $p_A^{(N)}$ and $p_B^{(N)}$ may be considered Gaussian random variables. $N \approx 10$ should be sufficient for this purpose.) By contrast, when $p_B \approx p_A \approx 1/2$, typical errors are only small for large N. Determining that a configuration lies near the separatrix thus requires a large number of fleeting trajectories.

The dependence of fluctuations in $p_A^{(N)}$ and $p_B^{(N)}$ on p_A, p_B, and N suggests a general scheme for estimating committor values. A desired level of statistical accuracy, σ_{des}, is chosen in advance. This choice determines the number of fleeting trajectories required to screen a given configuration. A minimum number, $N_{\min}$, of trajectories is first generated to ensure Gaussian

1. Start with a time slice just outside state A.
2. Generate $N_{\min} \sim 10$ short fleeting trajectories from that time slice with momenta selected from the appropriate distribution.
3. Determine p_B from the fraction of paths that end in state B.
4. If the interval $[p_B^{(N)} - \alpha\sigma^{(N)}, p_B^{(N)} + \alpha\sigma^{(N)}]$ does not include $\frac{1}{2}$ the time slice is rejected as a member of the transition state ensemble, otherwise more fleeting trajectories are generated until either $\frac{1}{2}$ is in $[p_B^{(N)} - \alpha\sigma^{(N)}, p_B^{(N)} + \alpha\sigma^{(N)}]$ or an upper limit $N_{\max}$ of the number of fleeting trajectories if reached.
5. If the latter is the case, the time slice is accepted as a member of the transition state ensemble. Otherwise it is rejected.
6. Move to the next time slice on the path and repeat steps 2–4 until region B is reached.

Scheme 1.9. Algorithm for determining transition states on a transition path.

statistics. Additional trajectories are subsequently generated until the error estimate

$$\sigma^{(N)} = \sqrt{\frac{p_B^{(N)}[1 - p_B^{(N)}]}{N}} \quad (1.110)$$

is smaller than σ_{des}. At this stage, the configuration is excluded from the transition state ensemble if the interval $[p_B^{(N)} - \alpha\sigma^{(N)}, p_B^{(N)} + \alpha\sigma^{(N)}]$ does not include the value $\frac{1}{2}$. The constant α depends on the desired confidence level. For example, $\alpha = 2$ is necessary to obtain a confidence level of 95%. If $\frac{1}{2}$ is inside the interval, we continue generating trajectories until the value $\frac{1}{2}$ falls outside the confidence interval. If a maximum number, $N_{\max}$, of trajectories is reached, and the confidence interval still includes $\frac{1}{2}$, then the configuration is accepted as a member of the transition state ensemble. This procedure, summarized in Scheme 1.9, minimizes the effort of excluding configurations that are far from the separatrix, and yields a consistent level of statistical error. As a rule of thumb, an error of 5% requires that $N_{\max} \sim 100$ trajectories are generated for configurations in the transition state region.

Our definition of a committor in Eq. (1.107) is applicable to both stochastic and deterministic dynamics. In the case of deterministic dynamics, care must be taken that fleeting trajectories are initiated with momenta drawn from the appropriate distribution. As discussed in Section III.A.2, global constraints on the system may complicate this distribution considerably. The techniques described in Section III.A.2 and in the Appendix of [10] for shooting moves may be simply generalized to draw initial momenta at random from the proper equilibrium distribution.

D. Committor Distributions

By screening transition pathways for examples of the separatrix, as described in Section V.D, an ensemble of transition states may be collected. For complex systems, this ensemble is expected to be structurally diverse, reflecting the many ways a particular transition can occur. But the reduced dimensionality of the separatrix ensures that even seemingly dissimilar transition states will share certain patterns. One may begin to search for these patterns by examining the distribution of various order parameters within the transition state ensemble. Coordinates that are important to the dynamical bottleneck will be narrowly distributed about values distinct from those characterizing stable states. In other words, they will be correlated with the transition. Such a correlation does not, however, guarantee that an order parameter is a useful reaction coordinate. Indeed, it is common for many coordinates to follow a transition adiabatically. These coordinates undergo a discernible change on the course from reactants to products, but do not play a significant role in driving the transition.

Distributions of committor values are a powerful diagnostic for differentiating coordinates that drive a transition from those that are simply correlated with it. Consider an order parameter q whose potential of mean force $w(q)$ has a maximum at $q = q^*$. If q serves as a reaction coordinate, then the ensemble of configurations with $q = q^*$ coincides with the separatrix [see Fig. 1.18(a)]. The committor distribution for this ensemble,

$$P(\tilde{p}_B) = \frac{\langle \delta[\tilde{p}_B - p_B(r, t_s)]\delta[q^* - q(r)]\rangle}{\langle \delta[q^* - q(r)]\rangle} \tag{1.111}$$

is thus sharply peaked at $p_B = \frac{1}{2}$. If, on the other hand, q does not characterize the dynamical bottleneck, most configurations in the ensemble with $q = q^*$ lie within the basins of attraction of states A and B [see Fig. 1.18(b)]. In this case, the distribution of p_B is dominated by peaks at $p_B = 0$ and $p_B = 1$. These two scenarios are distinguished by markedly different committor distributions. Figure 1.18 illustrates scenarios for three different schematic free energy surfaces.

In practice, determining informative committor distributions is a two-step process. The first step consists in generating configurations with the constraint $q = q^*$. Various Monte Carlo and molecular dynamics methods have been designed for this purpose [44, 68]. In the second step, the committor p_B is determined for each of the collected configurations, using the methods described in Section V.C. From these committor values a histogram $P(p_B)$ can then be constructed. The algorithm for determining committor distributions is summarized in Scheme 1.10.

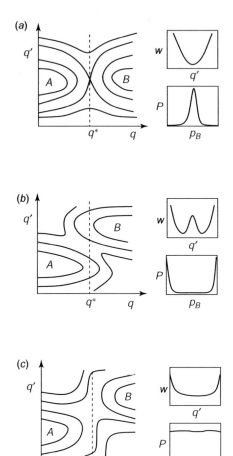

Figure 1.18. Three different free energy landscapes $w(q, q')$, the free energy $w(q^*, q')$ for $q = q^*$ and its corresponding committor distribution $P(p_B)$. (a) The reaction is correctly described by q and the committor distribution of the constrained ensemble with $q = q^*$ peaks at $p_B = 0.5$. (b) q' plays a significant additional role as a reaction coordinate, indicated by the additional barrier in $w(q^*, q')$ and the bimodal shape of $P(p_B)$. (c) Similar to case (b), but now the committor distribution is flat, suggesting diffusive barrier crossing along q'.

As an example, consider again proton transfer in the protonated water trimer as depicted in Figure 1.2. Imagine that the difference $\Delta r \equiv r_1 - r_2$ in the distances of the transferring proton to the oxygens of the donating and accepting water molecule has been postulated as the reaction coordinate for the proton transfer (r_1 and r_2 are defined as indicated in Fig. 1.2). Calculation of the reversible work necessary to control Δr yields a free

1. Compute the free energy $w(q)$ as a function of the proposed reaction coordinate q.
2. Determine the position q^* of the free energy maximum.
3. Collect a series of independent configurations at the top of the barrier, that is, $q = q^*$.
4. Calculate the committor p_B for every configuration.
5. Construct a histogram $P(p_B)$.

Scheme 1.10. Algorithm for determining a committor distribution.

energy profile with two minima separated by a barrier centered at $\Delta r = 0$. The minima of this free energy are located at values of Δr characteristic for the two stable states in each of which a different oxygen atom holds the excess proton. The committor distribution for configurations with $\Delta r = 0$, that is, configurations atop the barrier, is sharply peaked at 0 and 1 indicating that essentially all of these configurations clearly belong to the basin of attraction of one of the two stable states. Thus, the variable Δr is unable to capture the essential dynamics of the transition as the system passes through the dynamical bottleneck. Indeed, as discussed earlier, we found that a functioning reaction coordinate, such as the angular difference $\Delta \varphi$ defined earlier, must include degrees of freedom capable of describing the reorganization of the oxygen ring. The variable Δr merely follows this reorganization adiabatically without having any significant correlation with the transition dynamics. Committor distributions computed for configurations with $\Delta \varphi = 0$ peak at $p_B = \frac{1}{2}$ confirming the efficacy of $\Delta \varphi$ as a reaction coordinate.

This methodology has been also used to extract subtle mechanistic details of the dissociation mechanism of an NaCl ion pair in liquid water [12]. In this case, committor distributions revealed that rearrangement of solvent molecules around the dissociating ion pair significantly contributes to the free energy barrier separating the contact state from the dissociated state.

E. Path Quenching

Scrutinizing path ensembles for common factors can be a difficult task for several reasons. First, the paths are all different on a molecular scale due to thermal motion. Second, there might be different global pathways that are hard to distinguish. It would be helpful if the path ensemble could be somehow smoothed by removing all irrelevant dynamics orthogonal to the reaction coordinate, and reveal the essential dynamics of the optimal pathway. This smoothing can be done by path quenching. The idea behind this technique is that the path sampling method performs a random walk in trajectory space and is usually confined to a basin of attraction within this trajectory space. (This basin of attraction must be distinguished from the basis of attraction in configuration space, that is, the stable regions A and B.) Each basin of attraction in path space can be characterized by a minimum value of the action as defined in Section III.E. Quenching a path then amounts to minimizing this path action. In this way, different global pathways then correspond to different local minima in the action. As defined in Section III.E, the action $S_{AB}[x(\mathcal{T})]$ is related to the probability of finding a path by

$$P_{AB}[x(\mathcal{T})] = \exp(-S_{AB}[x(\mathcal{T})]) \qquad (1.112)$$

The analogy to the canonical Boltzmann distribution $\exp[-\beta V(x)]$ allows sampling of the path ensemble using the dynamical algorithms of Section III.E. By exploiting the same principle, one can use this dynamical algorithm to find the minimum action by sliding downhill in trajectory space. This amounts to quenching the "path temperature" to zero, which leads to a local minimum in path space. There might be many local minima, and in that case annealing the paths to find the nearest global minimum is probably a better option. For Newtonian dynamics with a canonical distribution of initial conditions, quenching the path action corresponds to finding the transition path with the lowest energy.

VI. OUTLOOK

Transition path sampling reduces the computational cost of harvesting reactive trajectories to that of generating dynamical paths of microscopic duration. Because its general methodology makes no reference to specific details of a dynamical system, we expect to see it widely applied. Indeed, examples of successful applications already include chemical reactions in the gas phase and in solution, nucleation of phase transitions, and diffusion of hydrocarbons in zeolites. We anticipate many more.

Some classes of applications will be made challenging by subtle difficulties in the implementation of transition path sampling. One such challenge is the identification of order parameters that unequivocally distinguish between basins of attraction. As discussed in Section II.F, failure to discriminate between stable states may lead to sampling of pathways that are not truly reactive. In many cases, physical intuition, combined with trial and error, is sufficient to determine a reasonable order parameter. But in general, characterizing stable states with only a few variables is a significant problem of pattern recognition. A similar problem arises in characterizing transition state surfaces. Even when many examples of the separatrix can be collected with the methods of Section V.B, identifying a successful reaction coordinate remains a challenge. A systematic approach to pattern recognition is lacking. Generalizations of standard methods for linear analysis [69] may offer progress in this area.

Sampling long trajectories is also a challenge. When the minimum duration of reactive trajectories (described in Section IV.F) is longer than the timescales characterizing chaos in a deterministic system, shooting moves will fail to generate useful trial trajectories. Due to divergence of neighboring phase-space elements, a typical trial path will differ markedly from the original path, even for the smallest possible momentum displacements. Most trial paths will therefore not be reactive, and thus will be rejected. Analogous difficulties hinder the sampling of strongly damped stochastic paths. When inertia is essentially inconsequential, only shooting points near the separatrix will generate reactive trajectories with a reasonable probability. In this case as well, sampling efficiency is severely degraded by the rejection of nearly all trial paths. It seems that a new genre of trial moves is necessary to confront these problems. One might imagine introducing a bias that imposes correlations between an existing path and trial trajectories generated from it.

For systems that are extremely large or complex, generating trajectories of even microscopic duration is computationally unfeasible. Clearly, harvesting true dynamical pathways of these systems is not possible. One might still hope, however, to construct chains of states that qualitatively resemble reactive trajectories. The stochastic path approach of Elber and co-workers [29, 70] focuses on such pathways. This method utilizes a weight functional for large time step trajectories that would normally be grossly unstable. The instability is artificially suppressed by the requirement that paths connect two given stable states. Although the stochastic path approach greatly extends the range of systems that can be considered, the pathways it generates are in some ways fundamentally dissimilar from natural trajectories: The effective dynamical propagation rule violates the fluctuation–dissipation theorem, and does not conserve an equilibrium distribution. It is not clear how to remedy these serious shortcomings while preserving the computational frugality that makes the method appealing. Transition path sampling may offer a more controlled context for this kind of approach.

In the course of future applications of our methods, we expect that many of the above difficulties will be overcome. Others' experience and insight will be essential for these advances. We also expect to see the perspectives underlying transition path sampling to be exploited in novel ways. An ensemble view of trajectories is especially promising for the studying dynamical structures like those of supercooled liquids [71, 72]. One can imagine that changes in relaxation patterns as a system approaches the glass transition are best captured using a "thermodynamics" of trajectories. While it is not obvious how to obtain simplifying insight into this thermodynamics (analogous to, say, van der Waals's picture of fluids), the ideas and formalism we have presented here would seem to be a first step.

ACKNOWLEDGEMENTS

We are very grateful to David Chandler, who inspired and directed the work described in this chapter. We acknowledge his guidance and support throughout our stay at UC Berkeley. This work also profited from enlightening and stimulating discussions with Hans Andersen, Christian Bartels, Felix Csajka, Gavin Crooks, Daan Frenkel, Daniel Laria, Jin Lee, Ka Lum, Jordi Marti, Tom McCormick, Vijay Pande, Javier Rodriguez, Dan Rokhsar, Udo Schmitt, Berend Smit, Sean Sun, Pieter-Rein Ten Wolde, and Thijs Vlugt.

REFERENCES

1. C. J. Cerjan and W. H. Miller, *J. Chem. Phys.* **75**, 2800 (1981).
2. J. P. K. Doye and D. J. Wales, *Z. Phys. D* **40**, 194 (1997).
3. J. B. Anderson, *J. Chem. Phys.* **58**, 4684 (1973); C. H. Bennett, in *Algorithms for Chemical Computations*, edited by R. E. Christoffersen, (American Chemical Society, Washington, D.C. 1977) pp. 63–97; D. Chandler, *J. Chem. Phys.* **68**, 2959 (1978).
4. L. P. Kadanoff, *Phys. Today*, **58**(5), 34 (August 2001).
5. C. Dellago, P. G. Bolhuis, F. S. Csajka, and D. Chandler, *J. Chem. Phys.* **108**, 1964 (1998).
6. C. Dellago, P. G. Bolhuis, and D. Chandler, *J. Chem. Phys.* **108**, 9263 (1998).
7. P. G. Bolhuis, C. Dellago, and D. Chandler, *Faraday Discuss.* **110**, 421 (1998).
8. F. S. Csajka and D. Chandler, *J. Chem. Phys.* **109**, 1125 (1998).
9. D. Chandler, in *Classical and Quantum Dynamics in Condensed Phas Simulations*, edited by B. J. Berne, G. Ciccotti, and D. F. Coker (World Scientific, Singapore, 1998), p. 51.
10. P. L. Geissler, C. Dellago, and D. Chandler, *Phys. Chem. Chem. Phys.* **1**, 1317 (1999).
11. C. Dellago, P. G. Bolhuis, and D. Chandler, *J. Chem. Phys.* **110**, 6617 (1999).
12. P. L. Geissler, C. Dellago, and D. Chandler, *J. Phys. Chem. B* **103**, 3706 (1999).
13. T. J. H. Vlugt, C. Dellago, and B. Smit, *J. Chem. Phys.* **113**, 8791 (2000).
14. P. L. Geissler, T. v. Voorhis, and C. Dellago, *Chem. Phys. Lett.* **324**, 149 (2000); E. P. F. Lee and J. M. Dyke, *Mol. Phys.* **73**, 375 (1991);
15. P. L. Geissler, C. Dellago, D. Chandler, J. Hutter, and M. Parrinello, *Chem. Phys. Lett.* **321**, 225 (2000).
16. P. L. Geissler and D. Chandler, *J. Chem. Phys.* **113**, 9759 (2000).
17. P. G. Bolhuis, C. Dellago, and D. Chandler, *Proc. Natl. Acad. USA* **97**, 5877 (2000).
18. P. G. Bolhuis, D. Chandler, C. Dellago, and P. L. Geissler, *Annu. Rev. Phys. Chem.*, in print (2000).
19. P. G. Bolhuis, C. Dellago, P. L. Geissler, and D. Chandler, *J. Phys.: Condens. Matter* **12**, A147 (2000).
20. J. Marti and F. S. Csajka, *J. Chem. Phys.* **113**, 1154 (2000).
21. J. Marti, F. S. Csajka, and D. Chandler, *Chem. Phys. Lett.* **328**, 169 (2000).

22. P. G. Bolhuis and D. Chandler, *J. Chem. Phys.* **113**, 8154 (2000).
23. T. J. H. Vlugt and B. Smit, *Phys. Chem. Comm.* **2**, 1 (2001).
24. P. L. Geissler, C. Dellago, D. Chandler, J. Hutter, and M. Parrinello, *Science* **291**, 2121 (2001).
25. D. Laria, J. Rodriguez, C. Dellago, and D. Chandler, *J. Phys. Chem. A* **105**, 2646 (2001).
26. G. E. Crooks and D. Chandler, *Phys. Rev. E* **64**, 026109 (2001).
27. L. R. Pratt, *J. Chem. Phys.* **85**, 5045 (1986).
28. R. Elber and M. Karplus, *Chem. Phys. Lett.* **139**, 375 (1987).
29. R. Elber, J. Meller, and R. Olender, *J. Phys. Chem. B* **103**, 899 (1999).
30. G. Henkelman, G. Johannesson, and H. Jonsson, in *Progress on Theoretical Chemistry and Physics*, edited by S. D. Schwartz, (Kluwer Academic Publishers, Dordrecht, 2000).
31. D. Passerone and M. Parrinello, *Phys. Rev. Lett.* **87**, 108302 (2001).
32. E. M. Sevick, A. T. Bell, and D. N. Theodorou, *J. Chem. Phys.* **98**, 3196 (1993).
33. A. F. Voter, *Phys. Rev. Lett.* **78**, 3908 (1997).
34. R. E. Gillilan and K. R. Wilson, *J. Chem. Phys.* **105**, 9299 (1996).
35. P. Eastman, N. Gronbech-Jensen, and S. Doniach, *J. Chem. Phys.* **114**, 3823 (2001).
36. R. Car and M. Parrinello, *Phys. Rev. Lett.* **55**, 2471 (1985).
37. W. G. Hoover, *Computational Statistical Mechanics* (Elsevier, Amsterdam, 1991).
38. S. Chandrasekhar, *Hydrodynamics and Hydromagnetic Stability* (Oxford University Press, Oxford, 1961).
39. M. Allen and D. J. Tildesley, *Computer Simulation of Liquids* (Oxford University Press, New York, 1987).
40. S. Chandrasekhar, *Rev. Mod. Phys.* **15**, 1 (1943).
41. D. P. Landau and K. Binder, *A Guide to Monte Carlo Simulation in Statistical Physics* (Cambridge University Press, Cambridge, 2000).
42. K. Lum and A. Luzar, *Phys. Rev. E* **56**, R6283 (1997).
43. E. I. Shakhnovich, *Curr. Opin. Struct. Biol.* **7**, 29 (1997); J. Shimada, E. L. Kussell, and E. I. Shakhnovich, *J. Mol. Bio.* **308**, 79 (2001).
44. D. Frenkel and B. Smit, *Understanding Molecular Simulation* (Academic Press, Boston, 1996).
45. N. Metropolis, A. W. Metropolis, M. N. Rosenbluth, A. H. Teller, and E. Teller, *J. Chem. Phys.* **21**, 1087 (1953).
46. H. Goldstein, *Classical Mechanics* (Addison-Wesley, Reading, MA, 1980).
47. H. C. Andersen, *J. Comp. Phys.* **52**, 24 (1983).
48. H. A. Posch and W. G. Hoover, *Phys. Rev. A* **39**, 2175 (1989).
49. P. G. de Gennes, *J. Phys. Chem.* **55**, 572 (1971).
50.. M. I. Dykman, D. G. Luchinsky, P. V. E. McClintock, and V. N. Smelyanskiy, *Phys. Rev. Lett.* **77** 5529 (1996).
51.. R. S. Maier and D. L. Stein, *Phys. Rev. Lett.* **69**, 3691 (1992).
52.. D. M. Ceperley, *Rev. Mod. Phys.* **67**, 279 (1995).
53.. H. C. Andersen, *J. Chem. Phys.* **72**, 2384 (1980).
54.. M. N. Rosenbluth and A. W. Rosenbluth, *J. Chem. Phys.* **23**, 356 (1955); J. I. Siepmann and D. Frenkel, *Mol. Phys.* **75**, 59 (1992); J. J. de Pablo, M. Laso, and U. W. Suter, *J. Chem. Phys.* **96**, 2395 (1992); D. Frenkel, G. C. A. M. Mooij, and B. Smit, *J. Phys. Condensed Matter* **4**, 3053 (1992).

55. M. Sprik, M. L. Klein, and D. Chandler, *Phys. Rev. B* **31**, 4234 (1985).
56. D. D. Frantz, D. L. Freeman, J. D. Doll, *J. Chem. Phys.* **93**, 2769 (1990).
57. B. A. Berg and T. Neuhaus, *Phys. Rev. Lett.* **68**, 9 (1992).
58. C. J. Geyer and E. A. Thompson, *J. Am. Stat. Assoc.* **90**, 909 (1995).
59. E. Marinari and G. Parisi, *Europhys. Lett.* **19**, 451 (1992).
60. U. H. E. Hansmann, *Chem. Phys. Lett.* **281**, 140 (1997).
61. T. Lazaridis and M. Karplus, *Science* **278**, 1928 (1997).
62. CPMD Version 3.3; developed by J. Hutter, A. Alavi, T. Deutsch, M. Bernasconi, S. Goedecker, D. Marx, M. Tuckerman, and M. Parrinello, Max-Planck–Institut für Festkörperforschung and IBM Zurich Research Laboraory (1995–1999).
63. D. Chandler, *Introduction to Modern Statistical Mechanics* (Oxford University Press, New York, 1987).
64. U. W. Schmitt and G. A. Voth, *J. Chem. Phys* **111**, 9361 (1999).
65. P. R. ten Wolde,, M. J. Ruiz-Montero, and D. Frenkel, *Discuss. Faraday Soc.* **104**, 93(1996).
66. U. Gasser, E. R. Weeks, A. Schofield, P. N. Nursey, and D. A. Weitz, *Science* **292**, 258 (2000).
67. M. M. Klosek, B. J. Matkowsky, and Z. Schuss, *Ber. Bunsenges. Phys. Chem.* **95**, 331 (1991); V. Pande, A. Y. Grosberg, T. Tanaka, and E. I. Shakhnovich, *J. Chem. Phys.* **108**, 334 (1998).
68. M. Sprik and G. Ciccotti, *J. Chem. Phys.* **109**, 7737 (1998).
69. J. B. Tenenbaum, V. de Silva, and J. C. Langford, *Science* **290**, 2319 (2000); S. T. Roweis and L. K. Saul, *Science* **290**, 2323 (2000).
70. V. Zaloj and R. Elber, *Comp. Phys. Comm.* **128**, 118 (2000).
71. D. N. Perera and P. Harrowell, *J. Non-Cryst. Solids* **235**, 314 (1998)
72. C. Donati, S. C. Glotzer, P. H. Poole, W. Kob, and S. J. Plimpton, *Phys. Rev. E* **60**, 3107 (1999).

CHEMICAL REACTION DYNAMICS: MANY-BODY CHAOS AND REGULARITY

TAMIKI KOMATSUZAKI

Department of Earth and Planetary Sciences, Faculty of Science, Kobe University, Kobe 657-8501 Japan

R. STEPHEN BERRY

Department of Chemistry and The James Franck Institute, The University of Chicago, Chicago, IL 60637

CONTENTS

I. Introduction
II. Canonical Perturbation Theory
    A. Lie Canonical Perturbation Theory
    B. Regional Hamiltonian
    C. Algebraic Quantization
III. Regularity in Chaotic Transitions
    A. Invariancy of Actions in Transition States
    B. "See" Trajectories Recrossing the Configurational Dividing Surface in Phase Space
    C. "Unify" Transition State Theory and Kramers–Grote–Hynes Theory
    D. "Visualize" a Reaction Bottleneck in Many-Body Phase Space
    E. "Extract" Invariant Stable and Unstable Manifolds in Transition States
    F. A Brief Comment on Semiclassical Theories
IV. Hierarchical Regularity in Transition States
V. Concluding Remarks and Future Prospects
VI. Acknowledgments
VII. Appendix
    A. The Proof of Eqs. (2.8) and (2.9)
    B. Lie Transforms
        1. Autonomous Cases
        2. Nonautonomous Cases
    C. Perturbation Theory Based on Lie Transforms
    D. A simple Illustration of Algebraic Quantization
    E. LCPT with One Imaginary Frequency Mode
References

Advances in Chemical Physics, Volume 123, Edited by I. Prigogine and Stuart A. Rice.
ISBN 0-471-21453-1 © 2002 John Wiley & Sons, Inc.

I. INTRODUCTION

Chemists have long envisioned reactions as passages from an initial, reactant state, locally stable, through an unstable transition state, to a final, stable product state. Described in terms of a multidimensional surface of internal energy as a function of the locations of the atomic nuclei, this model has the reacting system go from one local minimum across a saddle in the landscape to another local minimum. The questions, How fast does a system actually traverse the saddles? and What kinds of trajectories carry the system through?, have been among the most intriguing subjects in chemical reaction theories over the past several decades [1–15]. The introduction of the concept of "transition state" by Eyring and Wigner in 1930s [3–5] had great successes in the understanding of the kinetics of chemical reactions: It led to the definition of a hypersurface (generally in phase space) through which a reacting system should pass only once on the way from reactants to products. It has also provided us with a magnifying glass to understand the kinetics by decomposing the evolution of the reactions into, first, how a reacting system reaches into the transition state from the reactant state by getting a certain amount of thermal or light energy, and, second, how the system leaves the transition state after its arrival there, for example, its passage velocity and pattern of crossings.

A widespread assumption in a common class of chemical reaction theories [3–11] is the existence of such a hypersurface in phase space dividing the space into reactant and product regions that a chemical species crosses it only once on its path to reaction. However, many formulations of chemical reaction rate theories have had to allow this probability, the "transmission coefficient κ," to be less than unity. Toward resolving the recrossing problem that spoils this "no-return" hypothesis, researchers have so far tried to interpret the reaction rates by using either variational transition state theory (TST) [9–11] which optimizes a *configurational* dividing surface by minimizing the recrossings, or the (generalized) Langevin formalism developed by Kramers [14] and Grote and Hynes [15], which regards the recrossings as arising from "(molecular) friction" by the "bath" degrees of freedom, which retards any of the reactive trajectories. Neither of these approaches, however, could have clarified the actual mechanics of the systems' passage through a transition state.

Several findings, both theoretical [16–29] and experimental [30, 31], during the last decades have shed light on the mechanics of passage through the reaction bottlenecks, and on the concept of transition state, especially in systems with only a few degrees of freedom (dof). The striking experimental studies by Lovejoy et al. [30] "see" this transition state via the photofragment excitation spectra for unimolecular dissociation of highly vibrationally excited ketene. These spectra revealed that the rate of this reaction is controlled by the flux through *quantized thresholds* within a certain energy range above the barrier. The observability of the quantized thresholds in the transition state was first discussed by Chatfield et al. [32]. Marcus [33] pointed out that this indicates that the transverse vibrational quantum numbers might indeed be approximate constants of motion, presumably in the saddle region.

In the same period, Berry and his co-workers explored the nonuniformity of dynamical properties of Hamiltonian systems of several N-atom clusters, with N from 3 to 13; in particular, they explored how regular and chaotic behavior may vary locally with the topography of the potential energy surfaces (PESs) [16–23]. They revealed, by analyses of local Liapunov functions and Kolmogorov entropies, that when systems have just enough energy to pass through the transition state, the systems' trajectories become collimated and regularized through the transition state, developing approximate *local* invariants of motion different from those in the potential well. This occurs even though the dynamics in the potential well is fully chaotic under these conditions. It was also shown that at higher energies above the threshold, emerging mode–mode mixing wipes out these approximate invariants of motions even in the transition state.

Davis and Gray [24] first showed that in Hamiltonian systems with two dof, the transition state defined as the separatrix in the phase space is always free from barrier recrossings, so the transmission coefficient for such systems is unity. They also showed the existence of the dynamical bottlenecks to intramolecular energy transfer in the region of potential well, that is, cantori (in a two-dof system), which form partial barriers between irregular regions of phase space [24–26]. Zhao and Rice [26] developed a convenient approximation for the rate expression for the intermolecular energy transfer. However, their inference depends crucially on the Poincaré section having only two dimensions, and no general theory exists yet for systems of higher dimensionality [27, 34–36].

By focusing on the transition state periodic orbits in the vicinity of the unstable saddle points, Pechukas et al. [37] first showed in the late 1970s, for two-dimensional (2D) Hamiltonian systems such as the collinear $H + H_2$ reaction, that, within a suitable energy range just above the saddle, the reaction bottleneck over which no recrossings occur with a minimal flux of the system, can be uniquely identified as one periodic orbit dividing surface (PODS), a dividing surface $S(q_1 = 0)$. (Here q_1 is the hyperbolic normal coordinate about the saddle point.) Moreover, as the energy increases, pairs of the PODSs appearing on each reactant and product side migrate outward, toward reactant and product state, and the outermost PODS become identified as the reaction bottleneck. De Leon [28] developed a so-called reactive island theory; the reactive islands are the phase-space areas surrounded by the periodic orbits in the transition state, and reactions are interpreted as occurring along cylindrical invariant manifolds through the islands. Fair et al. [29] also found in their two- and three-dof models of the dissociation reaction of hydrazoic acid that a similar cylinder-like structure emerges in the phase space as it leaves the transition state. However, these are crucially based on the findings and the existence of (pure) periodic orbits for all the dof, at least in the transition states. Hence, some questions remain unresolved, for example, "How can one extract these periodic orbits from many-body dof phase space?" and "How can the periodic orbits persist at high energies above the saddle point, where chaos may wipe out any of them?

Recently, we developed a new method to look more deeply into these local regularities about the transition state of N-particle Hamiltonian systems [38–44]. The crux of the method is the application of Lie canonical perturbation theory (LCPT) [45–53], combined with microcanonical molecular dynamics (MD) simulation of a region around a saddle point. This theory constructs the nonlinear transformation to a hyperbolic coordinate system, which "rotates away" the recrossings and nonregular behavior, especially of the motion along the reaction coordinate. We showed by using intramolecular proton-transfer reaction of malonaldehyde [38, 39] and isomerization reactions in a simple cluster of six argon atoms [40–44] that, even to energies so high that the transition state becomes manifestly chaotic, at least one action associated with the reaction coordinate remains an approximate invariant of motion throughout the transition state. Moreover, it is possible to choose a multidimensional phase-space dividing surface through which the transmission coefficient for the classical reaction path is unity [40]. We "visualized" the dividing hypersurface in the phase space by constructing the projections onto subspaces of a very few coordinates and momenta, revealing how the "shape" of the reaction bottleneck depends on energy of the system and the passage velocity through the transition state, and how the complexity of the recrossings emerges over the saddle in the configurational space [41, 42]. (The dividing hypersurface migrates, depending on the passage velocity, just as PODS do.) We showed that this also makes it possible to visualize the stable and unstable invariant manifolds leading to and from the hyperbolic point of the transition state, like those of the one-dimensional (1D), integrable pendulum, and how this regularity turns to chaos with increasing total energy of the system. This, in turn, illuminates a new type of phase-space bottleneck in a transition state that emerges as the total energy, which keeps a reacting system increasingly trapped in that state, irrespective of any coordinate system one might see the dynamical events [43, 44].

This chapter is a review of our recent theoretical developments [38–44], in which we address two fundamental questions:

1. How can we extract a dividing hypersurface as free as possible from recrossings between the reactant and product states? And what is the physical foundation of why the reacting system can climb through the saddles?

2. How do the topographical features of a potential surface transform the dynamics of saddle crossings as the energy of the reacting system increases from threshold to much higher values? As a corollary, what role do saddles, including those of rank >1, play in the system's transition from regular to chaotic dynamics?

The outline of this chapter is as follows. In Section II, we briefly review canonical perturbation theory and an efficient technique, so-called algebraic quantization, for applying this method to a regional Hamiltonian about any stationary point. In Section III, we show the approximate local invariant of motion buried in the complexity of the original Hamiltonian $H(\mathbf{p}, \mathbf{q})$, without invoking any explicit assumption of its integrability. We use, as an illustrative vehicle, the isomerization of an Ar_6 cluster and show that the invariants associated with a reaction coordinate in the phase space—whose reactive trajectories are all "no-return" trajectories—densely distribute in the sea of chaotic dof in the regions of (first-rank) saddles. We also show how the invariants locate in the phase space and how they depend on the total energy of the system and other physical quantities, and discuss its implication for reaction dynamics, especially for many dof systems. In Section IV, we examine a universal consequence that holds in the regions of any first-rank saddle, that is, a hierarchical regularity in the transition states. In Section V, we give some concluding remarks and future prospects. In the appendices, we present the detailed description of Lie transform-based canonical perturbation theory with algebraic quantization, LCPT–AQ, and apply it there to a simple 2D system.

II. CANONICAL PERTURBATION THEORY

To begin, let us see what all the several forms of Canonical Perturbation Theories (CPT) provide. All the CPTs [45–53], including normal form theories [54, 55], require that an M-dimensional Hamiltonian $H(\mathbf{p}, \mathbf{q})$ in question be expandable as a series in powers of "ε," where the zeroth-order Hamiltonian H_0 is integrable as a function of the action variables $\mathbf{J}$ only

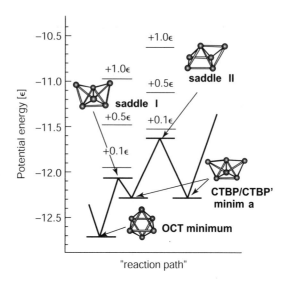

Figure 2.1. Potential energy profile of an Ar_6 atom cluster: ϵ is the unit of energy ($=121$ K).

and does not depend on the conjugate angle variables Θ,

$$H(\mathbf{p}, \mathbf{q}) = \sum_{n=0} \epsilon^n H_n(\mathbf{p}, \mathbf{q}) = H_0(\mathbf{J}) + \sum_{n=1} \epsilon^n H_n(\mathbf{J}, \Theta) \quad (2.1)$$

where $\mathbf{p}$ and $\mathbf{q}$ represent momenta and the conjugate coordinates of the system, respectively.

Furthermore, the canonical transformation W of the coordinate system minimizes the angular dependencies of the new Hamiltonian $\bar{H}$, thereby making the new action variables $\bar{\mathbf{J}}$ as nearly constant as possible. If $\bar{H}$ can be obtained altogether independent of the angle $\bar{\Theta}$ (at least, at the order of the perturbative calculation performed), then

$$H(\mathbf{p}, \mathbf{q}) \xrightarrow{W} \bar{H}(\bar{\mathbf{p}}, \bar{\mathbf{q}}) = \bar{H}(\bar{\mathbf{J}}) = \sum_{n=0} \epsilon^n \bar{H}_n(\bar{\mathbf{J}}) \quad (2.2)$$

so the new action and angle variables for mode k are expressed as

$$\frac{d\bar{J}_k}{dt} = \dot{\bar{J}}_k = -\frac{\partial \bar{H}(\bar{\mathbf{J}})}{\partial \bar{\Theta}_k} = 0 \quad (2.3)$$

$$\bar{J}_k = \text{constant} \quad (2.4)$$

and

$$\dot{\bar{\Theta}}_k = \frac{\partial \bar{H}(\bar{\mathbf{J}})}{\partial \bar{J}_k} \equiv \bar{\omega}_k(\bar{\mathbf{J}}) = \text{constant} \tag{2.5}$$

$$\bar{\Theta}_k = \bar{\omega}_k(\bar{\mathbf{J}})t + \beta_k \tag{2.6}$$

where β_k is the arbitrary initial phase factor of mode k, and $\bar{\mathbf{p}}(=\bar{p}(\mathbf{p}, \mathbf{q}))$ and $\bar{\mathbf{q}}(=\bar{q}(\mathbf{p}, \mathbf{q}))$ are canonically transformed new momenta and the conjugate coordinates, respectively.

If the zeroth-order Hamiltonian $H_0(\mathbf{p}, \mathbf{q})$ is a system of harmonic oscillators,

$$H_0(\mathbf{p}, \mathbf{q}) = \sum_{k=1}^{M} \frac{1}{2}(p_k^2 + \omega_k^2 q_k^2) = \sum_{k=1}^{M} \omega_k J_k \tag{2.7}$$

where ω_k is the fundamental frequency of mode k, these yield the equations of motion for $\bar{q}_k$ and $\bar{p}_k$, to obey the $\bar{H}$:

$$\frac{d^2 \bar{q}_k(\mathbf{p}, \mathbf{q})}{dt^2} + \bar{\omega}_k^2 \bar{q}_k(\mathbf{p}, \mathbf{q}) = 0 \tag{2.8}$$

and

$$\bar{p}_k(\mathbf{p}, \mathbf{q}) = \frac{\omega_k}{\bar{\omega}_k} \frac{d \bar{q}_k(\mathbf{p}, \mathbf{q})}{dt} \tag{2.9}$$

where $\bar{\omega}_k(=\bar{\omega}_k(\bar{\mathbf{J}}) = \bar{\omega}_k(\bar{\mathbf{p}}, \bar{\mathbf{q}}))$ is independent of time t because $\bar{\mathbf{J}}$ are constant (see its derivation in Appendix A). The general form of the solution can be represented as

$$\bar{q}_k(\mathbf{p}, \mathbf{q}) = \alpha e^{i\bar{\omega}_k(\bar{\mathbf{J}})t} + \beta e^{-i\bar{\omega}_k(\bar{\mathbf{J}})t} \tag{2.10}$$

$$\bar{p}_k(\mathbf{p}, \mathbf{q}) = \alpha \omega_k e^{i\bar{\omega}_k(\bar{\mathbf{J}})t} - \beta \omega_k e^{-i\bar{\omega}_k(\bar{\mathbf{J}})t} \tag{2.11}$$

Here, α and β are arbitrary constants depending on the initial value of $\bar{q}_k$ and $\bar{p}_k$.

The action, canonical momenta in action-angle coordinate system, is of fundamental importance for understanding of regularity in dynamics, that is, the constancy or invariancy of the action implies how separable the mode is. The advantage of any of the several forms of CPT is the reduction of dimensionality needed to describe the Hamiltonian. For example, Eqs. (2.8) and (2.9) tell us that even though the motions look quite complicated in the old coordinate system, they could be followed as simple decoupled periodic orbits in the phase space. For realistic many-body nonlinear systems, Eqs. (2.8) and (2.9) might not be retained through the dynamical evolution of the system because the (near-)commensurable conditions may densely distribute in typical regions throughout the phase space; that is, any integer linear combination of frequencies that vanishes identically at some order, ε^n;

$$\sum_{k=1}^{M} n_k \omega_k \leq \mathcal{O}(\varepsilon^n) \tag{2.12}$$

(n_k is arbitrary integer), makes the corresponding new Hamiltonian diverge and destroys invariants of motion. If the system satisfies any such (near-)commensurable condition, the new Hamiltonian might have to include the corresponding angle variables to avoid divergence [48–50, 55]. Otherwise the CPT calculation would have to be performed to infinite order in near-commensurable cases.

Until now, most studies based on the CPTs have focused on transforming the new Hamiltonian itself to as simple a form as possible, to avoid divergence, and to obtain this form through specific CPT calculations of low finite order. In other terms, CPT imputes the responsibility of determining the integrability of the Hamiltonian to the inclusion of some specific angle variables and/or the convergence of the perturbation calculation. Another, potentially powerful usage of CPT, especially for many-body chemical reaction systems, should be its application as a detector to monitor occurrence of local invariance, by use of the new action $\bar{J}_k(\mathbf{p}, \mathbf{q})$ and the new frequency $\bar{\omega}_k(\mathbf{p}, \mathbf{q})$ along classical trajectories obeying equations of motion of the *original* Hamiltonian $H(\mathbf{p}, \mathbf{q})$. That is, it is quite likely that the more dof in the system, the more the global invariants through the whole phase space become spoiled; nevertheless the invariants of motion might survive within a *certain locality*, that is, for a certain finite duration, a region of phase space or in a certain limited subset of dof. The standard resonance Hamiltonian [55] constructed to avoid the near-commensurability might also prevent the possibility of detecting such a limited, approximate invariant of motion retained in a certain locality. Note that the strength of local invariants could not be detected in use of (traditional) Liapunov analysis because *local* Liapunov exponents, characterized as the finite time averages of the rate of exponential growth of an infinitesimal deviation, are affected both by the well-known horseshoe mechanism and by the degrees of noncompactness or local hyperbolicity of potential energy topographies, for example, any Liapunov exponent becomes a positive definite for an integrable Hamiltonian system composed of negatively curved harmonic oscillators.

A. Lie Canonical Perturbation Theory

The traditional Poincaré–Von Zeipel approach [53] of CPT is based on mixed-variable generating functions F:

$$\bar{\mathbf{q}} = \frac{\partial F(\bar{\mathbf{p}}, \mathbf{q})}{\partial \bar{\mathbf{p}}} \qquad \mathbf{p} = \frac{\partial F(\bar{\mathbf{p}}, \mathbf{q})}{\partial \mathbf{q}} \qquad (2.13)$$

which requires functional inversion to obtain explicit formulas for $(\mathbf{p}, \mathbf{q})$ in terms of $(\bar{\mathbf{p}}, \bar{\mathbf{q}})$ and vice versa, at each order of the perturbative calculation. This imposes a major impediment to implementing higher order perturbations and to treating systems with many-degrees of freedom.

With the mixed-variable generating functions, after Birkoff [54], Gustavson [55] developed an elegant technique to extract the new Hamiltonian to avoid divergence by assuming that the new Hamiltonian is expandable in normal form; if complete inversion of the variables is not required, the procedure to calculate the new Hamiltonian can be rather straightforward.

LCPT [46, 47, 51, 52, 56] first developed by a Japanese astrophysicist, Hori [51, 52], is superior to all the most traditional methods, in that no cumbersome functions of mixed variables appear and all the terms in the series are repeating Poisson brackets. The crux is the use of Lie transforms, which is regarded as a "virtual" time evolution of phase-space variables $\mathbf{z}(=(\mathbf{p}, \mathbf{q}))$ along the "time" ε driven by a "Hamiltonian" W; that is,

$$\frac{d\mathbf{z}}{d\varepsilon} = \{\mathbf{z}, W(\mathbf{z})\} \equiv -L_W \mathbf{z} \qquad (2.14)$$

Here, $\{\ \}$ denotes the Poisson bracket. The formal solution can be represented as

$$\mathbf{z}(\varepsilon) = \exp\left[-\int^\varepsilon L_{W(\varepsilon')} d\varepsilon'\right] \mathbf{z}(0) \qquad (2.15)$$

As shown in Appendix B, it can easily be proved for any transforms described by the functional form of Eq. (2.15), that if $\mathbf{z}(0)$ are canonical, $\mathbf{z}(\varepsilon)$ are also canonical (and vice versa), as the time evolution of any Hamiltonian system is regarded as a canonical transformation from canonical variables at an initial time to those at another time, maintaining the structure of Hamilton's equations.

For any function f evaluated at "a point" $\mathbf{z}(0)$, the evolution operator T, defined as below, yields a new function g represented as a function of $\mathbf{z}(0)$ and ε, whose functional *value* is equal to $f(\mathbf{z}(\varepsilon))$:

$$f(\mathbf{z}(\varepsilon)) = Tf(\mathbf{z}(0)) = \exp\left[-\int^{\varepsilon} L_{W(\mathbf{z}(0);\varepsilon')}\,d\varepsilon'\right] f(\mathbf{z}(0)) = g(\mathbf{z}(0);\varepsilon) \quad (2.16)$$

The Lie transforms of an autonomous Hamiltonian H to a new Hamiltonian $\bar{H}$ can be brought about by

$$\bar{H}(\mathbf{z}(\varepsilon)) = T^{-1}H(\mathbf{z}(\varepsilon)) = H(\mathbf{z}(0)) \quad (2.17)$$

by determining W (also assumed to be expandable in powers of ε as H and $\bar{H}$ are) so as to make the new Hamiltonian as free from the new angle variables $\bar{\Theta}$ as possible, at each order in ε. Here, the *inverse* evolution operator T^{-1} brings the system dwelling at a "time" backward to a past in ε from *that* "time" along the dynamical evolution $\mathbf{z}$, yielding $H(\mathbf{z}(0))$ (see Appendix C in detail). We shall hereafter designate the initial values of $\mathbf{z}$, $\mathbf{z}(0)$, by $(\mathbf{p},\mathbf{q})$, and those at "time" ε by $(\bar{\mathbf{p}},\bar{\mathbf{q}})$. Then, one can see that Eq. (2.17) corresponds to a well-known relation between the old and new Hamiltonians hold under any canonical transformation for autonomous systems:

$$\bar{H}(\bar{\mathbf{p}},\bar{\mathbf{q}}) = H(\mathbf{p},\mathbf{q}) \quad (2.18)$$

The great advantage of LCPT in comparison with the Birkoff–Gustavson's normal form [54, 55] is that, after W is once established through each order, the new transformed physical quantities, for example, new action $\bar{J}_k$, frequency $\bar{\omega}_k$, momentum $\bar{p}_k$, and coordinate $\bar{q}_k$ of mode k, can be expressed straightforwardly as functions of the original momenta and coordinates $(\mathbf{p},\mathbf{q})$ by using the evolution operator T

$$\bar{J}_k(\mathbf{p},\mathbf{q}) = TJ_k(\mathbf{p},\mathbf{q}) = T\left(\frac{p_k^2 + \omega_k^2 q_k^2}{2\omega_k}\right) \quad (2.19)$$

$$\bar{\omega}_k(\mathbf{p},\mathbf{q}) = T\frac{\partial \bar{H}(\mathbf{J})}{\partial J_k} \quad (2.20)$$

$$\bar{p}_k(\mathbf{p},\mathbf{q}) = Tp_k \quad (2.21)$$

$$\bar{q}_k(\mathbf{p},\mathbf{q}) = Tq_k \quad (2.22)$$

For convenience, we denote hereafter the transformed quantities f in terms of original $(\mathbf{p},\mathbf{q})$ (and ε) by $\bar{f}(\mathbf{p},\mathbf{q})$, for example, $\bar{J}_k(\mathbf{p},\mathbf{q})$ when $f = J_k$, and we use the notation $J_k(\mathbf{p},\mathbf{q})$ to represent the action of H_0;

$$J_k(\mathbf{p},\mathbf{q}) = \frac{p_k^2 + \omega_k^2 q_k^2}{2\omega_k} = \frac{1}{2\pi}\oint_{E=H_0(\mathbf{p},\mathbf{q})} p_k\, dq_k \qquad (2.23)$$

Note that the *original* coordinate system $(\mathbf{p},\mathbf{q})$ are, in other terms, regarded as the canonical variables to represent harmonic motions of H_0, but $(\bar{\mathbf{p}}(\mathbf{p},\mathbf{q}), \bar{\mathbf{q}}(\mathbf{p},\mathbf{q}))$ correspond to the canonical variables, which represent periodic/hyperbolic regular motions in the phase space for the nonlinear $H(\mathbf{p},\mathbf{q})$ if $\bar{H}(\bar{\mathbf{p}},\bar{\mathbf{q}})$ actually exists.

For example, $\bar{p}_k^{ith}(\mathbf{p},\mathbf{q})$ and $\bar{q}_k^{ith}(\mathbf{p},\mathbf{q})$ have the following forms, respectively,

$$\bar{p}_k^{ith}(\mathbf{p},\mathbf{q}) = \sum_{n=0}^{i} \varepsilon^n \sum_j c_{nj}\mathbf{p}^{2\mathbf{s}_{nj}+1}\mathbf{q}^{\mathbf{t}_{nj}} \qquad (2.24)$$

$$\bar{q}_k^{ith}(\mathbf{p},\mathbf{q}) = \sum_{n=0}^{i} \varepsilon^n \sum_j c'_{nj}\mathbf{p}^{2\mathbf{s}'_{nj}}\mathbf{q}^{\mathbf{t}'_{nj}} \qquad (2.25)$$

where, for example,

$$\mathbf{p}^{2\mathbf{s}_{nj}+1} \equiv \prod_{l=1}^{M} p_l^{s_{njl}} \left(\sum_{l=1}^{M} s_{njl} = |2\mathbf{s}_{nj}+1| \right) \qquad (2.26)$$

and

$$\mathbf{q}^{\mathbf{t}_{nj}} \equiv \prod_{l=1}^{M} q_l^{t_{njl}} \left(\sum_{l=1}^{M} t_{njl} = |\mathbf{t}_{nj}| \right) \qquad (2.27)$$

Each coefficient depends on the original Hamiltonian and the order of CPT: for example, c_{nj} and c'_{nj} denote the (real) coefficients of the jth term at the nth order in $\bar{p}_k^{ith}(\mathbf{p},\mathbf{q})$ and $\bar{q}_k^{ith}(\mathbf{p},\mathbf{q})$, respectively; $\mathbf{s}_{nj}$ and $\mathbf{t}_{nj}$ are arbitrary positive integer vectors where $|\mathbf{s}_{nj}|, |\mathbf{t}_{nj}| \geq 0$ [s_{njl} and t_{njl}, arbitrary positive integers (≥ 0)] associated with the jth term at the nth order in $\bar{p}_k^{ith}(\mathbf{p},\mathbf{q})$. The new $\bar{p}_k^{ith}(\mathbf{p},\mathbf{q})$ and $\bar{q}_k^{ith}(\mathbf{p},\mathbf{q})$ maintain time reversibility. We showed in the online supplement [40] the expressions through second order for $\bar{p}_1(\mathbf{p},\mathbf{q})$ and $\bar{q}_1(\mathbf{p},\mathbf{q})$ at saddle I, defined below, of Ar_6. The contributions of the original p_1 and q_1 in $\bar{p}_1^{ith}(\mathbf{p},\mathbf{q})$ and $\bar{q}_1^{ith}(\mathbf{p},\mathbf{q})$ are not necessarily large and almost all modes contribute to $\bar{p}_1^{ith}(\mathbf{p},\mathbf{q})$ and $\bar{q}_1^{ith}(\mathbf{p},\mathbf{q})$ for $i \geq 1$.

Despite its versatility, CPT has rarely been applied to many dof realistic molecular systems. The main reason is twofold: First is the cumbersome task of the *analytical* derivative and integral calculations that appear successively in all kinds of CPT procedures. Second is the near-impossibility of obtaining even moderately simple *analytical* expressions to describe the accurate (e.g., ab initio) PESs in full. Two prescriptions for these obstacles have been developed, that is, the construction of an approximate Hamiltonian focusing on not global, but rather regional feature of dynamics in the vicinity of an arbitrary stationary point in question [38, 39], and a so-called "algebraic quantization" [38, 48–50], which replaces the cumbersome analytical differentiations and integrations carried out by computing directly with symbolic operations based on simple Poisson bracket rules.

B. Regional Hamiltonian

We first expand the full $3N$-dof potential energy surface about a chosen stationary point, that is, minimum, saddle, or higher rank saddle. By taking the zeroth-order Hamiltonian as a harmonic oscillator system, which might include some negatively curved modes, that is, reactive modes, we establish the higher order perturbation terms to consist of nonlinear couplings expressed in arbitrary combinations of coordinates.

$$H = H_0 + \sum_{n=1}^{\infty} \varepsilon^n H_n \qquad (2.28)$$

where

$$H_0 = \frac{1}{2} \sum_j (p_j^2 + \omega_j^2 q_j^2) \qquad (2.29)$$

$$\sum_{n=1}^{\infty} \varepsilon^n H_n = \varepsilon \sum_{j,k,l} C_{jkl} q_j q_k q_l + \varepsilon^2 \sum_{j,k,l,m} C_{jklm} q_j q_k q_l q_m + \cdots \qquad (2.30)$$

Here, q_j and p_j are the jth normal coordinate and its conjugate momentum, respectively; ω_j and C_{jkl}, C_{jklm}, ... are, respectively, the frequency of the jth mode, the coupling coefficient among q_j, q_k, and q_l and that among q_j, q_k, q_l, and q_m, and so forth. The frequency associated with an unstable reactive mode F and those of the other stable modes B are pure-imaginary and real, respectively. At any stationary point there are six zero-frequency modes corresponding to the total translational and infinitesimal rotational motions, and the normal coordinates of the infinitesimal rotational motions appear in the perturbation terms $H_n(\mathbf{q})$ ($n > 0$). The contribution of the total translational motion is simply separated.

We make no more mention of this. If one deals with a system whose total angular momentum is zero, one could eliminate the contributions of the total rotational motions from $H_n(\mathbf{q})$ ($n > 0$) by operating with a suitable projection operator [57]; at the stationary point it corresponds to putting to zero each normal coordinate and corresponding conjugate momentum representing the infinitesimal total rotational motion [58]. If the total angular momentum is not zero, the coupling elements among the rotational and vibrational modes must be taken into account. For the sake of simplicity, we focus on a $(3N-6)$-dof Hamiltonian system with total linear and angular momenta of zero, so that the kinetic and potential energies are purely vibrational [59]. For such a zeroth-order Hamiltonian $\omega_k \neq 0$ for all $k(=1, 2, \ldots, 3N-6)$, the associated action-angle variables of the stable modes B ($\omega_B \in \mathfrak{R}$:*real*) and the unstable mode F ($\omega_F \in \mathfrak{F}$:*imaginary*) are expressed as

$$J_B = \frac{1}{2\pi} \oint p_B \, dq_B = \frac{1}{2}\left(\frac{p_B^2}{\omega_B} + \omega_B q_B^2\right) \quad (2.31)$$

$$\Theta_B = \tan^{-1}\left(\frac{\omega_B q_B}{p_B}\right) \quad (2.32)$$

and

$$J_F = \frac{1}{2\pi} \operatorname{Im} \int_{\text{barrier}} p_F \, dq_F \quad (2.33)$$

$$= \frac{i}{2}\left(\frac{p_F^2}{|\omega_F|} - |\omega_F| q_F^2\right) \quad (2.34)$$

$$\Theta_F = -i \tanh^{-1}\left(\frac{|\omega_F| q_F}{p_F}\right) \quad \omega_F \equiv -|\omega_F| i \quad (2.35)$$

Here, the action associated with the reactive mode F has first been postulated in semiclassical transition state theory by Miller [12, 13, 60], and it is easily verified

$$\{\Theta_j, J_k\} = \delta_{jk} \quad \{\Theta_j, \Theta_k\} = \{J_j, J_k\} = 0 \quad (2.36)$$

that any set of variables $\mathbf{J}$ and $\mathbf{\Theta}$ is canonical, including those associated with the unbound mode F.

C. Algebraic Quantization

The CPT calculation requires decomposing the functions appearing at each order in ε, called $\xi(\mathbf{p}, \mathbf{q})$ for brevity, usually represented by a sum of an arbitrary combination of arbitrary power series in $\mathbf{p}$ and $\mathbf{q}$,

$$\xi(\mathbf{p}, \mathbf{q}) = \sum_j c_j \mathbf{p}^{\mathbf{s}_j} \mathbf{q}^{\mathbf{t}_j} \qquad (2.37)$$

into

$$\langle \xi \rangle = \frac{1}{2\pi} \oint \xi(\mathbf{p}, \mathbf{q}) \, d\Theta \quad \text{and} \quad \{\xi\} = \xi - \langle \xi \rangle \qquad (2.38)$$

where

$$\mathbf{p}^{\mathbf{s}_j} \equiv \prod_{l=1}^{M} p_l^{s_{jl}} \left(\sum_{l=1}^{M} s_{jl} = |\mathbf{s}_j| \right) \qquad (2.39)$$

$$\mathbf{q}^{\mathbf{t}_j} \equiv \prod_{l=1}^{M} q_l^{t_{jl}} \left(\sum_{l=1}^{M} t_{jl} = |\mathbf{t}_j| \right) \qquad (2.40)$$

Here c_j, the coefficient of the jth term of ξ, is not only real but also imaginary for the CPT calculations, if they include imaginary frequency mode(s); $\mathbf{s}_j$ and $\mathbf{t}_j$ are arbitrary positive integer vectors where $|\mathbf{s}_j|, |\mathbf{t}_j| \geq 0$ (their components s_{jl} and t_{jl}, arbitrary positive integers (≥ 0)) associated with $\mathbf{p}$ and $\mathbf{q}$ (p_l and q_l) of the jth term.

For a wide class of Hamiltonians described in Section II.B, a quite efficient technique, called algebraic quantization (AQ), was developed [38, 48–50]. This method first formally transforms $\xi(\mathbf{p}, \mathbf{q})$ to $\xi(\mathbf{a}^*, \mathbf{a})$ in terms of $(\mathbf{a}^*, \mathbf{a})$, which may correspond to customary creation and annihilation operators in quantum field theory, that is,

$$\xi(\mathbf{p}, \mathbf{q}) \to \xi(\mathbf{a}^*, \mathbf{a}) = \sum_s d_s \mathbf{a}^{*\mathbf{v}_s} \mathbf{a}^{\mathbf{u}_s} \qquad (2.41)$$

where

$$\mathbf{a}^{*\mathbf{v}_s} \equiv \prod_{l=1}^{M} (a_l^*)^{v_{sl}} \left(\sum_{l=1}^{M} v_{sl} = |\mathbf{v}_s| \right) \qquad (2.42)$$

$$\mathbf{a}^{\mathbf{u}_s} \equiv \prod_{l=1}^{M} a_l^{u_{sl}} \left(\sum_{l=1}^{M} u_{sl} = |\mathbf{u}_s| \right) \qquad (2.43)$$

and

$$a_k^* = \frac{1}{\sqrt{2}} (p_k + i\omega_k q_k) \qquad a_k = \frac{1}{\sqrt{2}} (p_k - i\omega_k q_k) \qquad (2.44)$$

Here, d_s, and $\mathbf{v}_s$ and $\mathbf{u}_s$, where $|\mathbf{v}_s|$, $|\mathbf{u}_s| \geq 0$ (their components v_{sl} and u_{sl}) depend on c_j, and $\mathbf{s}_j$ and $\mathbf{t}_j$ (s_{jl} and t_{jl}) in Eq. (2.37). By solving the following equation of motion,

$$\frac{da_k^*}{d\tau} = \{a_k^*, H_0\} \qquad \frac{da_k}{d\tau} = \{a_k, H_0\} \qquad (2.45)$$

a_k^* and a_k can be expressed in terms of action J_k, frequency ω_k, and time τ obeying Hamiltonian H_0:

$$a_k^*(\tau) = \sqrt{\omega_k J_k}\, e^{i\Theta_k} = \sqrt{\omega_k J_k}\, e^{i(\omega_k \tau + \beta_k)} \qquad (2.46)$$

$$a_k(\tau) = \sqrt{\omega_k J_k}\, e^{-i\Theta_k} = \sqrt{\omega_k J_k}\, e^{-i(\omega_k \tau + \beta_k)} \qquad (2.47)$$

Then, one can rewrite Eq. (2.41) thanks to these equations as

$$\xi(\tau) = \sum_s \text{constant} \times d_s \exp[i(\mathbf{v}_s - \mathbf{u}_s) \cdot \boldsymbol{\omega}\tau] \qquad (2.48)$$

which enables us to identify $\{\xi\}$ and $\langle\xi\rangle$ by simply checking the strength of the quantity associated with sth term,

$$|(\mathbf{v}_s - \mathbf{u}_s) \cdot \boldsymbol{\omega}| = \left|\sum_{l=1} (v_{sl} - u_{sl})\omega_l\right| \qquad (2.49)$$

that is, all the terms in the summation of Eq. (2.48), which are regarded as free from and depending on time τ, are those of $\langle\xi\rangle$ and $\{\xi\}$. Furthermore, the cumbersome analytical calculations of the convolutions by Poisson bracket are also replaced by symbolic operations with no special mathematical manipulators, thanks to the simple Poisson bracket rules,

$$\{a_j^*, a_k^*\} = \{a_j, a_k\} = 0 \qquad \{a_j^*, a_k\} = i\omega_k \delta_{jk} \qquad (2.50)$$

where δ is Kronecker delta. (See an illustrative example in Appendix D.)

III. REGULARITY IN CHAOTIC TRANSITIONS

We have applied this method to saddle crossing dynamics in intramolecular proton-transfer reaction of malonaldehyde [38, 39] and isomerization reaction of Ar_6 [40–44]. The former is a reacting system, involving a typical chemical bond breaking-and-forming; the latter is the smallest inert gas cluster in which no saddle dynamics more regular than the dynamics within the local wells was revealed by the local K entropy analysis by Hinde and Berry [21].

In this chapter, we will show our recent analyses of Ar_6 isomerization, as an illustrative vehicle, with no peculiar or specific mode(s), that offers well representable, generalizable situations. The potential energy of Ar_6 is represented by the sum of pairwise Lennard–Jones potentials,

$$V(\mathbf{r}) = 4\epsilon \sum_{i>j} \left[\left(\frac{\sigma}{r_{ij}}\right)^{12} - \left(\frac{\sigma}{r_{ij}}\right)^{6} \right] \quad (2.51)$$

Here, we assigned laboratory scales of energy and length appropriate for argon, that is, $\epsilon = 121$ K and $\sigma = 3.4$ Å with the atomic mass $m = 39.948$ amu, and the total linear and angular momenta are set to zero [40]. This cluster has two kinds of potential energy minima (see Figure 2.1). The global minimum corresponds to an octahedral arrangement of the atoms (OCT), with energy $E = -12.712\epsilon$, and the other, higher minimum, to a trigonal bipyramid structure of five atoms, capped on one face by the sixth atom (CTBP), with energy $E = -12.303\epsilon$. There are two distinct kinds of first-rank saddles. One, saddle I, at energy $E = -12.079\epsilon$ joins the OCT and the CTBP minima. The other higher saddle, saddle II, at energy $E = -11.630\epsilon$, joins two permutationally distinct CTBP structures. Saddle II is slightly flatter than the lower saddle (Table II.1). We analyzed the invariants of motion during the course of isomerization reaction at total energies $E = 0.1, 0.5,$ and 1.0ϵ above each saddle point energy at each saddle, for example, 16(45), 79(223), and 158(446)% of the barrier height of OCT → CTBP (OCT ← CTBP). The computational recipe for constructing the $3N-6(=12)$-dof regional Hamiltonian was described elsewhere [40]. The three- and four-body coupling terms for both the saddles were determined by introducing an appropriate cut-off value; the total numbers of terms were 106 three-, and 365 four-body couplings for saddle I, and 189 and 674 for saddle II.

Here, we analyzed the Lie-transformed physical quantities, for example, $\bar{J}_k(\mathbf{p},\mathbf{q})$, $\bar{\omega}_k(\mathbf{p},\mathbf{q})$, $\bar{p}_k(\mathbf{p},\mathbf{q})$, $\bar{q}_k(\mathbf{p},\mathbf{q})$, up to second order, through which no (exact) commensurable conditions were encountered.

Throughout this chapter the parabolic barrier, the reaction coordinate in the original $(\mathbf{p},\mathbf{q})$ space [and in the new $(\bar{\mathbf{p}},\bar{\mathbf{q}})$ space] is denoted as q_1 ($\bar{q}_1$) and the other nonreactive coordinates, as $q_2, q_3, \ldots, q_{12}$ ($\bar{q}_2, \bar{q}_3, \ldots, \bar{q}_{12}$) in order of increasing frequency, $\omega_2 \leqslant \omega_3 \leqslant, \ldots, \leqslant \omega_{12}$ ($\bar{\omega}_2 \leqslant \bar{\omega}_3 \leqslant, \ldots, \leqslant \bar{\omega}_{12}$). The units of energy, distance, momentum, action, frequency, temperature, mass and time are, respectively, ϵ, $m^{1/2}\sigma$, $m^{1/2}\sigma\text{ps}^{-1}$, kps, ps^{-1}, K, argon atomic mass and ps, unless otherwise noted.

TABLE II.1
The Fundamental Frequencies ω_k for Saddles I and II (cm^{-1})[a]

Mode	Saddle I	Saddle II
1	−49.29i(−1.477i)	−35.04i(−1.050i)
2	96.06 (2.880)	89.01 (2.668)
3	113.04 (3.389)	98.54 (2.954)
4	122.91 (3.684)	124.37 (3.729)
5	149.40 (4.478)	138.24 (4.144)
6	149.71 (4.488)	153.00 (4.587)
7	173.06 (5.188)	171.13 (5.130)
8	197.81 (5.930)	178.34 (5.346)
9	200.03 (5.996)	198.44 (5.949)
10	206.19 (6.181)	206.63 (6.195)
11	235.10 (7.047)	229.78 (6.889)
12	238.47 (7.148)	241.04 (7.226)

[a]The values in the parentheses are in the unit of reciprocal picoseconds (ps^{-1}).

For analyses of the infrequent saddle crossings, we employed a modified Keck–Anderson method [40] to generate the microcanonical ensemble of well-saddle-well trajectories. We generated 10,000 well-saddle-well trajectories for both the saddles, which were found to be enough to yield statistical convergence in calculating the transmission coefficients in terms of manybody phase-space dividing hypersurfaces $S(\bar{q}_1^{ith}(\mathbf{p},\mathbf{q}) = 0)$ $(i = 0, 1, 2)$ at $E = 0.1, 0.5, 1.0\epsilon$ above both the saddles. For the trajectory calculations, we used a fourth-order Runge–Kutta method with adaptive step-size control [61], and the total energies in their MD calculations were conserved within $\pm 1 \times 10^{-6}\epsilon$.

A. Invariancy of Actions in Transition States

To begin, we look into the new action variables $\bar{J}_k(\mathbf{p},\mathbf{q})$ along the saddle crossing trajectories over saddle I, linking the global minimum OCT and the higher minimum CTBP, obeying equations of motion of the original Hamiltonian $H(\mathbf{p},\mathbf{q})$. The trajectory of an isolated bound oscillator retraces the same points during each oscillation and the associated action is a constant of the motion. The extent to which the new kth action $\bar{J}_k(\mathbf{p},\mathbf{q})$ mimics this behavior indicates how separable the new kth mode, described by $\bar{p}_k$ and $\bar{q}_k$, is. Figure 2.2 shows a representative trajectory projected onto the (q_1, q_{12}) plane for each total energy, 0.05, and 0.5ϵ. Here the trajectory at 0.05ϵ is shown specifically because intuition tells us it should be less

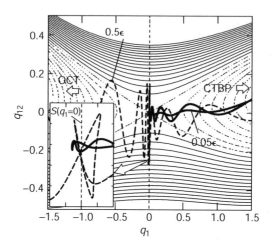

Figure 2.2. A representative saddle-recrossing trajectory at $E = 0.05$, and 0.5ϵ over the dividing surface $S(q_1 = 0)$, projected onto the (q_1, q_{12}) plane and the PES contour plot in this plane. The window in this figure is scaled to $-0.01 \leq q_1 \leq 0.01$ and $-0.3 \leq q_{12} \leq 0.2$. The trajectories from the left to the right side correspond to crossing from the OCT to the CTBP minimum. While the sampled trajectory at 0.05ϵ (bold solid line) is nonreactive, that is, coming from the CTBP and returning to the same CTBP, that at $E = 0.5\epsilon$ (dashed line) is reactive. The PES contour is plotted with an energy step 0.03ϵ, whose solid and dashed lines represent positive and negative values, respectively.

chaotic than those at $0.1 - 1.0\epsilon$, because at such an energy, in the vicinity of the saddle point, the system has insufficient kinetic energy to reach regions where nonlinearity and mode–mode mixing are significant.

Figure 2.3 shows the zeroth-, first-, and second-order new actions along the trajectory at 0.05ϵ. At even $E = 0.05\epsilon$, only slightly above the saddle point energy, almost of all the zeroth-order actions $\bar{\mathbf{J}}^{0\text{th}}(\mathbf{p}, \mathbf{q})$ do not maintain constancy of motion at all. This result implies that the system's trajectory reflects even very small nonlinearities on the PES and deviates from a simple normal mode picture. As we extend the order of LCPT, some but not all LCPT actions $\bar{J}_k$ tend to be conserved in the saddle region.

Figure 2.3. The time dependencies of $\bar{J}_k(\mathbf{p}, \mathbf{q})$ for the saddle-crossing trajectory at 0.05ϵ in Figure 2.2: (a) zeroth-, (b) first-, and (c) second-order actions. The units of action for mode 1 must be multiplied by a factor of an imaginary number i throughout the following figures. In $1.5 < t < 7.5$ ps, the system trajectories remain in a region $\sim -0.01 < q_1 < 0.2$ $[m^{1/2}\sigma]$. The solid, dashed, diamond, circle, square, doted, long-dashed, triangle, x, +, dot–dashed, and bold-solid lines throughout this chapter will denote modes 1, 2, 3, 4, 5, 6, 7, 8, 9, 10, 11, and 12, respectively.

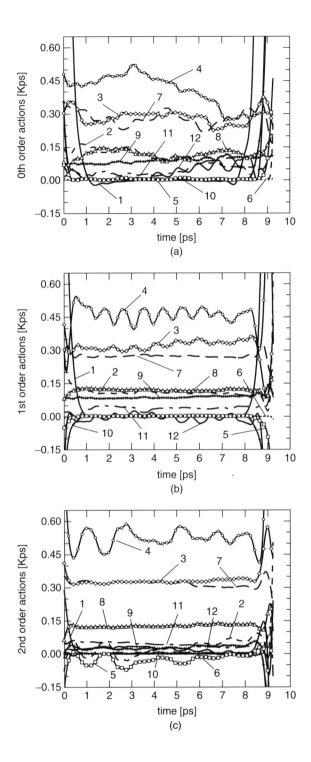

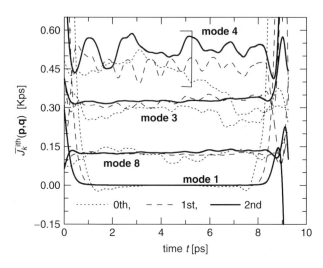

Figure 2.4. $\bar{J}_k^{ith}(\mathbf{p}(t), \mathbf{q}(t))$ ($i = 0, 1,$ and 2) at $E = 0.05\epsilon$ in Figure 2.3: thin, dot, and bold lines, respectively, denote the zeroth-, first-, and second-order LCPT action.

Figure 2.4 shows each order $\bar{J}_k^{ith}(\mathbf{p}, \mathbf{q})$ for $k = 1, 3, 4,$ and 8 on the same figure axes. Even at such an energy, only slightly above the saddle point energy; just 8% of the activation energy 0.633ϵ from the OCT minimum, almost none of the zeroth-order actions $\bar{\mathbf{J}}^{0th}(\mathbf{p}, \mathbf{q})$ maintain any constancy of motion at all; that is, even there, most modes violate a simple normal mode picture. The higher the order to which the LCPT is carried, the more *some* of the actions $\bar{J}_k$ tend to be well conserved, and to persist as nearly conserved quantities for longer times. The initial drop and/or rise observed in $\bar{J}_k(\mathbf{p}, \mathbf{q})$ at short times (e.g., 0–0.5 ps in Fig. 2.4) to the flat region implies that initially, the system is just entering a "regular region" near the saddle point, outside of which the system is subject to considerable nonlinearities of the PES. That is, in the saddle region, some modes are well decoupled and follow periodic orbits in phase space; examples are those of Eqs. (2.8) and (2.9), while the others are coupled at least within coupled mode-subsets in the $(\bar{\mathbf{p}}^{2nd}, \bar{\mathbf{q}}^{2nd})$ coordinate system.

How does the crossing dynamics change as the energy of the system increases? Intuition suggests that at higher total energies, the nonlinearities of the PES cannot be considered as a "sufficiently weak perturbation", and the number of approximate local invariants of motion becomes smaller and smaller, going to zero at sufficiently high energy, that is, causing the transition from quasiregular to chaotic dynamics. This is actually a universal picture assumed in almost of all chemical reaction theories, in the vicinity

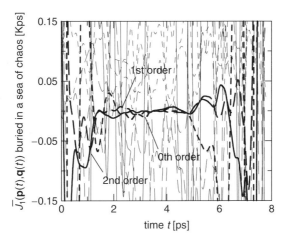

Figure 2.5. An example of $\bar{J}_k^{\text{ith}}(\mathbf{p}(t), \mathbf{q}(t))$ ($i = 0, 1, 2$) along a representative trajectory recrossed over $S(q_1 = 0)$ at saddle I ($E = 0.5\epsilon$): all the lines denote all the actions except bold ones $\bar{J}_1^{\text{ith}}$.

of potential energy minima, which validates the assumption of local vibrational equilibrium in the reactant well. In Figure 2.5, we show that at $E = 0.5\epsilon$, 79% above of the activation energy from the OCT minimum, while none of the action variables for the nonreactive modes, $\bar{q}_k$ ($2 \leq k \leq 3N - 6$), are conserved even through $\mathcal{O}(\epsilon^2)$, the reactive mode $\bar{q}_1$ stands out among all the rest: Its action is more conserved as the order of LCPT increases. (We rigorously demonstrated [43, 44] that these findings, that the phase-space reaction coordinate $\bar{q}_1(\mathbf{p}, \mathbf{q})$ persists its action even in the sea of chaos, is quite generic, irrespective of these sampled trajectories.)

However, how do these trajectories, observed as recrossing over the configurational dividing surface $S(q_1 = 0)$, look in the new $(\bar{\mathbf{p}}, \bar{\mathbf{q}})$ space?

B. "See" Trajectories Recrossing the Configurational Dividing Surface in Phase Space

Let us see these two trajectories, recrossing the conventional, configurational dividing surface $S(q_1 = 0)$, by projecting these onto the zeroth-, first-, and second-order new coordinate planes of $\bar{q}_j(\mathbf{p}, \mathbf{q})$ and $\bar{q}_k(\mathbf{p}, \mathbf{q})$. Remember that the zeroth-order coordinate system $(\bar{q}_j^{\text{0th}}, \bar{q}_k^{\text{0th}})$ is the original normal coordinate system (q_j, q_k). Figure 2.6 shows the projections onto a 2D subspace chosen from the nonreactive coordinates, $(\bar{q}_3^{\text{ith}}(\mathbf{p}, \mathbf{q}), \bar{q}_8^{\text{ith}}(\mathbf{p}, \mathbf{q}))$, at both the energies.

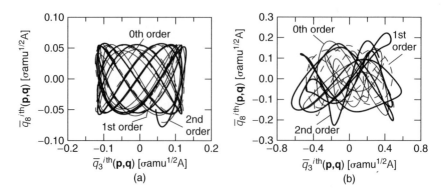

Figure 2.6. The viewpoint from ($\bar{q}_3^{ith}(\mathbf{p},\mathbf{q})$, $\bar{q}_8^{ith}(\mathbf{p},\mathbf{q})$) (saddle I). (a) $E = 0.05\epsilon$, (b) $E = 0.5\epsilon$.

At the lower energy, 0.05ϵ [see Fig. 2.6(a)], in all the orders one can see the approximate Lissajous figures, which implies that in the subspace of these two nonreactive dof, the motions are composed of two approximately decoupled, simple harmonic oscillations. As the total energy increases to 10 times higher, $\sim 0.5\epsilon$, as one may anticipate from Figure 2.5, that no approximate invariants of motion survive in the nonreactive subspace, and the nonreactive modes change from regular to fully chaotic dynamics [see Fig. 2.6(b)].

An even more striking consequence of the LCPT transformation appears in the behavior of the reactive degrees of freedom. Figure 2.7 shows the projections of the recrossing trajectories onto the ($\bar{q}_1^{ith}(\mathbf{p},\mathbf{q})$, $\bar{q}_4^{ith}(\mathbf{p},\mathbf{q})$). The abscissas in the figure correspond to a reaction coordinate, that is, $\bar{\omega}_1 \in \mathfrak{F}$, and the ordinates, to the nonreactive coordinates, that is, $\bar{\omega}_4 \in \mathfrak{R}$, in each order's coordinate system. To do this, we first examine the nonreactive recrossing trajectory at 0.05ϵ in Figure 2.2, which has returned to the original state after recrossing over the naive dividing surface $S(q_1 = 0)$. As shown in Figure 2.7, this trajectory never cross any dividing surface $S(\bar{q}_1^{1st}(\mathbf{p},\mathbf{q}) = 0)$ and $S(\bar{q}_1^{2nd}(\mathbf{p},\mathbf{q}) = 0)$ from the CTBP minimum where the trajectory originates. The trajectory is simply not that of a reaction. We can deduce one important feature from this: If the local invariants of motion persist along the higher order reactive coordinate, for example, $\bar{q}_1^{2nd}(\mathbf{p},\mathbf{q})$, all nonreactive recrossing trajectories observed over any configurational dividing surface, for example, $S(q_1 = 0)$, is transformed to trajectories that do not cross the dividing hypersurface $S(\bar{q}_1^{2nd}(\mathbf{p},\mathbf{q}) = 0)$ in the phase space. The reason is that decoupling the motion along the reactive LCPT coordinate removes all forces that would return the system back across the dividing

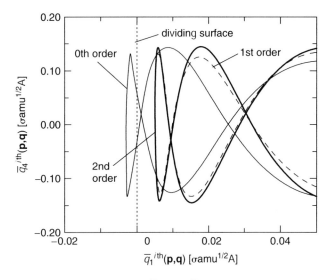

Figure 2.7. The projections onto $(\bar{q}_1^{ith}(\mathbf{p,q}), \bar{q}_4^{ith}(\mathbf{p,q}))$ at $E = 0.05\epsilon$ over saddle I.

hypersurface. Such nonreactive trajectories are those with insufficient incident momentum in the reactive coordinate $\bar{p}_1(\mathbf{p}(0), \mathbf{q}(0))$ to climb over the saddle. In other words, all the trajectories should react, whose incident momentum $\bar{p}_1(\mathbf{p}(0), \mathbf{q}(0))$ is larger than a certain threshold to carry the system through it.

Next, turn to the behavior of the reactive recrossing trajectory at 0.5ϵ, at which the transition state is almost chaotic (see Fig. 2.5). Here, as seen in Figure 2.8, the recrossings that occur over the naive dividing surface $S(q_1 = 0)$ in zeroth order are eliminated; they occur as no-return crossing motions over the second-order dividing surface $S(\bar{q}_1^{2nd}(\mathbf{p,q})=0)$. Furthermore, the system's trajectories along the second-order reactive coordinate $\bar{q}_1^{2nd}(\mathbf{p,q})$ are not forced to return to the dividing surface $S(\bar{q}_1^{2nd}(\mathbf{p,q}) = 0)$ over the (saddle) region, $-0.04 < \bar{q}_1^{2nd}(\mathbf{p,q}) < 0.04$. On the other hand, the zeroth- and first-order reactive coordinates are not decoupled from the other modes in the regions either near or more distant from the dividing surface. For example, the cyclic motion around $0.025 < \bar{q}_1^{1st}(\mathbf{p,q}) < 0.030$ implies the existence of some couplings between $\bar{q}_1^{1st}(\mathbf{p,q})$ and $\bar{q}_4^{1st}(\mathbf{p,q})$. Up to such moderately high energies, "apparent" recrossing reactive trajectories, observed in a low-order phase-space coordinate system $(\bar{\mathbf{p}}, \bar{\mathbf{q}})$ or the configurational space, can be rotated away to the ballistic, single crossing motions in the higher orders.

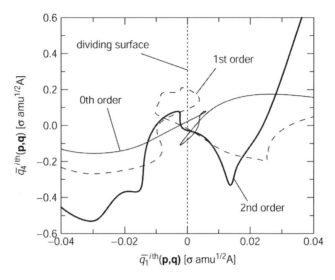

Figure 2.8. The projections onto $(\bar{q}_1^{ith}(\mathbf{p},\mathbf{q}), \bar{q}_4^{ith}(\mathbf{p},\mathbf{q}))$ at $E = 0.5\epsilon$ over saddle I.

C. "Unify" Transition State Theory and Kramers–Grote–Hynes Theory

In all the cases that saddle crossings have approximate local invariants of motion associated with the "reactive mode F" in a short time interval but long enough to determine the final state of the crossings, $S(\bar{q}_F(\mathbf{p},\mathbf{q}) = 0)$ can be identified as a "no-return" dividing hypersurface, free from the long-standing ambiguity in reaction rate theories, the recrossing problem. Recall that it is because there is no means or force returning the system to $S(\bar{q}_F(\mathbf{p},\mathbf{q}) = 0)$ even though the system may recross any of the configurational surfaces. The reformulated microcanonical (classical) transition state theory (TST) rate constant $\bar{k}^{TST}(E)$ is obtained [38] as an average of the one-way fluxes $j_+(=\dot{\bar{q}}_F(\mathbf{p},\mathbf{q})h(\dot{\bar{q}}_F(\mathbf{p},\mathbf{q})))$ across $S(\bar{q}_F(\mathbf{p},\mathbf{q}) = 0)$ over microcanonical ensembles constructed over a range of energies E.

$$\bar{k}^{TST}(E) = \langle j_+ \rangle_E = \langle \dot{\bar{q}}_F(\mathbf{p},\mathbf{q})\delta[\bar{q}_F(\mathbf{p},\mathbf{q})]h[\dot{\bar{q}}_F(\mathbf{p},\mathbf{q})] \rangle_E$$

$$= \int_1 dq_1 dp_1 \cdots \int_N dq_N dp_N$$

$$\times \delta[E - H(\mathbf{p},\mathbf{q})]\dot{\bar{q}}_F(\mathbf{p},\mathbf{q})\delta[\bar{q}_F(\mathbf{p},\mathbf{q})]h[\dot{\bar{q}}_F(\mathbf{p},\mathbf{q})] \qquad (2.52)$$

where $h(x)$ and $\delta(x)$, respectively, denote the Heaviside function and Dirac's delta function of x.

The deviation of $\bar{k}^{TST}(E)$ from the experimental rate coefficient $k^{exp}(E)$ is defined as a new transmission coefficient κ_c:

$$\kappa_c = \frac{k^{exp}(E)}{\bar{k}^{TST}(E)} \tag{2.53}$$

If the vibrational energy relaxation is fast enough to let us assume quasi-equilibration in the well, and the tunneling effect is negligible, one may use κ_c to estimate the extent of the *true* recrossing effect independent of the viewpoint or coordinate along which one observe its reaction events. This would be a means to measure the extent to which the action associated with the reactive mode cease to be approximate invariants; their nonconstancy reflects the degree of closeness to *fully developed* chaos in which no invariant of motion exists.

In order to focus on how the recrossings over a given dividing surface contribute to κ_c, we estimated the quantities $\kappa_c^{ith}(t)$

$$\kappa_c^{ith}(t) \equiv \kappa_c(t; S(\bar{q}_F^{ith}(\mathbf{p}, \mathbf{q}) = 0)) = \frac{\langle j(t=0)h[\bar{q}_F^{ith}(\mathbf{p}, \mathbf{q})]\rangle_E}{\langle j_+(t=0)\rangle_E} \tag{2.54}$$

using the 10,000 well-saddle-well classical trajectories across both the saddles at these three distinct energies above the threshold of Ar_6. Here $j(t=0)$ and $j_+(t=0)$, respectively, denote the initial total, and initial positive fluxes crossing the ith-order phase-space dividing surface $S(\bar{q}_F^{ith}(\mathbf{p}, \mathbf{q}) = 0)$, and the origin of time t was set when the system trajectory first crosses the given dividing surface.

The $\kappa_c^{ith}(t)$ in the regions of saddle I and saddle II are shown at $E = 0.1$, 0.5, and 1.0ε in Figure 2.9. The zeroth-order transmission coefficient $\kappa_c^{0th}(t)$, using the conventional configurational dividing surface $S(q_1 = 0)$, deviates significantly from unity (except at a very short times) and these deviations increase with increasing total energy. The plateau in $\kappa_c^{ith}(t)$ apparent in the figures implies that the recrossing trajectories eventually go into their final state and never again cross the given dividing surface within some interval long compared with the transit time. The plateau value of the $\kappa_c^{0th}(t)$ may be identified as the conventional transmission coefficient κ. All these κ's smaller for saddle I than for saddle II show that the traditional reaction coordinate q_1 is more coupled to the other nonreactive dof in the region of saddle I than it is near saddle II [20–22]. We showed, however, that in terms of the phase-space dividing hypersurface $S(\bar{q}_1^{ith}(\mathbf{p}, \mathbf{q}) = 0)$, for low and moderately high energies, ~ 0.1–0.5ε, for both the saddles, the higher is the order of the perturbative calculation, the closer the κ_c^{ith} is to unity. Even at

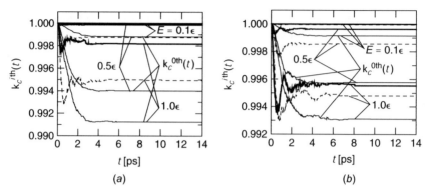

Figure 2.9. The new transmission coefficient $\kappa_c^{ith}(t)$ ($i = 0, 1, 2$) ($E = 0.1, 0.5,$ and 1.0ϵ); (a) saddle I; (b) saddle II. The solid, dot, and bold lines denote the zeroth-, first-, and second-order $\kappa_c^{ith}(t)$, respectively. The converged values at saddle I are 0.9988(0), 0.99996(1), 1.00000(2) ($E = 0.1\epsilon$); 0.9940(0), 0.9987(1), 0.9999(2) ($E = 0.5\epsilon$); 0.9912(0), 0.9949(1), 0.9982(2) ($E = 1.0\epsilon$), and those at saddle II 0.9991(0), 0.99995(1), 1.000000(2) ($E = 0.1\epsilon$); 0.9958(0), 0.9986(1), 0.9996(2) ($E = 0.5\epsilon$); 0.9931(0), 0.9948(1), 0.9955(2) ($E = 1.0\epsilon$). The number in parentheses denotes the order of LCPT.

energies $\sim 0.5\epsilon$, the plateau values of $\kappa_c^{2nd}(t)$ are almost unity, for example, 0.9999 (for saddle I) and 0.9996 (for saddle II). As the total energy becomes much higher, $\sim 1.0\epsilon$, even κ_c^{2nd} becomes deviate significantly from unity as the conventional κ_c^{0th} does. This "deviation from unity at the order $\mathcal{O}(\varepsilon^2)$" represents the degrees of

1. The difficulty of exposing the approximate invariants of motion associated with $\bar{q}_1$ with only a finite orders, $\sim \mathcal{O}(\varepsilon^2)$, to decouple the reactive $\bar{q}_1$ from the others.

2. Encroaching into a sufficiently high-energy region where the *length* of the path where the reactive mode is separable diminishes (i.e., even though it would be possible to carry CPT through an infinite order, it is anticipated, especially for nonlinear many-body systems, that there exist a certain energy region so high that the convergence radius of CPT becomes negligibly small).

The fact that the deviation is smaller for saddle I than that for saddle II implies that the crossing dynamics over saddle I should exhibit better approximate invariants of motion with $\bar{q}_1$ than that over saddle II at $\mathcal{O}(\varepsilon^2)$. Note that less chaotic saddle-crossing dynamics need not imply a better approximate invariant of motion associated with $\bar{q}_1$ at any specific order of the perturbative calculation. Furthermore, the strength of chaos in the regions of saddles, as characterized by the local Kolmogorov entropy, the sum of all the positive local Liapunov exponents λ^{sad}, is only weakly affected by a single (small) positive exponent λ_1^{sad} associated with the somewhat-regularized $\bar{q}_1(\mathbf{p},\mathbf{q})$ motion. This "burial" of a few locally regular modes in a sea of chaotic modes is apparent in the results of Hinde and Berry [21].

All these results indicate that, even in the region where the system is almost chaotic, an approximate analytical expression for (first-rank) saddle-crossing dynamics may nonetheless exist along a negatively curved mode in the phase space coordinate system $(\bar{\mathbf{p}}, \bar{\mathbf{q}})$;

$$\bar{q}_1(\mathbf{p}(t), \mathbf{q}(t)) \simeq \alpha e^{|\bar{\omega}_1|t} + \beta e^{-|\bar{\omega}_1|t} \tag{2.55}$$

$$\alpha = \frac{1}{2}\left(\bar{q}_1(t_0) + \frac{\bar{p}_1(t_0)}{|\bar{\omega}_1|}\right) \tag{2.56}$$

$$\beta = \frac{1}{2}\left(\bar{q}_1(t_0) - \frac{\bar{p}_1(t_0)}{|\bar{\omega}_1|}\right) \tag{2.57}$$

Here, t_0 is arbitrary time in t in the vicinity of saddles. These expressions imply that, even though almost all degrees of freedom of the system are chaotic, the final state (and initial state) may have been determined a priori. For example, if the trajectories that have initiated from $S(q_1 = 0)$ at time t_0 have $\alpha > 0$, the final state has already been determined at "the time t_0 when the system has just left the $S(q_1 = 0)$" to be a stable state directed by $\bar{q}_1 > 0$. Similarly, from only the phase-space information at $t = t_0$ (the sign of β), one can grasp whether the system on $S(q_1 = 0)$ at time t_0 has climbed from either stable state, that is, reactant or product, without calculating any time-reversed trajectory [62].

To address the recrossing problem, which spoils the "no-return" hypothesis, one has tried to interpret the reaction rates either by variational TST [9–11] or by (generalized) Langevin formalism by Kramers [14] and Grote and Hynes [15]. van der Zwan and Hynes [63] proved that, for a class of Hamiltonians representing (first-rank) saddles by parabolic barrier–harmonic oscillator systems linearly coupled with one another (i.e., an integrable system!), the TST rate constant is equivalent to that of the Grote–Hynes formulation with the parabolic mode as the reactive dof, if the reaction coordinate is chosen as an unstable normal coordinate composed of the total system (=reacting system + bath), under the condition that the vibrational relaxation is fast enough to attain "quasiequilibration" in the reactant well.

Our findings suggest that their equivalence is much more generic and applicable to a wider range of Hamiltonian classes, even when the system is almost chaotic. This stimulates us to reconsider a fundamental question of what constitutes the "thermal bath" for reacting systems. One may anticipate that reactions take place along a *ballistic* path composed of the *total* system in the sea of chaos, at least, in the region of saddles. The "thermal bath" for reactions, simply defined thus far as all the rest of the atoms or molecules except the reacting system, does not necessarily retard the reactive trajectories; rather, such a bath may control and assist the reactants to climb and go through the saddles.

Most reactions take place not on *smooth* but *rugged* PESs, for which the (generalized) Langevin formalism was originally developed. Nevertheless, a picture similar to that of a simple reacting system may emerge even on a rugged PES. In cases in which a "coarse-grained" landscape connecting one basin with another can be well approximated as parabolic at the zeroth order, we may elucidate how the coarse-grained action persists along the (coarse-grained) reactive degree of freedom during the dynamical evolution obeying the original Hamiltonian $H(\mathbf{p}, \mathbf{q})$. A technique or a scheme [64] to extract such a coarse-grained regularity or cooperativity from "random thermal" motions is quite demanding, especially to establish the equivalence or differences of descriptions by different representations or coordinate systems.

D. "Visualize" a Reaction Bottleneck in Many-Body Phase Space

The dividing surface in this representation is analogous to the conventional dividing surface in the sense that it is the point set for which the reaction coordinate has the constant value it has at the saddle-point singularity. However the nonlinear, full-phase-space character of the transformation makes the new crossing surface a complicated, abstract object. We proposed [41, 42] a visualization scheme of $S(\bar{q}_1(\mathbf{p}, \mathbf{q}) = 0)$ by projections into spaces of a few dimensions, for example, the (q_j, q_k) plane:

$$\bar{S}(q_j, q_k; E) = \langle \delta[\bar{q}_1(\mathbf{p}', \mathbf{q}')]\delta(q'_j - q_j)\delta(q'_k - q_k)\rangle_E$$

$$= \int_1 dq'_1 dp'_1 \cdots \int_N dq'_N dp'_N$$

$$\times \delta[E - H(\mathbf{p}', \mathbf{q}')]\delta[\bar{q}_1(\mathbf{p}', \mathbf{q}')]\delta(q'_j - q_j)\delta(q'_k - q_k) \quad (2.58)$$

For example, the projection onto the configurational reaction coordinate q_1 is an important device to reveal how $S(\bar{q}_1(\mathbf{p}, \mathbf{q}) = 0)$ differs from the conventional dividing surface $S(q_1 = 0)$. Remember that in an energy range close to the threshold energy in which the normal mode picture is approximately valid, the phase-space $S(\bar{q}_1 = 0)$ collapses onto the traditional configuration-space surface where $q_1 = 0$. Similarly, the projection onto p_1

TABLE II.2
Pattern Classification (N_{+-}, N_{-+}, σ) of Nonreactive, Recrossing Trajectories Over the Configurational Dividing Surface $S(q_1 = 0)$ at Saddle I

$E =$	$0.1\epsilon\ (=-11.979\epsilon)$	$0.5\epsilon\ (=-11.579\epsilon)$	$1.0\epsilon\ (=-11.079\epsilon)$
(1, 1, −)	161	329	369
(2, 2, −)	7	23	22
(3, 3, −)	0	1	3
(4, 4, −)	0	0	0
(1, 1, +)	26	67	86
(2, 2, +)	1	6	12
(3, 3, +)	0	0	0
(4, 4, +)	0	1	0

is an important device to reveal how the passage velocity through the saddles affects the $S(\bar{q}_1(\mathbf{p}, \mathbf{q}) = 0)$.

One should be careful about the implication of $S(q_1 = 0)$ in the phase space. Accessible values of $\mathbf{p}$ and $\mathbf{q}$ are restricted by total energy E. Thus, in stating that $S(\bar{q}_1(\mathbf{p}, \mathbf{q}) = 0)$ partially collapses onto $S(q_1 = 0)$, the relevant regions of $S(q_1 = 0)$ are not all of this surface determined by only q_1 irrespective of the other variables, but only parts of a hypersurface where $q_1 = 0$, generally determined by $\mathbf{p}$ and $\mathbf{q}$.

Next, we ask, What could one learn from these projections of the complicated, energy and momenta-dependent abstract object, $S(\bar{q}_1(\mathbf{p}, \mathbf{q}) = 0)$?

First, we look into the complicated nonreactive, recrossing behavior of trajectories over the conventional dividing surface $S(q_1 = 0)$ at saddle I. Table II.2 shows the numbers of nonreactive crossings in 10,000 well-saddle-well trajectories over $S(q_1 = 0)$ as the energy increases from $E = 0.1$ to 1.0ϵ. Here, (N_{+-}, N_{-+}, σ) represents the number of times each crosses the dividing surface in a specific direction: if a crossing trajectory, whose sign of the flux at the first crossing is σ, crosses a given dividing surface N_{+-} times from positive to negative, and N_{-+} times from negative to positive along the reactive coordinate, the trajectory is classified into the (N_{+-}, N_{-+}, σ)-type crossing, for example, for saddle I a trajectory that crosses the dividing surface two times and the first crossing is from the OCT to the CTBP minimum is (1, 1, +).

The table tells us the trajectories climbing from the CTBP minimum are more likely to return after crossing $S(q_1 = 0)$, than the trajectories from the OCT global minimum, for example, 329(1, 1, −) ≫ 67(1, 1, +) at 0.5ϵ (the corresponding figures are essentially the same for the symmetrical saddle II within the statistical error [41]).

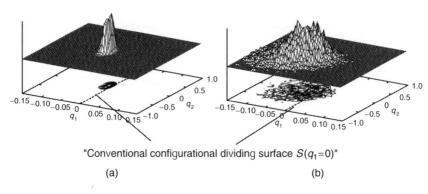

"Conventional configurational dividing surface $S(q_1=0)$"

(a)   (b)

Figure 2.10. $\bar{S}^{2nd}(q_1, q_2; E)$ at saddle I, (a) $E = 0.1\epsilon$; (b) $E = 0.5\epsilon$.

So far, there has been no means to address why, as the system passes through the transition state, there is such a distinct dependence of probability on the direction of climbing. The visualization scheme of the phase-space dividing surface lets us probe deeper into such questions. Figures 2.10 and 2.11 show projections of the ith order dividing surface $S(\bar{q}_1^{ith}(\mathbf{p}, \mathbf{q}) = 0)$ onto the two-dimensional (q_1, q_2) subspace at $E = 0.1$ and 0.5ϵ for saddle I and saddle II, respectively. As the total energy increases, the projections of the phase-space dividing surfaces, $S(\bar{q}_1^{ith}(\mathbf{p}, \mathbf{q}) = 0)$ ($i = 1, 2$), broaden and extend to regions more removed from the conventional dividing surface $S(q_1 = 0)$. Note for saddle I that these $S(\bar{q}_1(\mathbf{p}, \mathbf{q}) = 0)$ are more heavily distributed on the minus side (to the OCT minimum) than on the plus side (to the CTBP minimum) in the q_1 axis. This asymmetrical feature of the $S(\bar{q}_1(\mathbf{p}, \mathbf{q}) = 0)$ explains the higher frequencies found for $(n, n, -)$ than for $(n, n, +)$ type

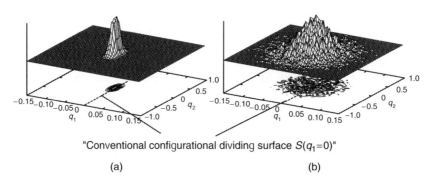

"Conventional configurational dividing surface $S(q_1=0)$"

(a)   (b)

Figure 2.11. $\bar{S}^{2nd}(q_1, q_2; E)$ at saddle II, (a) $E = 0.1\epsilon$; (b) $E = 0.5\epsilon$.

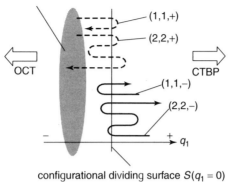

Figure 2.12. The schematic pictures of the $(1, 1, +)$, $(1, 1, -)$, $(2, 2, +)$, and $(2, 2, -)$ crossing patterns.

crossings over $S(q_1 = 0)$ of saddle I (see also Fig. 2.12):

- $- \rightarrow +$: If the system once crossed the *naive* dividing surface $S(q_1 = 0)$ from minus to plus, the system rarely returns to $S(q_1 = 0)$ because of the small driving force to make the system go back to $S(q_1 = 0)$ after it has passed the greater part of the distribution constituting the real $S(\bar{q}_1 = 0)$.

- $+ \rightarrow -$: Even if the system crossed $S(q_1 = 0)$ from plus to minus, the system has not necessarily passed the surface $S(\bar{q}_1 = 0)$. The system will recross $S(q_1 = 0)$ if the system does not possess sufficient incident (reactive) momentum $\bar{p}_1$ to pass through the $S(\bar{q}_1 = 0)$.

In other terms, almost nonreactive recrossings initiated from the CTBP state occur because the *real* dividing surface mainly distributes outwards to the OCT side from the $S(q_1 = 0)$, while the less frequent nonreactive recrossings from the other OCT state occur when the system finds an edge of the reaction bottleneck, that is, a tiny part of the dividing hypersurface in the phase space.

We also found [41] that besides total energy, the velocity across the transition state plays a major role in many-dof systems to migrate the reaction bottleneck outward from the naive dividing surface $S(q_1 = 0)$. A similar picture has been observed by Pechukas and co-workers [37] in 2D Hamiltonian systems, that is, as energy increases, pairs of the periodic orbit dividing surfaces (PODSs) appearing on each reactant and product side migrate outwards, toward reactant and product state, and the outermost

PODS can be identified as the reaction bottleneck. However, their crucial idea is based on the findings of *pure* periodic orbits in the saddle region. In many-body systems with more than three dof, the saddle-crossing dynamics often exhibits chaos due to resonance among nonreactive stable modes, extinguishing any periodic orbit in that region. As far as we know, this is the first example to picture reaction bottlenecks for such many-dof systems.

E. "Extract" Invariant Stable and Unstable Manifolds in Transition States

To identify those parts of space (either configurational, or phase space) in which invariants of motion "actually" survive or break during the course of dynamical evolution, obeying the exact Hamiltonian, we proposed [43, 44] "local invariancy analysis", in terms of a new concept of "duration of regularity (τ)", for each mode of the system, at each ith order of perturbation; these are the residence times each mode remains close to its near-constant values of the variables, as determined by a chosen bound on the fluctuation $\Delta \bar{J}$ or $\Delta \bar{\omega}$; for example, $\Delta \bar{J}$ for mode k, is

$$|\bar{J}_k^{\text{ith}}(\mathbf{p}(t+\tau), \mathbf{q}(t+\tau)) - \bar{J}_k^{\text{ith}}(\mathbf{p}(t), \mathbf{q}(t))| \leq \Delta \bar{J} \tag{2.59}$$

By transforming a time series of the variables, denoted hereinafter as $x(t)$, to a sequence of stationary points, $\ldots \min[i] - \max[i+1] - \min[i+2] - \cdots$ along $x(t)$ with the corresponding times $t[i]$, and choosing all the possible combinations of $\max[i]$ and $\min[j]$, one can calculate each residence time τ for which $x(t)$ traverses each fluctuation window Δx defined as $\max[i] - \min[j]$. For a bundle of $x(t)$, we can calculate how frequently $x(t)$ traverses the region of a certain fluctuation window Δx for a certain τ, that is, residence probability, say, $P_2(\Delta x; \tau)$, and also several distinct forms of joint probabilities, $P_{h+1}(\xi_1, \xi_2, \ldots, \xi_h; \tau)$ where ξ_i is either Δx, x, $\Delta x'$, or x' of any other variable $x'(t)$, x' and x are the short-term averages of $x'(t)$ and $x(t)$ for a certain period τ, say, from t' to $t' + \tau$, for example,

$$x' \equiv \frac{1}{\tau} \int_{t'}^{t'+\tau} x'(t)dt \tag{2.60}$$

This enables us to extract and visualize the stable and unstable invariant manifolds along the reaction coordinate in the phase space, to and from the hyperbolic point of the transition state of a many-body nonlinear system. $P_4(\Delta \bar{J}_1^{\text{2nd}}, \bar{p}_1, \bar{q}_1; \tau)$ and $P_4(\Delta \bar{J}_1^{\text{2nd}}, p_1, q_1; \tau)$ shown in Figure 2.13 can tell us how the system distributes in the two-dimensional $(\bar{p}_1(\mathbf{p}, \mathbf{q}), \bar{q}_1(\mathbf{p}, \mathbf{q}))$ and (p_1, q_1) spaces while it retains its local, approximate invariant of action $\bar{J}_1^{\text{2nd}}(\mathbf{p}, \mathbf{q})$ for a certain locality, $\sim \Delta \bar{J} = 0.05$ and $\tau \geq 0.5$, in the vicinity of

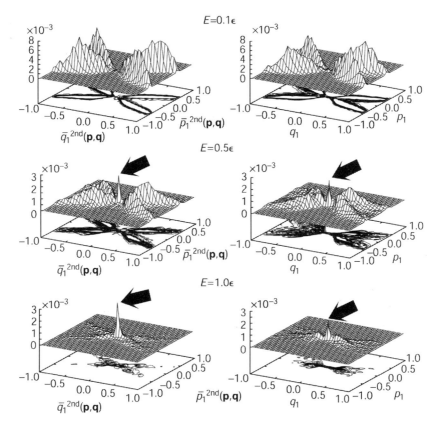

Figure 2.13. Probability distributions of approximate invariants of action $\bar{J}_1^{2nd}(\mathbf{p},\mathbf{q})$ on $(\bar{p}_1^{2nd}(\mathbf{p},\mathbf{q}), \bar{q}_1^{2nd}(\mathbf{p},\mathbf{q}))$ and (p_1, q_1) at $E = 0.1$, 0.5, and 1.0ε in the region of saddle I: $\Delta \bar{J} = 0.05$ and $\tau \geq 0.5$.

saddle I at the three distinct energies. The figures clearly capture the existence of a local near invariant of motion in the reactive coordinate $\bar{q}_1(\mathbf{p},\mathbf{q})$ even at moderately high energy, $\sim E = 0.5\varepsilon$ (such invariant regions in the nonreactive dof totally disappear at $E \geq 0.5\varepsilon$). The clear "X" shapes of the 2D contour maps on $(\bar{p}_1^{2nd}(\mathbf{p},\mathbf{q}), \bar{q}_1^{2nd}(\mathbf{p},\mathbf{q}))$ at all the three energies indicate that, without any explicit assumption of the separation of time scales associated with individual modes, as expected from Eqs. (2.8) and (2.9), one can extract and visualize the stable and unstable invariant manifolds, at least in the region of the first-rank saddle, along the 1D reaction coordinate $\bar{q}_1(\mathbf{p},\mathbf{q})$ in many-body nonlinear systems, just like that of a 1D, integrable pendulum.

These figures should be contrasted with the corresponding diffuse contour maps in the conventional (p_1, q_1) plane, especially, at $E \geqslant 0.5\varepsilon$. The more striking and significant consequence is this; the rather "long-lived" approximate invariant $\bar{J}_1(\mathbf{p}, \mathbf{q})$, around the origin, $\bar{p}_1(\mathbf{p}, \mathbf{q}) \cong \bar{q}_1(\mathbf{p}, \mathbf{q}) \cong 0$, emerges with an increase of energy, even surviving at 1.0ε, despite the consequent high passage velocity through the saddle (indicated by the arrows in the figures). Such a long-persistent invariant around that point could not be observed in the quasiregular region, up to 0.1ε. As shown in Eqs. (2.8) and (2.9), if approximate invariants of action $\bar{J}_1$ and frequency $\bar{\omega}_1$ survive, the entire phase-space flow $(\bar{p}_1, \bar{q}_1)$ should be just like those of a 1D integrable pendulum, and hence no sharp spike should appear in the region where such invariants exist.

The sharp spike around the origin $\bar{p}_1 \cong \bar{q}_1 \cong 0$ in the probability distribution of approximate constant of action $\bar{J}_1(\mathbf{p}, \mathbf{q})$ implies that slow passages through the reaction bottleneck tend to spoil the approximate invariant of frequency, and the system's reactive dof $(\bar{p}_1, \bar{q}_1)$ couples with the other nonreactive dof throughout the small region $\bar{p}_1 \cong \bar{q}_1 \cong 0$. The rather long residence in the region of constant $\bar{J}_1$ implies that the system is transiently trapped in the nonreactive space during the course of the reaction due to the mode–mode couplings that emerge with increasing total energy. In such an intermediate regime between these two energy regions, any simplistic picture, ballistic or diffusive, of the system's passage through transition states may be spoiled.

We also pointed out that with a residence time τ much shorter than that in Figure 2.13 (with the same fluctuation bound), for example, $\tau \leqslant 0.2$, a similar sharp spike exists at $E = 0.1\varepsilon$ around the origin, $\bar{p}_1^{2\text{nd}}(\mathbf{p}, \mathbf{q}) \cong \bar{q}_1^{2\text{nd}}(\mathbf{p}, \mathbf{q}) \cong 0$. This finding implies that the original Hamiltonian cannot completely be transformed to an exact, integrable Hamiltonian at second order in the LCPT calculation in the real situation at $E = 0.1\varepsilon$. However, as inferred [40] from the analysis of the transmission coefficients κ_c using the phase-space dividing hypersurface $S(\bar{q}_1^{2\text{nd}}(\mathbf{p}, \mathbf{q}) = 0)$, the system could be regarded as "fully" separable at 0.1ε in the transition state because the transmission coefficient κ_c's value was evaluated to be 1.00000, which suggests that the different rates of energy exchange between the nonreactive dof and the reactive dof make the nonreactive *near-integrable* subset of modes contribute far less, and with less influence on the kinetics, than those modes in the *chaotic* subset. Recall that the Rice–Ramsperger–Kassel–Marcus (RRKM) theory [6–8] postulates that the greater the number of dof that couple with the reactive mode, the slower is the process of a specific mode gathering the energy required to react. If some nonreactive modes remain very regular, we might expect them to contribute nothing at all toward trapping the trajectory.

F. A Brief Comment on Semiclassical Theories

At the end of this section, we want to make some comments on the applicability of LCPT to semiclassical theories. In this discussion, we neglected tunneling through the potential barriers for the sake of simplicity, but rather have focused on the (classical) physical foundations of transitions buried in the complexity of reactions. However, CPT, including normal form theories, also provides us with a versatile means to address multidimensional tunneling problems [65–67]. In the scope of the semiclassical WKB approach, Keshavamurthy and Miller [65] first developed a semiclassical theory to extract the 1D phase-space path along which the tunneling action is conserved locally, obeying the original nonintegrable Hamiltonian. By using the following two-mode nonlinear reacting system:

$$H(\mathbf{p},\mathbf{q}) = \frac{1}{2m}(p_1^2 + p_2^2) + \frac{1}{2}aq_1^2 - \frac{1}{3}bq_1^3 + \frac{1}{2}m\omega^2\left(q_2 - \frac{cq_1}{m\omega^2}\right)^2 \quad (2.61)$$

where the parameters a and b were chosen to correspond to a barrier height of 7.4 kcal/mol, the mass m of a hydrogen atom, the harmonic frequency ω to be 300 cm^{-1}, and c a coupling constant (typical values for hydrogen-atom transfer reactions), they showed that all the actions are well conserved irrespective of the coupling strength c. Moreover, the value of the new Hamiltonian truncated at the second-order E_{pert} is as close as possible to the exact total energy, in the *local* saddle region. The unimolecular decay rate evaluated by the tunneling time, chosen to be that when $\bar{J}_F$ is locally conserved (stationary) in time, coincides well with the exact quantum rate, even in a case in which strong coupling spoils any (apparent) mode-specificity of the reaction.

We infer here that this is always true irrespective of the kind of system, if the number of degrees-of-freedom N is two. It is because there is no source to yield "resonance" by a single imaginary- and a single real-frequency mode. If N is >2, it is no longer integrable in the saddles except at just above the threshold energy because of resonance arising from nonreactive–nonreactive modes' nonlinear couplings. As discussed in Section IV, there should be at least three distinct energy ranges in terms of regularity of transitions. In other terms, we can classify energy ranges according to the extent of the system to possess "good" approximate, local quantum numbers. In the *semichaotic* region, E_{pert} totally differs from the exact total energy, and the conventional semiclassical TST breaks down completely. Nevertheless, the tunneling action along the *phase-space* path is quite likely conserved during the events. One may thus anticipate that their semiclassical approach [65] is quite applicable through a wide range of total energies above the saddle point, in which the system is manifestly chaotic (even more than they might have expected!).

IV. HIERARCHICAL REGULARITY IN TRANSITION STATES

As seen in Sections I–III, there are at least three distinct energy regions above the saddle point energy that can be classified in terms of the regularity of saddle-crossing dynamics. We articulated the distinctions among them as follows:

Quasiregular Region: All or almost all the degrees of freedom of the system *locally* maintain approximate constants of motion in the transition state. The saddle crossing dynamics from well to well is fully deterministic, obeying M-analytical solutions [see Eqs. (2.8) and (2.9)] for systems of M degrees of freedom. The dynamical correlation between incoming and outgoing trajectories from and to the transition state is quite strong, and the dimensionality of saddle crossings is essentially one, corresponding to the reactive mode $\bar{q}_F$ in the $(\bar{\mathbf{p}}, \bar{\mathbf{q}})$ space. Barrier recrossing motions observed over a naive dividing

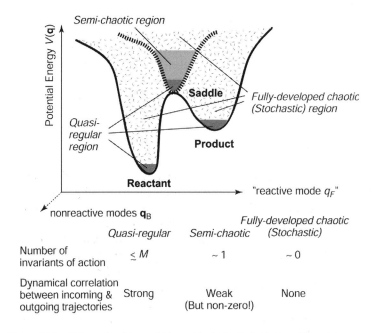

Figure 2.14. Schematic picture of hierarchical regularity in transition states.

surface defined in the configurational space are *all* rotated away to no-return single crossing motions across a phase-space dividing surface $S(\bar{q}_F(\mathbf{p}, \mathbf{q}) = 0)$. If the vibrational energy relaxes fast enough to let us assume quasiequilibration in the wells, the initial conditions $(\bar{\mathbf{p}}(0), \bar{\mathbf{q}}(0))$ of the system as it enters the transition state from either of the stable states can be simply sampled from microcanonical ensembles. One may then evaluate the (classical) exact rate constant, free from the recrossing problem. The staircase energy dependence observed by Lovejoy et al. [30, 31] for highly vibrationally excited ketene indicates that the transverse vibrational modes might indeed be approximately invariants of motion [33]. We classify such a range of energy, in which the rate coefficient shows staircase structure, as corresponding to this quasiregular region. Note that the incommensurable situations likely happen for systems having peculiar or specific mode(s), that is, modes *effectively* separable in time. We found in the proton-transfer reaction of malonaldehyde [38, 39], involving the proton movement, the O—O stretching, the out-of-plane wagging, and C=O stretching motions, that dynamics over the saddle is well regularized through all the dof, even at a moderately high total energy of the system, for example, 1.0 kcal/mol above the activation barrier whose height is 5.0 kcal/mol.

Intermediate, Semichaotic Region: Due to significant (near-)resonance emerging at these intermediate energies, *almost* all the approximate invariants of motion disappear, consequently inducing a topological change in dynamics from quasiregular-to-chaotic saddle crossings. However, at least one approximate invariant of motion survives during the saddle crossings, associated with the reactive coordinate $\bar{q}_F(\mathbf{p}, \mathbf{q})$. This is due to the fact that an arbitrary combination of modes cannot satisfy the resonance conditions of Eq. (2.12) if one mode has an imaginary frequency, the reactive mode in this case, is included in the combination. The other frequencies associated with nonreactive modes fall on the real axis, orthogonal to the imaginary axis in the complex ω plane. That is,

$$\left| \sum_{k=1}^{M} {}^{\dagger} n_k \omega_k \right| \geq |\omega_F| > O(\varepsilon^n) \tag{2.62}$$

for arbitrary integers n_k with $n_F \neq 0$, where $\Sigma^{\dagger}$ denotes the combination including the reactive mode. This finding was first pointed out by Hernandez and Miller in their semiclassical TST studies [60]. In this region, the dynamical correlation between incoming and outgoing trajectories to and from the transition state becomes weak (but nonzero!), and the saddle crossings' dimensionality is $\simeq M - 1$, excluding the 1D of $\bar{q}_F$, in this region. If the associated imaginary frequency $\bar{\omega}_F(\mathbf{p}, \mathbf{q})$ is approximately constant during a saddle crossing as the action $\bar{J}_F(\mathbf{p}, \mathbf{q})$ is, the reaction coordinate $\bar{q}_F$ decouples from a subspace composed of the other nonreactive dof, in which the system dynamics is manifestly chaotic. The $\bar{q}_F$ dynamics is then represented analytically during saddle crossings, and a dividing surface $S(\bar{q}_F(\mathbf{p}, \mathbf{q}) = 0)$ can still be extracted free from the recrossing problem, even for saddle crossings chaotic in the nonreactive modes.

We may expect that various kinds of resonance zones occur in the transition state, densely distributed, associated with very complicated patterns of level crossings in phase space, in a so-called "Arnold web" [53]. The transport among the states in such a web in many-dof systems raises many interesting and unresolved questions [24, 68, 69]. By using their local frequency analysis, Martens et al. [68] showed in a three-dof model for intramolecular energy flow in the OCS molecule that, although the motion is chaotic, some local frequencies are often fairly constant over times corresponding to many vibrational periods when the system moves along resonance zones, and long time-correlations are often observed near the junctions of resonance zones. As shown in Figures 7 [40] and 6 and 7 [44], one can see that in addition to the reactive mode frequency $\bar{\omega}_1^{2\text{nd}}(\mathbf{p}, \mathbf{q})$, some other frequencies are also fairly constant through the saddle region, although the corresponding actions do not maintain constancy at all. We may expect that the LCPT frequency analysis will provide us with a versatile tool to analyze the resonance mechanism in chaotic motions.

Stochastic (=Fully Developed Chaotic) Region: The system becomes subject to considerable nonlinearities of the PES at much higher energies, and the convergence radius becomes negligibly small for the LCPT near the fixed (saddle) point for the invariant of motion associated with the reactive coordinate $\bar{q}_F$. In this energy region, no approximate invariant of motion can be expected to exist, even in the passage over the saddle between wells.

The saddle-crossing dynamics is entirely stochastic, with dimensionality essentially equal to the number of degrees of freedom of the system. It may not be possible to extract a dividing surface free from barrier recrossings. Going from semichaotic into fully developed chaotic regions, a new type of phase-space bottleneck emerges, that makes a reacting system increasingly trapped in the transition state as the total energy of the system increases. This emergence of a new bottleneck challenges the simple pictures one has supposed for a reacting system crossing a transition state, that is, both the ballistic (or separable) and diffusive transitions, and sheds light on the nature of trapping of a reacting system in the transition state; one should distinguish "apparent" and "true" trappings. In the former, although the system looks transiently trapped around a saddle, its reaction proceeds, independently, along the *nearly separable* reactive degree of freedom in *phase* space, but in the latter, the system not only looks trapped but also resists proceeding along any reactive degree of freedom one might choose, due to the nonvanishing mode-couplings appearing between semichaotic and fully developed chaotic energy regimes. To describe reaction dynamics in this region, it will probably be more convenient to go back to the conventional reaction path approach in the configuration space **q**. In such a case, the variational TST approach, to choose the dividing surface to minimize the reactive flux, becomes one reasonable means to address the problem [9–11].

V. CONCLUDING REMARKS AND FUTURE PROSPECTS

So far, one has conventionally taken a reaction coordinate in the *configurational* space, for example, a distance between atoms to form or break their bond, a reactive normal coordinate, or (configurational) steepest descent path [70, 71]. Our results have shown the persistence of the approximate invariants of motion associated with the reaction coordinate $\bar{q}_1(\mathbf{p},\mathbf{q})$, at least, in the region of the (first-rank) saddle even in a "sea" of high-dimensional chaos, and strongly support the use of the concept of a single, nearly-separable reactive degree of freedom in the system's *phase space*, a dof that is as free as possible from coupling to all the rest of the dof. This result immediately tells us that the observed deviations from unity of the *conventional* transmission coefficient κ should be due to the choice of the reaction coordinate q_1 of the system along which one might want to see the reaction event, whenever the κ arises from the recrossing problem.

The remaining ambiguity in reaction rate theories is the assumption of local vibrational equilibrium. Recall that reactions involving (a) chemical bond formation and/or cleavage may have, as a typical activation energy, a few tens of kilocalories per mole while an average thermal energy associated with a single dof is ~0.6 kcal/mol at room temperature. Therefore, we may anticipate that such chemical reactions are regarded as very rare events, and any (strongly chaotic) many-dimensional reacting system moves through all the accessible phase space in the reactant domain before finding the transition state.

However, in some unimolecular reactions of a few degrees of freedom, [24–26], there is a dynamical bottleneck to intramolecular energy transfer even in the reactant well. That is, cantori appear to form partial barriers between irregular regions of phase space. As yet, there is no general answer or analytical tool to determine whether such a dynamical bottleneck even exists for larger nonlinear reacting systems, say, >10 dof.

Along this direction, although we only argued how the invariant of frequency arises, varying the ratios of frequencies $\bar{\omega}_k$ among the modes [68, 72, 73] should shed light on what kinds of energy flows take place among the modes of $\bar{q}_k(\mathbf{p},\mathbf{q})$ space, elucidated about potential minima. Obviously, the more the dof, the more possible combinations emerge to make the system very complicated. Another possible diagnostic method to look into this in many-body systems would be to execute the backward trajectory calculation, starting on the phase-space dividing hypersurface $S(\bar{q}_1(\mathbf{p},\mathbf{q}) = 0)$, sampled from the microcanonical ensemble. If the system exhibits an invariant of motion for a certain time in the reactant phase space, that is, if the system is trapped in a certain limited region for some period, this should imply how the local equilibrium is suppressed in the reaction. The backward calculations initiated with large momenta $\bar{p}_1(\mathbf{p},\mathbf{q})$ on that dividing hypersurface, that is, the bundle of the fast transitions from the reactant to product if one inverts the time, would reveal how any mode-specific nature of a reaction relates to the local topography of the phase space in the reactant state.

People have usually supposed that the elusive transition state is localized somewhere in the vicinity of a first-rank saddle linking reactant and product states, and that the evolution of the reactions can be decomposed into the two distinct parts, that is, how the system passes into the transition state from the reactant state, and then how it leaves there after its arrival. These have enabled us to argue the physical (classical) foundation of the deviation from the theories [as represented by a conventional $\kappa(<1)$], in terms of how violated "local equilibrium", and "no-return" assumptions may be. However, if, for example, the energy barrier of the reacting systems become comparable with the average thermal energy of a single dof, such a common scenario implicitly assumed so far should no longer be valid. If a dividing hypersurface were still to exist even in such cases, it might not be localized near the saddles but somehow delocalized throughout the whole accessible phase space as the separatrix theories [24–26] indicate for a few-dof systems. One may anticipate that the stable and unstable invariant manifolds we could extract in the region of the saddles even in a sea of *many-body* chaos can be connected through the rest of the phase space, and this provides us with an essential clue to generalize separatrix theories to *multidimensional* systems. Recently, Wiggins et al. [74] just started their research along this scenario.

Biological reactions take place on complex energy landscapes, involving sequential crossings over multiple saddles, which, however, result in specific *robust* functions in living organisms. The question arising now is this: Can we understand why such robust functions or efficiencies exist in biological systems? We start from a viewpoint that each reaction crossing over each saddle takes place in dynamically independent fashion, that is, the local equilibration is considered to be attained quickly in each basin before the system goes to the next saddle, so the system loses all *dynamical* memories. As yet, of course, there is no answer as to whether dynamical connectivity or non-Markovian nature along the sequential multiple saddle crossing dynamics plays a significant role in maintaining robust functions.

One of the relevant interesting papers on protein foldings is that by García et al [75]. They showed how non-Brownian strange kinetics emerge in multi-basin dynamics trajectories generated by all-atom MD simulations of cytochrome c in aqueous solution at a wide range of temperature. They used a so-called molecule optimal dynamic coordinates (MODC) derived by a linear transformation of the Cartesian coordinates of the protein system, which best represent the configurational protein fluctuations (in a least square sense). They found that some slow MODC manifestly exhibits non-Brownian dynamics, that is, protein motions are more suppressed and cover less configurational space than a normal Brownian process on a short time scale, but they become more enhanced as a faster, well-concerted motion on a long time scale between a temperature at which the protein is in the native state and a temperature above melting (see also [76–79]).

The Lie techniques may provide us with the physical footings or analytical means to elucidate dynamical correlations among successive saddle crossings by enabling us to scrutinize "connectivity" of manifolds from and to the sequential saddle points and "extent of volume" of the region of a junction of manifolds in terms of the backward system trajectories initiated from $S(\bar{q}_1(\mathbf{p}, \mathbf{q}) = 0)$ at one saddle point and the forward from the other $S(\bar{q}_1(\mathbf{p}, \mathbf{q}) = 0)$ at the previous saddle point, through which the system has passed before reaching the first [80].

At high energies above the lowest, presumably (but not necessarily) first-rank saddle, the system trajectories may pass over higher rank saddles of the PES. These provides us with a new, untouched, exciting problem, that is, what is the role of resonance in the imaginary ω-plane for the bifurcation? (This even arises in the degenerate bending modes for a linear transition state of a triatomic molecule.) This is one of the most exciting questions, especially for relaxation dynamics on a rugged PES, if the system finds higher rank saddles, which may be densely distributed in the regions of high potential energies, and would pass through such complicated regions at least as frequently as through the lowest, first-rank transition states. This will require going back to the fundamental question of what the transition state is, that is, whether a dividing hypersurface could still exist or be definable, in terms of separating the space of the system into regions identifiable with individual stable states.

Acknowledgments

We would like to acknowledge Professor Mikito Toda and Professor Stuart A. Rice for their continuous, stimulating discussions. We also thank Professor William Miller for his helpful comments and for sending us relevant articles [65]. T. K. also thanks Y. and Y. S. Komatsuzaki for their continuous encouragements. We thank the Aspen Center for Physics for its hospitality during the completion of this work. This research was supported by the National Science Foundation, the Japan Society for the Promotion of Science, and Grant-in-Aid for Research on Priority Areas (A) "Molecular Physical Chemistry" and (C) "Genome Information Science" of the Ministry of Education, Science, Sports and Culture of Japan.

APPENDIX

A. The Proof of Eqs. (2.8) and (2.9)

The equation of motion of new coordinates $\bar{\mathbf{q}}$ and momenta $\bar{\mathbf{p}}$ can straightforwardly be solved, with the new Hamiltonian $\bar{H}$ independent of the new angle variable $\bar{\Theta}$.

$$\frac{d\bar{p}_k}{dt} = -\frac{\partial \bar{H}(\bar{\mathbf{J}})}{\partial \bar{q}_k} = -\frac{\partial \bar{H}(\bar{\mathbf{J}})}{\partial \bar{J}_k} \frac{\partial \bar{J}_k}{\partial \bar{q}_k} \tag{A.1}$$

$$= -\bar{\omega}_k(\bar{\mathbf{J}})\omega_k \bar{q}_k \tag{A.2}$$

$$\frac{d\bar{q}_k}{dt} = \frac{\partial \bar{H}(\bar{\mathbf{J}})}{\partial \bar{p}_k} = \frac{\partial \bar{H}(\bar{\mathbf{J}})}{\partial \bar{J}_k} \frac{\partial \bar{J}_k}{\partial \bar{p}_k} \tag{A.3}$$

$$= \frac{\bar{\omega}_k(\bar{\mathbf{J}})}{\omega_k} \bar{p}_k \tag{A.4}$$

By differentiating Eq. (A.4) in time t and combining Eq. (A.2), one can obtain

$$\frac{d^2 \bar{q}_k}{dt^2} = \frac{\bar{\omega}_k(\bar{\mathbf{J}})}{\omega_k} \frac{d\bar{p}_k}{dt} = -\bar{\omega}_k^2(\bar{\mathbf{J}}) \bar{q}_k \tag{A.5}$$

Here, the last equal signs of Eqs. (A.2) and (A.4) follow from Eq. (2.5) and

$$\bar{J}_k = \frac{1}{2\pi} \oint \bar{p}_k \, d\bar{q}_k = \frac{\bar{p}_k^2 + \omega_k^2 \bar{q}_k^2}{2\omega_k} \tag{A.6}$$

One can easily see Eq. (A.6) as follows: as described in Appendix C, the new Hamiltonian $\bar{H}_0$ at zeroth-order corresponds to replacing the canonical variables, for example, $(\mathbf{p}, \mathbf{q})$, $(\mathbf{J}, \Theta)$, ..., in the counterpart of old Hamiltonian H_0 by the corresponding, new canonical variables, $(\bar{\mathbf{p}}, \bar{\mathbf{q}})$, $(\bar{\mathbf{J}}, \bar{\mathbf{\Theta}})$, ...

$$\bar{H}_0(\bar{\mathbf{p}}, \bar{\mathbf{q}}) = H_0(\bar{\mathbf{p}}, \bar{\mathbf{q}}) \tag{A.7}$$

$$= \sum_k \frac{1}{2}(\bar{p}_k^2 + \omega_k^2 \bar{q}_k^2) \tag{A.8}$$

$$= \sum_k \omega_k \bar{J}_k \tag{A.9}$$

[one can also verify the second equal sign in Eq. (A.6) by inserting Eqs. (2.10) and (2.11) into Eq. (A.6)].

B. Lie Transforms

Let us suppose the following Hamilton's equations of motion:

$$\frac{d\bar{p}_i}{d\varepsilon} = -\frac{\partial W}{\partial \bar{q}_i} \tag{A.10}$$

$$\frac{d\bar{q}_i}{d\varepsilon} = \frac{\partial W}{\partial \bar{p}_i} \tag{A.11}$$

where ε and W are "time" and "Hamiltonian", a function of canonical coordinates $\bar{\mathbf{q}}$ and its conjugate canonical momenta $\bar{\mathbf{p}}$. These can be represented collectively in the notation $z_i = (\bar{p}_i, \bar{q}_i)$:

$$\frac{dz_i}{d\varepsilon} = \{z_i, W(\mathbf{z})\} \equiv -L_W z_i \tag{A.12}$$

Here $\{\}$ denotes Poisson bracket:

$$\{u, v\} \equiv \sum_i \left(\frac{\partial u}{\partial \bar{q}_i} \frac{\partial v}{\partial \bar{p}_i} - \frac{\partial v}{\partial \bar{q}_i} \frac{\partial u}{\partial \bar{p}_i} \right) \tag{A.13}$$

which has the following properties, for arbitrary differentiable functions u, v, and w,

$$\{u, v\} = -\{v, u\} \tag{A.14}$$

$$\{u, v + w\} = \{u, v\} + \{u, w\} \tag{A.15}$$

$$\{u, vw\} = \{u, v\}w + v\{u, w\} \tag{A.16}$$

and

$$\{u, \{v, w\}\} + \{v, \{w, u\}\} + \{w, \{u, v\}\} = 0 \tag{A.17}$$

Recall that for a set of any canonical variables, for example $(\bar{\mathbf{p}}, \bar{\mathbf{q}})$,

$$\{\bar{q}_i, \bar{p}_j\} = \delta_{ij} \qquad \{\bar{q}_i, \bar{q}_j\} = \{\bar{p}_i, \bar{p}_j\} = 0 \tag{A.18}$$

(δ_{ij} is Kronecker delta) and, in turn, a set of variables that satisfy Eq. (A.18) is canonical.

An operator, $L_W (\equiv \{W, \ \})$, called the *Lie derivative generated by W*, obeys the following properties easily derived from Eqs. (A.15)–(A.17):

$$L_W(\alpha u + \beta v) = \alpha L_W u + \beta L_W v \tag{A.19}$$

$$L_W uv = u L_W v + v L_W u \tag{A.20}$$

$$L_W\{u, v\} = \{u, L_W v\} + \{L_W u, v\} \tag{A.21}$$

$$L_V L_W = L_{\{V,W\}} + L_W L_V \tag{A.22}$$

where α, β, and V are any numbers and any differentiable function. The n times repeated operations of L_W to Eqs. (A.19)–(A.21) give

$$L_W^n (\alpha u + \beta v) = \alpha L_W^n u + \beta L_W^n v \tag{A.23}$$

$$L_W^n uv = \sum_{m=0}^{n} {}_nC_m (L_W^m u)(L_W^{n-m} v) \tag{A.24}$$

$$L_W^n \{u, v\} = \sum_{m=0}^{n} {}_nC_m \{L_W^m u, L_W^{n-m} v\} \tag{A.25}$$

and, hence,

$$e^{-\varepsilon L_W} uv = (e^{-\varepsilon L_W} u)(\varepsilon^{-\varepsilon L_W} v) \tag{A.26}$$

$$e^{-\varepsilon L_W} \{u, v\} = \{e^{-\varepsilon L_W} u, e^{-\varepsilon L_W} v\} \tag{A.27}$$

1. Autonomous Cases

The formal solution of Eq. (A.12) can be represented as

$$\mathbf{z}(\varepsilon) = e^{-\varepsilon L_W} \mathbf{z}(0) \equiv T \mathbf{z}(0) \tag{A.28}$$

for autonomous systems having no explicit dependence on "time" ε of

"Hamiltonian" W. Here, we introduce the *evolution* operator T for brevity.

It can be easily proved for any transforms described by this that if $\mathbf{z}(0)$ are canonical, $\mathbf{z}(\varepsilon)$ are also canonical (and vice versa) as follows: Let us designate the phase-space variables $\mathbf{z}$ at the "time" being zero as $(\mathbf{p}, \mathbf{q})$ and those at the "time" ε as $(\bar{\mathbf{p}}, \bar{\mathbf{q}})$. Then,

$$\bar{p}_i = Tp_i = e^{-\varepsilon L_{W(\mathbf{p},\mathbf{q})}} p_i = \bar{p}_i(\mathbf{p}, \mathbf{q}; \varepsilon) \tag{A.29}$$

$$\bar{q}_i = Tq_i = e^{-\varepsilon L_{W(\mathbf{p},\mathbf{q})}} q_i = \bar{q}_i(\mathbf{p}, \mathbf{q}; \varepsilon) \tag{A.30}$$

and

$$\{\bar{q}_i, \bar{p}_j\} = \{Tq_i, Tp_j\} = T\{q_i, p_j\} \tag{A.31}$$

$$\{\bar{q}_i, \bar{q}_j\} = \{Tq_i, Tq_j\} = T\{q_i, q_j\} \tag{A.32}$$

$$\{\bar{p}_i, \bar{p}_j\} = \{Tp_i, Tp_j\} = T\{p_i, p_j\} \tag{A.33}$$

where the second equal signs of these three equations are thanks to Eq. (A.27). Therefore, if $(\mathbf{p}, \mathbf{q})$ is canonical [i.e., Eq. (A.18) is satisfied for $(\mathbf{p}, \mathbf{q})$], $(\bar{\mathbf{p}}, \bar{\mathbf{q}})$ is also canonical because

$$\{\bar{q}_i, \bar{p}_j\} = T\delta_{ij} = e^{-\varepsilon L_W}\delta_{ij} = \delta_{ij} \tag{A.34}$$

$$\{\bar{q}_i, \bar{q}_j\} = \{\bar{p}_i, \bar{p}_j\} = T \cdot 0 = 0 \tag{A.35}$$

Note that the successive operations of $L_{W(\mathbf{p},\mathbf{q})}$ to p_i or q_i usually produce, complicated, nonlinear functions expressed explicitly in terms of $(\mathbf{p}, \mathbf{q})$. In Eqs. (A.29) and (A.30), $\bar{p}_i$ and $\bar{q}_i$ are represented as functions of the "time" ε and the initial condition $(\mathbf{p}, \mathbf{q})$ along the dynamical evolution obeying the "Hamiltonian" W, that is, $\bar{p}_i(\mathbf{p}, \mathbf{q}; 0) = p_i$, $\bar{q}_i(\mathbf{p}, \mathbf{q}; 0) = q_i$.

By assuming the inverse transformation from ε to 0 in the "time", that is,

$$\mathbf{z}(0) = T^{-1}\mathbf{z}(\varepsilon) \tag{A.36}$$

in other terms,

$$p_i = T^{-1}\bar{p}_i = e^{\varepsilon L_{W(\mathbf{p},\mathbf{q})}}\bar{p}_i = p_i(\bar{\mathbf{p}}, \bar{\mathbf{q}}; \varepsilon) \tag{A.37}$$

$$q_i = T^{-1}\bar{q}_i = e^{\varepsilon L_{W(\mathbf{p},\mathbf{q})}}\bar{q}_i = q_i(\bar{\mathbf{p}}, \bar{\mathbf{q}}; \varepsilon) \tag{A.38}$$

where $TT^{-1} = T^{-1}T = 1$, it can also be proved straightforwardly by the premultiplication of Eqs. (A.31)–(A.32) by *inverse* evolution operator T^{-1} that, if $(\bar{\mathbf{p}}, \bar{\mathbf{q}})$ is canonical, $(\mathbf{p}, \mathbf{q})$ is also canonical.

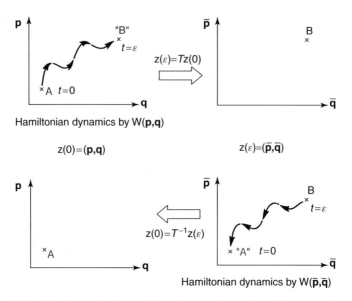

Figure 2.15. The schematic picture of the Lie transforms: If W is free from the "virtual" time t explicitly, the *functional* structure of W is preserved through the "time" evolution.

Figure 2.15 shows pictorially what the Lie transforms and its inversion perform: $T\mathbf{z}(0)$ transforms, for example, a point **A** in the $(\mathbf{p},\mathbf{q})$ coordinate system to a point **B** in the other $(\bar{\mathbf{p}},\bar{\mathbf{q}})$ system, which corresponds to a "virtual" time evolution of the phase-space variable (p_i, q_i) from "time" 0 to ε driven by a "Hamiltonian" $W(\mathbf{p},\mathbf{q})$. In turn, $T^{-1}\mathbf{z}(\varepsilon)$ transforms the point **B** in $(\bar{\mathbf{p}},\bar{\mathbf{q}})$ to the point **A** in $(\mathbf{p},\mathbf{q})$, the reversed-time evolution of $(\bar{p}_i, \bar{q}_i)$ from ε to 0 driven by a "Hamiltonian" $W(\bar{\mathbf{p}},\bar{\mathbf{q}})$, just replaced the "symbol" p_i and q_i in $W(\mathbf{p},\mathbf{q})$ by $\bar{p}_i$ and $\bar{q}_i$, whose functional *form* is unchanged.

How Does a Lie Transform Operate on Functions? Given an arbitrary differentiable function $f(\mathbf{p},\mathbf{q})$, let us consider the Lie transforms of the function, $Tf(\mathbf{p},\mathbf{q})$.

$$Tf(\mathbf{p},\mathbf{q}) = e^{-\varepsilon L_W} f(\mathbf{p},\mathbf{q}) = \sum_{n=0}^{\infty} \frac{\varepsilon^n}{n!} (-L_W)^n f(\mathbf{p},\mathbf{q}) \equiv g(\mathbf{p},\mathbf{q};\varepsilon) \quad (A.39)$$

where the recursive operations of the Lie derivatives by $W(\mathbf{p},\mathbf{q})$ on the function f yield a new function, denoted hereafter g, represented as a nonlinear function of $\mathbf{p}$ and $\mathbf{q}$, and ε. The resultant, transformed new function $g(\mathbf{p}, \mathbf{q}; \varepsilon)$ can be interpreted as follows:

$$L_W f = \{W, f\} = \sum_i \left(\frac{\partial W}{\partial \bar{q}_i} \frac{\partial f}{\partial \bar{p}_i} - \frac{\partial W}{\partial \bar{p}_i} \frac{\partial f}{\partial \bar{q}_i} \right) \tag{A.40}$$

$$= -\sum_i \left(\frac{d\bar{p}_i}{d\varepsilon} \frac{\partial f}{\partial \bar{p}_i} + \frac{d\bar{q}_i}{d\varepsilon} \frac{\partial f}{\partial \bar{q}_i} \right) \tag{A.41}$$

$$= -\frac{df}{d\varepsilon} \tag{A.42}$$

and, for example, the operation twice of the Lie derivatives on f, $L_W^2 f$, implies

$$L_W^2 f = L_W(L_W f) = L_W\left(-\frac{df}{d\varepsilon}\right) = \frac{d^2 f}{d\varepsilon^2} \tag{A.43}$$

and thus by recurrence over n,

$$(-L_W)^n f = \frac{d^n f}{d\varepsilon^n} \tag{A.44}$$

Therefore, the new function $g(\mathbf{z}(0); \varepsilon)$ is represented as

$$g(\mathbf{z}(0); \varepsilon) = Tf(\mathbf{z}(0)) = \sum_{n=0}^{\infty} \frac{\varepsilon^n}{n!} (-L_{W(\mathbf{z}(0))})^n f(\mathbf{z}(0)) = \sum_{n=0}^{\infty} \frac{\varepsilon^n}{n!} \frac{d^n f}{d\varepsilon^n}\bigg|_{\varepsilon=0} \tag{A.45}$$

The last term corresponds to a Taylor series in "time" ε (about the origin) of the function f, which does not depend explicitly on ε but implicitly through $\mathbf{z}$. Thus, we can lead

$$g(\mathbf{z}(0); \varepsilon) = f(\mathbf{z}(\varepsilon)) = f(T\mathbf{z}(0)) \tag{A.46}$$

One can see that the new function $g(\mathbf{z}(0); \varepsilon)$ represents the functional *value* f at *the point* $\mathbf{z}(\varepsilon)$ as a function of the *initial point* $\mathbf{z}(0)$ and the "*time*" ε, where the phase space variables $\mathbf{z}$ evolve in the "time" obeying "Hamiltonian" W.

One can interpret schematically what the Lie transforms and its inversion perform on *function f* in Figure 2.16: $Tf(\mathbf{p}, \mathbf{q})$ (as indicated by an arrow in the top of the figure) transforms f evaluated at a point $\mathbf{A}$, $(\mathbf{p}_A, \mathbf{q}_A)$ in the $(\mathbf{p}, \mathbf{q})$ system to a new function, say, $g(\mathbf{p}, \mathbf{q}; \varepsilon)$, represented at the *same* point $\mathbf{A}$, whose functional *value* is equivalent to $f(\bar{\mathbf{p}}_B, \bar{\mathbf{q}}_B)$ evaluated at a point $\mathbf{B}$

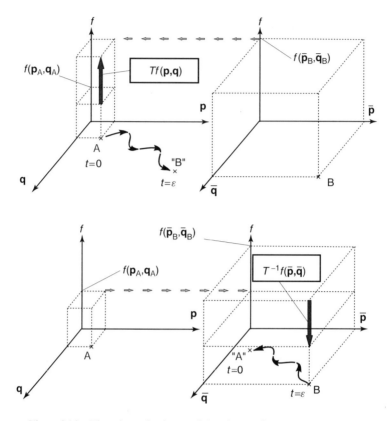

Figure 2.16. The schematic picture of the Lie transforms on a function f.

in the $(\bar{\mathbf{p}}, \bar{\mathbf{q}})$ system, which corresponds to a "virtual" time evolution driven by $W(\mathbf{p}, \mathbf{q})$, $(T\mathbf{p}_A, T\mathbf{q}_A)$. In turn, $T^{-1}f(\bar{\mathbf{p}}, \bar{\mathbf{q}})$ (as indicated by an another arrow in the bottom of the figure) transforms f evaluated at the point **B** in $(\bar{\mathbf{p}}, \bar{\mathbf{q}})$ to a new function, say, $g'(\bar{\mathbf{p}}, \bar{\mathbf{q}}; \varepsilon)$, at the *same* point **B**, whose functional *value* is equivalent to $f(\mathbf{p}_A, \mathbf{q}_A)$ evaluated at the point **A** in the $(\mathbf{p}, \mathbf{q})$ system.

Finally, the Lie transforms on functions f generated by "Hamiltonian" W is represented as

$$Tf(\mathbf{z}(0)) = f(\mathbf{z}(\varepsilon)) = f(T\mathbf{z}(0)) = g(\mathbf{z}(0); \varepsilon) \tag{A.47}$$

or

$$Tf(\mathbf{p}, \mathbf{q}) = f(\bar{\mathbf{p}}, \bar{\mathbf{q}}) = f(T\mathbf{p}, T\mathbf{q}) = g(\mathbf{p}, \mathbf{q}; \varepsilon) \tag{A.48}$$

It implies that for an arbitrary differentiable function f evaluated at $(\mathbf{p}, \mathbf{q})$, the Lie transforms provides us with a new function $g(\mathbf{p}, \mathbf{q}; \varepsilon)$, representing the functional *value* of f at $(\bar{\mathbf{p}}, \bar{\mathbf{q}})$, whose all components $\bar{p}_i$ and $\bar{q}_i$ obey $\bar{p}_i = Tp_i$ and $\bar{q}_i = Tq_i$.

For example, if f is the "Hamiltonian" W,

$$W(\bar{\mathbf{p}}, \bar{\mathbf{q}}) = W(e^{-\varepsilon L_W}\mathbf{p}, e^{-\varepsilon L_W}\mathbf{q}) \tag{A.49}$$

$$= e^{-\varepsilon L_W}W(\mathbf{p}, \mathbf{q}) \tag{A.50}$$

$$= W(\mathbf{p}, \mathbf{q}) \tag{A.51}$$

because $L_W^n W = 0$ except $n = 0$. That is, the Lie transforms generated by W, which does not depend on ε explicitly, preserves the functional form of W itself along the dynamical evolution in "time" ε.

Differentiating the equation $Tf(\mathbf{p}, \mathbf{q}) = f(\bar{\mathbf{p}}, \bar{\mathbf{q}})$ with respect to ε,

$$\frac{dT(\varepsilon)}{d\varepsilon} f(\mathbf{p}, \mathbf{q}) = \sum_i \left(\frac{\partial f(\bar{\mathbf{p}}, \bar{\mathbf{q}})}{\partial \bar{q}_i} \frac{d\bar{q}_i}{d\varepsilon} + \frac{\partial f(\bar{\mathbf{p}}, \bar{\mathbf{q}})}{\partial \bar{p}_i} \frac{d\bar{p}_i}{d\varepsilon} \right)$$

$$= \sum_i \left(\frac{\partial f(\bar{\mathbf{p}}, \bar{\mathbf{q}})}{\partial \bar{q}_i} \frac{\partial W}{\partial \bar{p}_i} - \frac{\partial f(\bar{\mathbf{p}}, \bar{\mathbf{q}})}{\partial \bar{p}_i} \frac{\partial W}{\partial \bar{q}_i} \right)$$

$$= \{f(\bar{\mathbf{p}}, \bar{\mathbf{q}}), W(\bar{\mathbf{p}}, \bar{\mathbf{q}})\}$$

$$= \{T(\varepsilon)f(\mathbf{p}, \mathbf{q}), T(\varepsilon)W(\mathbf{p}, \mathbf{q})\}$$

$$= T(\varepsilon)\{f(\mathbf{p}, \mathbf{q}), W(\mathbf{p}, \mathbf{q})\}$$

$$= -T(\varepsilon)L_W f(\mathbf{p}, \mathbf{q})$$

which holds for all $f(\mathbf{p}, \mathbf{q})$. Thus, one can have

$$\frac{dT}{d\varepsilon} = -TL_W \tag{A.52}$$

Similarly, differentiating the equation $TT^{-1} = 1$ with respect to ε yields

$$\frac{dT^{-1}}{d\varepsilon} = L_W T^{-1} \tag{A.53}$$

because

$$0 = \frac{dT(\varepsilon)T^{-1}(\varepsilon)}{d\varepsilon} = \frac{dT}{d\varepsilon}T^{-1} + T\frac{dT^{-1}}{d\varepsilon}$$

$$= -TL_W T^{-1} + T\frac{dT^{-1}}{d\varepsilon}$$

one can have Eq. (A.53) by premultiplying the final equation by T^{-1}.

2. Nonautonomous Cases

The Lie transforms for which the "Hamiltonian" W explicitly depends on "time" ε preserve the formal properties of the autonomous $W(\bar{\mathbf{p}}, \bar{\mathbf{q}})$. Hamilton's equations of motion do not change form for nonautonomous systems:

$$\frac{d\bar{p}_i}{d\varepsilon} = \{\bar{p}_i, W\} = -\frac{\partial W(\bar{\mathbf{p}}, \bar{\mathbf{q}}, \varepsilon)}{\partial \bar{q}_i}$$

$$\frac{d\bar{q}_i}{d\varepsilon} = \{\bar{q}_i, W\} = \frac{\partial W(\bar{\mathbf{p}}, \bar{\mathbf{q}}, \varepsilon)}{\partial \bar{p}_i}$$

and collectively,

$$\frac{dz_i}{d\varepsilon} = \{z_i, W(\mathbf{z}, \varepsilon)\} = -L_W z_i$$

The second derivatives of $\bar{p}_i$ over "time" ε becomes,

$$\frac{d^2\bar{p}_i}{d\varepsilon^2} = \frac{d}{d\varepsilon}\{\bar{p}_i, W\}$$

$$= \sum_j \left(\frac{\partial}{\partial \bar{q}_j}\{\bar{p}_i, W\}\frac{d\bar{q}_j}{d\varepsilon} + \frac{\partial}{\partial \bar{p}_j}\{\bar{p}_i, W\}\frac{d\bar{p}_j}{d\varepsilon}\right) + \frac{\partial}{\partial \varepsilon}\{\bar{p}_i, W\}$$

$$= \sum_j \left(\frac{\partial}{\partial \bar{q}_j}\{\bar{p}_i, W\}\frac{\partial W}{\partial \bar{p}_j} - \frac{\partial}{\partial \bar{p}_j}\{\bar{p}_i, W\}\frac{\partial W}{\partial \bar{q}_j}\right) + \left\{\bar{p}_i, \frac{\partial W}{\partial \varepsilon}\right\}$$

$$= \{\{\bar{p}_i, W\}, W\} + \left\{\bar{p}_i, \frac{\partial W}{\partial \varepsilon}\right\} \tag{A.54}$$

and similarly,

$$\frac{d^2\bar{q}_i}{d\varepsilon^2} = \{\{\bar{q}_i, W\}, W\} + \left\{\bar{q}_i, \frac{\partial W}{\partial \varepsilon}\right\} \quad \text{(A.55)}$$

One can combine these two expressions collectively in terms of z_i:

$$\frac{d^2 z_i}{d\varepsilon^2} = \frac{d}{d\varepsilon}\{z_i, W\} = \{\{z_i, W\}, W\} + \left\{z_i, \frac{\partial W}{\partial \varepsilon}\right\} \quad \text{(A.56)}$$

$$= \frac{d}{d\varepsilon}(\Delta_{-W}) z_i = \Delta^2_{-W} z_i \quad \text{(A.57)}$$

where a newly introduced operator Δ_W is defined by

$$\Delta_W \equiv \{W, \ \} + \frac{\partial}{\partial \varepsilon} = L_W + \frac{\partial}{\partial \varepsilon} \quad \text{(A.58)}$$

and easily verify that the nth derivatives of z_i in "time" ε is given, in terms of Δ_{-W}, by

$$\frac{d^n z_i}{d\varepsilon^n} = (\Delta_{-W})^n z_i \quad \text{(A.59)}$$

Thus, the expansion of the solution $\mathbf{z}(\varepsilon)$ in power series of ε around the origin result in the expressions using Δ_{-W}:

$$z_i(\varepsilon) = \sum_{n=0}^{\infty} \frac{\varepsilon^n}{n!} \frac{d^n z_i}{d\varepsilon^n}\bigg|_{\varepsilon=0} = \sum_{n=0}^{\infty} \frac{\varepsilon^n}{n!} (\Delta_{-W})^n z_i|_{\varepsilon=0} \quad \text{(A.60)}$$

$$= \exp\left[-\int^{\varepsilon} L_{W(\varepsilon')} d\varepsilon'\right] z_i(0) = T z_i(0) \quad \text{(A.61)}$$

$$\bar{p}_i = T p_i = \exp\left[-\int^{\varepsilon} L_{W(\mathbf{p},\mathbf{q};\varepsilon')} d\varepsilon'\right] p_i = \bar{p}_i(\mathbf{p}, \mathbf{q}; \varepsilon) \quad \text{(A.29a)}$$

$$\bar{q}_i = T q_i = \exp\left[-\int^{\varepsilon} L_{W(\mathbf{p},\mathbf{q};\varepsilon')} d\varepsilon'\right] q_i = \bar{q}_i(\mathbf{p}, \mathbf{q}; \varepsilon) \quad \text{(A.30a)}$$

just the same as the Lie transforms generated by an autonomous $W(\mathbf{p}, \mathbf{q})$. It can easily be verified that Δ_W satisfy similar properties of L_W that just replace L_W by Δ_W in Eqs. (A.19)–(A.25). The only exception is that, instead of Eq. (A.22), Δ_W holds

$$\Delta_V\Delta_W = \Delta_W\Delta_V + L_{\{V,W\}} + L_{\frac{\partial W}{\partial \varepsilon} - \frac{\partial V}{\partial \varepsilon}} \qquad (A.62)$$

Hence, one sees immediately that,

$$Tuv = (Tu)(Tv) \qquad (A.26a)$$

$$T\{u, v\} = \{Tu, Tv\} \qquad (A.27a)$$

and that Eqs. (A.29a) and (A.30a) yield a completely canonical mapping like those by an autonomous $W(\bar{\mathbf{p}}, \bar{\mathbf{q}})$.

In general, for an arbitrary differentiable function $f(\bar{\mathbf{p}}, \bar{\mathbf{q}})$, the nth derivatives of f are given by

$$\frac{d^n}{d\varepsilon^n} f = (\Delta_{-W})^n f$$

because

$$\frac{df}{d\varepsilon} = \sum_j \left(\frac{\partial f}{\partial \bar{q}_j} \frac{d\bar{q}_j}{d\varepsilon} + \frac{\partial f}{\partial \bar{p}_j} \frac{d\bar{p}_j}{d\varepsilon} \right)$$

$$= \sum_j \left(\frac{\partial f}{\partial \bar{q}_j} \frac{\partial W}{\partial \bar{p}_j} - \frac{\partial f}{\partial \bar{p}_j} \frac{\partial W}{\partial \bar{q}_j} \right) = \{f, W\} = \Delta_{-W} f$$

$$\frac{d^2 f}{d\varepsilon^2} = \frac{d}{d\varepsilon} (\Delta_{-W} f) = \{\Delta_{-W} f, W\} + \frac{\partial(\Delta_{-W} f)}{\partial \varepsilon}$$

$$= \Delta_{-W}^2 f$$

and so forth.

Consequently,

$$f(\bar{\mathbf{p}}, \bar{\mathbf{q}}) = \sum_{n=0}^{\infty} \frac{\varepsilon^n}{n!} \frac{d^n f}{d\varepsilon^n}\bigg|_{\varepsilon=0}$$

$$= \sum_{n=0}^{\infty} \frac{\varepsilon^n}{n!} (\Delta_{-W})^n f |_{\varepsilon=0}$$

$$= \exp\left[-\int_0^\varepsilon L_{W(\mathbf{p},\mathbf{q};\varepsilon')} \, d\varepsilon'\right] f(\mathbf{p}, \mathbf{q}) = Tf(\mathbf{p}, \mathbf{q})$$

$$= g(\mathbf{p}, \mathbf{q}; \varepsilon) \qquad (A.45a)$$

As argued in autonomous cases, $g(\mathbf{p},\mathbf{q};\varepsilon)$ implies that while the functional *value* of $Tf(\mathbf{p},\mathbf{q})$ is exactly equal to $f(\bar{\mathbf{p}},\bar{\mathbf{q}})$, the transformed, usually complicated, nonlinear function becomes represented as a function of $\mathbf{p}$ and $\mathbf{q}$, and ε, whose functional *form* differs from the original f.

Finally, the Lie transforms on functions f generated by a nonautonomous "Hamiltonian" W can be represented as

$$Tf(\mathbf{z}(0)) = f(\mathbf{z}(\varepsilon)) = f(T\mathbf{z}(0)) = g(\mathbf{z}(0);\varepsilon) \qquad (A.47a)$$

or

$$Tf(\mathbf{p},\mathbf{q}) = f(\bar{\mathbf{p}},\bar{\mathbf{q}}) = f(T\mathbf{p},T\mathbf{q}) = g(\mathbf{p},\mathbf{q};\varepsilon) \qquad (A.48a)$$

Here, one can see that we have the same formulations, Eqs. (A.47) and (A.48), derived for autonomous W. One can easily see that Eqs. (A.52) and (A.53), and the schematic interpretations of the Lie transforms, Figures 2.15 and 2.16 also hold for nonautonomous Lie transforms $W(\bar{\mathbf{p}},\bar{\mathbf{q}},\varepsilon)$, except the "Hamiltonian" W changes its form depending on the "time" ε along the "time" evolution.

C. Perturbation Theory Based on Lie Transforms

It is well known that for any canonical transforms $(\mathbf{p},\mathbf{q}) \to (\bar{\mathbf{p}},\bar{\mathbf{q}})$ of an autonomous system, the new Hamiltonian, denoted hereinafter $\bar{H}$, is related to the old Hamiltonian H by

$$\bar{H}(\bar{\mathbf{p}},\bar{\mathbf{q}}) = H(\mathbf{p},\mathbf{q}) \qquad (A.63)$$

Now let us suppose that the canonical transformation is brought about by "Lie transforms", and hence, in the $\mathbf{z}$ representation, this becomes

$$\bar{H}(\mathbf{z}(\varepsilon)) = H(\mathbf{z}(0)) \qquad (A.64)$$

that is, one can see that the *new* Hamiltonian $\bar{H}$ evaluated at the *new point* $\mathbf{z}(\varepsilon)$ along the dynamical evolution obeying W is equal to the *old* Hamiltonian H at the *old point* $\mathbf{z}(0)$.

By comparing this to the equation

$$f(\mathbf{z}(\varepsilon)) = g(\mathbf{z}(0)) \qquad (A.65)$$

[e.g., Eq. (A.47)] and applying $f = T^{-1}g$, one has

$$\bar{H}(\mathbf{z}(\varepsilon)) = T^{-1}H(\mathbf{z}(\varepsilon)) = H(\mathbf{z}(0)) \tag{A.66}$$

or,

$$\bar{H}(\bar{\mathbf{p}}, \bar{\mathbf{q}}) = T^{-1}H(\bar{\mathbf{p}}, \bar{\mathbf{q}}) = H(\mathbf{p}, \mathbf{q}) \tag{A.67}$$

Here, the *inverse* evolution operator T^{-1} brings the system dwelling at a "time" backward to a past in ε from *that* "time" along the dynamical evolution $\mathbf{z}$; yielding $H(\mathbf{z}(0))$, that is, $H(\mathbf{p}, \mathbf{q})$.

In cases that H, $\bar{H}$, and W (hence, also L_W, T, T^{-1}) are expandable as power series in ε, one can obtain a recursive, perturbation series to yield a desired new Hamiltonian.

$$H = \sum_{n=0} \varepsilon^n H_n \tag{A.68}$$

$$\bar{H} = \sum_{n=0} \varepsilon^n \bar{H}_n \tag{A.69}$$

$$W = \sum_{n=0} \varepsilon^n W_{n+1} \tag{A.70}$$

$$L_W = \sum_{n=0} \varepsilon^n L_{n+1} \tag{A.71}$$

$$T = \sum_{n=0} \varepsilon^n T_n \tag{A.72}$$

$$T^{-1} = \sum_{n=0} \varepsilon^n T_n^{-1} \tag{A.73}$$

where

$$L_n = \{W_n, \ \} \tag{A.74}$$

First, we derive a versatile perturbation series among the relations between T_n and T_n^{-1} and L_n. Inserting Eqs. (A.71) and (A.72) into Eq. (A.52), and Eqs. (A.71) and (A.73) into Eq. (A.53), and equating like powers of ε, one can obtain a recursive relation for T_n and T_n^{-1} ($n > 0$):

$$T_n = -\frac{1}{n} \sum_{m=0}^{n-1} T_m L_{n-m} \tag{A.75}$$

$$T_n^{-1} = \frac{1}{n} \sum_{m=0}^{n-1} L_{n-m} T_m^{-1} \tag{A.76}$$

where $T_0 = T_0^{-1} = 1$ because $T_0(T_0^{-1})$ corresponds to $T(T^{-1})$ at $\varepsilon = 0$. For example, to third order,

$$T_1 = -L_1 \tag{A.77}$$

$$T_2 = -\frac{1}{2}L_2 + \frac{1}{2}L_1^2 \tag{A.78}$$

$$T_3 = -\frac{1}{3}L_3 + \frac{1}{6}L_2L_1 + \frac{1}{3}L_1L_2 - \frac{1}{6}L_1^3 \tag{A.79}$$

$$T_1^{-1} = L_1 \tag{A.80}$$

$$T_2^{-1} = \frac{1}{2}L_2 + \frac{1}{2}L_1^2 \tag{A.81}$$

$$T_3^{-1} = \frac{1}{3}L_3 + \frac{1}{6}L_1L_2 + \frac{1}{3}L_2L_1 + \frac{1}{6}L_1^3 \tag{A.82}$$

(note that the T's and L's do not generally commute, e.g., $L_iL_j \neq L_jL_i$).

By inserting the series expansions of $\bar{H}$, T^{-1}, and H into $\bar{H} = T^{-1}H$ and equating like powers of ε, one can recursively solve the CPT with the Lie transforms: to second order, the new Hamiltonian $\bar{H}$ relates to the old Hamiltonian H as follows:

$$\varepsilon^0 : \bar{H}_0 = H_0 \tag{A.83}$$

$$\varepsilon^1 : \bar{H}_1 = T_0^{-1}H_1 + T_1^{-1}H_0 = H_1 + L_1H_0 \tag{A.84}$$

$$= H_1 + \frac{dW_1}{d\tau} \tag{A.85}$$

$$\varepsilon^2 : \bar{H}_2 = T_0^{-1}H_2 + T_1^{-1}H_1 + T_2^{-1}H_0 \tag{A.86}$$

$$= H_2 + \frac{1}{2}L_1(\bar{H}_1 + H_1) + \frac{1}{2}L_2H_0 \tag{A.87}$$

$$= H_2 + \frac{1}{2}\{W_1, \bar{H}_1 + H_1\} + \frac{1}{2}\frac{dW_2}{d\tau} \tag{A.88}$$

Here,

$$L_n\bar{H}_0 = L_nH_0 = \{W_n, H_0\} = \frac{dW_n}{d\tau} \tag{A.89}$$

where τ is the time along the orbits obeying the "unperturbed" H_0. In the representations of Eqs. (A.68)–(A.89), we *intentionally* omitted the arguments, that is, which kinds of forms of canonical variables we may use. The above equations hold for all kinds of the forms because the Poisson bracket calculation is canonically invariant, that is, the arguments themselves of the functions in the Lie transforms are really dummy variables. Hereafter, we use the arguments $(\bar{\mathbf{p}}, \bar{\mathbf{q}})$ as the new canonical variables in Eqs. (A.83)–(A.88).

From Eq. (A.83) it is straightforward at $\mathcal{O}(\varepsilon^0)$ that $\bar{H}_0(\bar{\mathbf{p}}, \bar{\mathbf{q}})$ is a function that just replaces the phase-space variables $(\mathbf{p}, \mathbf{q})$ in H_0 with those $(\bar{\mathbf{p}}, \bar{\mathbf{q}})$. However this is only the exception; at each order beyond $\mathcal{O}(\varepsilon^0)$, for each condition, there successively appear two unknown quantities, for example, $\bar{H}_1$ and W_1 at $\mathcal{O}(\varepsilon^1)$, and $\bar{H}_2$ and W_2 at $\mathcal{O}(\varepsilon^2)$ after $\bar{H}_1$ and W_1 have been determined in any fashion. In other words, there is flexibility to establish the new Hamiltonian $\bar{H}(\bar{\mathbf{p}}, \bar{\mathbf{q}})$ as one wishes.

Now let us suppose that the zeroth-order Hamiltonian H_0 can be represented as a function of only the action variables of H_0, that is, an integrable form, and the perturbation terms $H_n (n \geq 1)$ as functions of both the action and the angle variables:

$$H = \sum_{n=0} \varepsilon^n H_n = H_0(\mathbf{J}) + \sum_{n=1} \varepsilon^n H_n(\mathbf{J}, \mathbf{\Theta}) \tag{A.90}$$

and determine the unknown W_n so as to make the new Hamiltonian $\bar{H}_n$ as free from the new angle variables $\bar{\mathbf{\Theta}}$ as possible, at each order in ε^n; $\bar{H}_1$ and W_1 are determined at $\mathcal{O}(\varepsilon^1)$ as follows:

$$\bar{H}_1 = H_1 + \frac{dW_1}{d\tau} \tag{A.91}$$

$$= \langle H_1 \rangle + \{H_1\} + \frac{dW_1}{d\tau} \tag{A.92}$$

where

$$\langle H_1 \rangle = \lim_{\tau \to \infty} \frac{1}{\tau} \int_0^\tau H_1(\bar{\mathbf{p}}(\tau'), \bar{\mathbf{q}}(\tau')) \, d\tau' \tag{A.93}$$

$$\{H_1\} = H_1 - \langle H_1 \rangle \tag{A.94}$$

Here, the time τ obeying H_0 correlates with the new angle variables $\bar{\mathbf{\Theta}}$

linearly:

$$\frac{d\bar{\Theta}_k}{d\tau} = \{\bar{\Theta}_k, H_0\} = \frac{\partial H_0(\bar{\mathbf{J}})}{\partial \bar{J}_k} \equiv \omega_k(\bar{\mathbf{J}}) \qquad (A.95)$$

$$\bar{\Theta}_k = \omega_k(\bar{\mathbf{J}})\tau + \beta_k \quad (\beta_k: \text{initial phase factor}) \qquad (A.96)$$

Therefore, $\langle f \rangle$ and $\{f\}$ representing the average and oscillating parts over τ in f correspond to the free and dependent parts in all the $\bar{\Theta}_k$, respectively. The new Hamiltonian at first order $\bar{H}_1$ can be determined to be free from all the $\bar{\Theta}_k$ by

$$\bar{H}_1 = \langle H_1 \rangle \qquad (A.97)$$

$$W_1 = -\int \{H_1\} d\tau \qquad (A.98)$$

Similarly, at the second order $\mathcal{O}(\varepsilon^2)$,

$$\bar{H}_2 = \left\langle H_2 + \frac{1}{2}\{W_1, \bar{H}_1 + H_1\} \right\rangle \qquad (A.99)$$

$$W_2 = -\int \{2H_2 + \{W_1, \bar{H}_1 + H_1\}\} d\tau \qquad (A.100)$$

If the new Hamiltonian $\bar{H}(\bar{\mathbf{J}})$ can be shown to be independent of the time τ, we immediately obtain an invariant of motion $\bar{H}_0(\bar{\mathbf{p}}, \bar{\mathbf{q}})$, since

$$0 = \frac{d\bar{H}}{d\tau} = \{\bar{H}, \bar{H}_0\} = -\{\bar{H}_0, \bar{H}\} = -\frac{d\bar{H}_0}{dt} \qquad (A.101)$$

As seen in Appendix B, for any function f composed of $(\bar{\mathbf{p}}, \bar{\mathbf{q}})$, one can easily find an functional expression $\bar{f}$ in terms of the original $(\mathbf{p}, \mathbf{q})$ and ε by

$$f(\bar{\mathbf{p}}, \bar{\mathbf{q}}) = Tf(\mathbf{p}, \mathbf{q}) = \bar{f}(\mathbf{p}, \mathbf{q}; \varepsilon)$$

Hence, after W has once been established, one can find the functional expression $\bar{f}(\mathbf{p}, \mathbf{q}; \varepsilon)$ of any f at each order in ε thanks to Eq. (A.75):

$$\bar{f}(\mathbf{p}, \mathbf{q}; \varepsilon) = \sum_{n=0}^{\infty} \varepsilon^n \bar{f}_n(\mathbf{p}, \mathbf{q}) = \sum_{n=0}^{\infty} \varepsilon^n T_n f(\mathbf{p}, \mathbf{q}) \qquad (A.102)$$

For example, up to $\mathcal{O}(\varepsilon^2)$,

$$\varepsilon^0: \bar{f}_0(\mathbf{p}, \mathbf{q}) = f(\mathbf{p}, \mathbf{q}) \tag{A.103}$$

$$\varepsilon^1: \bar{f}_1(\mathbf{p}, \mathbf{q}) = -L_1(\mathbf{p}, \mathbf{q})f(\mathbf{p}, \mathbf{q}) = -\{W_1(\mathbf{p}, \mathbf{q}), f(\mathbf{p}, \mathbf{q})\} \tag{A.104}$$

$$\varepsilon^2: \bar{f}_2(\mathbf{p}, \mathbf{q}) = \frac{1}{2}(-L_2 + L_1^2)f \tag{A.105}$$

$$= \frac{1}{2}(\{W_1, \{W_1, f\}\} - \{W_2, f\}) \tag{A.106}$$

Note again that one might follow this by putting "$\mathbf{p}$" and "$\mathbf{q}$" into the "Hamiltonian" W as its arguments in the present case, while one might use "$\bar{\mathbf{p}}$" and "$\bar{\mathbf{q}}$" in an another case, for example, Eq. (A.67). A similar situation was seen in the original derivation for which the arguments of W in the transformations, for example, Eqs. (A.29) and (A.37), are either $(\mathbf{p}, \mathbf{q})$ or $(\bar{\mathbf{p}}, \bar{\mathbf{q}})$. The arguments themselves of the functions in the Lie transforms are really dummy variables and the Lie techniques involve operations on functions, rather than on variables.

An Illustrative Example. We apply LCPT to the following, simple 2D Hamiltonian, first demonstrated by Hori [51, 52],

$$H(\mathbf{p}, \mathbf{q}) = \frac{1}{2}(p_1^2 + p_2^2 + \omega_1^2 q_1^2 + \omega_2^2 q_2^2) - \varepsilon q_1 q_2^2 \tag{A.107}$$

$$= (\omega_1 J_1 + \omega_2 J_2) - \varepsilon \left(\frac{2J_1}{\omega_1}\right)^{1/2} \frac{2J_2}{\omega_2} \cos \Theta_1 \cos^2 \Theta_2 \tag{A.108}$$

where the Hamiltonian is integrable at $\mathcal{O}(\varepsilon^0)$.

According to the recipe of LCPT, Eq. (A.83), the new Hamiltonian $\bar{H}_0(\bar{\mathbf{p}}, \bar{\mathbf{q}})$ at $\mathcal{O}(\varepsilon^0)$ is given by

$$\bar{H}_0(\bar{\mathbf{p}}, \bar{\mathbf{q}}) = H_0(\bar{\mathbf{p}}, \bar{\mathbf{q}}) \tag{A.109}$$

$$= \frac{1}{2}(\bar{p}_1^2 + \bar{p}_2^2 + \omega_1^2 \bar{q}_1^2 + \omega_2^2 \bar{q}_2^2) = \omega_1 \bar{J}_1 + \omega_2 \bar{J}_2 \tag{A.110}$$

CHEMICAL REACTION DYNAMICS: MANY-BODY CHAOS AND REGULARITY 137

By using the general solution of the equation of motion obeying a system of $\bar{H}_0(\bar{\mathbf{p}}, \bar{\mathbf{q}})$;

$$\frac{d\bar{p}_k}{d\tau} = -\frac{\partial \bar{H}_0}{\partial \bar{q}_k} \qquad \frac{d\bar{q}_k}{d\tau} = \frac{\partial \bar{H}_0}{\partial \bar{p}_k} \qquad (A.111)$$

$$\bar{q}_k(\tau) = \sqrt{\frac{2\bar{J}_k}{\omega_k}} \cos \bar{\Theta}_k = \sqrt{\frac{2\bar{J}_k}{\omega_k}} \cos(\omega_k \tau + \beta_k) \qquad (A.112)$$

$$\bar{p}_k(\tau) = -\sqrt{2\omega_k \bar{J}_k} \sin \bar{\Theta}_k = -\sqrt{2\omega_k \bar{J}_k} \sin(\omega_k \tau + \beta_k) \qquad (k=1,2) \quad (A.113)$$

Now, we have H_1 as a function of $\bar{\mathbf{J}}$ and $\bar{\mathbf{\Theta}}$,

$$H_1(\bar{\mathbf{J}}, \bar{\mathbf{\Theta}}) = -\frac{1}{4}\left(\frac{2\bar{J}_1}{\omega_1}\right)^{1/2} \frac{2\bar{J}_2}{\omega_2} (2\cos\bar{\Theta}_1 + \cos(\bar{\Theta}_1 + 2\bar{\Theta}_2) + \cos(\bar{\Theta}_1 - 2\bar{\Theta}_2))$$

(A.114)

We first decompose this to $\langle H_1 \rangle$ and $\{H_1\}$, and try to establish the new Hamiltonian as free from the angle variables $\bar{\mathbf{\Theta}}$ as possible.

1. Non-(near) Resonant Case: $n_1\omega_1 + n_2\omega_2 \neq 0$

   In the cases that ω_1 is not commensurable with ω_2; in general, $n_1\omega_1 + n_2\omega_2 \neq 0$ (n_1, n_2: arbitrary integers), from Eq. (A.114), we have

$$\bar{H}_1 = 0$$

$$W_1 = -\int \{H_1\} d\tau = \frac{1}{4}\left(\frac{2\bar{J}_1}{\omega_1}\right)^{1/2} \frac{2\bar{J}_2}{\omega_2}$$

$$\times \left[\frac{2}{\omega_1}\sin\bar{\Theta}_1 + \frac{1}{\omega_1 + 2\omega_2}\sin(\bar{\Theta}_1 + 2\bar{\Theta}_2) + \frac{1}{\omega_1 - 2\omega_2}\sin(\bar{\Theta}_1 - 2\bar{\Theta}_2)\right]$$

(A.115)

or, in terms of $(\bar{\mathbf{p}}, \bar{\mathbf{q}})$,

$$W_1 = \frac{(2\omega_2^2 - \omega_1^2)\bar{p}_1\bar{q}_2^2 + 2\omega_1^2\bar{q}_1\bar{q}_2\bar{p}_2 + 2\bar{p}_1\bar{p}_2^2}{\omega_1^2(\omega_1 + 2\omega_2)(\omega_1 - 2\omega_2)} \qquad (A.116)$$

Hence, by using $H_1(\bar{\mathbf{p}}, \bar{\mathbf{q}}) = -\bar{q}_1\bar{q}_2^2$,

$$\{W_1, H_1\} = \frac{(2\omega_2^2 - \omega_1^2)\bar{q}_2^4 + 2\bar{q}_2^2\bar{p}_2^2 + 4\omega_1^2\bar{q}_1^2\bar{q}_2^2 + 8\bar{q}_1\bar{p}_1\bar{q}_2\bar{p}_2}{\omega_1^2(\omega_1 + 2\omega_2)(\omega_1 - 2\omega_2)} \qquad (A.117)$$

and, in the $(\bar{\mathbf{J}}, \bar{\mathbf{\Theta}})$ representation, this becomes

$$\{W_1, H_1\} = \frac{4}{\omega_1^2(\omega_1 + 2\omega_2)(\omega_1 - 2\omega_2)} \left[-\frac{1}{8}\left(\frac{\omega_1}{\omega_2}\right)^2 \right.$$

$$\times \bar{J}_2^2(\cos 4\bar{\Theta}_2 + 4\cos 2\bar{\Theta}_2 + 3) + \bar{J}_2^2(1 + \cos 2\bar{\Theta}_2)$$

$$+ \frac{\omega_1}{\omega_2} \bar{J}_1 \bar{J}_2 \left(1 + \cos 2\bar{\Theta}_1 + \cos 2\bar{\Theta}_2 \right.$$

$$+ \frac{1}{2}(\cos 2(\bar{\Theta}_1 + \bar{\Theta}_2) + \cos 2(\bar{\Theta}_1 - \bar{\Theta}_2))\bigg)$$

$$\left. - \bar{J}_1 \bar{J}_2 (\cos 2(\bar{\Theta}_1 + \bar{\Theta}_2) - \cos 2(\bar{\Theta}_1 - \bar{\Theta}_2)) \right]$$

Thus, up to $\mathcal{O}(\varepsilon^2)$, one can have the new Hamiltonian $\bar{H}$, given by

$$\bar{H} = \varepsilon^0(\omega_1 \bar{J}_1 + \omega_2 \bar{J}_2) + \varepsilon^2 \frac{2}{\omega_1^2(\omega_1 + 2\omega_2)(\omega_1 - 2\omega_2)}$$

$$\times \left[\left(1 - \frac{3}{8}\left(\frac{\omega_1}{\omega_2}\right)^2\right) \bar{J}_2^2 + \frac{\omega_1}{\omega_2} \bar{J}_1 \bar{J}_2 \right] \quad (A.118)$$

The equation of motion obeying the new Hamiltonian $\bar{H}(\bar{\mathbf{p}}, \bar{\mathbf{q}})$ represented up to the second-order $\mathcal{O}(\varepsilon^2)$,

$$\frac{d\bar{p}_k}{dt} = -\frac{\partial \bar{H}}{\partial \bar{q}_k} \quad \frac{d\bar{q}_k}{dt} = \frac{\partial \bar{H}}{\partial \bar{p}_k} \quad (A.119)$$

yields the general solution obeying $\bar{H}$;

$$\bar{q}_k(t) = \sqrt{\frac{2\bar{J}_k}{\omega_k}} \cos(\bar{\omega}_k(\bar{\mathbf{J}})t + \gamma_k) \quad (A.120)$$

$$\bar{p}_k(t) = -\sqrt{2\omega_k \bar{J}_k} \sin(\bar{\omega}_k(\bar{\mathbf{J}})t + \gamma_k) \quad (A.121)$$

where γ_k is the initial phase factor of mode k, and

$$\bar{\omega}_1(\bar{\mathbf{J}}) = \omega_1 + \varepsilon^2 \frac{2\bar{J}_2}{\omega_1 \omega_2 (\omega_1 + 2\omega_2)(\omega_1 - 2\omega_2)} \quad (A.122)$$

$$\bar{\omega}_2(\bar{\mathbf{J}}) = \omega_2 + \varepsilon^2 \frac{2}{\omega_1^2(\omega_1 + 2\omega_2)(\omega_1 - 2\omega_2)}$$

$$\times \left[\left(2 - \frac{3}{4}\left(\frac{\omega_1}{\omega_2}\right)^2 \right) \bar{J}_2 + \frac{\omega_1}{\omega_2} \bar{J}_1 \right] \quad (A.123)$$

As described in the preceding paragraphs, one can have new functional expressions in terms of the original variables $(\mathbf{p}, \mathbf{q})$, whose functional values are equivalent to $f(\bar{\mathbf{p}}, \bar{\mathbf{q}})$. Here, we shall denote the new functions as $\bar{f}(\mathbf{p}, \mathbf{q})$;

$$f(\bar{\mathbf{p}}, \bar{\mathbf{q}}) = Tf(\mathbf{p}, \mathbf{q})$$

$$= \varepsilon^0 f(\mathbf{p}, \mathbf{q}) - \varepsilon^1 \{W_1, f\} + \frac{\varepsilon^2}{2}(\{W_1, \{W_1, f\}\} - \{W_2, f\}) + \cdots$$

$$= \bar{f}(\mathbf{p}, \mathbf{q})$$

For example, up to $\mathcal{O}(\varepsilon^1)$,

$$\bar{q}_1(\mathbf{p}, \mathbf{q}) = q_1 - \varepsilon^1 \frac{(\omega_1^2 - 2\omega_2^2)q_2^2 - 2p_2^2}{\omega_1^2(\omega_1 + 2\omega_2)(\omega_1 - 2\omega_2)} + \mathcal{O}(\varepsilon^2) \quad (A.124)$$

$$\bar{p}_1(\mathbf{p}, \mathbf{q}) = p_1 - \varepsilon^1 \frac{2q_2 p_2}{(\omega_1 + 2\omega_2)(\omega_1 - 2\omega_2)} + \mathcal{O}(\varepsilon^2) \quad (A.125)$$

$$\bar{J}_1(\mathbf{p}, \mathbf{q}) = \frac{1}{2\omega_1}(p_1^2 + \omega_1^2 q_1^2) - \varepsilon^1 \frac{(\omega_1^2 - 2\omega_2^2)q_1 q_2^2 - 2q_1 p_2^2 + 2p_1 q_2 p_2}{\omega_1(\omega_1 + 2\omega_2)(\omega_1 - 2\omega_2)} + \mathcal{O}(\varepsilon^2)$$

$$(A.126)$$

and, for $\bar{\omega}_1(\mathbf{p}, \mathbf{q})$ up to $\mathcal{O}(\varepsilon^3)$,

$$\bar{\omega}_1(\mathbf{p}, \mathbf{q}) = \omega_1 + \varepsilon^2 \frac{2}{\omega_1 \omega_2(\omega_1 + 2\omega_2)(\omega_1 - 2\omega_2)}$$

$$\times \left[\frac{1}{2\omega_2}(p_2^2 + \omega_2^2 q_2^2) + \varepsilon^1 \frac{2\omega_2^2 q_1 q_2^2 - 2q_1 p_2^2 + 2p_1 q_2 p_2}{\omega_2(\omega_1 + 2\omega_2)(\omega_1 - 2\omega_2)} \right]$$

$$(A.127)$$

Note here that one can straightforwardly obtain all the inverse representations, for example, $q_1(\bar{\mathbf{p}}, \bar{\mathbf{q}})$, as

$$q_1(\bar{\mathbf{p}}, \bar{\mathbf{q}}) = \bar{q}_1 + \varepsilon^1 \frac{(\omega_1^2 - 2\omega_2^2)\bar{q}_2^2 - 2\bar{p}_2^2}{\omega_1^2(\omega_1 + 2\omega_2)(\omega_1 - 2\omega_2)} + \mathcal{O}(\varepsilon^2) \quad \text{(A.128)}$$

since

$$\bar{f}(\mathbf{p}, \mathbf{q}) = T^{-1} \bar{f}(\bar{\mathbf{p}}, \bar{\mathbf{q}})$$

$$= \varepsilon^0 \bar{f}(\bar{\mathbf{p}}, \bar{\mathbf{q}}) + \varepsilon^1 \{W_1, \bar{f}\} + \frac{\varepsilon^2}{2} (\{W_1, \{W_1, \bar{f}\}\} + \{W_2, \bar{f}\}) + \cdots$$

$$= f(\bar{\mathbf{p}}, \bar{\mathbf{q}})$$

2. (Near) Resonant Case: $n_1 \omega_1 + n_2 \omega_2 \simeq 0$
   The expression for $W_1(\bar{\mathbf{J}}, \bar{\mathbf{\Theta}})$ becomes divergent if $\omega_1 - 2\omega_2$ vanishes or becomes as small as $\mathcal{O}(\varepsilon^1)$, that is,

$$\omega_1 - 2\omega_2 \leq \mathcal{O}(\varepsilon^1) \quad \text{(A.129)}$$

One should regard the third term in the right-hand side of $H_1(\bar{\mathbf{J}}, \bar{\mathbf{\Theta}})$ [Eq. (A.114)] as free from τ and include it in $\bar{H}_1$;

$$\bar{H}_1 = -\frac{1}{4} \left(\frac{2\bar{J}_1}{\omega_1}\right)^{1/2} \frac{2\bar{J}_2}{\omega_2} \cos(\bar{\Theta}_1 - 2\bar{\Theta}_2)$$

$$W_1 = -\int (H_1 - \bar{H}_1) d\tau$$

$$= \frac{1}{4} \left(\frac{2\bar{J}_1}{\omega_1}\right)^{1/2} \frac{2\bar{J}_2}{\omega_2} \left[\frac{2}{\omega_1} \sin \bar{\Theta}_1 + \frac{1}{\omega_1 + 2\omega_2} \sin(\bar{\Theta}_1 + 2\bar{\Theta}_2)\right]$$

and so forth (see Hori [52] in detail).

Note here that in order to avoid such a small-denominator divergence, one might have to include the corresponding $\bar{\Theta}$ into the new Hamiltonian, or in the case of *near*-resonance, perform the CPT to infinite order $\mathcal{O}(\varepsilon^\infty)$.

At the end, we derive a versatile recursive series, first derived by Deprit [46], hold for an arbitrary order $\mathcal{O}(\varepsilon^n)$ of autonomous Hamiltonian systems that are expandable in power series in the perturbation strength ε. The following explanation relies heavily on a tutorial article by Cary [45, 53], which modified the original derivation by Deprit [46]:

First, we premultiply $\bar{H} = T^{-1}H$ by T and differentiate with respect to ε:

$$\frac{dT}{d\varepsilon}\bar{H} + T\frac{d\bar{H}}{d\varepsilon} = \frac{dH}{d\varepsilon} \tag{A.130}$$

By using Eq. (A.52) to eliminate $dT/d\varepsilon$ and premultiplying T^{-1}, we obtain

$$-L_W\bar{H} + \frac{d\bar{H}}{d\varepsilon} = T^{-1}\frac{dH}{d\varepsilon} \tag{A.131}$$

By inserting the series expansions and equating like powers of ε, one can have in each order of ε^n $(n > 0)$:

$$-\sum_{m=0}^{n-1} L_{n-m}\bar{H}_m + n\bar{H}_n = \sum_{m=1}^{n} mT_{n-m}^{-1}H_m \tag{A.132}$$

By writing out the first term in the first sum,

$$L_n\bar{H}_0 = L_nH_0 = \{W_n, H_0\} = \frac{dW_n}{d\tau} \tag{A.133}$$

From the last term in the last sum, and for $n > 0$, one can obtain a versatile perturbation series

$$\bar{H}_n = H_n + \frac{1}{n}\sum_{m=1}^{n-1}(L_{n-m}\bar{H}_m + mT_{n-m}^{-1}H_m) + \frac{1}{n}\frac{dW_n}{d\tau} \tag{A.134}$$

For nonautonomous systems, an additional term involving the time derivatives of $W(\mathbf{p}, \mathbf{q}; \varepsilon)$ must be included in Eq. (A.66) [45, 46, 53]. In this Appendix, we have described how Lie transforms provide us with an important breakthrough in the CPT free from any cumbersome mixed-variable generating function as one encounters in the traditional Poincaré–Von Zeipel approach. After the breakthrough in CPT by the introduction of the Lie transforms, a few modifications have been established in the late 1970s by Dewar [56] and Dragt and Finn [47]. Dewar established the general formulation of Lie canonical perturbation theories for systems in which the transformation is not expandable in a power series. Dragt and Finn developed a technique, particularly effective for high-order calculations more than, say, $\mathcal{O}(\varepsilon^5)$: It rewrites the evolution operator T as

$$T(\varepsilon) = e^{-\varepsilon L_1} e^{-\varepsilon L_2} e^{-\varepsilon L_3} \cdots \quad (A.135)$$

[which is validated for a wide class of Hamiltonians, e.g., Eqs. (2.28)–(2.30). The higher the order $\mathcal{O}(\varepsilon^n)$ at which one may want to perform the CPT, say $n \geq 5$, the smaller the number of terms are needed to represent the new Hamiltonian $\bar{H}$ in Dragt and Finn's technique [45], compared with those by Hori and Deprit. Note, however, that high dimensionality of the systems, to which one may want to apply CPT based on the Lie transforms, for example, six-atom cluster with 12 internal degrees of freedom, would make the total number of terms to be elucidated increase very quickly beyond a few orders, and make the direct applications very difficult, irrespective of kinds of (at least, existent) techniques in the Lie transforms-based CPTs.

D. A Simple Illustration of Algebraic Quantization

In general, a given Hamiltonian, Eqs. (2.28)–(2.30), can be rewritten at each order $\mathcal{O}(\varepsilon^n)$ in terms of $(\mathbf{a}^*, \mathbf{a})$: for example,

$$H_0 = \sum_i a_i^* a_i$$

$$H_1 = \sum_{j,k,l} B_{jkl}(a_j^* a_k^* a_l^* - a_j a_k a_l - 3(a_j a_k^* a_l^* - a_j^* a_k a_l))$$

$$H_2 = \sum_{j,k,l,m} B_{jklm}(a_j^* a_k^* a_l^* a_m^* + a_j a_k a_l a_m + 6a_j^* a_k^* a_l a_m - 4(a_j a_k^* a_l^* a_m^* + a_j^* a_k a_l a_m))$$

where

$$B_{jkl} = \frac{C_{jkl}}{(\sqrt{2}\, i)^3 \omega_j \omega_k \omega_l} \qquad B_{jklm} = \frac{C_{jklm}}{(\sqrt{2}\, i)^4 \omega_j \omega_k \omega_l \omega_m} \quad (A.136)$$

To see how the AQ simplifies the cumbersome analytical calculations, let us apply the AQ to a 2D system far from resonance, of Eq. (A.108),

$$H(\mathbf{a}^*, \mathbf{a}) = a_1^* a_1 + a_2^* a_2 - \varepsilon \frac{i}{2\sqrt{2}} \frac{1}{\omega_1 \omega_2^2}$$

$$\times (a_1^* a_2^{*2} - 2a_1^* a_2^* a_2 + a_1^* a_2^2 - a_1 a_2^{*2} + 2a_1 a_2^* a_2 - a_1 a_2^2) \quad (A.137)$$

Thanks to Eqs. (2.46)–(2.47), it is straightforward to establish that

$$\bar{H}_0(\bar{a}^*, \bar{a}) = H_0(\bar{a}^*, \bar{a}) = \bar{a}_1^* \bar{a}_1 + \bar{a}_2^* \bar{a}_2 \qquad (A.138)$$

$$\bar{H}_1(\bar{a}^*, \bar{a}) = 0 \qquad (A.139)$$

and

$$W_1(\bar{a}^*, \bar{a}) = -\int H_1(\bar{a}^*, \bar{a}) \, d\tau = \frac{1}{2\sqrt{2}} \frac{1}{\omega_1 \omega_2^2}$$

$$\times \left[\frac{\bar{a}_1^* \bar{a}_2^{*2} + \bar{a}_1 \bar{a}_2^2}{\omega_1 + 2\omega_2} - \frac{2}{\omega_1}(\bar{a}_1^* \bar{a}_2^* \bar{a}_2 + \bar{a}_1 \bar{a}_2 \bar{a}_2^*) + \frac{\bar{a}_1^* \bar{a}_2^2 + \bar{a}_1 \bar{a}_2^{*2}}{\omega_1 - 2\omega_2} \right]$$

(A.140)

$\{W_1, H_1\}$ can be solved *symbolically*, thanks to Eq. (2.50) and well-known general properties of Poisson bracket, Eqs. (A.14)–(A.17).

$$\{W_1, H_1\} = \{W_1(\bar{a}^*, \bar{a}), H_1(\bar{a}^*, \bar{a})\} = \frac{1}{\omega_1^2 \omega_2^2 (\omega_1 + 2\omega_2)(\omega_1 - 2\omega_2)}$$

$$\times \left[\left(4 - \frac{3}{2}\left(\frac{\omega_1}{\omega_2}\right)^2\right)(\bar{a}_2^* \bar{a}_2)^2 + 4\bar{a}_1^* \bar{a}_1 \bar{a}_2^* \bar{a}_2 \right.$$

$$- 2(\bar{a}_1^* \bar{a}_1 (\bar{a}_2^{*2} + \bar{a}_2^2) + \bar{a}_2^* \bar{a}_2 (\bar{a}_1^{*2} + \bar{a}_1^2))$$

$$+ \frac{\omega_1 - 2\omega_2}{\omega_1}((\bar{a}_1^* \bar{a}_2^*)^2 + (\bar{a}_1 \bar{a}_2)^2) + \frac{\omega_1 + 2\omega_2}{\omega_1}((\bar{a}_1^* \bar{a}_2)^2 + (\bar{a}_1 \bar{a}_2^*)^2)$$

$$\left. + \frac{\omega_1^2 - 2\omega_2^2}{\omega_2^2}(\bar{a}_2^{*3} \bar{a}_2 + \bar{a}_2^* \bar{a}_2^3) - \frac{\omega_1^2}{4\omega_2^2}(\bar{a}_2^{*4} + \bar{a}_2^4) \right] \qquad (A.141)$$

By using Eqs. (2.46) and (2.47), one can immediately establish the $\bar{\Theta}$-free terms in $\{W_1(\bar{a}^*, \bar{a}), H_1(\bar{a}^*, \bar{a})\}$, yielding

$$\bar{H}_2 = \frac{2}{\omega_1^2 \omega_2^2 (\omega_1 + 2\omega_2)(\omega_1 - 2\omega_2)} \left[\left(1 - \frac{3}{8}\left(\frac{\omega_1}{\omega_2}\right)^2\right)(\bar{a}_2^* \bar{a}_2)^2 + \bar{a}_1^* \bar{a}_1 \bar{a}_2^* \bar{a}_2 \right]$$

(A.142)

Here, one can see this being equal to $\bar{H}_2$, since $\bar{a}_k^* \bar{a}_k = \omega_k \bar{J}_k$ [see Eq. (A.118)].

All the functions ξ appearing through the LCPT–AQ procedure as Eq. (2.41),

$$\xi(\bar{\mathbf{a}}^*, \bar{\mathbf{a}}) = \sum_s d_s \bar{\mathbf{a}}^{*\mathbf{v}_s} \bar{\mathbf{a}}^{\mathbf{u}_s} \tag{A.143}$$

can be characterized by a set of parameters, $\{d_s, \mathbf{v}^s, \mathbf{u}^s\}_\xi$ for all the s in $\xi(\bar{\mathbf{a}}^*, \bar{\mathbf{a}})$, where coefficient d_s for the s-term can be real or imaginary, for systems involving imaginary frequency mode(s); $\mathbf{v}^s$ and $\mathbf{u}^s$ denote the vectors $\{v_1^s, v_2^s, \ldots\}$ and $\{u_1^s, u_2^s, \ldots\}$ (v_k^s, u_k^s: integers ≥ 0). The integrations of ξ over τ, for example, Eq. (A.140), can symbolically be carried out using the mathematical properties of exponential functions, for example,

$$\{d_s, \mathbf{v}^s, \mathbf{u}^s\}_{W_1} \leftarrow \left\{ \frac{id_s}{(\mathbf{v}^s - \mathbf{u}^s) \cdot \boldsymbol{\omega}}, \mathbf{v}^s, \mathbf{u}^s \right\}_{H_1} \tag{A.144}$$

The Poisson bracket calculations, for example, $\{\xi(\bar{\mathbf{a}}^*, \bar{\mathbf{a}}), \eta(\bar{\mathbf{a}}^*, \bar{\mathbf{a}})\}$ where $\eta(\bar{\mathbf{a}}^*, \bar{\mathbf{a}})$ is an arbitrary function of $\bar{\mathbf{a}}^*$ and $\bar{\mathbf{a}}$ as represented by Eq. (A.143), can also be established symbolically through Eq. (2.50), which replaces the cumbersome analytical derivations to searching the combinations $\{\cdots(\bar{a}_k^*)^n \cdots, \cdots(\bar{a}_k)^m \cdots\}$ and $\{\cdots(\bar{a}_k)^{n'} \cdots, \cdots(\bar{a}_k^*)^{m'} \cdots\}$ (n, m, n', m': arbitrary positive integers): for example,

$$\{\xi, \eta\} = \sum_{s,t} d_{s(\xi)} d_{t(\eta)} \{\xi_s, \eta_t\}$$

$$= \sum_{s,t}{}' d_{s(\xi)} d_{t(\eta)} \{\xi_{s'}(\bar{a}_k^*)^n \xi_{s''}, \eta_{t'}(\bar{a}_k)^m \eta_{t''}\}$$

$$+ \sum_{s,t}{}'' d_{s(\xi)} d_{t(\eta)} \{\xi_{s'}(\bar{a}_k)^{n'} \xi_{s''}, \eta_{t'}(\bar{a}_k^*)^{m'} \eta_{t''}\}$$

$$= \sum_{s,t}{}' d_{s(\xi)} d_{t(\eta)} (\cdots + inm\omega_k \xi_{s'}(\bar{a}_k^*)^{n-1} \xi_{s''} \eta_{t'}(\bar{a}_k)^{m-1} \eta_{t''} + \cdots)$$

$$+ \sum_{s,t}{}'' d_{s(\xi)} d_{t(\eta)} (\cdots - in'm'\omega_k \xi_{s'}(\bar{a}_k)^{n'-1} \xi_{s''} \eta_{t'}(\bar{a}_k^*)^{m'-1} \eta_{t''} + \cdots)$$

where $\xi_{s'}(\xi_{s''})$ and $\eta_{t'}(\eta_{t''})$ denote arbitrary multiplications over $\bar{a}_j^*$ and $\bar{a}_j$ involved in $\xi_s(\bar{\mathbf{a}}^*, \bar{\mathbf{a}})$ and $\eta_t(\bar{\mathbf{a}}^*, \bar{\mathbf{a}})$, respectively; Σ' and Σ'' mean that the summations are taken over all the terms which have the combinations $\{\cdots(\bar{a}_k^*)^n \cdots, \cdots(\bar{a}_k)^m \cdots\}$ and $\{\cdots(\bar{a}_k)^{n'} \cdots, \cdots(\bar{a}_k^*)^{m'} \cdots\}$ [all the other terms simply vanish because of Eq. (2.50)].

E. LCPT with One Imaginary Frequency Mode

In general, we may write the Hamiltonian in terms of N-dimensional actions $\mathbf{J}$ and angles $\mathbf{\Theta}$ of H_0:

$$H(\mathbf{J}, \mathbf{\Theta}) = H_0(\mathbf{J}) + \sum_n \varepsilon^n H_n(\mathbf{J}, \mathbf{\Theta})$$

$$= H_0(\mathbf{J})$$
$$+ \varepsilon \sum_{\mathbf{m}}^{\infty} H_{1\mathbf{m}}(\mathbf{J}) e^{i\mathbf{m} \cdot \mathbf{\Theta}} + \varepsilon^2 \sum_{\mathbf{m}}^{\infty} H_{2\mathbf{m}}(\mathbf{J}) e^{i\mathbf{m} \cdot \mathbf{\Theta}} + \cdots$$

for example,

$$H_{1\mathbf{m}}(\mathbf{J}) e^{i\mathbf{m} \cdot \mathbf{\Theta}} \equiv \prod_{k=1}^{N} H_{1m_k}(J_k) e^{im_k \Theta_k} \qquad (A.145)$$

where m_k and $H_{1m_k}(J_k)$ are integers and Fourier coefficients depending on the action of mode k, J_k. The frequencies of the unperturbed $H_0(\mathbf{J})$ are given by

$$\omega(\mathbf{J}) = \frac{\partial H_0(\mathbf{J})}{\partial \mathbf{J}} \qquad (A.146)$$

Here, we assume that ω_k may vary depending on the point $(\mathbf{p}, \mathbf{q})$ in the phase space as a function of the action J_k, for example, N-dimensional Morse oscillators,

$$H_0(\mathbf{p}, \mathbf{q}) = \sum_{k=1}^{N} \left[\frac{p_k^2}{2m_k} + D_k(e^{-2a_k(q_k - q_k^{eq})} - 2e^{-a_k(q_k - q_k^{eq})}) \right] \qquad (A.147)$$

$$\to H_0(\mathbf{J}) = -\sum_{k=1}^{N} D_k \left(1 - \frac{a_k J_k}{\sqrt{2m_k D_k}}\right)^2 \qquad (A.148)$$

where

$$\omega_k = \frac{\partial H_0(\mathbf{J})}{\partial J_k} = a_k \sqrt{\frac{2D_k}{m_k}} \left(1 - \frac{a_k J_k}{\sqrt{2m_k D_k}}\right) \qquad (A.149)$$

According to the recipe of LCPT, we have at zeroth order,

$$\bar{H}_0(\bar{\mathbf{J}}) = H_0(\bar{\mathbf{J}}) \tag{A.150}$$

and at first order,

$$\bar{H}_1 = H_1(\bar{\mathbf{J}}, \bar{\mathbf{\Theta}}) + \frac{dW_1(\bar{\mathbf{J}}, \bar{\mathbf{\Theta}})}{d\tau} \tag{A.151}$$

Extracting the $\bar{\mathbf{\Theta}}$-free part from H_1 by averaging it over τ yields

$$\bar{H}_1(\bar{\mathbf{J}}) = \langle H_1(\bar{\mathbf{J}}, \bar{\mathbf{\Theta}}) \rangle \tag{A.152}$$

and

$$W_1 = -\int^{\tau} d\tau' \{H_1(\bar{\mathbf{J}}, \bar{\mathbf{\Theta}}(\tau'))\} \tag{A.153}$$

Thus, one can solve W_1 by integrating the Fourier series for the oscillating part of H_1,

$$W_1(\bar{\mathbf{J}}, \bar{\mathbf{\Theta}}) = i \sum_{\mathbf{m} \neq 0} \frac{H_{1\mathbf{m}}(\bar{\mathbf{J}})}{\mathbf{m} \cdot \boldsymbol{\omega}(\bar{\mathbf{J}})} e^{i\mathbf{m} \cdot \bar{\mathbf{\Theta}}} \tag{A.154}$$

Suppose that in the vicinity of a first-rank saddle, the system is composed of one unstable mode F associated with an imaginary frequency $\omega_F(\bar{\mathbf{J}})(\in \mathfrak{F})$ and the other stable modes B with real frequencies $\omega_B(\bar{\mathbf{J}})(\in \mathfrak{R})$. Then, one can see that all the terms involving $\omega_F(\bar{\mathbf{J}})$ in Eq. (A.154) are prevented from diverging regardless of where the system dwells in the phase space, even though all the other terms excluding that mode may be ill-behaved. This is because, so long as the summation includes $\omega_F(\bar{\mathbf{J}})$, any arbitrary combination of a single imaginary, and the other real frequencies cannot be arbitrarily close to zero, that is, $|\mathbf{m} \cdot \boldsymbol{\omega}(\bar{\mathbf{J}})| \geq |\omega_F(\bar{\mathbf{J}})| > \mathcal{O}(\varepsilon^n)$. However, for all the other terms involved in the Eq. (A.154), that is, those excluding $\omega_F(\bar{\mathbf{J}})$, one may find any $\bar{\mathbf{J}}$ and $\mathbf{m}$ that assures that $\mathbf{m} \cdot \boldsymbol{\omega}(\bar{\mathbf{J}})$ is arbitrarily close to zero, threatening to cause a divergence of the parts of the summation.

How does the new Hamiltonian $\bar{H}(\bar{\mathbf{J}})$ become ruined or not, in the event of such a resonance? Suppose, for example, that $\bar{H}_2$ can be obtained by

$$\bar{H}_2 = \left\langle H_2 + \frac{1}{2} \{W_1, H_1 + \bar{H}_1\} \right\rangle \tag{A.155}$$

with

$$\langle\{W_1, H_1 + \bar{H}_1\}\rangle = \langle\{W_1, H_1\}\rangle$$
$$= \sum_{\mathbf{m}\neq 0}\sum_{\mathbf{m}'=0}\left\langle\left\{i\frac{H_{1\mathbf{m}}(\bar{\mathbf{J}})}{\mathbf{m}\cdot\omega(\bar{\mathbf{J}})}e^{i\mathbf{m}\cdot\bar{\Theta}}, H_{1\mathbf{m}'}(\bar{\mathbf{J}})e^{i\mathbf{m}'\cdot\bar{\Theta}}\right\}\right\rangle$$
$$= -\sum_{\mathbf{m}\neq 0}\sum_{\mathbf{m}'\neq 0}\left\langle\frac{e^{i(\mathbf{m}+\mathbf{m}')\cdot\bar{\Theta}}}{\mathbf{m}\cdot\omega(\bar{\mathbf{J}})}\right.$$
$$\times\sum_k\left[m_k\frac{\partial(H_{1\mathbf{m}}H_{1\mathbf{m}'})}{\partial\bar{J}_k} - (m_k + m'_k)H_{1\mathbf{m}'}\frac{\partial H_{1\mathbf{m}}}{\partial\bar{J}_k}\right.$$
$$\left.\left. + m'_k\frac{H_{1\mathbf{m}}H_{1\mathbf{m}'}}{\mathbf{m}\cdot\omega(\bar{\mathbf{J}})}\frac{\mathbf{m}\cdot\partial\omega(\bar{\mathbf{J}})}{\partial\bar{J}_k}\right]\right\rangle \quad (A.156)$$

Here, $\{W_1, \bar{H}_1\}$ has no average part free from τ because W_1 is oscillatory and $\bar{H}_1$ is averaged. Likewise, the second summation has no average part to emerge from a condition that $\mathbf{m}' = \mathbf{0}$. The average part of $\{W_1, H_1\}$ consists of all the terms such that

$$(\mathbf{m} + \mathbf{m}')\cdot\omega(\bar{\mathbf{J}}) = (m_F + m'_F)\omega_F(\bar{\mathbf{J}}) + (\mathbf{m}_B + \mathbf{m}'_B)\cdot\omega_B(\bar{\mathbf{J}})$$
$$= -(m_F + m'_F)|\omega_F(\bar{\mathbf{J}})|i + \sum_{k\in B}^N (m_k + m'_k)\omega_k(\bar{\mathbf{J}}) \leqslant \mathcal{O}(\varepsilon^2)$$

that is,

$$m_F = -m'_F \quad\text{and}\quad \sum_{k\in B}^N (m_k + m'_k)\omega_k(\bar{\mathbf{J}}) \simeq 0 \quad (A.157)$$

Then, Eq. (A.156) becomes

$$\langle\{W_1, H_1\}\rangle = \sum_{\mathbf{m}\neq 0}\sum_{\mathbf{m}'\neq 0}{}'\frac{1}{m_F\omega_F(\bar{\mathbf{J}}) + \mathbf{m}_B\cdot\omega_B(\bar{\mathbf{J}})}$$
$$\times\left[\sum_k\left(m_k\frac{\partial}{\partial\bar{J}_k}(G_{m_F,-m_F}(\bar{J}_F)G_{\mathbf{m}_B,\mathbf{m}'_B}(\bar{\mathbf{J}}_B))\right.\right.$$
$$\left. + m'_k\frac{G_{m_F,-m_F}(\bar{J}_F)G_{\mathbf{m}_B,\mathbf{m}'_B}(\bar{\mathbf{J}}_B)}{m_F\omega_F(\bar{\mathbf{J}}) + \mathbf{m}_B\cdot\omega_B(\bar{\mathbf{J}})}\frac{\mathbf{m}\cdot\partial\omega(\bar{\mathbf{J}})}{\partial\bar{J}_k}\right)$$
$$\left. - \sum_{k\in B}\left(G_{m_F,-m_F}(\bar{J}_F)H_{1\mathbf{m}'_B}(\bar{\mathbf{J}}_B)(m_k + m'_k)\frac{\partial H_{1\mathbf{m}_B}(\bar{\mathbf{J}}_B)}{\partial\bar{J}_k}\right)\right] \quad (A.158)$$

where

$$G_{m_k,m'_k} = H_{1m_k}H_{1m'_k} \quad\text{and}\quad G_{\mathbf{m},\mathbf{m}'} = \prod_k G_{m_k m'_k} \quad (A.159)$$

and the symbol Σ' denotes that the summation is taken under the condition of Eq. (A.157).

Note that all the terms in $\langle\{W_1, H_1\}\rangle$ involving the contribution of mode F, that is, $m_F = -m'_F \neq 0$, provide us with the *nondivergent* part of $\bar{H}_2(\bar{\mathbf{J}})$ (irrespective of where the system dwells in the phase space) as a function of $\bar{J}_F$ and $\bar{\mathbf{J}}_B$ because $\mathbf{m} \cdot \boldsymbol{\omega}(\bar{\mathbf{J}})$ involving one imaginary frequency $\omega_F(\bar{\mathbf{J}})$ always removes the small-denominator problem. The other terms in $\langle\{W_1, H_1\}\rangle$ excluding the F's, for which we have $m_F = m'_F = 0$ and $G_{m_F, -m_F}(\bar{J}_F) = 1$ in Eq. (A.158), that is, functions of $\bar{\mathbf{J}}_B$, may bring about divergence in such phase space regions that $\mathbf{m} \cdot \boldsymbol{\omega}_B(\bar{\mathbf{J}})$ becomes close to zero. This might require that we include the corresponding angle variables in the new Hamiltonian to avoid the divergence [48–50, 55]. Thus, one can deduce a generic feature inherent in the region of (first-rank) saddles, irrespective of the systems, that a negatively curved, reactive mode F tends *more* to preserve its invariant of action, than all the other stable, nonreactive modes B.

In turn, How does such a resonance occurring in the vicinity of first-rank saddles affect the associated local frequencies $\bar{\omega}_k(\bar{\mathbf{J}})$? One might anticipate naively that, if the system is not in the quasiregular regime where all or almost of all the actions are "good" approximate invariants, the invariants of all $\bar{\omega}_k(\bar{\mathbf{J}})$ are spoiled, including that of the reactive mode F, since $\bar{\omega}_F(\bar{\mathbf{J}})$ depends not only on the invariant $\bar{J}_F$ but also on the other $\bar{\mathbf{J}}_B$. Now, let us look into $\bar{\omega}_k(\bar{\mathbf{J}})$ at second order:

$$\bar{\omega}_k(\bar{\mathbf{J}}) = \frac{\partial \bar{H}}{\partial \bar{J}_k} = \omega_k(\bar{\mathbf{J}}) + \varepsilon \frac{\partial \langle H_1 \rangle}{\partial \bar{J}_k} + \varepsilon^2 \left(\frac{\partial \langle H_2 \rangle}{\partial \bar{J}_k} + \frac{1}{2} \frac{\partial \langle \{W_1, H_1\}\rangle}{\partial \bar{J}_k} \right) + \mathcal{O}(\varepsilon^3) \tag{A.160}$$

From Eq. (A.158),

$$\frac{\partial \langle \{W_1, H_1\}\rangle}{\partial \bar{J}_k} \sim \frac{\partial}{\partial \bar{J}_k} \left(\frac{h(\bar{J}_F, \bar{\mathbf{J}}_B)}{(m_F \omega_F(\bar{\mathbf{J}}) + \mathbf{m}_B \cdot \boldsymbol{\omega}_B(\bar{\mathbf{J}}))^n} + \frac{h'(\bar{\mathbf{J}}_B)}{(\mathbf{m}_B \cdot \boldsymbol{\omega}_B(\bar{\mathbf{J}}))^n} \right) \tag{A.161}$$

where exponent n is 1 or 2, and h and h' are, respectively, functions of $\bar{J}_F$ and $\bar{\mathbf{J}}_B$, and of $\bar{\mathbf{J}}_B$ only. If one takes the partial derivative of $\langle\{W_1, H_1\}\rangle$ with respect to $\bar{J}_F$, one can see that $\bar{\omega}_F(\bar{\mathbf{J}})$ is not affected by the second term of the right-hand side, where any $\bar{\mathbf{J}}_B$ and $\mathbf{m}_B$ can be found such that $\mathbf{m}_B \cdot \boldsymbol{\omega}_B(\bar{\mathbf{J}})$ is arbitrarily close to zero. Even if the implicit contributions from some fluctuating $\bar{\mathbf{J}}_B$ exist in the $\bar{\omega}_F(\bar{\mathbf{J}})$ at second order, they will be suppressed by the nonvanishing denominators that are always larger than $|\omega_F(\bar{\mathbf{J}})|$. On the contrary, if one takes the partial derivatives with respect to $\bar{\mathbf{J}}_B$, they are affected by both the first and second terms. Especially the second term may enhance the contributions from the fluctuating $\bar{\mathbf{J}}_B$ to the $\bar{\omega}_B(\bar{\mathbf{J}})$ through the small denominators occurring in some phase-space regions. Thus, one can deduce a generic consequence that the local frequency $\bar{\omega}_F(\bar{\mathbf{J}})$ is less influenced by the fluctuations of $\bar{\mathbf{J}}_B$ and more persistent as an approximate invariant than $\bar{\omega}_B(\bar{\mathbf{J}})$. [In fact, as shown in our recent numerical analysis using a bundle of well–saddle–well trajectories [44], even at a moderately high energy where almost all $\bar{\mathbf{J}}$ do not preserve their invariance—the exception being that $\bar{J}_F - \bar{\omega}_F^{2\text{nd}}(\bar{\mathbf{J}})$ tends to exhibit near-constancy with a much smaller fluctuation than those of $\bar{\omega}_B^{2\text{nd}}(\bar{\mathbf{J}})$].

It can easily be shown that the equation of motion of mode F obeying a Hamiltonian $\bar{H}(\bar{J}_F, \bar{\xi}_B)$ ($\bar{\xi}_B \equiv (\bar{\mathbf{J}}_B, \bar{\Theta}_B)$) is given by

$$\ddot{\bar{q}}_F(\mathbf{p}, \mathbf{q}) - \frac{\dot{\bar{\omega}}_F}{\bar{\omega}_F}\dot{\bar{q}}_F(\mathbf{p}, \mathbf{q}) + \bar{\omega}_F^2 \bar{q}_F(\mathbf{p}, \mathbf{q}) = 0 \qquad (A.162)$$

$$\bar{p}_F(\mathbf{p}, \mathbf{q}) = \frac{\omega_F}{\bar{\omega}_F}\dot{\bar{q}}_F(\mathbf{p}, \mathbf{q}) \qquad (A.163)$$

where

$$\bar{\omega}_F = \bar{\omega}_F(\bar{J}_F, \bar{\xi}_B) = \frac{\partial \bar{H}(\bar{J}_F, \bar{\xi}_B)}{\partial \bar{J}_F} \qquad (A.164)$$

Here, $\dot{x}$ and $\ddot{x}$ represent the first and second derivatives of x with respect to time t. $\bar{\omega}_F(\bar{J}_F, \bar{\xi}_B)$ depends on time only through nonreactive modes $\bar{\xi}_B$ because $\bar{J}_F$ is free from t; the $\bar{\xi}_B$ contributions to $\bar{\omega}_F$ usually arise from higher orders in ε; for example, it is second order in the vicinity of first-rank saddles, yielding [the second term of the left hand side of Eq. (A.162)] $\simeq \mathcal{O}(\varepsilon^2)\dot{\bar{q}}_F$, and, furthermore, these are suppressed due to nondivergent denominators involving one imaginary frequency ω_F.

Eq. (A.162) corresponds to a one-dimensional pendulum whose length will slowly change, being a well-used example to present the robust persistence of invariance of action under a small perturbation, referred to as "adiabatic invariance" (for example, see [53, 81]). By introducing the time dependencies of α, β, and $\bar{\omega}_F$ into the general solution of the "auxiliary equation" of Eq. (A.162) imposing $\dot{\bar{\omega}}_F = 0$, that is,

$$\bar{q}_F(t) = \alpha(t)l^{i\bar{\omega}_F(t)t} + \beta(t)e^{-i\bar{\omega}_F(t)t} \qquad (A.165)$$

and setting the supplementary condition

$$\dot{\alpha}e^{i\bar{\omega}_F t} + \dot{\beta}e^{-i\bar{\omega}_F t} + i\dot{\bar{\omega}}_F t(\alpha e^{i\bar{\omega}_F t} - \beta e^{-i\bar{\omega}_F t}) = 0 \qquad (A.166)$$

one can obtain [82] the solution of Eq. (A.162) with a slowly varying frequency by using the method of variation of constants:

$$\bar{q}_F(t) = \alpha(0)e^{i\int^t \bar{\omega}_F(t')dt'} + \beta(0)e^{-i\int^t \bar{\omega}_F(t')dt'} \qquad (A.167)$$

and

$$\bar{p}_F(t) = i\alpha(0)\omega_F e^{i\int^t \bar{\omega}_F(t')dt'} - i\beta(0)\omega_F e^{-i\int^t \bar{\omega}_F(t')dt'} \tag{A.168}$$

(One can easily verify that Eqs. (A.167) and (A.168) satisfy Eq. (A.6)). To pass through the dividing surface from the one side to the other, the crossing trajectories typically require, at most, only a half period of the reactive hyperbolic orbit $\sim \pi/|\bar{\omega}_F|$. Furthermore, in the regions of first-rank saddles, such passage time intervals are expected to be much shorter than a typical time of the systems to find the variation or modulation of the frequency $\bar{\omega}_F$ with respect to the curvature, that is, $\pi/|\bar{\omega}_F|(=\mathcal{O}(\varepsilon^0)) \ll |\bar{\omega}_F/\dot{\bar{\omega}}_F|(=\mathcal{O}(\varepsilon^{-2}))$. Thus, if $\bar{J}_F$ is conserved (i.e., the Hamiltonian is free from Θ_F), independent of the constancy or nonconstancy of $\bar{\omega}_F(\mathbf{p}, \mathbf{q})$, we can expect any nonconstancy of $\bar{\omega}_F$ to leave the separability of $\bar{q}_F$ as unaffected as $\bar{J}_F$ in the region of first-rank saddles. That is,

$$\bar{q}_F(t) \simeq \alpha(0)e^{|\bar{\omega}_F|t} + \beta(0)e^{-|\bar{\omega}_F|t} \tag{A.169}$$

and

$$\bar{p}_F(t) \simeq \alpha(0)|\omega_F|e^{|\bar{\omega}_F|t} - \beta(0)|\omega_F|e^{-|\bar{\omega}_F|t} \tag{A.170}$$

Note here that, even though the fluctuation of $\bar{\omega}_F(t)$ in a short passage time would be large enough to spoil the separability of $\bar{q}_F$ of Eqs. (A.169) and (A.170), one may still predict the final state of reactions *a priori* as far as the *sign* of $i\int^t \bar{\omega}_F(t')dt'$ will not change during the passage through the saddle. That is, for example, if the system leaving $S(q_F = 0)$ at time $t = 0$ have $\alpha(0) > 0$, the final state can be predicted at that time to be a stable state directed by $\bar{q}_F > 0$.

References

1. B. J. Berne, M. Borkovec, and J. E. Straub, *J. Phys. Chem.* **92**, 3711 (1988).
2. D. G. Truhlar, B. C. Garrett, and S. J. Klippenstein, *J. Phys. Chem.* **100**, 12771 (1996).
3. H. Eyring, *J. Chem. Phys.* **3**, 107 (1935).
4. M. G. Evans and M. Polanyi, *Trans. Faraday Soc.* **31**, 875 (1935).
5. E. Wigner, *J. Chem. Phys.* **5**, 720 (1938).
6. L. S. Kassel, *J. Phys. Chem.* **32**, 1065 (1928).
7. O. K. Rice and H. C. Ramsperger, *Am. Chem. Soc. Jpn.* **50**, 617 (1928).
8. R. A. Marcus, *J. Chem. Phys.* **20**, 359 (1952).
9. J. C. Keck, *Adv. Chem. Phys.* **13**, 85 (1967).
10. D. G. Truhlar and B. C. Garrett, *Acc. Chem. Res.* **13**, 440 (1980).
11. J. Villa and D. G. Truhlar, *Theor. Chem. Acc.* **97**, 317 (1997).
12. W. H. Miller, *Faraday Discuss. Chem. Soc.* **62**, 40 (1977).
13. T. Seideman and W. H. Miller, *J. Chem. Phys.* **95**, 1768 (1991).

14. H. A. Kramers, *Physica* **7**, 284 (1940).
15. J. T. Hynes, in *Theory of Chemical Reaction Dynamics*, edited by B. Baer (CRC Press, Boca Raton, FL, 1985), pp. 171–234.
16. C. Amitrano and R. S. Berry, *Phys. Rev. Lett.* **68**, 729 (1992); *Phys. Rev. E* **47**, 3158 (1993).
17. T. L. Beck, D. M. Leitner, and R. S. Berry, *J. Chem. Phys.* **89**, 1681 (1988).
18. R. S. Berry, *Chem. Rev.* **93**, 237 (1993).
19. R. S. Berry, *Int. J. Quantum Chem.* **58**, 657 (1996).
20. R. J. Hinde, R. S. Berry, and D. J. Wales, *J. Chem. Phys.* **96**, 1376 (1992).
21. R. J. Hinde and R. S. Berry, *J. Chem. Phys.* **99**, 2942 (1993).
22. D. J. Wales and R. S. Berry, *J. Phys. B: At. Mol. Opt. Phys.* **24**, L351 (1991).
23. S. K. Nayak, P. Jena, K. D. Ball, and R. S. Berry, *J. Chem. Phys.* **108**, 234 (1998).
24. M. J. Davis and S. K. Gray, *J. Chem. Phys.* **84**, 5389 (1986).
25. S. K. Gray and S. A. Rice, *J. Chem. Phys.* **87**, 2020 (1987).
26. M. Zhao and S. A. Rice, *J. Chem. Phys.* **96**, 6654 (1992).
27. R. E. Gillilan and G. S. Ezra, *J. Chem. Phys.* **94**, 2648 (1991).
28. N. De Leon, *J. Chem. Phys.* **96**, 285 (1992).
29. J. R. Fair, K. R. Wright, and J. S. Hutchinson, *J. Phys. Chem.* **99**, 14707 (1995).
30. E. R. Lovejoy, S. K. Kim, and C. B. Moore, *Science* **256**, 1541 (1992).
31. E. R. Lovejoy and C. B. Moore, *J. Chem. Phys.* **98**, 7846 (1993).
32. D. C. Chatfield, R. S. Friedman, D. G. Truhlar, B. C. Garrett, and D. W. Schwenke, *J. Am. Chem. Soc.* **113**, 486 (1991).
33. R. A. Marcus, *Science* **256**, 1523 (1992).
34. M. Toda, *Phys. Rev. Lett.* **74**, 2670 (1995).
35. M. Toda, *Phys. Lett. A* **227**, 232 (1997).
36. S. Wiggins, *Normally Hyperbolic Invariant Manifolds in Dynamical Systems* (Springer-Verlag, New York, 1991).
37. (a) P. Pechukas and E. Pollak, *J. Chem. Phys.* **67**, 5976 (1977); (b) E. Pollak and P. Pechukas, ibid. **69**, 1218 (1978); (c) ibid. **70**, 325 (1979); (d) E. Pollak, M. S. Child, and P. Pechukas, (e) ibid. **72**, 1669 (1980); (f) M. S. Child and E. Pollak, ibid. **73**, 4365 (1980).
38. T. Komatsuzaki and M. Nagaoka, *J. Chem. Phys.* **105**, 10838 (1996).
39. T. Komatsuzaki and M. Nagaoka, *Chem. Phys. Lett.* **265**, 91 (1997).
40. T. Komatsuzaki and R. S. Berry, *J. Chem. Phys.* **110**, 9160 (1999).
41. T. Komatsuzaki and R. S. Berry, *Phys. Chem. Chem. Phys.* **1**, 1387 (1999).
42. T. Komatsuzaki and R. S. Berry, *J. Mol. Struct.(THEOCHEM)* **506**, 55 (2000).
43. T. Komatsuzaki and R. S. Berry, *Proc. Natl. Acad. Sci. USA* **78**, 7666 (2001).
44. T. Komatsuzaki and R. S. Berry, *J. Chem. Phys.* **115**, 4105 (2001).
45. J. R. Cary, *Phys. Rep.* **79**, 130 (1981).
46. A. Deprit, *Celest. Mech.* **1**, 12 (1969).
47. A. J. Dragt and J. M. Finn, *J. Math. Phys.* **17**, 2215 (1976); **20**, 2649 (1979).
48. L. E. Fried and G. S. Ezra, *J. Chem. Phys.* **86**, 6270 (1987).
49. L. E. Fried and G. S. Ezra, *Comput. Phys. Commun.* **51**, 103 (1988).
50. L. E. Fried and G. S. Ezra, *J. Phys. Chem.* **92**, 3144 (1988).
51. G. Hori, *Pub. Astro. Soc. Jpn.* **18**, 287 (1966).

52. G. Hori, *Pub. Astro. Soc. Jpn.* **19**, 229 (1967).
53. A. J. Lichtenberg and M. A. Lieberman, *Regular and Chaotic Dynamics*, 2nd ed. (Springer, New York, 1992).
54. G. D. Birkhoff, *Dynamical Systems* (American Mathematical Society, New York, 1927).
55. F. Gustavson, *Astron. J.* **21**, 670 (1966).
56. R. L. Dewar, *J. Phys. A: Gen. Phys.* **9**, 2043 (1976).
57. M. Page and J. W. McIver, Jr., *J. Chem. Phys.* **88**, 922 (1988).
58. At a nonstationary point there exist only three zero-frequency modes corresponding to the total translational motions and the normal coordinates of the infinitesimal rotational motions appear in all the terms including H_0. The canonical perturbation theory can be applied to some non-zero total angular momentum systems by Dragt and Finn's two-step transformation [47].
59. This analysis has neglected any explicit treatment of the complex issue of systems that develop internal, vibrational angular momentum and, consequently, a counter-rotating "rigid body" angular momentum that just balances the vibrational angular momentum, leaving the total at its original, constant value. Such phenomena will require a higher level of analysis than is presented here.
60. R. Hernandez and W. H. Miller, *Chem. Phys. Lett.* **214**, 129 (1993).
61. W. H. Press, S. A. Teukolsky, W. T. Vetterling, and B. P. Flannery, *Numerical Recipes*, 2nd ed. (Cambridge University, New York, 1992).
62. T. Komatsuzaki and R. S. Berry, *J. Phys. Chem.* (submitted for publication).
63. G. van der Zwan and J. T. Hynes, *J. Chem. Phys.* **78**, 4174 (1983).
64. T. Komatsuzaki, Y. Matsunaga, M. Toda, S. A. Rice, and R. S. Berry (unpublished).
65. S. Keshavamurthy and W. H. Miller, *Chem. Phys. Lett.* **205**, 96 (1993); see His thesis at the University of California, Berkeley, CA, 1994, Chapt. 2, pp. 9–36.
66. G. V. Mil'nikov and A. J. C. Varandas, *Phys. Chem. Chem. Phys.* **1**, 1071 (1999).
67. S. Takada and H. Nakamura, *J. Chem. Phys.* **100**, 98 (1994).
68. C. C. Martens, M. J. Davis, and G. S. Ezra, *Chem. Phys. Lett.* **142**, 519 (1987).
69. M. J. Davis, *J. Chem. Phys.* **83**, 1016 (1985).
70. K. Fukui, *J. Phys. Chem.* **74**, 4161 (1970).
71. K. Fukui, S. Kato, and H. Fujimoto, *J. Am. Chem. Soc.* **97**, 1 (1975).
72. J. Laskar, *Icarus* **88**, 266 (1990).
73. J. C. Losada, J. M. Estebaranz, R. N. Benito, and F. Borondo, *J. Chem. Phys.* **108**, 63 (1998).
74. S. Wiggins, L. Wiesenfeld, C. Jaffé, and T. Uzer, *Phys. Rev. Lett.* **86**, 5478 (2001).
75. A. E. García and G. Hummer, *Proteins* **36**, 175 (1999).
76. A. E. García, R. Blumenfeld, G. Hummer, and J. A. Krumhansl, *Physica D*, **107**, 225 (1997).
77. B. Vekhter, K. D. Ball, J. Rose, and R. S. Berry, *J. Chem. Phys.* **106**, 4644 (1997).
78. Y. Matsunaga, K. S. Kostov, and T. Komatsuzaki, *J. Phys. Chem.* (submitted for publication).
79. S. S. Plotkin and P. G. Wolynes, *Phys. Rev. Lett.* **80**, 5015 (1998).
80. M. Toda, T. Komatsuzaki, R. S. Berry, and S. A. Rice (unpublished).
81. V. I. Arnold, *Mathematical Methods of Classical Mechanics*, 2nd Ed. (Springer, New York, 1978).
82. M. Toda, *Theory of Oscillations* (Japanese) (Baifukan, Tokyo, 1968).

DYNAMICS OF CHEMICAL REACTIONS AND CHAOS

MIKITO TODA

*Physics Department, Nara Women's University
Nara, 630-8506, Japan*

CONTENTS

I. Introduction
II. Transition State Theory Revisited
III. Recent Experimental Results
    A. Resonant Structures in the Rate
    B. Hierarchical Structures in IVR
    C. Breakdown of the Statistical Reaction Theory in Radicals
    D. Breakdown of Local Equilibrium in Reactions in Solution
IV. Toward Global Features of Dynamics
    A. Homoclinic Intersection
    B. Normally Hyperbolic Invariant Manifold
    C. Tangency
    D. Crisis
    E. Bifurcation in Reaction Paths
    F. Summary
V. Intramolecular Vibrational-Energy Redistribution
    A. Arnold Web
    B. Acetylene
    C. Hierarchy in Resonances
    D. Summary
VI. Toward Dynamical Correlation of Reaction Processes: Future Prospects
VII. Summary
Acknowledgments
References

I. INTRODUCTION

The interdisciplinary research field between nonlinear physics and chemistry is of much interest these days. The reasons for this are twofold: One is theoretical and the other is experimental. On the theoretical side, the development of the theory of chaos gives new approaches in understanding the dynamics of chemical reactions. In particular, the fundamental assumptions of statistical reaction theory have been examined from the viewpoint of dynamical systems. These studies not only offer the dynamical bases of the statistical reaction theory, but also suggest a new perspective that goes far beyond traditional theory. On the other hand, the progress of experimental studies has revealed new phenomena that do not fit into traditional theory, which is made possible by specifying the initial conditions with much finer precision and observation of the following dynamical processes within a shorter timescale. Therefore, these experiments lie outside of the applicability of the present reaction theory, thus indicating the necessity of a new framework based on the dynamical theory of chemical reactions.

The most fundamental assumption of the present reaction theory is the separation of timescales, that is, the characteristic timescale for the bath to recover equilibrium is much shorter than that of the reaction to take place. This assumption means that chemical reactions are among those statistical phenomena where local equilibrium of the environment is alway maintained. This makes it possible to apply the methods of equilibrium statistical physics to chemical reactions.

Based on this assumption, most of the features of reaction dynamics become irrelevant and can be replaced by statistical description, which enables us to focus our attention on local structures of phase space, that is, transition states. This assumption is the basic strategy of transition state theory. Thus in the traditional theory of chemical reactions, study of the transition states plays the key role.

However, some of the recent experiments cast doubt on the applicability of this assumption. First, experiments done in the gas phase are few-body problems where taking the thermodynamic limit is not always appropriate. In other words, we have to take into account the fact that the size of the environment is finite. Second, initial states prepared by laser are so highly excited that the timescale for the energy redistribution would be comparable to that of the reaction. Third, the timescale for observing reactions can be much shorter than that for relaxation. Therefore, dynamical behavior of reactions should be studied without assuming local equilibrium.

When the separation of the timescales does not hold, we have to study the mechanism of energy flow within the molecule after it is excited, which is the problem of intramolecular vibrational-energy redistribution (IVR).

The knowledge of IVR is expected to be substantial in those areas such as laser control of reactions, functional behavior of enzymes, and so on.

In short, recent experiments are observing phenomena that are far from equilibrium. Therefore, the bases of reaction theory has to be sought in their dynamical origin, which is the basic aim in interdisciplinary research between nonlinear science and chemical reactions.

From the standpoint of nonlinear physics, two main themes exist at present in studying chemical reactions: One is to reexamine the concept of transition states, the other is to understand the mechanism of IVR. Before studying these topics, let us briefly summarize the present status of nonlinear physics that would be relevant to understanding chemical reactions.

The dynamical origin of statistical physics has been one of the main topics in the study of chaos. The very existence of statistical averages, the concept of equilibrium and the mechanism of approaching equilibrium are among the most important. Here, the idea of ergodicity plays a crucial role. However, the current study of chaos is not limited to these topics related to equilibrium statistical physics. It also extends to entail dynamical behavior in systems far from equilibrium and the properties of those systems that do not necessarily show ergodicity.

Ergodicity has been taken for granted by many physicists even though it is very difficult to prove. However, the situation has started to change since the work of Fermi, Pasta, and Ulam became well known. In the numerical investigation of a nonlinear system that is now called the Fermi–Pasta–Ulam (FPU) model, they found that the system returns to its initial condition after some dynamical evolution. This finding is the first example where nonlinearly coupled systems do not necessarily show statistical behavior. Furthermore, the celebrated theorem of the Kolmogorov–Arnold–Moser (KAM) theorem proves the existence of tori (KAM tori) in Hamiltonian systems when perturbations from integrable systems are sufficiently small. This makes ergodicity problematic even to a physicist.

In Hamiltonian systems of two degrees of freedom, the existence of tori means their non-ergodicity, because a two-dimensional (2D) torus divides the three-dimensional (3D) equi-energy surface into two separate parts. Therefore, orbits starting from one side of the KAM torus cannot enter the other even when chaos exists. For systems of more than two degrees of freedom, however, the KAM tori do not divide the equi-energy surface, which means that KAM tori do not keep orbits from wondering around to cover the entire region of the equi-energy surface. Actually, a Russian mathematician Arnold gave an example where orbits move between KAM tori. Such a movement is now called Arnold diffusion after his name. Then, one may think that the existence of Arnold diffusion would enable the system to exhibit ergodicity at least to a physicist. However, it is not automatically so, because Arnold diffusion is supposed to be extremely slow so that its observation could take a much longer time than is possible in reality.

Arnold diffusion is known to take place on the network of nonlinear resonances. (Such a network is called the Arnold web.) Recent numerical simulations indicate that, in those regions where multiple resonances meet, chaotic diffusion can be faster than is expected from the theory of Arnold diffusion [1]. This result suggests that the global features of the Arnold web, especially how nonlinear resonances intersect in the web, would be much more important than Arnold diffusion along a single resonance, which implies that it is informative to characterize the connectivity of the web without going into the details of the system. If the web of the resonances covers the equi-energy surface uniformly and densely, the system would exhibit ergodicity within a physically relevant timescale. These ideas would lead us to formulate the dynamical origin of statistical behavior on the bases of the Arnold web.

When the web is not dense enough to guarantee that the timescale for ergodicity is physically meaningful, more detailed features of the phase space would matter. One of such features is heteroclinic (or homoclinic) intersection between stable and unstable manifolds, which would give the information on the connections among phase-space regions, thus enabling us to predict dynamical correlations between reaction processes. The study into this direction is an important step in going beyond the present theory, which is based on local information and statistical assumptions.

Having summarized the present status of nonlinear physics, we now go back to chemical reactions.

The first subject is to replace the traditional idea of transition states by a new concept that takes into account more global features of dynamics. The pioneering work by Davis and Gray [2] started an attempt to construct the dividing surface in phase space using homoclinic intersection. Their ideas work successfully in characterizing a nonstatistical nature of reaction dynamics for systems of two degrees of freedom. However, for systems of more than two degrees of freedom, Gillilan and Ezra [3] found that the phenomenon of homoclinic tangency hampers the construction of the dividing surface, because the simplectic condition of the Poincare map does not mean the conservation of the phase-space area for these systems. Thus, it turns out that the ideas of Davis and Gray critically depend on systems being two degrees of freedom.

Homoclinic (or heteroclinic) tangency would be a common phenomenon for systems of more than two degrees of freedom. Therefore, its role in reaction dynamics has to be taken seriously. Toda [4] noticed that the homoclinic tangency would lead to crisis where a transition between chaos of different natures takes place. Thus, the study of the tangency would enable us to understand the rich structures of the phase space, and lead us to go beyond the traditional statistical theory.

The second subject is to analyze the mechanism of IVR. In chemical reactions, IVR is a phenomenon where the Arnold web would be of interest. The reason is the following. Molecular systems consist of various degrees of freedom; some of them are movements of individual atoms, and some of them are collective. Therefore, their frequencies lie in a wide range of timescales. When a resonant condition is satisfied among some degrees of freedom, energy flow among them would be significant. However, contrary to linear resonances, infinite energy flow does not take place for nonlinear resonances, which is due to the change of frequencies that is caused by nonlinearity. After a certain amount of the energy flow has occurred, the frequencies of these modes would vary so that they become off-resonant. Thus, in order for a continuous energy flow to take place, the network of resonant conditions is important.

Recent experiments make it possible to actually analyze the Arnold web of specific molecules. A typical example is vibrationally highly excited acetylene. These analyses indicate that ergodicity is doubtful for this molecule, which opens the possibility of studying selective IVR, and would even lead to laser control of reactions. As a preliminary investigation toward this direction, we will report our study on the behavior of acetylene under external fields.

In the following sections, we will give more detailed explanations of the recent studies mentioned above. The outline of their contents is as follows. In Section II, we will discuss the fundamental assumptions of statistical reaction theory. In particular, we will focus our attention on the dynamical bases of the theory and its limitations.

In Section III, we will review some of the recent experiments and discuss their implications for reaction theory. In particular, we will point out that their results lie outside of the applicability of the present theory. In other words, these experiments present the necessity to question the separation of timescales.

In Section IV, we begin by discussing pioneering attempts to take into account global features of reaction dynamics. The concept of transition states is replaced by a more general one, that is, normally hyperbolic invariant manifolds. Then, intersections of their stable and unstable manifolds give a clue to characterize global features of reactions. In particular, we will show that new aspects of multidimensional Hamilton chaos play a crucial role, that is, tangency of stable and unstable manifolds and crisis in chaos. Their implications in reaction dynamics are discussed where multiexponential decay and bifurcation in reaction paths are revealed.

In Section V, we discuss the relationship between IVR and the network of nonlinear resonances (the Arnold web). By studying the web for vibrationally highly excited acetylene, we show that a nonstatistical nature of the IVR is explained by a sparse feature of the web, which comes from the selection rules imposed by the symmetry of the molecule. Furthermore, in studying the dynamics of the molecule under external fields, we show that the effects of the external field and the resonances are combined to produce a new behavior that each of the two cannot reveal. We suggest that combining the effects of external fields and resonances would enable us to manipulate reactions.

In Section VI, we discuss future prospects of the study on reactions from multidimensional Hamilton chaos. We point out the possibility that dynamical correlation among processes of crossing over barriers can be revealed by synthesizing the above two ideas, that is, intersections of stable and unstable manifolds and Arnold webs. We suggest that these studies would shed a new light on the mechanism of protein folding and molecular functions.

II. TRANSITION STATE THEORY REVISITED

In the statistical theory of chemical reactions, the concept of transition states plays a key role. In traditional theory, transition states are considered to be the saddle points in configuration space. As will be discussed later, however, their role in reactions can be more clearly understood when studied in phase space. So far, they are the only phase-space structure that is taken into account in reaction theory. In this section, we will focus attention on their role and the underlying assumptions of the statistical reaction theory.

The fundamental assumptions of the transition state theory are threefold.

First, it is supposed that chemical reactions can be described by a one-dimensional (1D) coordinate called the reaction coordinate. Then, the barrier for the reaction is characterized by an energy barrier of the potential function of the reaction coordinate. The top of the potential is the transition state (see Fig. 3.1 for a schematic picture of a transition state). It is actually a saddle point of the potential function of the whole degrees of freedom.

Second, it is assumed that, once the orbits cross over the transition state, they never come back. In other words, the reaction rate is supposedly given by the flux over the energy barrier from the reactant to the product. In fact, the rate thus estimated gives an upper bound of the actual reaction rate because some of the orbits may come back after they have gone over the transition state. Their existence results in the overestimate of the reaction rate. To solve this discrepancy is called the recrossing problem.

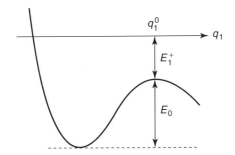

Figure 3.1. A schematic picture of a transition state.

Third, the characteristic time of the energy relaxation within the potential well is assumed to be much faster than that of the reaction. We call this assumption the separation of timescales. Then, we can treat those degrees of freedom other than the reaction coordinate as a heat bath, which means that local equilibrium within the well is always maintained during the reaction processes. It enables us to regard the whole dynamics taking place within the well as fluctuations in equilibrium.

Based on the above assumptions, the derivation of the reaction rate is well established so that we will not repeat it here [5]. When the total energy of the system is E, the reaction rate $k(E)$ is given by

$$k(E) = \frac{G^+(E - E_0)}{h\rho(E)} \qquad (3.1)$$

where $\rho(E)$ is the density of states in the reactant side, $G^+(E - E_0)$ is the sum of the number of the states from 0 to $E - E_0$ at the transition state, and h is the Planck constant.

Equation (3.1) indicates that there is no selectivity in reactions. In other words, each state in the reactant side has an equal opportunity of taking part in the reaction, that is, ergodicity within the potential well. In classical mechanics, this means that the phase space should be sufficiently chaotic so that there exist no tori or their remains that may work as dynamical barriers for the processes within the well. Then, we expect that ergodicity would be guaranteed for the reaction processes.

However, we have to notice the following difference between ergodicity here and ergodicity in statistical physics. While the ergodicity in statistical physics means that an orbit densely covers the whole phase space within the *infinite* timescale, ergodicity here means that an orbit visits virtually almost all parts of phase space within a *finite* timescale, that is, before it reaches the transition state. In other words, while the ergodicity in statistical physics is meant for a closed phase space, ergodicity here is for an open phase space. By taking into account the characteristic time for ergodicity (we will call it the ergodicity time for short), the criterion for statistical theory to be applicable is that the ergodicity time should be shorter than the reaction time.

The ergodicity time can be quite long as the size of the system becomes large, which presents the following difficulty for the traditional ideas of statistical reaction theory. In traditional theory, it is tacitly supposed that the statistical theory becomes more valid as the size of the system becomes larger. However, this is not necessarily so, since the ergodicity time would become longer than the reaction time, implying that, in the reaction theory, we cannot simply rely on conventional ideas for the foundation of statistical physics.

In quantum mechanics, we can interpret Eq. (3.1) in the following way [5]. A quantum state in the reactant side is not a stationary state but a resonant scattering state. The average interval δE of their resonant energies is given by $\delta E = 1/\rho(E)$. On the other hand, the mean reaction rate for *each* state at the transition state is $k(E)/G^+(E - E_0)$. This quantity is considered to be equal to the inverse of the average lifetime τ of the resonant scattering states, that is, $\tau = G^+(E - E_0)/k(E)$. By expressing Eq. (3.1) by these quantities, we obtain

$$\tau \delta E = h \qquad (3.2)$$

that is, the average energy interval δE is equal to the average width h/τ of the resonant states. In other words, Eq. (3.2) means that the resonant scattering states are overlapping with each other in the energy scale.

Suppose that the quantum system is ergodic in a physically suitable sense. For such systems, the timescale $h/\delta E$ would give an upper limit for their ergodicity times. Then, Eq. (3.2) means that the ergodicity time is shorter than the reaction time. Thus, interpretations of Eq. (3.1) are consistent between classical and quantum systems.

In order to make this argument more precise, we need a mathematical definition for quantum ergodicity. There exist mathematical studies concerning quantum ergodicity for a *closed* phase space (for a recent development, see [6, 7]). However, for *open* quantum systems, no study into this problem has been made so far. Therefore, we have limited our discussion for quantum systems to a physically intuitive level.

Going back to the underlying assumptions of statistical reaction theory, the third one is the most important. This means that the detail of the dynamical processes within the well is irrelevant for the reaction rate. Thus, it allows us to focus our attention on the local aspects of the potential function, that is, the transition states. Then, the recent work by Komatsuzaki and Nagaoka [8] shows that, for the saddles of index 1 in general, it is possible to choose, at least locally *in the phase space*, the reaction coordinate so that the recrossing problem does not occur [8]. In other words, the orbits that go over the transition state are ballistic along this coordinate.

When the separation of timescales breaks down, the local aspects are not sufficient to understand the reaction processes. Until recently, such possibilities have been completely neglected. However, some of the recent experiments indicate that those cases actually occur. Then, the problem we face is the following. How can we characterize those global features of the dynamics that are crucial in understanding reaction processes? This theme is in later sections. Before going to them, we will briefly review those experiments that revealed breakdown of the separation of timescales.

III. RECENT EXPERIMENTAL RESULTS

In this section, we will review the results of four experiments. The first three are done in the gas phase, and the last one is in the liquid. They imply that ergodicity of the dynamics in the reactant side is problematic. The discussion on these results would motivate the following theoretical studies into dynamical aspects of reaction processes.

A. Resonant Structures in the Rate

Experiments to observe transition states have been attempted by several groups. The one we discuss here was done by Moore and co-workers [9, 10]. In the dissociation of ketene, CH_2CO, they studied how the dissociation rate depends on the excitation energy.

The potential as a function of the bond length between C—C is shown in Figure 3.2. After it is excited to the upper electronic state, the bond breaks up leading to the formation of CO when the excess energy is large enough to go over the barrier. By measuring the amount of CO thus formed, we can estimate the rate $k(E)$ as a function of the excitation energy E. The results are shown by circles in Figure 3.3 where the calculations based on the statistical reaction theory are also indicated for comparison.

We can see in Figure 3.3 quantization of the rate, that is, the rate varies in steps as a function of the excitation energy, and is to be expected from Eq. (3.1) because the rate increases each time a quantum state on the transition

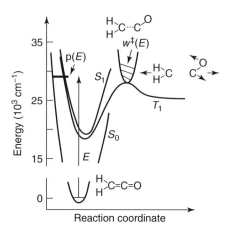

Figure 3.2. The potential as a function of the bond length between C—C in the dissociation of ketene, CH$_2$CO. [Reprinted with permission from E. R. Lovejoy, S. K. Kim, and C. B. Moore, *Science*, Vol. 256 (1992), p. 1541. Copyright © 1992, American Association for the Advancement of Science.]

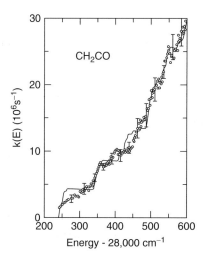

Figure 3.3. The dissociation rate $k(E)$ versus the excitation energy E for the dissociation of ketene. Comparisons are made between the experiment and the calculation. [Reprinted with permission from S. K. Kim, E. R. Lovejoy, and C. B. Moore, *J. Chem. Phys.*, Vol. 102 (1995), p. 3202. Copyright © 1995, American Institute of Physics.]

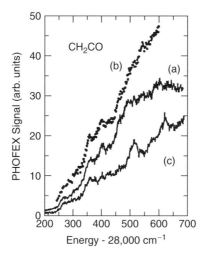

Figure 3.4. Breakdown of the amount of CO into final quantum states. [Reprinted with permission from S. K. Kim, E. R. Lovejoy, and C. B. Moore, *J. Chem. Phys.*, 102 (1995), p. 3202. Copyright © 1995, American Institute of Physics.]

state starts to contribute. Since quantum states on the transition state are discrete, the increase of the rate is stepwise.

Comparison between experiment and theory indicates overall agreement. However, a more detailed experiment reveals a new aspect that goes beyond conventional statistical theory. In Figure 3.4(*c*), breakdown of the amount of CO into final quantum states is indicated. There, we can see resonant structures in the dissociation rate, which implies the following two possibilities. One is that the reaction coordinate is not actually 1D but has couplings with other degrees of freedom. These couplings would create resonances in the transition region, which means that, among the basic assumptions of the statistical reaction theory, the first does not hold. Then, the second inevitably breaks down.

The other possibility is that there exist isolated resonant states in the reactant side, that is, some of the resonant states do not overlap in the energy scale with other resonant states. This would imply that the relaxation processes within the well are not statistical enough to guarantee the application of Eq. (3.1). Thus, the separation of timescales does not hold so that the dynamical processes within the well matters in the reaction.

In either cases, the experiments by Moore et al. [9, 10] imply necessity of going over the traditional transition state theory.

B. Hierarchical Structures in IVR

When the separation of timescales does not hold, we have to pay attention to the dynamics taking place within the well, which is the problem of IVR. For vibrationally highly excited molecules, their behavior can be chaotic in the classical sense because of nonlinear couplings. However, this does not directly result in its ergodicity. Vibrationally highly excited acetylene offers such a representative case. In particular, experiments by Yamanouchi, Field, and others reveal a hierarchical structure in its spectra, indicating a hierarchy of timescales in the IVR of acetylene (see Fig. 3.5).

Hierarchical structures in the timescales of IVR imply existence of intramolecular bottlenecks, which result in breakdown of ergodicity. In other words, the phase space consists of several regions that are connected by narrow pathways. Moreover, these regions themselves would consist of hierarchical structures of their own. As we will discuss in Section V.C, the network of nonlinear resonances provides one possibility for such hierarchical structures of phase space. Then, the existence of hierarchy in timescales of IVR would not be specific to acetylene. It can be quite generic under the condition that ergodicity of IVR does not hold.

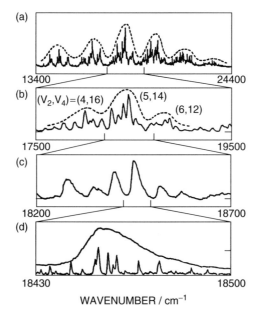

Figure 3.5. A hierarchical structure in the spectra for a vibrationally highly excited acetylene. [Reprinted with permission from K. Yamanouchi et al., *J. Chem. Phys.*, 95 (1991), p. 6330. Copyright © 1991, American Institute of Physics.]

C. Breakdown of the Statistical Reaction Theory in Radicals

In order for ergodicity of IVR to hold, energy exchange among modes in the well should be strongly chaotic in the classical sense, which requires that the network of nonlinear resonances be dense and uniform. On the other hand, gaps in the frequencies of modes make nonlinear resonances much more sparse than would be required for ergodicity. Such cases would be widely seen in van der Waals complexes, because gaps of timescales can be wide between the dynamics of chemical bonds and that involving van der Waals forces. For example, in HeI_2, studied by Davis and Gray [2], the frequency of the chemical bond of I_2 is about five times larger than that involving the wan der Waals force between He and I. Such gaps make it difficult for van der Waals complexes to satisfy resonant conditions between these two kinds of dynamical variables.

Another example for breakdown of the statistical theory is the dissociation of a radical that was recently studied by Suzuki's group [11]. They found that the rate of the dissociation is two orders of magnitude lower than that expected by statistical theory (see Figs. 3.6 and 3.7). According to their conjecture, this discrepancy results from the following mechanism. Because of the instability of radicals, some modes are strongly coupled with the reaction coordinate, thus leading to the fast dissociation of the radical. On the other hand, this means that the other modes are only loosely coupled with the reaction coordinate. If the initially excited modes are among the loosely coupled ones, the dissociation rate would be much lower than that estimated by the theory.

In general, this indicates that, when couplings with the reaction coordinate are not uniform, the rate would deviate from that estimated by statistical theory. For those modes that loosely couple with the reaction coordinate, the actual rate would be much lower than the theoretical value. In other words, weak couplings with the reaction coordinate play the role of reaction barriers, which means that the energy barriers in the traditional theory are not the only type of reaction barriers. Thus, we need to extend the concept of reaction barriers so that it also entails such barriers as are seen in the radical.

D. Breakdown of Local Equilibrium in Reactions in Solution

In the previous sections, we discussed breakdown of the statistical theory for reactions in the gas phase. For these reactions, the breakdown would seem understandable since the reasoning is based on the thermodynamic limit, which is doubtful for few-body systems. On the other hand, one might think that statistical theory holds for reactions in solution.

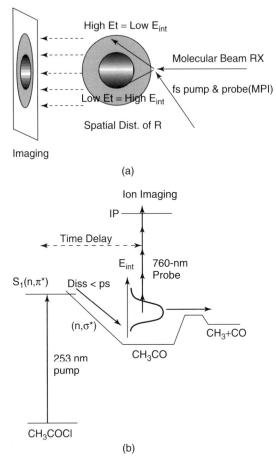

Figure 3.6. The experimental procedure for the dissociation of the radical CH_3CO. [Reprinted with permission from T. Shibata, H. Lai, H. Katayanagi, and T. Suzuki, *J. Phys. Chem.*, A102 (1998), p. 3643. Copyright © 1998, American Chemical Society.]

In this section, we introduce a recent experiment that shows this is not necessarily so.

The basic assumption for statistical theory is that local equilibrium within the well be maintained during the reaction. Resonant Raman spectroscopy offers an experimental method to see if this is true [12]. In particular, measurement of anti-Stokes shifts enables us to selectively observe the probability of vibrational states, which makes it possible to see if Boltzmann distribution is established during the reaction. In other words,

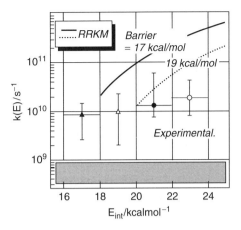

Figure 3.7. Comparison between the experiment and the theory for the dissociation rate of the radical CH_3CO. [Reprinted with permission from T. Shibata, H. Lai, H. Katayanagi, and T. Suzuki, *J. Phys. Chem.*, A102 (1998), p. 3643. Copyright © 1998. American Chemical Society.]

we can experimentally decide whether local equilibrium is maintained or not.

In the experiment by Jean et al. [12], they used *trans*-stilben as a solute and hexane as a solvent. The temperature of the solution is kept at 295 ± 2 K. After *trans*-stilben is electronically excited by a laser pulse, the intensities of the anti-Stokes Raman shifts are measured by changing the time interval after the pulse. Thus, they observed the time dependence of the vibrational energy distribution of the solute molecule. Figures 3.8 and 3.9 display their results.

In Fig. 3.8, they chose two vibrational modes and displayed the intensities of their Raman scattering. According to the experimentalists, the time dependence consists of two exponentials with different characteristic times (see the insert of Fig. 3.8). While the shorter one is ~ 2 ps, the longer one is ~ 12 ps. The first is supposed to characterize the IVR within the solute molecule, and the second is considered to describe the energy flow from the solute to the solvent.

The separation of the above two timescales may seem to indicate that the local equilibrium is established within the solute molecule. However, Figure 3.9 shows the opposite. In Figure 3.9, comparison is made between the slower change of the intensity shown in Figure 3.8 (the solid line) and the calculational result that assumes local equilibrium within the solute (the broken line).

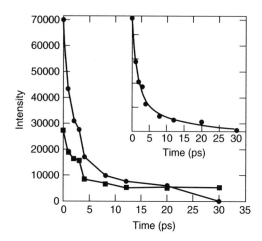

Figure 3.8. The time dependence of the intensities of Raman scattering for vibrational modes of *trans*-stilben. The figure insert displays a fitting. [Reprinted with permission from S. L. Schultz, J. Qian, and J. M. Jean, *J. Phys. Chem.*, A101 (1997), p. 1000. Copyright © 1997, American Chemical Society.]

As seen in Figure 3.9, the intensity observed in the experiment is always larger than the calculational result based on local equilibrium, which means that, in the process of the IVR, the solute molecule does not reach local equilibrium. Thus, even in reactions in solution, the assumption of local equilibrium does not always hold.

IV. TOWARD GLOBAL FEATURES OF DYNAMICS

In Section III, we briefly discussed some of the recent experiments that imply the breakdown of the separation of timescales. These results make us reconsider the underlying assumptions of the statistical reaction theory. Furthermore, they lead us to utilize the recent development in the field of dynamical systems for seeking clues for new ideas. In this section, we will discuss results of such attempts in the study of reaction processes.

A. Homoclinic Intersection

In the traditional theory of reactions, the separation of timescales makes it possible to focus our attention on local features of the potential, that is, the locations of its saddle points. When this hypothesis does not hold, more global aspects of the dynamics come into play in the theory of reactions.

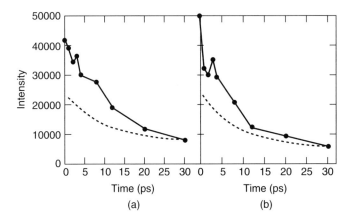

Figure 3.9. The slower change of the intensity of Raman scattering. Comparison is made between the experiment (the solid line) and the calculation that assumes local equilibrium (the broken line). [Reprinted with permission from S. L. Schultz, J. Qian, and J. M. Jean, *J. Phys. Chem.*, A101 (1997), p. 1000. Copyright © 1997, American Chemical Society.]

One such property is the heteroclinic (or homoclinic) intersection between stable and unstable manifolds.

Let us give a brief explanation on what this intersection means for chemical reactions. In Figure 3.10(a), a schematic picture is shown for a potential curve of a 1D reaction coordinate. This model potential is of the dissociation processes. Because of the interaction with other degrees of freedom, the potential curve is time dependent. Suppose that the interaction is periodic in time and take the Poincaré map of its movement. Then the flow in the phase space of the reaction coordinate looks like that shown in Figure 3.10(b).

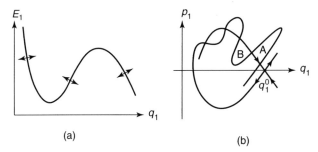

Figure 3.10. (a) A schematic picture of a time-dependent potential. (b) The stroboscopic map of the flow on the phase space under the potential.

In Figure 3.10(b), the point ($q_1 = q_1^0$, $p_1 = 0$) is a saddle point of the potential. While those orbits that asymptotically approach the saddle point constitute the stable manifold, those that asymptotically leave the saddle form the unstable manifold. These two manifolds do not in general coincide with each other. For systems of one degree of freedom under a periodic external force and those of two degrees of freedom, these two manifolds have intersections. When the two manifolds have the same saddle point in common as shown Figure 3.10(b), their intersections are called homoclinic. When they do not, their intersections are called heteroclinic.

In their pioneering work, Davis and Gray [2] tried to incorporate these intersections into the theory of reactions. Their ideas are the following. For systems of two degrees of freedom in general, a dividing curve can be constructed in the phase space of the Poincaré map using segments of the stable and unstable manifolds. We will call this curve the separatrix. For example, in Figure 3.10(b), we can take a separatrix consisting of the segments of the two manifolds each of which lies between the saddle point and one of the intersections. Before the dissociation takes place, orbits wonder around within the area enclosed by the separatrix. When dissociation takes place, the orbit crosses the separatrix. In Figure 3.10(b), the region denoted by A indicates a set of those orbits that are about to dissociate. On the other hand, the region denoted by B shows a set of those that are to be trapped. Because of the simplectic condition of the Poincaré map, these two regions are of the same area for systems of two degrees of freedom. Thus, the separatrix can unambiguously divide the phase space into the two separate parts, which means that there exists no recrossing problem in this method.

Based on the separatrix, the reaction rate can be defined as follows. Each time the Poincaré map is applied, new regions similar to the one denoted by A are created. Then, those orbits included in these regions will dissociate. Therefore, the reaction rate is proportional to the area of the region A multiplied by the number of regions created by a single application of the map. Thus, the concept of reaction rate is founded on the basis of dynamical behavior.

In order to estimate the rate itself, the analysis of the dynamics within the potential well is necessary, which is the problem of IVR, and is postponed to Section V.A.

B. Normally Hyperbolic Invariant Manifold

In Section IV.C, we will discuss the extension and limitations of the ideas of Davis and Gray for systems of n degrees of freedom in general. These arguments are based on dimensional counting of various manifolds in phase space. In order to understand them, more precise formulation of the ideas would be useful.

The location of the saddle point in phase space is specified by $q_1 = q_1^0$ and $p_1 = 0$, where q_1 is the reaction coordinate. On top of the saddle point, the reaction coordinate is completely separated from the rest of the degrees of freedom. Therefore, a set of orbits where (q_1, p_1) is fixed on the saddle point while the rest are arbitrary is invariant under dynamical evolution. Its dimension in phase space is $2n - 2$. Such invariant manifolds are considered as the phase-space structure corresponding to transition states, and will play a crucial role in the following discussion.

These manifolds are examples of normally hyperbolic invariant manifolds. Wiggins suggested that the concept of normally hyperbolic invariant manifolds would be useful in the study of chemical reactions [30]. A normally hyperbolic invariant manifold is a manifold where instability (either in a forward or a backward direction of time) along its normal directions is much stronger than that along its tangential directions. For example, consider the evolution equation of small deviations (Jacobi equation) from the saddle point and its eigenvalues. While pure imaginary eigenvalues correspond to stable degrees of freedom, those with none-zero real parts represent unstable directions (either in a forward or a backward direction of time). Divide the unstable directions into following two sets; the first one consists of those directions with smaller absolute values of the real parts than those belonging to the second. Then a normally hyperbolic invariant manifold can be constructed by choosing stable degrees of freedom and unstable ones belonging to the first set.

In chemical terms, normally hyperbolic invariant manifolds play the role of an extension of the concept of transition states. The reason why it is an extension is as follows. As already explained, transition states in the traditional sense are regarded as normally hyperbolic invariant manifolds in phase space. In addition to them, those saddle points with more than two unstable directions can be considered as normally hyperbolic invariant manifolds. Such saddle points are shown to play an important role in the dynamical phase transition of clusters [14]. Furthermore, as is already mentioned, a normally hyperbolic invariant manifold with unstable degrees of freedom along its tangential directions can be constructed as far as instability of its normal directions is stronger than its tangential ones. For either of the above cases, the reaction paths in the phase space correspond to the normal directions of these manifolds and constitute their stable or unstable manifolds.

The normally hyperbolic invariant manifolds are structurally stable under perturbations. The wider the gap of instability is between the normal and tangential directions, the more stable it is. The existence of this gap can be interpreted as an adiabatic condition between the reaction paths and the rest of the degrees of freedom.

For systems of n degrees of freedom, take a normally hyperbolic invariant manifold with $2r$ normal directions in phase space. Note that for Hamiltonian systems, the dimension of the normal directions in phase space is alway even, because the eigenvalues of the variational equation (Jacobi equation) is symmetric around the value 0. Thus, the dimension of the normally hyperbolic invariant manifold is $2n - 2r$ and, for its stable and unstable manifolds, their dimensions are $2n - r$, respectively. The dimension of their homoclinic intersection, if it exists, is $2n - 2r$ in the $2n$-dimensional phase space. When we consider the intersection manifold on the equi-energy surface, its dimension on the surface is $2n - 2r - 1$. Thus, the dimension d of the intersection on the Poincaré section is $d = 2n - 2r - 2$.

Let us consider saddle points with $r = 1$, that is, the cases that the traditional reaction theory treats. For $n = 2$ and $r = 1$, $d = 0$, that is, the intersection on the Poincare section consists of points. On the other hand, for $n = 3$ and $r = 1$, $d = 2$. Therefore, a continuous family of orbits connect the transition state.

This reasoning is not limited to homoclinic intersections. This argument can be easily extended to include heteroclinic intersections where stable and unstable manifolds may have different dimensions.

By using the above terminology and the argument of dimensional counting, we will show how things can be different for systems of more than two degrees of freedom.

C. Tangency

An attempt to apply the ideas of Davis and Gray to a system of three degrees of freedom is done by Gillilan and Ezra [3]. Their model is the predissociation of HeI_2 including the freedom of the internal rotation. In the following, r stands for the distance between the He atom and the center of mass of the I_2 molecule, R is the distance between the two I atoms, and γ is the angle between the directions of r and R. Their canonically conjugate momenta are represented as p_r, p_R, and p_γ, respectively.

In this system, the normally hyperbolic invariant manifold is the dissociation limit defined by $r = \infty$ and $p_r = 0$, thus forming a 3D manifold on the equi-energy surface. After taking the Poincaré section defined by $p_R = 0$, its stable and unstable manifolds are 3D on the equi-energy surface, respectively. Gillilan and Ezra [3] studied how the stable and unstable manifolds intersect in the four-dimensional (4D) Poincaré section. Since the Poincaré section cannot be directly displayed, they took additional sections of these manifolds by scanning the values of γ and p_γ.

In Figure 3.11, a typical case of homoclinic tangency is indicated by showing those sections of the stable and unstable manifolds. Here, we can

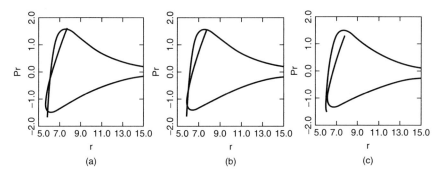

Figure 3.11. An example of homoclinic tangency in a model of HeI_2 that includes the freedom of internal rotation. [Reprinted with permission from R. E. Gillilan and G. S. Ezra, *J. Chem. Phys.*, Vol. 94 (1991), p. 2648. Copyright © 1991, American Institute of Physics.]

see that, after homoclinic tangency takes place, the stable and unstable manifolds do not divide the phase space any more, thus invalidating the ideas of Davis and Gray.

The difference between systems of two and three degrees of freedom can be easily understood based on the simplectic property of Hamilton's equations of motion and Poincaré sections. The simplectic property on $2n$-dimensional phase space says that the simplectic form

$$dw = \sum_{i=1}^{n} dp_i \wedge dq_i \qquad (3.3)$$

is conserved under time evolution. Here, $dp_i \wedge dq_i$ means the area on the phase space (q_i, p_i). Thus, this condition means that the sum of the areas is conserved under time evolution. For a one degree of freedom Hamiltonian and Poincaré sections for two degrees of freedom, this means that the area on the phase space itself is conserved. This fact makes such a case as shown in Figure 3.12(a) impossible because here the area inside the stable and unstable manifolds increases. To the contrary, for Poincaré sections of more than two degrees of freedom, the area on each of the phase space (q_i, p_i) is not necessarily kept constant. This reason shows why homoclinic tangency is possible in systems of more than two degrees of freedom.

The above reasoning is not limited to homoclinic tangency, but also can be applied to heteroclinic tangency such as that shown in Figure 3.12(b).

Considering the above argument, the tangency is not specific to the predissociation of HeI_2, but would be common in systems of more than two

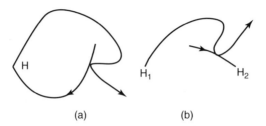

Figure 3.12. (a) An example of homoclinic tangency. (b) An example of heteroclinic tangency. Here, H, H_1, and H_2 indicate saddle points.

degrees of freedom. Therefore, the ideas of Davis and Gray needs serious reconsideration. Before doing so, we will discuss some of the outcomes of the tangency in reaction processes.

D. Crisis

In this section, we will show that the tangency triggers a transition between chaotic dynamics of different features. This idea stems from the study of crisis by Lai et al. [15]. Crisis is a phenomenon where tangency between stable and unstable manifolds leads to a transition in dynamics such as a merge of attracting basins, a sudden disappearance of chaotic behavior, and so on. Since crisis is a global bifurcation, local analyses of the dynamics cannot reveal its existence. In order to detect its occurrence and to see its outcomes, we need to perform global analyses by following stable and unstable manifolds.

Crisis can be seen both in dissipative and Hamiltonian systems. However, most of the studies so far are limited to lower dimensional systems. In these systems, crisis takes place when we scan one of the parameters of the system. On the other hand, for systems of multiple degrees of freedom, some degrees of freedom would play the role of these parameters in the sense that movement of these degrees of freedom triggers crisis. Then, depending on the movement of these variables, the system would dynamically go back and forth crossing the crisis. In this sense, crisis in systems of multiple degrees of freedom are dynamical compared to lower dimensional cases where they are static.

Recent studies on dissipative systems have shown that crisis in multiple degrees of freedom lead to a new type of behavior called chaotic itinerary [16]. On the other hand, for Hamiltonian systems of multiple degrees of freedom, crisis is not yet thoroughly studied, which is one of the reasons why the study of crisis for Hamilton chaos is urgent.

In order to capture the transition in the nature of chaos, we need some characteristic quantities that detect these changes. In the following, we will first discuss whether those quantities such as the Lyapunov exponents are effective for this purpose. Furthermore, we will propose to use new characteristics to reveal the crisis.

One of the most frequently used characteristics of chaos is Lyapunov exponents. They are defined as the time averages of the separation rates between neighboring orbits. These quantities reveal hyperbolic nature of chaos averaged in time and phase space, and give useful information for systems with homogeneous phase-space structure.

However, they are not necessarily effective in capturing dynamical behavior of chemical reactions because of the following three reasons. First, the averaged quantities over the whole phase space would not be of much value in reaction dynamics, because the potential function in reaction dynamics generally consists of multiple wells, and we are more interested in dynamical behavior within each of the wells rather than quantities averaged over the whole space. Second, the separation rates do not give any information on which degrees of freedom contributes the reaction dynamics in what ways. In studying the mechanism of reactions, the questions we ask are what coordinate is appropriate in describing the reaction, which degree of freedom couples strongly with this coordinate, and so on. However, Lyapunov exponents do not give any information on the roles each degree of freedom plays in the reaction. Third, local properties of the orbit like the Lyapunov exponents do not answer those questions related to global aspects of phase space, such as how large chaotic regions are in phase space, whether these chaotic regions are connected or not, how wide the paths are that connect chaotic regions, and so on.

In particular, a method to capture dynamical changes that take place during *a finite* time evolution is necessary, because crisis in multidimensional systems is a dynamical transition, that is, a transition taking place while an orbit wonders around phase space. Therefore, quantities averaged over the infinite time interval cannot be informative for this purpose.

In order to characterize chaos for *a finite* timescale, a concept of local Lyapunov exponents is proposed. Rather than taking the average over the infinite time interval, local Lyapunov exponents are defined as finite time averages of the separation rates. As far as appropriate intervals are selected for time average, these quantities would be useful in estimating instability of the dynamics within a certain area of the phase space. However, the criterion for the choice of time intervals is arbitrary. Moreover, since separation rates of the orbit only capture local features of the dynamics, they are not effective to make manifest those changes taking place in global aspects (e.g., how phase-space regions are connected with each other).

In order to make more direct correspondence between tangency and global changes in the dynamical behavior, we propose to use different methods to characterize chaos. The first one focuses attention on how normally hyperbolic invariant manifolds are connected with each other by their stable and unstable manifolds. Then, crisis would lead to a transition in their connections. The second one is to characterize chaos based on how unstable manifolds are folded as they approach normally hyperbolic invariant manifolds. Then, crisis would manifest itself as a change in their folding patterns. Let us explain these ideas in more detail.

The first method comes from the idea that the connections among normally hyperbolic invariant manifolds would form a network, which means that one manifold would be connected with multiple manifolds through homoclinic or heteroclinic intersections. Then, a tangency would signify a location in the phase space where their connections change. This idea offers a clue to understand, based on dynamics, those reactions where one transition state is connected with multiple transition states. In these reaction processes, the branching points of the reaction paths and the reaction rates to each of them are important. We expect that analysis of the network is the first step toward this direction.

In the second method, the number of the degrees of freedom that couple with the reaction coordinate is revealed. In chaotic dynamics, the unstable manifolds are folded each time they approach normally hyperbolic invariant manifolds. These folding processes result from the interactions between the reaction coordinate and the rest of the system. When the number of the interacting modes are smaller, the folding pattern of the unstable manifolds would look simpler. On the other hand, when their number is larger, the pattern would become more complicated. Thus, folding patterns of the unstable manifolds reveal information on the number of interacting modes that contribute to the reaction. In the following, the classification of caustics (Lagrangian singularity) will be used to estimate the number of interacting modes. These caustics will show up when the folded manifolds are projected onto the configuration space.

In [4], the above ideas are tested in the predissociation of HeI_2 with three degrees of freedom. For systems of more than two degrees of freedom, however, we cannot directly display the intersections between stable and unstable manifolds. Therefore, the following plot is used to display the homoclinic intersection and the folding patterns in 4D Poincaré sections.

As an illustrative example to explain the plot, see Figure 3.13(b) for 2D Poincaré sections. In the upper part of this figure, the folding pattern of the unstable manifold is displayed on the 2D phase space. There, the dots are intersection points, and the crosses indicate those points corresponding to caustics. In the lower part of the figure, the locations of these points are plotted on the 1D unstable manifold. From their locations on the unstable manifold, we can obtain information on the intersection between stable and unstable manifolds, and on the folding pattern of the unstable manifold. A similar plot is used to see 4D Poincare sections.

For the predissociation of HeI_2 with three degrees of freedom, the 4D Poincare section $p_R = 0$ is displayed in Figure 3.13(a) for a 2D slice of the unstable manifold. The slice is defined by the condition $p_\gamma \to 0$ as $r \to \infty$. In other words, the internal angular momentum asymptotically goes to zero as the distance between the He atom and I_2 becomes infinite. Here, the bold lines indicate the homoclinic intersection on the 2D slice of the unstable manifold, and the thin lines show the caustics there. Their locations on the unstable manifold are specified by (r, γ).

We can see in Figure 3.13(a) that, for the ranges $\gamma \sim 0$ and $\gamma \sim \pi/2$, the lines of the intersection points and those of the caustics appear alternately along the direction of r. This fact indicates that the unstable manifold is twofold along this direction in the same way as in 2D Poincare sections [cf. this with Fig. 13(b)]. This means that the number of degrees of

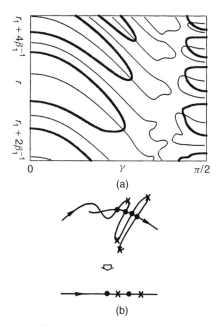

Figure 3.13. (a) The homoclinic intersection and the folding pattern for the predissociation of HeI_2, which includes the freedom of the internal rotation. (b) Intersection and folding are plotted on the unstable manifold for 2D Poincaré map. [Reprinted with permission from M. Toda, *Phys. Rev. Lett.*, Vol. 74 (1995), p. 2670. Copyright © 1995, American Institute of Physics.]

freedom involved in chaos in these regions is 2, that is, that the reaction coordinate r is interacting with another degree of freedom. We will see later that the degree of freedom interacting with the reaction coordinate is R, the distance between the two I atoms.

On the other hand, at around $\gamma \sim 0.3\pi$ and $\gamma \sim 0.4\pi$, the lines of the intersection points merge, respectively. There is no intersection between $\gamma \sim 0.3\pi$ and $\gamma \sim 0.4\pi$, which shows that the homoclinic tangency takes place as the variable γ crosses these values.

In order to see further the evolution of the unstable manifold, Figure 3.14 shows, on configuration space (r, γ), how parts of the unstable manifold evolve under the Poincaré map. The parts of the unstable manifold are specified by values of γ in the dissociation limit. In Figure 3.14, four representative cases are plotted. There, the thin lines indicate the equi-energy lines of the potential function for $R = R_{max}$, where R_{max} is the maximum value of R at this energy. This figure indicates how the topography of the potential gives rise to the crisis.

For $\gamma \sim 0$ and $\gamma \sim \pi/2$, the unstable manifold directly approach the dissociation limit under successive applications of the Poincaré map. On the

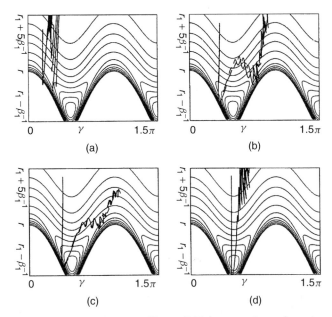

Figure 3.14. The evolution of the unstable manifold shown on the configuration space (r, γ). Thin lines indicate the equi-energy lines of the potential with $R = R_{max}$. (a) $\gamma \sim 0$. (b) $\gamma \sim 0.3\pi$. (c) $\gamma \sim 0.4\pi$. (d) $\gamma \sim 0.5\pi$. [Reprinted with permission from M. Toda, *Phys. Rev. Lett.*, Vol. 74 (1995), p. 2670. Copyright © 1995, American Institute of Physics.]

other hand, for $\gamma \sim 0.3\pi$ and $\gamma \sim 0.4\pi$, the unstable manifold does not. We can see in Figure 3.14 that, for these values of γ, the internal angular momentum is excited by collisions against the tilted parts of the potential wall. These collisions lead the unstable manifold in this region toward the saddle point at $\gamma = \pi$, that is, the transition state for the hindered rotation.

The above analysis shows the following three facts. First, Figure 3.14 shows how the connections of the unstable manifold change across the homoclinic tangency, that is, the change from the direct connection of the dissociation limit to the indirect one that goes via the transition state of the hindered rotation. In other words, it is the transition from the homoclinic intersection of the dissociation to the heteroclinic intersection between the dissociation limit and the transition state of the rotation. Second, it implies a transition in the number of degrees of freedom involved in chaos. Before the crisis, the number is 2, that is, the energy exchange takes place mainly between r and R. After the crisis, the number is 3, that is, the energy exchange occurs among the whole degrees of freedom r, R, and γ. Third, the topography of the potential function gives a clue to understanding the crisis.

The transition in the number of degrees of freedom involved in chaos can be more directly seen in the folding pattern. In Figure 3.15, the folding

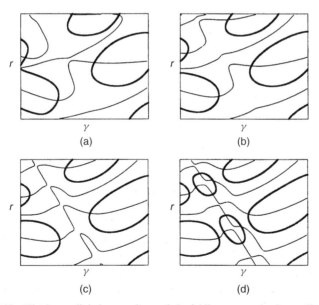

Figure 3.15. The homoclinic intersection and the folding pattern in the predissociation of HeI$_2$ in a latter stage of time evolution for $0.3\pi < \gamma < 0.4\pi$. (a) T = 16, (b) T = 17, (c) T = 18, (d) T = 19, where the unit of time is given by the Poincaré map. [Reprinted with permission from M. Toda, *Phys. Rev. Lett.*, Vol. 74 (1995), p. 2670. Copyright © 1995, American Institute of Physics.]

pattern of the unstable manifold for $0.3\pi < \gamma < 0.4\pi$ is plotted for a later stage of its time evolution. Here, the unstable manifold is fourfold, that is, twofold along two different directions, which indicates that two modes are interacting with the reaction coordinate and contrasts with the folding pattern shown in Figure 3.13. There, the folding takes place only along one direction, which clearly shows that, through the crisis, the numbers of the interacting modes change.

Thus, the tangency in the predissociation signals the transition between chaos of the different features. This transition is revealed by the changes in the connections and the folding patterns. As will be shown in Section IV.E, the first change results in bifurcation in the reaction paths. The second detects the transition in the number of modes interacting with the reaction coordinate.

E. Bifurcation in Reaction Paths

In Section IV.D, the crisis is studied in classical mechanics. In this section, we study how it manifests itself in quantum mechanics. There are two phenomena where the outcomes of the crisis can be observed. The first is bifurcation in reaction paths, which directly corresponds to the changes of the connections. The second is multiexponential decay in the predissociation processes, which reflects the existence of multiple dissociation paths with different decay times. We will further discuss the possibility of observing phenomena related to the crisis.

In Figure 3.16, the time evolution of a wave packet is shown for $\gamma \sim 0.3\pi$ [17]. Here, the wave packet splits into two parts; one directly approaches the dissociation limit and the other starts to take the path to internal rotation. These two paths correspond to the homoclinic intersection of the dissociation limit and the heteroclinic intersection, respectively. The superposition of these paths results in the split of wave packets in quantum mechanics.

When there exist multiple paths with different decay times, the dissociation processes would exhibit multiexponential decay. In Figure 3.17, the time dependences of the remaining probabilities of wave packets are shown for different values of γ [17]. This finding indicates that the dissociation consists of two processes with different decay times. The faster one corresponds to the homoclinic intersection, and the slower one to the heteroclinic intersection.

There can be other phenomena that reflect the crisis in classical mechanics. In Section IV.D, we showed that the folding patterns show the number of degrees of freedom involved in chaos. These patterns correspond, in quantum mechanics, to caustics. Therefore, the crisis would show up in the transition of the types of caustics. Thus, the difference between twofold and

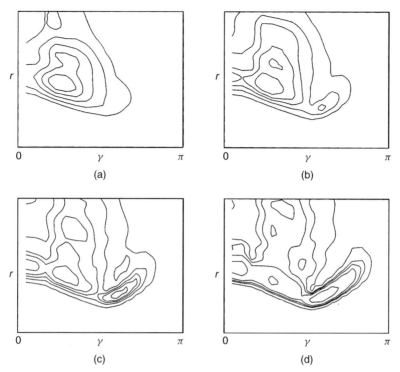

Figure 3.16. Time evolution of a wave packets for the predissociation of HeI_2 with the internal rotation. (a) T = 4, (b) T = 5, (c), T = 6, (d) T = 7, where the unit of time is the period of the vibrational motion. [Reprinted with permission from M. Toda, *Phys. Lett.*, Vol. A227 (1997), p. 232. Copyright © 1997, Elsevier Science.]

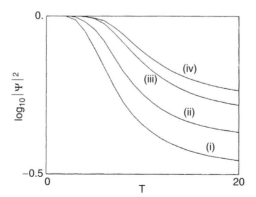

Figure 3.17. Multiexponential decay for the predissociation of HeI_2 with the internal rotation. (i) $\gamma = 0, 1\pi$, (ii) $\gamma = 0.2\pi$, (iii) $\gamma = 0.3\pi$, (iv) $\gamma = 0.4\pi$. [Reprinted with permission from M. Toda, *Phys. Lett.*, Vol. A227 (1997), p. 232. Copyright © 1997, Elsevier Science.]

fourfold manifolds would result in the transition in quantum interference. These phenomena will be studied elsewhere.

F. Summary

In summarizing the contents of this section, we focused our attention on global aspects of reaction dynamics. First, we replace the concept of transition states by a more general one, that is, normally hyperbolic invariant manifolds. Then, global aspects of the reaction processes are represented by the intersections between their stable and unstable manifolds. Furthermore, for systems of more than two degrees of freedom, new features emerge in their global aspects, that is, tangency and crisis in chaos. These aspects become manifest in the bifurcation of reaction paths, in multiexponential decay, and possibly in quantum interference.

In studying reaction processes from the viewpoint of Hamilton chaos of more than two degrees of freedom, our interest has shifted from the original purpose of defining the reaction rate to the network of reaction paths, which is revealed by the intersections of stable and unstable manifolds. For those cases where tangency does not take place, the original idea of Davis and Gray is applicable. However, for the situations where tangency and resulting crisis occurs, we should abandon the direct extension of the original idea. As an alternative, the main question is how reaction paths are organized in such a way as to form a network through the homoclinic and heteroclinic intersections. In doing so, the question related to the foundation of the statistical reaction theory, and the mechanism of selectivity in reactions are postponed until the argument of IVR has been done.

After discussing the dynamics taking place within the potential wells, we will come back in Section VI to the question how to understand the statistical–selective nature of reaction processes from the viewpoint of Hamilton chaos.

V. INTRAMOLECULAR VIBRATIONAL-ENERGY REDISTRIBUTION

A. Arnold Web

In the traditional theory of reactions, IVR in a potential well is supposed to be statistical. This idea is based on the separation of timescales between the relaxation and the reaction. Here, the relaxation within the well is assumed to take place much faster than the reaction. Then, thermal equilibrium within the well is supposed to be maintained during the whole processes of the reaction. In this framework, the reaction dynamics is often modeled as the dynamics of the reaction coordinate interacting with the heat bath composed of harmonic oscillators.

However, this framework is not applicable to the reactions taking place in nonequilibrium conditions. In particular, the heat bath consisting of harmonic oscillators does not exhibit any features characteristic in far from equilibrium. It is because nonlinearity plays a crucial role in the processes of energy exchange there, which is where the theory of dynamical systems comes into play. In Hamilton chaos, the mechanism of energy exchange lies in nonlinear resonances among dynamical variables. In the following, the dynamical processes of IVR will be analyzed based on this concept.

Davis and Gray were the first to study IVR from the point of view of nonlinear resonances [2]. They noticed that cantori created by nonlinear resonances constitute bottlenecks for the processes of IVR. As is seen in Figure 3.18, cantori trap orbits for a finite timescale. Contrary to KAM tori, cantori do not completely keep orbits from passing through them. Nevertheless, their existence results in breakdown of the separation of timescales, thus invalidating the statistical reaction theory.

For systems of more than two degrees of freedom, however, their role is not so obvious because their dimension is not large enough to work as barriers. This problem is closely related to the phenomenon of Arnold diffusion. In other words, dynamical processes along nonlinear resonances may create a way to go around these tori. In particular, intersections of resonances would play a dominant role in IVR since chaotic diffusion

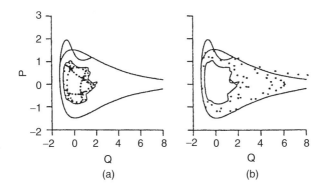

Figure 3.18. Bottleneck in the predissociation of HeI_2 (*a*) orbits trapped within the cantori created by the nonlinear resonance and (*b*) orbits trapped between the cantori and the separatrix. [Reprinted with permission from M. J. Davis and S. K. Gray, *J. Chem. Phys.*, Vol. 84 (1986), p. 5389. Copyright © 1986, American Institute of Physics.]

around these regions seems faster than that along a single isolated resonance [1]. Therefore, the network of nonlinear resonances as a whole should be of interest rather than the Arnold diffusion along a specific single resonance. The study of the Arnold web for planar OCS by Davis and co-workers [18], is an important example toward this direction. However, only a few studies have been done since then.

B. Acetylene

Acetylene is among those molecules where detailed analysis of the IVR is going on both experimentally and theoretically [19]. One of its interesting features is a hierarchical structure in spectra for vibrationally highly excited states [20]. Such a hierarchical structure in spectra implies a hierarchy in bottlenecks for the IVR, where each bottleneck would hinder the relaxation within the potential well for a finite duration. This idea can be contrasted with the one behind traditional theory, where relaxation is supposed to take place exponentially fast with a single time constant.

Nonstatistical features of IVR in acetylene is clearly revealed by recent experiments from the Field group [21], who showed that, for certain highly excited states, the IVR is unexpectedly regular. In other words, it indicates that the equi-energy surface of acetylene at this energy is non-ergodic. This result contradicts the naive expectation that IVR for highly excited molecules would be strongly chaotic.

In order to understand these features, we will analyze the Arnold web of acetylene using the effective Hamiltonian that is constructed by the Dunham expansion [22]. In our study, our main focus is to obtain a universal understanding for the mechanism of IVR based on the general properties of dynamical systems, rather than to study detailed aspects of specific molecules.

In the electronic ground state, acetylene is a linear molecule. By using the harmonic approximation, vibrational energy levels for small amplitude oscillations can be labeled as $(v_1, v_2, v_3, v_4^{l_4}, v_5^{l_5})^l$, where v_1 is the quantum number of the symmetric CH stretch, v_2 the CC stretch, v_3 the antisymmetric CH stretch, v_4 the trans-bend with the vibrational angular momentum l_4, v_5 the cis-bend with the vibrational angular momentum l_5, and $l = l_4 + l_5$.

As the molecule is vibrationally excited, couplings among these harmonic modes become important. These couplings are taken into account by the Dunham expansion. The Dunham expansion for the vibration–rotation energy of a linear polyatomic molecule above the zero-point level is given by

$$H_0 = \sum_i \omega_i v_i + \sum_{i \leq j} x_{ij} v_i v_j + \sum_{t \leq s} g_{ts} l_t l_s - B_v l^2 \qquad (3.4)$$

DYNAMICS OF CHEMICAL REACTIONS AND CHAOS 185

and

$$B_v = B_0 - \sum_i \alpha_i v_i + \sum_{i \leq j} \gamma_{ij} v_i v_j + \sum_{t \leq s} \gamma_{ts} l_t l_s \quad (3.5)$$

In this expression, we put the total angular momentum $J = 0$ for simplicity. The values of the constants appearing in the above expression are taken from Table I of [20].

In Eq. (3.4), the effective Hamiltonian H_0 is expressed solely by the quantum numbers $v_i (i = 1 \cdots 5)$ and $l_s (s = 4, 5)$, which means that these quantities are its conserved quantities. The eigenstates of H_0, which are in general anharmonic, are specified by the conserved quantities. In other words, Eq. (3.4) represents the diagonal part of the true Hamiltonian H.

IVR results from nondiagonal couplings among the anharmonic modes of the effective Hamiltonian H_0. The effects of these couplings are most significant when resonances among the modes take place. The locations where nonlinear resonances can occur are given by

$$\boldsymbol{n} \cdot \frac{\partial H_0}{\partial \boldsymbol{v}} = 0 \quad (3.6)$$

where $\boldsymbol{n} = (n_1, n_2, n_3, n_4, n_5)$ and $\boldsymbol{v} = (v_1, v_2, v_3, v_4, v_5)$, and $n_i (i = 1 \cdots 5)$ are integers. Here, we limit our argument to those resonances with $l_4 = 0$ and $l_5 = 0$ for simplicity. We will call $\Sigma_{i=1}^{5} |n_i|$ the order of a resonances $\boldsymbol{n}$. In these locations, the perturbation series are doomed to break down.

For acetylene, we calculate these locations where resonances can take place. This calculation does not mean that those nondiagonal terms actually exist. Its purpose is to indicate those locations where the effects of those couplings are most magnificent if they exist. In other words, when experiments are planned to look for the effects of certain coupling terms, these calculations show where to search.

In Figure 3.19, a section of the Arnold web of acetylene is displayed where nonlinear resonances up to seventh order are estimated. The value of the energy is chosen to be 18,797 cm^{-1}. Since, for systems of more than three degrees of freedom, we cannot directly display the whole web, we have to take a section. Here, the section is defined by $(v_1 = 0, v_2, v_3 = 0, v_4, v_5)$ and $(l_4 = 0, l_5 = 0)$. Although these sections offer only partial information on how the web is connected, we can derive some useful consequences as will be shown next.

The first thing we should notice in Figure 3.19 is that the distribution of the resonances is very sparse. There, only three major resonances exist: $\boldsymbol{n}_{44/55} = (0, 0, 0, 2, -2)$, $\boldsymbol{n}_{1/244} = (1, -1, 0, -2, 0)$ and $\boldsymbol{n}_{1/255} =$

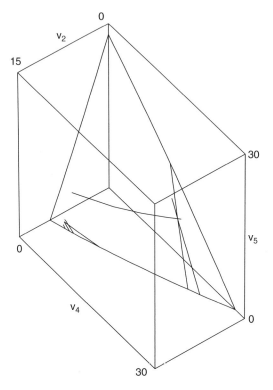

Figure 3.19. A section of the Arnold web for acetylene.

$(1, -1, 0, 0, -2)$, because the selection rules of the symmetry prohibit a large part of coupling terms. Thus, the symmetry of the molecule plays a crucial role in studying the Arnold web.

The second point to notice is correspondence between these calculations and experiments. All of the three major terms in Figure 3.19 actually exist in the IVR processes of acetylene, suggesting that these calculations can become a model for guidelines of future experiments, although they do not directly indicate the existence of these terms.

Furthermore, analysis of the resonances explains the recent experimental results from Field and co-workers [23]. They noticed that, contrary to the naive expectation, IVR from pure bending dynamics becomes simpler as the energy increases from 10,000 cm^{-1} and can be explained by the fact that, as the energy increases, the locations of the initial states actually leave the resonant region of $\boldsymbol{n}_{44/55} = (0, 0, 0, 2, -2)$ [24].

Up to now, the following coupling terms have already been identified in the experiments: $n = (-2, 0, 2, 0, 0)$, $(0, 0, 0, -2, 2)$, $(0, -1, 1, -1, -1)$, $(-1, 1, 0, 0, 2)$, $(-1, 1, 0, 2, 0)$, $(-1, 0, 1 -1, 1)$, and $(-1, -1, 2, -2, 0)$ [22]. In these experiments, only the SEP (Stimulated Emission Pumping) bright states $(0, v_2, 0, v_4, 0)$ are accessible as the initial conditions. One remarkable aspect about these terms is that the following quantities:

$$P = 5v_1 + 3v_2 + 5v_3 + v_4 + v_5 \quad (3.7)$$

$$R = v_1 + v_2 + v_3 \quad (3.8)$$

$$L_z = l_4 + l_5 \quad (3.9)$$

are conserved [25]. We call these quantities Kellman's constants.

By enumerating up to seventh order all of the resonances that satisfy the selection rules, we confirm that almost all of them conserve the Kellman's constants. We also notice the possibility of resonances that violate the constants. These results will be published elsewhere [24].

Existence of conserved quantities generally means that the IVR does not take place on the whole equi-energy surface, but is limited within a lower dimensional region specified by these quantities. For acetylene, the existence of the Kellman's constants indicates that the following limitation exists for the IVR starting from the SEP bright states.

By expressing Eqs. (3.7) and (3.8) in the following way:

$$v_1 + v_3 = R - v_2 \quad (3.10)$$

$$v_4 + v_5 = P - 5R + 2v_2 \quad (3.11)$$

we see that the sum $v_4 + v_5$ cannot increase in IVR from the SEP initial states. In order for the value of the sum to increase, the value of v_2 should increase because of Eq. (3.11). In order for the value of v_2 to increase, the value of the sum $v_1 + v_3$ should decrease because of Eq. (3.10). For the SEP initial states, however, the value of the sum $v_1 + v_3$ cannot decrease, because it already attains the lowest possible value, that is, 0.

Thus, the existence of Kellman's constants means that no IVR takes place from the SEP bright states that increases the value of the sum $v_4 + v_5$. This limitation has the following implications for a vibrationally highly excited acetylene.

For a vibrationally highly excited acetylene, the isomerization process from acetylene to vinylidene is of interest (see Fig. 3.20). In this process, both of the bending modes v_4 and v_5 should be excited where one of the hydrogen atoms goes half-way round the two carbon atoms. Therefore, the reaction coordinate of the isomerization is chosen to be the sum $v_4 + v_5$. However,

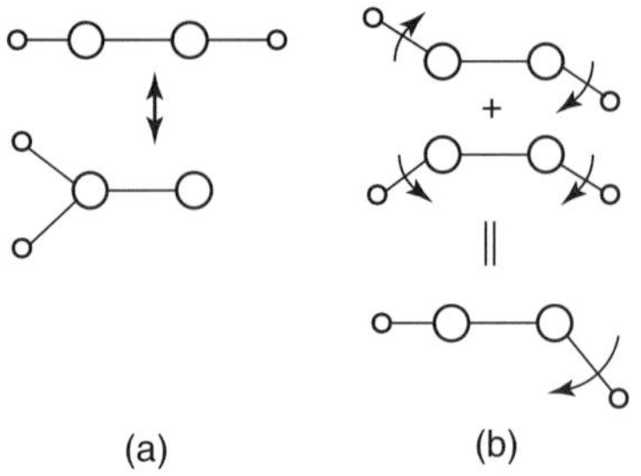

Figure 3.20. (a) Isomerization from acetylene to vinylidene and (b) the reaction coordinate for the isomerization.

the limitation imposed by Kellman's constants means that no IVR occurs from the SEP bright states that are directed toward the isomerization to vinylidene.

Thus, in order for the isomerization to take place from the SEP bright states, we have to seek for the following two possibilities. The first is to find coupling terms that violate the conservation of Kellman's constants. The second is to put the molecule under external fields that do not conserve the constants. Next, we pursue the second possibility.

Coupling with external electric fields is possible only for v_3 and v_5 because of the selection rules. Here, we put the molecule under the electric field that is resonant with the mode v_5, thereby violating the conservation of Kellman's constants. However, coupling the mode v_5 with the electric field is not sufficient, because both v_4 and v_5 need to be excited for the present purpose. Then, in order to make the energy flow from v_5 to v_4, we utilize the resonance $n_{44/55}$ by locating the SEP initial conditions near its resonant region. A schematic explanation of our ideas is shown in Figure 3.21.

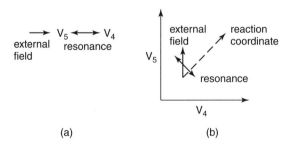

Figure 3.21. A schematic explanation for leading the molecule toward isomerization: (a) the processes of the energy flow and (b) the mechanism to lead the molecule toward isomerization.

Figure 3.21(a) displays the processes of the energy flow that consists of these two steps, that is, excitation of mode v_5 by the external field and that of mode v_4 by resonance $n_{44/55}$. In Figure 3.21(b), the mechanism to lead the molecule toward isomerization is shown. There, the two arrows with solid lines show the movements caused by the resonance $n_{44/55}$ and the external field, respectively. The arrow with broken lines indicates the behavior leading toward the isomerization. Our aim is to induce the behavior indicated by the broken arrow by combining the two movements shown by the solid arrows.

Figure 3.22 shows numerical results where the molecule is put under the external field. Its frequency ω_e is chosen to be 700 cm^{-1} so that the field becomes resonant with mode v_5. The amplitude E_e of the field is 30 cm^{-1}. Calculations are done by classical mechanics for the Hamiltonian that includes the diagonal part H_0 and the nondiagonal couplings identified experimentally so far. Initial conditions are taken from a uniform distribution that corresponds to the quantum state $(0, 3, 0, 22^0, 0^0)$. As can be seen in Figure 3.19, this state is located near the resonance $n_{44/55}$. Here, Figure 3.22(a) shows the time evolution of the mean of the sum

$$N = \langle v_4 + v_5 \rangle \tag{3.12}$$

and Figure 3.22(b) indicates that of its standard mean deviation

$$S = \sqrt{\langle (v_4 + v_5 - N)^2 \rangle} \tag{3.13}$$

where the bracket means taking the average concerning the distribution of the initial conditions.

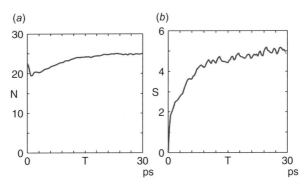

Figure 3.22. Numerical results to excite $v_4 + v_5$ by putting the molecule under the external fields. (a) T versus N. (b) T versus S.

From Figure 3.22, we can see that the diffusive behavior is induced with the modest increase of N. We can confirm that the values of $v_4 + v_5$ are increasing for some of the orbits where both the modes v_4 and v_5 are excited by an almost equal amount [26]. Thus, they are directed toward the isomerization.

Comparing the time evolution shown in Figure 3.22 with the Arnold web displayed in Figure 3.19, we can obtain more detailed understanding on the behavior of the molecule under the external field [26]. For example, the initial *decrease* of N is shown to be caused by the resonance $\boldsymbol{n}_{1/244} = (1, -1, 0, -2, 0)$. Since the resonance $\boldsymbol{n}_{1/244} = (1, -1, 0, -2, 0)$ lies very close to the resonance $\boldsymbol{n}_{44/55}$, the energy flow from v_4 to v_1 is strongly enhanced. This energy flow works adversely for the isomerization.

In this study, we have not made any attempt to design the external field to increase the rate of isomerization. In future studies, we will investigate the possibility of designing the field, and the results will be published elsewhere.

In summary, we have studied the Arnold web of acetylene, which is constructed utilizing the experimentally obtained parameters. We noticed that the symmetry of the molecule plays a crucial role where a large part of nondiagonal terms are prohibited. This results in a sparse feature of the web, leading to non-ergodicity of the IVR. We have also pointed out that the study of the web indicates those locations where the effects of nondiagonal terms are profoundly revealed, which would work as a guideline for future experiments that attempt to find unidentified coupling terms.

We have also studied the behavior of the molecule under external fields. By combining the effects of external fields and resonances, we can induce the energy flows that would not be possible for each of them. Then, comparison with the Arnold web is useful to understand the behavior of the molecule under external fields. We further point out the possibility that the study of this direction would enable us to manipulate the fields to control the reactions.

C. Hierarchy in Resonances

In general, detailed analyses of Arnold webs would be intractable for systems of more than three degrees of freedom. Then, our question is the following: What aspects of the Arnold webs would determine whether the reactions are statistical or nonstatistical? In other words, we need some methods to extract coarse grained features of the Arnold webs. In this section, we suggest focusing our attention on following two properties of the Arnold webs. The first is a hierarchy in the web. The second is whether the web is uniformly dense or not.

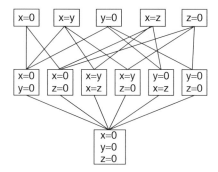

Figure 3.23. Hasse diagram of the hierarchy of nonlinear resonances.

Let us first focus attention on the relationship among resonance conditions, and temporarily neglect the widths of those resonances. Then, we note that there exists a hierarchy in the Arnold web. In general, for systems of n degrees of freedom, s independent resonances can intersect where s runs from 1 to $n - 1$. These intersections constitute the nodes in the network of nonlinear resonances. Moreover, every intersection of $s + 1$ independent resonances is included by some of the intersections of s independent resonances. When we order these intersections according to the number s of independent resonances, a hierarchical structure of the web becomes manifest. A diagram shown in Figure 3.23 is a schematic illustration of this idea. Here, a box in the uppermost alignment represents each of the nonlinear resonances, a box in the secondary alignment does an intersection of two independent resonances, a box in the third does an intersection of three independent resonances, and so on. The lines connecting boxes of successive alignments indicate the relation of inclusion between the intersections represented by the boxes. Such a diagram is in general called a Hasse diagram [27].

Around the regions in phase space where s independent resonances intersect, $n - s$ quantities other than the energy are conserved. Therefore, the more independent resonances intersect, the more degrees of freedom are involved in chaos there. Thus, the hierarchy of the web represents the hierarchy of chaotic regions in phase space.

Now, let us take into account the widths of resonances, and see how chaos can be different according to the characteristics of the webs. In Figure 3.24, schematic pictures of the webs are displayed for systems of three degrees of freedom, where the webs have different degrees of intersection and density.

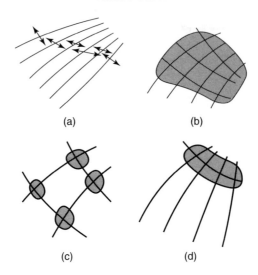

Figure 3.24 Schematic picture of Arnold webs with different degree of uniformity.

Figure 3.24(a) shows a web where no resonance intersection exists. Instead of intersections, resonant regions overlap with each other to cover the whole equi-energy surface. Then, global chaos occurs where orbits wonder around across the resonances. However, the number of degrees of freedom involved in chaos is small, which results in the fact that chaotic diffusion is not uniform, but occurs mainly in the direction across the resonances. In other words, the IVR only takes place along a limited direction although chaos may be strong.

Figure 3.24(b) shows the opposite to Figure 3.24(a) in the sense that a web has dense intersections. Moreover, those intersections overlap in phase space creating global chaos. Here, chaotic diffusion takes place in all the directions on the equi-energy surface. Then, the IVR takes place uniformly involving all the degrees of freedom, which is the case where the statistical reaction theory can be applied.

Figure 3.24(c) shows the case where resonance intersections exist but do not overlap. Then, the number of degrees of freedom involved in chaos would vary as orbits wonder around. It would increase around the intersections. However, between the intersections, it would decrease. Thus, the IVR in this case would be nonstationary and vary depending on where the orbit is wondering.

Figure 3.24(d) shows the intermediate case between (b) and (c). There exist intersections overlapping with each other. However, those overlapping intersections do not cover the whole equi-energy surface. Therefore, the IVR would depend on locations in phase space. While the IVR in overlapping intersections would be uniformly diffusive, it would take place only along limited directions outside the overlapping intersections.

The above discussions indicate that whether the webs are uniformly dense or not would play a crucial role in the processes of IVR, which is the second important point in the coarse-grained picture of the webs.

We can easily extend these discussions to systems of more than three degrees of freedom. In doing so, note that the dimension of resonance intersections in the action space is $n - s - 1$. Therefore, diffusion along resonance intersections would be possible for $s \leqslant n - 2$ even when they do not overlap, which extends the idea of Arnold diffusion into systems of more than three degrees of freedom. At present, the role of Arnold diffusion in the processes of IVR is unclear. However, as Lasker suggested [1], chaos in resonance intersections can be more remarkable than Arnold diffusion in the original sense. Then, Arnold diffusion in the extended sense is of interest for future study.

In this section, we have discussed two aspects of the Arnold webs. The first is the hierarchy in the web and the second is whether they are uniformly dense or not. How these two factors influence the spectra would be the next subject. In particular, the relationship between the hierarchy in the web and that in the spectra is important. We are currently studying this problem, and the results will be published in the near future.

D. Summary

To summarize this section, we have pointed out the relationship between IVR and the Arnold web. As an example, we have studied a vibrationally highly excited acetylene. We have found that, because of the symmetry of the molecule, a large part of nondiagonal couplings are prohibited, which results in non-ergodicity of the IVR in acetylene. We have also investigated the dynamics of the molecule under external fields. Then, the Arnold web turns out to be helpful in analyzing the dynamics. Furthermore, we suggest that the hierarchical structure and uniform/nonuniform nature of the Arnold web would give an important clue to understand the origin of statistical/selective features of the reaction.

VI. TOWARD DYNAMICAL CORRELATION OF REACTION PROCESSES: FUTURE PROSPECTS

Chemical reactions are composed of two dynamical processes: The first is crossing over potential barriers and the other is IVR within potential wells.

In Sections IV and V, we discussed these two processes from the viewpoint of chaos. In these discussions, the following points are of importance. As for the barrier crossing, intersection (either homoclinic or heteroclinic) between stable and unstable manifolds offers global information on the reaction paths. It not only includes how transition states are connected with each other, but also reveals how reaction paths bifurcate. Here, the concepts of normally hyperbolic invariant manifolds and crisis are essential. With regard to IVR, the concept of Arnold web is crucial. Then, we suggest that coarse-grained features of the Arnold webs should be studied. In particular, the hierarchy of the web and whether the web is uniformly dense or not would play an important role.

By combining these two points, we propose to capture global aspects of reaction processes. In particular, dynamical correlation among successive crossings over barriers is of interest. In other words, we pay attention to deviation from purely statistical description. Its reason is the following.

In the statistical reaction theory, each process of crossing over a barrier takes place independently with each other. In other words, the energy required for the crossing is completely thermalized after passing the barrier. Then, the next process of crossing a neighboring one starts only when the required energy happens to concentrate on the reaction coordinate. Its probability is independent from that of crossing the previous one. Therefore, the probability of crossing the two is given by their multiplication. Thus, for reactions with many barriers, the probability of going through the whole processes would become extremely small. If we can arrange the reaction so that successive crossings over barriers have dynamical correlation, we can increase the reaction rate.

Based on the discussions in Sections IV and V, we can envisage a possible mechanism for dynamical correlation. See Figure 3.25 for schematic explanations for this. Here, we suggest that there exist three levels of dynamical correlation.

Figure 3.25(*a*) shows a case where a strong correlation exists between processes of crossing two neighboring barriers. This correlation results from direct intersection between the stable and unstable manifolds of the normally hyperbolic invariant manifolds located on the tops of neighboring barriers. Then, some of the orbits starting from one of the normally hyperbolic invariant manifolds directly reach the other one without falling into the potential well. Some other orbits may fall into the well and would take some time to reach the other manifold. The ratio of these two types of orbits depends on how steep the intersection is.

Figure 3.25(*b*) displays a situation where dynamical correlation exists but is weak. Here, direct intersection does not exist between the stable and unstable manifolds of the normally hyperbolic invariant manifolds of the

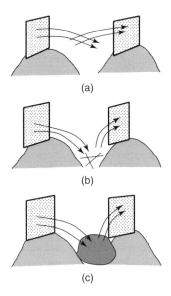

Figure 3.25. Schematic picture of dynamical correlation between two potential barriers.

neighboring barriers. Instead, the stable and unstable manifolds fall into the potential well and encounter the Arnold web. Then, complicated movement within the web takes place, thereby resulting in less correlation. However, when the Arnold web is sparse, some orbits from one of the normally hyperbolic invariant manifolds would soon find a way out to reach the stable manifold of the other. Thus, we can expect that dynamical correlation would, although weak, still exist in this situation.

Figure 3.25(c) exhibits a case where no dynamical correlation exists. Here, the stable and unstable manifolds fall into the potential well, and encounter densely populated resonances. Then, the movement within the web takes so much time that dynamical correlation would be completely lost there, which is the case where the statistical reaction theory is applicable.

The above explanations offer a unified point of view for various chemical reactions ranging from selective to statistical. The study on which level of dynamical correlation the reaction has would give a new insight to the origin of statistical or selective features of its dynamics. Next, we propose some of the topics that can be studied from this point of view.

One of the reactions where this approach can be fruitful is the folding of proteins. Recently, it was proposed that the energy landscape of the folding processes had a funnel-like structure [28, 29]. Usually, the funnel-like structure is interpreted based on the concept of free energy, that is, the concept of equilibrium statistical mechanics. However, if the folding processes take place far from equilibrium, the concept of equilibrium statistical mechanics cannot be applied. On the other hand, from our point of view, the funnel-like structure would give a typical example where the strong dynamical correlation exists. If we can analyze how stable and unstable manifolds intersect in the folding processes, we can have an alternative explanation concerning the funnel, which is a future project from our approach.

Another example is the mechanism where efficiency of reactions is affected by the surrounding molecules. Suppose that, depending on the surrounding molecules, intersection of the stable and unstable manifolds of a reaction changes between (a) and (b) or between (b) and (c). Then, these molecules can enhance or decrease the reaction rate by switching the level of dynamical correlation. Moreover, when the surrounding molecules induce bifurcation of reaction paths, their existence can trigger new reactions. These possibilities would offer a clue to understand molecular functions from a dynamical point of view.

In order to proceed further, we need a method to compute stable and unstable manifolds, and see if they intersect. We are currently studying this problem by extending the Lie perturbation methods. The results will be published elsewhere [30].

VII. SUMMARY

Summarizing this chapter, we have discussed, by examining recent experiments, the necessity of going over statistical reaction theory. Among the underlying assumptions of the traditional theory, we especially question the separation of timescales. Then, we showed how the study of chaos can contribute to this problem. First, the importance of taking into account global features of reaction dynamics is stressed. These features are revealed by the intersection of stable and unstable manifolds of normally hyperbolic invariant manifolds. Then, for multidimensional Hamilton chaos, new features emerge, that is, tangency and crisis in chaos. Second, the relationship between IVR and the Arnold web is pointed out. Then, non-ergodicity of IVR becomes manifest in sparse features of the web. As for future prospects of these studies, we point out that dynamical correlation among processes of successively crossing over barriers would be revealed by studying the intersections of stable and unstable manifolds. We expect that research toward this direction can shed new light on the mechanism of protein folding and molecular functions.

Acknowledgments

The author would like to thank Professor K. Yamanouchi, Professor M. Nagaoka, and Professor T. Komatsuzaki for fruitful discussions. This work is supported by the Grand-in-Aid for Scientific Research from the Ministry of Education, Science, Sports, and Culture.

References

1. L. Lasker, *Physica* **D67**, 257 (1993).
2. M. J. Davis and S. K. Gray, *J. Chem. Phys.*, **84**, 5389 (1986).
3. R. E. Gillilan and G. S. Ezra, *J. Chem. Phys.*, **94**, 2648 (1991).
4. M. Toda, *Phys. Rev. Lett.*, **74**, 2670 (1995).
5. T. Baer and W. L. Hase, In *Unimolecular Reaction Dynamics: Theory and Experiments* (Oxford, New York, 1996).
6. T. Sunada, *Trend Math.*, Birkhäuser, 175 (1997).
7. T. Tate, *J. Math. Soc. Jpn.*, **51**, 867, (1999).
8. T. Komatsuzaki and M. Nagaoka, *J. Chem. Phys.*, **105**, 10838 (1996)
9. E. R. Lovejoy, S. K. Kim, and C. B. Moore, *Science*, **256**, 1541 (1992).
10. S. K. Kim, E. R. Lovejoy, and C. B. Moore, *J. Chem. Phys.*, **102**, 3202 (1995).
11. T. Shibata, H. Lai, H. Katayanagi, and T. Suzuki, *J. Phys. Chem.*, **A102**, 3643 (1998).
12. S. L. Schultz, J. Qian, and J. M. Jean, *J. Phys. Chem.*, **A101**, 1000 (1997).
13. S. Wiggins, *Physica*, **D44**, 471 (1990).
14. N. Shida, *Generation and Degeneration in the Dynamics of Molecular Clusters*, Suurikagaku (in Japanese), Special Issue on Complex Systems (1996) pp. 50.
15. Y. C. Lai, C. Grebogi, R. Blümel, and I. Kan, *Phys. Rev. Lett.*, **71**, 2212 (1993).
16. M. Komuro, *A Mechanism of Chaotic Itinerary on Globally Coupled Maps* (in Japanese), RIMS Kokyuroku Vol. 1118 (1999), Singular Phenomena of Dynamical Systems, pp. 96–114.
17. M. Toda, *Phys. Lett.*, **A227**, 232 (1997).
18. C. C. Martens, M. J. Davis, and G. S. Ezra, *Chem. Phys. Lett.*, **142**, 519 (1987).
19. M. Herman, J. Lievin, J. V. Auwera, and A. Campargue, *Adv. Chem. Phys.*, **108**, 1 (1999).
20. K. Yamanouchi, N. Ikeda, S. Tsuchiya, D. M. Jonas, J. K. Lundberg, G. W. Adamson, and R. W. Field, *J. Chem. Phys.*, **95**, 6330 (1991).
21. M. P. Jacobson and R. W. Field, *J. Phys. Chem.*, **A104**, 3073 (2000).
22. M. A. Temsamani and M. Herman, *J. Chem. Phys.*, **102**, 6371 (1995).
23. M. P. Jacobson, J. P. O'Brien, R. J. Silbey, and R. W. Field, *J. Chem. Phys.*, **109**, 121 (1998).
24. M. Toda, *Arnold Web of Vibrationally Highly Excited Acetylene*, to be published.
25. M. E. Kellman, *J. Chem. Phys.*, **93**, 6630 (1990).
26. M. Toda, *Dynamical Behavior of Acetylene under External Fields*, in progress.
27. P. Orlik and H. Terao, *Arrangement of Hyperplanes* (Springer, New York, 1991).
28. J. N. Onuchic, P. G. Wolynes, Z. Luthey-Schulten, and N. D. Socci, *Proc. Natl. Acad. Sci., U.S.A.* **92**, 3626 (1995).

29. N. D. Socci, J. N. Onuchic, and P. G. Wolynes, *Proteins: Structure, Function, and Genetics*, **32**, 136 (1998).
30. M. Toda and T. Komatsuzaki, *Lie Perturbation Method Applied to Study Dynamical Correlation*, in progress.

A STOCHASTIC THEORY OF SINGLE MOLECULE SPECTROSCOPY

YOUNJOON JUNG, ELI BARKAI, and ROBERT J. SILBEY

Department of Chemistry, Massachusetts Institute of Technology, Cambridge, MA 02139

CONTENTS

I. Introduction
II. Theory
   A. Stochastic Optical Bloch Equation
   B. Classical Shot Noise
III. Simulation
IV. Q and the Three-Time Correlation Function
V. Two-State Jump Model: Exact Solution
   A. Model
   B. Solution
VI. Analysis of the Exact Solution: Limiting Cases
   A. Slow Modulation Regime: $R \ll v, \Gamma$
      1. Strong Modulation: Case 1
      2. Weak Modulation: Case 2
      3. Time Dependence
   B. Fast Modulation Regime: $v, \Gamma \ll R$
      1. Strong Modulation: Case 3
      2. Weak Modulation: Case 4
   C. Intermediate Modulation Regime
      1. Case 5
      2. Case 6
   D. Phase Diagram
VII. Connection to Experiments
VIII. Further Discussions on Validity of Present Theory
IX. Concluding Remarks
Acknowledgments
Appendix
   A. Calculation of Line Shape via Marginal Averaging Approach
   B. Perturbation Expansion
      1. First-Order Terms

Advances in Chemical Physics, Volume 123, Edited by I. Prigogine and Stuart A. Rice.
ISBN 0-471-21453-1 © 2002 John Wiley & Sons, Inc.

2. Second-Order Terms
3. Classical Lorentz Model
C. Exact Calculations of $\langle W \rangle$ and $\langle W^2 \rangle$
D. Q in the Long Time Limit
E. Q in the Slow Modulation Regime
F. Q in the Fast Modulation Regime
References

I. INTRODUCTION

In recent years, a new approach to condensed-phase spectroscopy has emerged, that focuses on the spectral properties of *a single molecule* (SM) embedded in a condensed phase [1–4]. Thanks to experimental advances made in optics and microscopy [5], it is now possible to perform single molecule spectroscopy (SMS) in many different systems. Motivations for SMS arise from a fundamental point of view (e.g., the investigation of the field–matter interaction at the level of a SM and the verification of statistical assumptions made in ensemble spectroscopy) and from the possibility of applications (e.g., the use of SMS as a probe for large biomolecules for which a SM is attached as a fluorescent marker).

In general, the spectral properties of each individual molecule vary from molecule to molecule due to differences in the local environments with which each SM is interacting [6–16]. With its unique ability to detect dynamical phenomena occurring at the level of an individual molecule surrounded by its local environment, SMS has uncovered the statistical distributions of microscopic quantities of the environment that are hidden in traditional ensemble averaged spectroscopy. In particular, a single molecule spectrum measured for a finite time necessarily "sees" the temporal fluctuations of the host environment that occur on timescales comparable to the measurement timescale, and therefore lead, in many cases, to a stochastically fluctuating single molecule spectrum. Time-dependent fluctuation phenomena in SMS occur in many ways, such as spectral diffusion [6–9, 12, 17] and fluorescence intermittency [18–23]. The physical mechanisms causing these fluctuation phenomena vary depending on the dynamical processes a SM is undergoing, including: triplet state dynamics [18–20], energy-transfer processes [24–26], exciton-transfer processes [17], chemical reactions [27–29, 86, 89], conformational changes [30–34], rotational dynamics [35, 36], and diffusion processes [37]. Thus SMS provides a unique microscopic tool to investigate the dynamical processes that a SM and its environment undergo during the measurement time.

One important process responsible for time-dependent fluctuations in SMS is spectral diffusion, that is, perturbations or excitations in the environment of the SM produce random changes in the transition frequency of the SM [6–9, 12, 17, 38–42], leading to a time-dependent spectrum. Spectral diffusion processes have been observed in various systems including dye molecules in a molecular crystal [6, 7] and in a low-temperature glass [8, 9], quantum dots [43], light harvesting systems [44, 45], and dendrimers [46]. Since the spectral diffusion process directly reflects both (a) the interaction between the SM and its environment and (b) the local dynamics of the latter, careful analysis of the time-dependent fluctuations of SMS illuminates the interplay between various dynamical processes in the condensed phase. In this work, we formulate a stochastic theory of SMS undergoing a spectral diffusion process. In particular, we address the issue of the counting statistics of emitted photons produced by a SM undergoing a spectral diffusion process. Our studies show how the fluctuations in SMS can be used to probe the dynamics of SM and its interaction with the excitations of the environment. A summary of our results was published in [79].

Previously, the photon counting statistics of an ensemble of molecules, studied by various methods, for example, the fluorescence correlation spectroscopy [47–51], has proved useful for investigating dynamical processes of different systems. The photon statistics of a SM is clearly different from that of the ensemble of molecules due both to the absence of inhomogeneous broadening and to the correlation between fluorescence photons that exists only on the SM level. In some SMS experiments, the measurement time is limited due to photobleaching, where the emission of a SM is quenched suddenly because of various reasons, for example, reaction with oxygen. Thus it is not an easy task in general to collect a sufficient number of photon counts to have good statistics. However, many SM–host systems have been found to remain stable for a long enough time to measure photon statistics [37, 52–57]. In view of these recent experimental activities, a theoretical investigation of the counting statistics of photons produced by single molecules, in particular *when there is a spectral diffusion process*, is timely and important.

We will analyze the SM spectra and their fluctuations semiclassically using the stochastic Bloch equation in the limit of a weak laser field. The Kubo–Anderson sudden jump approach [58–61] is used to describe the spectral diffusion process. For several decades, this model has been a useful tool for understanding line shape phenomena, namely, of the average number of counts $\langle n \rangle$ per measurement time T, and has found many applications mostly in ensemble measurements, for example, NMR [60], and nonlinear spectroscopy [62]. More recently, it was applied to model SMS in low-temperature glass systems in order to describe the static properties of line shapes [14–16, 63] and also to model the time-dependent fluctuations of SMS [64–66].

Mandel's Q parameter quantitatively describes the deviation of the photon statistics from the Poissonian case [67, 68],

$$Q = \frac{\langle n^2 \rangle - \langle n \rangle^2}{\langle n \rangle} - 1 \qquad (4.1)$$

where n is the random number of photon counts, and the average is taken over stochastic processes involved. In the case of Poisson counting statistics, $Q = 0$ while our semiclassical results show super-Poissonian behavior ($Q > 0$) for a SM undergoing a spectral diffusion process. For short enough times, fluorescent photons emitted by a SM show antibunching phenomena ($-1 < Q < 0$), a sub-Poissonian nonclassical effect [53, 54, 69–72]. Our semiclassical approach is valid when the number of photon counts is large. Further discussion of the validity of our approach is given in the text.

One of the other useful quantities to characterize dynamical processes in SMS is the fluorescence intensity correlation function, also called the second-order correlation function, $g^{(2)}(t)$, defined by [73, 74]

$$g^{(2)}(\tau) = \frac{\langle I(t+\tau)I(t) \rangle}{\langle I \rangle^2} \qquad (4.2)$$

This correlation function has been used to analyze dynamical processes involved in many SMS experiments [8, 26, 75–77]. Here, $I(t)$ is the random fluorescence intensity observed at time t. It is well known that for a stationary process there is a simple relation between $g^{(2)}(t)$ and $Q(T)$ [53, 71, 78]

$$Q(T) = \frac{2\langle I \rangle}{T} \int_0^T dt_1 \int_0^{t_1} dt_2 \, g^{(2)}(t_2) - \langle I \rangle T \qquad (4.3)$$

where T is the measurement time.

The essential quantity in the present formulation is a three-time correlation function, $C_3(\tau_1, \tau_2, \tau_3)$, which is similar to the nonlinear response function investigated in the context of four wave mixing processes [62]. The three-time correlation function contains all the microscopic information relevant for the calculation of the line shape fluctuations described by Q. It has appeared as well in a recent paper of Plakhotnik [66] in the context of intensity–time–frequency–correlation technique. In the present work, important time-ordering properties of this function are fully investigated, and an analytical expression for Q is found. The relation between $C_3(\tau_1, \tau_2, \tau_3)$

and line shape fluctuations described by Q generalizes the Wiener–Khintchine theorem, which gives the relation between the one-time correlation function and the averaged line shape.

The timescale of the bath fluctuations is an important issue in SMS. Bath fluctuations are typically characterized as being in either fast or slow modulation regimes (to be defined later) [60]. If the bath is very slow a simple adiabatic approximation is made based on the steady-state solution of the time-independent Bloch equation. Several studies have considered this simple limit in the context of SMS [8, 9, 38, 39]. From a theoretical and also experimental point of view it is interesting to go beyond the slow modulation case. In the fast modulation case, it is shown that a factorization approximation for the three-time correlation function yields a simple limiting solution. In this limit, the line shape exhibits the well-known behavior of motional narrowing (as the timescale of the bath becomes short, the line is narrowed). By considering a simple spectral diffusion process, we show that Q exhibits a more complicated behavior than the line shape does. When the timescale of the bath dynamics goes to zero, we find Poissonian photon statistics. Our exact results can be evaluated for an arbitrary timescale of the bath and are shown to interpolate between the fast and slow modulation regimes.

This chapter is organized as follows. In Section II.A, the stochastic Bloch equation is presented and a brief discussion of its physical interpretation is given. In Section II.B, the prescription for the relation between the solution of the optical Bloch equation and the discrete photon counts is described. We briefly review several results on counting statistics, which will later clarify the meaning of some of our results. Section III presents simple simulation results of SM spectra in the presence of spectral diffusion. These demonstrate a generic physical situation to which the present theory is applicable. In Section IV, an important relationship between Q and the three-time correlation function is found, and the general properties of the latter are investigated. An exact solution for a simple spectral diffusion process is found in Section V. In Section VI, we analyze the exact solution in various limiting cases so that the physical meaning of our results becomes clear. Connection of the present theory to experiments is made in Section VII. In Section VIII, we further discuss the validity of the present model in connection with other approaches. We conclude in Section IX. Many of the mathematical derivations are relegated to the Appendices.

II. THEORY

Our theory presented in this section consists of two parts. First, we model the time evolution of a SM in a dynamic environment by the stochastic optical Bloch equation. Second, we introduce the photon counting statistics of a SM by considering the shot noise process due to the discreteness of photons.

A. Stochastic Optical Bloch Equation

We assume a simple nondegenerate two level SM in an external classical laser field. The electronic excited state $|e\rangle$ is located at energy ω_0 above the ground state $|g\rangle$. We consider the time-dependent SM Hamiltonian

$$H = \frac{1}{2}\hbar\omega_0\sigma_z + \sum_{j=1}^{J}\frac{1}{2}\hbar\Delta\omega_j(t)\sigma_z - \mathbf{d}\cdot\mathbf{E}_0\cos(\omega_L t) \qquad (4.4)$$

where σ_z is the Pauli matrix. The second term reflects the effect of the fluctuation of the environment on the absorption frequency of the SM coupled to J perturbers. The stochastic frequency shifts $\Delta\omega_j(t)$ (i.e., the spectral diffusion) are random functions whose properties will be specified later. The last term in Eq. (4.4) describes the interaction between the SM and the laser field (frequency ω_L), where $\mathbf{d} \equiv \mathbf{d}_{eg}\sigma_x$ is the dipole operator with the real matrix element $\mathbf{d}_{eg} = \langle e|\mathbf{d}|g\rangle$. We assume that the molecule does not have any permanent dipole moment either in the ground or in the excited state, $\langle g|\mathbf{d}|g\rangle = \langle e|\mathbf{d}|e\rangle = 0$.

In the limit of a weak external field the model Hamiltonian describes the well-known Kubo–Anderson random frequency modulation process whose properties are specified by statistics of $\Delta\omega_j(t)$ [58–60]. When the fluctuating part of the optical frequency $\Delta\omega_j$ is a two-state random telegraph process, the Hamiltonian describes a SM (or spin of type A) coupled to J bath molecules (or spins of type B), these being two-level systems. Under certain conditions, this Hamiltonian describes a SM interacting with many two-level systems in low-temperature glasses that has been used to analyze SM line shapes [14–16, 63, 65, 66].

The molecule is described by 2×2 density matrix ρ whose elements are ρ_{gg}, ρ_{ee}, ρ_{ge}, and ρ_{eg}. Let us define

$$u \equiv \frac{1}{2}(\rho_{ge}e^{-i\omega_L t} + \rho_{eg}e^{i\omega_L t})$$

$$v \equiv \frac{1}{2i}(\rho_{ge}e^{-i\omega_L t} - \rho_{eg}e^{i\omega_L t})$$

$$w \equiv \frac{1}{2}(\rho_{ee} - \rho_{gg}) \qquad (4.5)$$

and note that from the normalization condition $\rho_{ee} + \rho_{gg} = 1$, we have

$\rho_{ee} = w + \frac{1}{2}$. By using Eq. (4.4), the stochastic Bloch equations in the rotating wave approximation are given by [80, 81]

$$\dot{u} = \delta_L(t)v - \frac{\Gamma u}{2}$$

$$\dot{v} = -\delta_L(t)u - \frac{\Gamma v}{2} - \Omega w$$

$$\dot{w} = \Omega v - \Gamma w - \frac{\Gamma}{2} \qquad (4.6)$$

$1/\Gamma$ is the radiative lifetime of the molecule added phenomenologically to describe spontaneous emission, $\Omega = -\mathbf{d}_{eg} \cdot \mathbf{E}_0/\hbar$ is the Rabi frequency, and the detuning frequency is defined by

$$\delta_L(t) = \omega_L - \omega_0 - \Delta\omega(t)$$

$$\Delta\omega(t) = \sum_{j=1}^{J} \Delta\omega_j(t) \qquad (4.7)$$

Besides the natural relaxation process described by Γ, other T_1 and T_2 processes can easily be included in the present theory. The parameter w represents one-half of the difference between the populations of the states $|e\rangle$ and $|g\rangle$, while u and v give the mean value of the dipole moment $\mathbf{d}$,

$$\text{Tr}(\rho \mathbf{d}) = 2\mathbf{d}_{ge}[u\cos(\omega_L t) - v\sin(\omega_L t)] \qquad (4.8)$$

In recent studies [82, 83] it has been demonstrated that the *deterministic* two-level optical Bloch equation approach captures the essential features of SMS in condensed phases, which further justifies our assumptions.

The physical interpretation of the optical Bloch equation in the absence of time-dependent fluctuations is well known [62, 73]. Now that the stochastic fluctuations are included in our theory we briefly discuss the additional assumptions needed for standard interpretation to hold. The time-dependent power absorbed by the SM due to work of the driving field is,

$$\frac{d\mathcal{W}}{dt} = \cos(\omega_L t)\mathbf{E}_0 \cdot \frac{d}{dt}\text{Tr}(\rho \mathbf{d}) \qquad (4.9)$$

As usual, additional averaging (denoted with overbar) of Eq. (4.9) over the optical period of the laser is made. This averaging is clearly justified for an ensemble of molecules each being out of phase. For a SM, such an additional averaging is meaningful when the laser timescale, $1/\omega_L$, is much shorter than any other timescale in the problem (besides $1/\omega_0$, of course). By using Eq. (4.8) this means,

$$\overline{v\cos^2(\omega_L t)} \approx \overline{v}\,\overline{\cos^2(\omega_L t)} \tag{4.10}$$

under the conditions $|\dot{u}(t)| \ll \omega_L |v(t)|$ etc., and hence we have

$$\frac{d\overline{\mathscr{W}}}{dt} \simeq \hbar\Omega\omega_L v(t) \tag{4.11}$$

The absorption photon current (unit 1/[time]) is [73]

$$I(t) = \frac{1}{\hbar\omega_L}\frac{d\overline{\mathscr{W}}}{dt} = \Omega v(t) \tag{4.12}$$

By neglecting photon shot noise (soon to be considered), $\int_0^T I(t)\,dt$ has the meaning of the number of absorbed photons in the time interval $(0, T)$ [i.e., since $\int_0^T (d\overline{\mathscr{W}}/dt)\,dt$ is the total work and each photon carries energy $\hbar\omega_L$]. By using Eq. (4.6), we have

$$\dot{\rho}_{ee} = \Omega v - \Gamma \rho_{ee} \tag{4.13}$$

In the steady state, $\dot{\rho}_{ee} = 0$, we have $\Omega v = \Gamma \rho_{ee}$, and since Ωv has a meaning of absorbed photon current, $\Gamma \rho_{ee}$ has the meaning of photon emission current. For the stochastic Bloch equation, a steady photon flux is never reached; however, integrating Eq. (4.13) over the counting time interval T,

$$[\rho_{ee}(T) - \rho_{ee}(0)] + \Gamma \int_0^T \rho_{ee}(t')\,dt' = \Omega \int_0^T v(t')\,dt', \tag{4.14}$$

and using $|\rho_{ee}(T) - \rho_{ee}(0)| \leq 1$ we find for large T that the absorption and emission photon counts are approximately equal,

$$\underbrace{\Gamma \int_0^T \rho_{ee}(t)\,dt}_{\text{emitted photons}} \simeq \underbrace{\int_0^T I(t)\,dt}_{\text{absorbed photons}}, \tag{4.15}$$

provided that $\int_0^T I(t)\,dt \gg 1$. Equation (4.15) is a necessary condition for the present theory to hold, and it means that the large number of absorbed photons is approximately equal to the large number of emitted photons (i.e., we have neglected any nonradiative decay channels). When there are nonradiative decay channels involved, one may modify Eq. (4.15) approximately by taking into account the fluorescence quantum yield, ϕ, the ratio of the number of emitted photons to the number of absorbed photons,

$$\Gamma \int_0^T \rho_{ee}(t)\,dt \simeq \phi \int_0^T I(t)\,dt \qquad (4.16)$$

B. Classical Shot Noise

Time-dependent fluctuations are not only produced by the fluctuating environment in SMS. In addition, an important source of fluctuations is the discreteness of the photon, that is, shot noise. By assuming a *classical* photon emission process, the probability of having a single photon emission event in time interval $(t, t + dt)$ is [68]

$$\text{Prob}(t, t + dt) = \Gamma \rho_{ee}(t)\,dt \qquad (4.17)$$

While this equation is certainly valid for ensemble of molecules all subjected to a hypothetical identical time-dependent environment, the validity of this equation for a SM is far from being obvious. In fact, as we discuss below, we can expect this equation to be valid only under certain conditions. By using Eq. (4.17) the probability of recording n photons in time interval $(0, T)$ is given by the classical counting formula [68]

$$p(n, T) = \frac{W^n}{n!} \exp(-W) \qquad (4.18)$$

with

$$W = \eta \int_0^T I(t)\,dt \qquad (4.19)$$

where η is a suitable constant depending on the detection efficiency. For simplicity, we set $\eta = 1$ here, but will reintroduce it later in our final expressions. Here, $W = \eta \bar{W}/\hbar\omega_L$ is η times the work done by the driving laser field whose frequency is ω_L divided by the energy of one photon $\hbar\omega_L$

[see Eq. (4.12)]. It is a dimensionless time-dependent random variable, described by a probability density function $P(W, T)$, which at least in principle can be evaluated based on the statistical properties of the spectral diffusion process and the stochastic Bloch equations. From Eq. (4.18) and for a specific realization of the stochastic process $\Delta\omega(t)$, the averaged number of photons counted in time interval $(0, T)$ is given by W,

$$\langle n \rangle_s = \sum_{n=0}^{\infty} n p(n, T) = W \tag{4.20}$$

where the shot noise average is $\langle \cdots \rangle_s = \sum_{n=0}^{\infty} \cdots p(n, T)$. Since W is random, additional averaging over the stochastic process $\Delta\omega(t)$ is necessary and statistical properties of the photon count are determined by $\langle p(n, T) \rangle$, where $\langle \cdots \rangle$ denotes averaging with respect to the spectral diffusion (i.e., not including the shot noise),

$$\langle p(n, T) \rangle = \left\langle \frac{W^n}{n!} \exp(-W) \right\rangle \tag{4.21}$$

Generally, the calculation of $\langle p(n, T) \rangle$ is nontrivial; however, in some cases simple behavior can be found. Assuming temporal fluctuations of W occur on the timescale τ_c, then we have the following:

1. For counting intervals $T \gg \tau_c$ and for ergodic systems,

$$\lim_{T \to \infty} \frac{\langle W \rangle}{T} = \lim_{T \to \infty} \frac{1}{T} \int_0^T I(t) \, dt \equiv \langle I(\omega_L) \rangle \tag{4.22}$$

which is the fluorescence line shape of the molecule, that is, the averaged number of photon counts per unit time when the excitation laser frequency is ω_L [later we suppress ω_L in $\langle I(\omega_L) \rangle$]. Several authors [51, 68, 84] argued quite generally (though not in the SM context) that in the long measurement time limit we may use the approximation $W \simeq \langle I \rangle T$ (i.e., neglect the fluctuations), and hence photon statistics becomes Poissonian [68],

$$\langle p(n, T) \rangle \simeq \frac{(\langle I \rangle T)^n}{n!} \exp(-\langle I \rangle T) \tag{4.23}$$

At least in principle, $\langle I \rangle$ can be calculated based on standard line shape theories (e.g., in Appendix A we calculate $\langle I \rangle$ for our working example considered in Section V). Equation (4.23) implies that a single measurement of the line shape (i.e., averaged number of emitted photons as a function of laser frequency) determines the statistics of the photon count in the limit of long measurement time. In fact, it tells us that in this case, counting statistics beyond the average will not reveal any new information on the SM interacting with its dynamical environment. Mathematically, this means that the distribution $P(W, T)$ satisfying

$$\langle p(n, T) \rangle = \int_0^\infty dW P(W, T) \frac{W^n}{n!} \exp(-W) \quad (4.24)$$

converges to

$$P(W, T) \to \delta(W - \langle I \rangle T) \quad (4.25)$$

when $T \to \infty$. We argue below, however, using the central limit theorem, for cases relevant for SMS, $P(W, T)$ is better described by a Gaussian distribution. The transformation Eq. (4.24) is called the Poisson transform of $P(W, T)$ [67].

2. In the opposite limit, $T \ll \tau_c$, we may use the approximation [85]

$$W \simeq I_0 T \quad (4.26)$$

where $I_0 = I(T = 0)$. In a steady state, $\langle p(n, T) \rangle$ can be calculated if the distribution of intensity (i.e., photon current) is known

$$\langle p(n, T) \rangle \simeq \int_0^\infty dI_0 P(I_0) e^{-I_0 T} \frac{(I_0 T)^n}{n!} \quad (4.27)$$

For example, assume that $I(t)$ is a two-state process, i.e. the case when a SM is coupled to a single slow two-level system in a glass, then

$$P(I) = p_1 \delta(I - I_1) + p_2 \delta(I - I_2) \quad (4.28)$$

and

$$\langle p(n, T) \rangle = \sum_{i=1,2} p_i \frac{(I_i T)^n}{n!} \exp(-I_i T) \quad (4.29)$$

and if, for example, $I_2 \sim 0$, the SM is either "on" or "off", a case encountered in several experiments [30, 43, 86].

3. A more challenging case is when $T \sim \tau_c$; later, we address this case in some detail.

We would like to emphasize that the photon statistics we consider is classical, while the Bloch equation describing dynamics of the SM has quantum mechanical elements in it (i.e., the coherence). In the weak laser intensity case, the Bloch equation approach allows a classical interpretation based on the Lorentz oscillator model as presented in Appendix B.3.

III. SIMULATION

To illustrate combined effects of the spectral diffusion and the shot noise on the fluorescence spectra of a SM, we present simulation results of spectral trails of a SM, where the fluorescence intensity of a SM is measured as a function of the laser frequency as the spectral diffusion proceeds [7, 12].

First, we present a simple algorithm for generating random fluorescence based on the theory presented in Section II, by using the stochastic Bloch equation, Eq. (4.6) and the classical photon counting distribution, Eq. (4.18). A measurement of the spectral trail is performed from $t = 0$ to $t = t_{end}$. As in the experimental situation, we divide t_{end} into N time bins each of which has a length of time T. For each bin time T, a random number of photon counts is recorded. Simulations are performed following the steps described below:

Step 1. Generate a spectral diffusion process $\Delta\omega(t)$ from $t = 0$ to $t = t_{end}$.

Step 2. Solve the stochastic Bloch equation, Eq. (4.6), for a random realization of the spectral diffusion process generated in Step 1 for a given value of ω_L during the time period $0 < t < t_{end}$.

Step 3. Determine $W(t_k, \omega_L)$ during the kth time bin ($k = 1, 2, 3, \ldots, N$), $(k-1)T < t < kT$, with a measurement time T according to Eq. (4.19),

$$W(t_k, \omega_L) = \int_{(k-1)T}^{kT} dt\, I(t) \tag{4.30}$$

Step 4. Generate a random number $0 < x < 1$ using a uniform random number generator, and then the random count n is found using the criterion,

$$\sum_{j=0}^{n-1} p(j, W) < x \leqslant \sum_{j=0}^{n} p(j, W) \tag{4.31}$$

According to Eq. (4.18), we find

$$\frac{\Gamma(n, W)}{(n-1)!} < x \leqslant \frac{\Gamma(n+1, W)}{n!} \qquad (4.32)$$

where $\Gamma(a, z)$ is the incomplete gamma function. Steps 1–4 must be repeated many times to get good statistics.

For an illustrative purpose, we choose a simple model of the spectral diffusion, which is called a two-state jump process or a dichotomic process. We assume that the frequency modulation can be either $\Delta\omega(t) = v$ or $\Delta\omega(t) = -v$, and the flipping rate between these two frequency modulations is given by R. This model will be used as a working example for which an analytical solution is obtained later in Section V.

In Figure 4.1, we present a simulation result of one realization of a spectral diffusion process when the fluctuation rate R is much smaller than Γ and v (slow modulation regime to be defined later). Parameters are given in the figure caption. Figure 4.1(a) shows $W(t_k, \omega_L)$, demonstrating the effects of the spectral diffusion process on the fluorescence spectra. Note that $W(t_k, \omega_L)$ has been defined *without the shot noise*. One can clearly see that the resonance frequency of a SM is jumping between two values as time goes on, and W shows its maximum values either at $\omega_L - \omega_0 = v$ or at $\omega_L - \omega_0 = -v$. Since shot noise is not considered in Figure 4.1(a), W appears smooth and regular between the flipping events. In Figure 4.1(b), we have taken into account the effects of shot noise as described in Step 4 and plotted the random counts n as a function of ω_L and t. Compared to Figure 4.1(a), the spectral trail shown in Figure 4.1(b) appears more fuzzy and noisy due to the shot noise effect. It looks similar to the experimentally observed spectral trails (see, e.g., [12]).

In Figure 4.2, we show the evolution of the photon counting distribution $P_k(n)$, Eq. (4.18), at a fixed laser frequency, chosen here as $\omega_L - \omega_0 = -v = -5\Gamma$. The spectral diffusion process is identical to the one shown in Figure 4.1. Here, k denotes the measurement performed during the kth time bin as described in Step 3. Notice that two distinct forms of the photon counting distributions appear. During the dark period at the chosen frequency $\omega_L - \omega_0 = -v$ in Figure 4.1 (e.g., $2.1 \times 10^4 < \Gamma t < 3.5 \times 10^4$ corresponding to $21 < k < 35$), $P_k(n)$ reaches its maximum at $n = 0$ with $P_k(n = 0) \simeq 1$ and $P_k(n > 1) \ll 1$, meaning that the probability for a SM not to emit any photon during each time bin in the dark period is almost one. However, during the bright period (e.g., $3.5 \times 10^4 < \Gamma t < 4.6 \times 10^4$ corresponding to $35 < k < 46$), $P_k(n)$ shows a wide distribution with $\langle n \rangle_s \simeq 35$, meaning that on the average $\langle n \rangle_s \simeq 35$ number of photons are emitted per bin during this period. As the spectral diffusion proceeds, one can see the

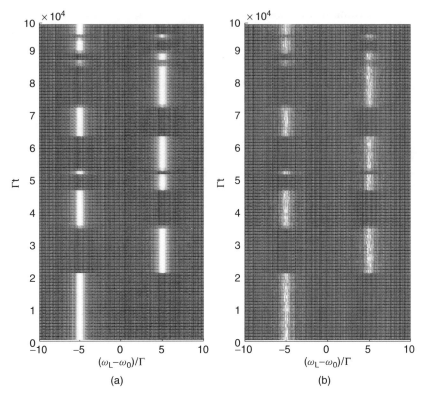

Figure 4.1. Spectral trails of a SM undergoing a very slow spectral diffusion process described by the two-state jump model are shown for (a) W (without the shot noise) and (b) n (with the shot noise). Parameters are chosen as $\Omega = 0.2\Gamma$, $R = 10^{-4}\Gamma$, $\nu = 5\Gamma$, $T = 10^3\Gamma^{-1}$, $t_{end} = 10^5\Gamma^{-1} = 100T$, and $\Gamma = 1$.

corresponding changes in the photon counting distribution, $P_k(n)$, typically among these two characteristic forms.

In Figure 4.3, we present a simulation result of a spectral trail for the case when the resonance frequency of a SM fluctuates very quickly compared with Γ and ν (i.e., $R \gg \nu \gg \Gamma$). In this case, since the frequency modulation is so fast compared with the spontaneous emission rate, a large number of frequency modulations are realized during the time $1/\Gamma$, and the frequency of the SM where the maximum photon counts are observed is dynamically averaged between $\omega_0 - \nu$ and $\omega_0 + \nu$ (i.e., a motional narrowing phenomenon) [60, 87]. The width of the spectral trail is $\sim \Gamma$, and no splitting is observed even though the frequency modulation ν is larger than the spontaneous decay rate Γ ($\nu = 5\Gamma$ in this case). This result is very different

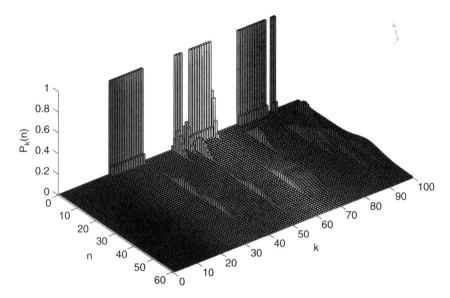

Figure 4.2. Time evolution of photon counting distributions $P_k(n)$ for the slow modulation case at $\omega_L - \omega_0 = -v$ in Figure 4.1 is shown. Other parameters are given in Figure 4.1.

from the slow modulation case shown in Figure 4.1, where two separate trails appear at $\omega_L = \omega_0 \pm v$.

In Figure 4.4, we also show the evolution of the photon counting distribution $P_k(n)$ during a spectral diffusion process at a fixed frequency $\omega_L = \omega_0$, where the line shape reaches its maximum in the fast fluctuation case shown in Figure 4.3. Unlike the slow modulation case in Figure 4.2, where one can see large fluctuations of the photon counting distributions, the fluctuations of the photon counting distributions are much smaller in the fast modulation case, and $P_k(n)$ shows a broad Gaussian-like behavior, centered at $\langle n \rangle_s \simeq 12$.

IV. Q AND THE THREE-TIME CORRELATION FUNCTION

Having observed the interplay between spectral diffusion and shot noise on the fluorescence spectra of a SM in the simple simulation results of Section III, it is natural to ask how one can analyze theoretically the photon counting statistics of a SM in the presence of a spectral diffusion process. The probability density of the number of photon counts $\langle p(n, T) \rangle$, or equivalently $P(W, T)$ in Eq. (4.24), would give complete information of the

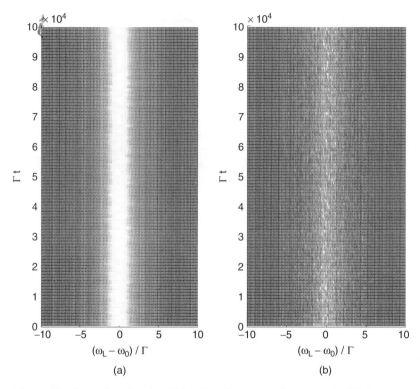

Figure 4.3. Spectral trails of a SM undergoing a very fast spectral diffusion process described by the two-state jump model are shown for (a) W (without the shot noise) and (b) n (with the shot noise). Parameters are chosen as $\Omega = 0.2\Gamma$, $R = 10\Gamma$, $v = 5\Gamma$, $T = 10^3\Gamma^{-1}$, $t_{end} = 10^5\Gamma^{-1} = 100T$, and $\Gamma = 1$.

dynamical processes of a SM undergoing a spectral diffusion process, but is difficult to calculate in general. In order to obtain dynamical information, we will consider the mean $\langle W \rangle$ and the second moment $\langle W^2 \rangle$ of the random photon counts.

It is easy to show that the average number of photons counted in time interval $(0, T)$ is given from Eq. (4.21),

$$\langle\langle n \rangle_s\rangle = \sum_{n=0}^{\infty} n \langle p(n, T) \rangle = \langle W \rangle \qquad (4.33)$$

and the second factorial moment of the photon counts in time interval $(0, T)$

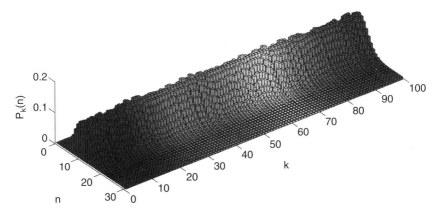

Figure 4.4. Time evolution of photon counting distributions $P_k(n)$ for the fast modulation case at $\omega_L = \omega_0$ in Figure 4.3 is shown. Other parameters are given in Figure 4.3.

is given by

$$\langle\langle n(n-1)\rangle_s\rangle = \sum_{n=0}^{\infty} n(n-1)\langle p(n,T)\rangle = \langle W^2 \rangle \quad (4.34)$$

The Mandel Q parameter is now introduced to characterize the fluctuations [68],

$$\frac{\langle\langle n^2\rangle_s\rangle - \langle\langle n\rangle_s\rangle^2}{\langle\langle n\rangle_s\rangle} = 1 + Q \quad (4.35)$$

and it is straightforward to show that [68]

$$Q = \frac{\langle W^2\rangle - \langle W\rangle^2}{\langle W\rangle} \quad (4.36)$$

This equation is important relating Q to the variance of the stochastic variable W. We see that $Q \geq 0$, indicating that photon statistics is super-Poissonian. For our classical case we anticipate:

1. For an ergodic system, when $T/\tau_c \to \infty$, and *if Eq. (4.23) is strictly valid*, $Q \to 0$ (i.e., Poissonian statistics). However, below we find an analytical expression for Q that is nonzero and in some cases large even in the limit of $T \to \infty$. We will discuss this subtle issue later.

2. In the opposite limit, $T \ll \tau_c$,

$$Q = \frac{\langle I_0^2 \rangle - \langle I_0 \rangle^2}{\langle I_0 \rangle} T \qquad (4.37)$$

3. If $I(t) = I$, independent of time, $Q = 0$, as expected.
4. It is easy to see that $Q \propto \eta$, hence when $\eta \to 0$, counts recorded in the measuring device tend to follow the Poissonian counting statistics.

Now, we consider the important limit of weak laser intensity. In this limit, the Wiener–Khintchine theorem relating the line shape to the one-time correlation function holds. As we shall show now, a three-time correlation function is the central ingredient of the theory of fluctuations of SMS in this limit. In Appendix B, we perform a straightforward perturbation expansion with respect to the Rabi frequency Ω in the Bloch equation, Eq. (4.6), to find

$$v(t) = \frac{\Omega}{2} \operatorname{Re} \left\{ \int_0^t dt_1 \exp\left[-i \int_{t_1}^t dt' \delta_L(t') - \frac{\Gamma}{2}(t - t_1) \right] \right\} \qquad (4.38)$$

According to the discussion in Section II, the random number of photons absorbed in time interval $(0, T)$ is determined by $W = \Omega \int_0^T v(t)\, dt$ [see Eqs. (4.12) and (4.19)], and from Eq. (4.38) we find

$$W = \frac{\Omega^2}{2} \operatorname{Re} \left[\int_0^T dt_2 \int_0^{t_2} dt_1 e^{-i\omega_L(t_2-t_1) - \Gamma(t_2-t_1)/2 + i \int_{t_1}^{t_2} dt' \Delta\omega(t')} \right] \qquad (4.39)$$

where we neglected terms of higher order than Ω^2. In standard line shape theories, Eq. (4.39) is averaged over the stationary stochastic process and the long time limit is taken, leading to the well-known result for the (unnormalized) line shape

$$\langle I(\omega_L) \rangle = \lim_{T \to \infty} \frac{\langle W \rangle}{T} = \frac{\Omega^2}{2} \operatorname{Re}\left[\int_0^\infty d\tau\, e^{-i\omega_L \tau - \Gamma \tau/2} C_1^{-1}(\tau) \right] \qquad (4.40)$$

where we have set $\omega_0 = 0$. The one-time correlation function $C_1^l(\tau)$ is defined by

$$C_1^l(\tau) = \langle e^{-il \int_0^\tau \Delta\omega(t')\, dt'} \rangle \qquad (4.41)$$

where $l = \pm 1$ and $\langle \cdots \rangle$ is an average over the stochastic trajectory $\Delta\omega$. Equation (4.40) is the celebrated Wiener–Khintchine formula relating the one-time correlation function to the average number of photon counts, i.e. the averaged line shape of a SM. We now investigate line shape fluctuation by considering the statistical properties of W.

By using Eq. (4.38), we show in Appendix B that

$$\langle W^2 \rangle = \frac{\Omega^4}{16} \int_0^T \int_0^T \int_0^T \int_0^T dt_1\, dt_2\, dt_3\, dt_4\, e^{-i\omega_L(t_2 - t_1 + t_3 - t_4) - \Gamma(|t_1 - t_2| + |t_3 - t_4|)/2}$$

$$\times \left\langle \exp\left[i \int_{t_1}^{t_2} \Delta\omega(t')\, dt' - i \int_{t_3}^{t_4} \Delta\omega(t')\, dt'\right] \right\rangle \quad (4.42)$$

As can be seen from Eq. (4.42), the key quantity of the theory of line shape fluctuation is the three-time correlation function,

$$C_3(\tau_1, \tau_2, \tau_3) = \left\langle \exp\left[i \int_{t_1}^{t_2} \Delta\omega(t')\, dt' - i \int_{t_3}^{t_4} \Delta\omega(t')\, dt'\right] \right\rangle \quad (4.43)$$

which depends on the time ordering of t_1, t_2, t_3, t_4. In Eq. (4.43), we have defined the time ordered set of $\{t_1, t_2, t_3, t_4\}$ as $\{t_I, t_{II}, t_{III}, t_{IV}\}$ such that $t_I < t_{II} < t_{III} < t_{IV}$, and $\tau_1 = t_{II} - t_I$, $\tau_2 = t_{III} - t_{II}$ and $\tau_3 = t_{IV} - t_{III}$. Due to the stationarity of the process, $C_3(\tau_1, \tau_2, \tau_3)$ does not depend on the time elapsing between start of observation $t = 0$ and t_I. It has a similar mathematical structure to that of the nonlinear response function used to describe four-wave mixing spectroscopies such as photon echo or hole burning [62].

In Eq. (4.43), there are $4! = 24$ options for the time ordering of (t_1, t_2, t_3, t_4); however, as we show below, only three of them (plus their complex conjugates) are needed. It is convenient to rewrite the three-time correlation function as a characteristic functional,

$$C_3^m(\tau_1, \tau_2, \tau_3) = \left\langle \exp\left[-i \int_{t_I}^{t_{IV}} S_m(t')\Delta\omega(t')\, dt'\right] \right\rangle \quad (4.44)$$

TABLE IV.1
Three-time correlation functions C_3^m

m	$S_m(t)$	$C_3^m(\tau_1, \tau_2, \tau_3)$	Time Ordering
1	1, 0, −1 (pulse shape); $t_I, t_{II}, t_{III}, t_{IV}$	$\frac{1}{2}\sum P_{ij}^{-1}(\tau_1)P_{jk}^{0}(\tau_2)P_{kl}^{1}(\tau_3)$	$t_1 < t_2 < t_3 < t_4$
2	1, 0, −1	$\frac{1}{2}\sum P_{ij}^{1}(\tau_1)P_{jk}^{0}(\tau_2)P_{kl}^{-1}(\tau_3)$	$t_2 < t_1 < t_4 < t_3$
3	1, 0	$\frac{1}{2}\sum P_{ij}^{1}(\tau_1)P_{jk}^{0}(\tau_2)P_{kl}^{1}(\tau_3)$	$t_2 < t_1 < t_3 < t_4$
4	0, −1	$\frac{1}{2}\sum P_{ij}^{-1}(\tau_1)P_{jk}^{0}(\tau_2)P_{kl}^{-1}(\tau_3)$	$t_1 < t_2 < t_4 < t_3$
5	2, 1	$\frac{1}{2}\sum P_{ij}^{1}(\tau_1)P_{jk}^{2}(\tau_2)P_{kl}^{1}(\tau_3)$	$t_2 < t_3 < t_1 < t_4$
6	−1, −2	$\frac{1}{2}\sum P_{ij}^{-1}(\tau_1)P_{jk}^{-2}(\tau_2)P_{kl}^{-1}(\tau_3)$	$t_1 < t_4 < t_2 < t_3$

^aThree-time correlation functions C_3^m are represented by six different pulse shape functions $S_m(t)$ for 24 time-ordering schemes. Values of $S_m(t)$ for each time interval are shown to the left of the pulse shape. Only one representative time ordering scheme for each class is shown in the third column. Three other time-ordering schemes belonging to the same class are obtained by exchanging $t_1 \leftrightarrow t_4$ and/or $t_2 \leftrightarrow t_3$. For example, three other time orderings belonging to $m = 1$ are $t_4 < t_2 < t_3 < t_1$, $t_1 < t_3 < t_2 < t_4$, and $t_4 < t_3 < t_2 < t_1$. In the second column, expressions of the three-time correlation functions C_3^m are shown in terms of the weight functions considered in Section V. Note that $C_3^{2n-1}(\tau_1, \tau_2, \tau_3)$ and $C_3^{2n}(\tau_1, \tau_2, \tau_3)$ ($n = 1, 2, 3$) are complex conjugates (C.C.) of each other.

where $S_m(t)$ ($m = 1, 2, \ldots, 6$) is defined in Table IV.1 as the pulse shape function corresponding to the mth time ordering. Let us consider as an example the case $m = 1$, $t_1 < t_2 < t_3 < t_4$ (for which $t_1 = t_I$, $t_2 = t_{II}, \ldots$, and $\tau_1 = t_2 - t_1$, $\tau_2 = t_3 - t_2$, and $\tau_3 = t_4 - t_3$). Then the pulse shape function is given by

$$S_1(t) = \begin{cases} -1 & t_I < t < t_{II} \\ 0 & t_{II} < t < t_{III} \\ 1 & t_{III} < t < t_{IV} \end{cases} \quad (4.45)$$

and the shape of this pulse is shown in the first line of Table IV.1. Similarly, other pulse shapes describe the other time orderings.

The four-dimensional (4D) integration in Eq. (4.42) is over 24 time orderings. We note, however, that

$$e^{-i\omega_L(t_2-t_1+t_3-t_4)-\Gamma(|t_1-t_2|+|t_3-t_4|)/2}\left\langle \exp\left[i\int_{t_1}^{t_2}dt'\Delta\omega(t') - i\int_{t_3}^{t_4}dt'\Delta\omega(t')\right]\right\rangle$$

in Eq. (4.42) has two important properties: (a) the expression is invariant when we replace t_1 with t_4 and t_2 with t_3, and (b) the replacement of t_1 with t_2 (or t_3) and of t_3 with t_4 (or t_1) has a meaning of taking the complex conjugate. Hence, it is easy to see that only three types of time orderings (plus their complex conjugates) must be considered. Each time ordering corresponds to a different pulse shape function, $S_m(t)$. The corresponding pulse shape functions are presented in Table IV.1 for all six time-ordering schemes. We also give expressions of $C_3^m(t)$ for the working example to be considered soon in Section V.

We note that if $\tau_1 = \tau_3$ the pulse in Eq. (4.45) is identical to that in the three-time photon echo experiments. The important relation between line shape fluctuations and nonlinear spectroscopy has been pointed out by Plakhotnik in [66] in the context of intensity–time–frequency–correlation measurement technique.

V. TWO-STATE JUMP MODEL: EXACT SOLUTION

A. Model

In order to investigate basic properties of line shape fluctuations, we consider a simple case as a working example. Note, however, that our results in Section IV have a more general validity. We assume $J = 1$ and $\Delta\omega_1(t) = vh(t)$, where v is the magnitude of the frequency shift and measures the interaction of the chromophore with the bath, and $h(t)$ describes a random telegraph process $h(t) = 1$ or $h(t) = -1$ depending on the bath state, up or down, respectively. For simplicity, the transition rate from up to down and vice versa are assumed to be R. The generalization to the case of different up and down transition rates is important, but are not considered here. A schematic representation of the spectral diffusion process is given in Figure 4.5.

The above model describes a single molecule coupled to a single two-level system in low temperature glasses as explained in Section VII. In this case, v depends on the distance between the SM and the two level system [88]. Another physical example of this model is the following: consider a chromophore that is attached to a macromolecule, and assume that conformational fluctuations exist between two conformations of the macro-

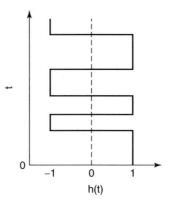

Figure 4.5. A schematic representation of the spectral diffusion process modeled by two-state jump process (random telegraph noise). The fluctuating transition frequency is given by $\Delta\omega(t) = vh(t)$.

molecule. Depending on the conformation of the macromolecule, the transition frequency of the chromophore is either $\omega_0 - v$ or $\omega_0 + v$ [89].

B. Solution

By using a method of Suárez and Silbey [90], developed in the context of photon echo experiments, we now analyze the properties of the three-time correlation function. We first define the weight functions,

$$P^a_{if}(t) = \left\langle \exp\left[-ia \int_0^t \Delta\omega(t') \, dt'\right] \right\rangle_{if} \quad (4.46)$$

where the initial (final) state of the stochastic process $\Delta\omega$ is $i(f)$ and $a = 0$ or $a = \pm 1$ or $a = \pm 2$. For example, $P^{-1}_{++}(t)$ is the value of $\langle e^{i\int_0^t \Delta\omega(t') \, dt'} \rangle$ for a path restricted to have $\Delta\omega(0) = v$ and $\Delta\omega(t) = v$. The one-time correlation function $C^l_1(\tau)$ defined in Eq. (4.41) can be written as the sum of these weights,

$$C^l_1(\tau) = \frac{1}{2} \sum_{i,j} P^l_{ij}(\tau) \quad (4.47)$$

where a prefactor $\frac{1}{2}$ is due to the symmetric initial condition and the summation is over all the possible paths during time τ (i.e., $i = \pm, j = \pm$). Also, by using the Markovian property of the process, we can express all the C^m_3 functions in terms of the weights. For example, for the pulse shape in Eq. (4.45)

$$C_3^1(\tau_1, \tau_2, \tau_3) = \frac{1}{2} \sum_{i,j,k,l} P_{ij}^{-1}(\tau_1) P_{jk}^0(\tau_2) P_{kl}^1(\tau_3) \tag{4.48}$$

where the summations are over all possible values of $i = \pm$, $j = \pm$, $k = \pm$, and $l = \pm$. The other C_3^m functions are expressed in terms of weights in Table IV.1. Explicit expressions of the weights for the working model are given by

$$P_{if}^0(t) = (\tfrac{1}{2})[1 + (-1)^{\delta_{if}+1} \exp(-2Rt)]$$

$$P_{+-}^1(t) = P_{-+}^1(t) = R \exp(-Rt) \sin(Y_1 t)/Y_1$$

$$P_{\pm\pm}^1(t) = \exp(-Rt)\left[\cos(Y_1 t) \mp iv \frac{\sin(Y_1 t)}{Y_1}\right] \tag{4.49}$$

$P_{if}^{-1}(t) = \text{C.C. } [P_{if}^1(t)]$ and $Y_1 = \sqrt{v^2 - R^2}$. $P_{if}^{\pm 2}(t)$ is given by the same expressions as the corresponding $P_{if}^{\pm 1}(t)$ with v replaced by $2v$.

Now, we can evaluate $\langle W \rangle$ and $\langle W^2 \rangle$ explicitly. First, we consider $\langle W^2 \rangle$ and, in particular, focus on the case $m = 1$, $0 < t_1 < t_2 < t_3 < t_4 < T$. By using Table IV.1, the contribution of $\langle W^2 \rangle$ to Eq. (4.42) is

$$\langle W^2 \rangle_{1234} = \frac{\Omega^4}{16} \int_0^T dt_4 \int_0^{t_4} dt_3 \int_0^{t_3} dt_2 \int_0^{t_2} dt_1 e^{-i\omega_L(t_2-t_1+t_3-t_4) - \Gamma(|t_1-t_2|+|t_3-t_4|)/2}$$

$$\times \frac{1}{2} \sum_{i,j,k,l} P_{ij}^{-1}(t_2 - t_1) P_{jk}^0(t_3 - t_2) P_{kl}^1(t_4 - t_3) \tag{4.50}$$

We use the convolution theorem of Laplace transform four times and find

$$\langle W^2 \rangle_{1234} = \frac{\Omega^4}{32} \mathscr{L}^{-1} \left\{ \frac{1}{s^2} \sum_{i,j,k,l} \hat{P}_{ij}^{-1}(s + \Gamma/2 + i\omega_L) \hat{P}_{jk}^0(s) \hat{P}_{kl}^1(s + \Gamma/2 - i\omega_L) \right\} \tag{4.51}$$

where $\mathscr{L}^{-1}$ denotes the inverse Laplace transform, where the Laplace $T \to s$ transform is defined by

$$\hat{f}(s) = \int_0^\infty f(T) e^{-sT} dT \tag{4.52}$$

and the Laplace transforms of the functions $\hat{P}_{ij}^a(s)$ are listed in Eq. (A.33).

By considering the other 23 time orderings, we find

$$\langle W^2 \rangle = \frac{\Omega^4}{16} \mathscr{L}^{-1} \left\{ \frac{1}{s^2} \sum_{i,j,k,l} [\hat{P}_{ij}^{-1}(s+\Gamma/2+i\omega_L)\hat{P}_{jk}^{0}(s)\hat{P}_{kl}^{+1}(s+\Gamma/2-i\omega_L) \right.$$
$$+ \hat{P}_{ij}^{-1}(s + \Gamma/2 + i\omega_L)\hat{P}_{jk}^{0}(s + \Gamma)\hat{P}_{kl}^{+1}(s + \Gamma/2 - i\omega_L)$$
$$+ \hat{P}_{ij}^{+1}(s + \Gamma/2 - i\omega_L)\hat{P}_{jk}^{0}(s)\hat{P}_{kl}^{+1}(s + \Gamma/2 - i\omega_L)$$
$$+ \hat{P}_{ij}^{+1}(s + \Gamma/2 - i\omega_L)\hat{P}_{jk}^{0}(s + \Gamma)\hat{P}_{kl}^{+1}(s + \Gamma/2 - i\omega_L)$$
$$\left. + 2\hat{P}_{ij}^{+1}(s+\Gamma/2-i\omega_L)\hat{P}_{jk}^{+2}(s+\Gamma-2i\omega_L)\hat{P}_{kl}^{+1}(s+\Gamma/2-i\omega_L) + \text{C.C.}] \right\}. $$

(4.53)

Eq. (4.53), which can be used to describe the line shape fluctuations, is our main result so far. In Appendix C, we invert this equation from the Laplace s domain to the time T domain using straightforward complex analysis. Our goal is to investigate Mandel's Q parameter, Eq. (4.36); it is calculated using Eq. (4.53) and

$$\langle W \rangle = \frac{\Omega^2}{8} \mathscr{L}^{-1} \left\{ \frac{1}{s^2} \sum_{i,j} \hat{P}_{ij}^{-}(s + \Gamma/2 + i\omega_L) + \text{C.C.} \right\} \quad (4.54)$$

which is also evaluated in Appendix C. As mentioned, Eq. (4.54) is the celebrated Wiener–Khintchine formula for the line shape (in the limit of $T \to \infty$) while Eq. (4.53) describes the fluctuations of the line shape within linear response theory. Note that $\langle W^2 \rangle$ and $\langle W \rangle$ in Eqs. (4.53) and (4.54) are time dependent, and these time dependences are of interest only when the dynamics of the environment is slow (see more details below). Exact time-dependent results of $\langle W^2 \rangle$ and $\langle W \rangle$, and thus Q, are obtained in Appendix C. The limit $T \to \infty$ has, of course, special interest since it is used in standard line shape theories, and does not depend on an assumption of whether the frequency modulations are slow. The exact expression for Q in the limit of $T \to \infty$ is given in Eqs. (A.47) and (A.48). These equations are one of the main results of this chapter. It turns out that Q is not a simple function of the model parameters; however, as we show below, in certain limits, simple behaviors are found.

VI. ANALYSIS OF THE EXACT SOLUTION: LIMITING CASES

In this section, we investigate the behavior of $Q(\omega_L, T)$ for several physically important cases. In the two-state model considered in Section V, in addition to two control parameters ω_L and T, we have three model parameters that depend on the chromophore and the bath: Γ, R, and v. Depending on their relative magnitudes we consider six different limiting cases:

1. $R \ll \Gamma \ll v$

2. $R \ll v \ll \Gamma$

3. $\Gamma \ll v \ll R$

4. $v \ll \Gamma \ll R$

5. $v \ll R \ll \Gamma$

6. $\Gamma \ll R \ll v$

We discuss these various limits in this section.

A. Slow Modulation Regime: $R \ll v, \Gamma$

First, we consider the slow modulation regime, $R \ll v, \Gamma$, where the bath fluctuation process is very slow compared with the radiative decay rate and the frequency fluctuation amplitude. In this case, the foregoing Eqs. (4.53) and (4.54) can be simplified. This case is similar to situations in several SM experiments in condensed phases [6–9, 12, 17].

Within the slow modulation regime, we can have two distinct behaviors of the line shape depending on the magnitude of the frequency modulation, v. When the frequency modulation is slow but strong such that $R \ll \Gamma \ll v$ [case (1)], the line shape exhibits a splitting with the two peaks centered at $\omega_L = \pm v$ [see Fig. 4.6(a)]. On the other hand, when the frequency modulation is slow and weak, $R \ll v \ll \Gamma$ [case (2)] a single peak centered at $\omega_L = 0$ appears in the line shape [see Fig. 4.7(a)]. From now on, we will term the case $\Gamma \ll v$ as the strong modulation limit and the other case $\Gamma \gg v$ as the weak modulation limit. The same distinction can also be applied to the fast modulation regime considered later.

In the slow modulation regime, we can find $Q(T)$ using random walk theory and compare it with the exact result obtained in Appendix C. In this regime, the molecule can be found either in the up $(+)$ or in the down $(-)$ state, if the transition times [i.e., typically $\mathcal{O}(1/R)$] between these two states are long; the rate of photon emission in these two states is determined by the *stationary* solution of the time-independent Bloch equation in the limit

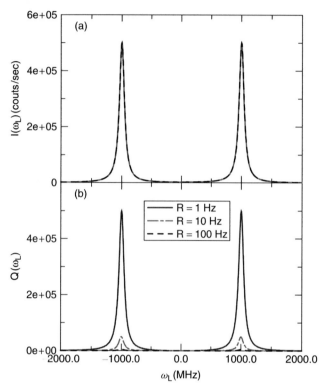

Figure 4.6. Case 1 ($R \ll \Gamma \ll v$) in the steady-state limit. Exact results for line shape, Eq. (A.3), and for Q, Eqs. (A.47), (A.48), are shown for the strong, slow modulation case as functions of ω_L. Parameters are chosen as $v = 1$ GHz, $\Gamma = 100$ MHz, $\Omega = \Gamma/10$, $R = 1$–100 Hz, and $T \to \infty$. They mimic a SM coupled to a single slow two level system in a glass. Notice that line shape does not change with R while a does.

of weak laser intensity [73] [see also Eqs. (A.4) and (A.5)]

$$I_{\pm}(\omega_L) = \frac{\Omega^2 \Gamma}{4\left[(\omega_L \mp v)^2 + \frac{\Gamma^2}{4}\right]} \qquad (4.55)$$

Now the stochastic variable $W = \int_0^T I(t)\, dt$ must be considered, where $I(t)$ follows a two state process, $I(t) = I_+$ or $I(t) = I_-$ with transitions $+ \to -$ and $- \to +$ described by the rate R. One can map this problem onto a simple two-state random walk problem [91], where a particle moves with a "velocity" either I_+ or I_- and the "coordinate" of the particle is W. Then, from the random walk theory, it is easy to see that for long

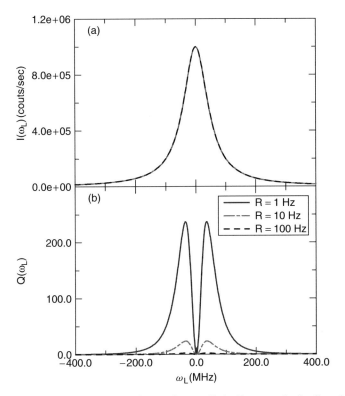

Figure 4.7. Case 2 ($R \ll v \ll \Gamma$) in the steady-state limit. Exact results for line shape, Eq. (A.3), and for Q, Eqs. (A.47), (A.48), are shown for the weak, slow modulation case as functions of ω_L. Parameters are chosen as $v = 1$ MHz, $\Gamma = 100$ MHz, $\Omega = \Gamma/10$, $R = 1$ Hz–100 Hz, and $T \to \infty$.

times ($T \gg 1/R$),

$$\langle W \rangle \simeq \left(\frac{I_+ + I_-}{2}\right) T = \langle I \rangle T \qquad (4.56)$$

meaning that the line is composed of two Lorentzians centered at $\pm v$ with a width determined by the lifetime of the molecule,

$$\langle I \rangle = \frac{1}{2}(I_+ + I_-) \qquad (4.57)$$

and the "mean square displacement" is given by

$$\langle W^2 \rangle - \langle W \rangle^2 \simeq \frac{(I_+ - I_-)^2}{4R} T \qquad (4.58)$$

After straightforward algebra Eqs. (4.56), (4.58), and (4.36) yield

$$Q = \frac{\Omega^2 \Gamma v^2 \omega_L^2}{R\left(\omega_L^2 + v^2 + \frac{\Gamma^2}{4}\right)\left((\omega_L - v)^2 + \frac{\Gamma^2}{4}\right)\left((\omega_L + v)^2 + \frac{\Gamma^2}{4}\right)} \quad (4.59)$$

in the limit $T \gg 1/R$. The full-time dependent behaviors of $\langle W \rangle$ and $\langle W^2 \rangle$ are calculated in Appendix E using the two-state random walk model [91]. From Eq. (A.55), we have Q as a function of the measurement time in the slow modulation regime

$$Q(T) \simeq \frac{\langle (\Delta I)^2 \rangle}{\langle I \rangle R}\left(1 + \frac{e^{-2RT} - 1}{2RT}\right) \quad (4.60)$$

where the "variance" of the line shape is defined by

$$\langle (\Delta I)^2 \rangle = \frac{1}{2}[(I_+ - \langle I \rangle)^2 + (I_- - \langle I \rangle)^2] \quad (4.61)$$

Eq. (4.60) shows that Q is factorized as a product of frequency-dependent part and the time-dependent part in the slow modulation regime. It is important to note that Eq. (4.60) could also have been derived from the exact result presented in Appendix C by considering the slow modulation conditions. Briefly, from the exact expressions of $\langle W \rangle$ and $\langle W^2 \rangle$ in the Laplace domain given in Appendix C, by keeping only the pole s that satisfies $|\text{Re}(s)| \sim \mathcal{O}(\mathcal{R})$ and neglecting other poles such that $|\text{Re}(s)| \gg R$, we can recover the result given in Eq. (4.60), thus confirming the validity of the random walk model in the slow modulation regime. In the limit of short and long times, we have

$$Q \simeq \begin{cases} \dfrac{\langle (\Delta I)^2 \rangle}{\langle I \rangle} R^{-1} & T \gg 1/R \\ \dfrac{\langle (\Delta I)^2 \rangle}{\langle I \rangle} T & T \ll 1/R \end{cases} \quad (4.62)$$

We recover Eq. (4.59) from Eq. (4.62) in the limit of $T \to \infty$. Equation (4.62) for $T \ll 1/R$ is a special case of Eq. (4.37). Note that the results in this subsection can be easily generalized to the case of strong external fields using Eqs. (A.4) and (A.5).

1. Strong Modulation: Case 1

When $\Gamma \ll v$, case 1, the two intensities $I_{\pm}$ in Eq. (4.55) are well separated by the amplitude of the frequency modulation, $2v$, and therefore we can approximate $\langle I \rangle \simeq I_+/2$ when $I_- \simeq 0$ and vice versa, which leads to

$$\langle (\Delta I)^2 \rangle \simeq \langle I \rangle^2 \qquad \Gamma \ll v \qquad (4.63)$$

In this case, Q is given by

$$Q \simeq \begin{cases} \langle I \rangle R^{-1} & T \gg 1/R \\ \langle I \rangle T & T \ll 1/R \end{cases} \qquad (4.64)$$

We show the line shapes and Q as functions of the laser frequency ω_L at the steady-state limit ($T \to \infty$) for case 1 in Figure 4.6. In all the calculations shown in the figures in this chapter, we have assumed an ideal measurement, $\eta = 1$. Values of parameters are given in the figure caption, which is relevant for the case that a chromophore is strongly coupled to a single two level system in a low temperature glass [8, 9]. Since $Q \sim \langle I \rangle / R$ in this case, both the line shape [Fig. 4.6(a)] and Q [Fig. 4.6(b)] are similar to each other; two Lorentzian peaks located at $\omega_L = \pm v$ with widths Γ. Note that in this limit the value of Q is very large compared with that in the fast modulation regime considered later. While the line shape is independent of R in this limit, the magnitude of Q decreases with R, hence it is Q not $\langle I \rangle$, which yields information on the dynamics of the environment.

2. Weak Modulation: Case 2

On the other hand, in case 2, where the fluctuation is very weak ($v \ll \Gamma$), we notice that $I_+ \simeq I_-$ from Eq. (4.55) (note that when there is no spectral diffusion, $v = 0$, $I_+ = I_-$, thus $Q = 0$). The line shape is given by a single Lorentzian centered at $\omega_L = 0$,

$$\langle I \rangle \simeq \frac{\Omega^2 \Gamma/4}{\omega_L^2 + \Gamma^2/4} \qquad (4.65)$$

Since $v \ll \Gamma$ in this case we expand $I_{\pm}$ in terms of v to find

$$\langle (\Delta I)^2 \rangle \simeq v^2 \left(\frac{d\langle I \rangle}{d\omega_L} \right)^2 = \frac{v^2 \Omega^4 \Gamma^2 \omega_L^2}{4(\omega_L^2 + \Gamma^2/4)^4} \qquad v \ll \Gamma \qquad (4.66)$$

Therefore, in case 2, Q is given by

$$Q \simeq \begin{cases} \dfrac{v^2}{\langle I \rangle}\left(\dfrac{d\langle I \rangle}{d\omega_L}\right)^2 R^{-1} & T \gg 1/R \\ \dfrac{v^2}{\langle I \rangle}\left(\dfrac{d\langle I \rangle}{d\omega_L}\right)^2 T & T \ll 1/R \end{cases} \qquad (4.67)$$

Note that $Q \propto (1/\langle I \rangle)(d\langle I \rangle/d\omega_L)^2$ in the weak, slow modulation case while $Q \propto \langle I \rangle$ in the strong, slow modulation case.

In Figure 4.7, we show the line shape and Q for the weak, slow modulation limit, case 2. The line shape [Fig. 4.7(a)] is a Lorentzian with a width Γ, to a good approximation, thus the features of the line shape do not depend on the properties of the coupling of the SM to an environment such as v and R. On the other hand, Q [Fig. 4.7(b)], shows a richer behavior. Recalling Eq. (4.67) for $T \gg 1/R$ in Eq. (4.66), it exhibits doublet peaks separated by $\sim \mathcal{O}(\Gamma)$ with the dip located at $\omega_L = 0$, and its magnitude is proportional to $1/R$. We will later show that this kind of a doublet and a dip in Q is a generic feature of the weak coupling case, found not only in the slow but also in the fast modulation case considered in Section VI.A.3.

Both in the strong (Fig. 4.6) and the weak (Fig. 4.7) cases, we find $Q(\omega_L = 0) \simeq 0$, which is expected in the slow modulation case considered here. Physically, when the laser detuning frequency is exactly in the middle of two frequency shifts, $\pm v$, the rate of photon emissions is identical whether the molecule is in the up state $(+v)$ or in the down state $(-v)$. Therefore, the effect of bath fluctuation on the photon counting statistics is negligible, which leads to Poissonian counting statistics at $\omega_L = 0$.

3. Time Dependence

Additional information on the environmental fluctuations can be gained by measuring the time dependence of Q in the slow modulation regime. In Figure 4.8, we show Q versus the measurement time T for the strong, slow modulation limit, case 1, both for the exact result calculated in Appendix C and for the approximate result, Eq. (4.60). We choose the resonance condition $\omega_L = v$, and used parameters relevant to SMS in glass systems. The approximate result based on the two-state random walk model, Eq. (4.60) shows a perfect agreement with the exact result in Figure 4.8(a) as expected. When $T \ll 1/R$, Q increases linearly with T as predicted from Eq. (4.62) for $RT \ll 1$, and it reaches the steady-state value given by Eq. (4.62) when T becomes $RT \gg 1$. We also notice that even in the long measurement time limit the value of Q is large: $Q = 5 \times 10^5 \eta$ in the example given in

A STOCHASTIC THEORY OF SINGLE MOLECULE SPECTROSCOPY 229

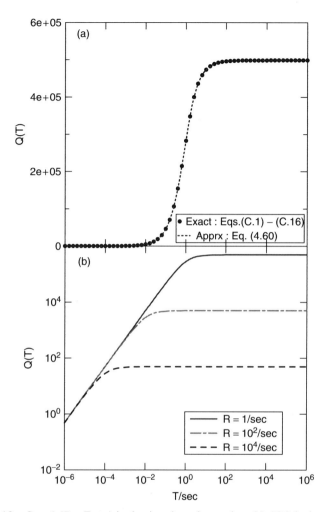

Figure 4.8. Case 1 ($R \ll \Gamma \ll v$) in the time-dependent regime. (a) $Q(T)$ is shown both for the exact result and for the approximate result given in the linear-log scale. This figure demonstrates that our exact results given in Appendix C are well reproduced by the approximation, Eq. (4.60). The same parameters are used as in Figure 4.6 except $\omega_L = v = 1$ GHz, $R = 1$ Hz, and T (varied). (b) $Q(T)$ for different values of R in the slow modulation regime in the log-log scale. All the parameters except for R are chosen the same as those in (a).

Figure 4.8(a) (including the detection efficiency). Therefore, even if we consider the imperfect detection of the photon counting device (e.g., $\eta = 15\%$ has been reported recently [92]), deviation of the photon statistics from Poissonian is observable in the slow modulation regime of SMS. This is seemingly contrary to propositions made in the literature [51, 68, 84], in which the claim is made that the Poissonian distribution is achieved in the long time limit. We defer discussion of this issue to the end of this section. We note that it has been shown that $Q(T)$ can be very large ($Q \sim 10^4$) at long times in the atomic three-level system with a metastable state but without a spectral diffusion process [78]. Figure 8(b) illustrates that the steady-state value of $Q(T)$ is reached when $RT \gg 1$, and the magnitude of Q in the steady state decreases as $1/R$ as predicted from Eq. (4.62) for $RT \gg 1$. This therefore illustrates that valuable information on the bath fluctuation timescales can be obtained by measuring the fluctuation of the line shape as a function of measurement time.

Figure 4.9 shows a two-dimensional plot of Q as a function of the laser frequency and the measurement time for the parameters chosen in Figure 4.8(a). We observe that two Lorentzian peaks located at resonance frequencies become noticeable when $T \simeq 1/R$. Similar time-dependent behavior can be also found for the weak, slow modulation limit, case 2, however, then Q along the ω_L axis shows doublet peaks separated by $\sim \mathcal{O}(\Gamma)$ as shown in Figure 4.7.

We can extend our result and describe photon statistics beyond the second moment. Based on the central limit theorem, the probability density

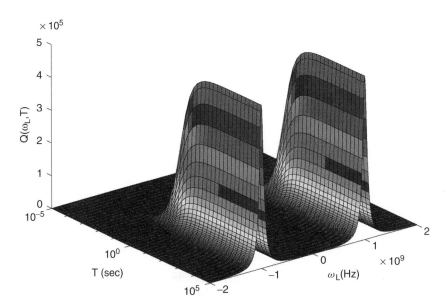

Figure 4.9. Case 1 ($R \ll \Gamma \ll v$) in the time dependent case. Two-dimensional plot of $Q(\omega_L, T)$ for the slow modulation as a function of ω_L and T. The same parameters are used as in Figure 4.8(a) except ω_L (varied).

function of the two-state random walk variable W in the long time limit is described by

$$P(W, T) \xrightarrow{RT \gg 1} G(W, T) \equiv \frac{1}{\sqrt{4\pi DT}} \exp\left[-\frac{(W - VT)^2}{4DT}\right] \quad (4.68)$$

with $D = \langle(\Delta I)^2\rangle/(2R)$ and $V = \langle I \rangle$ when $\eta = 1$. We also note that $Q \simeq 2D/V$ in the long time limit. By using Eq. (4.24),

$$\langle p(n, T) \rangle \simeq \int_0^\infty dW\, G(W, T)\, \frac{W^n}{n!} \exp(-W) \quad (4.69)$$

Equation (4.69) shows that in the long time limit the photon statistics is the Poisson transform [67] of a Gaussian. This transformation can be found explicitly [one can slightly improve this approximation by considering a normalized truncated Gaussian with $G(W, T) = 0$ for $W < 0$; we expect that our approximation will work well when $V^2 T \gg D$]. In contrast to Eq. (4.68), the proposed Eq. (4.25) suggests that $G(W, T)$ be replaced with a delta function, for which a single parameter V controls the photon statistics, while according to our approach both V and D, or equivalently $\langle W \rangle$ and Q, are important. Mathematically, when $T \to \infty$, the Gaussian distribution Eq. (4.68) may be said to "converge" to the delta function distribution, Eq. (4.25), in a sense that the mean is $\langle W \rangle \sim \mathcal{O}(T)$ while the standard deviation is $\Delta W = (\langle W^2 \rangle - \langle W \rangle^2)^{1/2} \sim \mathcal{O}(T^{1/2})$. This argument corresponds to the proposition made in the literature, Eq. (4.25), and then the photon statistics is only determined by the mean $\langle W \rangle \simeq VT$. However, physically it is more informative and meaningful to consider not only the mean $\langle W \rangle$ but also the variance $(\Delta W)^2 \simeq 2DT \simeq Q\langle W \rangle$ since it contains information on the bath fluctuation. In a sense, the delta function approximation might be misleading since it implies that $Q = 0$ in the long time limit, which is clearly *not* true in general.

B. Fast Modulation Regime: $v, \Gamma \ll R$

In this section, we consider the fast modulation regime, $v, \Gamma \ll R$. Usually, in this fast modulation regime the dynamics of the bath (here modeled with R) is so fast that only the long time limit of our solution should be considered [i.e., Eqs. (A.47) and (A.48)]. Hence, all our results below are derived in the limit of $T \to \infty$, since the time dependence of Q is irrelevant. The fast modulation regime considered in this section includes case 3 ($\Gamma \ll v \ll R$) as the strong, fast modulation case and case 4 ($v \ll \Gamma \ll R$) as the weak, fast modulation case.

It is well known that the line shape is Lorentzian [62] in the fast modulation regime (soon to be defined precisely),

$$\langle I(\omega_L) \rangle = \frac{\Omega^2(\Gamma + \Gamma_f)}{4\omega_L^2 + (\Gamma + \Gamma_f)^2} \tag{4.70}$$

where Γ_f is the line width due to the dephasing induced by the bath *fluctuations* and given by

$$\Gamma_f = \frac{v^2}{R} \tag{4.71}$$

The line shape given in Eq. (4.70) exhibits motional narrowing, namely, as R is increased the line becomes narrower and in the limit $R \to \infty$ the width of the line is simply Γ.

Now, we define the fast modulation regime considered in this work. When $R \to \infty$ (and other parameters fixed) $Q \to 0$, so fluctuations become Poissonian, which is expected and in a sense trivial because the molecule cannot respond to the very fast bath, so the two limits $R \to \infty$ and $v = 0$ (i.e., no interaction with the bath) are equivalent. It is physically more interesting to consider the case that $R \to \infty$ with Γ_f/Γ remaining finite, which is the standard definition of the fast modulation regime [60]. In this fast modulation regime, the well-known line shape is Lorentzian as given in Eq. (4.70) with a width $\Gamma + \Gamma_f$.

Here, we present a simple calculation of Q in the fast modulation regime based on physically motivated approximations. We justify our approximations by comparing the resulting expression with our exact result, Eqs. (A.47) and (A.48). When the dynamics of the bath is very fast, the correlation between the state of the molecule during one pulse interval with that of the following pulse interval is not significant in the stochastic averaging in Eq. (4.44). Therefore in the fast modulation regime, we can approximately factorize the three-time correlation function $C_3^m(\tau_1, \tau_2, \tau_3)$ in Eq. (4.44) in terms of one-time correlation functions $C_1^l(\tau)$ as was done by Mukamel and Loring [93] in the context of four-wave mixing spectroscopy,

$$\begin{aligned} C_3^m(\tau_1, \tau_2, \tau_3) &= \langle e^{-ia\int_{t_I}^{t_{II}} \Delta\omega(t')\,dt' - ib\int_{t_{II}}^{t_{III}} \Delta\omega(t')\,dt' - ic\int_{t_{III}}^{t_{IV}} \Delta\omega(t')\,dt'} \rangle \\ &\simeq \langle e^{-ia\int_{t_I}^{t_{II}} \Delta\omega(t')\,dt'} \rangle \langle e^{-ib\int_{t_{II}}^{t_{III}} \Delta\omega(t')\,dt'} \rangle \langle e^{-ic\int_{t_{III}}^{t_{IV}} \Delta\omega(t')\,dt'} \rangle \\ &= C_1^a(\tau_1) C_1^b(\tau_2) C_1^c(\tau_3) \end{aligned} \tag{4.72}$$

where a, b, and c are the values of $S_m(t)$ in time intervals, $t_I < t < t_{II}$,

$t_{II} < t < t_{III}$, and $t_{III} < t < t_{IV}$, respectively. For example, $a = -1$, $b = 0$, and $c = 1$ for the case $m = 1$ (see Table IV.1). The one-time correlation function is evaluated using the second-order cumulant approximation, which is also valid for the fast modulation regime [93],

$$C_1^l(t) \simeq \exp\left(-l^2 \int_0^t dt' \int_0^{t'} dt'' \langle \Delta\omega(t')\Delta\omega(t'') \rangle\right) \qquad (4.73)$$

Note that these two approximations, Eqs. (4.72) and (4.73), are not limited to the two-state jump model but are generally valid in the fast modulation regime. The frequency correlation function is given by

$$\langle \Delta\omega(t')\Delta\omega(t'') \rangle = v^2 \exp(-2R|t' - t''|) \qquad (4.74)$$

for the two-state jump model. By substituting Eq. (4.74) into Eq. (4.73), we have

$$C_1^l(t) \simeq \exp\left[-l^2 v^2 \left(\frac{t}{2R} + \frac{e^{-2Rt} - 1}{(2R)^2}\right)\right] \qquad (4.75)$$

In the limit of $R \to \infty$, we can approximate $C_1^l(t)$ further

$$C_1^l(t) \simeq \exp(-l^2 \Gamma_f t/2) \qquad (4.76)$$

and in turn evaluate all the three-time correlation functions corresponding to different time orderings within these approximations using Eqs. (4.72) and (4.76). Note that the line shape shows a Lorentzian behavior with the width $\Gamma + \Gamma_f$ in Eq. (4.70) because of the exponential decay of the one-time correlation function in Eq. (4.76). Since we have simple analytical expressions of the three-time correlation functions, it is straightforward to calculate Q in the fast modulation regime. After straightforward algebra described in Appendix F, Q in the fast modulation regime is

$$Q = \frac{\Omega^2 \Gamma_f [8\Gamma_f \omega_L^4 + 2(\Gamma + \Gamma_f)(8\Gamma^2 + 15\Gamma\Gamma_f + 5\Gamma_f^2)\omega_L^2 + \Gamma_f(\Gamma + 2\Gamma_f)(\Gamma + \Gamma_f)^3]}{16\Gamma(\Gamma + \Gamma_f)\left(\omega_L^2 + \frac{(\Gamma + \Gamma_f)^2}{4}\right)^2 \left(\omega_L^2 + \frac{(\Gamma + 2\Gamma_f)^2}{4}\right)}$$

(4.77)

Note that the exact expression for Q obtained in Appendix D yields the same expression as Eq. (4.77) when $R \to \infty$ but Γ_f remains finite, which justifies the approximations introduced.

Let us now estimate the magnitude of these fluctuation. We consider $\omega_L = 0$ since the photon current is strongest for this case (i.e., the line shape has a maximum at $\omega_L = 0$), then

$$Q = \frac{4\eta\Omega^2\Gamma_f^2}{\Gamma(\Gamma + \Gamma_f)^2(\Gamma + 2\Gamma_f)} \quad (4.78)$$

where the detection efficiency η has been restored. The maximum of Q is found when $\Gamma_f = \Gamma(1 + \sqrt{5})/2$ and then $Q \simeq 0.361\eta\Omega^2/\Gamma^2$. Even if we take $|\Omega|/\Gamma = 1/10$ and $\eta = 5 \times 10^{-2}$ as reasonable estimates for a weak laser field and detection efficiency we find $Q \simeq 2 \times 10^{-4} \ll 1$, which shows the difficulties of measuring deviations from Poissonian statistics in this limit. We note, however, that values of Q as small as $|Q| \leq 10^{-4}$ have been measured recently (although in the short time regime, not in the steady-state case) [53]. Therefore it might also be possible to observe the deviation from Poissonian photon statistics in the fast modulation limit under appropriate experimental situations.

In Figure 4.10, we show the results of exact steady-state calculations for the line shape and Q [Eqs. (A.47) and (A.48)] for different values of the fluctuation rate R in the fast modulation regime. We have chosen the parameters as $v = 100$ MHz, $\Gamma = 1$ MHz, and $\Omega = \Gamma/10$, and R is varied from 1–100 GHz, corresponding to case 3. For this parameter set, we have checked that the fast modulation approximation for Q given in Eq. (4.77) agrees well with the exact calculation, Eqs. (A.47) and (A.48). The line shape shown in Figure 4.10(a) shows the well-known motional narrowing behavior. When $R = 1$ GHz, the line shape is a broad Lorentzian with the width $\Gamma + \Gamma_f \simeq \Gamma_f$ (note that $\Gamma_f \gg \Gamma$ in this case). As R is increased further the line becomes narrowed and finally its width is given by Γ as $R \to \infty$.

Compared with the line shape, Q in Figure 4.10(b) shows richer behavior. The most obvious feature is that when $R \to \infty$, $Q \to 0$, which is expected since when the bath is very fast the molecule cannot respond to it, hence fluctuations are Poissonian and $Q \to 0$. It is noticeable that, unlike $I(\omega_L)$, Q shows a type of narrowing behavior *with splitting* as R is increased. The line shape remains Lorentzian regardless of Γ_f in the fast modulation case, while Q changes from a broad Lorentzian line with the width Γ_f when $\Gamma_f \gg \Gamma$ to doublet peaks separated approximately by Γ when $\Gamma_f \ll \Gamma$, which will be analyzed in the following [see, e.g., Eq. (4.79)]. Therefore, although the value of Q is small in the fast modulation regime, it yields additional information on the relative contributions of Γ and Γ_f, which are not differentiated in the line shape measurement.

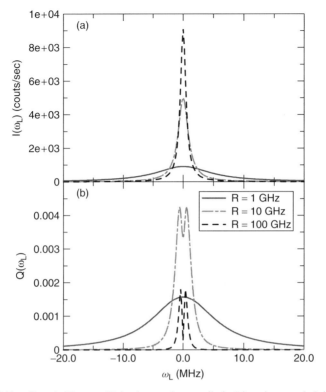

Figure 4.10. Case 3 ($\Gamma \ll v \ll R$) in the steady-state limit. Line shape and Q in the fast modulation regime are shown as functions of ω_L. Parameters are chosen as $v = 100$ MHz, $\Gamma = 1$ MHz, $\Omega = \Gamma/10$, $R = 1\text{--}100$ GHz, and $T \to \infty$.

1. Strong Modulation: Case 3

We further analyze the case that the bath fluctuation is both strong and fast, case 3, $\Gamma \ll v \ll R$. The results shown in Figure 4.10 correspond to this case. We find that Eq. (4.77) is further simplified in two different limits, $\Gamma_f \gg \Gamma$ and $\Gamma_f \ll \Gamma$,

$$Q \simeq \begin{cases} \dfrac{\Omega^2 \Gamma_f}{2\Gamma(\omega_L^2 + \Gamma_f^2/4)} & \Gamma_f \gg \Gamma \\ \dfrac{\Omega^2 \Gamma_f \Gamma \omega_L^2}{(\omega_L^2 + \Gamma^2/4)^3} & \Gamma_f \ll \Gamma \end{cases} \quad (4.79)$$

When $\Gamma_f \gg \Gamma$, both Q and the line shape are a Lorentzian located at $\omega_L = 0$

with a width Γ_f, which yields the relation $Q \simeq 2\langle I(\omega_L)\rangle/\Gamma$. Both exhibit motional narrowing behaviors. In the other limit $\Gamma_f \ll \Gamma$, we have neglected $\mathcal{O}((\Gamma_f/\Gamma)^2)$ terms with additional conditions for $\omega_L^2, \Gamma_f/\Gamma \ll (\omega_L/\Gamma)^2 \ll \Gamma/\Gamma_f$. In this case, Q shows a splitting behavior at $|\omega_L| \sim \mathcal{O}(\Gamma)$. Note that Eq. (4.79) for $\Gamma_f \ll \Gamma$ is exactly the same as Eq. (4.67) although the parameters have different relative magnitudes. This is the case because the very fast frequency modulation corresponds to the weak modulation case, if we recall that the dephasing rate due to the bath fluctuation is given by $\Gamma_f = v^2/R$ in the fast modulation regime.

In Fig. 4.11, we checked the validity of the limiting expressions of Q for the Lorentzian and the splitting cases [Eq. (4.79)] by comparing them to the exact results, Eqs. (A.47), (A.48). In Figure 4.11(a) the parameters are chosen such that $\Gamma_f = 100\Gamma$, while in Figure 4.11(b) $\Gamma_f = \Gamma/100$. Approximate expressions (dashed line) show a good agreement with exact expressions (solid line) in each case.

2. Weak Modulation: Case 4

Now, we consider the weak, fast modulation case, case 4 ($v \ll \Gamma \ll R$). In this limit, the line shape is simply a single Lorentzian peak given by Eq. (4.70) with $\Gamma + \Gamma_f \simeq \Gamma$. We obtain the following limiting expression for Q from

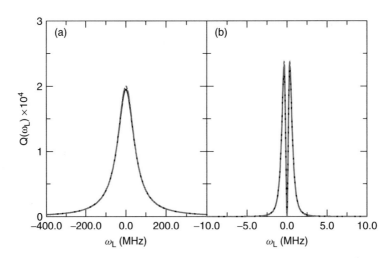

Figure 4.11. Case 3 ($\Gamma \ll v \ll R$) in the steady-state limit. Exact calculations (solid line) of Q, Eqs. (A.47) and (A.48) are compared with the approximate expressions, Eq. (4.79), (dashed line) for (a) the Lorentzian ($\Gamma_f \gg \Gamma$), and (b) the splitting ($\Gamma_f \ll \Gamma$) cases in the fast modulation case. Parameters are chosen as $v = 1$ GHz, $\Gamma = 1$ MHz, $\Omega = \Gamma/10$, $T \to \infty$, and $R = 10$ GHz in (a) while $R = 10^5$ GHz in (b).

A STOCHASTIC THEORY OF SINGLE MOLECULE SPECTROSCOPY 237

TABLE IV.2
Summary of parameter regimes investigated in this work

	Slow	Intermediate	Fast
	$R \ll v \ll \Gamma$: Case 2	$v \ll R \ll \Gamma$: Case 5	$v \ll \Gamma \ll R$: Case 4
Weak		$\langle I \rangle \sim \dfrac{\Omega^2 \Gamma}{4[\omega_L^2 + \Gamma^2/4]}$	
		$Q \sim \dfrac{\Gamma_f}{\langle I \rangle}\left(\dfrac{d\langle I \rangle}{d\omega_L}\right)^2$	
	$R \ll \Gamma \ll v$: Case 1	$\Gamma \ll R \ll v$: Case 6	$\Gamma \ll v \ll R$: Case 3
Strong	$\langle I \rangle \sim \dfrac{1}{2}(I_+ + I_-)$	$\langle I \rangle \sim \dfrac{\Omega^2 v^2 R}{(\omega_L^2 - v^2)^2 + 4R^2\omega_L^2}$	$\langle I \rangle \sim \dfrac{\Omega^2(\Gamma + \Gamma_f)}{4[\omega_L^2 + (\Gamma + \Gamma_f)^2/4]}$
	$Q \sim \dfrac{\langle I \rangle}{R}$	$Q \sim \dfrac{2\langle I \rangle}{\Gamma}$	$Q \sim \begin{cases} \dfrac{2\langle I \rangle}{\Gamma} & \Gamma_f \gg \Gamma \\ \dfrac{\Gamma_f}{\langle I \rangle}\left(\dfrac{d\langle I \rangle}{d\omega_L}\right)^2 & \Gamma_f \ll \Gamma \end{cases}$

$^a I_\pm$ have been defined in Eq. (4.55).

Eq. (4.77), noting $\Gamma_f = v^2/R$,

$$Q \simeq \frac{v^2 \Omega^2 \Gamma \omega_L^2}{R(\omega_L^2 + \Gamma^2/4)^3} \quad (4.80)$$

Also, here Q shows splitting behavior. It is given by the same expression as that in the strong, fast modulation case with $\Gamma_f \ll \Gamma$ [see Eq. (4.79)] and also as that in the weak, slow modulation case with $T \gg 1/R$ [see Eq. (4.67)] (see also Table IV.2 in Section VI.D). When Q is plotted as a function of ω_L, it would look similar to Figure 4.11(b).

C. Intermediate Modulation Regime

So far, we have considered four limiting cases: (1) strong, slow, (2) weak, slow, (3) strong, fast, and (4) weak, fast case. Now, we consider cases 5 ($v \ll R \ll \Gamma$) and 6 ($\Gamma \ll R \ll v$). They are neither in the slow nor in the fast modulation regime according to our definition.

1. Case 5

In case 5, since $v \ll R \ll \Gamma$ we can approximate the exact results by considering small v limit in Eqs. (A.47) and (A.48) for Q, and Eq. (A.3) for the line shape. By taking this limit, we find that the line shape is well described by a single Lorentzian given by Eq. (4.65), and Q by Eq. (4.67). Note that for all weak modulation cases, cases 2, 4, and 5, both the slow and fast modulation approximate results are valid (see Table IV.2). Also, a simple relation between $\langle I \rangle$ and Q holds in the limit, $v \to 0$,

$$\lim_{v \to 0} Q = \frac{v^2}{R} \lim_{v \to 0} \frac{1}{\langle I \rangle} \left(\frac{d\langle I \rangle}{d\omega_L} \right)^2 \tag{4.81}$$

The exact results of line shape and Q in Figure 4.12 show good agreement with the approximate results, Eqs. (4.65) and (4.67).

2. Case 6

The only remaining case is case 6 ($\Gamma \ll R \ll v$), where the bath fluctuation is fast compared with the radiative decay rate but not compared with the fluctuation amplitude. Because $\Gamma \ll R, v$ in this case we can approximate the exact results for $\langle I \rangle$ and Q by their limiting expressions corresponding to $\Gamma \to 0$, yielding Eqs. (A.8) and (A.49), and an important relation holds in this limit,

$$\lim_{\Gamma \to 0} Q = \frac{2}{\Gamma} \lim_{\Gamma \to 0} \langle I \rangle \tag{4.82}$$

Note that the same relation between Q and $\langle I \rangle$ was also found to be valid in one of the fast modulation regimes, Eq. (4.79) with $\Gamma_f \gg \Gamma$. Figure 4.13 shows that in this case the limiting expressions approximate well the exact results.

D. Phase Diagram

We investigate the overall effect of the bath fluctuation on the photon statistics for the steady-state case as the fluctuation rate R is varied from slow to fast modulation regime. To characterize the overall fluctuation behavior of the photon statistics, we define an order parameter q,

$$q \equiv \frac{\Gamma}{\pi \Omega^2} \int_{-\infty}^{\infty} Q(\omega_L) \, d\omega_L \tag{4.83}$$

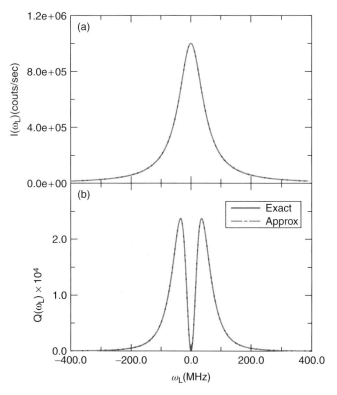

Figure 4.12. Case 6 ($v \ll R \ll \Gamma$) in the steady-state limit. Exact results for line shape, Eq. (A.3), and for Q, Eqs. (A.47), (A.48), are compared with approximations corresponding to $v \to 0$ limit, Eqs. (4.65) and (4.67), respectively. Parameters are chosen as $v = 1$ MHz, $\Gamma = 100$ MHz, $\Omega = \Gamma/10$, $R = 10$ MHz, and $T \to \infty$.

where Q in the steady state is given in Eqs. (A.47) and (A.48). Before we discuss the behavior of q, it is worthwhile mentioning that the line shape is normalized to a constant regardless of R and Γ,

$$\int_{-\infty}^{\infty} \langle I(\omega_L) \rangle \, d\omega_L = \frac{\pi \Omega^2}{2} \tag{4.84}$$

which can be easily verified from Eq. (4.40). In contrast, q exhibits nontrivial behavior reminiscent of a phase transition. In Figures 4.14 and 4.15, we show q versus R/Γ. The figures clearly demonstrate how the photon statistics of SMS in the presence of the spectral diffusion becomes Poissonian as $R \to \infty$ or $v \to 0$ (i.e., $q \to 0$ when $R \to \infty$ or $v \to 0$).

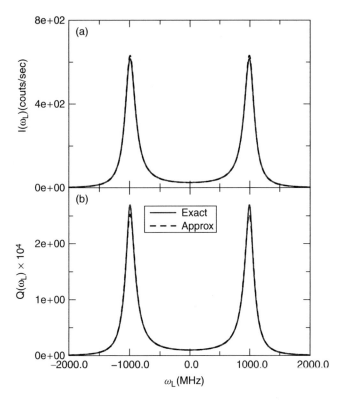

Figure 4.13. Case 5 ($\Gamma \ll R \ll v$) in the steady-state limit. Exact results for line shape, Eq. (A.3), and for Q, Eqs. (A.47), (A.48), are compared with approximations corresponding to $\Gamma \to 0$ limit, Eqs. (A.8) and (A.49), respectively. Parameters are chosen as $v = 1$ GHz, $\Gamma = 5$ MHz, $\Omega = \Gamma/10$, $R = 100$ MHz, and $T \to \infty$.

First, we discuss the *strong modulation regime*, $v \gg \Gamma$, shown in Figure 4.14 with $v/\Gamma = 100$ in this case. Depending on the fluctuation rate, there are three distinct regimes:

1. In the slow modulation regime, $R \ll \Gamma$, q decreases as $1/R$. The approximate calculation (dot–dashed line) based on the slow modulation approximation, Eq. (4.64) shows good agreement with the exact calculation.

2. When R is such that $\Gamma \ll R \ll v$ (case 6), the intermediate regime is achieved, and q starts to show a plateau behavior. The plateau behavior is found whenever $Q = 2\langle I(\omega_L)\rangle/\Gamma$, which yields $q = 1$ as can be easily seen from Eqs. (4.83) and (4.84). As R is increased further

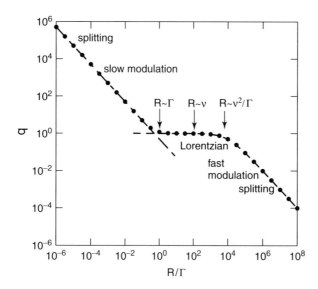

Figure 4.14. The parameter q versus R/Γ in the strong modulation case. Filled circles are the result of exact calculation based on Eqs. (A.47) and (A.48) when $v/\Gamma = 100$. The dot–dashed and dashed curves are calculations based on approximate expressions of Q, Eq. (4.64) (slow modulation case) and Eq. (4.77) (fast modulation case), respectively.

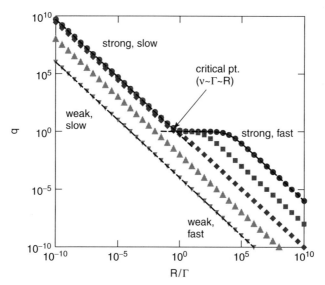

Figure 4.15. Phase diagram of q versus R/Γ. Symbols are the results of exact calculations based on Eqs. (A.47) and (A.48) as v/Γ is varied from the lower left to the upper right cases: 0.01 (downward triangle), 0.1 (upward triangle), 1 (diamond), 10 (square), and 100 (circle). The dashed curves are calculations based on appropriate approximate expressions of Q for each case, Eqs. (4.80) (weak modulation), (4.64), (strong, slow modulation), and (4.77) (strong, fast modulation).

241

such that $v \ll R$, the fast modulation regime is reached, and q still shows a plateau until $R \simeq v^2/\Gamma$. In this regime the Lorentzian behavior of Q is observed in Eq. (4.79) for $\Gamma_f \gg \Gamma$.

3. When the bath fluctuation becomes extremely fast such that $R \gg v^2/\Gamma$, the splitting behavior of Q is observed, as discussed in Eq. (4.79) for $\Gamma_f \ll \Gamma$, and then $q \propto 1/R$ similar to the slow modulation regime. The approximate value of q based on fast modulation approximation, Eq. (4.77) (dotted line) shows good agreement with the exact calculation found using Eqs. (A.47) and (A.48). Finally, when $R \to \infty$, $Q = 0$. As mentioned before, this is expected since the molecule cannot interact with a very fast bath hence the photon statistics becomes Poissonian.

Now, we discuss the effect of magnitude of frequency fluctuation, v, on q in Figure 4.15. We have calculated q as a function of $1/R$ as v is varied from $v/\Gamma = 100$ (circles) to $v/\Gamma = 0.01$ (downward triangles). We see in Figure 4.15 that the $(q, R/\Gamma)$ diagram exhibits a behavior similar to a phase transition as v is varied. In the strong modulation regime ($v/\Gamma \gg 1$), three distinct regimes appear in the $(q, R/\Gamma)$ diagram ($R \leqslant \Gamma$, $\Gamma \leqslant R \leqslant v^2/\Gamma$, and $R \geqslant v^2/\Gamma$), while in the weak modulation regime ($v/\Gamma \ll 1$) q always decreases as $1/R$. When three parameters, v, Γ, and R, have similar values, $v \sim \Gamma \sim R$, there appears a "critical point" in the "phase diagram".

Table IV.2 summarizes various expressions for the line shape and Q found in the limiting cases of the two-state jump model investigated in this work. For simplicity, we have set $\eta = 1$. We see that although the fluctuation model itself is a simple one, rich behaviors are found. We believe that these behaviors are generic (although we do not have a mathematical proof). In the weak modulation cases, both $\langle I \rangle$ and Q can be described by a single expression, irrespective of the fluctuation rate R. However, in the strong modulation cases, $\langle I \rangle$ and Q change their qualitative features as R changes.

VII. CONNECTION TO EXPERIMENTS

Single molecule spectroscopy has begun to reveal the microscopic nature of low-temperature glasses [9, 12, 14–16]. An important question is whether the standard tunneling model of low-temperature glass developed by Anderson et al. [94] and Phillips [95] is valid or not. As far as macroscopic measurements of acoustic, thermal, and optical properties are concerned, this model has proved to be compatible with experimental results [88]. However, on a more microscopic level we do not have much experimental or theoretical proof (or disproof) that the model is valid. At the heart of the standard tunneling model is the concept of a two-level system(TLS). At very low temperatures, the complicated multidimensional potential energy surface of the glass system is presumed to reduce to a multitude of noninteracting double well potentials whose two local minima correspond to reorientations of clusters of atoms or molecules [96]. Hence, the complicated behavior of glasses is reduced to a simple picture of many nonidentical and noninteracting TLSs. For a different perspective on the nature of low-temperature excitation in glasses, see [97].

Geva and Skinner [14] provided a theoretical interpretation of the static line shape properties in a glass (i.e., $\langle W \rangle$). The theory relied on the standard tunneling model and the Kubo–Anderson approach as means to quantify the line shape behavior (i.e., the time-dependent fluctuations of W are neglected). In [16], the distribution of static line shapes in a glass was found analytically and the relation of this problem to Lévy statistics was demonstrated.

Orrit and co-workers [8, 9] measured spectral trails as well as the line shapes and the fluorescent intensity correlation function $g^{(2)}(t)$. In [12], spectral trails of 70 molecules were investigated and 22 exhibited behaviors that seemed incompatible with the standard tunneling model. While the number of molecules investigated is not sufficient to determine whether the standard tunneling model is valid, the experiments are approaching a direct verification of this model.

Our theory in the slow modulation limit describes SMS experiments in glasses. The TLSs in the glass flip between their up and down states with a rate R determined by the coupling of TLS to phonons, the energy asymmetry, and the tunneling matrix element of the TLS [94, 95]. When a TLS makes a transition from the up to the down state, or vice versa, a frequency shift $v \propto 1/r^3$ occurs in the absorption frequency of a SM, where r is the distance between the SM and the TLS. The $1/r^3$ dependence is due to an elastic dipole interaction between the SM and the TLS. In a low-temperature glass, the density of TLSs is very low, hence one finds in experiment that the SM is coupled strongly to only a few TLSs. In some cases, when one TLS is in the vicinity of the SM, it is a reasonable approximation to neglect all the background TLSs. In this case, our theory describes SMS for chromophores in glasses with a single TLS strongly coupled to SM. Extension of our work to coupling of SM to many TLSs is important, and can be done in a straightforward manner provided that the TLSs are not interacting with each other.

Fleury et al. [9] measured $g^{(2)}(t)$ for a single terrylene molecule coupled to single TLS in polyethylene matrix. They showed that their experimental results are well described by

$$g^{(2)}(\tau) = 1 + \frac{R_+ R_- (I_+ - I_-)^2}{(R_+ I_+ + R_- I_-)^2} e^{-(R_+ + R_-)\tau} \tag{4.85}$$

where R_+ and R_- are the upward and downward transition rates, respectively. These two rates, R_+ and R_-, are due to the asymmetry of the TLS. This result is compatible with our result for Q in the slow modulation limit. When Eq. (4.85) is used in Eq. (4.13) with $R_+ = R_- = R$ for the symmetric transition case considered in this work, we exactly reproduce the result of $Q(T)$ for the slow modulation given in Eq. (4.60). [The asymmetric rate case $R_+ \neq R_-$ is also readily formulated for Q in the slow modulation limit, and again leads to a result compatible with Eq. (4.85)]. Hence, at least in this limit, our results are in an agreement with the experiment.

As far as we are aware, however, measurements of photon counting statistics for SMS in fast modulation regimes have not been made yet. The theory presented here suggests that even in the fast modulation regime, the deviation from Poisson statistics due to a spectral diffusion process might be observed under suitable experimental situations, for example, when the contributions of other mechanisms such as the quantum mechanical antibunching process and the blinking process due to the triplet state are known a priori. In this case, it can give more information on the distribution of the fluctuation rates, the strength of the chromophore–environment interaction, and the bath dynamics than the line shape measurement alone.

VIII. FURTHER DISCUSSIONS ON VALIDITY OF PRESENT THEORY

All along this work we have specified the conditions under which the present model is valid. Here, we will emphasize the validity and the physical limitations of our model. We will also discuss other possible approaches to the problem at hand.

In the present work, we have used classical photon counting statistics in the weak laser field limit. In the case of strong laser intensity, quantum mechanical effects on the photon counting statistics are expected to be important. From theories developed to describe two-level atoms interacting with a photon field *in the absence of environmental fluctuations*, it is known that, for strong field cases, deviations from classical Poissonian statistics can become significant [68, 99]. One of the well-known quantum mechanical effects on the counting statistics is the photon antibunching effect [53, 54, 69–72, 98]. In this case, a sub-Poissonian behavior is obtained, $-1 < Q_{qm} < 0$, where the subscript qm stands for quantum mechanical contribution. It is clear that when the spectral diffusion process is significant (i.e., cases 1 and 2) any quantum mechanical correction to Q is negligibly small. However, in the fast modulation case where we typically found a small value of Q, for example, $Q \sim 10^{-4}$, quantum mechanical corrections due to antibunching phenomenon might be important unless experiments under extremely weak fields can be performed such that $|Q_{qm}| < Q_{sd}$, where the subscript sd stands for spectral diffusion. The interplay between truly quantum mechanical effects and the fast dynamics of the bath is left for future work.

In this context, it is worthwhile to recall the quantum jump approach developed in the quantum optics community. In this approach, an emission of a photon corresponds to a quantum jump from the excited to the ground state. For a molecule with two levels, this means that right after each emission event, $\rho_{ee} = 0$ (i.e., the system is in the ground state). Within the classical approach this type of wave function collapse never occurs. Instead, the emission event is described with the probability of emission per unit time being $\Gamma \rho_{ee}(t)$, where $\rho_{ee}(t)$ is described by the stochastic Bloch equation. At least in principle, the quantum jump approach, also known as the Monte Carlo wave function approach [98–103], can be adapted to calculate the photon statistics of a SM in the presence of spectral diffusion.

Another important source of fluctuation in SMS is due to the triplet state dynamics. Indeed, one of our basic assumptions was the description of the electronic transition of the molecule in terms of a two-state model. Blinking behavior is found in many SMS experiments [18–20, 25, 26]. Due to the existence of metastable triplet states (usually long lived) the molecule switches from the bright to the dark states (i.e., when molecule is shelved in the triplet state, no fluorescence is recorded). Kim and Knight [78] pointed out that Q can become very large in the case of the metastable three level system *in the absence of spectral diffusion*, which is especially the case when the lifetime of the metastable state is long. Molski et al. have considered the effect of the triple state blinking on the photon counting statistics of SMS [56, 109].

Therefore at least three sources of fluctuations can contribute to the measured value of Q in SMS: (1) Q_{qm}, well investigated in quantum optics community; (2) $Q_{triplet}$, which can be described using the approach of [78, 56, 109], and now we have calculated the third contribution to Q; (3) Q_{sd}. Our approach is designed to describe a situation for which the spectral diffusion process is dominant over the others.

It is interesting to see if one can experimentally distinguish Q_{sd} and $Q_{triplet}$ for a SM in a condensed environment. One may think of the following *gedanken experiment*: Consider a case where the SM jumps between the bright to the dark state, and assume that we can identify the dark state when the SM is in the metastable triplet state. Further, let us assume that dark state is long lived compared to the time between emission events in the bright state. Then, at least in principle, one may filter out the effect of the dark triplet state on Q especially when the time scale of the spectral diffusion process is short compared with the dark period by measuring the photon statistics during the bright period.

IX. CONCLUDING REMARKS

In this chapter, we developed a stochastic theory of single molecule fluorescence spectroscopy. Fluctuations described by Q are evaluated in terms of a three-time correlation function $C_3(\tau_1, \tau_2, \tau_3)$ related to the response function in nonlinear spectroscopy. This function depends on the characteristics of the spectral diffusion process. Important time-ordering properties of the three-time correlation function were investigated here in detail. Since the fluctuations (i.e., Q) depend on the three-time correlation function, necessarily they contain more information than the line shape that depends on the one-time correlation function $C_1(\tau_1)$ via the Wiener–Khintchine theorem.

We have evaluated the three-time correlation function and Q for a stochastic model for a bath with an arbitrary timescale. The exact results for Q permit a better understanding of the non-Poissonian photon statistics of the single molecule induced by spectral diffusion. Depending on the bath timescale, different time orderings contribute to the line shape fluctuations and in the fast modulation regime all time orderings contribute. The theory predicts that Q is small in the fast modulation regime, increasing as the timescale of the bath (i.e., $1/R$) is increased. We have found nontrivial behavior of Q as the bath fluctuation becomes fast. Results obtained in this work are applicable to the experiment in the slow-to-intermediate modulation regime (provided that detection efficiency is high), and our results in the more challenging fast modulation regime give the theoretical limitations on the measurement accuracy needed to detect $Q > 0$.

The model system considered in this work is simple enough to allow an exact solution, but still complicated enough to exhibit nontrivial behavior. Extensions of the present work are certainly possible in several important aspects. It is worthwhile to consider photon counting statistics for a more complicated chromophore-bath model, for example, the case of many TLSs coupled to the chromophore, to see to what extent the results obtained in this work would remain generic. Also the effects of a triplet bottleneck state on the photon counting statistics can be investigated as a generalization of the theory presented here. Another direction for the extension of the present theory is to formulate the theory of SMS starting from the microscopic model of bath dynamics (e.g., the harmonic oscillator bath model). Effects of the interplay between the bath fluctuation and the quantum mechanical photon statistics on SMS is also left for future work.

The standard assumption of Markovian processes (e.g., the Poissonian Kubo–Anderson processes considered here) fails to explain the statistical properties of emission for certain single "molecular" systems such as quantum dots [21–23]. Instead of the usual Poissonian processes, a power-law process has been found in those systems. For such highly non-Markovian dynamics stationarity is never reached, and hence our approach as well as the Wiener–Khintchine theorem does not apply. This behavior is the topic of our recent work in [104].

Acknowledgments

This work was supported by the NSF. YJ is grateful for a partial financial support from the Korea Foundation for Advanced Studies. EB thank Michel Orrit and Gert Zumofen for helpful discussions.

APPENDIX

A. Calculation of Line Shape via Marginal Averaging Approach

In this Appendix, we calculate the line shape for our working example. We set $J = 1$ and $\omega_0 = 0$, and as mentioned the stochastic frequency modulation follows $\Delta\omega(t) = vh(t)$, where $h(t)$ describes a two-state telegraph process with $h(t) = +1$ (up) or $h(t) = -1$ (down). Transitions from state up to down and down to up are described by the rate R.

We use the marginal averaging approach [80, 105, 106] to calculate the average line shape. Briefly, the method gives a general prescription for the calculation of averages $\langle y \rangle$ (y is a vector) described by stochastic equation $\dot{y} = M(t)y$, where $M(t)$ is a matrix whose elements are fluctuating according to a Poissonian process (see [80, 106] for details). We define the marginal averages, $\langle v(t) \rangle_x$, $\langle u(t) \rangle_x$, and $\langle w(t) \rangle_x$, where $x = +$ or $x = -$ denotes the state of the two-state process at time t. For the stochastic Bloch equation, the evolution equation for the marginal averages are

$$\begin{pmatrix} \langle \dot{u} \rangle_+ \\ \langle \dot{v} \rangle_+ \\ \langle \dot{w} \rangle_+ \\ \langle \dot{u} \rangle_- \\ \langle \dot{v} \rangle_- \\ \langle \dot{w} \rangle_- \end{pmatrix} = \begin{pmatrix} -R-\frac{\Gamma}{2} & \delta_L^+ & 0 & R & 0 & 0 \\ -\delta_L^+ & -R-\frac{\Gamma}{2} & -\Omega & 0 & R & 0 \\ 0 & \Omega & -\Gamma-R & 0 & 0 & R \\ R & 0 & 0 & -R-\frac{\Gamma}{2} & \delta_L^- & 0 \\ 0 & R & 0 & -\delta_L^- & -R-\frac{\Gamma}{2} & -\Omega \\ 0 & 0 & R & 0 & \Omega & -R-\Gamma \end{pmatrix}$$

$$\times \begin{pmatrix} \langle u \rangle_+ \\ \langle v \rangle_+ \\ \langle w \rangle_+ \\ \langle u \rangle_- \\ \langle v \rangle_- \\ \langle w \rangle_- \end{pmatrix} + \begin{pmatrix} 0 \\ 0 \\ -\Gamma/2 \\ 0 \\ 0 \\ -\Gamma/2 \end{pmatrix}$$

where $\delta_L^\pm = \delta_L \mp v$. The steady-state solution is found by using a symbolic program such as *Mathematica* [107],

$$\langle I(\omega_L) \rangle = \frac{1}{2}\Omega(\langle v_{st} \rangle_- + \langle v_{st} \rangle_+) = \frac{1}{2}\Gamma(\langle w_{st} \rangle_+ + \langle w_{st} \rangle_- + 1) \quad (A.2)$$

where $(\langle w_{st} \rangle_+ + \langle w_{st} \rangle_- + 1 \rangle_-)/2$ represents the steady-state occupation of the excited level. We find

$$\langle I(\omega_L) \rangle = \frac{A}{B} \quad (A.3)$$

$A = \Gamma\Omega^2[(\Gamma + 2R)(\Gamma(4\omega_L^2 + (\Gamma + 4R)^2) + 4v^2(\Gamma + 4R)) + 2\Omega^2\Gamma(\Gamma + 4R)]$

$B = \Gamma(\Gamma + 2R)((4(\omega_L^2 - v^2) - \Gamma(\Gamma + 4R))^2 + 16\omega_L^2(\Gamma + 2R))$

$\quad + 4\Omega^2(4\omega_L^2\Gamma(\Gamma + 3R) + (\Gamma^2 + 4\Gamma R + 4v^2)(\Gamma^2 + 3\Gamma R + 4R^2))$

$\quad + 4\Omega^4\Gamma(\Gamma + 4R)$

Remark 1: When $R \to 0$, it is easy to show that $\langle I(\omega_L) \rangle$ is a sum of two Lorentzians centered at $\pm v$,

$$\langle I(\omega_L) \rangle = \frac{\Omega}{2}(v_+ + v_-) \tag{A.4}$$

where $v_\pm$ are the steady-state solution of the Bloch equation for two-level atom (see [73])

$$v_\pm = \frac{\Omega}{2} \frac{\Gamma/2}{(\delta_L^\pm)^2 + \Gamma^2/4 + \Omega^2/2} \tag{A.5}$$

Remark 2: If $\Omega \to 0$, the solution can also be found based on the Wiener–Khintchine formula, using the weights $\hat{P}_{ij}^{-1}(s)$ defined in Appendix C

$$\langle I(\omega_L) \rangle = \frac{\Omega^2}{4} \operatorname{Re}\left[\sum_{i=\pm, j=\pm} \hat{P}_{ij}^{-1}(i\omega_L + \Gamma/2)\right] \tag{A.6}$$

which gives

$$\langle I(\omega_L) \rangle = \frac{\Omega^2(4\Gamma\omega_L^2 + (\Gamma + 4R)(\Gamma^2 + 4\Gamma R + 4v^2))}{(4(\omega_L^2 - v^2) - \Gamma(\Gamma + 4R))^2 + 16\omega_L^2(\Gamma + 2R)^2} \tag{A.7}$$

Now, if $\Gamma \to 0$ we get the well-known result of Kubo

$$\langle I(\omega_L) \rangle = \frac{\Omega^2 v^2 R}{(\omega_L^2 - v^2)^2 + 4R^2 \omega_L^2} \tag{A.8}$$

then in the slow modulation limit, $v \gg R$ the line $\langle I(\omega_L) \rangle$ exhibits splitting (i.e., two peaks at $\pm v$), while for the fast modulation regime $v \ll R$ the line is a Lorentzian centered at $\omega_L = 0$ and motional narrowing is observed.

Remark 3: If $v \to 0$ the solution reduces to the well-known Bloch equation solution of a stable two-level atom, which is independent of R.

Remark 4: We have assumed that occupation of state $+$ and state $-$ are equal. A more general case, limited to weak laser intensity regime, was considered in [39].

B. Perturbation Expansion

In Appendix B, we find expressions for the photon current using the perturbation expansion. In Section B.3, we use the Lorentz oscillator model to derive similar results based on a classical picture.

We use the stochastic Bloch equation [80, 81] to investigate $\rho_{ee}(t)$ in the limit of weak external laser field when we expect $\rho_{ee} \simeq 0$, $\rho_{gg} \simeq 1$ for times $\Gamma T \gg 1$. We rewrite Eq. (4.6)

$$\frac{d\hat{\rho}_{ee}}{dt} = -\Gamma\hat{\rho}_{ee} + \frac{i\Omega}{2}(\hat{\rho}_{eg} - \hat{\rho}_{ge})$$

$$\frac{d\hat{\rho}_{ge}}{dt} = -[i\delta_L(t) + \Gamma/2]\hat{\rho}_{ge} - \frac{i\Omega}{2}(\hat{\rho}_{ee} - \hat{\rho}_{gg}), \quad \text{(A.9)}$$

where $\hat{\rho}_{eg} = \rho_{eg}e^{i\omega_L t}$, $\hat{\rho}_{ge} = \rho_{ge}e^{-i\omega_L t}$, $\hat{\rho}_{ee} = \rho_{ee}$, and $\hat{\rho}_{gg} = \rho_{gg}$. By using Eq. (A.9), the normalization condition $\hat{\rho}_{ee} + \hat{\rho}_{gg} = 1$, and $\hat{\rho}_{ge} = \text{C.C.}[\hat{\rho}_{eg}]$, the four matrix elements of the density matrix can be determined in principle when the initial conditions and the stochastic trajectory $\Delta\omega(t)$ are specified. We use the perturbation expansion

$$\hat{\rho}_{ee}(t) = \hat{\rho}_{ee}^{(0)}(t) + \Omega\hat{\rho}_{ee}^{(1)}(t) + \Omega^2\hat{\rho}_{ee}^{(2)}(t) + \cdots$$

$$\hat{\rho}_{ge}(t) = \hat{\rho}_{ge}^{(0)}(t) + \Omega\hat{\rho}_{ge}^{(1)}(t) + \Omega^2\hat{\rho}_{ge}^{(2)}(t) + \cdots \quad \text{(A.10)}$$

and initially $\hat{\rho}_{ee}^{(i)}(t=0) = \hat{\rho}_{eg}^{(i)}(t=0) = 0$ for $i = 1, 2, 3, \ldots$. We insert Eq. (A.10) in (A.9) and first consider only the zeroth-order terms in Ω. We find $\hat{\rho}_{ee}^{(0)}(t) = \hat{\rho}_{ee}(0)\exp(-\Gamma t)$, which is expected since when the laser field is absent, population in the excited state is decreasing due to spontaneous emission. The off-diagonal term is $\hat{\rho}_{ge}^{(0)}(t) = \hat{\rho}_{ge}^{(0)}(0)\exp[-\Gamma t/2 - i\int_0^t dt'\delta_L(t')]$. This term is described by the dynamics of a Kubo–Anderson classical oscillator, $\dot{x} = [-i\delta_L(t) - \Gamma/2]x$ [61].

1. First-Order Terms

The first-order term is described by the equation

$$\frac{d\hat{\rho}_{ee}^{(1)}}{dt} = -\Gamma\hat{\rho}_{ee}^{(1)} + \frac{i}{2}(\hat{\rho}_{eg}^{(0)} - \hat{\rho}_{ge}^{(0)}) \quad \text{(A.11)}$$

This equations yields the solution

$$\hat{\rho}_{ee}^{(1)}(t) = e^{-\Gamma t} \int_0^t e^{\Gamma t'} dt' \, \text{Im}[\hat{\rho}_{ge}^{(0)}(t')] \quad (A.12)$$

One can show that $\rho_{ee}^{(1)}(t)$ unimportant for times $\Gamma t \gg 1$, like all the other terms that depend on the initial condition.

For the off-diagonal term, we find

$$\frac{d\hat{\rho}_{ge}^{(1)}}{dt} = -[i\delta_L(t) + \Gamma/2]\hat{\rho}_{ge}^{(1)} - \frac{i}{2}(2\hat{\rho}_{ee}^{(0)} - 1) \quad (A.13)$$

The transient term $\hat{\rho}_{ee}^{(0)} = \rho_{ee}(0)\exp(-\Gamma t)$ is unimportant, and Eq. (A.13) yields

$$\hat{\rho}_{ge}^{(1)}(t) = \frac{i}{2}\int_0^t dt_1 \exp\left[-i\int_{t_1}^t dt'\delta_L(t') - \Gamma(t-t_1)/2\right] \quad (A.14)$$

By using $v(t) = \Omega \, \text{Im}[\hat{\rho}_{ge}^{(1)}(t)]$, we find Eq. (4.38).

According to the discussion in the text, the number of photons absorbed in time interval $(0, T)$ is determined by time integration of the photon current $W \simeq \Omega \int_0^T dt v(t)$. By using Eq. (4.38) and the definition of $\delta_L(t)$, we obtain Eq. (4.39). In addition, it is convenient to rewrite Eq. (4.39) in the following form:

$$W = \frac{\Omega^2}{4}\int_0^T dt_2 \int_0^T dt_1 e^{-i\omega_L(t_2-t_1)-\Gamma|t_2-t_1|/2+i\int_{t_1}^{t_2}\Delta w(t')dt'} \quad (A.15)$$

We calculate the average number of counts $\langle W \rangle$ using Eq. (4.39). The integration variables are changed to $\tau = t_2 - t_1$ and t_1, and for such a transformation the Jacobian is unity. Integration of t_1 is carried out from 0 to $T - \tau$, resulting in

$$\langle W \rangle = \frac{\Omega^2 T}{2} \text{Re}\left[\int_0^T d\tau \left(1 - \frac{\tau}{T}\right) e^{-i\omega_L\tau - \Gamma\tau/2} C_1^{-1}(\tau)\right] \quad (A.16)$$

and $C_1^l(\tau) = \langle e^{-il\int_0^\tau \Delta w(t')dt'}\rangle$ is the one-time correlation function. In the limit of $T \to \infty$ we find Eq. (4.10).

By using Eq. (A.15), the fluctuations are determined by

$$\langle W^2 \rangle = \frac{\Omega^4}{16} \int_0^T \int_0^T \int_0^T \int_0^T dt_1\, dt_2\, dt_3\, dt_4\, e^{-i\omega_L(t_2 - t_1 + t_4 - t_3) - \Gamma(|t_1 - t_2| + |t_3 - t_4|)/2}$$

$$\times \left\langle \exp\left[i \int_{t_1}^{t_2} dt' \Delta\omega(t') + i \int_{t_3}^{t_4} dt' \Delta\omega(t') \right] \right\rangle \quad \text{(A.17)}$$

and changing integration variables, $t_3 \to t_4$ and $t_4 \to t_3$, yields Eq. (4.42). We replaced t_3 and t_4 to get a pulse shape similar to that in the three-time photon echo experiment (when $t_1 < t_2 < t_3 < t_4$). Note that the derivation did not assume a specific type of random process $\Delta\omega(t)$ and our results are not limited to the two-state telegraph process we analyze in the text.

2. Second-Order Terms

According to Eqs. (4.12) and (4.15) $\Gamma \int_0^T \rho_{ee}(t)\, dt = \Omega \int_0^T v(t)\, dt \gg 1$. We now show that this equation is valid within perturbation theory. For this purpose we must consider second-order perturbation theory. The equation for the second-order term

$$\frac{d\hat{\rho}_{ee}^{(2)}(t)}{dt} = -\Gamma \hat{\rho}_{ee}^{(2)} + \text{Im}[\hat{\rho}_{ge}^{(1)}(t)] \quad \text{(A.18)}$$

yields

$$\hat{\rho}_{ee}^{(2)}(t) = \frac{1}{2} e^{-\Gamma t} \text{Re}\left[\int_0^t dt_2 \int_0^{t_2} dt_1 e^{\Gamma(t_1 + t_2)/2 - i\omega_L(t_2 - t_1) + i \int_{t_1}^{t_2} \Delta\omega(t')\, dt'} \right] \quad \text{(A.19)}$$

where we have neglected terms depending on initial condition. Since $\hat{\rho}_{ee}(t) \simeq \Omega^2 \hat{\rho}_{ee}^{(2)}(t)$, we see that the response is quadratic with respect to the Rabi frequency as we expect from symmetry (population in excited state does not depend on sign of E_0). By using Eq. (A.19), we find

$$\hat{\rho}_{ee}(t) = \frac{1}{4} \Omega^2 e^{-\Gamma t} \int_0^t dt_2 \int_0^t dt_1 e^{\Gamma(t_1 + t_2)/2 - i\omega_L(t_2 - t_1) + i \int_{t_1}^{t_2} \Delta\omega(t')\, dt'} \quad \text{(A.20)}$$

Now, we consider the standard ensemble measurement and average $\rho_{ee}^{(2)}(t)$ with respect to history of the process $\Delta\omega(t')$. By assuming stationarity and changing variables, we find in the limit of $T \to \infty$

$$\langle \rho_{ee}^{(2)}(\infty) \rangle = \frac{\Omega^2}{2\Gamma} \text{Re} \left[\int_0^\infty d\tau C_1^{-1}(\tau) e^{-\Gamma\tau/2 - i\omega_L \tau} \right] \qquad (A.21)$$

which is the line shape. From Eq. (4.40) we see that $\Gamma \langle \rho_{ee}^{(2)}(\infty) \rangle = \lim_{T \to \infty} \langle W/T \rangle = \langle I(\omega_L) \rangle$. Thus the theory of the averaged line shape can be based on either v (first-order perturbation theory) or on ρ_{ee} (second-order perturbation theory).

Instead of the averages, let us consider the stochastic variables $\Gamma \Omega^2 \int_0^T \hat{\rho}_{ee}^{(2)}(t) \, dt$ and $\Omega^2 \int_0^T v^{(1)}(t) \, dt$, where $v^{(1)} = \text{Im}[\hat{\rho}_{eg}^{(1)}(t)]$. By using Eq. (A.19), and the Laplace $T \to s$ transform

$$\Omega^2 \int_0^T dT e^{-sT} \left[\Gamma \int_0^T dt \hat{\rho}_{ee}^{(2)}(t) \right] = \frac{\Omega^2}{2s} \frac{\Gamma}{\Gamma + s} \hat{f}_{\text{sto}}(s) \qquad (A.22)$$

where

$$\hat{f}_{\text{sto}}(s) = \int_0^\infty dt e^{-st} \text{Re} \left[\int_0^t dt_1 e^{-\Gamma(t - t_1)/2 - i\omega_L(t - t_1) + i \int_{t_1}^t \Delta\omega(t') \, dt'} \right] \qquad (A.23)$$

is a functional of the stochastic function $\Delta\omega(t)$. By using Eq. (4.39), we find

$$\int_0^\infty dT e^{-sT} \left[\Omega^2 \int_0^T dt v^{(1)}(t) \right] = \frac{\Omega^2}{2s} \hat{f}_{\text{sto}}(s) \qquad (A.24)$$

By comparing Eqs. (A.22) and (A.24), we see that a theory based on ρ_{ee} or on v are not entirely identical. However, for long times $\Gamma T \gg 1$, we may use small s behavior (i.e., $\Gamma/(s + \Gamma) \simeq 1$), and $\Gamma \int_0^T \rho_{ee}(t) \, dt = \Omega \int_0^T v(t) \, dt$ as expected.

3. Classical Lorentz Model

The stochastic Bloch equation is a semiphenomenological equation with some elements of quantum mechanics in it. To understand better whether our results are quantum mechanical in origin, we analyze a classical model. Lorentz invented the theory of classical, linear interaction of light with matter. Here, we investigate a stochastic Lorentz oscillator model. We follow Allen and Eberley [108] who considered the deterministic model in detail. The classical model is also helpful because its physical interpretation is clear. We show that for weak laser intensity, the stochastic Bloch equations are equivalent to classical Lorentz approach.

We consider the equation for harmonic dipole $|e|x(t)$ in the driving field, $\mathcal{E}(t) = E\cos(\omega_L t)$,

$$\ddot{x} + \Gamma\dot{x} + \omega_0^2(t)x = \frac{|e|}{m}\mathcal{E}(t) \quad \text{(A.25)}$$

where all symbols have their usual meanings and $\omega_0(t) = \omega_0 + vh(t)$ is a stochastic time-dependent frequency and $|vh(t)| \ll \omega_0$. All along this section we use symbols that appear also in the Bloch formalism since their meanings in the Bloch and in the Lorentz models are identical, as we show in Eqs. (A.29) and (A.30).

We decompose x into two parts.

$$x(t) = x_0[u\cos(\omega_L t) - v\sin(\omega_L t)] \quad \text{(A.26)}$$

x_0 is a time-independent constant, while u and v vary slowly in time. The work done by the laser force $F = |e|E\cos(\omega_L t)$ on the particle is $d\mathcal{W} = -F\,dx$, hence $d\mathcal{W}/dt = -F\dot{x}$. By using Eq. (A.26), we find

$$\frac{\overline{d\mathcal{W}}}{dt} = \frac{1}{2}|e|Ex_0\omega_L v \quad \text{(A.27)}$$

and the overbar denotes average over rapid laser oscillations [e.g., we assume that the noise term $h(t)$ evolves slowly if compared with the laser period $2\pi/\omega_L$]. Since $h(t)$ is stochastic so is $v(t)$, hence the power $d\overline{\mathcal{W}}/dt$ is also a stochastic function.

As in [108], we assume that u and v vary slowly in time such that

$$|\dot{u}| \ll \omega_L|v|, \quad |\ddot{u}| \ll \omega_L^2|u|, \quad |\dot{v}| \ll \omega_L|u|, \quad |\ddot{v}| \ll \omega_L^2|v| \quad \text{(A.28)}$$

then insert Eq. (A.26) into Eq. (A.25) and find two equations for the envelopes u and v,

$$\dot{u} = [\omega_L - \omega_0(t)]v - \frac{\Gamma u}{2}$$

$$\dot{v} = -[\omega_L - \omega_0(t)]u - \frac{\Gamma v}{2} + \frac{|e|E}{2m\omega_L x_0} \quad \text{(A.29)}$$

where the relation $[\omega_0^2(t) - \omega_L^2]/(2\omega_L) \simeq \omega_0(t) - \omega_L$ was used. By comparing Eqs. (A.27), (A.29) with Eqs. (4.11), (4.6), we will now show that in the weak laser intensity limit the Bloch equation describes the dynamics described by the Lorentz model. To see this clearly, note that when $\Omega \to 0$, $\rho_{ee} \simeq 0$, and hence $w \simeq -\frac{1}{2}$. Therefore, if we replace $-\Omega w$ in the Bloch equation, Eq. (4.6) with $\Omega/2$, the Bloch equations for u and v become uncoupled from that for w. By using this approximation, we find

$$\dot{u} = \delta_L(t)v - \frac{\Gamma u}{2}$$

$$\dot{v} = -\delta_L(t)u - \frac{\Gamma v}{2} + \frac{\Omega}{2} \quad \text{(A.30)}$$

It is clearly seen that the Bloch equation in the weak intensity limit, Eq. (A.30), has the same structure as the Lorentz equation, Eq. (A.29). Note that two parameters, x_0 and m, only appear in the Lorentz model while two other parameters, $\Omega = -\mathbf{d}_{eg} \cdot \mathbf{E}/\hbar$ and $\hbar$, appear in the Bloch equation. To make the equivalence between these two approaches complete, the following relations can be deduced by comparing Eq. (A.29) with Eq. (A.30), and Eq. (A.27) with Eq. (4.11),

$$\Omega \leftrightarrow \frac{|e|E}{m\omega_L x_0} \qquad \hbar\Omega \leftrightarrow \frac{|e|x_0 E}{2}$$

or

$$d_{eg} \leftrightarrow -\frac{1}{2}|e|x_0, \qquad \hbar\omega_L \leftrightarrow \frac{1}{2}m\omega_L^2 x_0^2$$

In conclusion, when the laser intensity is not strong the stochastic phenomenological Bloch equation describes classical behavior.

C. Exact Calculations of $\langle W \rangle$ and $\langle W^2 \rangle$

In this Appendix, we use straightforward complex analysis and find

$$\langle W^2 \rangle = \frac{\Omega^4}{16} \mathscr{L}^{-1}\left\{\sum_{i=1}^{5} \hat{\xi}_i(s) + \text{C.C.}\right\} \quad \text{(A.31)}$$

where

$$\hat{\xi}_1(s) = \frac{1}{s^2} \sum_{i,j,k,l} \hat{P}_{ij}^{-1}(s+s_+)\hat{P}_{jk}^{0}(s)\hat{P}_{kl}^{1}(s+s_-)$$

$$\hat{\xi}_2(s) = \frac{1}{s^2} \sum_{i,j,k,l} \hat{P}_{ij}^{-1}(s+s_+)\hat{P}_{jk}^0(s+\Gamma)\hat{P}_{kl}^1(s+s_-)$$

$$\hat{\xi}_3(s) = \frac{1}{s^2} \sum_{i,j,k,l} \hat{P}_{ij}^1(s+s_-)\hat{P}_{jk}^0(s)\hat{P}_{kl}^1(s+s_-)$$

$$\hat{\xi}_4(s) = \frac{1}{s^2} \sum_{i,j,k,l} \hat{P}_{ij}^1(s+s_-)\hat{P}_{jk}^0(s+\Gamma)\hat{P}_{kl}^1(s+s_-)$$

$$\hat{\xi}_5(s) = \frac{2}{s^2} \sum_{i,j,k,l} \hat{P}_{ij}^1(s+s_-)\hat{P}_{jk}^2(s+2s_-)\hat{P}_{kl}^1(s+s_-) \quad (A.32)$$

and $s_\pm = \Gamma/2 \pm i\omega_L$. The Laplace transforms of the weights $P_{ij}^a(\tau)$ are

$$\hat{P}_{++}^0(s) = \hat{P}_{--}^0(s) = \frac{R+s}{s(s+2R)}$$

$$\hat{P}_{+-}^0(s) = \hat{P}_{-+}^0(s) = \frac{R}{s(s+2R)}$$

$$\hat{P}_{++}^1(s) = \hat{P}_{--}^{-1}(s) = \frac{R+s-iv}{(s-s_1)(s-s_2)}$$

$$\hat{P}_{+-}^1(s) = \hat{P}_{-+}^{-1}(s) = \hat{P}_{-+}^1(s) = \hat{P}_{+-}^{-1}(s) = \frac{R}{(s-s_1)(s-s_2)}$$

$$\hat{P}_{--}^1(s) = \hat{P}_{++}^{-1}(s) = \frac{R+s+iv}{(s-s_1)(s-s_2)}$$

$$\hat{P}_{ij}^2(s) = \hat{P}_{ij}^1(s)|_{v \to 2v} \quad (A.33)$$

where

$$s_{1,2} = -R \pm \sqrt{R^2 - v^2}$$

The inverse Laplace transforms of $\hat{\xi}_i(s)$ in Eq. (A.35) are calculated using standard methods of complex analysis to yield $\xi_i(T)$:

1. $$\xi_1(T) = \sum_{m=1}^{5} [s^3 \hat{\xi}_1(s)(s-z_m)]|_{s=z_m} f_1(z_m, T) \quad (A.34)$$

where the simple poles z_i are given by

$$z_1 = -s_+ + s_1 = -\frac{\Gamma}{2} - i\omega_L - R + \sqrt{R^2 - v^2}$$

$$z_2 = -s_+ + s_2 = -\frac{\Gamma}{2} - i\omega_L - R - \sqrt{R^2 - v^2}$$

$$z_3 = -s_- + s_1 = -\frac{\Gamma}{2} + i\omega_L - R + \sqrt{R^2 - v^2}$$

$$z_4 = -s_- + s_2 = -\frac{\Gamma}{2} + i\omega_L - R - \sqrt{R^2 - v^2}$$

$$z_5 = -2R \quad \text{(A.35)}$$

and $f_1(z, T)$ is

$$f_1(z, T) = -\frac{T^2}{2z} - \frac{T}{z^2} - \frac{1 - e^{zT}}{z^3} \quad \text{(A.36)}$$

Note that if $\omega_L = 0$, $z_1 = z_3$, and $z_2 = z_4$, then the poles become second order. Also, we can neglect exponential decays $\exp(z_i T)$ for $i = 1\text{-}4$ since $\Gamma T \gg 1$, and the term $\exp(z_5 T)$ is important when $RT \leqslant 1$.

2.
$$\xi_2(T) = \sum_{m=1}^{6} [s^2 \hat{\xi}_2(s)(s - z_m)]|_{s=z_m} f_2(z_m, T) \quad \text{(A.37)}$$

where $z_1 = -s_+ + s_1$, $z_2 = -s_+ + s_2$, $z_3 = -s_- + s_1$, $z_4 = -s_- + s_2$, $z_5 = -\Gamma$, $z_6 = -2R - \Gamma$ and

$$f_2(z, T) = -\frac{T}{z} - \frac{1 - e^{zT}}{z^2} \quad \text{(A.38)}$$

The expression $s^2 \hat{\xi}_2(s)(s - z_m)|_{s=z_m}$ in Eq. (A.37) is the residue of $s^2 \hat{\xi}_2(s)$ when $s = z_m$.

3.
$$\xi_3(T) = [s^3(s - z_3)\hat{\xi}_3(s)]|_{s=z_3} f_1(z_3, T)$$

$$+ \sum_{m=1}^{2} [s^3(s - z_m)^2 \hat{\xi}_3(s)]|_{s=z_m} f_3(z_m, T)$$

$$+ \sum_{m=1}^{2} \left[\frac{d}{ds} s^3(s - z_m)^2 \hat{\xi}_3(s)\right]\bigg|_{s=z_m} f_1(z_m, T) \quad \text{(A.39)}$$

where $z_1 = -s_- + s_1$, $z_2 = -s_- + s_2$, $z_3 = -2R$ and

$$f_3(z, T) = \frac{T^2}{2z^2} + \frac{2T}{z^3} + \frac{Te^{zT}}{z^3} + \frac{3(1 - e^{zT})}{z^4} \tag{A.40}$$

4.
$$\xi_4(T) = \sum_{m=1}^{2} [s^2(s - z_m)^2 \hat{\xi}_4(s)]|_{s=z_m} f_4(z_m, T)$$
$$+ \sum_{m=1}^{2} \left[\frac{d}{ds} s^2(s - z_m)^2 \hat{\xi}_4(s)\right]\bigg|_{s=z_m} f_2(z_m, T)$$
$$+ \sum_{m=3}^{4} [s^2(s - z_m) \hat{\xi}_4(s)]|_{s=z_m} f_2(z_m, T) \tag{A.41}$$

where $z_1 = -s_- + s_1$, $z_2 = -s_- + s_2$, $z_3 = -\Gamma$, $z_4 = -2R - \Gamma$ and

$$f_4(z, T) = \frac{T}{z^2} + \frac{Te^{zT}}{z^2} + \frac{2(1 - e^{zT})}{z^3} \tag{A.42}$$

Finally,

5.
$$\xi_5(T) = \sum_{m=1}^{2} [s^2(s - z_m)^2 \hat{\xi}_5(s)]|_{s=z_m} f_4(z, T)$$
$$+ \sum_{m=1}^{2} \left[\frac{d}{ds} s^2(s - z_m)^2 \hat{\xi}_5(s)\right]\bigg|_{s=z_m} f_2(z_m, T)$$
$$+ \sum_{m=3}^{4} [s^2(s - z_m) \hat{\xi}_5(s)]|_{s=z_m} f_2(z_m, T) \tag{A.43}$$

where $z_1 = -s_- + s_1$, $z_2 = -s_- + s_2$, $z_3 = -2s_- + \tilde{s}_1$ and $z_4 = -2s_- + \tilde{s}_2$ and $\tilde{s}_{1,2} = -R + \sqrt{R^2 - v^2}$.

From Eq. (A.16), the average counting number can be written as

$$\langle W \rangle = \frac{\Omega^2}{4} \mathscr{L}^{-1} \left\{ \frac{1}{s^2} [\hat{C}_1^{-1}(s + s_+) + \text{C.C.}] \right\} \tag{A.44}$$

which leads to Eq. (4.54) with Eq. (4.47). It is also easy to show that

$$\langle W \rangle = \frac{\Omega^2}{4} \mathscr{L}^{-1} \left\{ \frac{1}{s_1 - s_2} [-s_2 f_2(s_1 - s_+, T) + s_1 f_2(s_2 - s_+, T)] + \text{C.C.} \right\} \tag{A.45}$$

where $f_2(z, T)$ was defined in Eq. (A.38). In the limit of large T, we have

$$\langle W \rangle \simeq \frac{\Omega^2}{8} T \left[\sum_{i,j} \hat{P}_{ij}^-(s_+) + \text{C.C.} \right] \quad (A.46)$$

D. Q in the Long Time Limit

The exact expression for the Q parameter in the long time limit is given by $Q = \text{Numerator}[Q]/\text{Denominator}[Q]$, where

$$\begin{aligned}
\text{Denominator}[Q] = &\; \Gamma R [\Gamma^3 + 8\Gamma^2 R + 16v^2 R + 4\Gamma(v^2 + 4R^2 + \omega_L^2)] \\
&\times [\Gamma^4 + 4\Gamma^3 R + 16\Gamma R(v^2 + \omega_L^2) + 4\Gamma^2(2v^2 + R^2 + 2\omega_L^2) \\
&+ 16(v^4 - 2v^2\omega_L^2 + R^2\omega_L^2 + \omega_L^4)] \\
&\times [\Gamma^4 + 8\Gamma^3 R + 32\Gamma R(v^2 + \omega_L^2) + 8\Gamma^2(v^2 + 2R^2 + \omega_L^2) \\
&+ 16(v^4 - 2v^2\omega_L^2 + 4R^2\omega_L^2 + \omega_L^4)]^2 \quad (A.47)
\end{aligned}$$

and

$$\begin{aligned}
\text{Numerator}[Q] = &\; 64v^2\Omega^2 [\Gamma^{11}\omega_L^2 + 28\Gamma^{10} R \omega_L^2 \\
&+ 4\Gamma^9(85R^2\omega_L^2 + 4\omega_L^4 + v^2(R^2 + 4\omega_L^2)) \\
&+ 16\Gamma^8 R(146R^2\omega_L^2 + 21\omega_L^4 + v^2(4R^2 + 21\omega_L^2)) \\
&+ 512\Gamma^2 v^2 R^3(11v^6 + 92R^4\omega_L^2 + 96R^2\omega_L^4 + 3\omega_L^6 \\
&+ v^4(18R^2 - 19\omega_L^2) + 5v^2(18R^2\omega_L^2 + \omega_L^4)) \\
&+ 16\Gamma^7(620R^4\omega_L^2 + 183R^2\omega_L^4 + 6\omega_L^6 \\
&+ v^4(4R^2 + 6\omega_L^2) + v^2(25R^4 + 182R^2\omega_L^2 + 4\omega_L^4)) \\
&+ 32\Gamma^6 R(832R^4\omega_L^2 + 424R^2\omega_L^4 + 42\omega_L^6 \\
&+ v^4(25R^2 + 42\omega_L^2) + v^2(38R^4 + 437R^2\omega_L^2 + 28\omega_L^4)) \\
&+ 128\Gamma^4 R(v^6(27R^2 + 14\omega_L^2) + 2(10R^2 + 7\omega_L^2) \\
&\times (4R^2\omega_L + \omega_L^3)^2 + 2v^4(28R^4 + 65R^2\omega_L^2 - 7\omega_L^4) \\
&+ v^2(8R^6 + 625R^4\omega_L^2 + 207R^2\omega_L^4 - 14\omega_L^6)) \\
&+ 1024\Gamma v^2 R^2(v^8 + 8R^4\omega_L^4 + 3R^2\omega_L^6 + \omega_L^8 \\
&+ v^6(7R^2 - 4\omega_L^2) + v^4(-11R^2\omega_L^2 + 6\omega_L^4) \\
&+ v^2(40R^4\omega_L^2 + R^2\omega_L^4 - 4\omega_L^6))
\end{aligned}$$

$$+ 64\Gamma^5(688R^6\omega_L^2 + 552R^4\omega_L^4 + 111R^2\omega_L^6 + 4\omega_L^8$$
$$+ v^6(6R^2 + 4\omega_L^2) + v^4(57R^4 + 105R^2\omega_L^2 - 4\omega_L^4)$$
$$+ v^2(28R^6 + 653R^4\omega_L^2 + 102R^2\omega_L^4 - 4\omega_L^6))$$
$$+ 2048v^2R^3(v^8 - 4v^6\omega_L^2 + 4R^4\omega_L^4 + 5R^2\omega_L^6 + \omega_L^8$$
$$+ v^4(5R^2\omega_L^2 + 6\omega_L^4) - 2v^2(5R^2\omega_L^4 + 2\omega_L^6))$$
$$+ 256\Gamma^3(v^8(4R^2 + \omega_L^2) + \omega_L^2(R^2 + \omega_L^2)(4R^2 + \omega_L^2)^3$$
$$+ v^6(39R^4 + 4R^2\omega_L^2 - 4\omega_L^4)$$
$$+ v^4(20R^6 + 114R^4\omega_L^2 - 7R^2\omega_L^4 + 6\omega_L^6)$$
$$+ v^2(356R^6\omega_L^2 + 215R^4\omega_L^4 - 14R^2\omega_L^6 - 4\omega_L^8))] \quad (A.48)$$

To obtain these results, we find the small s expansion of the Laplace transforms $\langle \hat{W}^2 \rangle$ and $\langle \hat{W} \rangle$. These give in a standard way the long time behavior of the averages $\langle W^2 \rangle$ and $\langle W \rangle$ with which Q is found. These equations were derived using *Mathematica* [107], without which the calculation is cumbersome. Note that Q is an even function of ω_L and v, as expected from symmetry. For some special cases, discussed in this chapter, the exact results are much simplified.

Remark 1: In the limit of $\Gamma \to 0$, which corresponds to the limit of Kubo's line shape theory, we get

$$Q = \frac{2v^2\Omega^2 R}{\Gamma[(\omega_L^2 - v^2)^2 + 4R^2\omega_L^2]} \quad (A.49)$$

Remark 2: For $\omega_L = 0$, we get

$$Q = \frac{256\Omega^2 v^4 R(\Gamma + 2R)}{\Gamma(\Gamma + 4R)(\Gamma^2 + 4v^2 + 2\Gamma R)(\Gamma^2 + 4v^2 + 4\Gamma R)^2} \quad (A.50)$$

E. Q in the Slow Modulation Regime

In the two-state random walk model, where the "velocity" of the particle is either I_+ or I_-, we can conveniently calculate the first and second moments of the "coordinate" W by introducing the generating function $G(k, T)$,

$$G(k, T) = \left\langle \exp\left(-ik \int_0^T I(t)\, dt\right) \right\rangle = \langle e^{-ikW} \rangle \quad (A.51)$$

where k is an auxiliary variable. For the stochastic process where the "velocity" of the particle alternates between I_+ and I_- with transition rate R, we can easily evaluate $G(k, T)$ using the Laplace $s \to T$ transformation,

$$G(k, T) = \mathscr{L}^{-1}\left\{\frac{s + ik(I_+ + I_-)/2 + 2R}{(s + ikI_+ + R)(s + ikI_- + R) - R^2}\right\} \quad (A.52)$$

Now, from Eq. (A.51) we have $\langle W \rangle$ and $\langle W^2 \rangle$

$$\langle W \rangle = \left\langle \int_0^T I(t)\, dt \right\rangle = -\left.\frac{\partial G(k, T)}{\partial ik}\right|_{k=0}$$

$$\langle W^2 \rangle = \left\langle \left[\int_0^T I(t)\, dt\right]^2 \right\rangle = \left.\frac{\partial^2 G(k, T)}{\partial (ik)^2}\right|_{k=0} \quad (A.53)$$

By taking derivatives of $G(k, T)$, we find

$$-\left.\frac{\partial G(k, T)}{\partial ik}\right|_{k=0} = \mathscr{L}^{-1}\left\{\frac{I_+ + I_-}{2s^2}\right\}$$

$$\left.\frac{\partial^2 G(k, T)}{\partial (ik)^2}\right|_{k=0} = \mathscr{L}^{-1}\left\{\frac{(I_+^2 + I_-^2)}{s^2(s + 2R)} + \frac{(I_+ + I_-)^2 R}{s^3(s + 2R)}\right\} \quad (A.54)$$

which yield after the inverse Laplace transform,

$$\langle W \rangle = \left(\frac{I_+ + I_-}{2}\right) T$$

$$\langle W^2 \rangle = \frac{(I_+ + I_-)^2}{4} T^2 + \frac{(I_+ - I_-)^2}{4}\left(\frac{T}{R} + \frac{e^{-2RT} - 1}{2R^2}\right) \quad (A.55)$$

Q is given by Eq. (4.60) in the slow modulation regime.

F. Q in the Fast Modulation Regime

Based on the approximation introduced in the text we can calculate Q in the fast modulation regime. Once the factorization of the three-time correlation functions is made in Eq. (4.72), $\hat{\xi}_i(s)$, the functions determining $\langle W^2 \rangle$

[Eq. (A.32)] can be written as

$$\hat{\xi}_1(s) = \frac{2}{s^2} \hat{C}_1^{-1}(s + s_+) \hat{C}_1^1(s + s_-)$$

$$\hat{\xi}_2(s) = \frac{2}{s^2} \hat{C}_1^{-1}(s + s_+) \hat{C}_1^1(s + s_-)$$

$$\hat{\xi}_3(s) = \frac{2}{s^2} \hat{C}_1^1(s + s_-) \hat{C}_1^1(s + s_-)$$

$$\hat{\xi}_4(s) = \frac{2}{s^2} \hat{C}_1^1(s + s_-) \hat{C}_1^1(s + s_-)$$

$$\hat{\xi}_5(s) = \frac{4}{s^2} \hat{C}_1^1(s + s_-) \hat{C}_1^2(s + 2s_-) \hat{C}_1^1(s + s_-) \tag{A.56}$$

where $s_\pm = \Gamma/2 \pm i\omega_L$ as defined in the Appendix C. Then $\langle W \rangle$ and $\langle W^2 \rangle$ are calculated from Eqs. (4.53) and (4.54),

$$\langle W \rangle = \frac{\Omega^2}{4} \mathcal{L}^{-1} \left\{ \frac{1}{s^2} F_1(s) \right\} \tag{A.57}$$

$$\langle W^2 \rangle = \frac{\Omega^4}{8} \mathcal{L}^{-1} \left\{ \frac{1}{s^3} F_2(s) \right\} \tag{A.58}$$

where

$$F_1(s) = c_1(s) + \text{C.C.}$$

$$F_2(s) = \left(1 + \frac{s}{s + \Gamma}\right) F_1(s)^2 + 2s(c_1(s)^2 c_2(s) + \text{C.C.})$$

$$c_1(s) \equiv \hat{C}_1^1(s + s_-)$$

$$c_2(s) \equiv \hat{C}_1^2(s + 2s_-) \tag{A.59}$$

After the cumulant approximation is made for $C_1^l(t)$ and the long time limit is taken in Eq. (4.76) $c_1(z)$ and $c_2(z)$ are simply given by

$$c_1(s) = \frac{1}{s + s_- + \Gamma_f/2}$$

$$c_2(s) = \frac{1}{s + 2s_- + 2\Gamma_f} \tag{A.60}$$

Since only the long time limit is relevant for the calculation of Q in the fast modulation regime we make expansions of $\langle W \rangle$ and $\langle W^2 \rangle$ around $s = 0$, and find in the long time limit

$$Q = \frac{\Omega^2}{2} \frac{F'_2(0) - F_1(0)F'_1(0)}{F_1(0)} \quad \text{(A.61)}$$

Note that Eq. (A.61) is valid once the factorization approximation is made irrespective of the second-order cumulant approximation. After performing a lengthy but straightforward algebra using Eqs. (A.59)–(A.61), we obtain the result of Q in the fast modulation regime given as Eq. (4.77).

References

1. W. E. Moerner and M. Orrit, *Science* **283**, 1670 (1999).
2. W. E. Moerner and L. Kador, *Phys. Rev. Lett.* **62**, 2535 (1989).
3. M. Orrit and J. Bernard, *Phys. Rev. Lett.* **65**, 2716 (1990).
4. P. Tamarat, A. Maali, B. Lounis, and M. Orrit, *J. Phys. Chem. A* **104**, 1 (2000).
5. *Single-Molecule Optical Detection, Imaging and Spectroscopy*, edited by T. Basché, W. E. Moerner, M. Orrit, and U. P. Wild (VCH, Berlin, 1996).
6. W. P. Ambrose and W. E. Moerner, *Nature* (*London*) **349**, 225 (1991).
7. W. P. Ambrose, T. Basché, and W. E. Moerner, *J. Chem. Phys.* **95**, 7150 (1991).
8. A. Zumbusch, L. Fleury, R. Brown, J. Bernard, and M. Orrit, *Phy. Rev. Lett* **70**, 3584 (1993).
9. L. Fleury, A. Zumbusch, M. Orrit, R. Brown, and J. Bernard, *J. Lumin.* **56**, 15 (1993).
10. B. Kozankiewicz, J. Bernard, and M. Orrit, *J. Chem. Phys.* **101**, 9377 (1994).
11. M. Vacha, Y. Liu, H. Nakatsuka, and T. Tani, *J. Chem. Phys.* **106**, 8324 (1997).
12. A.-M. Boiron, Ph. Tamarat, B. Lounis, R. Brown, and M. Orrit, *Chem. Phys.* **247**, 119 (1999).
13. A. V. Naumov, Y. G. Vainer, M. Bauer, S. Zilker, and L. Kador, *Phys. Rev. B* **63**, 212302 (2001).
14. E. Geva and J. L. Skinner, *J. Phys. Chem. B* **101**, 8920 (1997).
15. F. L. H. Brown and R. J. Silbey, *J. Chem. Phys.* **108**, 7434 (1998).
16. E. Barkai, R. Silbey, and G. Zumofen, *Phys. Rev. Lett.* **84**, 5339 (2000).
17. H. Bach, A. Renn, G. Zumofen, and U. P. Wild, *Phys. Rev. Lett.* **82**, 2195 (1999).
18. J. Bernard, L. Fleury, H. Talon, and M. Orrit, *J. Chem. Phys.* **98**, 850 (1993).
19. T. Basché, S. Kummer, and C. Bräuchle, *Nature* (*London*) **373**, 132 (1995).
20. A. C. J. Brouwer, E. J. J. Groenen, and J. Schmidt, *Phys. Rev. Lett.* **80**, 3944 (1998).
21. M. Kuno, D. P. Fromm, H. F. Hamann, A. Gallagher, and D. J. Nesbitt, *J. Chem. Phys.* **112**, 3117 (2000).
22. R. G. Neuhauser, K. Shimizu, W. K. Woo, S. A. Empedocles, and M. G. Bawendi, *Phys. Rev. Lett.* **85**, 3301 (2000).
23. K. T. Shimizu, R. G. Neuhauser, C. A. Leatherdale, S. A. Empedocles, W. K. Woo, and M. G. Bawendi, *Phys. Rev. B* **63**, 205316 (2001).

24. T. Ha, T. Enderle, D. F. Ogletree, D. S. Chemla, P. R. Selvin, and S. Weiss, *Proc. Natl. Acad. Sci. U.S.A.* **93**, 6264 (1996).
25. D. A. vanden Bout, W. Yip, D. Hu, D. Fu, T. M. Swager, and P. F. Barbara, *Science* **277**, 1074 (1997).
26. W. Yip, D. Hu, J. Yu, D. A. vanden Bout, and P. F. Barbara, *J. Phys. Chem. A* **102**, 7564 (1998).
27. J. Wang and P. G. Wolynes, *Phys. Rev. Lett.* **74**, 4317 (1995).
28. V. Chernyak, M. Schulz, and S. Mukamel, *J. Chem. Phys.* **111**, 7416 (1999).
29. A. M. Berezhkovskii, A. Szabo, and G. H. Weiss, *J. Phys. Chem. B* **104**, 3776 (2000).
30. H. P. Lu, L. Y. Xun, and X. S. Xie, *Science* **282**, 1877 (1998).
31. G. K. Schenter, H. P. Lu, and X. S. Xie, *J. Phys. Chem. A* **103**, 10477 (1999).
32. N. Agmon, *J. Phys. Chem. B* **104**, 7830 (2000).
33. J. Cao, *Chem. Phys. Lett.* **327**, 38 (2000).
34. S. Yang and J. Cao, *J. Phys. Chem. B* **105**, 6536 (2001).
35. T. Ha, T. Enderle, D. S. Chemla, P. R. Selvin, and S. Weiss, *Phys. Rev. Lett.* **77**, 3979 (1996).
36. T. Ha, J. Glass, T. Enderle, D. S. Chemla, and S. Weiss, *Phys. Rev. Lett.* **80**, 2093 (1998).
37. L. Edman, *J. Phys. Chem. A* **104**, 6165 (2000).
38. P. D. Reilly and J. L. Skinner, *Phys. Rev. Lett.* **71**, 4257 (1993).
39. P. D. Reilly and J. L. Skinner, *J. Chem. Phys.* **101**, 959 (1994).
40. G. Zumofen and J. Klafter, *Chem. Phys. Lett.* **219**, 303 (1994).
41. Y. Tanimura, H. Takano, and J. Klafter, *J. Chem. Phys.* **108**, 1851 (1998).
42. I. S. Osad'ko and L. B. Yershova, *J. Chem. Phys.* **112**, 9645 (2000).
43. S. A. Empedocles, D. J. Norris, and M. G. Bawendi, *Phys. Rev. Lett.* **77**, 3873 (1996).
44. A. M. van Oijen, M. Ketelaars, J. Kohler, T. J. Aartsma, and J. Schmidt, *Science* **285**, 400 (1999).
45. M. A. Bopp, Y. W. Jia, L. Q. Li, R. J. Cogdell, and R. M. Hochstrasser, *Proc. Natl. Acad. Sci. U.S.A.* **94**, 10630 (1997).
46. J. Hofkens, M. Maus, T. Gensch, T. Vosch, M. Cotlet, F. Kohn, A. Herrmann, and K. Mullen, *J. Chem. Soc. Am.* **122**, 9278 (2000).
47. E. L. Elson and D. Magde, *Biopolymers* **13**, 1 (1974).
48. M. Ehrenberg and R. Rigler, *Chem. Phys.* **4**, 390 (1974).
49. D. E. Koppel, *Phys. Rev. A* **10**, 1938 (1974).
50. H. Qian, *Biophys. Chem.* **38**, 49 (1990).
51. Y. Chen, J. D. Müller, P. T. C. So, and E. Gratton, *Biophys. J.* **77**, 553 (1999).
52. C. Brunel, B. Lounis, P. Tamarat, and M. Orrit, *Phys. Rev. Lett.* **83**, 2722 (1999).
53. L. Fleury, J.-M. Segura, G. Zumofen, B. Hecht, and U. P. Wild, *Phys. Rev. Lett.* **84**, 1148 (2000).
54. B. Lounis and W. E. Moerner, *Nature (London)* **407**, 491 (2000).
55. T. Nonn and T. Plakhotnik, *Phys. Rev. Lett.* **85**, 1556 (2000).
56. A. Molski, *Chem. Phys. Lett.* **324**, 301 (2000).
57. E. Novikov and N. Boens, *J. Chem. Phys.* **114**, 1745 (2001).
58. P. W. Anderson, *J. Phys. Soc. Jpn.* **9**, 316 (1954).

59. R. Kubo, *J. Phys. Soc. Jpn.* **9**, 935 (1954).

60. R. Kubo, in *Fluctuation, Relaxation, and Resonance in Magnetic Systems*, edited by D. T. Haar (Oliver and Boyd, Edinburgh and London, 1962).

61. R. Kubo, M. Toda, and N. Hashitsume, *Statistical Physics*, 2nd ed. (Springer-Verlag, Berlin, 1991), Vol. 2.

62. S. Mukamel, *Principles of Nonlinear Optical Spectroscopy* (Oxford University Press, Oxford, 1995).

63. E. Barkai, R. Silbey, and G. Zumofen, *J. Chem. Phys.* **113**, 5853 (2000).

64. T. Plakhotnik and D. Walser, *Phys. Rev. Lett.* **80**, 4064 (1998).

65. T. Plakhotnik, *J. Lumin.* **83-4**, 221 (1999).

66. T. Plakhotnik, *Phys. Rev. B* **59**, 4658 (1999).

67. B. Saleh, *Photoelectron Statistics* (Springer-Verlag, Berlin, 1978).

68. L. Mandel and E. Wolf, *Optical Coherence and Quantum Optics* (Cambridge University Press, New York, 1995).

69. T. Basché, W. E. Moerner, M. Orrit, and H. Talon, *Phys. Rev. Lett.* **69**, 1516 (1992).

70. P. Michler, A. Imamoglu, M. D. Mason, P. J. Carson, G. F. Strouse, and S. K. Buratto, *Nature (London)* **406**, 968 (2000).

71. R. Short and L. Mandel, *Phys. Rev. Lett.* **51**, 384 (1983).

72. M. Schubert, I. Siemers, R. Blatt, W. Neuhauser, and P. E. Toschek, *Phys. Rev. Lett.* **68**, 3016 (1992).

73. C. Cohen-Tannoudji, J. Dupon-Roc, and G. Grynberg, *Atom Photon Interaction* (John Wiley & Sons, Inc., New York, 1993).

74. R. Loudon, *The Quantum Theory of Light*, 2nd ed. (Oxford University Press, Oxford, 1983).

75. K. D. Weston, P. J. Carson, H. Metiu, and S. K. Buratto, *J. Chem. Phys.* **109**, 7474 (1998).

76. I. S. Osad'ko, *J. Exp. Theo. Phys.* **86**, 875 (1998).

77. I. S. Osad'ko, *J. Exp. Theo. Phys.* **89**, 513 (1999).

78. M. S. Kim and P. L. Knight, *Phys. Rev. A* **36**, 5265 (1987).

79. E. Barkai, Y. Jung, and R. J. Silbey, *Phys. Rev. Lett.* **87**, 207403 (2001).

80. B. W. Shore, *J. Opt. Soc. Am. B* **1**, 176 (1984).

81. P. J. Colmenares, J. L. Paz, R. Almeida, and E. Squitieri, *J. Mol. Structure (THEOCHEM)* **390**, 33 (1997).

82. B. Lounis, F. Jelezko, and M. Orrit, *Phys. Rev. Lett.* **78**, 3673 (1997).

83. C. Brunel, B. Lounis, P. Tamarat, and M. Orrit, *Phys. Rev. Lett.* **81**, 2679 (1998).

84. A. Schenzle and R. G. Brewer, *Phys. Rev. A* **34**, 3127 (1986).

85. D. F. Walls and G. J. Milburn, *Quantum Optics* (Springer-Verlag, Berlin, 1994).

86. X. S. Xie, *Acc. Chem. Res.* **29**, 598 (1996).

87. H. Talon, L. Fleury, J. Bernard, and M. Orrit, *J. Opt. Soc. Am. B* **9**, 825 (1992). In this work, the motional narrowing phenomenon has been observed in the single molecule spectrum of pentacene in *p*-terphenyl crystal as the temperature of the system is increased.

88. *Tunneling systems in amorphous and crystalline solids*, edited by P. Esquinazi (Springer, Berlin, 1998).

89. Y. Jia, A. Sytnik, L. Li, S. Vladimirov, B. S. Cooperman, and R. M. Hochstrasser, *Proc. Natl. Acad. Sci. U.S.A.* **94**, 7932 (1997).
90. A. Suárez and R. Silbey, *Chem. Phys. Lett.* **218**, 445 (1994).
91. G. H. Weiss, *Aspects and Applications of the Random Walks* (North-Holland, Amsterdam, 1994).
92. W. P. Ambrose, P. M. Goodwin, J. Enderlein, D. J. Semin, J. C. Martin, and K. A. Keller, *Chem. Phys. Lett.* **269**, 365 (1997).
93. S. Mukamel and R. F. Loring, *J. Opt. Soc. Am. B* **3**, 595 (1986).
94. P. W. Anderson, B. I. Halperin, and C. M. Varma, *Philos. Mag.* **25**, 1 (1971).
95. W. A. Phillips, *J. Low Temp. Phys.* **7**, 351 (1972).
96. A. Heuer and R. Silbey, *Phys. Rev. Lett.* **70**, 3911 (1993).
97. V. Lubchenko and P. G. Wolynes, *Phys. Rev. Lett.* **87**, 195901 (2001).
98. R. Dum, A. S. Parkins, P. Zoller, and C. W. Gardiner, *Phys. Rev. A* **46**, 4382 (1992).
99. M. B. Plenio and P. L. Knight, *Rev. Mod. Phys.* **70**, 101 (1998).
100. J. Dalibard, Y. Castin, and K. Møllmer, *Phys. Rev. Lett.* **68**, 580 (1992).
101. K. Møllmer, Y. Castin, and J. Dalibard, *J. Opt. Soc. Am. B.* **10**, 524 (1993).
102. D. E. Makarov and H. Metiu, *J. Chem. Phys.* **111**, 10126 (1999).
103. D. E. Makarov and H. Metiu (2000), *J. Chem. Phys.* **115**, 5989 (2001).
104. Y. Jung, E. Barkai, and R. J. Silbey, *Chem. Phys.* (Strange Kinetics special issue), in press (2002).
105. A. I. Burshtein, *Sov. Phys. JETP* **22**, 939 (1966).
106. B. W. Shore, *The theory of coherent atomic excitation* (John Wiley & Sons, Inc., New York, 1990), Vol. 2.
107. S. Wolfram, *Mathematica*, Wolfram Research Inc., Champaign, IL.
108. L. Allen and J. H. Eberly, *Optical Resonance and Two-Level Atoms* (John Wiley & Sons, Inc., New York, 1975).
109. A. Molski, J. Hufkens, T. Gensch, N. Boens, and F. de Schryver, *Chem. Phys. Lett.*, **318**, 325 (2000).

THE ROLE OF SELF-SIMILARITY IN RENORMALIZATION GROUP THEORY

ANDRZEJ R. ALTENBERGER and JOHN S. DAHLER

*Department of Chemical Engineering and Materials Science
University of Minnesota, Minneapolis, MN 55455*

CONTENTS

I. Introduction
   A. The Self-Similarity of Ideal Systems
   B. Examples of Ideal, Strictly Self-Similar Systems
II. Nonideal Systems and the Hypothesis of Asymptotic Self-Similarity
   A. Functional Equations of Evolution for the Generalized Propagator Renormalization Group (GPRG)
   B. Direct Mapping Method for Solving the Equations of Evolution
   C. Lie Equations of the GPRG
   D. Callan–Symanzik Equations of the GPRG
   E. Average Squared Magnetization of an Open Chain of Ising Spins
III. Stretched-Exponential Scaling and the Positive Function Renormalization Group (PFRG)
   A. Iterative Solution of the PFRG Functional Equations
   B. Differential Equations of the PFRG
   C. PFRG Calculation of the Partition Function of the One-Dimensional Ising Model
IV. Linear, Nonideal Polymer Chain
   A. Calculation of Swelling Factor Using the GPRG Method
   B. Partition Function Calculation Using the PFRG Method
V. Compressibility Factors of Hard-Particle Fluids
   A. Self-Similarity of the Fluid Pressure
   B. Theory and Calculations
VI. Applications to Quantum Field Theory (QFT)
   A. Photon Propagator of Quantum Electrodynamics
   B. Screening in an Electron Gas
VII. Additive Formulation of the Renormalization Group Method
   A. Translational Self-Similarity
   B. Viscosity of a Hard-Sphere Suspension
VIII. Concluding Remarks
Acknowledgments
References

I. INTRODUCTION

During the past few decades, the concepts of scaling and self-similarity have become common fare in various fields of chemistry and physics. These concepts have provided the building blocks for a new calculational technique, commonly referred to as the renormalization group (RG) method. The fundamental ideas of scaling and self-similarity, upon which this technique is based, have long been recognized. However, while all calculational procedures identified by their originators as RG methods invariably involve scaling arguments of one sort or another, many make no explicit mention or use of the concept of self-similarity. Furthermore, it frequently happens that little evidence can be found of a connection between the computational algorithm used in a particular RG study and a mathematical group on which the procedure presumably is reliant. Some RG formulations are based on detailed theoretical analyses of the systems involved (e.g., quantum electrodynamics) while others rely on physical or abstract mathematical analogies and/or on field theoretic models. Thus, there is a wide variety of schemes to which the label of RG method has been affixed.

The procedure we present here is distinguished from others by our deliberate effort to exploit the notion of self-similarity and determine the extent to which this concept can be used to construct robust mathematical frameworks suitable for predicting various properties of physical systems that are deemed to be self-similar. In general terms, one identifies a particular system to be self-similar provided that its parts exhibit properties like those of the whole. Our RG method is not a theory specific to a single physical system or class thereof, but a systematic procedure designed to utilize theoretical calculations and/or computer simulations available for only limited values of a system parameter (e.g., number of particles, system density, strength of interactions) to generate reliable predictions of system properties over a much larger range of parameter values. There is no pretense that the predictions of this method are exact, for were that so we would have succeeded in solving incredibly difficult many-body problems. Thus, our self-similarity based RG method should be judged solely by its ability to produce accurate numerical approximations.

A. The Self-Similarity of Ideal Systems

A good example of a self-similar geometric object is the Koch curve [1], depicted in Figure 5.1, which is a deterministic fractal object generated by the following procedure: (1) A straight line segment of length r_0 is divided into three equal subsegments; (2) the middle of these is replaced with an equilateral triangle, thereby forming the "generator" of the transformation; (3) these two operations are reapplied to all subsegments of the generator

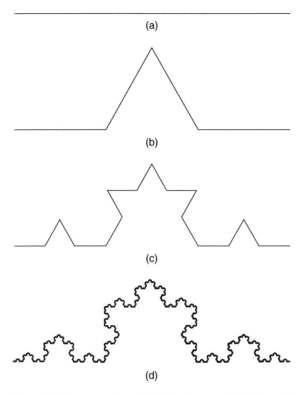

Figure 5.1. Stages in the construction of the Koch curve; (a) initiator; (b) generator; (c) reapplication of procedure and; (d) curve resulting from many repetitions of procedure.

and, finally; (4) the entire procedure is repeated n times. With N_n denoting the resulting number of subsegments and r_n the length of each, one identifies two sequences generated by this procedure:

$$
\begin{aligned}
n &= 0 & N_0 &= 1 & r_0 \\
n &= 1 & N_1 &= 4N_0 & r_1 &= r_0/3 \\
n &= 2 & N_2 &= 4^2 N_0 & r_2 &= r_0/3^2 \\
&\vdots & &\vdots & &\vdots \\
n &= k & N_k &= 4^k N_0 & r_k &= r_0/3^k
\end{aligned}
$$

The relationship between the number of subsegments N_k, produced by k repetitions and the length r_k, of the corresponding subsegments is found by solving the following pair of equations:

$$\ln \frac{N_k}{N_0} = k \ln 4$$

$$\ln \frac{r_k}{r_0} = -k \ln 3 \tag{5.1}$$

One thereby obtains the power-law formula

$$\frac{N_k}{N_0} = \left(\frac{r_k}{r_0}\right)^{-d_H} \tag{5.2}$$

involving a scaling exponent

$$d_H = \frac{\ln 4}{\ln 3} = \log_3 4 \tag{5.3}$$

known as the Hausdorf dimension of this geometric fractal object. The pure number d_H is independent from the arbitrary scale r_0, used as an initial measure of segment length. One can, in fact, use another scale such as $r = x$, without altering the number of segments N_k. Indeed, it is an immediate consequence of Eq. (5.2) that

$$\frac{N_k}{N_0} = \left(\frac{r_k}{x}\right)^{-d_H} \left(\frac{x}{r_0}\right)^{-d_H} \tag{5.4}$$

and this, in turn, can be rewritten as a functional equation

$$N_k\left(\frac{r_k}{r_0}; N_0\right) = N_k\left[\frac{r_k}{x}; N_k\left(\frac{x}{r_0}; N_0\right)\right] \tag{5.5}$$

for which the initial condition is

$$N_k(1; N_0) = N_0 \tag{5.6}$$

In these last two formulas, the number of segments generated by k repetitions of the subdivision procedure is treated as a function of the initial state N_0 and the relative length r_k/r_0 of the resulting subsegments. The latter of these plays the role of a generalized time, marking the number of repetitions. The power-law formula (5.2) and the functional equation (5.5) will appear later in several different contexts. However, as we shall see, the power law is only one member of a broad class of scaling relations that arise quite naturally from considerations of self-similarity.

The idea of functional similarity is basic to computer simulations, by means of which one seeks to extract the physical properties of many-body systems from studies conducted on similarly constituted systems composed of small numbers of interacting particles. The hope motivating such investigations is that properties obtained by examining a miniscule subsystem will differ insignificantly from those of a macroscopic system, the population of which may be of the order of Avogadro's number. For example, Monte Carlo simulations performed on samples consisting of a few hundreds or thousands of hard spheres often are designed to generate the equation of state $p = p(n, T)$, with $n = N/V$ denoting the particle density. The reason for conducting such an exercise does, of course, lie in the implicit assumption that precisely this same equation of state will be applicable to a macroscopic system of hard spheres: the equations of state of the microscopic and macroscopic systems are expected to exhibit functional self-similarity.

The assumption of self-similarity often is applied, almost as a matter of course, when one uses inductive reasoning to predict the properties of a large system on the basis of studies performed on relatively small subsystems thereof. This holds true not only in the physical sciences but for other areas of research as well. For example, statisticians obviously are only able to interview very limited samples of human populations. To generalize such findings to an entire population it is assumed that testimony elicited from the small sample accurately reflects opinions held by all members. Thus, the (response distribution of the) population is assumed to be self-similar, on the "scale" set by the sample size. This assumption can be tested by interviewing several samples, each consisting of a different number of participants. If the distributions of opinions are identical for all samples beyond a certain threshold size, it then is concluded that results based on the larger samples accurately represent the population as a whole, that is, the distribution is "asymptotically self-similar".

This example brings to the fore another important feature of the problem, namely, that prior to an investigation, the minimum or threshold size of a representative sample usually is unknown. We may have good reasons for expecting the scale to exhibit such a threshold value but in most cases its precise value must be determined empirically. In this respect, the behaviors of self-similar systems of physical interest are expected to be more complicated than those of the deterministic fractals, of which the previously mentioned Koch curve is one example.

Although the concept of functional self-similarity would appear to be intuitively clear, it is less apparent how this property should be defined mathematically. In most cases, interest is focused on the behavior of some physical quantity f that can be treated as a function of a single dimensionless real variable x, defined on the interval $1 \leq x < \infty$. And it usually is so that the "initial value" of this function at $x = 1$ is known to equal f_0. Thus, the physical quantity f can be represented by a function of the variable x (considered as a "generalized time" specific to the problem) and of the parameter f_0, namely, $f = f(x; f_0)$ with $f_0 \equiv f(1; f_0)$. Then, we say that f is self-similar with respect to the scaling transformation

$$x \to x' = x/t$$
$$f_0 \to f'_0 = f(t; f_0) \tag{5.7}$$

provided that it satisfies the functional equation

$$f(x; f_0) = f\left[\frac{x}{t}; f(t; f_0)\right] \tag{5.8}$$

According to this definition, the function $f(x; f_0)$ can be expressed as a (possibly quite complicated) function of its value at an earlier value t ($1 < t \leq x$) of the generalized time, now used as a revised value of the parameter f_0 and the rescaled, new time x/t. The defining equation (5.8) states that the value of the function at time x is unchanged if we shift the scale of time from $\tau = 1$ to $\tau = t$, *provided* that we simultaneously replace the initial condition $f_0 = f(1; f_0)$ with $f'_0 = f(t; f_0)$. The self-similar function f is said to be an invariant of the scaling transformation (5.7).

It easily is verified that, among the many functions that are self-similar in the sense of the "conservation law" (5.8), are the power law

$$f(x; f_0) = f_0 x^\sigma \tag{5.9}$$

the stretched exponential

$$f(x; f_0) = f_0^{x^\sigma} \tag{5.10}$$

and the function

$$f(x; f_0) = \frac{f_0}{1 + \sigma f_0 \ln x} \tag{5.11}$$

The symbol σ appearing in these three formulas denotes a numerical constant. In addition to these few examples, an entire class of self-similar functions can be generated by solving the differential equation

$$\frac{\partial f(t; f_0)}{\partial \ln t} = \beta[f(t; f_0)] \quad (5.12)$$

wherein

$$\beta(f_0) = \left.\frac{\partial f(t; f_0)}{\partial t}\right|_{t=1} \quad (5.13)$$

is a function called the "infinitesimal generator" of the scaling transformation.

Equation (5.12), known as the Lie equation of evolution for a dynamic system, is obtained directly from (5.8) by differentiating with respect to the scale. In quantum field theory [2], Eq. (5.12) often is called the Gell-Mann–Low equation or the equation for an invariant charge. The generator $\beta(f_0)$ frequently is referred to as the Gell-Mann function. The infinitesimal generators corresponding to the three functions of Eqs. (5.9)–(5.11) are

$$\beta(f_0) = \sigma f_0 \quad (5.14)$$

$$\beta(f_0) = \sigma f_0 \ln f_0 \quad (5.15)$$

and

$$\beta(f_0) = -\sigma f_0^2 \quad (5.16)$$

respectively.

Different from the Lie equation (5.12) is the Callan–Symanzik equation

$$\frac{\partial f(t; f_0)}{\partial \ln t} = \beta(f_0) \frac{\partial f(t; f_0)}{\partial f_0} \quad (5.17)$$

which can be obtained by differentiating the functional equation (5.8) with respect to the generalized time variable x. The functional equation (5.8), the Lie equation (5.12), and the Callan–Symanzik equation (5.17) are three essentially equivalent forms of the strict self-similarity condition for the object function $f(t; f_0)$.

An alternative interpretation of the functional equation (5.8) is to identify it as the defining relation of the group operation associated with the semigroup of scaling transformations (5.7). The variable x then plays the role of the group parameter, a discrete, or continuous index that labels elements of the group.

The functional equation formulation of RG theory is due to Bogoliubov and Shirkov [3–5]. As we shall see, this approach often is more convenient in practical, numerical calculations than the better known and more widely used formulations based on the Lie and Callan–Symanzik differential equations.

B. Examples of Ideal, Strictly Self-Similar Systems

A simple example of a physical quantity that is explicitly invariant with respect to the scaling transformation (5.7) is the average squared displacement associated with an unrestricted random walk. This quantity, $\overline{R^2}(N; \ell^2)$, depends on N, the number of steps and on ℓ^2, the squared length of a single step. The initial condition is $\overline{R^2}(1; \ell^2) = \ell^2$. For a random trajectory consisting of N uncorrelated steps [6]

$$\overline{R^2}(N; \ell^2) = N\ell^2 \tag{5.18}$$

This expression is an example of the power-law formula (5.9) with $f_0 = \ell^2$ and $x = N$.

The right-hand side of Eq. (5.18) can be manipulated as follows:

$$\overline{R^2}(N; \ell^2) = \frac{N}{K} K\ell^2 = \frac{N}{K} \overline{R^2}(K; \ell^2) = \overline{R^2}\left(\frac{N}{K}; \overline{R^2}(K; \ell^2)\right) \tag{5.19}$$

thus resulting in a "conservation equation" of the form (5.8). This equation (5.19) is a mathematical statement of a physically obvious fact: the average square of the distance traversed by an N-step random walk is unaltered (conserved) when one replaces the original squared step length ℓ^2 with the average squared displacement $\overline{R^2}(K; \ell^2)$ of a K-step subsequence and simultaneously reduces the number of steps from N to N/K. Consequently, the unrestricted random walk is a self-similar function of the number of steps and $\overline{R^2}(N; \ell^2)$ is an invariant of the scaling transformation

$$N \to N' = N/K$$
$$\ell^2 \to \ell'^2 = \overline{R^2}(K; \ell^2) \tag{5.20}$$

A second example of a self-similar object is the canonical partition function

$$Q(N; q(T)) = q(T)^N \tag{5.21}$$

specific to a system composed of N identical but distinguishable, noninteracting particles. This clearly is a highly nonlinear stretched-exponential function of the single-particle partition function $q(T)$, which also occurs in the initial condition $Q(1; q(T)) = q(T)$. From Eq. (5.21), one immediately obtains the relationship

$$Q(N; q(T)) = [q(T)^K]^{N/K} = [Q(K; q(T))]^{N/K}$$
$$= Q\left(\frac{N}{K}; Q(K; q(T))\right) \qquad (5.22)$$

which is a conservation equation identical in form to (5.8). It can be interpreted as a renormalization procedure: equivalent to N independent, single-particle subsystems each with a partition function $q(T)$ is a collection of N/K larger subsystems, each with a K-particle partition function $Q(K; q(T))$. The idea of clustering subelements to form larger, composite units is an essential part of the real-space RG method pioneered by Kadanoff and co-workers [7, 8] and de Gennes [9].

The examples of self-similar functions considered above fall into an especially simple category: all are functions of the "generalized time" or group parameter and functionals (either linear or nonlinear) of the initial condition. However, there are many self-similar physical properties and mathematical objects that depend on additional, unscaled variables. For example, the probability density for the distribution of end-to-end distances of a linear, ideal (phantom) polymer chain is given by the expression

$$G(\boldsymbol{R}; N, \psi) = \int \prod_{i=1}^{N} \{d\boldsymbol{q}_i \psi(\boldsymbol{q}_i)\} \delta\left[\boldsymbol{R} - \sum_{j=1}^{N} \boldsymbol{q}_j\right] \qquad (5.23)$$

Here, $\boldsymbol{R}$ denotes the vector connecting the two ends of the chain and $\boldsymbol{q}_i (i = 1, 2, \ldots, N)$ is the bond vector extending from site i to $i + 1$. The function $G(\boldsymbol{R}; N, \psi)$ depends parametrically on the bond number N, which for a given value of $\boldsymbol{R}$, can be identified as the generalized time appropriate to this problem. Furthermore, $G(\boldsymbol{R}; N, \psi)$ is a nonlinear functional of the unit-normalized bond-vector distribution function

$$\psi(\boldsymbol{q}) = Z(T)^{-1} \exp[-\beta u(|\boldsymbol{q}|)] \qquad (5.24a)$$

which depends, in turn, on the nearest-neighbor site interaction potential $u(|\boldsymbol{q}|)$. And finally, the initial condition obviously satisfied by this function is

$$G(\boldsymbol{R}; 1, \psi) = \psi(\boldsymbol{R}) \qquad (5.24b)$$

Alternatively, Eq. (5.23) can be recast as a Chapman–Kolmogorov equation,

$$G(\mathbf{R}; N, \psi) = \int d\mathbf{R}' \psi(\mathbf{R} - \mathbf{R}') G(\mathbf{R}'; N - 1, \psi) \tag{5.25}$$

for the Markov process associated with the transition probability $\psi(\mathbf{R})$. The corresponding initial condition is

$$G(\mathbf{R}; 0, \psi) = \delta(\mathbf{R}) \tag{5.26}$$

To show explicitly that the probability density $G(\mathbf{R}; N, \psi)$ is indeed a self-similar object, we introduce its Fourier transform (FT)

$$G_k(N; \psi) \equiv \int d\mathbf{R} e^{-i\mathbf{k}\cdot\mathbf{R}} G(\mathbf{R}; N, \psi) = \psi_k^N \tag{5.27}$$

wherein

$$\psi_k = \int d\mathbf{R} e^{i\mathbf{k}\cdot\mathbf{R}} \psi(\mathbf{R}) \tag{5.28}$$

Then, the relationship (5.27) can be rewritten as a functional equation

$$G_k(N; \psi_k) = G_k\left[\frac{N}{K}; G_k(K; \psi_k)\right] \tag{5.29}$$

for which the initial condition is

$$G_k(1; \psi_k) = \psi_k \tag{5.30}$$

According to (5.29) the (FT of the) probability densities for the end-to-end vectors of all K-bond macrosegments are functionally similar to the corresponding probability density for the entire chain. These macrosegments differ from the complete chain only by virtue of the number of monomers from which they are constituted. The fact that the segment is part of the larger chain does not affect its statistical properties. This is a feature common to all strictly self-similar, "ideal" systems.

In all the preceding examples, the generalized time (or group parameter) has been a discrete variable. However, the concept of self-similarity is equally applicable to physical objects or processes described by a continuous time-like variable. For example, the probability density that, during a time interval of duration $t > 0$, a Brownian particle will experience a vector displacement R is given by the formula

$$p(R; t) = [8(\pi Dt)^{3/2}]^{-1} \exp[-R^2/4Dt] \qquad (5.31)$$

Here, D denotes the particle diffusion coefficient. The FT of this function is

$$p_k(t) = \exp[-k^2 Dt] \qquad (5.32)$$

By introducing an arbitrarily selected unit of time, τ_0, and the corresponding dimensionless time variable $\tau = t/\tau_0$, we can rewrite Eq. (5.32) in the form

$$p_k(\tau; p_k^0) = [p_k^0]^\tau \qquad (5.33)$$

wherein

$$p_k^0 = \exp[-k^2 D\tau_0] \qquad (5.34)$$

With τ taken to be the *discrete* variable $\tau = n\tau_0$ (for $n = 1, 2, \ldots$), we obtain from (5.33) an expression

$$p_k(n\tau_0; p_k^0) = p_k^0 p_k[(n-1)\tau_0; p_k^0] \qquad (5.35)$$

which is the Fourier representation of the Chapman–Kolmogorov equation appropriate to the diffusion problem. Alternatively, particle diffusion can be identified as a self-similar process for which the conservation equation is

$$p_k(\tau; p_k^0) = p_k\left[\frac{\tau}{x}; p_k(x; p_k^0)\right] \qquad (5.36)$$

Strictly speaking, the size of the scale τ_0 [appearing in (5.34) and (5.35)] must exceed the threshold value $\tau_{th} \simeq \beta m D$. Here m denotes the mass of the diffusing particle and $\beta = 1/k_B T$ with k_B the Boltzmann constant. For intervals of time shorter than τ_{th}, the particle motions are not controlled by collisions with other particles of the medium. Instead, the particles move freely along "ballistic" trajectories, $R(t) = R(0) + (p/m)t$, as they do in the example of a dilute gas to which we now turn. Specifically, we consider the van Hove intermediate scattering function for a single thermalized particle.

This function is given by the formula

$$I_k(t) = \int d\boldsymbol{p}\varphi(\boldsymbol{p}) \exp[-i\boldsymbol{k}\cdot\{\boldsymbol{R}(t) - \boldsymbol{R}(0)\}] = \exp\left[\frac{-k^2 t^2}{2m\beta}\right] \quad (5.37)$$

with $\varphi(\boldsymbol{p})$ denoting the Maxwellian momentum distribution function. In terms of the scale unit τ_0 and the corresponding dimensionless time $\tau = t/\tau_0$, this formula can be rewritten as

$$I_k(\tau; I_k^0) = [I_k^0]^{\tau^2} \quad (5.38)$$

with

$$I_k^0 = \exp[-k^2\tau_0^2/2m\beta] \quad (5.39)$$

equal to the vanHove function evaluated for the scale interval of time. The expression (5.38) has the form of the extended exponential function (5.8). As such, it represents a quantity that satisfies the conservation equation

$$I_k(\tau; I_k^0) = I_k\left[\frac{\tau}{x}; I_k(x; I_k^0)\right] \quad (5.40)$$

expressing the condition of scaling invariance.

The examples provided above illustrate both the meaning and the ubiquitousness of functional self-similarity, thereby offering an alternative perspective from which to view physical problems for which mathematical solutions already may be well known. Generally speaking, strict functional self-similarity of a physical quantity with respect to a scaling transformation means that it can be represented mathematically as the solution of a functional equation of the form (5.8). However, one must look elsewhere to discover the utility of the self-similarity concept as the basis of a systematic approximation method applicable to more difficult problems. Systems that are strictly self-similar, such as those considered above, invariably are formed from elements that are uncorrelated. Whether one considers an unrestricted random walk or the partition function of a collection of noninteracting particles, their common feature is that any subdivision of the "generalized time" variable produces subsystems with properties described by the same function of time (and other parameters as well) as that pertaining to the entire system.

Things are very different when one considers systems for which correlations are of great importance. Thus, restricted (or self-avoiding) random walks and/or systems composed of interacting particles are not expected to be strictly self-similar for *any* choice of scale. For example, it is well known that small atomic or molecular clusters often exhibit properties that are quite different from those of the corresponding bulk materials. Experimental investigations as well as studies of solvable statistical models [10] provide information on the rates at which the macroscopic, bulk values of physical properties are approached as the number of particles grows. The concepts of a "thermodynamic limit" and of the asymptotic, "long-time" behavior of dynamical properties of many-body systems clearly are related to expectations that the values of certain system properties will become stabilized provided that the chosen scale exceeds some threshold value *and* that the system itself is much larger than this critical scale. In other words, even strongly correlated systems may prove to be approximately self-similar, provided that their behaviors are examined from the perspective of sufficiently coarse-grained scales. This hypothesis of the asymptotic self-similarity of large correlated systems lies at the heart of the RG technique outlined in Section II.

These ideas of asymptotic functional self-similarity and of a critical scale find their expression in the postulatory formulations of the thermodynamics of uniform systems due to Tisza [11] and Callen [12]. Thus, the internal energy of a single-component system is written in the form

$$U(S, V, N) = TS - pV + \mu N \tag{5.41}$$

of an Euler equation for a first-order homogeneous function of the extensive variables S, V, and N, the entropy, volume, and number of particles, respectively. The absolute temperature $T = (\partial U/\partial S)_{V,N}$, the pressure $p = -(\partial U/\partial V)_{S,N}$, and the chemical potential $\mu = (\partial U/\partial N)_{S,V}$ are the associated intensive variables and can be identified as zero-order homogeneous functions.

The Euler equation (5.41) also can be written in the form

$$U(t; U_0) = U(tS_0, tV_0, tN_0) = t(TS_0 - pV_0 + \mu N_0) \tag{5.42}$$

with t denoting the scaling factor

$$t = U/U_0 = S/S_0 = V/V_0 = N/N_0 \tag{5.43}$$

It is obvious from this that the internal energy is a self-similar function with respect to the scaling transformation that divides the system into t identical subsystems, each characterized by the variables S_0, V_0, and N_0. The associated functional relationship

$$U(t;\ U_0) = U[t/x;\ U(x;\ U_0)] \tag{5.44}$$

is expected to be valid provided that the values of the variables S_0, V_0, and N_0 are not too small, that is, provided that the reference system itself is macroscopic.

II. NONIDEAL SYSTEMS AND THE HYPOTHESIS OF ASYMPTOTIC SELF-SIMILARITY

The occurrence of correlations among the elementary units of a system, be they physical particles, steps of a random walk, or some more complex objects, usually can be viewed as a consequence of mutual interactions, to which some sort of potential energy can be assigned. This, in turn, introduces an additional parameter, g, called the "bare" coupling constant. The value of this parameter provides a measure of the strength of the specific interactions that are responsible for deviations of property values from those characteristic of the corresponding ideal system and for which the value of g is zero.

It often is convenient, and always possible, to separate the "ideal" and "excess" parts of any physical quantity $f(t; g, f_0)$ by writing it in the form

$$f(t; g, f_0) = f_{id}(t; f_0)\delta f(t; g) \tag{5.45}$$

with

$$f_{id}(t; f_0) = f(t; 0, f_0) \tag{5.46}$$

denoting the ideal part of the quantity f. The complementary excess part, δf, is independent from the initial condition f_0. Furthermore, because of the initial condition

$$f_0 = f(1; g, f_0) \tag{5.47}$$

it follows that

$$\delta f(1; g) = 1 \tag{5.48}$$

One example of a physical quantity exhibiting a separation such as (5.45) is the average squared end-to-end separation of a nonideal polymer chain (with nonnearest-neighbor site interactions). This system property can be written as the product

$$\overline{R^2}(N;g,\ell^2) = N\ell^2\alpha^2(N;g) \tag{5.49}$$

of a swelling factor α^2, serving as an "excess" corrective factor, with the average squared end-to-end separation $\overline{R^2_{id}}(N;\ell^2) = N\ell^2$ of the corresponding ideal chain [9]. For the latter, there are only nearest-neighbor site interactions. Here, it has been assumed that the nonnearest-neighbor, "excluded-volume" interactions can be characterized by a single scalar valued coupling parameter g. A second example is the canonical partition function of this same nonideal chain, written in the form

$$Q[N;g,q(T)] = q(T)^N \delta Q(N;g) \tag{5.50}$$

The quantity $q(T)$ is the partition function of a single segment, connecting two adjacent sites of the chain. The excess partition function δQ accounts for the correlations among these segments, produced by the excluded-volume interactions. A third and final example is the equation of state of an imperfect gas that often is written as the product

$$p(N;g,k_B T/V) = N(k_B T/V)\phi(N;g) \tag{5.51}$$

of the ideal contribution and a corrective factor $\phi(N;g)$ called the compressibility factor [13].

These three examples, as well as others, will be treated in greater detail below. Our intention here is simply to illustrate the practicability of separating the function representative of a physical quantity into an ideal part that easily can be evaluated and an excess part that almost never is susceptible to exact evaluation. The challenge, of course, is to devise a (necessarily approximate) method for estimating the excess part. It is in this context that the hypothesis of asymptotic self-similarity can be used to construct a nonperturbative calculational procedure.

The basic premise of the RG method presented here is that, under certain conditions, a nonideal system may display a self-similarity that closely mimics that shown by the corresponding ideal system. However, unlike the ideal, strictly self-similar systems for which scale size is essentially irrelevant, nonideal systems are expected to display self-similarity with respect to scaling only for scales in excess of some, initially unspecified threshold value. Indeed, the value of this threshold must be determined a posteriori, by comparing theoretical predictions with reference data that serve to gauge the effectiveness of the approximation scheme.

To provide mathematical implementation for these premises, we write the basic conservation equation for a quantity $f(t; g, f_0)$ as

$$f(t; g, f_0) = f\left[\frac{t}{x}; \bar{g}(x; g), f(x; g, f_0)\right] \quad (5.52)$$

and identify this equation as a generalized statement of invariance with respect to the scaling transformation

$$t \to t' = t/x$$
$$g \to g' = \bar{g}(x; g) \quad (5.53)$$
$$f_0 \to f'_0 = f(x; g, f_0)$$

The function $\bar{g}(t; g)$ is known as the effective (or running) coupling function.

According to Eq. (5.52), the function $f(t; g, f_0)$ is an invariant of a renormalization procedure, in the course of which the generalized time is divided into t/x steps, each of length, x, the bare coupling parameter is replaced by its effective counterpart and the initial value of the function at $t = 1$ is replaced with the function evaluated at the new scale length x. In the limit $g = 0$ the conservation equation (5.52) must reduce to the corresponding equation (5.6), appropriate to the ideal system. Therefore, the effective coupling function must satisfy the condition $\bar{g}(t; 0) = 0$, $\forall t$.

If the conservation equation (5.52) is to be satisfied, the as yet unspecified effective coupling function must be a functional of the object function $f(t; g, f_0)$. A self-consistent procedure for establishing this functional relationship is presented later in this section. Once this has been done, one can derive coupled functional equations of evolution for the excess part $\delta f(t; g)$ and the effective coupling function $\bar{g}(t, g)$. From these equations, estimates of $\delta f(t; g)$ can be constructed that are applicable to greater ranges of group parameter and bare coupling parameter values than is possible with a standard perturbation series such as

$$\delta f(t; g) = 1 + g\delta f_1(t) + \frac{1}{2}g^2 \delta f_2(t) + \cdots \quad (5.54)$$

A. Functional Equations of Evolution for the Generalized Propagator Renormalization Group (GPRG)

Problems of particular interest are those for which the quantity $f(t; g, f_0)$ is a linear function of the initial condition and for which the ideal part has a

THE ROLE OF SELF-SIMILARITY IN RENORMALIZATION GROUP THEORY 283

power-law dependence on the generalized time (characterized here by a numerical exponent σ), namely,

$$f(t; g, f_0) = \{f_0 t^\sigma\} \delta f(t; g) \tag{5.55}$$

The right-hand side of this expression can be rewritten as the sequence of equalities

$$f(t; g, f_0) = f_0 x^\sigma \left(\frac{t}{x}\right)^\sigma \delta f(t; g)$$

$$= f_0 x^\sigma \delta f(x; g) \left(\frac{t}{x}\right)^\sigma \frac{\delta f(t; g)}{\delta f(x; g)} = f(x; g, f_0) \left(\frac{t}{x}\right)^\sigma \frac{\delta f(t; g)}{\delta f(x; g)} \tag{5.56}$$

If one then is able to find an effective coupling function $\bar{g}(t; g)$ such that the excess part of f satisfies the nonlinear functional equation

$$\frac{\delta f(t; g)}{\delta f(x; g)} = \delta f\left[\frac{t}{x}; \bar{g}(x; g)\right] \tag{5.57}$$

Eq. (5.56) can be further manipulated into a form

$$f(t; g, f_0) = f(x; g, f_0) \left(\frac{t}{x}\right)^\sigma \delta f\left[\frac{t}{x}; \bar{g}(x; g)\right]$$

$$= f\left[\frac{t}{x}; \bar{g}(x; g), f(x; g, f_0)\right] \tag{5.58}$$

which is precisely that of the previously proposed, generalized self-similarity conservation equation (5.52). The quantity $f(t; g, f_0)$ will be self-similar with respect to the scaling transformation (5.53) only if the excess part of f satisfies Eq. (5.57).

The following, different rearrangement of (5.56) and (5.57)

$$f(t; g, f_0) = f_0 \left(\frac{t}{x}\right)^\sigma \delta f\left[\frac{t}{x}; \bar{g}(x, g)\right] x^\sigma \delta f(x; g)$$

$$= f\left[\frac{t}{x}; \bar{g}(x; g), f_0\right] x^\sigma \delta f(x; g) \tag{5.59}$$

provides an alternative and equivalent form of the conservation equation (5.52), namely,

$$f(t; g, f_0) = f\left[x; g, f\left(\frac{t}{x}; \bar{g}(x; g), f_0\right)\right] \qquad (5.60)$$

The effective coupling function thus far remains essentially unspecified. However, it is obvious that this function plays a central role in the RG theory and, as we now shall demonstrate, the proper choice of this function is not arbitrary, as sometimes has been suggested [14], but a natural consequence of the basic conservation equation (5.52). To obtain the functional equation of evolution for $\bar{g}(t; g)$ we rewrite Eq. (5.57) in the multiplicative form

$$\delta f(tx; g) = \delta f(x; g)\delta f[t; \bar{g}(x; g)]$$
$$= \delta f(t; g)\delta f[x; \bar{g}(t; g)] \qquad (5.61)$$

which clearly is invariant with respect to the transposition of indexes, $t' = tx = xt$. This equation can be considered as representative of a semigroup; it provides a prescription for creating (the representative of) a new group element with index $t' = tx$ from two others with group indexes t and x.

From the last two equalities of (5.61), we obtain the expression

$$\delta f[t; \bar{g}(x; g)] = \frac{\delta f(t; g)}{\delta f(x; g)} \delta f[x; \bar{g}(t; g)]$$
$$= \delta f\left[\frac{t}{x}; \bar{g}(x; g)\right]\delta f[x; \bar{g}(t; g)] \qquad (5.62)$$

or, equivalently,

$$\frac{\delta f[t; \bar{g}(x; g)]}{\delta f[t/x; \bar{g}(x; g)]} = \delta f[x; \bar{g}(t; g)] \qquad (5.63)$$

Next, by applying the rule (5.58) to the left-hand side of (5.63) it follows that

$$\delta f\left[x; \bar{g}\left(\frac{t}{x}; \bar{g}(x; g)\right)\right] = \delta f[x; \bar{g}(t; g)] \qquad (5.64)$$

Since this is an identity only if

$$\bar{g}(t; g) = \bar{g}\left[\frac{t}{x}; \bar{g}(x; g)\right] \tag{5.65}$$

we conclude that the effective coupling function is itself an invariant of the scaling transformation. Furthermore, it belongs to the class of strictly self-similar functions, defined by Eq. (5.8). And finally, this function obviously satisfies the initial condition

$$\bar{g}(1; g) = g \tag{5.66}$$

Although the equation of evolution for $\bar{g}(t; g)$ now has been identified, we still must specify the relationship between this function and the object function $f(t; g, f_0)$. We also must develop a practical, computational procedure for implementing the multiplicative RG method. These items will be treated in the following sections. However, before proceeding to these issues it is worthy of notice that in standard formulations of the theory the equations of evolution are derived not from the conservation equation (5.52), as given above, but by postulating a proportionality between the "original" and "renormalized" excess functions. This postulate can be written in the form

$$\delta f(t; g) = z(x; g)\delta f\left[\frac{t}{x}; \bar{g}(x; g)\right] \tag{5.67}$$

where $z(x; g)$ is an as yet unspecified factor that usually is called the renormalization constant (or amplitude) but which, in fact, is a function of the new "scale" x and the bare coupling constant g. These are treated as parameters of the RG transformation

$$t \to t' = t/x$$
$$g \to g' = \bar{g}(x; g) \tag{5.68}$$

in the course of which the scale of the generalized time is shifted (from $x = 1$ to some $x > 1$) and the "bare" coupling constant g is replaced with the "dressed" coupling function $\bar{g}(x; g)$. The value of the renormalization constant z can be determined from the normalization condition (5.48). Thus, by setting $t = x$ in Eq. (5.67) we obtain the result

$$z(x; g) = \delta f(x; g) \tag{5.69}$$

which, when substituted in (5.67), leads by a different route to Eq. (5.57). This second, alternative way of introducing the renormalization procedure has been used, for example, by Bogoliubov and Shirkov [3] and by Shirkov [4]. It has appeared in other expositions of quantum field theory, as well. The pair of evolution equations, (5.61) and (5.65), often arise in quantum field theory in connection with the determination of various Green functions by means of the RG method. For this reason, we call them the generalized propagator renormalization group (GPRG), although the same equations occur in different contexts in other areas of physics.

B. Direct Mapping Method for Solving the Functional Equations of Evolution

The initial conditions for the RG transformation require that one specify the threshold scale τ, above which the system is expected to exhibit the property of self-similarity. Once this scale has been selected we express the generalized time as a power of the scale, namely, $t = \tau^n$, so that (with $x = \tau$) the functional equation (5.57) can be written in the form

$$\delta f(\tau^n; g) = \delta f(\tau; g) \delta f[\tau^{n-1}; \bar{g}(\tau; g)] \tag{5.70}$$

Simple iteration then permits us to express the object function $\delta f(t = \tau^n; g)$, for $t > \tau$, as a functional of this same excess function evaluated at a time equal to the scale length. Thus, we find that

$$\delta f(\tau^n; g) = \delta f(\tau; g) \prod_{i=1}^{n-1} \delta f[\tau; \bar{g}^{(i)}(\tau; g)] \tag{5.71}$$

where $\bar{g}^{(i)}(\tau; g)$ denotes the i-fold nested iterate of $\bar{g}(\tau; g)$, defined as follows:

$$\bar{g}^{(1)}(\tau; g) = \bar{g}(\tau; g)$$
$$\bar{g}^{(2)}(\tau; g) = \bar{g}[\tau; \bar{g}(\tau; g)] \tag{5.72}$$
$$\vdots$$
$$\bar{g}^{(n)}(\tau; g) = \bar{g}[\tau; \bar{g}^{(n-1)}(\tau; g)]$$

Similarly, the evolution equation for the effective coupling function can be written in the form

$$\bar{g}(\tau^n; g) = \bar{g}[\tau; \bar{g}(\tau^{n-1}; g)] \tag{5.73}$$

THE ROLE OF SELF-SIMILARITY IN RENORMALIZATION GROUP THEORY   287

and the solution thereof as

$$\bar{g}(\tau^n; g) = \bar{g}^{(n)}(\tau; g) \tag{5.74}$$

The formula (5.71) allows one to compute the value of excess function $\delta f(t = \tau^n; g)$, provided that the value of this function is known for a time equal to the scale τ *and* that one is able to determine $\bar{g}(\tau; g)$, the value of the effective coupling function for this same scale. To obtain the latter, we have only to solve the equation [(5.71) for $n = 2$]

$$\frac{\delta f(\tau^2; g)}{\delta f(\tau; g)} = \delta f[\tau; \bar{g}(\tau; g)] \tag{5.75}$$

thus determining $\bar{g}(\tau; g)$ as a functional of the ratio $\delta f(\tau^2; g)/\delta f(\tau; g)$. This is the step that insures the self-consistency of the method, namely, that the effective coupling function $\bar{g}(t; g)$ be a functional of the object function $\delta f(t; g)$.

To initiate this computational algorithm it is necessary to know two values of the excess function, namely, $\delta f(\tau; g)$ and $\delta f(\tau^2; g)$, as functions of the bare coupling parameter g. The scale "length" τ should, in principle, be selected equal to the shortest interval of the generalized time beyond which the nonideal system is expected to display self-similarity. Once $\delta f(\tau; g)$ and $\delta f(\tau^2; g)$ have been evaluated, construction of the excess quantity $\delta f(t; g)$ for $t \gg \tau$ becomes a relatively easy task. This is the computational advantage that one gains by using the RG method of calculation based on the postulate of self-similarity.

The evolution equation (5.70) obviously is nonlinear and, as such, it can exhibit a variety of interesting properties. Of particular interest is the appearance of so-called "fixed points", by which we refer to values g^* of the bare coupling parameter that satisfy the condition

$$\bar{g}(t; g^*) = g^* \tag{5.76}$$

and/or

$$\lim_{t \to \infty} \bar{g}(t; g^*) = g^* \tag{5.77}$$

In other words, g^* is a horizontal asymptote of the function $\bar{g}(t; g)$, in the limit $t \to \infty$. The fixed points characteristic of the dynamic system $\delta f(t; g)$

and for the scale τ can be determined by solving the equation

$$\delta f(\tau^2; g^*) = [\delta f(\tau; g^*)]^2 \tag{5.78}$$

which follows directly from (5.75). The roots of this equation constitute the fixed-point spectrum of $\delta f(t; g)$ associated with τ, the selected value of the scale. If this system is indeed asymptotically self-similar, the spectrum will be independent of scale for values of τ in excess of the threshold.

For $g = g^*(\tau)$ the formula (5.71) reduces to the expression

$$\delta f[t; g^*(\tau)] = [\delta f(\tau; g^*(\tau))]^{\ln t/\ln \tau} = t^{\ln \delta f[\tau; g^*(\tau)]/\ln \tau} \equiv t^{\log_\tau \delta f[\tau; g^*(\tau)]} \tag{5.79}$$

which is a power law with a scale-dependent exponent

$$\Lambda(\tau) = \log_\tau \delta f[\tau; g^*(\tau)] \tag{5.80}$$

Because this exponent is a functional of the excess quantity δf, it depends on the microscopic model involved in the definition of the physical property f. Thus, the scaling exponents of nonideal, correlated systems are not simple numbers as were those of the uncorrelated, fractal objects considered in Section I. Instead, they depend on the threshold value of the scale, above which the system is expected to exhibit self-similarity.

For values of g that differ from a particular fixed point $g^*(\tau)$, but lie within its domain of attraction (values that tend to g^* at $t \to \infty$), the excess quantity $\delta f(t; g)$ can be written as a product,

$$\delta f(t; g) = t^{\Lambda(\tau)} \Phi[t; \tau, g, g^*(\tau)] \tag{5.81}$$

of the dominant ($t \to \infty$) factor given by (5.79) and the "scaling function" or "subdominant part" Φ. The latter is defined by the formula

$$\Phi = \frac{\delta f(\tau; g)}{\delta f(\tau; g^*)} = \prod_{i=1}^{n-1} \frac{\delta f[\tau; \bar{g}^{(i)}(\tau; g)]}{\delta f[\tau; g^*]} \tag{5.82}$$

with $n = \log_\tau t$. This scaling function is expected to approach a constant value for $t \to \infty$ and $g \neq g^*$.

As may be evident to some readers, the method described above is excellently suited to the task of calculating the swelling factor for a nonideal polymer chain [15]. In this case, the threshold scale for self-similarity can be identified with the length of the Kuhn segment, a statistically uncorrelated macrosegment of the chain. The idea of dividing a chain into a collection of uncorrelated "blobs" first was used by de Gennes [9] in his formulation of the RG method. We later shall examine this system in more detail, as one of several examples of the general RG procedure.

C. Lie Equations of the GPRG

While the functional equations (5.61) and (5.65), are very well suited to the above mentioned, iterative method for determining the (generalized) time evolution of the dynamic system $f(t; g, f_0)$, most presentations of the RG method have instead used an approach based on differential equations. These differential equations can be written directly as equations of evolution for the object function $\delta f(t; g)$ and the related effective coupling function $\bar{g}(t; g)$ [3–5, 16] or as a pair of partial differential equations known as the Callan–Symanzik equations [3–5, 17]. These three forms of the RG theory are essentially equivalent. However, we personally favor the functional equation approach, not only from a computational point of view but because it provides better insight into the workings of the postulates of the self-similarity based RG technique.

The Lie equations of the generalized propagator renormalization group can be derived directly from the functional equations (5.57) and (5.65). Thus, by differentiating the first of these with respect to the group parameter t we obtain an expression

$$\frac{\partial \delta f(t; g)}{\partial t} = \delta f(x; g) \frac{\partial \delta f[t/x; \bar{g}(x; g)]}{\partial(t/x)} \frac{1}{x} \quad (5.83)$$

which is satisfied for all times t greater than or equal to the fixed scale x. With t set equal to x we obtain from (5.82) the Lie equation

$$\frac{\partial \ln \delta f(t; g)}{\partial \ln t} = \gamma[\bar{g}(t; g)] \quad (5.84)$$

with $\gamma(g)$, the so-called infinitesimal generator of the RG transformation, defined as follows:

$$\gamma(g) = \frac{\partial \delta f(t; g)}{\partial t}\bigg|_{t=1} \quad (5.85)$$

In similar fashion, one obtains from (5.65) a relationship

$$\frac{\partial \bar{g}(t; g)}{\partial t} = \frac{1}{x} \frac{\partial \bar{g}(t/x; \bar{g}(x; g))}{\partial(t/x)} \quad (5.86)$$

which, in the limit $t \to x$, reduces to the differential Lie equation of evolution for the effective coupling function:

$$\frac{\partial \bar{g}(t; g)}{\partial \ln t} = \beta[\bar{g}(t; g)] \quad (5.87)$$

The function $\beta(g)$, defined by the expression

$$\beta(g) = \frac{\partial \bar{g}(t; g)}{\partial t}\bigg|_{t=1} \quad (5.88)$$

and known as the Gell–Mann–Low function, also can be identified as the infinitesimal generator of the effective coupling transformation.

The pair of Lie equations (5.84) and (5.87) govern the evolution of the dynamical system $\delta f(t; g)$. As we previously have emphasized, the effective coupling function must be determined self-consistently, as a function of δf. This critical step in the procedure is accomplished by expressing the generator $\beta(g)$ as the following functional of the object function δf;

$$\beta(g) = \frac{\partial/\partial t[\partial \ln \delta f(t; g)/\partial \ln t]|_{t=1}}{d\gamma(g)/dg} = \frac{\gamma(g) - \gamma(g)^2 + (\partial^2 \delta f/\partial t^2)|_{t=1}}{d\gamma(g)/dg} \quad (5.89)$$

This important relationship is the result of differentiating both sides of Eq. (5.84) with respect to the generalized time t and then setting t equal to unity.

In this formulation of the theory, the fixed points of the GPRG are identified as zeros of the infinitesimal generator $\beta(g)$, that is, as roots of the equation

$$\beta(g^*) = 0 \quad (5.90)$$

Fixed points that are stable in the limit $t \to \infty$ are those for which $d\beta(g)/dg|_{g=g^*} < 0$. The approximate solutions obtained by solving the Lie equations are dependent on the initial estimates of the infinitesimal generators β and γ and these, in turn, depend functionally on the initial estimate that one has adopted for the excess quantity $\delta f(t; g)$.

When the bare coupling parameter precisely equals the fixed-point value g^*, the solution of equation (5.84) is the power-law formula

$$\delta f(t; g = g^*) = t^{\gamma(g^*)} \quad t > 1 \quad (5.91)$$

the exponent of which equals the value of the infinitesimal generator $\gamma(g)$, evaluated for $g = g^*$. The function defined by this formula, (5.91), obviously is an invariant of the scaling transformation, subject to the initial condition $\delta f(t = 1; g^*) = 1$. For values of the bare coupling parameter different from g^* (but within its domain of attraction), the expression (5.91) represents the asymptotic ($t \to \infty$) form of the function $\delta f(t; g)$. The exact solution of the Lie equation can be written in the form

$$\delta f(t; g) = t^{\gamma(g^*)} \exp\left\{\int_1^t dt'[\gamma(\bar{g}(t'; g)) - \gamma(g^*)]\right\}$$

$$= t^{\gamma(g^*)} \exp\left\{\int_{1/t}^1 dy \left[\gamma\left(\bar{g}\left(\frac{1}{y}; g\right)\right) - \gamma(g^*)\right]\right\} \quad (5.92)$$

In the long-time limit, the exponential factor (scaling function) appearing on the right-hand side of this expression approaches a constant value, dependent on both g and g^*.

A comparison of the formulas (5.91) and (5.79) might lead one to conclude that use of the Lie differential equation formalism has enabled us to eliminate the uncertainty associated with the choice of the threshold value τ for the self-similarity scale. This is not so. The self-similarity threshold, like the period of a periodic function, is an inherent feature of each physical object expected to exhibit self-similarity. By adopting the differential equation formulation of the GPRG theory one implicitly has selected a threshold scale equal to unity, that is, $\tau = 1$.

D. Callan–Symanzik Equations of the GPRG

Alternative to the Lie equations associated with a self-similar nonideal system is the pair of partial differential equations obtained by differentiating the functional equations (5.57) and (5.65) with respect to the scale x. Thus, the result of differentiating (5.57) is the relationship

$$0 = \frac{\partial \delta f(t/x; g)}{\partial(t/x)} \delta f[x; \bar{g}(t/x; g)](-t/x^2) + \delta f(t/x; g)$$

$$\times \left\{\frac{\partial \delta f[x; \bar{g}(t/x; g)]}{\partial x}\bigg|_{\bar{g}(t/x;g)} + \frac{\partial \delta f[x; \bar{g}(t/x; g)]}{\partial \bar{g}(t/x; g)} \frac{\partial \bar{g}(t/x; g)}{\partial(t/x)}(-t/x^2)\right\} \quad (5.93)$$

which, with x set equal to t, reduces to the Callan–Symanzik equation,

$$\frac{\partial \ln \delta f(t; g)}{\partial \ln t} = \gamma(g) + \beta(g)\frac{\partial \ln \delta f(x; g)}{\partial g} \quad (5.94)$$

for the excess function. This same procedure leads from (5.65) to the second Callan–Symanzik equation

$$\frac{\partial \bar{g}(t; g)}{\partial \ln t} = \beta(g) \frac{\partial \bar{g}(t; g)}{\partial g} \qquad (5.95)$$

The formulation of the GPRG method that has been presented here differs from earlier presentations of the "old renormalization group method" [3–5], not only by virtue of the manner in which the effective coupling function is introduced [see the derivation of Eq. (5.65)] but by the role this function plays in enforcing the validity of the self-similarity assumption [see the derivation of Eq. (5.57)]. The intimate relationship between the effective coupling function and the object function, as illustrated by Eqs. (5.75) and (5.89), is an essential part of the theory that is unique to our approach. The remainder of this section is devoted to two examples that are illustrative of how the method works in practice.

E. Average Squared Magnetization of an Open Chain of Ising Spins

Most problems involving nonideal systems of physical interest cannot be solved exactly. In these circumstances, the RG method can be used to overcome the intrinsic inadequacies of initially available approximate results, such as a truncated perturbation series developed in powers of the bare coupling parameter. A regrettable shortcoming of this procedure is the difficulty of assessing the effectiveness of the RG method. How reliable are the predictions it produces? The problem treated here is one for which the exact answer is known. Our examination of this specific case not only provides a simple illustration of how the formal theory is to be implemented computationally but also permits a comparison to be made between predictions generated by the GPRG approximation method and the corresponding exact answers. This test is something that rarely can be conducted for the more complex problems to which we later shall turn.

The average squared magnetization (ASM) of an open, one-dimensional (1D) chain of N spins (each interacting exclusively with its nearest neighbors) is given by the formula [18]

$$\overline{M^2}(N; g, m^2) = Nm^2 e^{2g} \left[1 + \frac{1}{N} (\tanh^N g - 1) \sinh(2g) \right] \qquad (5.96)$$

Here, m denotes the magnetic moment of a single spin so that

$$\overline{M^2}(1; g, m^2) = m^2$$

The bare coupling parameter, $g = J/k_B T$, is proportional to the nearest-neighbor exchange energy J. According to the formula (5.96) the ASM exhibits the two limits

$$\lim_{g \to 0} \overline{M^2} = Nm^2 \tag{5.97}$$

and

$$\lim_{g \to \infty} \overline{M^2} = N^2 m^2 \tag{5.98}$$

corresponding, respectively, to the high-temperature paramagnetic phase and the low-temperature ferromagnetic phase. The second-order transition between these two phases occurs at a characteristic "Curie temperature", the value of which depends on the number of spins, N, and approaches absolute zero for $N \to \infty$ [10].

By imagining the chain to be divided into N/K blocks, each consisting of K contiguous spins, we can express the assumption of self-similarity as a conservation equation

$$\begin{aligned} \overline{M^2}(N; g, m^2) &= \overline{M^2}\left[K; g, \overline{M^2}\left(\frac{N}{K}; \bar{g}(K; g), m^2\right)\right] \\ &= \overline{M^2}\left[\frac{N}{K}; \bar{g}(K; g), \overline{M^2}(K; g, m^2)\right] \end{aligned} \tag{5.99}$$

for the average squared magnetization. The first of these equalities states that the ASM of the N-spin system is identical to that of a system of K spins, each with a square magnetization equal to $m'^2 = \overline{M^2}(N/K; \bar{g}(K; g), m^2)$ that depends on the number of spin blocks N/K and on the modified coupling parameter $g' = \bar{g}(K; g)$. The interpretation of the second equality of (5.99) more closely resembles the Kadanoff idea. Thus, the individual spins are clustered into K-spin blocks from which is formed an "effective" 1D lattice, the N/K elements of which each has a square magnetization $m'^2(K; g, m^2) = \overline{M^2}(K; g, m^2)$ and whose interactions with its neighbors are characterized by an effective coupling parameter $\bar{g}(K; g)$.

The excess function appropriate to this problem is the reduced (R) zero-field magnetic susceptibility $\chi_R(N; g)$ defined by the expression

$$\overline{M^2}(N; g, m^2) = Nm^2 \chi_R(N; g) \tag{5.100}$$

By combining this definition with the conservation equation (5.99), we find immediately that the susceptibility satisfies a functional equation

$$\chi_R(N; g) = \chi_R\left(\frac{N}{K}; g\right)\chi_R\left[K; \bar{g}\left(\frac{N}{K}; g\right)\right] \quad (5.101)$$

of the form characteristic of the GPRG. The effective coupling function appearing in these equations is governed by Eq. (5.65). If Eq. (5.101) is to be solved by means of the discrete mapping technique of Section II.B, the reduced susceptibilities for systems of K and K^2 spins must be known functions of the bare coupling parameter g. In the calculations presented here, we use the exact formula (5.96) to generate these two functions for the scale block of $K = 5$ spins.

The effective coupling function $\bar{g}(K; g)$ then can be determined from the equation [(5.101) with $N = K^2$]

$$\chi_R(K^2; g) = \chi_R(K; g)\chi_R[K; \bar{g}(K; g)] \quad (5.102)$$

The solution of this equation can be written as

$$\bar{g}(K; g) = \text{Inv } \chi_R\left[K; \frac{\chi_R(K^2; g)}{\chi_R(K; g)}\right] \quad (5.103)$$

with Inv $\chi_R[K,\ldots]$ denoting the inverse of the function $\chi_R(K; g)$, specific to a given value of the block size K. In practice, we begin by solving the transcendental equation

$$0 = \chi_R - e^{2g}\left[1 + \frac{1}{K}(\tanh^K g - 1)\sinh(2g)\right] \quad (5.104)$$

and writing the solution in the form

$$g = g_K(\chi_R) \quad (5.105)$$

This solution then is used to generate the effective coupling function, according to the prescription [cf. (5.102) or (5.103)]

$$\bar{g}(K; g) = g_K[\chi_R(K^2; g)/\chi_R(K; g)] \quad (5.106)$$

The function $\bar{g}(K; g)$ constructed in this way is shown in Figure 5.2 for a $K = 5$ spin block. Since $g < \bar{g}(5; g)$ except for the trivial (stable) fixed point at $g^* = 0$, the system is "asymptotically free".

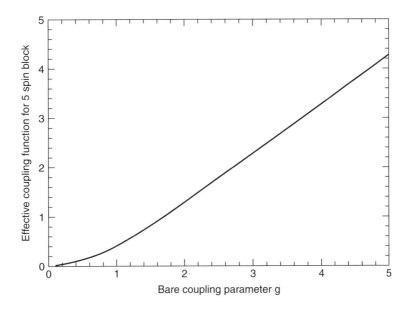

Figure 5.2. Effective coupling function for a 5-spin block plotted versus the bare coupling parameter $g = J/k_B T$.

The reduced magnetic susceptibility now can be calculated iteratively, using the discrete mapping method, that is, for systems composed of $N = 5^n$ spins with $n > 2$. For example, the susceptibility $\chi_R(N; g)$ for $N = 5^6 = 15{,}625$ is plotted in Figure 5.3 as a function of the reduced temperature $g^{-1} = k_B T/J$. The dependence of the susceptibility on the size of the spin system is shown in Figure 5.4. The points appearing in both of these figures indicate predictions produced by the GPRG method (with $K = 5$), whereas the continuous curves were obtained directly from the exact formula (5.96). The good agreement between the two shows that the GPRG method can indeed provide reliable predictions for large systems based solely on the basis of information about much smaller systems that are assumed to be functionally self-similar to the former.

III. STRETCHED-EXPONENTIAL SCALING AND THE POSITIVE FUNCTION RENORMALIZATION GROUP (PFRG)

The equations of evolution derived in Section II are specific to situations for which the object function $f(t; g, f_0)$ depends linearly on the initial value f_0. This linear dependence is so frequently encountered in practice that it is the

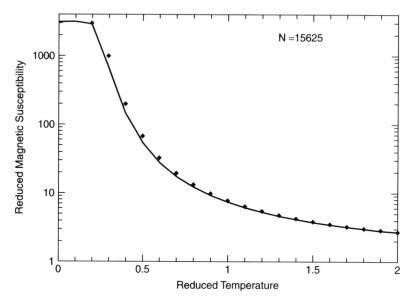

Figure 5.3. The reduced magnetic susceptibility $\chi_R(N; g)$ of Eq. (5.100) plotted versus the reduced temperature $g^{-1} = k_B T/J$. The continuous curve is a plot of the exact formula [cf. Eq. (5.96)]. The points indicate results of RG calculations based on the choice of $K = 5$, that is, on a 5-spin block.

only type considered in most, if not all, previous presentations of RG theory. However, there is at least one other type of initial value dependence that arises in problems of physical interest, namely, the stretched exponential

$$f(t; g, f_0) = f_0^{t^\sigma} \delta f(t; g) \tag{5.107}$$

with σ a numerical constant. The right-hand side of this relationship can be manipulated as follows;

$$\begin{aligned} f_0^{t^\sigma} \delta f(t; g) &= [f_0^{x^\sigma} \delta f(x; g)]^{(t/x)^\sigma} \frac{\delta f(t; g)}{[\delta f(x; g)]^{(t/x)^\sigma}} \\ &= [f(x; g, f_0)]^{(t/x)^\sigma} \delta f\left[\frac{t}{x}; \bar{g}(x; g)\right] \\ &= f\left[\frac{t}{x}; \bar{g}(x; g), f(x; g, f_0)\right] \end{aligned} \tag{5.108}$$

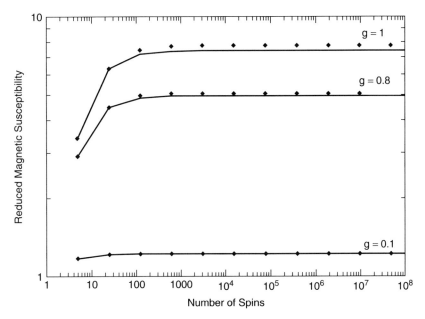

Figure 5.4. The dependence of the reduced magnetic susceptibility $\chi_R(N;g)$ on the number of spins N, for several values of the bare coupling parameter g. The solid curves are exact results obtained from Eq. (5.96). The points are results of RG calculations based on a 5-spin block.

obtaining by the second step, a final result that is valid only if one can find an effective coupling function $\bar{g}(x;g)$, which ensures that the relationship

$$\frac{\delta f(t;g)}{[\delta f(x;g)]^{(t/x)^\sigma}} = \delta f\left[\frac{t}{x};\bar{g}(x;g)\right] \quad (5.109)$$

is a mathematical equality.

From (5.107) and (5.108), we obtain a conservation equation

$$f(t;g,f_0) = f\left[\frac{t}{x};\bar{g}(x;g), f(x;g,f_0)\right] \quad (5.110)$$

that is identical with Eq. (5.52) and that previously has been identified in Section II as an expression of scaling invariance with respect to the RG transformation (5.53). The terms of (5.108) can be manipulated in a different order, namely,

$$\begin{aligned}f_0^{t^\sigma}\delta f(t;g) &= \left[f_0^{(t/x)^\sigma}\delta f\left(\frac{t}{x};\bar{g}(x;g)\right)\right]^{x^\sigma}\frac{\delta f(t;g)}{\left[\delta f\left(\frac{t}{x};\bar{g}(x;g)\right)\right]^{x^\sigma}}\\ &= \left[f\left(\frac{t}{x};\bar{g}(x;g),f_0\right)\right]^{x^\sigma}\delta f(t;g)\\ &= f\left[x;g,f\left(\frac{t}{x};\bar{g}(x;g),f_0\right)\right]\end{aligned} \tag{5.111}$$

provided that one can construct an effective coupling function $\bar{g}(x;g)$ that causes the relationship

$$\frac{\delta f(t;g)}{\left[\delta f\left(\frac{t}{x};\bar{g}(x;g)\right)\right]^{x^\sigma}} = \delta f(x;g) \tag{5.112}$$

to be a mathematical equality. The two equations (5.107) and (5.111) lead directly to a conservation equation

$$f(t;g,f_0) = f\left[x;g,f\left(\frac{t}{x};\bar{g}(x;g),f_0\right)\right] \tag{5.113}$$

which is the same as (5.60).

It now will be shown that Eq. (5.109) requires the effective coupling function to be an invariant of the scaling transformation and thus a solution of Eq. (5.65). We begin by rewriting (5.109) in the form

$$\begin{aligned}\delta f[tx;g] &= [\delta f(x;g)]^{t^\sigma}\delta f[t;\bar{g}(x;g)]\\ &= [\delta f(t;g)]^{x^\sigma}\delta f[x;\bar{g}(t;g)]\end{aligned} \tag{5.114}$$

the second line of which follows from the requirement of invariance with respect to transposition of the variables t and x. From (5.114), it can be seen that

$$\frac{\delta f(t;\bar{g}(x;g))}{\delta f(x;\bar{g}(t;g))} = \left[\frac{\delta f(x;g)^{(x/t)^\sigma}}{\delta f(t;g)}\right]^{t^\sigma} = \left[\frac{1}{\delta f\left(\frac{x}{t};\bar{g}(t;g)\right)}\right]^{t^\sigma}$$

or

$$\frac{\delta f[x; \bar{g}(t; x)]}{\left[\delta f\left(\frac{x}{t}; \bar{g}(t; x)\right)\right]^{t^\sigma}} = \delta f[t; \bar{g}(x; g)] \tag{5.115}$$

Finally, by using (5.112) to reexpress the left-hand side of (5.115) we obtain the equality

$$\delta f\left[t; \bar{g}\left(\frac{x}{t}; \bar{g}(t; g)\right)\right] = \delta f[t; \bar{g}(x; g)] \tag{5.116}$$

The proof is now complete since this can be so only if the effective coupling function satisfies Eq. (5.65). We leave it to the reader to verify that this same requirement of invariance follows from (5.112) as well.

An alternative derivation of Eq. (5.112) can be established using the "classical" method of renormalization constants. To illustrate this, we note first that the quantity $f_{id} = f_0^{t^\sigma}$ for the corresponding ideal system satisfies the relationship

$$f_{id}(t; f_0) = \left[f_{id}\left(\frac{t}{x}; f_0\right)\right]^{x^\sigma} \tag{5.117}$$

Therefore, if the behavior of the nonideal system is to resemble that of its ideal counterpart, the corresponding excess part should behave as follows:

$$\delta f(t; g) = z(x; g)\left[\delta f\left(\frac{t}{x}; \bar{g}(x; g)\right)\right]^{x^\sigma} \tag{5.118}$$

when the generalized time is divided into t/x intervals, each with a length equal to the new scale x. The "renormalization constant" $z(x; g)$ now can be evaluated by requiring that the function be normalized for the new scale. Thus, with t set equal to x, we conclude from (5.118) that

$$z(x; g) = \delta f(x; g) \tag{5.119}$$

The result of substituting this into (5.118) is the previously obtained Eq. (5.112).

It is important to recognize that the excess function associated with an object function exhibiting a stretched-exponential dependence on the initial value satisfies an evolution equation different from that considered in

Section II. Because this evolution equation (5.112) also can be derived by requiring the excess function to be positive definite [19], we call this form of the theory the positive (or partition) function renormalization group (PFRG).

A. Iterative Solution of the PFRG Functional Equations

The discrete mapping method presented in Section II.B also can be used to solve Eq. (5.112). Thus, with the generalized time expressed as a power of the scale x (i.e., $t = x^n$) we find that

$$\delta f(x^n; g) = \delta f(x; g) \prod_{j=1}^{n-1} \{\delta f[x; \bar{g}^{(j)}(x; g)]\}^{x^{j\sigma}} \tag{5.120}$$

The objects $\bar{g}^{(j)}(x; g)$ appearing here are the nested iterates of the effective coupling function defined previously by Eqs. (5.72)–(5.74).

The effective coupling function specific to this scale is found by solving the equation

$$\frac{\delta f(x^2; g)}{\delta f(x; g)} = \{\delta f[x; \bar{g}(x; g)]\}^{x^\sigma} \tag{5.121}$$

To construct $g(x; g)$, one must be provided with $\delta f(x; g)$ and $\delta f(x^2; g)$ as functions of the bare coupling constant g.

The fixed points for the scale x are solutions of the equation

$$\delta f(x^2; g^*) = [\delta f(x; g^*)]^{x^\sigma + 1} \tag{5.122}$$

When the bare coupling parameter g precisely equals the fixed point value $g^* = g^*(x)$, Eq. (5.120) reduces to

$$\delta f(t; g^*(x)) = [\delta f(x; g^*(x))]^{(t^\sigma - 1)/(x^\sigma - 1)} \tag{5.123}$$

and the function $\delta f(t; g^*(x))$ is asymptotically self-similar in the limit $t \to \infty$ (for $\sigma > 0$). In this same limit,

$$\lim_{t \to \infty} f(t; g^*, f_0) = \{f_0[\delta f(x; g^*(x))]^{(1)/(x^\sigma - 1)}\}^{t^\sigma} \tag{5.124}$$

For values of the bare coupling parameter that differ from $g^*(x)$, but lie within its domain of convergence, the formula (5.123) provides an asymptotic ($t \to \infty$) limit,

Here again we can introduce a "scaling function" $\Phi(t; g)$, defined by the relationship

$$\delta f(t; g) = \delta f(t; g^*)\Phi(t; g) \tag{5.125}$$

Substitution of this expression into the equation of evolution (5.112) results in the formula

$$\Phi(t; g) = \Phi(x; g)\left[\Phi\left(\frac{t}{x}; \bar{g}(x; g)\right)\right]^{x^\sigma} \tag{5.126}$$

thus indicating that the scaling function satisfies the same functional equation as does the excess function $\delta f(t; g)$. It additionally is subject to the condition $\Phi(t; g^*) = 1$.

B. Differential Equations of the PFRG

The Lie and Callan–Symanzik equations for the PFRG are readily obtained from the relationship

$$\ln \delta f(t; g) = \ln \delta f(x; g) + x^\sigma \ln \delta f\left[\frac{t}{x}; \bar{g}(x; g)\right] \tag{5.127}$$

which follows directly from Eq. (5.112). Thus, by differentiating (5.127) with respect to the generalized time t and subsequently setting t equal to x we obtain the Lie equation of evolution

$$\frac{\partial \ln \delta f(t; g)}{\partial \ln t} = t^\sigma \gamma[\bar{g}(t; g)] \tag{5.128}$$

with $\gamma(g)$ defined by the formula (5.84).

The equation of evolution for the effective coupling function again is given by (5.86). However, the infinitesimal generator $\beta(g)$ now is related to the excess function by the formula

$$\beta(g) = \frac{\partial^2 \ln \delta f(t; g)/\partial t^2|_{t=1} - (\sigma - 1)\gamma(g)}{d\gamma/dg} \tag{5.129}$$

As before, the fixed points of the RG transformation (if such do indeed exist) are zeros of the function $\beta(g)$. A particular fixed point is stable if the derivative of this function is negative definite, that is, if $d\beta(g)/dg|_{g=g^*} < 0$.

It follows from (5.128) that, when the bare coupling parameter equals the fixed point value g^* [and $\bar{g}(t; g^*)$ therefore is equal to g^*],

$$\delta f(t; g^*) = \begin{cases} t^{\gamma(g^*)}; & \sigma = 2 \\ \exp\left[\dfrac{\gamma(g^*)}{\sigma}(t^\sigma - 1)\right]; & \sigma \neq 2 \end{cases} \tag{5.130}$$

If g^* is the only nontrivial and stable fixed point of the transformation, then the formula (5.130) gives the asymptotic, long-time limit of the function $\delta f(t; g)$ for $g \neq g^*$. In this case, the excess function can be written in the form

$$\delta f(t; g) = \delta f(t; g^*)\phi(t; g) \tag{5.131}$$

with

$$\phi(t; g) = \exp\left\{\int_{1/t}^{1} dy \, \frac{1}{y^{\sigma+1}} \left[\gamma\left(\bar{g}\left(\frac{1}{y}; g\right)\right) - \gamma(g^*)\right]\right\} \tag{5.132}$$

In the limit as $t \to \infty$, this scaling function reduces to a constant, dependent on both g and g^*.

Also derivable from the PFRG relationship (5.127) are linear partial differential equations of the Callan–Symanzik type. By differentiating this relationship with respect to the scale x and then taking the limit $x \to 1$, one obtains the equation

$$\frac{\partial \ln \delta f(t; g)}{\partial \ln t} = \gamma(g) + \beta(g) \frac{\delta \ln \delta f(t; g)}{\partial g} + \sigma \ln \delta f(t; g) \tag{5.133}$$

The Callan–Symanzik equation for the effective coupling function is the same as before, namely, Eq. (5.95).

C. PFRG Calculation of the Partition Function of the One-Dimensional Ising Model

Here, we demonstrate the PFRG method by applying it to the very simple task of calculating the partition function of an open-ended, 1D Ising lattice of interacting spins. The partition function for the corresponding "ideal" system of *non*interacting spins is given by the product of single-spin partition functions

$$Q_{id}(N; q) = q^N \tag{5.134}$$

with $q = 2$.

THE ROLE OF SELF-SIMILARITY IN RENORMALIZATION GROUP THEORY   303

Each spin of the nonideal system is coupled to its nearest neighbors by an exchange energy J and an associated coupling constant $g = J/k_B T$. The partition function of this system then can be written as

$$Q(N; g, q) = q^N \delta Q(N; g) \tag{5.135}$$

which is of the form (5.107) with the number of spins identified as the generalized time and $\sigma = 1$. We now divide the lattice of N spins into N/K blocks, each consisting of K contiguous spins. Next, we invoke the physically well-founded requirement that the measure provided by the number of configurations $Q(N; g, q)$ be conserved. According to (5.113) this conservation condition is expressed by the functional relationship

$$Q(N; g, q) = Q\left[\frac{N}{K}; g, Q\left(K; \bar{g}\left(\frac{N}{K}; g\right), q\right)\right] \tag{5.136}$$

To solve this equation, we use the discrete mapping method and select the simplest scale value of $K = 2$, corresponding to a block of two interacting spins. The excess partition functions for blocks of $K = 2$ and $K^2 = 2^2 = 4$ spins easily can be calculated exactly by direct summation:

$$\delta Q(2; g) = \cosh g$$
$$\delta Q(4; g) = \cosh^3 g \tag{5.137}$$

For the scale $K = 2$, Eq. (5.121) then takes the simple form

$$\frac{\cosh^3 g}{\cosh g} = \cosh^2[\bar{g}(2; g)] \tag{5.138}$$

from which it follows that

$$\bar{g}(2; g) = g \tag{5.139}$$

The system therefore has a *continuous* spectrum of fixed points, which is rather unusual and happens here because of the extreme simplicity of the system.

The general solution of the basic functional equation, as given by (5.120), now can be written as

$$dQ(2^n; g) = \delta Q(2; g) \prod_{j=1}^{n-1} [\delta Q(2; \bar{g}^{(j)}(2; g))]^{2^j} \qquad (5.140)$$

where, for this special case, $\bar{g}^{(j)}(2; g) = g$, $\forall j$. It then follows that

$$\delta Q(2^n; g) = [\delta Q(2; g)]^{1 + 2 + 2^2 + \cdots + 2^{n-1}} = [\delta Q(2; g)]^{2^n - 1} \qquad (5.141)$$

Finally, by recalling that $N = 2^n$ we can rewrite this last result as

$$\delta Q(N; g) = [\delta Q(2; g)]^{N-1} = \cosh^{N-1} g \qquad (5.142)$$

which is the exact result, usually obtained by the transfer matrix method [18].

This simple example of the PFRG together with the GPRG example of Section II.E illustrate that an assortment of physical properties can be calculated successfully using the self-similarity hypothesis, *provided* that the appropriate choice of renormalization group can be identified. Furthermore, these examples suggest that to identify the appropriate group one should examine the properties of the corresponding ideal systems.

In the sections that follow, we consider less trivial applications of the GPRG and PFRG techniques to properties of systems for which exact results are not available.

IV. LINEAR, NONIDEAL POLYMER CHAINS

The notions of functional self-similarity and scaling arise very naturally in polymer science and have found applications in this field for many years. Thus, a linear polymer chain with excluded-volume interactions is a perfect example of a physical object to which scaling should be applicable, as a subdivision of the entire chain into a collection of "blobs" or "Kuhn macrosegments" [9, 14]. One of these segments may be envisioned as a fragment of sufficient length to ensure that its statistical properties are effectively independent from the remainder of the chain.

Here, we limit our considerations to the simple "pearl necklace" (site-and-bond) model of a linear polymeric macromolecule, the identical sites of which are permanently connected to one another. The Hamiltonian function for such a chain is given by the sum

$$H_N = H_N^0 + V_N \tag{5.143}$$

of the Hamiltonian

$$H_N^0 = \sum_{i=1}^{N+1} \frac{m}{2} \dot{r}_i^2 + \sum_{i=1}^{N} u(|r_{i+1} - r_i|) \tag{5.144}$$

for the corresponding ideal chain, consisting of $N + 1$ permanently connected interaction sites (N bonds) linked to one another by a nearest-neighbor bond potential $u(|r_{i+1} - r_i|) = u(|q_i|)$, and

$$\begin{aligned}V_N &= \sum_{i=1}^{N-1} \sum_{j=i+2}^{N-1} v(|r_i - r_j|) \\ &= \sum_{i=1}^{N-1} v(|q_i + q_{i+1}|) + \sum_{i=1}^{N-2} v(|q_i + q_{i+1} + q_{i+2}|) + \cdots \\ &\quad + \sum_{i=1}^{2} v(|q_i + q_{i+1} + \cdots + q_{N-2+i}|) + v(|q_1 + q_2 + \cdots + q_N|)\end{aligned} \tag{5.145}$$

The latter is the potential energy associated with the (nonbonding) pair interactions among nonnearest neighboring sites. The symbols r_i indicate position vectors of the sites and $q_i = r_{i+1} - r_i$ denotes the bond vector extending from site i to site $i + 1$.

The Gibbs canonical partition function of this chain is given by the expression

$$Q_N = Q_0 Q_N^0 \delta Q_N \tag{5.146}$$

where

$$Q_0 = [2\pi m k_B T \sigma^2 / h^2]^{d/2} (\Omega/\sigma^d) \tag{5.147}$$

is the partition function of an end site of mass m. Here, σ denotes a length characteristic of the excluded-volume potential v and d is the dimension of the space in which the chain is embedded. The function Q_N^0 is related by the formula

$$Q_N^0 = [q(T)]^N \tag{5.148}$$

to the partition function

$$q(T) = [2\pi m k_B T \sigma^2/h^2]^{d/2} Z(T)/\sigma^d \qquad (5.149)$$

of a single chain link. The quantity $Z(T)$ is the bond configuration integral defined by

$$Z(T) = \int d\mathbf{q} e^{-\beta u(|q|)} = \frac{2\pi^{d/2}}{\Gamma(d/2)} \int_0^\infty dq\, q^{d-1} e^{-\beta u(q)} \qquad (5.150)$$

Finally, the excess partition function

$$\delta Q_N = \langle e^{-\beta V_N} \rangle \qquad (5.151)$$

measures the degree by which the statistical properties of the nonideal, self-repelling chain deviate from those of the corresponding ideal chain. The pointed bracket $\langle \cdots \rangle$ appearing in (5.151) indicates an average over the bond vector probability distribution for the ideal chain, namely,

$$\langle \chi \rangle = \int \chi \prod_{i=1}^N \{\psi(\mathbf{q}_i) d\mathbf{q}_i\} \qquad (5.152)$$

with

$$\psi(\mathbf{q}) = Z(T)^{-1} e^{-\beta u(|q|)} \qquad (5.153)$$

denoting the bond vector probability density.

A remarkable feature of the nonideal polymer chain is that both types of scalings (GPRG and PFRG) arise in the analysis of its statistical properties. It is clear from Eq. (5.148) that the ideal chain partition function Q_N^0 has a nonlinear, power-law dependence on its "initial value", the single-particle partition function $Q_1^0 = q(T)$. Consequently, we expect the partition function of the corresponding nonideal chain to behave in a similar manner and so to scale in accordance with the PFRG. On the other hand, the average squared end-to-end length of an ideal chain is given by the formula

$$\overline{\mathbf{R}_{id}^2}(N; \ell^2) = N\ell^2 \qquad (5.154)$$

and so depends linearly on its initial value

$$\overline{R_{id}^2}(1;\ell^2) = \ell^2 = \int d\mathbf{q}\, q^2 \psi(\mathbf{q}) \tag{5.155}$$

Because of this, the average squared end-to-end length of a nonideal, self-avoiding chain is expected to scale according to the GPRG of Section II.A.

Both of these characteristic properties of the nonideal chain, its partition function and its average squared end-to-end length, can be calculated using the self-similarity based RG techniques described in the previous sections but, because these two properties depend differently on their initial values, they are subject to two different renormalization groups. The situation just described closely resembles our previous treatments in Sections II.E and III.C of two different properties of the 1D Ising model, namely, its average squared magnetization and its partition function.

The RG method, as presented here, allows one to calculate the partition function and swelling factor for a broad range of chain contour lengths, that is, monomer numbers N. Its capability of predicting the unmeasurable asymptotic ($N \to \infty$) properties of chains is an issue of secondary importance but one that nevertheless must be addressed since these asymptotic properties are the foci of the conventional RG investigations with which ours inevitably will be compared.

A. Calculation of the Swelling Factor Using the GPRG Method

The swelling factor $\alpha^2(N;g)$, defined by Eq. (5.49), is a correction factor or "excess function" that takes into account the interactions among non-nearest-neighbor chain sites. These are not included in the model for the corresponding ideal (or phantom) chain. The quantity g, on which the swelling factor depends, is the "bare" coupling parameter characteristic of the excluded-volume interaction potential $v(r)$, [cf. Eq. (5.145)].

Our previous considerations lead us to expect that the average squared end-to-end length of a nonideal chain will be subject to the conservation condition

$$\overline{R^2}(N;g,\ell^2) = \overline{R^2}\left[\frac{N}{K};g,\overline{R^2}\left(K;\bar{g}\left(\frac{N}{K};g\right),\ell^2\right)\right] \tag{5.156}$$

The physical basis for this equation is the plausible expectation that the value of the average squared end-to-end length will remain unchanged when the chain of N segments is divided into N/K equal parts, each a Kuhn macrosegment consisting of K monomer units characterized by an average squared end-to-end length of $\overline{R^2}[K; \bar{g}(N/K; g), \ell^2]$. Because these macrosegments are not autonomous chains but subdivisions of the larger chain, the bare coupling parameter g is replaced with the effective coupling function $\bar{g}(N/K; g)$. The renormalization procedure consists of scaling (dividing) the chain length N by the scale K and replacing the average squared end-to-end length of a monomer by that of the Kuhn segment, that is,

$$N \to N' = N/K$$
$$\ell^2 \to \ell'^2 = \overline{R^2}[K; \bar{g}(N/K; g), \ell^2] \qquad (5.157)$$

Once the Kuhn length K has been selected, the only remaining unknown in Eq. (5.156) is the effective coupling function. It is determined by a self-consistent procedure to ensure that (5.156) is a mathematical identity.

The swelling factor satisfies the equation

$$\alpha^2(N; g) = \alpha^2\left(\frac{N}{K}; g\right) \alpha^2\left[K; \bar{g}\left(\frac{N}{K}; g\right)\right] \qquad (5.158)$$

which is of the form (5.61), appropriate to the excess function of a property governed by the GPRG. This equation is supplemented by the relationship [cf. (5.65)]

$$\bar{g}(N; g) = \bar{g}\left[\frac{N}{K}; \bar{g}(K; g)\right] \qquad (5.159)$$

which states the scaling invariance of the effective coupling function.

Kuhn's idea of the asymptotic self-similarity of a nonideal chain now can be expressed as two propositions: (1) there exists a value g^* of the bare coupling parameter for which Eq. (5.156) reduces to the form

$$\overline{R^2}(N; g^*, \ell^2) = \overline{R^2}\left[\frac{N}{K}; g^*, \overline{R^2}(K; g^*, \ell^2)\right] \qquad (5.160)$$

that is, for which $\overline{R^2}$ becomes *strictly* invariant with respect to the scaling transformation, and; (2) that

$$\lim_{N \to \infty} \bar{g}(N; g) = g^* \qquad (5.161)$$

so that the scaling transformation of the effective coupling function exhibits a stable fixed point g^* (horizontal asymptote) in the long-chain limit.

We now select a value $K \geq 2$ for the Kuhn length, such that $N = K^n$, and so conclude from (5.71) that

$$\alpha^2(K^n; g) = \alpha^2(K; g) \prod_{i=1}^{n-1} \alpha^2[K; \bar{g}^{(i)}(K; g)] \qquad (5.162)$$

This formula expresses the swelling factor of the entire chain as a functional of the swelling factor for a much shorter Kuhn segment of length $K \ll N$. The latter obviously presents a less formidable calculational task than the former.

The effective coupling function for the Kuhn segment is obtained by solving the equation

$$\alpha^2(K^2; g) = \alpha^2(K; g)\alpha^2[K; \bar{g}(K; g)] \qquad (5.163)$$

and this, in turn, requires as input data the functions $\alpha^2(K; g)$ and $\alpha^2(K^2; g)$. These must be constructed either by direct calculation or by conducting computer simulations for nonideal chains of length K and K^2.

The asymptotic $(N \to \infty)$ behavior of the swelling factor is determined by the fixed-point spectrum of the effective coupling function, defined as the roots $g^*(K)$ of the nonlinear equation

$$\alpha^2(K^2; g^*) = [\alpha^2(K; g^*)]^2 \qquad (5.164)$$

If there is but one stable fixed point, the asymptotic form of the swelling factor is then given by the power-law formula

$$\alpha^2(N; g^*) \approx [\alpha^2(K; g^*)]^{\ln N/\ln K} = N^{2\nu(K)-1} \qquad (5.165)$$

This formula involves a scaling exponent $\nu(K)$ defined as follows:

$$2\nu(K) - 1 = \log_K \alpha^2[K; g^*(K)] \qquad (5.166)$$

and dependent, in principle, on the selected Kuhn length K and on the specific microscopic, mechanical model used to describe the chain.

The assumption that use of a sufficiently large scale K will eliminate both of these dependences is known as the universality hypothesis. If it is valid, there then will exist a limit ν_∞ to $\nu(K)$ that is common to all chain models. In the context of the self-similarity based RG method, as presented here, it should be possible to test the validity of this hypothesis by conducting numerical studies similar to those described below for a variety of chain models.

We recently [15] performed calculations of the swelling factor for a randomly jointed chain (fixed nearest-neighbor separation, ℓ) with rigid-sphere excluded volume interactions. The bare coupling coefficient for this model was taken to be $g = (\sigma/\ell)^2$ with σ the hard-sphere diameter and ℓ the bond length. Depending on the numerical value of g, a chain can be identified as belonging to one of three classes. For $0 < g < \frac{1}{2}$, the chain contour is a partially restricted random walk; sites cannot overlap but the range of the repulsive excluded-volume interactions is too small to protect the entire contour. Consequently, bonds linking distant pairs can touch. For the chain to be truly self-avoiding it is necessary that $g > \frac{1}{2}$. Chains for which $\frac{1}{2} < g < 2$ are flexible and self-repelling. Bonds cannot touch one another. As the value of g increases so also does the degree of chain stiffness. Chains with $2 < g < 4$ may be described as flexible rods, with the limit $g = 4$ corresponding to a stiff, rod-like macromolecule.

Some results of our calculations, based on a Kuhn length of $K = 4$, are shown in Figures 5.5–5.7 The first of these, Figure 5.5, is a plot of the effective coupling function $\bar{g}(4; g)$, which should be compared with the effective coupling function for the 1D Ising model presented in Figure 5.2. The basic evolution equations for these two systems [Eqs. (5.101) and (5.163)] are closely analogous. However, the 1D Ising chain was found to

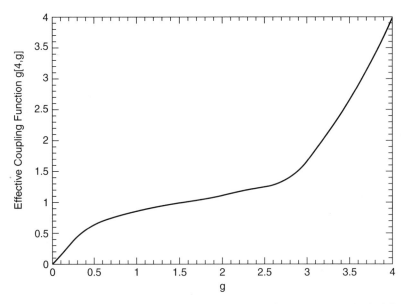

Figure 5.5. The GPRG effective coupling function $\bar{g}(4; g)$ for a 4-bond randomly jointed polymeric macrosegment plotted versus the bare coupling parameter g.

THE ROLE OF SELF-SIMILARITY IN RENORMALIZATION GROUP THEORY 311

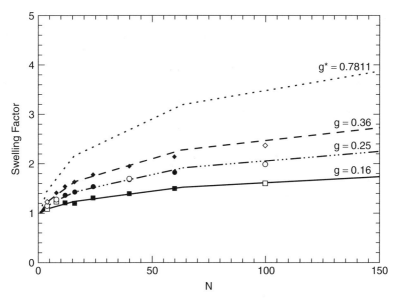

Figure 5.6. A comparison of swelling factors predicted by the GPRG method with $K = 4$ and the results of computer simulations. Filled and open circles indicate data of Kremer et al. [19] and of Altenberger et al. [15], respectively.

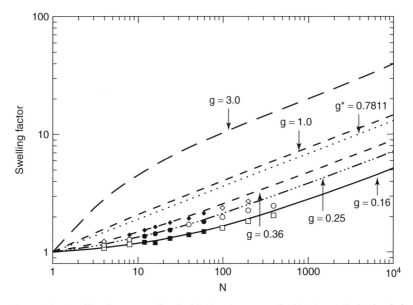

Figure 5.7. Swelling factors predicted with $K = 4$, compared with Monte Carlo simulation data. The latter are labeled as they were in Figure 5.6

be asymptotically free, whereas the average squared end-to-end polymeric length is not; for $K = 4$ the latter exhibits a nontrivial, stable fixed point at $g^* = 0.7811$. In Figures 5.6 and 5.7, we compare values of the swelling factor calculated according to the GPRG method with the results of computer simulations conducted both by Kremer et al. [20] (filled circles) and ourselves [15] (open circles). Calculations were performed for Kuhn segments of $K = 2, 4$, and 8. The spectrum of fixed points was found to be discrete with two stable fixed points. As the value of K increased, the values of both diminished. The lower point tended to zero and the upper approached an extrapolated $(K \to N)$ limit of $g^* = 0.3595$, for which the corresponding scaling exponent is $v_\infty(D = 3) = 0.5916$. Calculations for the same chain model in two and four dimensions produced the exponents $v(D = 2) = 0.760$ and $v(D = 4) = 0.500$. These three results are in close agreement with previous estimates of the scaling coefficients for self-avoiding chains.

B. Partition Function Calculation Using the PFRG Method

The task of evaluating the excess part of the partition function for a self-repelling chain is in many ways similar to that of calculating the partition function for the 1D Ising model, (cf. Section III.C). As previously noted, the partition function of an ideal polymer chain is a strictly self-similar quantity. We expect the partition function of the corresponding nonideal chain to exhibit self similarity provided that the chain itself is sufficiently long and that a large enough value is assigned to the Kuhn length. In particular, the total partition function for a nonideal chain is expected to satisfy the general conservation equation (5.110) and the associated excess partition function is governed by a functional equation of the PFRG type (5.114). When the chain length N and the Kuhn length K are related in the manner $N = K^n$, this excess function can be expressed as a functional of the excess partition function of a Kuhn segment. According to the general formula (5.120), this relationship is given by

$$\delta Q(K^n; g) = \delta Q(K; g) \prod_{j=1}^{n-1} [\delta Q(K; \bar{g}^{(j)}(K; g))]^{K^j} \qquad (5.167)$$

and, in this case, the effective coupling function $\bar{g}(K; g)$ for the Kuhn segment is determined by the equation [cf. (5.121)]

$$\left[\frac{\delta Q(K^2; g)}{\delta Q(K; g)} \right]^{1/K} = \delta Q[K; \bar{g}(K; g)] \qquad (5.168)$$

As before, the prerequisite to solving this equation is a knowledge of the two excess functions $\delta Q(K; g)$ and $\delta Q(K^2; g)$ as functions of the bare coupling parameter g.

If there is only one stable and nontrivial fixed point g^* of the effective coupling function $\bar{g}(K; g)$, it then follows directly from Eq. (5.167) that

$$\delta Q(K^n; g^*) = [\delta Q(K; g^*)]^{(K^n - 1)/(K - 1)} \tag{5.169}$$

or, equivalently

$$\delta Q(N; g^*) = [\delta Q(K; g^*)]^{(N - 1)/(K - 1)} \equiv \exp\{-\lambda(N - 1)\} \tag{5.170}$$

The quantity $\lambda(K; g^*)$, defined by the formula

$$\lambda(K; g^*) = |\ln \delta Q(K; g^*)|/(K - 1) \tag{5.171}$$

is called the "attrition coefficient" of the self-avoiding walk associated with the configurations of the nonideal chain. The numerical value of this dimensionless coefficient depends on the fixed point of the scaling transformation and this, in turn, is dependent on the choice of K, the length of the Kuhn segment. There are severe limitations on the scale sizes K that lie within the range of practical numerical calculations. However, if one were able to determine the limit

$$\lambda_\infty = \lim_{K \to \infty} \lambda(K; g^*) \tag{5.172}$$

by using some method of extrapolation, λ_∞ then could be identified as a "universal" attrition constant characteristic of a prescribed model for the polymer chain.

We recently performed calculations of the partition function for a randomly jointed chain with hard-sphere excluded-volume interactions [21], namely, the same model for which the swelling factor was calculated in Section IV.A. The effective coupling function for the two-bond ($K = 2$) Kuhn segment is displayed in Figure 5.8. It is apparent that the spectrum of fixed points becomes quasicontinuous for $g > 2$. (An examination of the roots of the equation $\delta Q(K^n; g) - [\delta Q(K; g)]^{(K^n - 1)/(K - 1)} = 0$ confirms this conclusion.) This behavior is not particularly surprising since for $g > 2$ the chains more closely resemble flexible rods than coils. For values of g in this range the excess partition function is indistinguishable from the exponential form of (5.170) while for $g < 2$ it is asymptotically exponential.

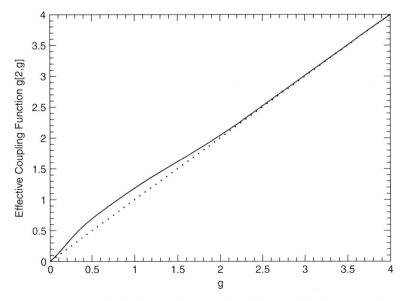

Figure 5.8. The PFRG effective coupling function $\bar{g}(2; g)$ for a 2-bond randomly jointed polymeric macrosegment plotted versus the bare coupling parameter g.

Figure 5.9 shows the bare coupling parameter dependence of the excess partition function, calculated as a function of chain length for the Kuhn length $K = 2$. The continuous, dashed and dotted curves are predictions of the RG theory. Circles, squares, triangles, and so on indicate the results of Monte Carlo computer simulations. The two-bond Kuhn segment is too short to produce good agreement between PFRG predictions and Monte Carlo results for the longer chains. Discrepancies are particularly evident for small values of the bare coupling parameter g. Significant improvements were obtained by shifting to the larger Kuhn segment with $K = 4$.

Figures (5.10) and (5.11) illustrate how the excess partition function depends on chain length for $K = 4$ and for various values of g. Table V.1 contains values of the fixed points, attrition coefficients and "effective coordination numbers" $z = e^{-\lambda}$, obtained by using Kuhn macrosegments of several lengths.

It is important to bear in mind that the asymptotic, long-chain behavior of the partition function differs significantly from that of the chain's average squared end-to-end length. In the long-chain limit, the former is an asymptotically exponential function of chain length for $g \neq g^*$ and strictly exponential for $g = g^*$. Deviations from this behavior occur only for short chains. This is consistent with thermodynamics, according to which the

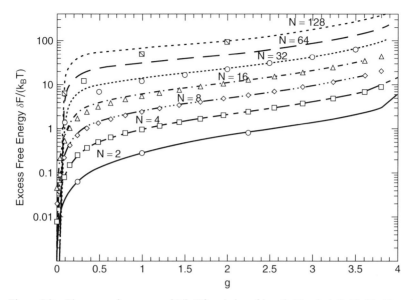

Figure 5.9. The excess free energy $\delta F/k_B T$ for chains of length $N = 2, 4, 8, 16, 32, 64$, and 128 plotted versus the bare coupling parameter g. The curves are PFRG predictions based on the Kuhn length $K = 2$. The points, indicated by various symbols, are results of Monte Carlo simulations.

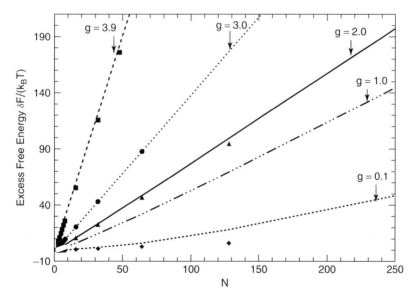

Figure 5.10. The chain length dependence of the excess free energy for various values of the bare coupling parameters. Curves indicate PFRG predictions for $K = 2$. Simulation results for $g = 0.1, 1.0, 2.0, 3.0$, and 3.9 are indicated by diamonds, open circles, triangles, filled circles, and squares, respectively.

TABLE V.1

Values of the Stable Fixed Point g_2^* and the Associated Attrition Constant λ and Effective Coordination Number z, Calculated for Several Choices of the Kuhn Length K

Macrosegment Size, K	Fixed Point g_2^*	Attrition Constant λ	Coordination Number z
2	3.783	2.915	0.0542
3	3.300	1.739	0.176
4	3.159	1.560	0.210
8	3.164	1.567	0.209

free energy of a long chain should be a first-order homogeneous function of chain length and, indeed, the number of monomers (or bonds) is the only extensive variable associated with the system.

The differences between the average squared end-to-end distance $\alpha^2(N, g)$ and the excess partition function $\delta Q(N, g)$ also are evident from the associated Callan–Symanzik equations [see Eqs. (5.94) and (5.133)]

$$\gamma(g) + \beta(g)\frac{\partial}{\partial g}\ln \alpha^2(N, g) - \frac{\partial}{\partial \ln N}\ln \alpha^2(N, g) = 0 \qquad (5.173)$$

and

$$\gamma(g) + \beta(g)\frac{\partial}{\partial g}\ln \delta Q(N, g) - \frac{\partial}{\partial \ln N}\ln \delta Q(N, g) + \ln \delta Q(N, g) = 0 \quad (5.174)$$

the solutions of which are given by the formulas

$$\ln \alpha^2(N, g) = \gamma(g^*)\ln N + \int_0^{\ln N} dt\left\{\gamma\left[g - \int_{s(g)}^{s(g)-\ln N+t} ds' B^{-1}(s')\right] - \gamma[g^*]\right\} \qquad (5.175)$$

and

$$\ln \delta Q(N, g) = \gamma(g^*)(N - 1)$$
$$+ N\int_0^{\ln N} dt\, e^{-t}\left\{\gamma\left[g - \int_{s(g)}^{s(g)-\ln N+t} ds' B^{-1}(s')\right] - \gamma[g^*]\right\} \quad (5.176)$$

respectively. Here,

$$s(g) = -\int_0^g dg'\beta^{-1}(g'), \quad g(s) = -\int_0^s ds' B^{-1}(s') \quad \text{and} \quad B(s) = \beta^{-1}(g)$$

In both cases, the solution of the second Callan–Symanzik equation [see Eq. (5.95)]

$$\beta(g)\frac{\partial}{\partial g}\bar{g}(N, g) - \frac{\partial}{\partial \ln N}\bar{g}(N, g) = 0 \tag{5.177}$$

is

$$\bar{g}(N, g) = g - \int_{s(g)}^{s(g)-\ln N} ds' B^{-1}(s') \tag{5.178}$$

Note, however, that the self-consistency relationships connecting the generators $\beta(g)$ and $\gamma(g)$ are different for the GPRG [Eq. (5.89)] and the PFRG [Eq. (5.129)].

V. COMPRESSIBILITY FACTORS OF HARD-PARTICLE FLUIDS

A system composed of N identical particles with hard-core mutual interactions is one of the most basic models considered in statistical mechanics. The finite sizes of the particle cores give rise to excluded volume effects and limit the number of particles that can be squeezed into a container of finite size. The upper limit of the packing fraction (or volume fraction) can be established from geometric considerations. Thus, it is known that a three-dimensional (3D) system of hard spheres can form two densely packed, ordered structures, hexagonal (face-centered A3) and cubic (face-centered A1), both with packing fractions equal to $\eta_{CP} = (\pi/6)\sqrt{2} \simeq 0.74$ and coordination numbers $z = 12$. Hard spheres also can form a less densely packed body-centered cubic structure A2 for which $\eta = (\pi/6)\sqrt{3} \simeq 0.68$ and $z = 8$ and a simple cubic structure with $\eta = (\pi/6) \simeq 0.524$ and $z = 6$. All of these are regular, periodic structures pertaining to spheres that are in static, direct contact with neighbors. The packing of hard spheres also has been studied experimentally by Berryman [22] who established the value of $\eta_{RCP} \simeq 0.64 \pm 0.02$ for random close packing. Here again the arrangement is static, with neighboring particles in direct contact. How these packings are affected by the finite motions characteristic of thermal equilibrium is a matter of conjecture.

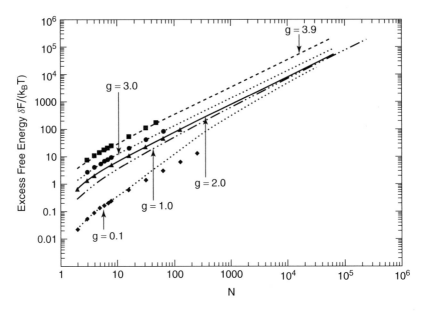

Figure 5.11. An expanded-scale comparison of PFRG predictions for $K = 4$ and the results of computer simulations.

The thermodynamic properties of a classical hard-sphere system are derivable from the Gibbs canonical partition function

$$Q(T, V, N) = [2\pi m k_B T/h^2]^{3N/2} \frac{1}{N!} \left(\frac{V}{\sigma^3}\right)^N \frac{1}{V^N} \int dr^N \prod_{1 \leqslant i < j \leqslant N} \theta(|\mathbf{r}_i - \mathbf{r}_j| - \sigma)$$

(5.179)

with $\theta(x)$ denoting the Heaviside step function [for which $\theta(x) = 1$ for $x \geqslant 1$ and $\theta(x) = 0$ for $x < 0$]. The symbol σ indicates the diameter of a sphere and $\mathbf{r}_i$ is the position vector of particle i. Due to our choice of hard-sphere particle interactions the configuration integral appearing in (5.179) is independent of temperature.

Because the excess free energy defined by the expression

$$\delta F = -k_B T \ln\left[\frac{1}{V^N} \int dr^N \prod_{1 \leqslant i < j \leqslant N} \theta(|\mathbf{r}_i - \mathbf{r}_j| - \sigma)\right] \quad (5.180)$$

cannot be calculated exactly for arbitrary values of N, one frequently uses the alternative virial series formula

$$\delta F = Nk_B T \int_V^\infty \left(\frac{pV}{Nk_B T} - 1\right)\frac{dV}{V} = Nk_B T \sum_{j \geq 1} \eta^j \bar{B}_{j+1}/j \quad (5.181)$$

Corresponding to this is the formula

$$\phi(\eta) = 1 + \sum_{j \geq 1} \bar{B}_{j+1} \eta^j \quad (5.182)$$

for the compressibility factor $\phi = pV/nk_B T$. Here $\eta = (\pi\sigma^3/6)(N/V)$ denotes the packing fraction of the spheres and $\bar{B}_j$ is a dimensionless virial coefficient that can be expressed in terms of the configuration integrals for systems of $j, j-1, \ldots, 2$ particles. Thus, the virial expansions provide formal representations of the excess free energy and pressure of an N-particle system in terms of thermodynamic quantities pertaining to systems of smaller size.

The coefficients $\bar{B}_j$ do, however, prove to be so difficult to evaluate that, even for hard spheres, the values of only eight presently are known [23], (cf. Table V.2). The corresponding *truncated* virial series (or virial polynomial) produces quite accurate estimates of the compressibility factor for values of η as large as 0.4. However, the series (5.182) probably is only asymptotic and so, even if more virial coefficients were available, it would be useful to construct an alternative approximation that is more suitable for high densities. One of the most successful of the various techniques proposed for this purpose is that of Carnahan and Starling [24] who replaced the exact hard-sphere virial coefficients of Table V.2 with the integer approximations 4, 10, 18, 28, 40, 56, and 70. These then could be represented by the simple formula $\bar{B}_j = j^2 + j - 2$ and the correspondingly modified virial series

TABLE V.2
Numerical Values of the Dimensionless Virial Coefficients $\bar{B}_{j+1} = B_{j+1}/(\pi\sigma^3/6)$ for Hard Spheres [23][a]

$\bar{B}_j$		$\bar{B}_j$	
$\bar{B}_2$	4	$\bar{B}_6$	39.5
$\bar{B}_3$	10	$\bar{B}_7$	56.5
$\bar{B}_4$	18.365	$\bar{B}_8$	70.779
$\bar{B}_5$	28.24		

[a] Here, σ denotes the diameter of an individual sphere.

summed over *all* values of j to produce the well-known Carnahan–Starling (CS) equation of state

$$\phi_{\text{CS}}(\eta) = \frac{1 + \eta + \eta^2 - \eta^3}{(1 - \eta)^3} \tag{5.183}$$

This formula produces estimates of the compressibility factor that agree much better at high densities with the results of computer simulations than those generated by the 7-term virial polynomial. It now will be demonstrated that an even better approximation can be obtained by using the GPRG.

A. Self-Similarity of the Fluid Pressure

We begin by observing that the pressure of an ideal gas can be written as the product of the number of particles N and $k_B T/V$, the pressure of a system consisting of a single particle;

$$p = p(N; k_B T/V) = N \frac{k_B T}{V} = Np(1; k_B T/V) \tag{5.184}$$

With N identified as the group parameter, it then follows that the ideal gas pressure is self-similar with respect to scaling of the particle number, namely,

$$p(N; k_B T/V) = \frac{N}{K} p(K; k_B T/V) = p\left[\frac{N}{K}; p(K; k_B T/N)\right] \tag{5.185}$$

This can be identified as a form of the Dalton law, pertaining to the additivity of the partial pressures exerted by N/K subsystems, each at the same temperature and consisting of K particles confined to a common container of volume V.

When attention is shifted to a *nonideal* gas of hard spheres, one must introduce a suitable bare coupling parameter to account for the finite sizes of the particles. We select this quantity to be $g = \pi\sigma^3/6V$ so that the packing fraction of spheres is given by the relationship $\eta = Ng$. The pressure then can be treated as a function of the three independent variables N, g and $k_B T/V$ and the basic conservation equation becomes

$$p(N; g, k_B T/V) = \frac{N}{K} \phi\left[\frac{N}{K}; \bar{g}(K; g)\right] p(K; g, k_B T/V)$$

$$= p\left[\frac{N}{K}; \bar{g}(K; g), p(K; g, k_B T/V)\right] \tag{5.186}$$

Here, $\bar{g}(K; g)$ denotes an effective coupling function that satisfies the scaling invariance relationship (5.159). Finally, the evolution of the compressibility factor is governed by the GPRG functional equation

$$\phi(N; g) = \phi(K; g)\phi\left(\frac{N}{K}; \bar{g}(K; g)\right) \qquad (5.187)$$

To interpret Eq. (5.186), we first note that this expression describes a scaling of the pressure. The system of N particles is (virtually) separated into N/K subsystems, every one of which contains K interacting particles. Each of these subsystems occupies the entire volume V and contributes a partial pressure $p(K; g, k_B T/V)$. If it were permissible to neglect interactions among particles assigned to different subsystems, the total pressure would equal $(N/K)p(K; g, k_B T/V)$. However, particles belonging to different subsystems do indeed interact with one another, so that it is necessary to introduce a corrective factor, namely, the renormalized compressibility factor $\phi(N/K; \bar{g}(K; g))$. Incorporation of this factor produces Eq. (5.186), which has been rewritten in the form of a functional equation, expressing the requirement that the value of the pressure be unaltered by the virtual subdivision of the system.

Before proceeding further we attend to a technical problem connected with the asymptotic nature of the virial series (5.182). This representation of the compressibility factor is appropriate in the so-called thermodynamic limit, for which $N \to \infty$, $V \to \infty$ with N/V finite. However, truncated virial series such as

$$\phi_M(N; g) = \sum_{j=1}^{M} \bar{B}_{j+1}(gN)^j \qquad (5.188)$$

do not satisfy the condition $\phi(1; g) = 1$ and so are not unit normalized with respect to the group parameter N. Instead of attempting to construct a modified virial series (a feasible but very complicated task) we introduce the "transfer function" or renormalized compressibility factor

$$\tilde{\phi}(N; g) = \phi(N; g)/\phi(1; g) \qquad (5.189)$$

The relationship $p(N; g, k_B T/V) = p(1; k_B T/V)N\phi(N; g)$ and the initial condition $p(1; k_B T/V) = k_B T/V$ then are replaced with

$$p\left[N; g, \frac{k_B T}{V}\phi(1; g)\right] = p\left[1; \frac{k_B T}{V}\phi(1; g)\right]N\tilde{\phi}(N; g) \qquad (5.190)$$

and

$$p\left[1; \frac{k_B T}{V} \phi(1; g)\right] = \frac{k_B T}{V} \phi(1; g) \qquad (5.191)$$

respectively, and the conservation equation (5.186) becomes

$$p\left[N; g, \frac{k_B T}{V} \phi(1; g)\right] = p\left\{\frac{N}{K}; g, p\left[K; \bar{g}\left(\frac{N}{K}; g\right), \frac{k_B T}{V} \phi(1; g)\right]\right\} \qquad (5.192)$$

Finally, the transfer function $\tilde{\phi}(N; g)$ satisfies the functional equation (5.187) with ϕ replaced by $\tilde{\phi}$. The function $\bar{g}(K; g)$ continues to satisfy (5.159).

The stage is now set to apply the GPRG Lie equations [see Section II.C]

$$\frac{\partial \ln \tilde{\phi}(N; g)}{\partial \ln N} = \gamma[\bar{g}(N; g)] \qquad (5.193)$$

and

$$\frac{\partial \bar{g}(N; g)}{\partial \ln N} = \beta[\bar{g}(N; g)] \qquad (5.194)$$

to the task of evaluating the effective coupling function and transfer function associated with the compressibility factor of a hard-sphere fluid.

The infinitesimal generator $\gamma(g)$ is obtained from the relationship

$$\gamma(g) = \left.\frac{\partial \tilde{\phi}(N; g)}{\partial N}\right|_{N=1} = \frac{g}{\phi(g)} \frac{d\phi(g)}{dg} \qquad (5.195)$$

using a virial polynomial as the initial $(N \to 1)$ approximation for the compressibility factor. As before, the effective coupling function must be determined self-consistently in order to insure the validity of the self-similarity hypothesis. To accomplish this, Eq. (5.193) is first differentiated with respect to N, to produce the expression

$$\frac{\partial}{\partial N}\left[\frac{\partial \ln \tilde{\phi}}{\partial \ln N}\right] = \frac{\partial \gamma[\bar{g}(N; g)]}{\partial \bar{g}(N; g)} \frac{\partial \bar{g}(N; g)}{\partial N} \qquad (5.196)$$

We then take the limit $N \to 1$ to obtain the following, desired relationship:

$$\beta(g) = \frac{(\partial/\partial N)[\partial \ln \tilde{\phi}/\partial \ln N]_{N=1}}{d\gamma/dg} = \frac{\gamma(g) - \gamma^2(g) + (\partial^2 \tilde{\phi}/\partial N^2)|_{N=1}}{d\gamma/dg}$$

$$= \frac{\gamma(g) - \gamma^2(g) + [g^2/(\phi(g))](d^2\phi/(dg^2))}{d\gamma/dg} \qquad (5.197)$$

between the two generators $\beta(g)$ and $\gamma(g)$. The last equalities of (5.195) and (5.197) have been obtained by noticing that in the virial series $\phi(N; g) = \phi(Ng) = \phi(\eta)$. Thus, in the asymptotic limit the compressibility factor is, in fact, a function of the product of the variables g and N.

A simple calculation now reveals that $\beta(g) = g$ and so the proper solution of Eq. (5.194) is

$$\bar{g}(N; g) = Ng \qquad (5.198)$$

The Lie equation (5.193) then can be written in the form

$$\tilde{\phi}(N; g) = \exp\left\{ \int_1^N dN \gamma(Ng)/N \right\} \qquad (5.199)$$

Next, we select as an initial approximation a virial polynomial such as

$$\phi_M(g) = 1 + \sum_{j=1}^{M} \bar{B}_{j+1} g^j \qquad (5.200)$$

and so obtain from Eq. (5.195) and the corresponding approximation

$$\gamma_M(g) = \frac{\sum_{j=1}^{M} j \bar{B}_{j+1} g^j}{1 + \sum_{j=1}^{M} \bar{B}_{j+1} g^j} \equiv \sum_{j \geq 1} \Gamma_{j+1} g^j \qquad (5.201)$$

for the generator. The coefficients Γ_j are related to the virial coefficients $\bar{B}_j$ as follows:

$$\Gamma_2 = \bar{B}_2$$
$$\Gamma_3 = 2\bar{B}_3 - \bar{B}_2^2$$
$$\Gamma_4 = 3\bar{B}_4 - 3\bar{B}_2\bar{B}_3 + \bar{B}_2^3 \qquad (5.202)$$
$$\vdots$$

The transfer function $\tilde{\phi}(N; g)$ now can be evaluated by substituting the formula (5.201) into (5.199). The results generated by this procedure do, of course, depend on the polynomial $\phi_M(g)$ used as an initial approximation. For example, if only terms to second order in g are retained, it is found that

$$\tilde{\phi}(N; g) = \exp\left[\bar{B}_2 g(N-1) + \frac{1}{2}(2\bar{B}_3 - \bar{B}_2^2)g^2(N^2 - 1)\right] \quad (5.203)$$

In the limit $N \to \infty$, $g \to 0$ and, with $Ng = \eta$ finite, this produces an expression for the compressibility factor, namely,

$$\phi(\eta) = \exp\left[\bar{B}_2 \eta + \frac{1}{2}(2\bar{B}_3 - \bar{B}_2^2)\eta^2\right] \quad (5.204)$$

which is identical with that derived by Shinomoto [25].

The use of higher order polynomial approximations leads to significantly more accurate estimates of the equation of state. Thus, by incorporating all the presently known virial coefficients into the initial approximation we obtain the cumulant formula

$$\phi(\eta) = \exp[4\eta + 2\eta^2 - 0.3020\eta^3 + 0.7653\eta^4 + 1.8302\eta^5 - 0.2013\eta^6 + 0.7146\eta^7]$$
(5.205)

This formula appears to provide better agreement with the results of computer simulations than other, currently existing analytic expressions for the compressibility factor of a hard-sphere fluid. However, as indicated in Figure 5.12, the simulation results of several investigators are not in agreement for $\eta \geqslant 0.6$.

Also included in this figure are predictions based on the 8-term virial series and on the Shinomoto formula. On the scale of this figure the predictions of the Carnahan–Starling formula are indistinguishable from those of the RG theory. However, when presented in tabular form (see Table V.3), the predictions of the RG formula (5.205) exhibit an average deviation from the simulation data of Erpenbeck and Wood [26] that is only one-fourteenth of that associated with the Carnahan–Starling formula. The symbols $\eta_{RCP} \simeq 0.64$ and $\eta_{OCP} \simeq 0.74$ indicate the current best estimates [22] of the densities for random (RCP) and ordered (OCP) closest packing of hard spheres.

Our presentation here has been limited to the 3D case of a hard-sphere fluid. However, the GPRG method is equally successful in predicting the compressibility factor for a two-dimensional (2D) fluid of hard disks [29].

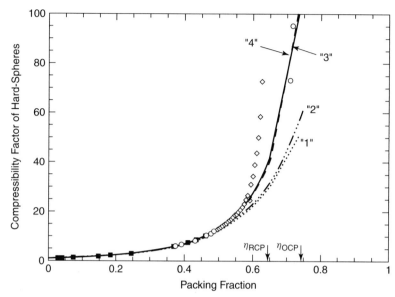

Figure 5.12. The compressibility factor of a hard-sphere fluid. The solid squares, circles, and diamonds indicate values obtained from the computer simulations of Erpenbeck and Wood [26], Rotenberg [27] and Woodcock [28], respectively. Curve "1" is a plot of the seventh-order virial polynomial. Curve "2" is the Shinomoto, second-order cumulant approximation of Eq. (5.204). Finally, "3" is a plot of the RG seventh-order cumulant approximation of Eq. (5.205) and of the Carnahan–Starling equation (5.183) as well, since the two are indistinguishable on the scale of this figure.

TABLE V.3
Compressibility Factors for Hard Spheres[a]

Packing Fraction η	Simulations MC	Virial Series	CS Eq. (5.183)	Shinomoto Eq. (5.204)	Padé Approximant	RG Theory Eq. (5.205)
0.0296	1.12777	1.12766	1.1277	1.12767	1.12766	1.12766
0.0411	1.18282	1.18265	1.1826	1.18267	1.18265	1.18265
0.0740	1.35939	1.35915	1.3590	1.35928	1.35915	1.35915
0.1481	1.88839	1.88848	1.8872	1.88941	1.88847	1.88850
0.1851	2.24356	2.24394	2.2416	2.24549	2.24394	2.24409
0.2468	3.03114	3.02952	3.0252	3.03139	3.03026	3.03129
0.3702	5.85016	5.79471	5.8305	5.78285	5.83311	5.85230
0.4114	7.43040	7.29440	7.4099	7.27250	7.39697	7.44104
0.4356	8.60034	8.37362	8.5806	8.34694	8.54838	8.61875
0.4628	10.19388	9.79982	10.1780	9.77246	10.10784	10.22531
100Δ	0	0.940	1.14	1.053	0.228	0.078

[a] The parameter η denotes the packing fraction and $\Delta = \frac{1}{10}\sum_{j=1}^{10} |\phi_{MC}(\eta_j) - \phi(\eta_j)|/\phi_{MC}(\eta_j)$ is the average fractional deviation from the Monte Carlo (MC), computer simulation data of [26].

The numerical values of eight virial coefficients also are available for this system, as are the results of several Monte Carlo computer simulation studies.

VI. APPLICATIONS TO QUANTUM FIELD THEORY (QFT)

In the previously considered, statistical mechanical applications of the RG method the number of particles belonging to the system was almost invariably the natural choice for the group parameter. However, other situations occur in which quite different quantities can be identified as the generalized time variable appropriate to the problem. Two such problems are considered here, both of which arise in QFT.

A. Photon Propagator of Quantum Electrodynamics

The first problem to which we turn is that of determining the asymptotic, high four-momentum behavior of the photon propagator. It was to solve this problem that the RG method originally was invented. An historical outline of the application of RG ideas to this and other problems in quantum field theory can be found in the well-known monograph by Bogoliubov and Shirkov [3], recent review articles by Shirkov [4, 5], and the monograph written by Itzykson and Zuber [30].

The photon propagator, defined by the formula [31]

$$D_{\mu\nu}(x - x') = i\langle 0|T\{A_\mu(x)A_\nu(x')\}|0\rangle \tag{5.206}$$

is one of the fundamental physical quantities of quantum electrodynamics. In the momentum representation this function can be written as a sum of longitudinal and transverse parts

$$D_{\mu\nu}(k) = D_\perp(k)\left(\delta_{\mu\nu} - \frac{k_\mu k_\nu}{k^2}\right) + D_{||}(k)\frac{k_\mu k_\nu}{k^2} \tag{5.207}$$

For low-energy photons ($k^2 \ll m^2$), where m is the mass of electron) radiative corrections are negligible, the photon is "ideal" (i.e., non-self-interacting) and

$$D_\perp^0(k) = D_{||}^0(k) = \frac{1}{k^2} \tag{5.208}$$

However, for photons with energies sufficient for pair creation ($k^2 > 4m^2$) the transverse part becomes dependent on the fine structure constant $\alpha = e^2/4\pi$ while the longitudinal part remains unchanged.

The transverse part of the propagator then usually is written as the product

$$D_\perp(k^2; \alpha) = D_\perp^0(k^2)\, d\left(\frac{k^2}{m^2}; \alpha\right) \qquad (5.209)$$

of the "ideal" part given by (5.208) and an "excess", correction factor of $d(k^2/m^2; \alpha)$ known as the photon propagator amplitude. The latter can be represented by the third-order perturbation polynomial

$$\tilde{d}(t; \alpha) = 1 + \alpha \frac{1}{3\pi} \ln t + \alpha^2 \frac{1}{9\pi^2}\left[\frac{7}{12} + \ln t\right] \ln t$$

$$+ \alpha^3 \frac{1}{3\pi^3}\left[[2\zeta(3)] + \frac{1}{9}\ln^2 t + \frac{55}{216}\ln t - \frac{4627}{2592}\right]\ln t + \cdots \qquad (5.210)$$

obtained from the formula given by Rafael and Rosner [32] for the asymptotic form of the proper photon self-energy. This polynomial suffers from the so-called "ultraviolet catastrophe", diverging at high values of the group parameter t (i.e., when $k^2 \gg m^2$). However, because of the relatively small value of the radiative coupling coefficient $[\alpha = (137.036)^{-1}]$ and the logarithmic t-dependence of the series (5.210) this perturbation series should, in fact, provide a very good approximation to the propagator amplitude for values of t as large as 10^6, that is, for photon energies $\leqslant 3 \times 10^5$ MeV. Furthermore, although the series cannot tell us much about the behavior of the amplitude in the limit $t \to \infty$, it can serve quite well as an initial approximation for RG method calculations.

Since experiments currently are limited to energies of the order 100 GeV (corresponding to $t \leqslant 4 \times 10^7$), questions concerning the asymptotic behavior of the photon propagator amplitude obviously are more of theoretical interest than practical importance. Within this experimentally accessible domain it is reasonable to expect that Landau's [33] "leading logarithm" formula

$$\tilde{d}(t; \alpha) = \frac{1}{1 - (\alpha/3\pi)\ln t} \qquad (5.211)$$

will provide an adequate approximation and, indeed, it is possible that this formula will be applicable even near the limiting value of $t = \exp(3\pi/\alpha) \simeq 10^{561}$ at which the famous Landau ghost pole occurs.

The Landau formula (5.211) can be obtained directly from the perturbation series (5.210) by assuming that terms of the form $[(\alpha/3\pi)\ln t]^n$ occur for all values of n and then constructing the sum of this geometric progression. However, it is clear that this formula cannot be used to determine the asymptotic behavior of the propagator because: (a) neglected terms of the series (5.210) surely contribute to the sum and; (b) conclusions concerning the limit $t \to \infty$ require information about the propagator amplitude for values of t in excess of the convergence limit, $t = \exp(3\pi/\alpha)$, of the geometric series. Additionally, the leading logarithm formula (5.211) produces unphysical consequences that caused Landau and Pomeranchuk [33] to question the basic premises of quantum electrodynamics.

We now identify the generalized time appropriate to this problem by the relationship $t = k^2/m^2$, using m^2 as the basic scale with respect to which the four momentum is measured. The quantity $1/m^2$ is the scale of the propagator.

It easily is demonstrated that the unperturbed (by radiative corrections) transverse propagator $D_\perp^0(k)$ is strictly self-similar with respect to scaling of the four momentum. Thus, with $D_\perp^0$ treated as a function of the group parameter t and the propagator amplitude $1/m^2$, namely,

$$D_\perp^0(t; 1/m^2) = t^{-1} \frac{1}{m^2} \qquad (5.212)$$

we see that

$$D_\perp^0(t; 1/m^2) = D_\perp^0\left[\frac{t}{x}; D_\perp^0(x; 1/m^2)\right] \qquad (5.213)$$

and so conclude that this function is an invariant of the scaling transformation

$$t \to t' = t/x$$
$$D_\perp^0(1; 1/m^2) \to D_\perp^0(x; 1/m^2) \qquad (5.214)$$

As the value of the four momentum increases the propagator acquires a dependence on the fine structure constant and so evolves from $D_\perp^0(t; 1/m^2)$ into a more complicated function $D_\perp(t; \alpha, 1/m^2)$. This equation we write in a form

$$D_\perp(t; \alpha, 1/m^2) = D_\perp^0(t; 1/m^2) d(t, \alpha) = t^{-1} \frac{1}{m^2} d(t; \alpha)$$

$$= \frac{x}{t} D_\perp(x; \alpha, 1/m^2) \frac{d(t; \alpha)}{d(x; \alpha)} \qquad (5.215)$$

THE ROLE OF SELF-SIMILARITY IN RENORMALIZATION GROUP THEORY 329

that is comparable to (5.209). The formula (5.215) can, in turn, be identified as a generalization of the self-similarity relation (5.213), *provided* that there exists an effective radiative constant $\bar{\alpha}(x; \alpha)$ such that

$$\frac{d(t; \alpha)}{d(x; \alpha)} = d\left[\frac{t}{x}; \bar{\alpha}(x; \alpha)\right] \qquad (5.216)$$

Thus, by substituting (5.216) into (5.215), we obtain a relationship

$$D_\perp(t; \alpha, 1/m^2) = \frac{x}{t} D_\perp(x; \alpha, 1/m^2) d\left[\frac{t}{x}; \bar{\alpha}(x; \alpha)\right]$$

$$= D_\perp\left[\frac{t}{x}; \bar{\alpha}(x; \alpha), D_\perp(x; , 1/m^2)\right] \qquad (5.217)$$

which is identical in form with the generalized conservation equation (5.58).

The functional equation (5.127) expresses the invariance of the propagator with respect to the RG transformation

$$t \to t' = t/x$$

$$\alpha \to \alpha' = \bar{\alpha}(x; \alpha)$$

$$\frac{1}{m^2} \to D_\perp(x; \alpha, 1/m^2) \qquad (5.218)$$

Implicit in Eq. (5.217) are the following basic assumptions of the RG method: (1) That the propagators defined on the larger interval t and the smaller interval t/x are functionally similar, provided that the bare coupling parameter α is replaced with a suitably defined effective coupling function $\bar{\alpha}(x; \alpha)$ and that the propagator scale $1/m^2$ at the value $t = 1$ is replaced with the value $D_\perp(x; \alpha, 1/m^2)$ appropriate to the new scale $t = x$ for the generalized time; (2) the self-similarity of the transformation (5.218) is enforced by relating the effective coupling function (effective radiative constant) to the ratio of propagator amplitudes according to Eq. (5.216). The mathematical implication of this second condition is that, for each value of the generalized time t, there is an inverse relation that expresses the bare coupling parameter α as a function of the propagator amplitude, that is, that associated with the relationship $d = d(t; \alpha)$ is an inverse function $\alpha = F(t; d)$. Consequently, for any value of the new scale $\tau = x$ the effective coupling function can be formally represented as follows:

$$\bar{\alpha}(x; \alpha) = F\left(\frac{t}{x}; \frac{d(t; \alpha)}{d(x; \alpha)}\right) \qquad (5.219)$$

with the actual value of the generalized time expressed as a multiple $t = (t/x)x$ of the new scale.

Equation (5.216), together with the supplementary equation (for the invariant charge)

$$\bar{\alpha}(x; \alpha) = \bar{\alpha}\left(\frac{x}{t}; \bar{\alpha}(t; \alpha)\right) \tag{5.220}$$

are the basic elements of the GPRG, as applied to this problem. These equations can be solved by the iterative procedure outlined in Section II.B. Thus, with the generalized time t expressed as an integral power of the scale τ ($t = \tau^n$), the solution of (5.216) may be written in the form

$$d(\tau^n, \alpha) = d(\tau; \alpha) \prod_{i=1}^{n-1} d[\tau; \bar{\alpha}^{(i)}(\tau; \alpha)] \tag{5.221}$$

with $\bar{\alpha}^{(i)}(\tau; \alpha)$ denoting the i-fold nested iterate of $\bar{\alpha}(\tau; \alpha)$. The corresponding solution of (5.220) is $\bar{\alpha}(\tau^n; \alpha) = \bar{\alpha}^{(n)}(\tau; \alpha)$. The relationship (5.221) expresses the propagator amplitude $d(\tau^n; \alpha)$ as a functional of this same amplitude, evaluated for the shorter interval of time τ *and* for the effective coupling function $\bar{\alpha}(\tau; \alpha)$ specific to this smaller interval. The latter is obtained by solving the equation

$$d[\tau; \bar{\alpha}(\tau; \alpha)] = d(\tau^2; \alpha)/d(\tau; \alpha) \tag{5.222}$$

gotten from (5.221) by setting $n = 2$.

Although the value of the scale length τ is not given by the theory, it necessarily must exceed the self-similarity threshold value τ_0 discussed at the beginning of Section II. If self-similarity exists for some scale $\tau_0 > 1$, the propagator amplitude $d(t; \alpha)$ then is a fractal object. However, should this condition not be met, the system will be only *asymptotically self-similar* and the scaling exponent will depend on the scale value. In this latter event, the photon propagator amplitude then joins the large class of physical objects that exhibit fractal behavior only for limited ranges of the physically significant scale [34]. It is important to recognize that the minimum scale value above which the system exhibits self-similarity is not known a priori. To determine this minimum, threshold value of the scale — and, indeed, to establish its very existence — one must rely on independent theoretical or experimental information. The situation here is closely analogous to that encountered in the polymer chain problems of Section IV, where the length of the repetitive unit, the Kuhn segment, served as an adjustable parameter comparable to the scale τ_0. In the polymer case, the minimum Kuhn segment length was determined by requiring agreement between RG theoretical predications and Monte Carlo simulations for long, but finite chains. Although no procedure strictly analogous to this is applicable to the photon propagator amplitude, possible alternatives will be identified below.

If the effective coupling function $\bar{\alpha}(t; \alpha)$ exhibits a stable fixed point $\alpha^*(\tau)$ for any given value τ of the scale, its asymptotic form will be given by the familiar scaling relation

$$\lim_{t \to \infty} d(t; \alpha) \simeq [d(\tau; \alpha^*(\tau))]^{\ln t/\ln \tau} = t^{\lambda(\tau)} \tag{5.223}$$

with

$$\lambda(\tau) = \frac{\ln d[\tau; \alpha^*(\tau)]}{\ln \tau} \tag{5.224}$$

denoting a scaling exponent that sometimes is referred to as an *anomalous dimension*. The value of this exponent depends on the length τ of the assumed threshold scale, which is the only adjustable parameter associated with the method.

Generally speaking, the results of RG Lie equation calculations are functionals of the initial approximation selected for the object function (here the propagator amplitude). The more accurate the initial approximation, the better are the expected results. For example, if we begin with the first-order (in powers of α) approximation [cf. (5.210)]

$$d(\tau; \alpha) = 1 + \frac{\alpha}{3\pi} \ln \tau$$

$$d(\tau^2; \alpha) = 1 + \frac{2\alpha}{3\pi} \ln \tau \tag{5.225}$$

Equation (5.222) then takes a form

$$1 + \bar{\alpha}(\tau; \alpha) \frac{1}{3\pi} \ln \tau = \frac{1 + (2\alpha/3\pi) \ln \tau}{1 + (\alpha/3\pi) \ln \tau} \tag{5.226}$$

from which one obtains for the effective coupling function the corresponding approximation

$$\bar{\alpha}(\tau; \alpha) = \frac{\alpha}{1 + (\alpha/3\pi) \ln \tau} \tag{5.227}$$

The associated nested iterates are given by the formula

$$\bar{\alpha}^{(k)}(\tau; \alpha) = \frac{\alpha}{1 + (k\alpha/3\pi) \ln \tau} \qquad k \geq 1 \quad (5.228)$$

Then, since $n = \ln t/\ln \tau$ we find that

$$\bar{\alpha}(t; \alpha) = \frac{\alpha}{1 + (\alpha/3\pi) \ln t} \quad (5.229)$$

and

$$d(t; \alpha) = \left[1 + \frac{\alpha}{3\pi} \ln \tau\right]\left[\frac{1 + (2\alpha/3\pi) \ln \tau}{1 + (\alpha/3\pi) \ln \tau}\right] \cdots \left[\frac{1 + (n\alpha/3\pi) \ln \tau}{1 + (n-1)\alpha/3\pi \ln \tau}\right] = 1 + \frac{\alpha}{3\pi} \ln t \quad (5.230)$$

Consequently, the chosen initial approximation is valid for all $t \geq \tau$. We furthermore conclude from (5.230) that when the linear approximation (5.225) is used as input to our RG calculations, the only fixed point of the self-similarity transformation is the trivial point $\alpha = 0$.

The situation changes dramatically when higher order terms are included in the initial polynomial approximation for the propagator amplitude [35]. Beginning with the initial, two-loop approximation that is quadratic in α, one finds that the dynamical system $\{d(t; \alpha), \bar{\alpha}(t; \alpha)\}$ exhibits a nontrivial, stable fixed point or "attractor" $\alpha^*(\tau)$. This indicates that the corresponding approximation to the photon propagator amplitude is an asymptotically self-similar fractal object.

As previously indicated [see (5.224)], the value of the exponent λ is dependent on the choice of the self-similarity threshold. However, this dependence is not at all dramatic for the values of t that presently are experimentally accessible. The curves of Figure 5.13 illustrate the group parameter dependence of the propagator amplitude for three values of the threshold scale, using the second-order polynomial initial approximation for $d(t; \alpha)$. Also included in this figure is the Landau leading-logarithm approximation (5.211), indicated by crosses, †, and the second- and third-order polynomials (indicated by circles) obtained by truncating the perturbation series (5.200). Our RG results are practically identical with predictions of the Landau formula for group parameter values up to $t \simeq 10^{651}$, regardless of the value selected for the self-similarity threshold τ. As t draws nearer and nearer to the Landau ghost pole, the values of $d(t; \alpha)$ predicted by the RG theory become progressively smaller than those given by the Landau formula.

THE ROLE OF SELF-SIMILARITY IN RENORMALIZATION GROUP THEORY 333

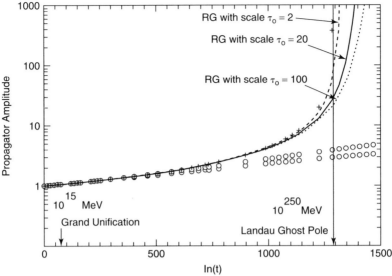

Figure 5.13. Group parameter dependence of the propagator amplitude $d_2(\tau; \alpha)$. RG calculations (based on the second-order polynomial initial data) are presented for three values of the scale τ. Also indicated are points (indicated by crosses, †) computed from the Landau formula (5.211) and second and third order polynomials (indicated by centered and open circles, ⊙ and ○, respectively).

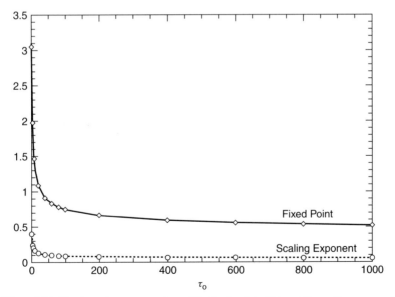

Figure 5.14. The dependences on scale τ of the fixed point $\alpha^*(\tau)$ and the scaling exponent or anomalous dimension $\lambda(\tau)$. Calculations based on second-order polynomial as initial data.

The curves of Figure 5.14 show how the values of the fixed point $\alpha^*(\tau)$ and the anomalous dimension $\lambda(\tau)$ depend on the choice of scale. Both $\alpha^*(\tau)$ and $\lambda(\tau)$ decrease monotonically as the value of the scale increases. To determine the proper threshold value τ, in excess of which the propagator amplitude is self-similar for all values of the group parameter t, we must acquire independent information about the expected asymptotic behavior of the effective coupling function such as, for example, the correct limiting value (at $t \to \infty$) of the coupling function. One possibility is that the onset of self-similarity will become apparent if one selects a sufficiently large value of the scale. Implementation of this notion requires an extrapolation of our calculated results to infinite values of τ, a procedure that leads to the results $\alpha^*(\tau \to \infty) \simeq 0.95$ and $\lambda(\tau \to \infty) \simeq 0.12$.

Alternative to this is the value $\alpha^*(\tau = 23.21) = \pi/3 \simeq 1.05$ which has been identified (by an analysis of spontaneous symmetry breaking in massless QED [36]) as the value of the short-ranged, high-momentum effective coupling function that is characteristic of the boundary between massless and massive phases. A detailed discussion of this aspect of quantum electrodynamics can be found in the review article by Fomin et al. [36]. For present purposes, this result simply allows us to establish a threshold scale τ and so also to calculate [using the second-order polynomial as the initial approximation for $d(t; \alpha)$] the corresponding anomalous dimension $\lambda(\tau = 23.21) = 0.1277$. Although the values of $\alpha^*(\tau)$ and $\lambda(\tau)$ produced by these two schemes are remarkably similar, the agreement may be accidental. A third possibility is to argue that the physically natural threshold value is $\tau = 4$, corresponding to the photon energy threshold $k^2 = 4m^2$ for pair production. By using this scale and the second-order polynomial as an initial approximation, it is found that $\alpha^*(\tau = 4) = 2$ and $\lambda(\tau = 4) = 0.25$.

Although these results fall short of fully resolving the issue of the high four-momentum behavior of the transverse photon propagator, they do establish that *if* this propagator is asymptotically self-similar it will exhibit a fractal-like asymptotic form

$$D_\perp(k) \underset{k^2 \to \infty}{\simeq} \text{const } k^{2(\lambda - 1)} \tag{5.231}$$

characterized by an anomalous dimension λ whose value falls within the interval $(0.12, 1)$.

B. Screening in an Electron Gas

The electron gas is a simple model, often used to gain insights into the many-body aspects of metals. According to this model, the electrons interact Coulombically with one another and the ionic charge is statically distributed so as to form a uniform positive background through which the electrons move. The interaction energy of two isolated electrons is given by the familiar Coulomb law, the Fourier transform of which is

$$V_0(k) = 4\pi e^2/k^2 \tag{5.232}$$

However, because two electrons belonging to the electron gas polarize the surrounding medium, their effective interaction energy, V_{eff}, differs from the "bare" Coulomb function of (5.232). In particular, V_{eff} is related to V_0 by the Dyson equation [37]

$$V_{\text{eff}}(k) = V_0(k) + V_{\text{eff}}(k)\Pi(k)V_0(k) \tag{5.233}$$

with $\Pi(k)$ denoting the so-called polarization function. This equation can be rewritten in the form

$$V_{\text{eff}}(k) = \varepsilon(k)^{-1}V_0(k) \tag{5.234}$$

with

$$\varepsilon(k) = 1 - \Pi(k)V_0(k) \tag{5.235}$$

identified as the Fourier transform of the local dielectric function. (Due to the assumed spatial isotropy of the system, the functions V_{eff}, Π, and ε depend only on the magnitude k of the wave vector $\boldsymbol{k}$.)

The "bare" interaction energy is invariant with respect to scaling of the wave vector. This can be illustrated by selecting an arbitrary scale k_0 and introducing the dimensionless group parameter

$$t = k_0/k \tag{5.236}$$

The bare interaction energy then can be treated as a function of this group parameter t and the "initial condition"

$$V_0(k_0) = 4\pi e^2/k_0^2 \tag{5.237}$$

Indeed, from the identity

$$V_0(k) \equiv V_0[t; V_0(k_0)] = \frac{4\pi e^2}{k_0^2} t^2 = V_0(k_0)t^2 \tag{5.238}$$

one immediately obtains the relationship,

$$V_0[t; V_0(k_0)] = V_0[t/x; V_0(x; V_0(k_0))] \tag{5.239}$$

characteristic of strict functional self-similarity.

The effective interaction V_{eff} is expected to depend not only on the wave vector but also on n, the number density of electrons in the medium. This, in turn, is related to the Fermi wave number k_F by the definition

$$k_F = (3\pi^2 n)^{1/3} \tag{5.240}$$

Henceforth, we identify this natural scale value k_F as the unit in which wave vectors are measured. It is connected by the formula $k_F = (\alpha r_s a_0)^{-1}$ to the constant $\alpha = (4/9\pi)^{1/3}$ and the ratio $r_s = r_0/a_0$ of $r_0 = (3/4\pi n)^{1/3}$, the average separation of neighboring electrons, and the Bohr radius $a_0 = \hbar^2/me^2$.

In the high-density limit, where the value of r_s is small, the dominant contribution to the polarization Π is the sum over all Feynman diagrams of the repeated pair-bubble or ring type [16]. The resulting, so-called random-phase approximation (r) for the polarization function is given by the formula [16]

$$\Pi_r(k) = -\frac{m k_F}{2\pi^2 \hbar^2}\left[1 - \frac{k_F}{k}\left(1 - \frac{k^2}{4k_F^2}\right)\ln\left|\frac{2k_F - k}{2k_F + k}\right|\right] \tag{5.241}$$

The first factor of this expression can be rewritten as $g(2/\pi)(k_F^2/4\pi e^2)$ with g denoting the dimensionless squared charge

$$g = \frac{e^2 m}{\hbar^2 k_F} = \frac{1}{k_F a_0} = \alpha r_s \tag{5.242}$$

which we identify as the "bare coupling variable" controlling the magnitude of the polarization. The dielectric function associated with this approximation then can be written in the form

$$\varepsilon_r(t; g) = 1 + g\phi(t) \tag{5.243}$$

with

$$\phi(t) = \frac{2}{\pi} t^2 \left[1 - t\left(1 - \frac{1}{4t^2}\right)\ln\left|\frac{2t - 1}{2t + 1}\right|\right] \tag{5.244}$$

Corresponding to (5.243) is the RPA approximation

$$V_{\text{eff},r}(k) = \frac{1}{\varepsilon_r(k)} V_0(k) \qquad (5.245)$$

for the effective, electron-pair interaction energy. By using the same integration procedure that Fetter and Walecka [37, Chap. 5] employed to calculate the charge density induced by an ionic impurity, we obtain the asymptotic formula

$$V_{\text{eff},r}(r) \underset{r\to\infty}{\sim} \frac{e^2}{r} \frac{\sin 2k_F r}{k_F r} \frac{4\xi}{(4+\xi)^2} \qquad (5.246)$$

for the Fourier transform of $V_{\text{eff},r}(k)$. Here $\xi = (2/\pi)g$. The term given explicitly by Eq. (5.246) dominates a Yukawa, shielded-Coulomb contribution that vanishes exponentially for large values of the electron separation r.

It is clear from (5.243) and the restrictive conditions that led to its derivation that the RPA formulas are strictly applicable only to the very high electron densities at which $r_s \ll 1$. For real metals, r_s lies in the range 1.8 to 5.6 [16]. To produce estimates of V_{eff} appropriate to the latter, less stringent conditions we invoke the RG method. The first step in this direction is to express the effective interaction energy in the form

$$V_{\text{eff}} = \frac{4\pi e^2}{\varepsilon(k)k^2} = \frac{\varepsilon(k_F)}{\varepsilon(k)} \frac{4\pi e^2}{\varepsilon(k_F)k_F^2} \frac{k_F^2}{k^2} = \frac{1}{\tilde{\varepsilon}(t;g)} V_{\text{eff}}(k_F) t^2 \equiv V_{\text{eff}}[t;g,V_{\text{eff}}(k_F)] \qquad (5.247)$$

with

$$\tilde{\varepsilon}(t;g) = \varepsilon(t;g)/\varepsilon(1;g) \qquad (5.248)$$

and where $\varepsilon(1;g) = \varepsilon(k_F)$. Next, we assume that $V_{\text{eff}}[t;g,V_{\text{eff}}(k_F)]$ can be treated as a functionally self-similar object and therefore satisfies the conservation equation

$$V_{\text{eff}}[t;g,V_{\text{eff}}(k_F)] = V_{\text{eff}}[t/x; \bar{g}(x;g), V_{\text{eff}}(x;g,V_{\text{eff}}(k_F))] \qquad (5.249)$$

together with the initial condition

$$V_{\text{eff}}[t=1; g, V_{\text{eff}}(k_F)] = V_{\text{eff}}(k_F) \qquad (5.250)$$

As always, the coupling function $\bar{g}$ satisfies the functional equation (5.65), characteristic of a strictly self-similar function.

According to Eq. (5.249) the value $V_{\text{eff}}[t; g, V_{\text{eff}}(k_F)]$ of the effective energy is independent of the choice of the initial condition, provided only that the latter lies on the dynamic trajectory of the system. Thus, we can recover the value $V_{\text{eff}}[t; g, V_{\text{eff}}(k_F)]$ after a "time" t, beginning with the value $V_{\text{eff}}(k_F)$ or by "traveling" a time t/x, starting from the initial value $V_{\text{eff}}[x; g; V_{\text{eff}}(k_F)]$. Furthermore, the equation of evolution (5.249) for the nonideal, polarizable system with $g \neq 0$ must reduce to Eq. (5.239) in the limit $g \to 0$: This is assured by the fact that $\bar{g}(t; 0) = 0$. Finally, as a consequence of (5.247) and (5.249) it follows that the reciprocal of the "normalized" (at $k = k_F$) dielectric constant $\tilde{\varepsilon}(t; g)$ satisfies the functional equation

$$\tilde{\varepsilon}^{-1}(t; g) = \tilde{\varepsilon}^{-1}(x; g)\tilde{\varepsilon}^{-1}[t/x; \bar{g}(x; g)] \quad (5.251)$$

which is characteristic of the GPRG and for which the corresponding Lie equations are

$$\frac{\partial \ln \tilde{\varepsilon}^{-1}(t; g)}{\partial \ln t} = \gamma[\bar{g}(t; g)] \quad (5.252)$$

$$\frac{\partial \bar{g}(t; g)}{\partial \ln t} = \beta[\bar{g}(t; g)] \quad (5.253)$$

with

$$\gamma(g) = \frac{\partial}{\partial t} \tilde{\varepsilon}^{-1}(t; g)\big|_{t=1} \quad (5.254)$$

$$\beta(g) = \frac{\partial}{\partial t} \bar{g}(t; g)\big|_{t=1} \quad (5.255)$$

We now use the RPA approximation for $\tilde{\varepsilon}^{-1}$, as given by the formula

$$\tilde{\varepsilon}_r^{-1}(t; g) = \frac{1 + g\phi(1)}{1 + g\phi(t)} \quad (5.256)$$

and the connection [cf. (5.89)]

$$\beta(g) = \frac{\gamma - \gamma^2 + \frac{\partial^2}{\partial t^2}\tilde{\varepsilon}^{-1}(t; g)\big|_{t=1}}{d\gamma/dg} \quad (5.257)$$

to calculate the infinitesimal generators γ and β. The results are

$$\gamma(g) = -\frac{g\phi'(1)}{1+g\phi(1)} = -\frac{2.5600g}{1+1.1612g} \tag{5.258}$$

and

$$\beta(g) = g\left[1+\frac{\phi''(1)}{\phi'(1)}\right] + g^2\left[\phi(1)-\phi'(1)+\frac{\phi(1)\phi''(1)}{\phi'(1)}\right] = 1.9761g - 0.2654g^2 \tag{5.259}$$

respectively. From the second of these we conclude that there is a stable and nontrivial fixed point at $g^* = 7.4457$ (corresponding to $r_s = 14.289$). Furthermore, by substituting (5.259) into (5.253) we find that

$$\bar{g}(t;g) = \frac{t^A}{\dfrac{1}{g}+\dfrac{1}{g^*}(t^A-1)} \tag{5.260}$$

with

$$A = 1 + \phi''(1)/\phi'(1) = -\gamma(g^*) = 1.9761 \tag{5.261}$$

Finally, with the formula (5.260) for $\bar{g}(t;g)$ and $\gamma(g)$ given by (5.258) the solution of the Lie equation (5.252) is given by the expression

$$\tilde{\varepsilon}^{-1}(t;g) = \frac{\chi(g)+1}{\chi(g)+t^A} \tag{5.262}$$

with

$$\chi(g) = \frac{g^*/g - 1}{1+g^*\phi(1)} = \frac{g^*/g - 1}{8.6457} \tag{5.263}$$

From this, it follows that in the long wavelength limit of $t \to \infty$ (or $k \to 0$) the reduced dielectric function exhibits the power-law behavior

$$\tilde{\varepsilon} \underset{t\to\infty}{\sim} t^A \tag{5.264}$$

With $\tilde{\varepsilon}^{-1}(t; g)$ given by Eq. (5.262) one find that

$$V_{\text{eff}}(k) = V_{\text{eff}}[t; g, V_{\text{eff}}(k_F)] = \frac{1}{\tilde{\varepsilon}(t; g)} \frac{4\pi e^2}{\varepsilon(1; g)k_F^2}$$

$$= \frac{4\pi e^2[1 + \chi(g)]}{\varepsilon(1; g)} \frac{1}{k^2\chi(g) + k^{2-A}k_F^{-A}} \tag{5.265}$$

wherein $2 - A = 0.0239$ and $\varepsilon(1; g) = \varepsilon(k_F)$. Since the effective interaction should reduce to the function $V_0(k)$ of (5.232) in the limit $k \to \infty$, we therefore identify $\varepsilon(k_F)$ with $1 + \chi(g)^{-1}$ and so conclude that

$$V_{\text{eff}}[t; g, V_{\text{eff}}(k_F)] \underset{t \to \infty}{=} V_{\text{eff}}(k) \underset{k \to 0}{\sim} \frac{4\pi e^2 \chi(g)}{k_F^{1.9761} k^{0.0239}} \tag{5.266}$$

With this expression treated as a mathematical distribution [38], it then follows that:

$$V_{\text{eff}}(\tau) \underset{r \to \infty}{\sim} \frac{e^2}{r(k_F r)^A} \left\{ \frac{4\pi\chi(g)\Gamma[(1+A)/2]}{2^{2-A}\pi^{3/2}\Gamma[(2-A)/2]} \right\} = 0.02366 \left(\frac{g^*}{g} - 1 \right) \frac{e^2}{r} \frac{1}{(k_F r)^{1.9761}} \tag{5.267}$$

The GPRG estimate for the effective electron-pair interaction therefore is significantly shorter ranged ($\sim r^{-2.9761}$) than the RPA estimate given by Eq. (5.246).

VII. ADDITIVE FORMULATION OF THE RENORMALIZATION GROUP METHOD

In the preceding sections, we focused exclusively on system properties that were expected (due either to physical or mathematical reasons) to exhibit functional self-similarity with respect to scaling. Implicit in this expectation of self-similarity is the assumption that the system under consideration can be parceled into subunits, the physical properties of which are proportional (at least approximately) to those of the entire system.

This proportionality is clearly evident, for example, in the scaling transformation for the averaged square end-to-end separation of a polymer chain, which for an ideal chain of N bonds (each with an averaged square length equal to ℓ^2) is given by the formula

$$\overline{R_{id}^2}(N; \ell^2) = N\ell^2 = \frac{N}{K} \overline{R_{id}^2}(K; \ell^2) \tag{5.268}$$

Thus, the averaged square end-to-end separation of the N-bond chain is related to that of a smaller K-bond fragment by the proportionality factor N/K. Bond correlations due to nonlocal (excluded-volume) interactions complicate the problem, but it is still reasonable, as we have seen, to assume that subunits of the system will, in some sense, be replicas of the entire chain. More generally, the many examples presented in the preceding sections illustrate that there are many properties of physically interesting systems that can be analyzed successfully from this point of view and that, in particular, the appropriate application of the assumption of self-similarity can be used to generate reliable approximations for physical quantities that we otherwise would be unable to calculate.

A. Translational Self-Similarity

Having thus argued in support of the idea of self-similarity with respect to scaling, we now draw attention to a separate class of problems to which a different type of self-similarity is appropriate, namely, self-similarity with respect to *translation* of the generalized time. A function $f(t; f_0)$, subject to the initial condition $f_0 = f(0; f_0)$, is said to be translationally invariant if it satisfies the functional relationship

$$f(t; f_0) = f(t - \tau; f(\tau; f_0)) \tag{5.269}$$

According to this equation, the value of the function $f(t; f_0)$ is unchanged when the value of the generalized time t and the initial condition f_0 are altered simultaneously by the transformation

$$t \to t' = t - \tau$$
$$f_0 \to f_0' = f(\tau; f_0) \tag{5.270}$$

A simple example of invariance with respect to the transformation (5.270) is provided by the decay of a collection of radioactive particles. The number of surviving particles is given by the familiar formula

$$N(t; N_0) = N_0 e^{-\lambda t} = N_0 e^{-\lambda \tau} e^{-\lambda(t - \tau)}$$
$$= N(\tau; N_0) e^{-\lambda(t - \tau)} = N[t - \tau; N(\tau; N_0)] \tag{5.271}$$

with λ equal to the specific rate of decay. A second example (due to Shirkov [4]) is a homogeneous, elastic rod that is bent by some external force such as gravity or fluid drag, as depicted in Figure 5.15. The shape of the rod is described by the angle θ (between the tangent to the rod and the direction of the force), considered to be a function of distance along the rod, measured

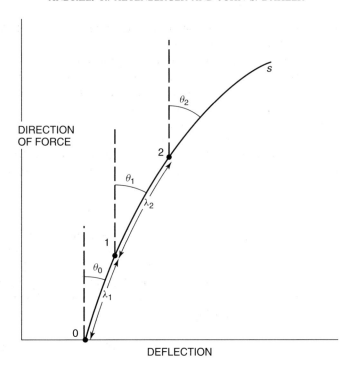

Figure 5.15. Schematic depiction of a bent rod. λ_1 and λ_2 are sequential segments of the rod measured along the rod axis s. The θ_i are angles between the direction of the bending force and the tangent to the rod.

from the "fixed point" at $s = 0$, that is, $\theta(s) = G(s; \theta_0)$ with $\theta_0 = \theta(s = 0)$. From this, we conclude that

$$\theta_1 = G(\lambda_1; \theta_0)$$
$$\theta_2 = G(\lambda_1 + \lambda_2; \theta_0) = G[\lambda_2; G(\lambda_1; \theta_0)]$$

and so, in general,

$$G(\lambda + \ell; \theta_0) = G[\ell; G(\lambda; \theta_0)] \tag{5.272}$$

Closely related to this is the "transfer problem" considered by Mnatsakamjan [39], according to which one seeks a relationship between the flux of radiation (or material particles) and the depth of penetration into a homogeneous medium.

The simplicity of the conservation equation (5.269) is due to the fact that the "ideal system" that it represents is composed of independent constituent parts. When this independence is compromised by interactions, characterized by a coupling parameter g, the system becomes nonideal. Then, we propose in place of (5.269) the following generalization

$$f(t; g, f_0) = f[t - \tau; g, f(\tau; \bar{g}(t - \tau; g), f_0)] \tag{5.273}$$

Here, $\bar{g}(t; g)$ is an effective coupling function that must be so chosen that $f(t; g, f_0)$ is indeed an invariant of the transformation

$$t \to t' = t - \tau$$
$$f_0 \to f_0' = f(\tau; \bar{g}(t - \tau; g), f_0) \tag{5.274}$$

The generalized time variable is restricted to the range $0 \leqslant t \leqslant \infty$ and the two functions $f(t; g, f_0)$ and $\bar{g}(t; g)$ are subject to the initial conditions

$$f(0; g, f_0) = f_0$$
$$\bar{g}(0, g) = g \tag{5.275}$$

It is natural (but not obligatory) to express the function $f(t; g, f_0)$ as the product of the initial value f_0, which can be interpreted as the "scale" of f, and a dimensionless "excess" function $\delta f(t; g)$, namely,

$$f(t; g, f_0) = f_0 \delta f(t; g) \tag{5.276}$$

with $\delta f(0, g) = 1$. By substituting this expression into the conservation equation (5.273), one then obtains for δf the functional equation of evolution

$$\delta f(t; g) = \delta f(t - x; g) \delta f[x; \bar{g}(t - x; g)] \tag{5.277}$$

This can be rewritten in the equivalent forms

$$\delta f(x + y; g) = \delta f[x; \bar{g}(y; g)] \delta f(y; g) = \delta f[y; \bar{g}(x; g)] \delta f(x; g) \tag{5.278}$$

from which one then obtains the identity

$$\delta f[x; \bar{g}(y; g)] = \delta f[y; \bar{g}(x; g)] \frac{\delta f(x; g)}{\delta f(y; g)} \tag{5.279}$$

This, in turn, can be transformed into the expression

$$\frac{\delta f[x; \bar{g}(y; g)]}{\delta f[x - y; \bar{g}(y; g)]} = \delta f[y; \bar{g}(x; g)] \qquad (5.280)$$

which subsequently reduces to the equality

$$\delta f[y; \bar{g}(x - y; \bar{g}(y; g))] = \delta f[y; \bar{g}(x; g)] \qquad (5.281)$$

This last relationship is satisfied only if the effective coupling function is a solution of the functional equation

$$\bar{g}(x; g) = \bar{g}[x - y; \bar{g}(y; g)] \qquad (5.282)$$

According to Eq. (5.269), this is the condition for $\bar{g}$ to be an invariant of the translational transformation.

The Lie equations for $\delta f(x; g)$ and $\bar{g}(x; g)$ are obtained by differentiating the corresponding functional equations (5.277) and (5.282) with respect to t and x, respectively, and then setting these variables equal to zero. One thereby obtains the equations

$$\frac{\delta \ln \delta f(y; g)}{\partial y} = \gamma[\bar{g}(y; g)] \qquad (5.283)$$

and

$$\frac{\partial \bar{g}(y; g)}{\partial y} = \beta[\bar{g}(y; g)] \qquad (5.284)$$

with $\beta(g)$ and $\gamma(g)$ denoting the infinitesimal generators defined as follows:

$$\beta(g) = \left.\frac{\partial \bar{g}(t; g)}{\partial t}\right|_{t=0} \qquad (5.285)$$

$$\gamma(g) = \left.\frac{\partial \delta f(t; g)}{\partial t}\right|_{t=0} \qquad (5.286)$$

In a similar way, one obtains partial differential equations of the Callan–Symanzik type, namely

$$\frac{\partial \ln \delta f(t; g)}{\partial t} = \beta(g) \frac{\partial \ln \delta f(t; g)}{\partial g} + \gamma(g) \qquad (5.287)$$

and

$$\frac{\partial \bar{g}(t; g)}{\partial t} = \beta(g) \frac{\partial \bar{g}(t; g)}{\partial g} \quad (5.288)$$

There is, of course, a direct connection between scaling and translational invariances, which we indicate as follows; $t' = \ln t$, $x' = \ln x$, $\bar{g}(t; x) = \bar{g}'(t'; g)$ and $\delta f(tx; g) = \delta f'(t' + x'; g)$.

B. Viscosity of a Hard-Sphere Suspension

As a simple example of the additive version of the RG technique we consider a suspension of hard spheres dispersed in a continuous, Newtonian solvent. The effective viscosity of this composite system is a function of the solvent viscosity, η_0, the number of hard spheres, N, and a coupling parameter, $g = \pi\sigma^3/6V$, defined as the ratio of the volume of a hard sphere of diameter σ to the total volume of the suspension. This suspension can be treated as an effectively continuous fluid with a viscosity $\eta = \eta(N; g, \eta_0)$ that approaches the value η_0 as the volume fraction of spheres, $\phi = Ng$, tends to zero.

In order to calculate this effective viscosity, we assume the suspension to be additively self-similar, in the sense that

$$\eta(N; g, \eta_0) = \eta[N - K; g, \eta(K; \bar{g}(N - K; g), \eta_0)] \quad (5.289)$$

According to this conservation equation, the viscosity of a suspension of N spheres dispersed in pure solvent, treated as a continuous fluid, is equal to the viscosity of a suspension of $N - K$ spheres dispersed in an effective medium consisting of the solvent and the K remaining spheres, to each of which is assigned an appropriately modified diameter.

It is immediately evident from (5.289) that the excess (or reduced) viscosity $\delta\eta$, defined by the relationship

$$\eta(N, g, \eta_0) = \eta_0 \delta\eta(N; g) \quad (5.290)$$

satisfies the functional equation

$$\delta\eta(N; g) = \delta\eta(N - K; g)\delta\eta[K; \bar{g}(N - K; g)] \quad (5.291)$$

and, as shown previously, the effective coupling function $\bar{g}(N; g)$ must satisfy Eq. (5.282). Finally, the demands of self-consistency require that $\bar{g}(N; g)$ be a functional of the excess viscosity $\delta\eta(N; g)$.

To proceed further, we shall use the Lie equations, and hence need an initial approximation for the excess viscosity (valid for small volume fractions) from which estimates of the infinitesimal generators can be constructed. At the present time, very little information of this type exists. Indeed, reliable values are available only for the second- and third-order coefficients v_2 and v_3 appearing in the viscosity virial expansion

$$\delta\eta(N; g) = 1 + v_2 N g + v_3 N^2 g^2 + \cdots \tag{5.292}$$

The numerical values of these coefficients depend on the fluid boundary condition (e.g., stick or slip) that is assumed to apply at the solvent–sphere interface.

According to (5.286) the infinitesimal generator γ is related to $\delta\eta$ by the formula $\gamma(g) = \partial\delta\eta(N; g)/\partial N|_{N=0}$ and, by differentiating the Lie equation for $\delta\eta$ [viz., $\partial \ln \delta\eta/\partial N = \gamma(\bar{g}(N; g))$] with respect to N and then taking the limit $N \to 0$, we obtain for $\beta(g)$ the expression

$$\beta(g) = \frac{(\partial^2 \delta\eta/\partial N^2)_{N=0} - (\partial\delta\eta/\partial N)^2_{N=0}}{d\gamma/dg} \tag{5.293}$$

By using the initial approximation (5.292), it then follows that

$$\gamma(g) = v_2 g \tag{5.294}$$

and

$$\beta(g) = g^2 \frac{2v_3 - v_2^2}{v_2} \tag{5.295}$$

The second of these two formulas provides the information needed to solve the Lie equation (5.284) and so to conclude that

$$\bar{g}(N; g) = \frac{g}{1 - gN(2v_3 - v_2^2)/v_2} \tag{5.296}$$

By inserting this result into the Lie equation for $\delta\eta$, we obtain the following approximate formula, expressing the excess viscosity as a function of the volume fraction;

$$\delta\eta(\phi) = \frac{1}{|1 - \phi/\phi^*|^{v_2 \phi^*}} \tag{5.297}$$

TABLE V.4
Viscosity Virial Coefficients [see Eq. (5.281) and Associated Values of the "Threshold Volume Fraction" ϕ^*, defined by Eq. (5.298), and the Exponent $v_2\phi^*$ of Eq. (5.297)]

Boundary Condition	v_2	v_3	ϕ^*	$v_2\phi^*$	Reference
Stick	2.5	5.9147	0.448	1.120	42
Slip	1.0	1.8954	0.358	0.358	42
Stick	2.5	4.32	1.05	2.62	43
Stick	2.5	6.2	0.41	1.02	44
Stick	2.5	4.84	0.73	1.82	45
Stick	2.7	5.9147	0.5948	1.606	

The quantity ϕ^* appearing in this expression and defined by the formula

$$\phi^* = \frac{v_2}{2v_3 - v_2^2} \tag{5.298}$$

is the value of the hard-sphere volume fraction at which the suspension viscosity becomes unbounded. An expression of the same form as (5.297) was derived previously by Dougherty [40] and Krieger [41], using a quite different type of reasoning.

Essentially exact values for v_2 and v_3 have been reported by Wajnryb and Dahler [42] for both stick and slip boundary conditions. These values are recorded in Table V.4, along with estimates obtained by several previous investigators. In Figure 5.16, predictions based on the formula (5.297) are compared with the available experimental data. The solid curve is based on the stick boundary condition values $v_2 = 2.5$ (Einstein) and $v_3 = 5.9147$ (Wajnryb and Dahler). To obtain the dashed curve, which agrees much better with (some of) the experimental data and for which $\phi^* = \phi_{RCP} = 0.64$, we have replaced the second of these viscosity virial coefficients with $v_3 = 5.0781$.

There are several possible sources for the discrepancies between theory and experiment: (1) experimental problems, including the difficulty of preparing and maintaining homogeneous, monodisperse suspensions (notice the lack of agreement among the several experimental studies); (2) uncertainties concerning the particle surfaces and the choice of an appropriate solvent–sphere boundary condition; (3) limitations intrinsic to the severely truncated, two-term virial series used to construct the infinitesimal generators of the RG transformation. Since none of these problems can be dealt

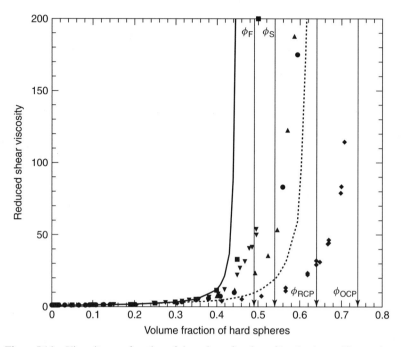

Figure 5.16. Viscosity as a function of the volume fraction of hard spheres. The continuous curve is of Eq. (5.297) with $v_2 = 2.5$ and $v_3 = 5.9147$. The dashed curve is for $v_2 = 2.5$ and $v_3 = 5.0781$, with the value of the latter so chosen that $\phi^* = \phi_{RCP} = 0.64$ [22]. The symbols (●, ▼, ▲, ◆, ■) indicate experimental data reported in references 46–50 by de Kruif, Segre, Jones, Eilers, and Vand, respectively. The symbols ϕ_{SCF} and ϕ_{RCP} are the volume fractions of ordered and random closest packing, respectively [22], whereas ϕ_F and ϕ_S denote values of the volume fraction reported for the coexisting fluid (F) and solid (S) the phases of a neat, hard sphere fluid [51, 52].

with at the present time, the most we can conclude is that the additive formulation of the RG method does provide an approximate formula for the excess viscosity that is in qualitative accord with experiment.

VIII. CONCLUDING REMARKS

The studies reported here have been addressed to problems in theoretical physics concerning physical properties about which one has only limited analytic or computer generated data. These data may be available only for small values of the system density as, for example, in the form of a "virial polynomial", or they may be limited to an analogous, truncated series in powers of some parameter gauging the strength of an interaction that causes the many-body system to behave nonideally. Alternatively, data characteristic of a many-body system may be available only for similarly constituted systems of much smaller size, for example, short segments differing only in length from the macromolecular chain of interest or small replicas of a much larger lattice-spin system. One wishes to use these available data to produce numerical estimates of the property value applicable to a much broader range of parameter values and/or system sizes.

Among the few tools available for this purpose are empirical procedures such as the asymptotic-ratio method, the method of Padé approximants and, as characterized by Robertson [53], "the physically mysterious rescaling process". The last of these has evolved over the past few decades into the powerful tool of renormalization, first introduced by researchers in quantum field theory and subsequently extended and refined by a host of outstanding scientists. Although the self-similarity-based RG method described in these pages differs in several respects from the mainstream of these RG theories, there are many overlaps in terminology, formalism, and objectives. What distinguishes our approach from most others is that we have sought to exploit the ubiquitousness in the characterization of many physically interesting situations of the concept of self-similarity. When this similarity is "strict", as defined and exemplified in Section II, the application of our formalism generally is quite obvious and the results are essentially trivial. The true power of the method emerges only when the self-similarity is at best "asymptotic" and therefore likely to become evident only when a critical scale size (range of values of the characteristic variable) has been exceeded. This asymptotic self-similarity seldom if ever can be established a priori and may not be unequivocally demonstrable even in situations where the numerical estimates generated by the method match quite well all the data available for comparison. By our standards, the method will have succeeded if it passes this last test and/or if it has produced physically reasonable predictions appropriate to conditions for which no demonstrably superior alternative already exists.

Our objective has not been to discover new and better fundamental approaches for solving the physical problems of concern, but to develop a general procedure by which information limited to small values of a characteristic variable (group parameter) can be used to produce estimates of the object function (quantity of interest) for a much larger range of characteristic variable values. The limited data with which one initially is supplied invariably have been obtained by the careful use of the appropriate classical or quantal mechanics, but practical, computational limitations prevent further exploitation of these conventional tools. Thus, for example, each successive term of a perturbation series invariably is more difficult to evaluate than its predecessor and the resulting series often exhibit very limited radii of convergence or, indeed, even may appear to be divergent.

The alternative approach adopted in the RG method is to seek an (almost necessarily approximate) equation of motion or evolution, governing the dependence of the object function on a variable characteristic of the "size" of the system, for example, the number of component parts (atoms, spins, etc.) or some other measure of its complexity. The derivation (or divination) of this evolution equation must rely on some argument or point of view that transcends the usual orthodoxy and therefore is unlikely to be as well founded as the conventional laws of physics. In the simplest of cases, namely, the idealized strictly self-similar examples considered in Section II, exact equations of evolution actually can be constructed and solved. However, when faced with a more complicated problem for which exact results neither are presently available nor apt to be in the future, one then searches for a representation of the object function to which a modified version of strict self-similarity may be applicable. This search is guided by examining related "ideal" object functions that exhibit strict self-similarity. The success of this search is vitally dependent on concocting a physically plausible expectation that the complex object function will in fact exhibit strict self-similarity *provided* that the scale on which it is treated exceeds some threshold value. The examples treated in Sections III–VII illustrate that this procedure can indeed be applied successfully to a rather broad spectrum of problems. One does not expect the results to be exact and, in fact, exact results often are unavailable for comparison. Success is measured by how well the approximate results of the RG method reproduce whatever data are available and by the reasonableness of the results. These criteria are, of course, the same as those used to judge all approximation schemes. A remarkable feature of our approach, not shared by most other formulations of the RG method is its capability of generating estimates of the object function for finite values of the group parameter (e.g., the number of constituent parts of the system) as well as the corresponding asymptotic limit.

Like many other RG methods, ours rests on the implicit assumption that the evolution of the object function, namely, its dependence on the selected group parameter or "generalized time", is autonomous, that is, fully characterized by the value of the group parameter and by initial data, generally restricted to small values thereof; equivalently, the parameters of the system are not time dependent. Confirmation of this hypothesis almost never can be extracted from the conventional laws of physics, for were it otherwise, analyses based on those laws already would have succeeded in solving the very same, incredibly complicated many-body problems that led us from the start to examine the RG method.

Finally, it should be mentioned that the particular form of the self-similarity-based RG method presented here certainly is not sufficiently flexible to treat all problems. Specifically, it is reliant on the assumption that

THE ROLE OF SELF-SIMILARITY IN RENORMALIZATION GROUP THEORY   351

there is a single bare coupling parameter g and that the rescaled problem can be treated in terms of a single modified coupling parameter $\bar{g}$. However, this deficiency would appear to be easily remedied since much of the formalism developed in Sections II and II is applicable to situations for which the single bare coupling parameter g is replaced with a set $\boldsymbol{g} = [g_1, g_2, \ldots, g_\ell]$. In place of the single effective coupling function $\bar{g}(x; g)$ there is then a corresponding set, $\bar{\boldsymbol{g}}(x; \bar{\boldsymbol{g}}) = [\bar{g}_1(x; \boldsymbol{g}), \bar{g}_2(x; \boldsymbol{g}), \ldots, \bar{g}_\ell(x; \boldsymbol{g})]$, and the relevant GPRG or PFRG excess function will satisfy an equation of the same form as (5.58) or (5.121), but with $\boldsymbol{g}$ and $\bar{\boldsymbol{g}}$ in place of g and $\bar{g}$, respectively. Finally, arguments identical with those presented previously establish that the members of the set $\bar{\boldsymbol{g}}$ are invariants of the scaling transformation

$$t \to t' = t/x$$
$$\boldsymbol{g} \to \boldsymbol{g}' = \bar{\boldsymbol{g}}(x; \boldsymbol{g}) \tag{5.299}$$

satisfying the equations

$$\bar{g}_k(t; \bar{\boldsymbol{g}}) = \bar{g}_k\left[\frac{t}{x}; \bar{\boldsymbol{g}}(x; \boldsymbol{g})\right] \quad k = 1, 2, \ldots, \ell \tag{5.300}$$

An illustration is provided by the case of two coupling parameters, g_1 and g_2, for which the GPRG excess function satisfies the equation

$$\frac{\delta f(t; g_1, g_2)}{\delta f(x; g_1, g_2)} = \delta f\left[\frac{t}{x}; \bar{g}_1(x; g_1, g_2), \bar{g}_2(x; g_1, g_2)\right] \tag{5.301}$$

With $t = \tau^2$, $x = \tau$ and $t = \tau^3$, $x = \tau$ and a selected value of τ we obtain from (5.301) the pair of coupled equations

$$\Psi(g_1, g_2) = \psi(\bar{g}_1, \bar{g}_2)$$
$$\Phi(g_1, g_2) = \phi(\bar{g}_1, \bar{g}_2) \tag{5.302}$$

wherein

$$\Psi(g_1, g_2) = \delta f(\tau^2; g_1, g_2)/\delta f(\tau; g_1, g_2)$$
$$\Phi(g_1, g_2) = \delta f(\tau^3; g_1, g_2)/\delta f(\tau; g_1, g_2) \tag{5.303}$$
$$\psi(\bar{g}_1, \bar{g}_2) = \delta f[\tau; \bar{g}_1(\tau; g_1, g_2), \bar{g}_2(\tau; g_1, g_2)]$$

and

$$\phi(\bar{g}_1, \bar{g}_2) = \delta f[\tau^2; \bar{g}_1(\tau; g_1, g_2), \bar{g}_2(\tau; g_1, g_2)]$$

Equation (5.302) then can be solved numerically for $\bar{g}_1$ and $\bar{g}_2$ as functions of g_1 and g_2 (for the chosen value of τ), *provided* that the three functions $\delta f(\tau; g_1, g_2)$, $\delta f(\tau^2; g_1, g_2)$, and $\delta f(\tau^3; g_1, g_2)$ are known, either analytically or from numerical simulations. Once this has been accomplished, the object function $\delta f(t; \mathbf{g})$ can be constructed, for all $t = \tau^n$, using the appropriate generalization of Eq. (5.71).

An example of a problem fitting into this category is the calculation of the mean-squared end-to-end separation, $\overline{R_N^2}$, for a linear polymeric chain with nonnearest-neighbor interactions characterized by a two-parameter potential energy function, for example, a Lennard–Jones 12-6 potential with length and energy parameters σ and ε. Here, the group parameter t is to be identified with the number of chain links N and $g_1 = (\sigma/\ell)^2$ and $g_2 = \varepsilon/k_B T$. The computational requirements for this task are significantly greater than those for the single-parameter problem described in Section IV.A. Not only is it necessary to solve the coupled equations (5.302), but one first must construct the three functions $\delta f(\tau; g) = \overline{R_K^2}(\mathbf{g})$, $\delta f(\tau^2; \mathbf{g}) = \overline{R_{K^2}^2}(\mathbf{g})$, and $\delta f(\tau^3; \mathbf{g}) = \overline{R_{K^3}^2}(\mathbf{g})$. It is natural to begin with the smallest possible choice for the group parameter, namely $K = 2$, and so to construct $\overline{R_N^2}(\mathbf{g})$ for $N = 2, 4$, and 8. The first of these involves a single quadrature but the remaining two must be generated by computer simulations.

The partition function for this same polymeric model can be constructed by a completely analogous procedure, thus providing the means for computing the characteristic θ-temperature at which the excess partition function $\delta Q(N; g_1, g_2)$ is equal to unity.

Since the two-parameter model already requires a significant computational effort, the investigation of more elaborate, multiparameter models would seem to be inadvisable. In fact, at the time of this writing even the two-parameter formalism remains untested.

Acknowledgments

The work reported here was begun with support from the National Science Foundation and completed with the aid of a grant from the Donors of the Petroleum Research Fund administered by the American Chemical Society.

References

1. M. Schroeder, *Fractals, Chaos, Power Laws*, (W. H. Freeman, New York 1991), Chap 1.
2. K. Huang, *Quantum Field Theory* (John Wiley & Sons, Inc., New York, 1948).
3. N. N. Bogoliubov and D. V. Shirkov, *Quantum Fields*, (Benjamin/Cummings, London, 1982).

4. D. V. Shirkov, *Int. J. Mod. Phys.* **73**, 1321 (1988).
5. D. V. Shirkov, *Theor. Math. Phys. (USSR)* **60**, 218 (1984).
6. M. N. Barber and B. W. Ninham, *Random and Restricted Walks* (Gordon and Breach, New York, 1970).
7. L. P. Kadanoff, *Physics* **2**, 263 (1966).
8. H. J. Maris and L. P. Kadanoff, *Am. J. Phys.* **46**, 652 (1978).
9. P. G. deGennes, *Scaling Concepts in Polymer Physics* (Cornell University Press, Ithaca, 1979).
10. A. R. Altenberger and J. S. Dahler, *Adv. Chem. Phys.* **112**, 337 (2000).
11. L. Tisza in *Phase Transformations in Solids*, edited by R. Smoluchowski and W. A. Weyl (John Wiley & Sons, Inc., New York, 1951).
12. H. B. Callen, *Thermodynamics and Introduction to Thermostatics*, (John Wiley & Sons, Inc., New York, 1965).
13. D. A. McQuarrie, *Statistical Mechanics* (Harper & Row, New York, 1976).
14. J. desCloizeaux and G. Jannink, *Polymers in Solution* (Clarendon Press, Oxford, 1990).
15. A. R. Altenberger, J. I. Siepmann, and J. S. Dahler, *Physica* **A272**, 22 (1999).
16. R. D. Mattuck, *A Guide to Feynman Diagrams in the Many-Body Problem* (Dover, New York, 1976).
17. M. le Bellac, *Quantum and Statistical Field Theory* (Clarendon Press, Oxford, 1991).
18. B. M. McCoy and T. T. Wu, *The Two-Dimensional Ising Model*, Chap. 1 (Harvard University Press, Cambridge, MA, 1973).
19. A. R. Altenberger and J. S. Dahler, *Phys. Rev. E* **54**, 6242 (1997).
20. K. Kremer, A. Baumgärtner, and K. Binder, *Z. Phys.* **B40**, 331 (1981) and *J. Phys.* **A15**, 2879 1981).
21. A. R. Altenberger, J. I. Siepmann, and J. S. Dahler, *Physica* **A289**, 107 (2001).
22. J. G. Berryman, *Phys. Rev.* **A27**, 1053 (1983).
23. J. E. J. van Rensburg, *J. Phys.* **A26**, 4805 (1993).
24. N. F. Carnahan and K. Starling, *J. Chem. Phys.* **51**, 635 (1969).
25. S. Shinomoto, *J. Statist. Phys.* **32**, 105 (1983).
26. J. J. Erpenbeck and W. W. Wood, *J. Statist. Phys.* **35**, 321 (1984).
27. A. Rotenberg, *J. Chem. Phys.* **42**, 1126 (1965).
28. L. V. Woodcock, *Ann. N. Y. Acad. Sci.* **371**, 274 (1981).
29. A. R. Altenberger and J. S. Dahler, *Polish J. Chem.* **75**, 601 (2001).
30. C. Itzykson and J. B. Zuber, *Quantum Field Theory* (McGraw-Hill, New York, 1980).
31. V. B. Berestetskii, E. M. Lifshitz, and L. P. Pitaevskii, *Quantum Electrodynamics*, §76 (Pergamon Press, Oxford, 1982), p. 300.
32. E. deRafael and J. Rosner, *Ann. Phys.* **82** (1974), 369. See also Chapter 13 of [29].
33. L. D. Landau and I. Pomeranchuk, *Dokl. Akad. Nauk CCCP* **102** (1955), 450.
34. A. Yu. Grosberg and A. R. Khokhlov, *Giant Molecules* (Academic Press, New York, 1997).
35. A. R. Altenberger and J. S. Dahler *Int. J. Mod. Phys.* **A73**, 279 (2002).
36. P. I. Fomin, V. P. Gusynin, V. A. Miransky, and Yu. A. Sitenko, *Riv. Nuovo Cimento* **6** (1983). See also, J. Bartholomew, S. H. Shenker, J. Sloan, J. Kogut, M. Stone, H. W. Wyld, J. Shigmetsu, and D. K. Sinclair, *Nucl. Phys. B* **230** [FS10] (1984), 222 and V. A. Miransky, *Il Nuovo Cimento* **90A** (1985), 149.

37. A. L. Fetter and J. D. Walecka, *Quantum Theory of Many-Particle Systems* (McGraw-Hill, New York, 1971), Chap. 3.
38. I. M. Gelfand and G. E. Shilov, *Generalized Functions*, Vol. 1 (Academic Press, New York, 1964).
39. M. A. Mnatsakamjan, *Sov. Phys. Dokl.* **27**, 123 (1987).
40. T. J. Dougherty, *Some Problems in the Theory of Colloids*, Ph.D. Thesis, Case Institute of Technology (1959).
41. I. M. Krieger, in *Surfaces and Coatings Related to Paper and Wood*, edited by R. Marcuessault and C. Skaar (Syracuse University Press, Syracuse, NY, 1967); *Adv. Colloid Interf. Sci.* **3**, 111 (1972).
42. E. Wajnryb and J. S. Dahler, *Adv. Chem. Phys.* **102**, 193 (1997).
43. M. Fixman and J. M. Peterson, *J. Chem. Phys.* **39**, 2516 (1963).
44. G. K. Batchelor and J. T. Green, *J. Fluid Mech.* **56**, 375, 401 (1972).
45. C. W. J. Beenakker, *Physica* **128A**, 48 (1984).
46. C. G. deKriuf, E. M. F. van Lersel, and A. Brij, *J. Chem. Phys.* **83**, 4717 (1985).
47. P. N. Segre, S. P. Meeker, P. N. Pusey, and W. C. K. Poon, *Phys. Rev. Lett.* **75**, 958 (1995).
48. D. A. R. Jones, B. Leary, and D. V. Boger, *J. Colloid Interf. Sci.* **147**, 479 (1991).
49. V. M. Eilers, *Kolloid Z.* **97**, 313 (1941).
50. V. Vand, *J. Phys. Chem. Colloid Chem.* **52**, 300 (1948).
51. W. G. Hoover and F. H. Ree, *J. Chem. Phys.* **49**, 3609 (1968).
52. L. V. Woodcock, *Nature (London)* **385**, 141 (1997).
53. H. S. Robertson, *Statistical Thermophysics* (Prentice Hall, Englewood Cliffs, New Jersey, 1993).

ELECTRON-CORRELATED APPROACHES FOR THE CALCULATION OF NMR CHEMICAL SHIFTS

JÜRGEN GAUSS

Institut für Physikalische Chemie,
Universität Mainz,
D-55099 Mainz, Germany

JOHN F. STANTON

Institute for Theoretical Chemistry,
Departments of Chemistry and Biochemistry,
The University of Texas at Austin,
Austin, TX 78712

CONTENTS

I. Introduction
II. Basic Aspects for the Calculation of NMR Chemical Shifts
    A. General Remarks
    B. Shielding Tensors as Second Derivatives of the Energy
    C. Molecular Hamiltonian in the Presence of a Magnetic Field
    D. Gauge Invariance and Gauge-Origin Independence
    E. Local Gauge-Origin Methods
    F. Analytic Second Derivatives
    G. General Recommendations
III. Quantum Chemical Methods for the Calculation of NMR Chemical Shifts
    A. Electron-Correlated Quantum Chemical Approaches
    B. Electron-Correlated Approaches for NMR Chemical Shift Calculations
    C. Description of the GIAO–MP2 Approach
    D. Further Methodological Developments
    E. DFT Approach to NMR Chemical Shifts

Advances in Chemical Physics, Volume 123, Edited by I. Prigogine and Stuart A. Rice.
ISBN 0-471-21453-1 © 2002 John Wiley & Sons, Inc.

IV. Illustrative Examples
   A. Benchmark Calculations and Absolute Nuclear Magnetic Shielding Constants
   B. Chemical Applications and Relative NMR Chemical Shifts
V. Summary and Outlook
Acknowledgments
References

I. INTRODUCTION

Quantum chemical nuclear magnetic resonance (NMR) chemical shift calculations enjoy great popularity since they facilitate interpretation of the spectroscopic technique that is most widely used in chemistry [1–11]. The reason that theory is so useful in this area is that there is no clear relationship between the experimentally measured NMR shifts and the structural parameters of interest. NMR chemical shift calculations can provide the missing connection and in this way have proved to be useful in many areas of chemistry. A large number of examples including the interpretation of NMR spectra of carbocations [12], boranes [10, 13], carboranes [10, 13–15], low-valent aluminum compounds [16–18], fullerenes [19–21] as well as the interpretation of solid-state NMR spectra [22–26] can be found in the literature.

An important prerequisite for the application of NMR chemical shift calculations is that accurate calculations can be performed routinely. Although the underlying theory for chemical shielding tensors has been well-understood since the classic work of Ramsay [27], quantum chemical calculations of NMR chemical shifts were very slow to develop. The main reason was the gauge problem connected with the theoretical treatment of magnetic properties. It was realized early on that the computation of magnetic properties employing standard quantum chemical techniques and approximations provide results that depend on the choice of origin, and hence are somewhat unreliable [28]. While the gauge problem can be dealt with in a brute force fashion by using extremely large atomic orbital basis sets [29], this cannot be considered a practical solution for anything but the smallest (and therefore not particularly interesting) molecules.

An important breakthrough concerning the origin dependence (gauge problem) was the work of Kutzelnigg [30] and Schindler and Kutzelnigg [31] who suggested and implemented a solution based on individual gauges for localized molecular orbitals (IGLO). Their IGLO approach was the first scheme suitable for routine application, and was exploited by many research groups, especially Schleyer and co-workers [12]. Nevertheless, other solutions to the gauge problem in NMR chemical shift calculations have also been suggested [32–37]. For several reasons, the gauge-including atomic orbital (GIAO) approach has become the de facto standard for this area of quantum chemical application. This solution to the gauge problem was first advocated long ago by London [38] in the context of magnetic susceptibilities and later used by Hameka [33], Ditchfield [34], and others [39, 40] for chemical shifts. For a long time, the potential of this approach was not appreciated, but the seminal work by Wolinski et al. [35] proved in a convincing manner the suitability and efficiency of the GIAO ansatz for computing chemical shifts. The GIAO approach has since then been implemented in essentially all major quantum chemistry packages (ACESII [41], DALTON [42], GAUSSIAN [43], TEXAS [44], and TURBOMOLE [45]). Furthermore, the GIAO approach facilitated the goal of making accurate NMR shift calculations via electron-correlated methods routine [46–62] and these are the focus of this chapter.

It is justified to conclude that the gauge problem in quantum chemical calculations of magnetic properties is now well understood and has been solved in a satisfactory manner in the sense that the use of local gauge origins is strongly recommended. Hence, the focus of method development in NMR chemical shift calculations shifted in the 1990s from remedies of the gauge problem to incorporation of electron correlation effects in these calculations [4–11]. In addition, relativistic effects (which can be important for compounds that contain heavy atoms) [63–65], rovibrational averaging [66–67] and environmental effects [68–71] on NMR chemical shifts have been active research areas in the last decade. However, since NMR spectroscopy is so vital to many areas of chemistry, it is not surprising that electron correlation has been the most intense focus among the above mentioned efforts. The range of methods that incorporate electron correlation for GIAO based NMR calculations include traditional approaches such as many-body perturbation (MBPT or MP) or coupled-cluster (CC) theory but also density functional theory (DFT), which evolved during the last decade to one of the standard quantum chemical approaches.

This chapter provides a report on the current status of research in the field of NMR chemical shift calculations. The theoretical background required for these calculations will be outlined in Section II, and schemes available for carrying out these calculations with electron correlation will be discussed in Section III. Finally, a few examples are given in Section IV that demonstrate the performance of the various schemes and emphasize the importance of electron correlation for accurate NMR chemical shift calculations.

II. BASIC ASPECTS FOR THE CALCULATION OF NMR CHEMICAL SHIFTS

A. General Remarks

The chemical shielding observed in NMR spectroscopy is an electronic effect, and therefore can be treated by quantum chemical calculations. The underlying physical mechanism is that the applied (external) magnetic field **B** induces an electronic current **j**, which according to the Biot–Savart law [72] generates an additional, induced magnetic field. The nuclear magnetic moments $\mathbf{m}_J$ thus experience an effective field given by the sum of applied external and induced field:

$$\mathbf{B}_{\text{eff}} = \mathbf{B} + \mathbf{B}_{\text{ind}} \tag{6.1}$$

To first order, the induced field $\mathbf{B}_{\text{ind}}$ is proportional to the applied field

$$\mathbf{B}_{\text{ind}} = -\boldsymbol{\sigma}\mathbf{B} \tag{6.2}$$

with the "proportionality" carried by the nuclear magnetic chemical shielding tensor $\boldsymbol{\sigma}$. Note that $\boldsymbol{\sigma}$ must be a tensor since the applied and induced fields are not necessarily parallel.

For the splitting of the nuclear spin energy levels in the presence of an external magnetic field, we then obtain

$$\Delta E = -\mathbf{m}_J \mathbf{B}_{\text{eff}}$$
$$= -\mathbf{m}_J (1 - \boldsymbol{\sigma})\mathbf{B} \tag{6.3}$$

This splitting is proportional to **B** but also to the shielding tensor $\boldsymbol{\sigma}$. By measuring transitions between different nuclear spin states, information about the chemical shielding and, thus, the underlying electronic structure of the molecule can be obtained. These are the transitions that are probed in NMR spectroscopy [73].

For practical purposes, one usually does not determine absolute chemical shifts in NMR spectroscopy, but rather shifts relative to those of some reference compound. Though absolute shieldings are in principle observable quantities, it turns out to be considerably easier to measure relative shifts in the laboratory. The latter can be determined easily from experimental NMR spectra with high precision (and do not require knowledge of the exact magnetic field strength), while absolute shieldings are typically determined in a rather involved procedure from experimental spin-rotation constants obtained from microwave spectra. In addition, one should note that in isotropic media (gases and liquids) only the trace of the shielding tensor, the so-called isotropic shielding constants

$$\sigma = \frac{1}{3} \operatorname{Tr} \boldsymbol{\sigma} \tag{6.4}$$

can be measured. Determination of individual elements of the tensor requires spectra to be taken in oriented media (solid state or liquid crystals, see, e.g., [74]).

Thus, although the shielding tensor $\boldsymbol{\sigma}$ is the most physically fundamental object in NMR theory, relative shifts δ are the quantities actually used in interpreting spectra. These relative shifts are defined by

$$\delta = \sigma_{\text{ref}} - \sigma \tag{6.5}$$

with σ_{ref} as the shielding of a suitably chosen reference compounds, e.g., tetramethylsilane (TMS) for ^{1}H, ^{13}C, and ^{29}Si NMR. In the same manner, experimentally determined tensor elements in solid-state investigations are reported with respect to a reference compound. However, as the actual tensor elements depend on the chosen coordinate system, one usually characterizes shielding tensors by the following (invariant) quantities: the chemical shift anisotropy defined by

$$\Delta\sigma = \sigma_{33} - \frac{1}{2}(\sigma_{11} + \sigma_{22}) \tag{6.6}$$

the asymmetry of the shielding defined by

$$\eta = \frac{\sigma_{22} - \sigma_{11}}{\sigma_{33} - \sigma} \tag{6.7}$$

and finally the principal components $\sigma_{33} \geqslant \sigma_{22} \geqslant \sigma_{11}$, which are eigenvalues of the symmetrized shielding tensors. It should be noted that all quantities above are dimensionless and except for the asymmetry parameter η usually given in parts per million (ppm).

B. Shielding Tensors as Second Derivatives of the Energy

For the quantum chemical calculation of NMR chemical shifts, it is essential to realize that the shielding tensor $\boldsymbol{\sigma}$ in Eq. (6.3) appears in a term bilinear in the external magnetic field $\mathbf{B}$ and the nuclear magnetic moment $\mathbf{m}_J$ of interest. This means that the shielding tensor is a second-order property (with respect to the two "perturbations" **B** and $\mathbf{m}_J$) and thus can be obtained via the corresponding sum-over-states expressions of second-order perturbation theory (as demonstrated by Ramsay [27]) or via the corresponding second derivative of the electronic energy [75]. Both possibilities are equivalent in the exact limit, but the derivative approach is more convenient for most quantum chemical calculations, as it avoids the explicit summation over all excited states (which in fact cannot even be done in a well-defined way when MBPT is used). The corresponding computational recipe for the quantum chemical calculation of shielding tensors is thus given by

$$\sigma = \left(\frac{dE_{\text{electronic}}}{d\mathbf{B}d\mathbf{m}_J}\right)_{\mathbf{B},\mathbf{m}_J=0} \qquad (6.8)$$

Nevertheless, it should be noted that the sum-over-states expression is sometimes also used for the computation of shieldings. It forms the basis, for example, for the so-called propagator ansätze [76, 77] and has been invoked in some empirical schemes for the calculation of shieldings within the DFT framework [78, 79].

C. Molecular Hamiltonian in the Presence of a Magnetic Field

For the computation of the second derivatives given in Eq. (6.8) for the shielding tensor, it is necessary to specify the molecular (electronic) Hamiltonian H in the presence of a magnetic field. The latter is obtained by replacing the canonical momentum **p** in the kinetic energy part of H by the kinetic momentum π

$$\mathbf{p} \rightarrow \pi = \mathbf{p} + e\mathbf{A} \qquad (6.9)$$

with e as elementary charge and **A** as the vector potential describing the external field as well as interactions with nuclear magnetic moments. The latter takes the form

$$\mathbf{A}(\mathbf{r}) = \frac{1}{2}\mathbf{B} \times (\mathbf{r} - \mathbf{R}_O) \qquad (6.10)$$

for the external magnetic field with **r** as the spatial coordinate, $\mathbf{R}_O$ as the (arbitrary) gauge origin (see next section), and

$$\mathbf{A}_J(\mathbf{r}) = \alpha^2 \frac{\mathbf{m}_J \times (\mathbf{r} - \mathbf{R}_J)}{|\mathbf{r} - \mathbf{R}_J|^3} \qquad (6.11)$$

for the nuclear magnetic moments, where α is the fine-structure constant and $\mathbf{R}_J$ is the position of nucleus J. Inserting Eqs. (6.10) and (6.11) for $\mathbf{A}$ in the Hamiltonian and expanding H in a Taylor series with respect to $\mathbf{B}$ and $\mathbf{m}_J$ leads to

$$H = H_0 + \sum_i \left(\left.\frac{\partial h(i)}{\partial \mathbf{B}}\right|_{\mathbf{B}=0} \mathbf{B} + \sum_J \left.\frac{\partial h(i)}{\partial \mathbf{m}_J}\right|_{\mathbf{m}_J=0} \mathbf{m}_J + \sum_J \left.\frac{\partial^2 h(i)}{\partial \mathbf{B} \partial \mathbf{m}_J}\right|_{\mathbf{B},\mathbf{m}_J=0} \mathbf{B}\,\mathbf{m}_J + \cdots \right)$$

(6.12)

that involves the following (one-electron) derivatives:

$$\frac{\partial h}{\partial \mathbf{B}} = \frac{1}{2}(\mathbf{r} - \mathbf{R}_O) \times \mathbf{p} \tag{6.13}$$

$$\frac{\partial h}{\partial \mathbf{m}_J} = \alpha^2 \frac{\mathbf{r} - \mathbf{R}_J) \times \mathbf{p}}{|\mathbf{r} - \mathbf{R}_J|^3} \tag{6.14}$$

$$\frac{\partial^2 h}{\partial \mathbf{B} \partial \mathbf{m}_J} = \frac{\alpha^2}{2} \frac{(\mathbf{r} - \mathbf{R}_O) \cdot (\mathbf{r} - \mathbf{R}_J)\mathbf{1} - (\mathbf{r} - \mathbf{R}_O)(\mathbf{r} - \mathbf{R}_J)}{|\mathbf{r} - \mathbf{R}_J|^3} \tag{6.15}$$

The sums in Eq. (6.12) run over all electrons (lower case indices) and nuclei (upper case). This form of the Hamiltonian is a particularly suitable starting point for the treatment of chemical shifts. We note that the molecular Hamiltonian in its original form does not explicitly include the magnetic field $\mathbf{B}$, which instead enters indirectly via the vector potential $\mathbf{A}$. This formulation of magnetic interactions is consistent with the preference of the Lagrangian and Hamiltonian formulations of mechanics for potentials as central quantities instead of forces and field strengths. In addition, the use of the vector potential $\mathbf{A}$ allows a straightforward discussion of the gauge problem in the (quantum chemical) calculation of magnetic properties (see Section II.D). Finally, it should be noted that the magnetic field $\mathbf{B}$ is completely determined by the specification of the vector potential $\mathbf{A}$ via

$$\mathbf{B} = \nabla \times \mathbf{A}(\mathbf{r}) \tag{6.16}$$

D. Gauge Invariance and Gauge-Origin Independence

While $\mathbf{A}$ uniquely determines the magnetic field $\mathbf{B}$, the opposite is not true and there is no unique choice for the vector potential corresponding to a given magnetic field. This introduces some arbitrariness in the description of magnetic interactions. For example, it is always possible to add a gradient of any scalar function to $\mathbf{A}$ without changing the magnetic field or the physical system under consideration [72]. (Note that this is due to the fact that for all scalar functions f the following is true: $\nabla \times \nabla f = 0$.) Therefore different possibilities exist for describing the same fundamental phenomenon that differ in the gauge, i.e., the choice of the vector potential. The most important example for our discussion is illustrated in Eq. (6.10), which specifies (an infinite number of) possible vector potentials for a static homogeneous magnetic field differing only in the (arbitrary) gauge origin $\mathbf{R}_O$.

The possibility of describing the same physical system with different choices for the gauge (or gauge origin) leads to a natural requirement, namely, that the computed values for all physically observable quantities must be independent of the chosen gauge (or gauge origin). This necessary requirement is usually called the principle of gauge invariance or, for the special case of Eq. (6.10), the principle of gauge-origin independence. Gauge invariance is trivially satisfied by exact solutions of the Schrödinger equation (for a proof see, e.g., [80]). However, gauge invariance is not guaranteed for approximate solutions. Of course, this is a serious problem for quantum chemistry which essentially always deals with approximate solutions. It can be easily shown that gauge invariance is indeed not satisfied for standard quantum chemical approaches for the calculation of energy, wave function, and molecular properties. A detailed analysis reveals that this is due to (a) the use of finite basis sets within the linear combination of atomic orbital (LCAO) approximation to the molecular orbitals; and (b) the fact that some quantum chemical schemes do not obey the hypervirial theorem. The first reason is typically considered to be the most serious; the latter feature is mentioned less frequently in the literature [81]. However, methods that do satisfy the hypervirial theorem — such as Hartree–Fock — can be shown to be gauge invariant in the limit of a complete basis set [82].

E. Local Gauge-Origin Methods

As mentioned in the introduction, the problem associated with gauge-origin dependence represented a serious obstacle to efficient and routine calculations of magnetic properties. The results for magnetic properties of such nongauge-invariant calculations depend on the chosen gauge (or gauge-orgin), and in this way on an arbitrary parameter, i.e., the gauge origin $\mathbf{R}_O$. To demonstrate this, we show in Table VI.1 the calculated proton shieldings for the hydrogen fluoride molecule as obtained in standard Hartree–Fock self-consistent-field (HF–SCF) calculations using different gauge origins. Differences in the results are due to the fact that the basis sets used in these calculations are not complete, and gauge-origin independence is not satisfied even though the underlying method satisfies the hypervirial theorem. A second problem, which of course is related to the gauge issue, is the slow

Table VI.1
Gauge-Origin Dependence of ^{1}H NMR Chemical Shifts
(in ppm) Calculated for Hydrogen Fluoride Using
Different Basis Sets at the HF–SCF Level

Basis Set	Gauge Origin		
	Center of Mass	Fluorine	Hydrogen
dz + d	29.3	27.6	60.1
tz + d	28.4	27.2	50.8
qz + 2d	27.7	27.0	40.4

convergence with respect to basis set exhibited by quantum chemical calculations of magnetic properties. This is illustrated in Figure 6.1, where the computed proton shielding of hydrogen fluoride is displayed as a function of basis set. It is clearly seen that standard basis sets of polarized double- or triple-zeta quality do not — as for many other properties — provide adequate results. Reliable results are obtained only when very large basis sets are used, thereby mitigating the gauge problem. However, this is

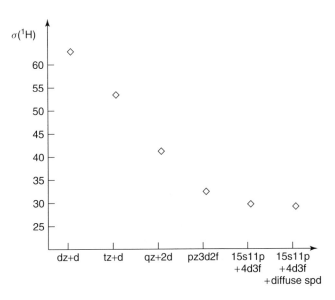

Figure 6.1. Basis set convergence in the calculation of ^{1}H shielding constant of hydrogen fluoride. All calculations have been carried out with the gauge origin at the hydrogen.

only feasible for very small molecules for which the gauge problem is somewhat less serious anyway. For recent examples of such calculations, the reader is referred to [29], but we must emphasize that this is not a practical scheme for magnetic property calculations in general.

The solution to the gauge problem lies in the use of so-called local gauge-origin approaches. To understand the motivation for these methods, it is important to realize that the use of a single gauge origin can never provide an adequate solution to the gauge problem in calculations for molecules. An analysis of the problem for atoms shows that there the best possible choice for the gauge origin is the nuclear position—in the sense that convergence with respect to the basis set is most rapid—and that the results deteriorate with increasing distance between the gauge origin and the nucleus. For molecules, the problem is that there is no optimal gauge origin since there is not just a single nucleus. In particular, for large molecules (those for which brute force techniques are totally inapplicable), not all parts of the molecule can be treated adequately and described in a balanced manner when using a single gauge origin. The introduction of a set of individual gauge origins for different local parts of the wave function resolves the problem; in this way, a balanced treatment for all parts of the molecule and molecular wave function can be achieved. As magnetic interactions are of the one-electron type [see, e.g., the Hamiltonian given in Eq. (6.12)], it is recommended to introduce local gauge origins for the one-electron constituents of the wave functions, that is, either the molecular orbitals or the atomic orbitals (basis functions). In case of the molecular orbitals, serious problems arise when canonical Hartree–Fock orbitals are used that unfortunately are almost always highly delocalized. This drawback can be remedied by using localized molecular orbitals as in the IGLO approach by Kutzelnigg [30] and Schindler and Kutzelnigg [31] and the localized orbital/local origin (LORG) scheme by Hansen and Bouman [32]. However, this leads to additional constraints for the wave functions that (at least in electron-correlated treatments) cause complications. Thus, we believe that the use of individual gauge origins for atomic orbitals is the more useful and elegant solution to this problem. Atomic orbitals are necessarily localized; therefore no additional constraint concerning the quantum chemical treatment is required. The use of local (nucleus-centered) origins for atomic orbitals is usually referred to as the GIAO approach [33–35].

Technically, local gauge origins are introduced by "non-canonical" gauge-origin transformations. A standard "canonical" gauge-origin transformation of the form

$$\Psi \to \Psi' = \exp(-\Lambda(\mathbf{r}))\Psi \qquad (6.17)$$

$$H \to H' = \exp(-\Lambda(\mathbf{r}))H\exp(\Lambda(\mathbf{r})) \qquad (6.18)$$

with

$$\Lambda(\mathbf{r}) = \frac{i}{2}((\mathbf{R}'_O - \mathbf{R}_O) \times \mathbf{B})\mathbf{r} \quad (6.19)$$

moves the gauge origin from its original position $\mathbf{R}_O$ to the new position $\mathbf{R}'_O$. The non-canonical gauge-origin transformation

$$\Psi \to \Psi' = \sum_\mu \exp(-\Lambda_\mu(\mathbf{r})) P_\mu \Psi \quad (6.20)$$

$$H \to H' = \sum_\mu P_\mu \exp(-\Lambda_\mu(\mathbf{r})) H \sum_\nu \exp(\Lambda_\nu(\mathbf{r})) P_\nu \quad (6.21)$$

with

$$\Lambda_\mu(\mathbf{r}) = \frac{i}{2}((\mathbf{R}_\mu - \mathbf{R}_O) \times \mathbf{B})\mathbf{r} \quad (6.22)$$

and P_μ as an appropriately chosen projection operator on "local" fragments of the wave function shifts the gauge origin for the part denoted by μ to the new position $\mathbf{R}_\mu$. For the GIAO approach in which local gauge origins are introduced for atomic orbitals, the projector P_μ is given by

$$P_\mu = \sum_\nu |\chi_\mu\rangle S_{\mu\nu}^{-1} \langle\chi_\nu| \quad (6.23)$$

where the $S_{\mu\nu}$ are elements of the atomic orbital overlap matrix and the new origin is placed at the corresponding nuclear position $\mathbf{R}_\mu$. The complicated form of the projector P_μ within the GIAO approach is due to the nonorthogonality of the atomic orbitals, but causes no computational difficulties.

What is gained by introducing local gauge origins? Certainly, we are *not* solving the formal problem of gauge invariance, as the local gauge-origin approaches are not introducing gauge invariance in a rigorous manner. Instead, they "solve" the gauge problem by fixing the gauge in a well-defined and optimal manner. In this way, unique results are guaranteed (which of course is highly desirable from a practical standpoint). One can at least say that these methods provide results that are no longer dependent of the global gauge origin and thus these approaches are necessarily independent of where the coordinate system origin is located. Furthermore, one should note that the choice of local gauge origins is physically motivated and in this manner it is not surprising that local gauge-origin approaches exhibit improved basis set convergence, and therefore resolve the second problem in the quantum chemical calculation of magnetic properties. This is also

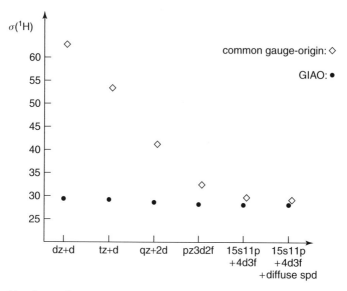

Figure 6.2. Comparison of the basis set convergence in common gauge-origin and GIAO calculations of the ^{1}H shielding constant of hydrogen fluoride.

illustrated in Figure 6.2 where the basis set convergence of GIAO calculations is compared with the slow convergence of calculations using a common gauge origin.

While the description of the GIAO approach above is perhaps most satisfying and insightful, a different perspective is actually more convenient in the context of actual calculations. Instead of viewing the method as involving the usual, magnetic-field independent basis functions χ_μ, it is equivalent to use the field-dependent functions

$$\chi_\mu(\mathbf{B}) = \exp\left(-\frac{i}{2}(\mathbf{B} \times (\mathbf{R}_\mu - \mathbf{R}_O)\mathbf{r}\right)\chi_\mu \qquad (6.24)$$

The latter are usually termed gauge-including atomic orbitals. [These basis functions have been originally called in a misleading manner gauge-invariant or gauge-independent atomic orbitals. The term "gauge-including atomic orbitals" has been suggested by Hansen and Bouman [32] in an effort to find a more appropriate name while keeping the original acronym. Note that the basis functions given in Eq. (6.24) are also known as London atomic orbitals (LAOs).] Both descriptions of the GIAO approach, that is, as a local gauge-origin method or as an approach with field-dependent basis functions, are equally valid. It is just a matter of taste whether one attaches the phase factors introduced by the gauge transformation in Eq. (6.20) to the Hamiltonian ($\rightarrow$ local gauge-origin approach) or to the wave functions ($\rightarrow$ field-dependent basis functions). The use of explicitly field-dependent basis functions somewhat complicates the discussion of gauge invariance, but facilitates rationalization of the fast basis set convergence. It is rather easy to show that the basis functions given in Eq. (6.24) provide correct first-order solutions in the presence of a magnetic field if the corresponding field-independent basis functions are the exact solutions in the field-free case [83].

F. Analytic Second Derivatives

The nuclear magnetic shielding tensor is formally given as a second derivative of the energy with respect to the external magnetic field and the nuclear magnetic moments, as shown in Eq. (6.8). Evaluating second derivatives of the energy is clearly a more difficult task than simply evaluating the energy and the wave function.

For the evaluation of energy derivatives, two principal possibilities exist in general. One can calculate derivatives using finite difference schemes (which involve some numerical differentiation) or by means of purely analytic techniques. For electric-field perturbations involved in the computation of electric properties such as dipole moments, polarizabilites, numerical differentiation schemes are rather easy to implement and these properties are thus easy to calculate for all schemes that are available for energy calculations, although the analytic approach — when available — is more efficient computationally and provides more precise results. However, the drawback of analytic approaches is that they are usually complicated and that the derivation and implementation of the equations is a rather foreboding prospect, even for experienced quantum chemists. Reflecting this, it is not surprising that analytic schemes for most highly correlated approaches have been reported only rather recently [57, 84, 85] (nearly a decade after the methods were first used for energy calculations) or still are unavailable. On the other hand, the advantages of the analytic approaches cannot be denied, as they often form the basis for the routine calculation of the corresponding properties. For example, the fact that nowadays molecular geometries can be routinely determined in the course of quantum chemical investigation is entirely due to the availability of analytically evaluated forces (first derivatives of the energy with respect to nuclear positions) [86].

For magnetic properties, there are some subtleties that are not encountered when calculating electric properties or simple derivatives of the energy with respect to nuclear displacements. Specifically, magnetic perturbations are—unlike electric perturbations—what we term formally imaginary perturbations. This means that matrix elements of the first-order perturbed Hamiltonian [as given in Eqs. (6.13) and (6.14)] over real basis functions are imaginary. Application of finite difference schemes for the computation of magnetic properties thus faces the serious problem that the wave function parameters, that is, MO coefficients, cluster amplitudes, and so on, become complex in the presence of the perturbation. As most quantum chemical programs are not capable of dealing with complex wave function parameters, numerical differentiation is not a viable means to calculate properties that depend on formally imaginary perturbations such as the magnetic field (however, for an example of the application of numerical differentiation to the calculation of shieldings, see [49]). Difficulties associated with numerical differentiation are the principal reason that analytic second derivatives are generally considered to be an important prerequisite for the calculation of NMR chemical shifts and why such calculations became possible only after such analytic methods were developed. The problem of complex wave function parameters is avoided in analytic calculations; all quantities encountered in these calculations are either real or (purely) imaginary, and real arithmetic can be used in all stages of the calculation. The imaginary character of the magnetic perturbations only shows up in the fact that some (first-order perturbed) matrices are antisymmetric instead of being symmetric and that some signs need to be changed. These issues are rather easily incorporated in a computer implementation that uses only real arithmetic.

In the following, a few general remarks concerning the analytic evaluation of shielding tensors are given. They are of relevance for the computational efficiency of shielding constant calculations and valid for all quantum chemical schemes.

Most of the following is motivated by the fact that the nuclear magnetic shieldings are—unlike polarizabilities and quadratic force constants—second-order properties that involve two distinct classes of perturbations. This fact can be exploited in the following manner.

Second derivative expressions [The following discussion strictly applies (for simplicity) to the use of a common gauge origin, though conclusions are also valid for the use of GIAOs] can be obtained for the shielding tensor in three different ways:

1. Differentiation first with respect to $\mathbf{m}_J$ followed by differentiation with respect to $\mathbf{B}$.
2. Differentiation first with respect to $\mathbf{B}$ followed by differentiation with respect to $\mathbf{m}_J$.

3. Differentiation as in (1) or (2) with subsequent symmetrization of the expressions so obtained.

As explained below, variants (1) and (2) lead to *asymmetric* expressions, while (3) obviously provides a *symmetric* derivative expression.

Differentiation of the energy with respect to a one-electron perturbation x yields (after suitable algebraic manipulations) the following derivative expressions [75]

$$\frac{dE}{dx} = \sum_{\mu\nu} D_{\mu\nu} \frac{\partial h_{\mu\nu}}{\partial x} \quad (6.25)$$

where the indices μ, ν represent atomic orbitals and $D_{\mu\nu}$ elements of the (effective) one-particle density matrix in this atomic orbital basis. The latter can always be computed from the unperturbed wave function parameters (MO coefficients, configuration interaction coefficients, coupled-cluster amplitudes, etc.), and therefore is independent of the perturbation x. For variational approaches such as HF or MCSCF the definition of the density matrix is straightforward [86], while for nonvariational approaches the expressions for $D_{\mu\nu}$ are more involved. For those approaches, it can be demonstrated (using either the interchange theorem of perturbation theory [87] or an approach based on the use of Lagrangian multipliers [88]) that the expression for the energy gradient takes the form given by Eq. (6.25) and does not require knowledge of any perturbed wave function parameters. However, it might be that for the construction of the effective density matrix additional (though perturbation-independent) equations need to be solved. Most relevant in this context are the Z-vector equations [89] that account for the relaxation of the molecular orbitals, and the Lambda equations in coupled-cluster theory [90].

Differentiation of these gradient expressions with respect to a second one-electron perturbation y yields

$$\frac{d^2E}{dx\,dy} = \sum_{\mu\nu} D_{\mu\nu} \frac{\partial^2 h_{\mu\nu}}{\partial x \, \partial y} + \sum_{\mu\nu} \frac{\partial D_{\mu\nu}}{\partial y} \frac{\partial h_{\mu\nu}}{\partial x} \quad (6.26)$$

where it should be noted that now the perturbed one-electron density matrix appears. This means that evaluation of second derivatives (in accordance with the $(2n + 1)$ rule of perturbation theory) requires the knowledge of the perturbed wave function in the form of the first derivatives of their parameters. However, from Eq. (6.26) it is also clear that the perturbed wave function parameters are required *only* for perturbation y, a fact that can be exploited in NMR chemical shift calculations.

Focusing on the possibilities sketched above for the derivation of second derivative expressions, we have the following three options:

Variant (1)

$$\sigma = \sum_{\mu\nu} D_{\mu\nu} \frac{\partial^2 h_{\mu\nu}}{\partial \mathbf{B} \partial \mathbf{m}_J} + \sum_{\mu\nu} \frac{\partial D_{\mu\nu}}{\partial \mathbf{B}} \frac{\partial h_{\mu\nu}}{\partial \mathbf{m}_J} \qquad (6.27)$$

requires determination of perturbed wave function parameters and perturbed density matrix for the three magnetic field components.

Variant (2)

$$\sigma = \sum_{\mu\nu} D_{\mu\nu} \frac{\partial^2 h_{\mu\nu}}{\partial \mathbf{B} \partial \mathbf{m}_J} + \sum_{\mu\nu} \frac{\partial D_{\mu\nu}}{\partial \mathbf{m}_J} \frac{\partial h_{\mu\nu}}{\partial \mathbf{B}} \qquad (6.28)$$

involves determination of perturbed wave function parameters and perturbed density matrix for the three times N_{atoms} nuclear magnetic moment perturbations with N_{atoms} as the number of nuclei in the molecule,

Variant (3)

$$\sigma = \sum_{\mu\nu} D_{\mu\nu} \frac{\partial^2 h_{\mu\nu}}{\partial \mathbf{B} \partial \mathbf{m}_J} + \frac{1}{2}\sum_{\mu\nu} \frac{\partial D_{\mu\nu}}{\partial \mathbf{B}} \frac{\partial h_{\mu\nu}}{\partial \mathbf{m}_J} + \frac{1}{2}\sum_{\mu\nu} \frac{\partial D_{\mu\nu}}{\partial \mathbf{m}_J} \frac{\partial h_{\mu\nu}}{\partial \mathbf{B}} \qquad (6.29)$$

requires determination of perturbed wave function parameters and perturbed density matrix for the three magnetic field components as well as all $3N_{atoms}$ nuclear magnetic moment perturbations.

Assuming—as it is actually the case—that the determination of the perturbed wave function is the time-limiting step, it is clear that variant (1) is the preferred choice and computationally most efficient. Specifically, there are only *three* magnetic field components, while there are N_{atoms} times this many nuclear magnetic moments. This conclusion holds for Hartree–Fock where the coupled-perturbed HF (CPHF) equations are only solved for **B** (see, e.g., [35]) as well as the correlated approaches discussed in Section III. We thus conclude that the second derivatives needed for the computation of NMR chemical shifts are best calculated using an *asymmetric* derivative expression.

G. General Recommendations

We summarize this general section on computing NMR chemical shifts by stating that NMR chemical shifts can be routinely calculated as the corresponding second derivatives of the energy with respect to the external magnetic field and nuclear magnetic moments [Eq. (6.8)] provided the gauge-origin problem is handled via a local gauge-origin approach. For the latter, the GIAO approach appears to be most convenient because (unlike other schemes) it does not introduce additional constraints (localization conditions) on the molecular wave function.

III. QUANTUM CHEMICAL METHODS FOR THE CALCULATION OF NMR CHEMICAL SHIFTS

The calculation of NMR chemical shifts at the uncorrelated HF–SCF level is nowadays routinely possible using CPHF theory. Efficient implementations have been reported in particular within the GIAO approach [35, 91] and (using integral-direct techniques) calculations on large molecules consisting of more than 100 atoms are routinely possible [91]. Recent examples include HF–SCF calculations on hexabenzocoronene derivatives [24, 25] as well on aluminum halide clusters [18]. The status here has been earlier reviewed by others and thus will not be repeated. Our focus will be instead on electron-correlated approaches, in particular, as consideration of electron-correlation effects is essential in many cases.

A. Electron-Correlated Quantum Chemical Approaches

Figure 6.3 gives an overview of existing methods for the treatment of electron correlation, which are organized into three categories. Those that

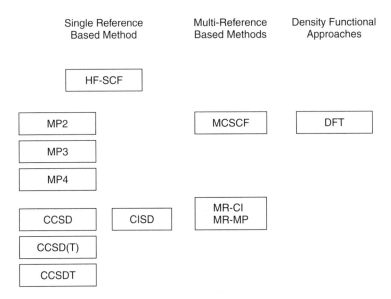

Figure 6.3. Overview over quantum chemical methods for the routine treatment of electron-correlation effects.

belong to the first category are termed single-reference methods. For these approaches, one starts from an HF–SCF reference determinant and treats electron correlation either by perturbation theory (Møller–Plesset or many-body perturbation theory [92]), coupled-cluster [93] or configuration–interaction (CI) methods [94]. Perhaps most popular among these are second-order perturbation theory [MP2 or MBPT(2)], which in many cases allows a low-cost treatment of electron correlation of reasonable accuracy, as well as the CCSD [95] and CCSD(T) approaches [96], the latter usually providing near-quantitative accuracy when the reference determinant is a reasonably good approximation to the exact wave function. Though very popular in the 1970s and 1980s, CI methods are less often used now, mostly because these approaches do not satisfy the property of "size-consistency". In cases where the single-reference approach is not appropriate because no single determinant makes the dominant contribution to the wave function, multiconfigurational approaches can be used. The HF–SCF ansatz is then replaced by an MCSCF treatment [97] in which the wave function consists of a linear combination of several Slater determinants or configuration state functions. In this way, static correlation effects resulting from the (near) degeneracies of several configurations can in principle be adequately described. Dynamic correlation effects are then introduced using either multireference perturbation theory [MR–MP, usually second-order perturbation theory based on a complete-active space SCF (CASSCF) wave function, i.e., CASPT2] [98], multireference configuration–interaction (MR-CI) [94], or multireference-coupled-pair methods such as MR–ACPF or MR–AQCC (which are nearly identical methods) [99]. These approaches—especially the last three—are designed to provide high-level descriptions, although the accuracy that can be achieved depends critically on the size of the MCSCF expansion or the active space in the CASSCF treatment.

Finally, there is an alternative and decidedly different way to incorporate electron correlation in quantum chemical calculations that is growing rapidly in importance: DFT [100]. By using the Kohn–Sham formulation, DFT methods have been used extensively in quantum chemistry during the last decade and yield results that are superior to HF–SCF calculations at essentially the same cost. A further advantage seems to be that DFT appears to hold promise in the treatment of transition metal compounds, which is an area where standard methods (except elaborate MCSCF and MR–CI treatments) often fail catastrophically. Concerning the treatment of electron correlation, it should be noted that DFT methods—unlike the more traditional methods discussed so far—are semiempirical in nature and therefore only provide an implicit treatment. Correlation effects are incorporated in DFT (via an adequate parametrization) through the exchange-correlation functional and not explicitly treated in the usual sense.

We also note that there are other, more advanced techniques for the treatment of electron correlation such as the r_{12} approaches [101], geminal techniques [101], or quantum Monte Carlo methods [102]. Although these approaches are undeniably very powerful treatments for obtaining accurate ground-state energies, they are still awaiting implementations of analytic derivatives and therefore are outside the scope of this chapter.

B. Electron-Correlated Approaches for NMR Chemical Shift Calculations

As described in Section II, NMR chemical shifts can be computed as second derivatives [Eq. (6.8)]. With the development of analytic second derivative techniques for electron-correlated schemes [84], calculation of NMR chemical shifts thus became possible. In the last decade, a number of correlated approaches have been applied (together with appropriate measures for treating the gauge-origin problem) to the calculation of chemical shielding tensors, which has allowed its applications to several chemically interesting problems.

Figure 6.4 gives an overview about the existing quantum chemical schemes for computing NMR chemical shifts, with the proviso that we have included only those methods that involve a satisfactory treatment of the

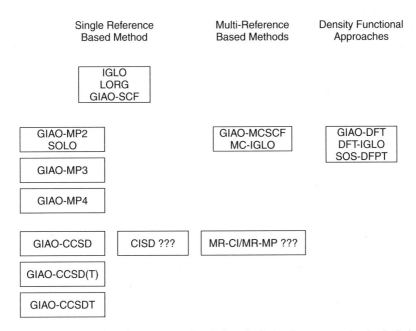

Figure 6.4. Overview over quantum chemical methods for electron-correlated calculation of NMR chemical shifts using local gauge-origin methods.

gauge problem. (Note that for some of the quantum chemical schemes listed in Figure 6.4, corresponding schemes for common gauge-origin calculations have been presented as well. We mention here the second-order polarization propagator approximation (SOPPA) approach advocated by Oddershede et al. [77] as well as the MP2, MP3 and linearized CCD schemes by Cybulski and Bishop [29]. However, we again emphasize that these schemes are at most useful for benchmark calculations but not intended for routine application.)

The first electron-correlated approach using local gauge origins was presented in 1990 by Bouman and Hansen [103]. However, their second-order LORG (SOLO) scheme does not exploit the derivative formula given in Eq. (6.8) for the shielding tensor. Rather, it is based on the sum-over-states ansatz used in propagator theory and correlation effects are introduced by considering additional second-order terms in the propagator expression (see [76] for a detailed discussion). The SOLO scheme was applied in [103] to the calculation of ^{31}P NMR chemical shifts, but otherwise has seen scarce application since other electron-correlated schemes providing increased accuracy at comparable costs have come on the scene since that time. In particular, the SOLO scheme appears to systematically underestimate the effect of electron correlation.

In 1992, Gauss presented the first rigorous scheme for calculating NMR chemical shifts at the MP2[MBPT(2)] level based on the GIAO ansatz [47]. (Note that an earlier report on GIAO–MP2 [104] has been shown to be erroneous, as the approach misses several terms as explained in [47].) Subsequent calculations using this GIAO–MP2 approach convincingly demonstrated the importance of electron-correlation effects in NMR chemical shift calculations. Substantial improvements are observed at this level of theory (relative to HF–SCF) for absolute shieldings [48], but also for relative shifts. In addition, a number of problems concerning the interpretation of experimental spectra of boranes [14], carboranes [14, 15], and carbocations [105–108] could be resolved. The development of GIAO–MP2 has indeed been useful, and is now included in the most popular quantum chemical program packages (i.e., ACESII [41], GAUSSIAN [43], and TURBOMOLE [45]).

The work on GIAO–MP2 was subsequently extended to third- and fourth-order perturbation theory [50] as well as to coupled-cluster methods [52–57]. All of these developments were based on use of the GIAO ansatz, analytic second-derivative techniques, and the asymmetric approach discussed in Section II.F. While GIAO–MP3 — as had already been known for other properties — turns out to be less useful than GIAO–MP2, GIAO–MP4(SDQ) (MP4 without triple excitations) calculations provide some indication that GIAO–MP2 tends to overestimate electron-correlation effects on the absolute shielding constants [50]. The GIAO-CC schemes — available at CC singles and doubles (CCSD) [52, 53], CCSD(T) [55], and most recently also for the full CC singles, doubles, and triples (CCSDT) level [57] — are intended to provide for the first time shieldings with quantitative accuracy. The GIAO–CCSD(T) scheme has proven particularly valuable and has been extensively used for benchmark calculations. These have focused both on establishing accurate NMR scales [66], and the resolution of some challenging chemical problems [109–111].

Parallel to the development of perturbation theory and coupled-cluster schemes for the calculation of NMR chemical shifts, MCSCF approaches have been formulated and implemented. In 1993, van Wüllen and Kutzelnigg presented a multiconfigurational extension of the original IGLO approach termed MC–IGLO [112–114], while Ruud et al. [51] reported an MCSCF implementation within the GIAO framework (GIAO–MCSCF). As demonstrated in several examples, MCSCF calculations turn out to be essential in cases with strong static electron correlation effects. On the other hand, they do not necessarily provide results of quantitative accuracy, which can only be achieved by including residual dynamic correlation effects by means of perturbation theory or configuration–interaction treatments. However, due to the lack of any corresponding analytic second derivative methods for these approaches, no implementations for the calculation of NMR chemical shifts is yet available.

Due to the increasing popularity of DFT schemes in the last decade, it is perhaps not surprising that DFT-based schemes have also been developed for the calculation of NMR chemical shifts. While we will discuss some of the formal problems of DFT calculations of magnetic properties in Section III.E, the available schemes will be summarized here largely without editorial comment. The DFT methods for the calculation of NMR chemical shifts within the IGLO framework have been mainly pursued by Malkin et al. [115], while Schreckenbach and Ziegler [58] (using Slater-type orbitals), Rauhut et al. [59], as well as Cheeseman et al. [60] reported implementations using GIAOs. In addition, effort has been made to improve upon standard DFT schemes either by considering current-dependent functionals [61] or by empirical modifications of the basic theoretical expressions [62, 78, 79, 116]. Despite undeniable problems, DFT calculations of NMR chemical shifts have nevertheless become popular in recent years and have evolved into a standard tool of computational chemistry.

C. Description of the GIAO–MP2 Approach

It is far beyond the scope of this chapter to discuss the theoretical details for all existing electron-correlated approaches. However, to give the reader an impression of what is needed for electron-correlated calculations of NMR chemical shifts, a detailed account of the GIAO–MP2 approach [47, 48] is presented in this section.

Second-order Møller-Plesset perturbation theory is the simplest approach to account for electron correlation when starting from a HF–SCF reference determinant. The correlation energy in MP2 is given by

$$E(MP2) = \frac{1}{4} \sum_{ij} \sum_{ab} t_{ij}^{ab*} \langle ab||ij\rangle \tag{6.30}$$

with $\langle pq||rs\rangle$ as the antisymmetrized two-electron integrals over spin orbitals

$$\langle pq||rs\rangle = \int \varphi_p^*(1)\varphi_q^*(2)\frac{1}{r_{12}}[\varphi_r(1)\varphi_s(2) - \varphi_s(1)\varphi_r(2)]\,d\tau_1\,d\tau_2 \tag{6.31}$$

and t_{ij}^{ab} as the first-order double excitation amplitudes

$$t_{ij}^{ab} = \langle ab||ij\rangle/(f_{ii} + f_{jj} - f_{aa} - f_{bb}) \tag{6.32}$$

The molecular orbitals φ_p are assumed to be eigenfunctions of the Fock operator with corresponding eigenvalues f_{pp}. The indices $i, j, k, l, \ldots$ refer to occupied spin orbitals and $a, b, c, d, \ldots$ to virtual (unoccupied) spin orbitals. The generic indices $p, q, r, s, \ldots$ are used to denote spin orbitals that might be occupied or virtual. In all equations, complex conjugate quantities are denoted by an asterisk. Computationally, it is important to realize that the expression for the correlation energy is given in terms of molecular orbitals. At the MP2 level, a transformation of the two-electron integrals from the atomic orbital (AO) to the molecular orbital (MO) basis is the most computationally demanding part of the energy evaluation.

Expressions for the MP2 correction to the shielding tensor are—as already discussed—best obtained by differentiating first Eq. (6.30) with respect to the nuclear magnetic moment $\mathbf{m}_J$ and then, in a second step, with respect to the magnetic field $\mathbf{B}$.

Following the general theory of gradients in MP theory [117], differentiation of the MP2 energy given in Eq. (6.30) with respect to the components m_i^J of $\mathbf{m}_J$ yields. (Note that this expression is valid only for one-electron perturbations and perturbation-independent basis functions. The more general case requires additional terms. (See [117].))

$$\frac{dE(MP2)}{dm_i^J} = \sum_{\mu\nu} D_{\mu\nu} \frac{\partial h_{\mu\nu}}{m_j^J} \tag{6.33}$$

where the $D_{\mu\nu}$ are elements of the effective MP2 one-particle density matrix (often referred to as "relaxed" or "response" density matrix [118–120]) and $h_{\mu\nu}$ are those of the one-electron Hamiltonian. Both quantities are given in the AO representation. According to Eq. (6.33), evaluation of the MP2 gradient requires knowledge of the perturbed integrals (which can be usually computed in a straightforward manner using standard techniques) as well as the unperturbed MP2 density matrix. Expressions for the latter are preferably given in the MO representation, where one obtains

$$D_{ij} = -\frac{1}{2}\sum_m \sum_{ef} t_{im}^{ef} t_{jm}^{ef*} \qquad (6.34)$$

for the occupied–occupied block and

$$D_{ab} = \frac{1}{2}\sum_{mn}\sum_e t_{mn}^{ae*} t_{mn}^{be} \qquad (6.35)$$

for the virtual–virtual block. The occupied–virtual block (and thus also the virtual–occupied block) depends solely on relaxation of the molecular orbitals (first-order changes of the orbitals with respect to the perturbation) and is determined as solution of the so-called Z-vector equations [89]

$$\sum_m \sum_e D_{em}[\langle ei||ma\rangle + \delta_{im}\delta_{ae}(f_{aa} - f_{ii})] + D_{em}^*\langle mi||ea\rangle = -X_{ai} \qquad (6.36)$$

The X_{ai} term on the right-hand side of the perturbation-independent Z-vector equations is defined as

$$X_{ai} = \frac{1}{2}\sum_m\sum_{ef}\langle ef||am\rangle t_{im}^{ef*} - \frac{1}{2}\sum_{mn}\sum_e\langle ie||mn\rangle t_{mn}^{ae*}$$
$$+ \sum_{mn} D_{mn}\langle im||an\rangle + \sum_{ef} D_{ef}\langle ie||af\rangle \qquad (6.37)$$

involving the antisymmetrized two-electron integrals as well as the double-excitation amplitudes t_{ij}^{ab}.

The AO density matrix in Eq. (6.33) is obtained via

$$D_{\mu\nu} = \sum_{pq} c_{\mu p}^* D_{pq} c_{\nu q} \qquad (6.38)$$

from the MO expressions given in Eqs. (6.34) to (6.36); the $c_{\mu p}$ denoting the MO coefficients (specifically, the contribution of AO μ to MO p).

The expression required for evaluation of shielding tensors at the MP2 level are obtained by differentiating Eq. (6.33) with respect to the magnetic field components B_i. While derivation of the gradient expression in Eq. (6.33) requires some algebraic manipulations in order to eliminate derivatives of the double excitation amplitudes and of the MO coefficients, the second step consists of a simple, straightforward differentiation without further manipulations, which leads to

$$\sigma_{ji}(\text{MP2}) = \sum_{\mu\nu} D_{\mu\nu} \frac{\partial^2 h_{\mu\nu}}{\partial B_i \partial m_j^J} + \sum_{\mu\nu} \frac{\partial D_{\mu\nu}}{\partial B_i} \frac{\partial h_{\mu\nu}}{\partial m_j^J} \quad (6.39)$$

The derivative of the AO relaxed density matrix is now related to the corresponding MO quantity via

$$\frac{\partial D_{\mu\nu}}{\partial B_i} = \sum_{pq} c_{\mu p}^* \frac{\partial D_{pq}}{\partial B_i} c_{\nu q} + \sum_{pqr} [U_{pr}^{B_i*} c_{\mu p}^* D_{rq} c_{\nu q} + c_{\mu p}^* D_{pr} U_{qr}^{B_i} c_{\nu q}] \quad (6.40)$$

and includes contributions due to the derivatives of the MO coefficients. The latter are usually parametrized using the coupled-perturbed HF coefficients $U_{pq}^{B_i}$ in terms of the unperturbed MO coefficients

$$\frac{\partial c_{\mu p}}{\partial B_i} = \sum_q U_{qp}^{B_i} c_{\mu q} \quad (6.41)$$

and obtained from the solution of the coupled-perturbed HF equations [121] and/or the orthonormality constraint for the perturbed orbitals. Expressions for the perturbed MO density matrix are obtained by straightforward differentiation of Eqs. (6.34) to (6.36). For the occupied–occupied and virtual–virtual blocks, the expressions

$$\frac{\partial D_{ij}}{\partial B_i} = -\frac{1}{2} \sum_m \sum_{ef} \left\{ \frac{\partial t_{im}^{ef}}{\partial B_i} t_{jm}^{ef*} + t_{im}^{ef} \frac{\partial t_{jm}^{ef*}}{\partial B_i} \right\} \quad (6.42)$$

and

$$\frac{\partial D_{ab}}{\partial B_i} = \frac{1}{2} \sum_{mn} \sum_e \left\{ \frac{\partial t_{mn}^{ae*}}{\partial B_i} t_{mn}^{be} + t_{mn}^{ae*} \frac{\partial t_{mn}^{be}}{\partial B_i} \right\} \quad (6.43)$$

are obtained. The differentiated occupied–virtual block is obtained by

CALCULATION OF NMR CHEMICAL SHIFTS 379

solving the first-order Z-vector equations [47]

$$\sum_m \sum_e \frac{\partial D_{em}}{\partial B_i} \{\langle ei||ma\rangle - \langle mi||ea\rangle + \delta_{im}\delta_{ea}(f_{aa} - f_{ii})\}$$
$$= -\frac{\partial X_{ai}}{\partial B_i} - \sum_m \sum_e D_{em} \left\{ \frac{\partial \langle ei||ma\rangle}{\partial B_i} + \frac{\partial \langle mi||ea\rangle}{\partial B_i} + \delta_{im}\frac{\partial f_{ea}}{\partial B_i} - \delta_{ea}\frac{\partial f_{im}}{\partial B_i} \right\}$$
(6.44)

which can be derived from the corresponding zeroth-order equations, Eq. (6.36). The derivative of the intermediate X_{ai} in Eq. (6.44) is given as

$$\frac{\partial X_{ai}}{\partial B_i} = \frac{1}{2}\sum_m \sum_{ef} \left\{ \frac{\partial t_{im}^{ef*}}{\partial B_i} \langle ef||am\rangle + t_{im}^{ef*}\frac{\partial \langle ef||am\rangle}{\partial B_i} \right\}$$
$$- \frac{1}{2}\sum_{mn} \sum_e \left\{ \frac{\partial t_{mn}^{ae*}}{\partial B_i} \langle ie||mn\rangle + t_{mn}^{ae*}\frac{\partial \langle ie||mn\rangle}{\partial B_i} \right\}$$
$$+ \sum_{mn} \left\{ \frac{\partial D_{mn}}{\partial B_i} \langle mi||na\rangle + D_{mn}\frac{\partial \langle mi||na\rangle}{\partial B_i} \right\}$$
$$+ \sum_{ef} \left\{ \frac{\partial D_{ef}}{\partial B_i} \langle ei||fa\rangle + D_{ef}\frac{\partial \langle ei||fa\rangle}{\partial B_i} \right\}$$
(6.45)

and involves perturbed antisymmetrized two-electron integrals as well as perturbed double-excitation amplitudes. The perturbed MO integrals depend as follows on perturbed AO integrals and CPHF coefficients

$$\frac{\partial \langle pq||rs\rangle}{\partial B_i} = \sum_{\mu\nu\rho\sigma} c_{\mu p}^* c_{\nu q}^* c_{\sigma r} c_{\rho s} \frac{\partial \langle \mu\nu||\sigma\rho\rangle}{\partial B_i}$$
$$+ \sum_t \{U_{tp}^{B_i^*}\langle tq||rs\rangle + U_{tq}^{B_i^*}\langle pt||rs\rangle + U_{tr}^{B_i}\langle pq||ts\rangle + U_{ts}^{B_i}\langle pq||rt\rangle\}$$
(6.46)

while the perturbed amplitudes are given by

$$\frac{\partial t_{ij}^{ab}}{\partial B_i} = \left\{ \frac{\partial \langle ab||ij\rangle}{\partial B_i} + \sum_e \left[\frac{\partial f_{ae}}{\partial B_i} t_{ij}^{eb} + \frac{\partial f_{be}}{\partial B_i} t_{ij}^{ae} \right] \right.$$
$$\left. - \sum_m \left[t_{mj}^{ab}\frac{\partial f_{mi}}{\partial B_i} + t_{im}^{ab}\frac{\partial f_{mj}}{\partial B_i} \right] \right\} \bigg/ (f_{ii} + f_{jj} - f_{aa} - f_{bb}) \quad (6.47)$$

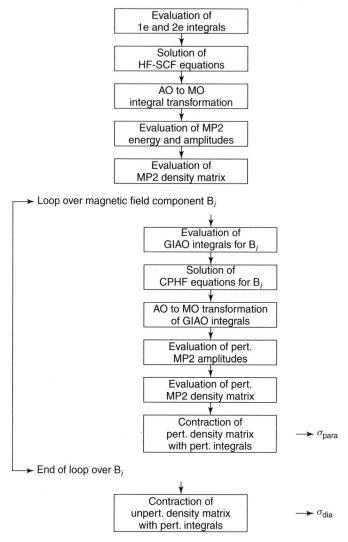

Figure 6.5. Flowchart of the various computational steps required in a GIAO–MP2 NMR chemical shift calculation.

Note that we do not assume in Eq. (6.47) that the perturbed orbitals are canonical. Indeed, it has long been recognized that skipping the canonical constraint is highly advantageous (for a detailed discussion of this issue, see [122]).

Equations (6.30) to (6.47) summarize the theoretical framework for the GIAO–MP2 method.

Figure 6.5 sketches the different computational steps required for a GIAO–MP2 calculation. The first part is identical to a normal MP2 gradient calculation and consists of the evaluation of AO integrals, the SCF step, as well as the AO to MO integral transformation and the computation of the unperturbed density matrix. The second part consists of a loop over the three magnetic field components. For each component, the required GIAO integrals are computed (see below) and the CPHF equations solved. These steps (which are actually also required for GIAO–HF–SCF calculations) are followed by an AO to MO transformation of the GIAO two-electron integrals [see Eq. (6.46)] as well as evaluation of perturbed amplitudes [Eq. (6.47)] and perturbed density matrices [Eqs. (6.42)–(6.44)].

From the equations above, it should be relatively obvious that calculation of the perturbed density matrices requires essentially the same steps as the calculation of the unperturbed density matrix. The only difference is that the unperturbed amplitudes and integrals in the latter are replaced sequentially (and individually) by their perturbed counterparts. As a result two, rather than one, matrix multiplications are required for their evaluation.

An additional comment is warranted about the additional (perturbed) integrals required in a GIAO calculation of NMR chemical shifts. The integrals needed consist of the following one-electron integrals:

$$\frac{\partial S_{\mu\nu}}{\partial B_i} = \left\langle \frac{\partial \chi_\mu}{\partial B_i} \Big| \chi_\nu \right\rangle + \left\langle \chi_\mu \Big| \frac{\partial \chi_\nu}{\partial B_i} \right\rangle \quad (6.48)$$

$$\frac{\partial h_{\mu\nu}}{\partial B_i} = \left\langle \chi_\mu \Big| \frac{\partial h}{\partial B_i} \Big| \chi_\nu \right\rangle + \left\langle \frac{\partial \chi_\mu}{\partial B_i} \Big| h \Big| \chi_\nu \right\rangle + \left\langle \chi_\mu | h | \frac{\partial \chi_\nu}{\partial B_i} \right\rangle \quad (6.49)$$

$$\frac{\partial h_{\mu\nu}}{\partial m_i^J} = \left\langle \chi_\mu \Big| \frac{\partial h}{\partial m_j^J} \Big| \chi_\nu \right\rangle \quad (6.50)$$

$$\frac{\partial^2 h_{\mu\nu}}{\partial B_i \partial m_j^J} = \left\langle \chi_\mu \Big| \frac{\partial^2 h}{\partial B_i \partial m_j^J} \Big| \chi_\nu \right\rangle + \left\langle \frac{\partial \chi_\mu}{\partial B_i} \Big| \frac{\partial h}{\partial m_j^J} \Big| \chi_\nu \right\rangle + \left\langle \chi_\mu \Big| \frac{\partial h}{\partial m_j^J} \Big| \frac{\partial \chi_\nu}{\partial B_i} \right\rangle \quad (6.51)$$

and the following two-electron integrals

$$\frac{\partial \langle \mu\nu|\sigma\rho \rangle}{\partial B_i} = \left\langle \frac{\partial \chi_\mu}{\partial B_i} \nu | \sigma\rho \right\rangle + \left\langle \mu \frac{\partial \chi_\nu}{\partial B_i} | \sigma\rho \right\rangle + \left\langle \mu\nu \Big| \frac{\partial \chi_\sigma}{\partial B_i} \rho \right\rangle + \left\langle \mu\nu | \sigma \frac{\partial \chi_\rho}{\partial B_i} \right\rangle$$

$$(6.52)$$

This should be seen in contrast to calculations based on a common gauge origin, where just the following one-electron integrals are needed

$$\frac{\partial h_{\mu\nu}}{\partial B_i} = \left\langle \chi_\mu \left| \frac{\partial h}{\partial B_i} \right| \chi_\nu \right\rangle \tag{6.53}$$

$$\frac{\partial h_{\mu\nu}}{\partial m_i^J} = \left\langle \chi_\mu \left| \frac{\partial h}{\partial m_j^J} \right| \chi_\nu \right\rangle \tag{6.54}$$

and

$$\frac{\partial^2 h_{\mu\nu}}{\partial B_i \partial m_j^J} = \left\langle \chi_\mu \left| \frac{\partial^2 h}{\partial B_i \partial m_j^J} \right| \chi_\nu \right\rangle \tag{6.55}$$

The handling of the perturbed two-electron GIAO integrals was originally considered to be a major obstacle, but Pulay and co-workers [35, 123] showed that they can be computed in a straightforward manner. In their work, the similarity of perturbed GIAO integrals to usual two-electron integral derivatives needed in gradient calculations was exploited, and they also demonstrated that no storage of the perturbed GIAO integrals is required in GIAO–HF–SCF calculations. Although storage of these integrals is required in electron-correlated calculations (see, however, Section III.D), the ease with which these integrals can be calculated is an important factor. To be more specific, Pulay exploited the similarity of AOs differentiated with respect to nuclear coordinates

$$\frac{\partial}{\partial \mathbf{R}_A} \exp(-\alpha_\mu (\mathbf{r} - \mathbf{R}_\mu)^2) = 2\alpha_\mu \delta_{A\mu} (\mathbf{r} - \mathbf{R}_\mu) \exp(-\alpha_\mu (\mathbf{r} - \mathbf{R}_\mu)^2) \tag{6.56}$$

with those obtained by differentiating GIAOs with respect to B_i

$$\frac{\partial}{\partial \mathbf{B}} \left\{ \exp\left(\frac{-i}{2} (\mathbf{B} \times (\mathbf{R}_\mu - \mathbf{R}_O)\mathbf{r}) \exp(-\alpha_\mu (\mathbf{r} - \mathbf{R}_\mu)^2) \right) \right\} \bigg|_{\mathbf{B}=0}$$

$$= \frac{-i}{2} [(\mathbf{R}_\mu - \mathbf{R}_O) \times \mathbf{r}] \exp(-\alpha_\mu (\mathbf{r} - \mathbf{R}_\mu)^2) \tag{6.57}$$

as both are (apart from factors) proportional to $\mathbf{r}$ times the original Gaussian. It is thus clear that GIAO integrals can be obtained from a standard integral derivative code with just a few slight modifications. For example, for the derivatives of the overlap integrals $S_{\mu\nu}$ with respect to $\mathbf{B}$, we obtain

$$\frac{\partial S_{\mu\nu}}{\partial \mathbf{B}} = -\frac{i}{2}(\mathbf{R}_\mu - \mathbf{R}_\nu) \times \langle \mu | \mathbf{r} | \nu \rangle \qquad (6.58)$$

while the corresponding derivative with respect to the coordinate $\mathbf{R}_A$ is

$$\frac{\partial S_{\mu\nu}}{\partial \mathbf{R}_A} = (2\alpha_\mu \mathbf{R}_\mu \delta_{A\mu} + 2\alpha_\nu \mathbf{R}_\nu \delta_{A\nu})S_{\mu\nu} + 2(\alpha_\mu \delta_{A\mu} + \alpha_\nu \delta_{A\nu})\langle \mu | \mathbf{r} | \nu \rangle \qquad (6.59)$$

Comparison of both formulas reveals that the basic integral, that is, $\langle \mu | \mathbf{r} | \nu \rangle$ is the same in both cases. Note that the relationships between derivatives of integrals with respect to magnetic field components and nuclear coordinates strictly hold only for uncontracted Gaussians. For contracted basis functions, these relationship can only be exploited at the level of the underlying primitive Gaussians. However, the main conclusion of our brief discussion is that all the perturbed integrals required in the GIAO approach can easily be computed using standard integral evaluation techniques, that is, either based on Rys polynomials [124], the McMurchie–Davidson scheme employing Hermite Gaussians [125], or the Obara–Saika recursion methods [126].

A further remark addresses the (permutational) symmetry relations of GIAO integrals. While for the unperturbed integrals the following relations hold:

$$h_{\mu\nu} = h_{\nu\mu} \qquad (6.60)$$

and

$$\langle \mu\sigma | \nu\rho \rangle = \langle \nu\sigma | \mu\rho \rangle = \langle \mu\rho | \nu\sigma \rangle = \langle \nu\rho | \mu\sigma \rangle = \langle \sigma\mu | \rho\nu \rangle = \langle \sigma\nu | \rho\mu \rangle$$
$$= \langle \rho\mu | \sigma\nu \rangle = \langle \rho\nu | \sigma\mu \rangle \qquad (6.61)$$

the following symmetry relations apply to the first-order perturbed integrals

$$\frac{\partial h_{\mu\nu}}{\partial B_i} = -\frac{\partial h_{\nu\mu}}{\partial B_i} \qquad (6.62)$$

and

$$\frac{\partial \langle \mu\sigma | \nu\rho \rangle}{\partial B_i} = -\frac{\partial \langle \nu\rho | \mu\sigma \rangle}{\partial B_i} = \frac{\partial \langle \sigma\mu | \rho\nu \rangle}{\partial B_i} = -\frac{\partial \langle \rho\nu | \sigma\mu \rangle}{\partial B_i} \qquad (6.63)$$

Besides the additional signs, the main difference is that the perturbed two-electron integrals possess fourfold permutation symmetry instead of the usual eightfold symmetry.

Finally, we note that for the double excitation amplitudes the following relations hold upon complex conjugation:

$$t_{ij}^{ab*} = t_{ij}^{ab} \tag{6.64}$$

in the case of the unperturbed amplitudes and

$$\frac{\partial t_{ij}^{ab*}}{\partial \mathbf{B}} = -\frac{\partial t_{ij}^{ab}}{\partial \mathbf{B}} \tag{6.65}$$

in case of the perturbed amplitudes.

The computational cost of GIAO–MP2 scales with N^5, as does a gradient calculation at the same level of theory. The rate-limiting steps are the AO to MO integral transformation of the unperturbed and perturbed two-electron integrals as well as calculation of the corresponding X_{ai} intermediates. The evaluation of the GIAO integrals in the AO basis scales at most with N^4 and thus is not a limiting factor. However, unlike for GIAO–HF–SCF calculations, it should be noted that a conventional implementation of the GIAO–MP2 approach (as described in [47] and [48]) requires storage of the GIAO two-electron integrals. This leads to increased disk space usage and turns out to be the main bottleneck in conventional GIAO–MP2 calculations. Currently, such calculations are limited — depending on the molecular point group — to molecules with up to 300–400 basis functions. However, it is very advantageous that the computational cost of a GIAO–MP2 calculation does not exhibit an additional dependence on the size of the molecule (apart from that associated with the increased number of basis functions), as the cost ratio between an MP2 gradient and an NMR chemical shift calculation is essentially constant (factor of 3 due to the magnetic field perturbations and the consideration that for the perturbed quantities two contractions are required when only one suffices in the unperturbed case.) Consequently, it should always be possible to perform a GIAO–MP2 calculation following a MP2 geometry optimization that uses analytic gradients.

The theoretical formulation of NMR chemical shifts at other electron-correlated levels follows very similar strategies to that used for GIAO–MP2; some exceptions nevertheless occur for MCSCF where different strategies are applicable. Typical is that all correlated approaches (in their conventional implementations) require the transformation of unperturbed/

perturbed AO integrals to the MO representation and storage of the corresponding integrals. The expressions for the amplitudes, however, are more involved for higher order MP theory as well as coupled-cluster methods, which also leads to an increased scaling of the computational cost that is found to be N^6 for MP3 and CCSD and N^7 for MP4 and CCSD(T). Consequently, the applicability of these schemes is much more limited than that associated with GIAO–MP2; GIAO–CCSD(T) calculations are currently possible for systems consisting of about 10 to 15 atoms. For GIAO–MCSCF calculations, the computational cost heavily depend on the size of the active space (or equivalently the length of the MCSCF expansion). Hence, these methods are applicable only to smaller systems, as well.

D. Further Methodological Developments

The main and immediate challenge for electron-correlated schemes for computing NMR chemical shifts is certainly the application of these methods to larger systems. The significance of this challenge is particularly apparent with the advent of DFT methods, since these methods are applicable to reasonably large systems and generally offer results (with inclusion of electron correlation) better than HF–SCF at essentially the same cost. However, developments reported in recent years [127] demonstrate that many possibilities exist to improve the computational efficiency of conventional approaches and thus to render them attractive in comparison to DFT. While most of these developments are still restricted to energy calculations, some efforts have also been undertaken concerning NMR chemical shift calculations [128–130]. These developments will be discussed forthwith, and include the application of integral-direct techniques in electron-correlated calculations of NMR chemical shifts [128, 129] and the use of the local correlation ansatz [131, 132] in the context of NMR chemical shift calculations.

Before discussing these developments in some detail, the reader is reminded that the bottlenecks in electron-correlated calculations are (a) the disk-space requirement due to the storage of unperturbed and perturbed AO and MO integrals as well as amplitudes and/or CI coefficients and (b) the steep increase of the CPU time requirement with the molecular size ranging from N^5 for MP2 up to N^7 in the case of CCSD(T).

Existing schemes designed to overcome these limitations can be classified as to whether they reduce the computational cost by using various numerical techniques without any approximation of the equations (these preserve the answers that would be obtained with the traditional approach if the computational resources were available), and those that introduce additional (often physically motivated) approximations in order to speed up calculations. Integral-direct techniques with prescreening [133] belong to the first class, while resolution-of-identity (RI) methods [134] and local correlation methods [131, 132] are examples of the second class.

Since disk-space bottlenecks present a more severe impediment than excessive CPU time requirements, initial efforts focused on reducing disk storage requirements. (This statement seems unjustified, as the CPU time increase rises so rapidly with the size of the system, and indeed more rapidly that the requirement of disk space. However, one should realize that if there is not enough disk space a calculation cannot be performed, while excessive CPU time requirement just means that the calculations run for a very long time and thus are still possible, in principle.) Kollwitz et al. [128, 129] thus reported an integral-direct GIAO–MP2 implementation (within the TURBOMOLE program package [45]), using concepts that were essentially the same as those exploited in previous integral-direct energy and gradient calculations at the MP2 level [135, 136].

The key step of the integral-direct GIAO–MP2 implementation is a direct AO to MO integral transformation. While these steps are also encountered in energy and gradient calculations, GIAO–MP2 has the additional requirement that this step needs to be carried out for the perturbed two-electron integrals. As seen from Eq. (6.46), this step actually involves five transformations involving either perturbed AO integrals or perturbed MO coefficients. The main advantage in the use of a direct AO to MO integral transformation is that no AO integrals need to be stored, but rather are computed as needed in the transformation step. A reduction in the storage requirements concerning the MO quantities, that is, unperturbed and perturbed MO integrals as well as double excitation amplitudes, is achieved by realizing that one loop over occupied orbitals (index i) can be moved to the outside. The integral transformation as well as all subsequent contractions are thus performed with one fixed index and in this way the storage requirements (e.g., $\langle ip||qr\rangle$ with fixed i and $p, q, r = 1, \ldots, N$ in the case of unperturbed integrals) are reduced from $\mathcal{O}(N^4)$ to $\mathcal{O}(N^3)$. For the fixed index, a subset of all unperturbed and perturbed MO integrals is then used to compute the corresponding subset of double-excitation amplitudes as well as all contributions to the density matrices, before the next subset of integrals with a different i is handled. Efficient use of molecular point group symmetry as well as coarse-grain parallelization further extends the range of applicability as discussed in [129]. NMR chemical shift calculations on the naphthalenium ($C_{10}H_9^+$) and anthracenium ($C_{14}H_{11}^+$) ion involving 232 and 288 basis functions, respectively, could be carried on modestly equipped workstations (128 MB memory and 2 GB disk) with these direct GIAO–MP2 methods. In addition, calculations were reported for the highly symmetric molecules Al_4Cp_4 (see Fig. 6.6) as well as B_4t-Bu_4 (see Fig. 6.7). These calculations, particularly the latter

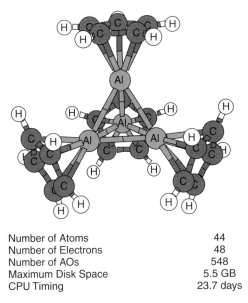

Number of Atoms	44
Number of Electrons	48
Number of AOs	548
Maximum Disk Space	5.5 GB
CPU Timing	23.7 days

Figure 6.6 Computational details of the GIAO–MP2/tzp/dz calculation [129] for Al$_4$Cp$_4$ (D_{2d} symmery). CPU timings are given for a SGI power challenge.

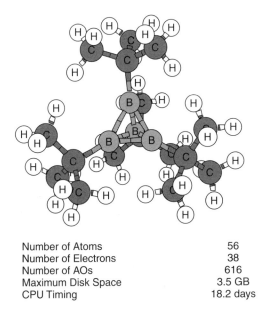

Number of Atoms	56
Number of Electrons	38
Number of AOs	616
Maximum Disk Space	3.5 GB
CPU Timing	18.2 days

Figure 6.7. Computational details of the GIAO–MP2/tzp calculation [129] for B$_4$t-Bu$_4$ (T_d symmery). CPU timings are given for an IBM RS6000/3CT.

examples, amply demonstrate that integral-direct schemes greatly extend the range of applicability over that associated with conventional implementations where full storage of AO and MO integrals is required.

While the integral-direct implementation significantly enhances the applicability of the GIAO–MP2 approach to about 500–1000 basis functions (depending on symmetry), it is clear that after elimination of the disk-space bottleneck, the $\mathcal{O}(N^5)$ dependence of the CPU requirement becomes the actual limiting factor. This rather obvious observation is reflected by the timings reported for Al_4Cp_4 and B_4tBu_4.

The first efforts to tackle the CPU problem in GIAO–MP2 calculations have been undertaken by Gauss and Werner [130] who tested the feasibility of local correlation schemes for the calculation of NMR chemical shifts. In these methods, one tries to exploit the local character of dynamical electron correlation from the onset. This idea is realized by considering only excitations from localized orbitals to projected AOs (i.e., AOs orthogonalized with respect to the occupied space) that reside on a set of well-defined closeby centers. Since a constraint is thereby introduced for the excitation space, it should be clear that energies from a local correlation scheme such as local MP2 (LMP2) are not equivalent to those obtained in corresponding conventional calculations. However, for chemical purposes the agreement is more than sufficient, and it has in fact been argued that part of the difference is due to so-called "basis set superposition errors" that are avoided in the local schemes. Local correlation methods have been implemented in the framework of MP2 [137, 138], CCSD [139, 140], and most recently also for CCSD(T) [141] and the iterative triples model known as CCSDT-1 [142]. Furthermore, it has been demonstrated that linear scaling of the computational cost is possible within local correlation schemes [138, 140]. These methods hold great promise for future applications, and should be a very visible part of method development in the coming years.

However, since local schemes include some constraints in the treatment of electron correlation, their suitability for property calculations is not entirely clear and the correspondence between traditional GIAO–MP2 and GIAO–LMP2 schemes needs to be checked. This was done by Gauss and Werner [130], who implemented a local GIAO–MP2 scheme (GIAO–LMP2) within a conventional GIAO–MP2 program. Before comparing some results, it should be mentioned that a GIAO–LMP2 calculation requires additional steps. The use of localized occupied orbitals necessitates an iterative solution for the MP2 double excitation amplitudes and the restriction to local excitations leads to a loss of invariance with respect to rotations among occupied orbitals. Hence, additional coupled-perturbed equations due to the localization conditions need to be considered. These

Table VI.2
Comparison of Calculated ^{13}C, ^{15}N, and ^{17}O Isotropic Nuclear Magnetic Shielding Constants (in ppm) Obtained at HF–SCF, MP2, and LMP2 Levels Using the GIAO Ansatz Together with a qz2p Basis[a]

Molecule	GIAO–HF–SCF	GIAO–MP2	GIAO–LMP2
(a) ^{13}C shieldings			
C_2H_6	184.0	188.0	187.9
$\underline{C}H_3OH$	143.8	142.2	142.7
$\underline{C}H_3NH_2$	163.8	164.9	165.1
$\underline{C}H_3CN$	190.8	193.5	193.8
$\underline{C}H_3CHO$	162.3	162.9	163.2
$CH_3\underline{C}HO$	−15.7	1.3	1.4
$CH_3\underline{C}N$	60.5	76.0	76.0
$\underline{C}H_2CCH_2$	114.0	120.8	121.3
$CH_2\underline{C}CH_2$	−44.3	−25.8	−26.1
(b) ^{17}O shieldings			
CH_3OH	338.9	350.4	351.1
CH_3CHO	−375.9	−292.1	−290.5
(c) ^{15}N shieldings			
CH_3NH_2	250.7	261.1	260.9
CH_3CN	−46.7	12.9	12.9

[a] For further details, see [130].

additions are already required for LMP2 gradient calculations and have been recently extended (within a pilot implementation) to chemical shift calculations.

Table VI.2 compares results obtained from GIAO–LMP2 calculations to those from standard GIAO–MP2 calculations. It is seen that differences are quite small (typically 1 ppm or less), so that one can conclude that GIAO–LMP2 holds great potential for the future. Efficient implementation of this method has not yet been achieved, but will open the way for the routine electron-correlated treatment of large molecules. Extensions to GIAO–LCCSD and GIAO–LCCSD(T) can also be envisioned and might then — when available — provide results of unprecedented accuracy for large molecules.

E. The DFT Approach to NMR Chemical Shifts

The calculation of NMR chemical shifts within the DFT framework is connected with some formal, theoretical problems. Of course, it is possible to compute shielding tensors in a straightforward manner via the corresponding second derivatives as described in Section II, and this is indeed done in practice. The schemes described in [58–60, 115] follow such a prescription and are successfully applied in many quantum chemical investigations. On the other hand, there are some formal problems associated with this approach (which are not widely appreciated) and such calculations are not particularly satisfying from a purely theoretical perspective.

The main problem in DFT calculations of magnetic properties is that the Hohenberg–Kohn theorems [100] do not hold in the presence of magnetic fields [143]. The first theorem establishes a one-to-one correspondence between the given external potential $v(\mathbf{r})$ and the electron density $\rho(\mathbf{r})$. The problem is that magnetic interactions are not described via a local potential and that the introduction of the so-called vector potential $\mathbf{A}(\mathbf{r})$ is necessary. This means that the usual assumptions of the Hohenberg–Kohn theorems are not satisfied and thus that the original formulation of the theorems is not valid in this case. We discuss later how this problem can be addressed by extending the Hohenberg–Kohn theorems.

Ignoring the aforementioned problems, the following second derivative expression for the chemical shielding tensor is obtained within the Kohn–Sham approach

$$\sigma_{ji} = 2 \sum_i \left\langle \frac{\partial \varphi_i}{\partial B_i} \middle| \frac{\partial h}{\partial m_j^J} \middle| \varphi_i \right\rangle + \sum_i \left\langle \varphi_i \middle| \frac{\partial^2 h}{\partial B_i \partial m_j^J} \middle| \varphi_i \right\rangle \qquad (6.66)$$

with

$$\frac{\partial \varphi_i}{\partial B_i} = -\sum_a \varphi_a \frac{\left\langle \varphi_a \middle| \frac{\partial h}{\partial B_i} \middle| \varphi_i \right\rangle}{\varepsilon_a - \varepsilon_i} \qquad (6.67)$$

In Eqs. (6.66) and (6.67), the orbitals φ_p are the Kohn–Sham orbitals with corresponding one-particle energies ε_p. Note that in order to simplify the discussion, we limit ourselves here and in the following to a common gauge origin, though the arguments hold in a similar manner for the local gauge-origin schemes.

It is easily realized that the expressions given in Eqs. (6.66) and (6.67) correspond to what is usually termed the "uncoupled approach". The argument is simply that there appears no coupling between the different perturbed orbitals that appear in Eq. (6.67). This finding also means that the expression given in Eq. (6.66) can be rewritten in the following sum-over-states form

$$\sigma_{ji} = -2\sum_i\sum_a \frac{\left\langle \varphi_i \left| \frac{\partial h}{\partial B_i} \right| \varphi_a \right\rangle \left\langle \varphi_a \left| \frac{\partial h}{\partial m_j^J} \right| \varphi_i \right\rangle}{\varepsilon_a - \varepsilon_i} + \sum_i \left\langle \varphi_i \left| \frac{\partial^2 h}{\partial B_i \partial m_j^J} \right| \varphi_i \right\rangle \quad (6.68)$$

where the sum runs over all single excitations *with the corresponding excitation energies given by the Kohn–Sham orbital energy differences.*

There is a simple physical explanation that explains why an uncoupled expression is obtained in DFT. Due to the imaginary character of the perturbation, the perturbed first-order density vanishes

$$\begin{aligned}\frac{\partial \rho}{\partial B_i} &= \sum_i \frac{\partial \varphi_i^*}{\partial B_i} \varphi_i + \varphi_i^* \frac{\partial \varphi_i}{\partial B_i} \\ &= 0 \end{aligned} \quad (6.69)$$

This is generally true and, for example, is the reason why in the CPHF equations for magnetic perturbations there is only an exchange contribution and no Coulomb contribution due to the perturbed orbitals. The first-order HF density vanishes, while the first-order HF density matrix is non-zero. In DFT, all contributions (Coulomb, exchange, and correlation) are described via the electron density with the (physically incorrect) consequence that there are no corresponding perturbed contributions in the case of magnetic perturbations. As there is no perturbed contribution, there is no coupling between the perturbed orbitals and we obtain an uncoupled approach. Computationally, the uncoupled approach is easy to implement, but it needs to be emphasized that it is not correct from a theoretical viewpoint. What are the solutions to the above mentioned problems?

1. The simplest (and perhaps the most popular) "solution" is to simply ignore the problem. The numerical results obtained in DFT calculations with the uncoupled approach (see Section IV) provide some pragmatic justification for this approach, but of course it is not satisfying.

2. Serious attempts at resolving the problem are based on extensions of the underlying theory. Vignale et al. [143] showed that the formal problems can be resolved by introducing a functional that depends on both electron density $\rho(\mathbf{r})$ as well as the paramagnetic part of current density $\mathbf{j}_p(\mathbf{r})$. The latter is defined by

$$\mathbf{j}_p(\mathbf{r}) = -i \sum_i (\varphi_i^* \nabla \varphi_i - \varphi_i \nabla \varphi_i^*) \quad (6.70)$$

The associated modified Hohenberg–Kohn theorems are then valid in the presence of magnetic fields. A problem with $\mathbf{j}_p(\mathbf{r})$ is the issue of gauge invariance, as $\mathbf{j}_p$ is not a gauge-invariant quantity. Vignale et al. [143] solved the problem by reformulating the theory in terms of the following gauge-invariant quantity v

$$v = \nabla \times \frac{\mathbf{j}_p}{\rho} \tag{6.71}$$

The current dependence of the functionals reintroduces the coupling between the perturbed Kohn–Sham orbitals and thus leads to a coupled DFT variant for the calculation of NMR chemical shifts. A first implementation of current-dependent functionals was reported in 1995 by Lee et al. [61]. They found that the changes in the calculated shifts due to additional consideration of the current are small, but also that incorporation of these effects tend to worsen the results (see also Section IV). These empirical findings are often used as a basis to argue that the current dependence can be neglected, but Lee et al. [61] pointed out that no adequate current dependent functionals are yet available and essentially nothing is known about their form.

Another modification of the underlying theory has been suggested by Salisbury and Harris [144], who introduce explicitly magnetic-field dependent functionals and therefore avoid the use of the current density as second variable.

3. Starting from the sum-over-states expressions, it has been argued that the problem is due to the incorrect excitation energies used in the denominator. Malkin et al. [78] thus suggested that we "improve" the excitation energies in the following (empirical) manner:

$$\Delta E = \varepsilon_a - \varepsilon_i + \delta E_{i \to a}^{xc} \tag{6.72}$$

with either

$$\Delta E_{i \to a}^{xc} = \frac{1}{2} \left(\frac{3}{4\pi} \right)^{1/3} \int \rho^{\uparrow}(\mathbf{r})^{-2/3} \rho_i(\mathbf{r}) \rho_a(\mathbf{r}) \, d^3r \quad \text{(Loc1 approximation)} \tag{6.73}$$

or

$$\Delta E_{i \to a}^{xc} = \frac{2}{3} \left(\frac{3}{4\pi} \right)^{1/3} \int \rho^{\uparrow}(\mathbf{r})^{-2/3} \rho_i(\mathbf{r}) \rho_a(\mathbf{r}) \, d^3r \quad \text{(Loc2 approximation)} \tag{6.74}$$

The numerical results suggest that this modification, known as sum-over-states density-functional perturbation theory (SOS–DFPT), leads in some cases to improved results compared to experiment [78, 116, 145].

Another empirical modification of the SOS ansatz has been proposed recently by Wilson and co-workers [62, 79]. They modify the SOS calculations such that they use (admittedly in a somewhat inconsistent way) Kohn–Sham orbitals from a hybrid functional calculation. Initial numerical results (see Section IV) again indicate some improvement, although this approach has been criticized in the literature [146].

IV. ILLUSTRATIVE EXAMPLES

In the following, we will illustrate the importance of electron correlation effects in NMR chemical shift calculations by giving several examples. Before relative NMR shifts — which are the quantities measured in virtually all experiments — are discussed, we will first focus on benchmark calculations of absolute shieldings. Several cases in which the inclusion of electron correlation is vital will be shown.

A. Benchmark Calculations and Absolute Nuclear Magnetic Shielding Constants

Many calculations reported in the literature demonstrate the importance of electron correlation for the accurate prediction of shielding tensors and absolute shielding constants. However, in order to make accurate theoretical predictions of shieldings, it is necessary to calibrate the various existing quantum chemical approaches. This can be done either by comparison with reliable experimental data or — from a theoretical point of view the preferred way — by comparison with full configuration–interaction (FCI) results since this allows us to focus entirely on the electron correlation problem. (Comparisons with FCI and experiment are equivalent only in the limit of a complete basis set, and then only when all other effects, e.g., relativistic effects and rovibrational contributions, are neglected.) Though FCI calculations have become possible for more and more interesting problems, FCI calculations for chemical shifts are still scarce. The only values that have been presented in the literature are for the trivial case of H_2 [147] (where FCI is of course equivalent to CCSD!) as well as BH [148].

Figure 6.8 shows the performance of MP and CC methods in the calculation of the ^{11}B shielding constant in boron monohydride in comparison with FCI. BH is a challenging case with rather large correlation effects of about 80 ppm. (The BH results have been reported in [55], [57], and [148]. Note that extensive MCSCF calculations for BH (within the MC–

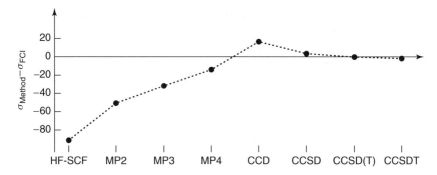

Figure 6.8. ^{11}B shielding constant of BH calculated at different levels of theory in comparison to the corresponding FCI result (for further details, see [148]).

IGLO framework) have been reported by van Wüllen [149].) The MP series exhibits a slow monotonic convergence in this case; MP2 recovers less than 50% of the correlation contribution to the ^{11}B shielding. On the other hand, the CC methods yields results very close to the exact value; residual deviations are 3.4 ppm (CCSD), 0.4 ppm (CCSD(T)), and 1.7 ppm (CCSDT). It is interesting that the full CCSDT result [57] disagrees more with the FCI shielding than does that obtained at the CCSD(T) level. Since CCSD(T) is an approximation to CCSDT, this clearly results from fortuitous error cancellation. Specifically, this involves overestimation of triple excitation contributions by CCSD(T) and its neglect of quadruple excitation. On the other hand, the somewhat inferior performance of CCD (16.4 ppm deviation from FCI) points to the importance of including single excitations in the cluster operator, as these are *essential* for the accurate description of the paramagnetic part of the shielding.

BH is actually atypical. For most other molecules, the MP series characteristically exhibits an oscillatory behavior [92], as found, for example, in a previous study of the NNO molecule [52]. Unfortunately, however, no other FCI shieldings (apart from H_2) are available and comparisons with experimental data are necessary for further validation of the methods. Such a comparison is shown in Figure 6.9 for N_2. Again, this is a case with large correlation effects of about 55 ppm. Second-order MP theory overestimates the correlation contribution by nearly 20 ppm (36%), while MP3 underestimates it by 15 ppm (27%). Fourth-order MP again overestimates the correlation corrections, in particular when triple excitations are included. On the other hand, the CC results are very accurate, thus lending support to the general belief that near quantitative accuracy is achieved with these methods.

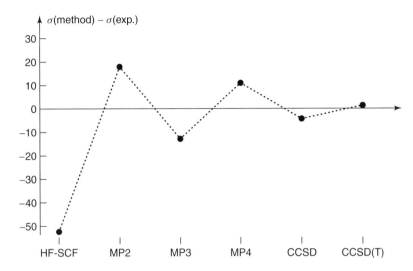

Figure 6.9. ^{15}N shielding constant of N_2 calculated at different levels of theory in comparison to the corresponding experimental equilibrium value.

Other electron-correlated methods do not perform as well for N_2. SOPPA calculations [150] significantly underestimate correlation effects (the values are in the range of -72 to -82 ppm), while MCSCF calculations [51, 113] give results that exhibit a rather strong dependence on the chosen active space.

A more comprehensive comparison of the different methods is given in Table VI.3 where shielding constants computed for HF, H_2O, NH_3, CH_4, N_2, CO, and F_2 are collected. The results support our previous conclusions. Oscillations are seen in the MP calculations, while the CCSD(T) calculations are in satisfactory agreement with experiment. The GIAO–MCSCF results in Table VI.3 show that for simple cases such as HF and H_2O good agreement between the different electron correlation treatments is obtained, while discrepancies exist for the more challenging cases N_2, CO, and F_2. In particular for F_2, the GIAO–MCSCF calculations in [51] reveal that convergence with respect to the chosen active space is extremely slow (cf. Table VI.4). As this problem is even more pronounced for larger molecules, it seems that results obtained in MCSCF calculations should be regarded with caution. Having said that, there are on the other hand cases where static correlation is important and where MCSCF methods thus are appropriate (see, e.g., calculations reported in [113]). One example is the ozone molecule. Both HF–SCF and MP calculations are an unmitigated disaster in this case, but MC–IGLO and GIAO–MCSCF yield qualitatively

Table VI.3
Comparison of Nuclear Magnetic Shielding Constants (in ppm) Calculated at HF–SCF, MP2, MP3, MP4, CCSD, CCSD(T), and MCSCF Levels of Theory Using the GIAO Ansatz[a]

Molecule	Nucleus	HF–SCF	MP2	MP3	MP4	CCSD	CCSD(T)	MCSCF	Experimental σ_e[b]
HF	^{19}F	413.6	424.2	417.8	419.9	418.1	418.6	419.6	420.0 ± 1.0
	^{1}H	28.4	28.9	29.1	29.2	29.1	29.2	28.5	28.90 ± 0.01
H$_2$O	^{17}O	328.1	346.1	336.7	339.7	336.9	337.9	335.3	357.6 ± 17.2
	^{1}H	30.7	30.7	30.9	30.9	30.9	30.9	30.2	
NH$_3$	^{15}N	262.3	276.5	270.1	271.8	269.7	270.7		273.3 ± 0.3
	^{1}H	31.7	31.4	31.6	31.6	31.6	31.6		
CH$_4$	^{13}C	194.8	201.0	198.8	199.5	198.7	199.1	198.2	198.4 ± 0.9
	^{1}H	31.7	31.4	31.5	31.5	31.5	31.5	31.3	
CO	^{13}C	−25.5	10.6	−4.2	12.7	0.8	5.6	8.2	3.29 ± 0.9
	^{17}O	−87.7	−46.5	−68.3	−44.0	−56.0	−52.9	−38.9	−38.7 ± 17.2
N$_2$	^{15}N	−112.4	−41.6	−72.2	−48.5	−53.9	−58.1	−53.1	−57.3
F$_2$	^{19}F	−167.9	−170.0	−176.9	−180.7	−171.1	−186.5	−136.6	−196.0 ± 1.0

[a] HF–SCF, MP2, CCSD, CCSD(T) results from [55], MP3 results from [50], MP4 results from [151], and MCSCF results from [51].
[b] Experimental equilibrium values for HF, CO, and N$_2$ derived from σ_0 values and rovibrational corrections given in [66], σ_e values for H$_2$O, NH$_3$, and CH$_4$ as compiled in [55].

Table VI.4
Convergence of GIAO–MCSCF Calculations of the ^{19}F Shielding (in ppm) for F_2 with Respect to the Chosen Active Space[a]

Active Space	^{19}F Shielding
HF	−167.3
Full-valence CAS	−205.5
CAS(32202220)	−145.1
CAS(32203220)	−136.6
Exp. σ_0	−231.6 ± 1.0
Exp. σ_e	−196.9 ± 1.0

[a]Computational results from [51], experimental values taken from [66].

correct descriptions [112, 152] in satisfactory agreement with experiment [153, 154] as do the CC treatments (see Table IV.5). Note that satisfactory results for O_3 are also obtained in DFT calculations [78].

The performance of DFT for the prediction of absolute shieldings is documented in Table VI.6, where we compare results from uncoupled and coupled DFT calculations with those obtained using the more empirical schemes of Malkin et al. [78] and Wilson et al. [79]. When comparing those results with the best available ab initio data and σ_e values derived from

Table VI.5
Comparison of Computed and Experimental ^{17}O Nuclear Magnetic Shielding Constants (in ppm) of Ozone

Method	$O_{terminal}$	$O_{central}$	Reference
IGLO	−2814	−2929	112
GIAO–HF	−2862	−2768	55
MC–IGLO	−1153	−658	112
GIAO–MCSCF	−1126	−703	152
GIAO–MP2	1248	2875	55
GIAO–CCSD	−1403	−968	55
GIAO–CCSD(T)	−1183	−724	55
GIAO–CCSDT-1	−927	−415	57
GIAO–CCSDT	−1261	−775	57
SOS–DFPT	−1230	−753	78
Experimental	−1290	−724	153
	−1289(170)	625(240)	154

Table VI.6
Comparison of Calculated Nuclear Magnetic Shielding Constants (in ppm) Using Various DFT Variants with the Best Available Ab Initio Values as Well as Experimental σ_e Values

Molecule	Nucleus	HF–SCF[a]	Uncoupled DFT[a]	Coupled DFT[a]	SOS–DFPT[b]	Modified DFT[c]	Best Ab Initio[d]	Experimental σ_e^e
HF	^{19}F	410.4	405.1	404.0	406.9	417.6	419.6	420.0 ± 1.0
H_2O	^{17}O	320.5	317.9	316.6	323.2	334.5	337.9	357.6 ± 17.2
CH_4	^{13}C	193.4	184.3	182.9	191.2	198.9	198.9	198.4 ± 0.9
CO	^{13}C	−23.7	−15.4	−20.2	3.2	4.8	5.6	3.29 ± 0.9
	^{17}O	−84.3	−77.1	−83.7	−51.6	−44.2	−52.9	−38.7 ± 17.2
N_2	^{15}N	−110.0	−84.8	−90.4	−66.0	−56.0	−58.1	−57.3
F_2	^{19}F	−167.3	271.7	−289.9	−227.7	−186.3	−189.3	−196.0 ± 1.0

[a]HF–SCF, uncoupled and coupled DFT results from [61].
[b]SOS–DFPT results from [116].
[c]Modified DFT (DFT–B3LYP$_{GGA}^{0.05}$) results from [79].
[d]CCSD(T) results from [55] and [66].
[e]Values for HF, CO, N_2, and F_2 obtained from σ_0 values and rovibrational corrections given in [66]; values for H_2O, NH_3, and CH_4 as compiled in [55].

experiment, several interesting observations are made. First, it appears that calculated DFT shielding constants are not particularly accurate and sometimes show large deviations compared to the given reference data. It further appears that the correlation corrections obtained in the DFT calculations often *have the wrong sign*. Nevertheless, for the more challenging cases, DFT results appear to be superior to corresponding HF–SCF results. Shieldings from uncoupled DFT and coupled DFT calculations differ only marginally, with the (unphysical) uncoupled approach somewhat better. Significant improvements, however, are achieved in the empirically corrected schemes, which is especially true for more difficult molecules such as CO, N_2, and F_2. However, it should be noted that these molecules were among those used to adjust the empirical parameters in the scheme by Wilson. Hence, the excellent performance of that scheme is to be expected.

To conclude the discussion on DFT NMR chemical shift calculations, we note that the currently available DFT schemes do not provide highly accurate shieldings; there is still a need for further improvements. These should include not just the usual development and testing of new exchange-correlation functionals but also a deeper understanding of the theoretical basis for calculating shieldings within the DFT framework. A recent study by Wilson and Tozer [155] about the possibility of predicting accurate shieldings directly from available electron densities might be of some relevance to this discussion.

While highly accurate calculations of NMR chemical shifts with explicit inclusion of electron correlation are essential for benchmark studies, such calculations are also important for the accurate determination of absolute NMR scales [66]. As mentioned earlier in this chapter, absolute NMR shielding constants are not determined by means of usual NMR measurements. Instead, the standard procedure (see Fig. 6.10) uses spin-rotation constants $\mathcal{M}$ determined from analysis of experimental microwave spectra and exploits the relationship between these constants and the paramagnetic part of the shielding [156, 157]. As seen in Fig. 6.10, the actual procedure is quite complicated. After determination of the spin-rotation constants, these have to be corrected for rovibrational corrections (preferably from calculations, see [158–160]). Then, the equilibrium values for the spin-rotation constants are converted to the paramagnetic part of the shielding (with the corresponding nucleus as gauge origin) using

$$\sigma_{\text{para}} = \frac{1}{2\gamma}(\mathcal{M} - \mathcal{M}_{\text{nuc}})\mathbf{I} \qquad (6.75)$$

where $\mathcal{M}_{\text{nuc}}$ is the nuclear contribution to $\mathcal{M}$, $\mathbf{I}$ is the inertia tensor of the molecule, and γ the gyromagnetic ratio for the nucleus under study. The

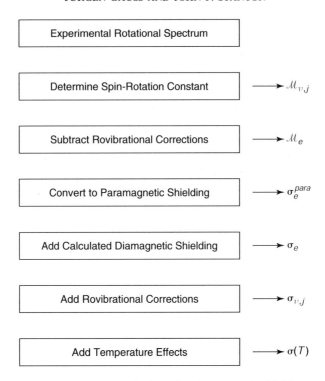

Figure 6.10. Flowchart for the derivation of nuclear magnetic shielding constants from experimental spin-rotation constants.

paramagnetic shielding is then combined with (calculated) values of the diamagnetic part of σ and augmented by rovibrational and temperature corrections to obtain values for the shielding constants at the given temperature. The calculation of the diamagnetic contribution via the corresponding expectation value is generally considered unproblematic and can easily be carried out with sufficient accuracy. The rovibrational corrections are also best determined by calculations, as they are not as easily extracted from the experimental data (usually via the temperature dependence of the shieldings). It thus turns out that the "experimental" determination of absolute shieldings involves a number of steps for which theoretical (computational) information is required. In addition, it is in our opinion important to verify the experimental NMR scales derived in the manner described above by strictly computational results. While low-level treatments are not sufficiently accurate for this purpose, the GIAO–CCSD(T) method provides a serviceable approach for such investigations. Table VI.7

summarizes efforts in this direction. The rovibrational corrections reported there have been obtained by solving the rovibrational Schrödinger equation using finite-element techniques and CCSD(T) potential curves followed by integration over the CCSD(T) shielding functions [66]. The temperature correction is then computed by standard Boltzmann summation. As seen in Table VI.7, the rovibrational corrections are significant ranging from a few tenths of a ppm for hydrogens to about several ppm for the other nuclei. A special case is F_2 for which a particularly large value of -30.9 ppm is found for the rovibrational correction. It is obvious that these effects must be considered when aiming for highly accurate values for absolute shielding constants. On the other hand, temperature effects are typically an order of magnitude smaller at 300 K. A special case is again F_2 where the largest thermal correction (-4.7 ppm) has been predicted. Temperature effects are thus essential for the explanation of the experimental NMR chemical shift of the fluorine molecule, but less essential in other cases and thus can be neglected in most cases.

Table VI.7 also summarizes the shielding constants derived in [66] from experimental spin-rotation constants using computed rovibrational corrections. These values are compared with earlier (less accurate) literature values as well as pure computational results from large basis-set CC calculations. The agreement between these newly determined "experimental" and the best available theoretical values is excellent and lends further support to the high accuracy provided by the CC treatment. A larger discrepancy, though within

Table VI.7
Calculated Rovibrational Contributions, Temperature Corrections, and Nuclear Magnetic Shielding Constants at 300 K [$\sigma(300\,\text{K})$] in Comparison with Corresponding Shielding Constants Derived from Experimental Spin-Rotation Constants[a,b]

Molecule	Nucleus	σ_e	$\sigma_0 - \sigma_e$	$\sigma(300\,\text{K}) - \sigma_0$	$\sigma(300\,\text{K},\text{Theor.})$	$\sigma(300\,\text{K},\text{Exp.})$
H_2	^{1}H	26.667	-0.355	-0.014	26.298	26.288 ± 0.002
HF	^{1}H	28.84	-0.323	-0.035	28.48	28.54 ± 0.01
	^{19}F	419.58	-10.00	-0.42	409.2	409.6 ± 1.0
CO	^{13}C	5.29	-2.24	-0.15	2.9	0.9 ± 2.0
	^{17}O	-53.26	-5.73	-0.35	-59.3	-44.8 ± 17.2
N_2	^{15}N	-58.42	-4.03	-0.24	-62.7	-104.8 ± 19.3^c -61.6^d
F_2	^{19}F	-190.08	-30.87	-4.69	-225.5	-233.0 ± 1.0

[a]All quantities are in parts per million (ppm).
[b]CCSD(T) calculations as described in [66]; experimental values as given in [66].
[c]Based on spin-rotation constants determined for N_2.
[d]Based on spin-rotation constants determined for NH_3.

the experimental error bar, is only noted for the ^{17}O scale for which the theoretical values are about 15 ppm lower than the experimental values. This discrepancy can be clearly attributed to the uncertaintity in the measured spin-rotation constant for ^{12}C^{17}O (obtained from radioastronomical data [161]) and is enlarged by the conversion to shieldings in Eq. (6.75). Thus, at least for ^{17}O, the use of improved absolute NMR scales based on calculations [66, 162] is strongly recommended and more and more accepted in the literature [163]. In addition, the calculations clearly reveal that the spin-rotation constants reported for N_2 is incorrect, as the derived shielding constants differ significantly from those obtained from calculations as well as those using a ^{15}N scale based on spin-rotation constants for ammonia. Furthermore, the data in [66] indicates that accurate shielding scales are preferably derived from spin-rotation constants for compounds containing hydrogens, as there the scaling factor in Eq. (6.75) is smaller due to the smaller moment of inertia and thus does not magnify uncertainties associated with the experimental spin-rotation constants.

We finally note that a more clear-cut connection with experiment can be established by directly calculating spin-rotation constants (via second derivatives with respect to the total angular momentum and the nuclear spin of interest) instead of shieldings. Such calculations have been reported in the literature [77, 158–160, 164, 165]. However, as demonstrated in [158], calculation of spin-rotation constants are hampered by a dependence on the choice of origin for the electronic angular momentum. In analogy to NMR chemical shift calculations, this problem can be cured by using explicitly perturbation-dependent basis functions of the following type [158]:

$$\chi_\mu(\mathbf{J}) = \exp(i\mathbf{I}^{-1}(\mathbf{J} \times \mathbf{R}_\mu) \cdot \mathbf{r})\chi_\mu \tag{6.76}$$

The latter ensure origin-independent results as well as fast convergence to the basis set limit in calculations of spin-rotation constants [158]. Table VI.8 compares calculated spin-rotation constants (including rovibrational corrections) with available experimental values for some diatomic molecules. We also note that the discussion of spin-rotation constants in [158] leads to well-defined expressions for para- and diamagnetic parts of the shieldings within the GIAO ansatz. In order to keep the close relationship between the paramagnetic part of the shielding and spin-rotation constants, the diamagnetic part is best defined in the same manner as in common gauge-origin calculations as the expectation value of the corresponding second-order part of the Hamiltonian. The paramagnetic part is then obtained as the difference between the total shielding (computed with GIAOs) and the so defined diamagnetic contribution [158].

Table VI.8
Calculated Spin-Rotation Constants Including Rovibrational Corrections (in kHz) for H_2, HF, CO, N_2, and F_2 as Obtained in Large Basis-Set CCSD(T) Calculations in Comparison with Corresponding Experimental Values[a]

Molecule	Nucleus	Calculation	Experiment
H_2	^1H	114.064	113.904 ± 0.030
HF	^1H	72.08	71.10 ± 0.02
	^{19}F	309.92	307.65 ± 0.02
CO	^{13}C	−32.44	−32.70 ± 0.12
	^{17}O	−31.32	−30.4 ± 1.2
N_2	^{15}N	19.85	22.0 ± 1.0
F_2	^{19}F	−154.80	−156.85 ± 1.0

[a] For further computational details as well as the used compilation of experimental values see [159].

B. Chemical Applications and Relative NMR Chemical Shifts

The quantities that are almost always measured in NMR experiments are relative chemical shifts, and it is consequently more relevant to calculate these than the absolute nuclear magnetic shielding constants. Though relative shifts of course differ from absolute shieldings only by sign and an additive factor, some important distinctions exist between the computational determination of these quantities. Relative shifts are somewhat simpler to calculate accurately, because they involve taking differences between shieldings calculated for both the compound of interest and the reference compound. Hence, as effects due to electron correlation, vibrational corrections, and basis set deficiencies tend to have the same sign and to be of similar magnitudes for different compounds, a partial cancellation of errors occurs. This cancellation of largely systematic errors also is the main reason why HF–SCF calculations of NMR chemical shifts have proven useful, even in cases where electron correlation contributions to the *absolute* shieldings are relatively pronounced. However, despite these successes, we will demonstrate the importance of electron correlation effects in the following and show that their inclusion is often essential in order to make a definitive analysis of the experimental spectrum.

We begin with a benchmark study of ^{13}C relative NMR chemical shifts in which calculated values are compared to experimental gas-phase results from the work of Jameson and Jameson [166]. The set of molecules here has been used in a number of studies [47, 52, 60, 116, 167] in order to calibrate the accuracy of different theoretical approaches. Figure 6.11

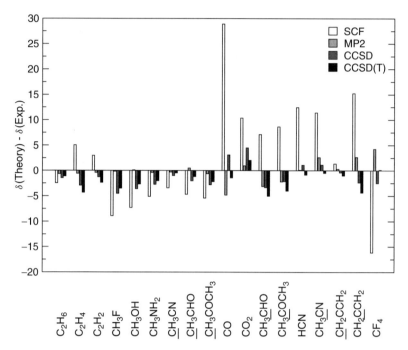

Figure 6.11. Calculated relative ^{13}C NMR shifts (in ppm, vs. CH$_4$) as obtained at HF–SCF, MP2, CCSD, and CCSD(T) level in comparison with experimental gas-phase data [166].

compares results obtained at the HF–SCF, MP2, CCSD, and CCSD(T) level using the GIAO ansatz together with a large polarized quadruple zeta basis. It is seen that the HF–SCF results follow a decided pattern; carbons within methyl groups that are in singly bonded environments have the smallest errors (5–10 ppm), while those involved in multiple bonds to oxygen (CO, CO$_2$, carbonyls) or nitrogen (cyano groups) have errors that are about three times larger. That deviations are primarily due to neglect of electron correlation is clearly seen from the remaining calculations. Deviations from experiment are significantly reduced at all of the correlated levels; standard deviations are 1.4 ppm (MP2), 2.4 ppm (CCSD), and 2.2 ppm (CCSD(T), while that given by HF–SCF is 8.7 ppm. Note that there is a curiosity: The MP2 results appear to be better than both CCSD and CCSD(T)! This suggests that MP2 results benefit from some sort of fortuitous error cancellation. Indeed, as shown in [167], use of large basis sets together with consideration of zero-point vibrational effects are required in order to discriminate between the intrinsic accuracy of these methods.

When these are included, the superiority of CCSD(T) to MP2 is apparent [167].

Figure 6.12 shows for the same set of molecules the performance of various DFT schemes for the prediction of chemical shifts [167]. As it is seen the DFT results are less satisfactory than those obtained with the more traditional schemes for treating electron correlation. Some improvements compared to HF-SCF are noted, especially for the challenging cases, but the DFT calculations do not reach the same accuracy as the corresponding MP2 calculations.

The first chemical applications to show the importance of electron correlation effects for NMR chemical shift calculations dealt with the chemistry of boranes and carboranes, which has often been a fertile area for theoretical insights. Since ^{11}B NMR is one of the main tools for establishing molecular structures in this area, there is accordingly great interest in the accurate prediction of chemical shifts for these compounds. Bühl and Schleyer [13] found that uncorrelated calculations are sufficient in most

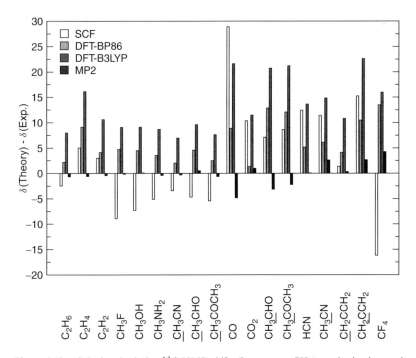

Figure 6.12. Calculated relative ^{13}C NMR shifts (in ppm, vs. CH_4) as obtained at various DFT levels in comparison with the corresponding MP2 values and experimental gas-phase data [166].

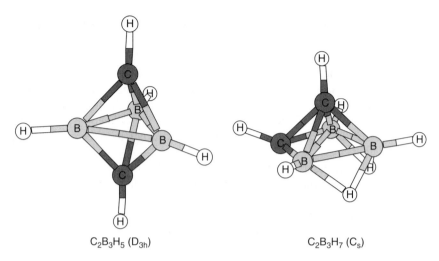

Figure 6.13. Optimized MP2/6-31G* structures (from [13]) for the carboranes $C_2B_3H_5$ and $C_2B_3H_7$.

cases in a study using the IGLO approach. The mean deviation between experiment and calculated shifts was less than 3 ppm for the boron shifts. However, significantly larger deviations have been observed in a few cases. For example, in $C_2B_3H_5$ (see Fig. 6.13) the discrepancy between theory and experiment was about 10 ppm, while the deviation was nearly 10 ppm for $C_2B_3H_7$ (see Fig. 6.13). As the structure for the former system had been well established, Bühl and Schleyer [13] speculated that the discrepancy (in both cases) was due to the neglect of electron correlation.

Subsequent GIAO–MP2 calculations [14] confirmed this conjecture in case of $C_2B_3H_5$ and gave excellent agreement with the experimental data for both the ^{11}B and ^{13}C NMR chemical shifts (see Table VI.9). However, for the second problematic case, the GIAO–MP2 calculations indicated that — at least for boron — electron correlation effects are small. Thus, a different explanation for the discrepancy was needed. Since a search for alternative isomers was unsuccessful, the accuracy of the experimental data came into question [14]. Greatrex and co-workers [15] subsequently resynthesized $C_2B_3H_7$ and measured the ^{11}B and ^{13}C NMR chemical spectra. Their spectra turned out to be in excellent agreement with the theoretical predictions from [13] and [14], but disagreed with the older NMR data from [169]. The superiority of the GIAO–MP2 calculations in the case of $C_2B_3H_7$ is seen in the ^{13}C NMR chemical shifts for which the HF–SCF results show the expected deviations of about 6–9 ppm.

Table VI.9

Comparison of Calculated and Experimental ^{11}B and ^{13}C NMR Chemical Shifts (in ppm, ^{11}B Shifts Relative to BF$_3$. OEt$_2$, ^{13}C Shifts Relative to Tetramethylsilane (TMS)) for the Carboranes C$_2$B$_3$H$_5$ (D_{3h} Symmetry) and C$_2$B$_3$H$_7$ (C_s) and the Tetraboranes(4) B$_4$R$_4$ with R = H, Cl, Methyl, and tert-Butyl

(a) C$_2$B$_3$H$_5$

	^{13}C	^{11}B
GIAO–HF–SCF[a]	95.4	11.8
GIAO–MP2[a]	104.2	1.9
Experiment	103.3	1.4[b], 3.5[c]

[a] Reference [14].
[b] Reference [168].
[c] R. Greatrex and M. A. Fox (1993), see [15].

(b) C$_2$B$_3$H$_7$

	^{13}C		^{11}B	
GIAO–HF–SCF[d]	−30.5	−51.7	−13.3	−14.7
GIAO–MP2[d]	−23.1	−57.4	−14.6	−16.0
Experiment 1972[e]			−21.7	−23.7
Experiment 1993[f]	−21.5	−57.9	−13.4	−15.1

[d] Reference [14].
[e] Reference [169].
[f] R. Greatrex and M. A. Fox (1993), see [15].

(c) B$_4$R$_4$ with R = H, Cl, Me, t-Bu

	GIAO–HF–SCF[g]	GIAO–MP2[g]	Experiment
R = H	200.9	151.2	
R = Methyl	170.0	133.9	
R = tert-Butyl	167.7	130.7	135.4[h]
R = Cl	116.6	81.4	85.5[i]

[g] Reference [129], B$_4$Cl$_4$ results from [14].
[h] Reference [170].
[i] Reference [171].

Unusually large correlation effects of 35–50 ppm are predicted for the tetrahedral tetraboranes(4) B$_4$R$_4$ with R = H, Me, t-Bu, and Cl. Comparison with experiment shows for these compounds that uncorrelated calculations are completely unreliable, while GIAO–MP2 calculations satisfactorily reproduce the experimental NMR chemical shifts [14, 129]. The GIAO–MP2 calculation for B$_4$t-Bu$_4$ (T_d symmetry) with 616 basis functions had been possible due to the integral-direct implementations [128, 129] discussed in Section III.D.

Substantial correlation effects are also observed in ^{13}C chemical shift calculations for carbocations. As pointed out by Lipscomb long ago, there is a formal similarity between carbocations and monoboranes since they are isoelectronic [172]. Prominent examples include the phenonium and benzenonium cations [105], allyl cations [107], as well as vinyl cations [109, 173–175]. In case of the phenonium ion, a decided improvement with respect to HF–SCF is achieved at the GIAO–MP2 level, thus providing further support for the proposed C_{2v} structure (see Fig. 6.14 for comparison of the corresponding GIAO–HF–SCF and GIAO–MP2 results with experimental data [177]). Significant electron correlation effects have also been found for the benzenonium cation ($C_6H_7^+$) [105] as well as the nahpthalenium ($C_{10}H_9^+$) and anthracenium ions ($C_{14}H_{11}^+$) [128].

In the case of the allyl cations, substantial deviations between the HF–SCF data and the available experimental ^{13}C shifts have been noted [107]. Again, GIAO–MP2 calculations provide satisfactory agreement with the experimental data (cf. Table VI.10) and for the first time enabled a reliable prediction of the NMR chemical shifts of the parent system. Indeed, a previous claim of experimental observation of this important prototype molecule [178] was shown to be incorrect. To date, it seems that the latter has been characterized only by infrared (IR) spectroscopy [107]; no NMR evidence for its existence has been found thus far. In Table VI.10, we also list some GIAO–LMP2 data from [130], which are in excellent agreement with the original GIAO–MP2 results, exemplifying the general behavior that was alluded to in Section III.

Dramatic correlation effects are observed in calculations for vinyl cations, which is not particularly surprising, since there is a low-lying empty p orbital on the carbocation center. While the parent system has only been detected by elaborate spectroscopic techniques [179], vinyl cations stabilized by appropriate substituents have been synthesized. These have been characterized by low-temperature ^{13}C NMR spectroscopy using special techniques [180]. A particularly interesting system in this regard is the 1-cyclopropylcyclopropylidenemethyl cation ($C_7H_9^+$, see Fig. 6.15), where the stabilization is conferred entirely by three-membered ring substituents. When the NMR spectrum of $C_7H_9^+$ was first observed [173], quantum chemical calculations were unable to satisfactorily reproduce the spectrum. GIAO–HF–SCF and GIAO–MP2 calculations both turned out insufficient (see Fig. 6.15). The substantial disagreement between calculated and observed NMR chemical shifts for the α carbon (carbocation center) was subsequently attributed to solvent effects [181]. The low-temperature NMR spectrum was obtained in frozen superacid solution, and the authors of

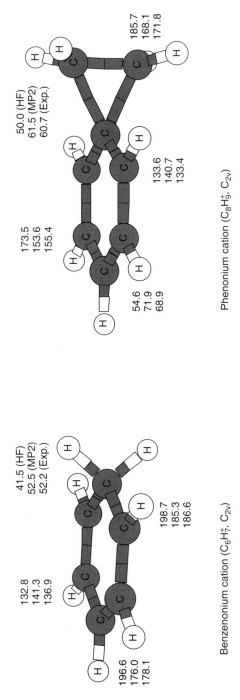

Figure 6.14. Electron correlation effects on the calculated ^{13}C NMR chemical shifts of the benzenonium and phenonium cation. Computational results from [105], experimental values from [176, 177].

409

Table VI.10
Calculated ^{13}C NMR Chemical Shifts (in ppm, with Respect to TMS) for Various Allyl Cations in Comparison to Experiment[a]

Nucleus	GIAO–HF–SCF	GIAO–MP2	GIAO–LMP2	Experimental
Allyl cation				
CH	236.5	227.4		
CH$_2$	142.8	152.8		
1-Methyl allyl cation				
CH	141.3	151.1	150.6	149.8
CH$_2$	217.5	205.5	205.9	201.5
CH(Me)	261.5	255.1	255.5	255.1
CH$_3$	31.1	38.2	38.4	36.5
1,1-Dimethyl allyl cation				
CH	137.3	147.1	146.4	146.0
CH$_2$	195.3	181.2	181.1	175.0
C(CH$_3$)$_2$	279.9	275.1	275.5	274.3
CH$_3$ (cis)	28.7	33.7	33.7	33.1
CH$_3$ (trans)	35.9	42.6	42.5	41.4
1,3-Dimethyl allyl cation				
CH	138.9	149.5	148.8	147.0
CH(Me)	244.2	234.9	235.1	231.3
CH$_3$	27.9	33.7	34.9	29.8
1,1,3,3-Tetramethyl allyl cation				
CH	132.2	145.4	144.1	141.4
C(CH$_3$)$_2$	249.7	239.1	239.6	233.7
CH$_3$ (cis)	29.0	34.5	34.4	
CH$_3$ (trans)	36.2	42.1	41.8	

[a]Computational results from [107] and [130], experimental results as compiled in [107].

[181] suggested that the optimized geometry used in the calculation (which features a nearly linear arrangement of the carbon atoms) differed qualitatively from that in solution. However, high-level coupled-cluster calculations (which were the first chemical application of these techniques) resolved the issue (see Fig. 6.15) [109]. At the GIAO–CCSD(T) level, deviations between experiment and theory for the α carbon are reduced to 1 ppm. Hence, this is a case where coupled-cluster methods are required to obtain satisfactory agreement with experiment.

A related example is the vinyl-substituted vinyl cation

$$H_2C=C^+-CH=CHCH_3$$

which has been recently prepared by Siehl et al. [175] in superacid solution at low temperature and characterized by NMR spectroscopy. The ^{13}C

CALCULATION OF NMR CHEMICAL SHIFTS

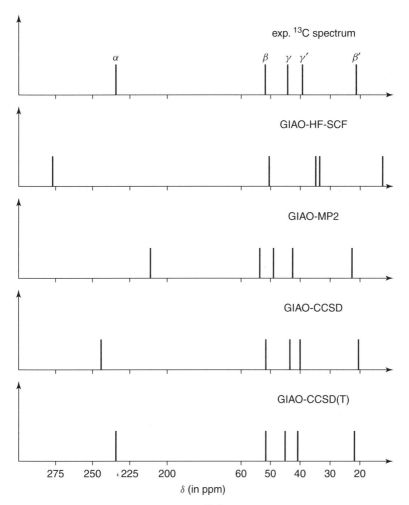

Figure 6.15. Calculated and experimental ^{13}C NMR spectra [109] for the 1-cyclopropyl-cyclopropylidenemethyl cation.

NMR spectrum (Fig. 6.16) shows that two geometrical isomers are formed, since the spectrum exhibits five *pairs* of signals. A challenging task of quantum chemistry is now to assign the peaks in the spectrum to the two isomers, a task that would probably be difficult from experimental alone. Figure 6.16 shows spectra calculated at HF–SCF, MP2, CCSD, and CCSD(T) levels along with the experimental spectrum. Since the HF–SCF calculations give rather poor agreement with experiment (the deviations are in the 40 ppm range), any efforts to distinguish between the two isomers of $H_2C=C^+—CH=CHCH_3$ would be ill-advised at this level of theory. On the other hand, the correlated calculations are in essentially perfect agreement with experiment. It is therefore straightforward to assign the signals to the (Z) and (E) isomers of this compound, as is done in Figure 6.16. Note that the MP2 and CCSD(T) results are somewhat superior to the CCSD values, which is qualitatively in accord with what was found in the ^{13}C benchmark study. The results for the vinyl cations clearly demonstrate the accuracy provided by state-of-the-art coupled-cluster calculations of NMR chemical shifts. They provide results that are sufficient for the unequivocal assignment of NMR signals that differ by only a few parts per million for specific geometrical isomers. In this way, theory can provide useful insights into molecular structure when used in concert with experimental NMR spectra.

Another interesting case is the 10-annulene molecule ($C_{10}H_{10}$). The cyclic form of this molecule has been the subject of considerable attention from theory [182–184] since it has 10 π electrons and is a potentially aromatic species (it is the prototype $n = 2$ molecule for the $4n + 2$ rule). However, calculations over the years have shown that the symmetric D_{10h} structure of this molecule is not geometrically stable. Calculations by Schleyer and co-workers [184] showed that although a large amount of "resonance energy" is indeed found for this hypothetical cyclic structure, it is more than offset by unfavorable ring strain (CCC bond angles in the regular decagonal structure would be 144°). Hence, less symmetric geometries are preferred for this interesting compound. Computational studies show that three structures of this compound are energetically most favorable. One of these is a twisted structure of C_2 symmetry, another is an interesting "heart" form in which one of the hydrogen atoms lies *inside* the ring, and a third is a "bicyclic" form that can be viewed as a dihydronaphthalene (see Fig. 6.17). Both DFT and MP2 calculations favor the heart isomer [183], but high-level CCSD(T) calculations indicate otherwise [182, 185]. These predict that the "twist" is actually the favored form. However, the magnitude of the relative stability found at this level of theory (all forms lie within 5 kcal/mol) is certainly not accurate enough to draw definitive conclusions. The only spectroscopic characterization of 10-annulene comes from a

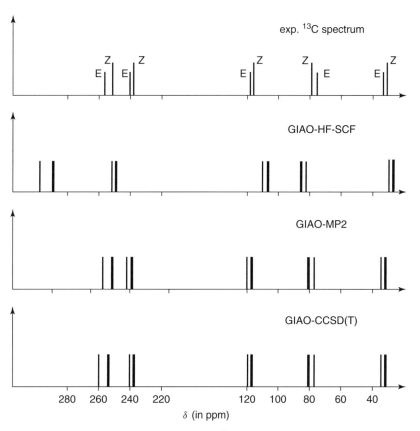

Figure 6.16. Assignment of the experimental ^{13}C NMR spectra [175] to the (Z) and (E) isomers of $CH_2=C^+-CH=CHCH_3$ on the basis of NMR chemical shift calculations. Thick lines correspond to the (Z) isomer in the calculations, thin lines to the (E) isomer.

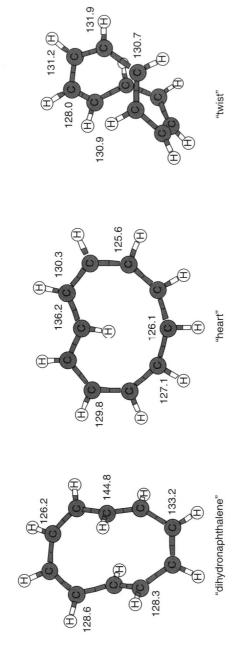

Figure 6.17. Structures and calculated ^{13}C NMR chemical shifts (GIAO–CCSD(T), in ppm) of the three isomers of 10-annulene [111]. Experimental ^{13}C NMR chemical shifts from [186].

exp. ^{13}C shifts: 128.4, 131.5, 131.6, 132.3, 132.5

low-temperature ^{13}C NMR experiment [186]. Here, five distinct transitions are observed, which is consistent with either the "napthalene" form or the "twist" form, assuming that there are no overlapping resonances in the "heart" structure. GIAO calculations for the NMR shifts at the HF–SCF level [184] were found to be most consistent with the "twist" form, but the errors associated with this level of theory for multiply bonded systems are such that only speculative conclusions can be drawn. However, high-level CCSD(T) calculations recently carried out give chemical shifts for the "twist" structure that agree with experiment to within 1.1 ppm for all signals; those calculated for the other two isomers differ from the experimental positions by as much as 12 ppm (dihydronaphthalene) and 4 ppm (heart). Hence, it seems clear that the "twist" structure is indeed the observed form of 10-annulene, a conclusion that would not be firm without the availability of electron-correlated calculations of NMR chemical shifts.

V. SUMMARY AND OUTLOOK

NMR spectroscopy is without doubt the most widely used spectroscopic technique in chemistry, both as a means to detect the presence of familiar molecules as well as to determine structural aspects of so far unknown species. While it has long been appreciated that theory can be a great aid for interpretating experimental spectroscopic data, almost all initial efforts in this direction focused on vibrational spectroscopy. However, the last two decades has seen tremendous advances in the theoretical treatment of other spectroscopic properties, for example, excitation energies and NMR chemical shifts, so that a vast range of spectroscopic investigations can now be reliably supported by quantum chemical calculations.

Since NMR spectroscopy occupies such a prominent position in most areas of chemistry, it is not surprising that the theoretical treatment of NMR parameters has attracted such a large interest in the last two decades. After the introduction of local gauge-origin methods (IGLO, LORG, and GIAO) and the seminal work by Pulay and co-workers [35] in demonstrating the suitability and high efficiency of the GIAO approach this area in quantum chemistry has grown explosively. The incorporation of electron correlation via single-reference approaches (MP and CC theory) or through multireference methods has improved significantly the accuracy of NMR chemical shift calculations. This has turned out essential, as the definitive interpretation of NMR spectra often is not possible at the HF–SCF level and electron-correlated treatments (see Section III) are required. The power of the electron-correlated schemes for chemical shift calculations such as GIAO–MP2 and GIAO–CCSD(T) has been amply demonstrated in many cases and renders them useful and valuable tools for the interpretation of experimental spectra (see Section IV). Focusing on larger molecules, the availability of integral-direct schemes as well as the prospect of an efficient GIAO–LMP2 implementation might soon allow the routine treatment of systems with a hundred carbon atoms or more in a routine fashion. We believe that this technique could be applicable to large molecules and that the applicability of electron-correlated techniques to larger molecules will be the most important challenge in quantum chemical method development during the next decade. Developments such as GIAO–LMP2 will have a significant impact on applied quantum chemistry and in this way will also impact the general chemistry community.

While this chapter has focused on the area of our expertise—the treatment of electron correlation in NMR calculations via MP and CC theories—it should also be mentioned that DFT-based calculations of magnetic properties offer great promise for the treatment of large molecules. However, there are formal problems associated with the way that these calculations are currently carried out, and further theoretical developments in this area are needed. In its present state, errors in NMR shielding constants obtained in DFT calculations can be quite substantial (see again Fig. 6.12) and render definitive interpretation of experimental spectra in some, though not all cases difficult.

Acknowledgments

The authors acknowledge fruitful collaborations as well as discussions with the following colleagues: R. Ahlrichs (Karlsruhe, Germany), A. A. Auer (Mainz, Germany), M. Bühl (Mülheim, Germany), O. Christiansen (Lund, Sweden), M. Häser (Karlsruhe, Germany), T. Helgaker (Oslo, Norway) M. Kollwitz (Karlsruhe, Germany), C. Ochsenfeld (Mainz, Germany), D. Price (Austin, USA), K. Ruud (Tromsø, Norway), P.v.R. Schleyer (Erlangen, Germany, and Athens, USA), H.-U. Siehl (Ulm, Germany), D. Sundholm (Helsinki, Finland), and H.-J. Werner (Stuttgart, Germany). Financial support from the Deutsche Forschungsgemeinschaft (JG), the Fonds der Chemischen Industrie (JG), the National Science Foundation (Young Investigator award to JFS), and the Robert A. Welch Foundation (JFS) is gratefully acknowledged.

References

1. W. Kutzelnigg, U. Fleischer, and M. Schindler, in *NMR Basic Principles and Progress*, edited by P. Diehl, E. Fluck, H. Günther, R. Kosfeld, and J. Seelig (Springer, Berlin, 1990), Vol. 23, p. 165.
2. See articles, in *Nuclear Magnetic Shieldings and Molecular Structure*, edited by J. A. Tossell (NATO ASI Series C 386, Kluwer, Dordrecht, 1993).
3. D. B. Chestnut, *Ann. Rep. NMR Spectrosc.* **29**, 71 (1994).
4. J. Gauss, *Ber. Bunsenges. Phys. Chem.* **99**, 1001 (1995).
5. D. B. Chestnut, in *Reviews in Computational Chemistry*, Vol. 8, edited by K. B. Lipkowitz, and D. B. Boyd (VCH Publishers, New York, 1996), p. 245.

6. U. Fleischer, W. Kutzelnigg, and C. v. Wüllen, in *Encyclopedia of Computational Chemistry*, edited by P.v.R. Schleyer, N. L. Allinger, T. Clark, J. Gasteiger, P. Kollman, H. F. Schaefer, and P. R. Schreiner (Wiley, Chichester, 1998), p. 1827.
7. J. C. Facelli, in *Encyclopedia of Nuclear Magnetic Resonance*, edited by D. M. Frant and R. K. Harris (Wiley, Chichester, 1996), Vol. 7, p. 4327.
8. P. Pulay and J. F. Hinton, in *Encyclopedia of Nuclear Magnetic Resonance*, edited by D. M. Frant and R. K. Harris (Wiley, Chichester, 1996), Vol. 7, p. 4334.
9. A. C. de Dios, *Prog. NMR Spectrosc.* **29**, 229 (1996).
10. M. Bühl, in *Encyclopedia of Computational Chemistry*, edited by P.v.R. Schleyer, N. L. Allinger, T. Clark, J. Gasteiger, P. Kollman, H. F. Schaefer, and P. R. Schreiner (Wiley, Chichester, 1998), p. 1835.
11. T. Helgaker, M. Jaszunski, and K. Ruud, *Chem. Rev.* **99**, 293 (1999).
12. P. Buzek, S. Sieber, and P.v.R. Schleyer, *Chem. Z.* **26**, 116 (1992).
13. M. Bühl and P.v.R. Schleyer, *J. Am. Chem. Soc.* **114**, 477 (1991).
14. M. Bühl, J. Gauss, M. Hofmann, and P.v.R. Schleyer, *J. Am. Chem. Soc.* **115**, 12385 (1993).
15. P.v.R. Schleyer, J. Gauss, M. Bühl, R. Greatrex, and M. Fox, *J. Chem. Soc. Chem. Commun.* 1766 (1993).
16. J. Gauss, U. Schneider, R. Ahlrichs, C. Dohmeier, and H. Schnöckel, *J. Am. Chem. Soc.* **115**, 2402 (1993).
17. C. Dohmeier, H. Schnöckel, C. Robl, U. Schneider, and R. Ahlrichs, *Angew. Chem.* **106**, 225 (1994).
18. C. Klemp, M. Bruns, J. Gauss, U. Häussermann, G. Stößer, L. v. Wüllen, M. Jansen, and H. Schnöckel, *J. Am. Chem. Soc.* **123**, 9099 (2001).
19. U. Schneider, S. Richard, M. M. Kappes, and R. Ahlrichs, *Chem. Phys. Lett.* **210**, 165 (1993).
20. F. H. Hennrich, R. H. Michel, A. Fischer, S. Richard-Schneider, S. Gilb, M. M. Kappes, D. Fuchs, M. Bürk, K. Kobayashi, and S. Nagase, *Angew. Chem. Int. Ed. Engl.* **35**, 1732 (1996).
21. M. Bühl, W. Thiel, H. Jiao, P.v.R. Schleyer, and M. Saunders, *J. Am. Chem. Soc.* **116**, 6005 (1994).
22. R. Born and H. W. Spiess, *Macromolecules*, **28**, 7785 (1995).
23. R. Born and H. W. Spiess, in *NMR Basic Principles and Progress*, edited by P. Diehl, E. Fluck, H. Günther, R. Kosfeld, and J. Seelig (Springer, Berlin, 1997), Vol. 35, p. 1.
24. C. Ochsenfeld, S. P. Brown, I. Schnell, J. Gauss, and H. W. Spiess, *J. Am. Chem. Soc.* **123**, 2597 (2001).
25. C. Ochsenfeld, *Phys. Chem. Chem. Phys.* **2**, 2153 (2000).
26. S. P. Brown, T. Schaller, U. P. Seelbach, F. Koziol, C. Ochsenfeld, F.-G. Klärner, and H. W. Spiess *Angew. Chem. Int. Ed. Engl.* **40**, 717 (2001).
27. N. F. Ramsay, *Phys. Rev.* **78**, 699 (1950).
28. See, for example, H. F. Hameka, *Advanced Quantum Chemistry* (Addison-Wesley, New York, 1963).
29. S. M. Cybulski and D. M. Bishop, *Chem. Phys. Lett.* **250**, 471 (1996); S. M. Cybulski and D. M. Bishop, *J. Chem. Phys.* **106**, 4082 (1996).
30. W. Kutzelnigg, *Isr. J. Chem.* **19**, 193 (1980).
31. M. Schindler and W. Kutzelnigg, *J. Chem. Phys.* **76**, 1919 (1982).
32. A. E. Hansen and T. D. Bouman, *J. Chem. Phys.* **82**, 5036 (1985).
33. H. Hameka, *Mol. Phys.* **1**, 203 (1958); H. Hameka, *Z. Naturforsch. A: Phys. Sci.* **14**, 599 (1959).
34. R. Ditchfield, *J. Chem. Phys.* **56**, 5688 (1972); R. Ditchfield, *Mol. Phys.* **27**, 789 (1974).
35. K. Wolinski, J. F. Hinton, and P. Pulay, *J. Am. Chem. Soc.* **112**, 8251 (1990).

36. T. A. Keith and R. F. W. Bader, *Chem. Phys. Lett.* **194**, 1 (1992).
37. T. A. Keith and R. F. W. Bader, *Chem. Phys. Lett.* **210**, 223 (1993); P. Lazzeretti, M. Malagoli, and R. Zanasi, *Chem. Phys. Lett.* **220**, 299 (1994).
38. F. London, *J. Phys. Radium* **8**, 397 (1937).
39. F. Ribas Prado, C. Giessner-Prettre, J.-P. Daudey, A. Pullmann, F. Young, J. F. Hinton, and D. Harpool, *J. Magn. Reson.* **37**, 431 (1980).
40. H. Fukui, K. Miura, H. Yamazaki, and T. Nosaka, *J. Chem. Phys.* **82**, 1410 (1985).
41. Aces2: J. F. Stanton, J. Gauss, J. D. Watts, W. J. Lauderdale, and R. J. Bartlett, *Int. J. Quantum Chem. Symp.* **26**, 879 (1992).
42. Dalton (Release 1.0): T. Helgaker, H. J. Aa. Jensen, P. Jørgensen, J. Olsen, K. Ruud, H. Ågren, T. Andersen, K. L. Bak, V. Bakken, O. Christiansen, P. Dahle, E. K. Dalskov, T. Enevoldsen, B. Fernandez, H. Heiberg, H. Hettema, D. Jonsson, S. Kirpekar, R. Kobayashi, H. Koch, K. V. Mikkelsen, P. Norman, M. J. Packer, T. Saue, P. R. Taylor, and O. Vahtras, 1997.
43. Gaussian 98: M. J. Frisch et al., Gaussian Inc., Pittsburgh, PA, 1999.
44. Texas: P. Pulay et al. University of Arkansas, Fayetteville, AR, 1990.
45. Turbomole: R. Ahlrichs, M. Bär, M. Häser, H. Horn, and C. Kölmel, *Chem. Phys. Lett.* **162**, 165 (1989).
46. T. Helgaker and P. Jørgensen, *J. Chem. Phys.* **95**, 2595 (1991).
47. J. Gauss, *Chem. Phys. Lett.* **191**, 614 (1992).
48. J. Gauss, *J. Chem. Phys.* **99**, 3629 (1993).
49. H. Fukui, T. Baba, H. Matsuda, and K. Miura, *J. Chem. Phys.* **100**, 6609 (1994).
50. J. Gauss, *Chem. Phys. Lett.* **229**, 198 (1994).
51. K. Ruud, T. Helgaker, R. Kobayashi, P. Jørgensen, K. L. Bak, and H. J. Aa. Jensen, *J. Chem. Phys.* **100**, 8178 (1994).
52. J. Gauss and J. F. Stanton, *J. Chem. Phys.* **102**, 251 (1995).
53. J. Gauss and J. F. Stanton, *J. Chem. Phys.* **103**, 3561 (1995).
54. O. Christiansen, J. Gauss, and J. F. Stanton, *Chem. Phys. Lett.* **266**, 53 (1997).
55. J. Gauss and J. F. Stanton, *J. Chem. Phys.* **104**, 2574 (1996).
56. J. Gauss and J. F. Stanton, *Phys. Chem. Chem. Phys.* **2**, 2047 (2000).
57. J. Gauss, *J. Chem. Phys.*, **116**, 4773 (2002).
58. G. Schreckenbach and T. Ziegler, *J. Phys. Chem.* **99**, 606 (1995).
59. G. Rauhut, S. Puyear, K. Wolinski, and P. Pulay, *J. Phys. Chem.* **100**, 6310 (1996).
60. J. R. Cheeseman, G. W. Trucks, T. A. Keith, and M. J. Frisch, *J. Chem. Phys.* **104**, 5497 (1996).
61. A. M. Lee, N. C. Handy, and S. M. Colwell, *J. Chem. Phys.* **103**, 10095 (1995).
62. T. Helgaker, P. J. Wilson, R. D. Amos, and N. C. Handy, *J. Chem. Phys.* **113**, 2983 (2000).
63. H. Nakatsuji, M. Takashima, and M. Hada, *Chem. Phys. Lett.* **233**, 95 (1995).
64. O. L. Malkina, B. Schimmelpfennig, M. Kaupp, B. A. Hess, P. Chandra, U. Wahlgren, and V. G. Malkin, *Chem. Phys. Lett.* **296**, 93 (1998).
65. J. Vaara, K. Ruud, O. Vahtras, H. Ågren, and J. Jokisaari, *J. Chem. Phys.* **109**, 1212 (1998); J. Vaara, K. Ruud, and O. Vahtras, *J. Chem. Phys.* **111**, 2900 (1999).
66. D. Sundholm, J. Gauss, and A. Schäfer, *J. Chem. Phys.* **105**, 11051 (1996).
67. K. Ruud, P.-O. Åstrand, and P. R. Taylor, *J. Chem. Phys.* **112**, 2668 (2000); K. Ruud, P.-O. Åstrand, and P. R. Taylor, *J. Am. Chem. Soc.* **123**, 4826 (2001).
68. D. Cremer, L. Olsson, F. Reichel, and E. Kraka, *Isr. J. Chem.* **33**, 369 (1993).
69. C. G. Zhan and D. M. Chipman, *J. Chem. Phys.* **110**, 1611 (1999).
70. K. V. Mikkelsen, P. Jørgensen, K. Ruud, and T. Helgaker, *J. Chem. Phys.* **106**, 1170 (1997).
71. R. Cammi, *J. Chem. Phys.* **109**, 3185 (1998); R. Cammi, B. Mennucci, and J. Tomasi, *J. Chem. Phys.* **110**, 7627 (1999).

72. See, for example, L. D. Landau and E. M. Lifshitz, *Lehrbuch der Theoretischen Physik*, Band II (Akademie Verlag, Berlin, 1976).
73. See, for example, A. Abragam, *Principles of Nuclear Magnetism* (Oxford University Press, Oxford, 1961).
74. See, for example, K. Schmidt-Rohr and H. W. Spiess, *Multidimensional Solid-State NMR and Polymers* (Academic Press, New York, 1994).
75. See, for example, the general discussion in: J. Gauss, in *Modern Methods and Algorithms of Quantum Chemistry*, edited by J. Grotendorst (John von Neumann Institute for Computing, Jülich, 2000), p. 509.
76. J. Oddershede, P. Jørgensen, and D. L. Yeager, *Comput. Phys. Rep.* **2**, 33 (1984).
77. J. Oddershede and J. Geertsen, *J. Chem. Phys.* **92**, 6036 (1990); I. Paiderova, J. Komasa, and J. Oddershede, *Mol. Phys.* **72**, 559 (1991).
78. V. G. Malkin, O. L. Malkina, M. E. Casida, and D. R. Salahub, *J. Am. Chem. Soc.* **116**, 5898 (1994).
79. P. J. Wilson, R. D. Amos, and N. C. Handy, *Chem. Phys. Lett.* **312**, 475 (1999).
80. W. Kutzelnigg, *J. Mol. Struct. (THEOCHEM)* **202**, 11 (1989).
81. For a corresponding discussion focusing on coupled-cluster theory, see T. B. Pedersen, H. Koch, and C. Hättig, *J. Chem. Phys.* **110**, 8318 (1999).
82. See, for example, S. T. Epstein, *J. Chem. Phys.* **42**, 2897 (1965).
83. P. Pulay, J. F. Hinton, and K. Wolinski, in [2], p. 243.
84. See, for example, J. F. Stanton and J. Gauss, *Int. Rev. Phys. Chem.* **19**, 61 (2000).
85. See, for example, J. Gauss and J. F. Stanton, *J. Chem. Phys.*, **116**, 1773 (2002).
86. See, for example, P. Pulay, in *Modern Electronic Structure Theory, Part II*, edited by D. R. Yarkony (World Scientific, Singapore, 1995) p. 1191.
87. See, for example, R. M. Sternheimer and H. M. Foley, *Phys. Rev.* **92**, 1460 (1952); A. M. Dalgarno and A. L. Stewart, *Proc. R. Soc. London Ser. A* **247**, 245 (1958).
88. P. Jørgensen and T. Helgaker, *J. Chem. Phys.* **89**, 1560 (1989); T. Helgaker and P. Jørgensen, *Theor. Chim. Acta* **75**, 111 (1989); P. G. Szalay, *Int J. Quantum Chem.* **55**, 151 (1995).
89. N. C. Handy and H. F. Schaefer, *J. Chem. Phys.* **81**, 5031 (1984).
90. L. Adamowicz, W. D. Laidig, and R. J. Bartlett, *Int. J. Quant. Chem. Symp.* **18**, 245 (1984).
91. M. Häser, R. Ahlrichs, H. P. Baron, P. Weis, and H. Horn, *Theor. Chim. Acta* **83**, 455 (1993).
92. For a recent review, see D. Cremer, in *Encyclopedia of Computational Chemistry*, edited by P.v.R. Schleyer, N. L. Allinger, T. Clark, J. Gasteiger, P. Kollman, H. F. Schaefer, and P. R. Schreiner (John Wiley & Sons, Ltd., Chichester, 1998), p. 1706.
93. For recent reviews, see R. J. Bartlett and J. F. Stanton, in *Reviews in Computational Chemistry*, edited by K. B. Lipkowitz and D. B. Boyd (VCH Publisher, New York, 1994), Vol. 5, p. 65; R. J. Bartlett, in *Modern Electronic Structure Theory, Part II*, edited by D. R. Yarkony (World Scientific, Singapore, 1995), p. 1047; T. J. Lee and G. E. Scuseria, in *Quantum Mechanical Electronic Structure Calculations with Chemical Accuracy*, edited by S. R. Langhoff (Kluwer Academic Publisher, Dordrecht, 1995), p. 47; J. Gauss, in *Encyclopedia of Computational Chemistry*, edited by P.v.R. Schleyer, N. L. Allinger, T. Clark, J. Gasteiger, P. Kollman, H. F. Schaefer, and P. R. Schreiner (John Wiley & Sons, Ltd., Chichester, 1998), p. 615.
94. For a review, see I. Shavitt, in *The Electronic Structure of Atoms and Molecules*, edited by H. F. Schaefer (Addison-Wesley, Reading, MA, 1972), p. 189.
95. G. D. Purvis and R. J. Bartlett, *J. Chem. Phys.* **76**, 1910 (1982).
96. K. Raghavachari, G. W. Trucks, J. A. Pople, and M. Head-Gordon, *Chem. Phys. Lett.* **157**, 479 (1989).

97. For reviews, see H.-J. Werner, in *Ab Initio Methods in Quantum Chemistry*, edited by K. P. Lawley (John Wiley & Sons, Inc., New York, 1987), Part II, p. 1: R. Shepard, in *Ab Initio Methods in Quantum Chemistry*, edited by K. P. Lawley (John Wiley & Sons, Inc., New York, 1987), Part II, p. 62: B. O. Roos, in *Ab Initio Methods in Quantum Chemistry*, edited by K. P. Lawley (John Wiley & Sons, Inc., New York, 1987), Part II, p. 399.
98. B. O. Roos and K. Andersson, in *Modern Electronic Structure Theory, Part I*, edited by D. R. Yarkony (World Scientific, Singapore, 1995), p. 55.
99. P. Szalay, in *Modern Ideas in Coupled-Cluster Methods*, edited by R. J. Bartlett (World Scientific, Singapore, 1997), p. 81.
100. R. G. Parr and W. Yang, *Density-Functional Theory of Atoms and Molecules* (Oxford University Press, New York, 1989); W. Koch and M. C. Holthausen, *A Chemist's Guide to Density Functional Theory* (VCH-Wiley, Weinheim, 2000).
101. W. Klopper, in *Encyclopedia of Computational Chemistry*, edited by P.v.R. Schleyer, N. L. Allinger, T. Clark, J. Gasteiger, P. Kollman, H. F. Schaefer, and P. R. Schreiner (John Wiley & Sons, Ltd., Chichester, 1998), p. 2351.
102. *Recent Advances in Quantum Monte Carlo Methods*, edited by W. A. Lester (World Scientific, Singapore, 1998).
103. T. D. Bouman and A. E. Hansen, *Chem. Phys. Lett.* **175**, 292 (1990).
104. E. C. Vauthier, M. Comeau, S. Odiot, and S. Fliszar, *Can. J. Chem.* **66**, 1781 (1988).
105. S. Sieber, P.v.R. Schleyer, and J. Gauss *J. Am. Chem. Soc.* **115**, 6987 (1993).
106. S. Sieber, P.v.R. Schleyer, A. H. Otto, J. Gauss, F. Reichel, and D. Cremer, *J. Phys. Org. Chem.* **6**, 445 (1993).
107. P. Buzek, P.v.R. Schleyer, H. Vancik, Z. Mihalic, and J. Gauss, *Angew. Chem.* **104**, 470 (1994).
108. H.-U. Siehl, M. Fuss, and J. Gauss, *J. Am. Chem. Soc.* **117**, 5983 (1995).
109. J. F. Stanton, J. Gauss, and H.-U. Siehl, *Chem. Phys. Lett.* **262**, 183 (1996).
110. K. Christe, W. W. Wilson, J. A. Sheehy, and J. A. Boatz, *Angew. Chem. Int. Ed. Engl.* **38**, 2004 (1999); the GIAO–CCSD(T) calculations reported in this reference have been carried out by J. F. Stanton (unpublished results).
111. D. R. Price and J. F. Stanton, to be submitted.
112. W. Kutzelnigg, C. v. Wüllen, U. Fleischer, R. Franke, and T. v. Mourik, in [2], p. 141.
113. C. v. Wüllen and W. Kutzelnigg, *Chem. Phys. Lett.* **205**, 563 (1993).
114. C. v. Wüllen and W. Kutzelnigg, *J. Chem. Phys.* **104**, 2330 (1996).
115. V. G. Malkin, O. L. Malkina, and D. R. Salahub, *Chem. Phys. Lett.* **221**, 91 (1994).
116. L. Olsson and D. Cremer, *J. Chem. Phys.* **105**, 8995 (1996).
117. See, for example, J. Gauss and D. Cremer, *Adv. Quant. Chem.* **23**, 205 (1992).
118. R. J. Bartlett, in *Geometrical Derivatives of Energy Surfaces and Molecular Properties*, edited by P. Jørgensen and J. Simons, (NATO ASI Series C 166, Reidel, Dordrecht, 1986), p. 35.
119. J. E. Rice and R. D. Amos, *Chem. Phys. Lett.* **122**, 585 (1985).
120. E. Kraka, J. Gauss, and D. Cremer, *J. Mol. Struct. (THEOCHEM)*, **234**, 95 (1991).
121. R. McWeeny, *Rev. Mod. Phys.* **32**, 335 (1960); *Phys. Rev.* **116**, 1028 (1962); R. M. Stevens, R. M. Pitzer, and W. N. Lipscomb, *J. Chem. Phys.* **38**, 550 (1963); J. Gerratt and I. M. Mills, *J. Chem. Phys.* **49**, 1719 (1968); J. A. Pople, R. Krishnan, H. B. Schlegel, and J. S. Binkley, *Int. J. Quantum. Chem. Symp.* **13**, 225 (1979).
122. N. C. Handy, R. D. Amos, J. F. Gaw, J. E. Rice, and E. D. Simandiras, *Chem. Phys. Lett.*, **120**, 151 (1985).
123. P. Pulay, *Adv. Chem. Phys.* **69**, 241 (1987).
124. M. Dupuis, J. Rys, and H. F. King, *J. Chem. Phys.* **65**, 111 (1976).
125. L. E. McMurchie and E. R. Davidson, *J. Comp. Phys.* **26**, 218 (1978).

126. S. Obara and A. Saika, *J. Chem. Phys.* **84**, 3963 (1986).
127. See, for example, R. Ahlrichs, S. D. Elliott, and U. Huniar, *Ber. Bunsen-Ges. Phys. Chem.* **102**, 795 (1998); M. Head-Gordon, *J. Phys. Chem.* **100**, 13213 (1996).
128. M. Kollwitz and J. Gauss, *Chem. Phys. Lett.* **260**, 639 (1996).
129. M. Kollwitz, M. Häser, and J. Gauss, *J. Chem. Phys.* **108**, 8295 (1998).
130. J. Gauss and H.-J. Werner, *Phys. Chem. Chem. Phys.* **2**, 2083 (2000).
131. S. Saebø and P. Pulay, *Ann. Rev. Phys. Chem.* **44**, 213 (1993) and references cited therein.
132. P. Knowles, H.-J. Werner, and M. Schütz, in *Modern Methods and Algorithms of Quantum Chemistry*, edited by J. Grotendorst (John von Neumann Institute for Computing, Jülich, 2000), p. 97.
133. See, for example, J. Almlöf, in *Modern Electronic Structure Theory, Part I*, edited by D. R. Yarkony (World Scientific, Singapore, 1995), p. 110; M. Schütz, R. Lindh, and H.-J. Werner, *Mol. Phys.* **96**, 719 (1999).
134. O. Vahtras, J. Almlöf, and M. W. Feyereisen, *Chem. Phys. Lett.* **213**, 514 (1993); F. Weigend and M. Häser, *Theor. Chim. Acta* **97**, 331 (1997); F. Weigend, M. Häser, H. Patzelt, and R. Ahlrichs, *Chem. Phys. Lett.* **294**, 143 (1998).
135. J. Almlöf and S. Saebø, *Chem. Phys. Lett.* **154**, 83 (1989); M. Head-Gordon, J. A. Pople, and M. J. Frisch, *Chem. Phys. Lett.* **153**, 503 (1988).
136. M. J. Frisch, M. Head-Gordon, and J. A. Pople, *Chem. Phys. Lett.* **166**, 281 (1990); F. Haase and R. Ahlrichs, *J. Comp. Chem.* **14**, 907 (1993).
137. P. Pulay, *Chem. Phys. Lett.* **100**, 151 (1983); P. Pulay and S. Saebø, *Theor. Chim. Acta* **69**, 357 (1986); S. Saebø and P. Pulay, *J. Chem. Phys.* **b86**, 914 (1987).
138. M. Schütz, G. Hetzer, and H.-J. Werner, *J. Chem. Phys.* **111**, 5691 (1999).
139. C. Hampel and H.-J. Werner, *J. Chem. Phys.* **104**, 6286 (1996).
140. M. Schütz and H.-J. Werner, *J. Chem. Phys.* **114**, 661 (2000).
141. M. Schütz and H.-J. Werner, *Chem. Phys. Lett.* **318**, 370 (2000); M. Schütz, *J. Chem. Phys.* **114**, 661 (2001).
142. M. Schütz, Habilitationsschrift, Universität Stuttgart, 2000.
143. G. Vignale, M. Rasolt, and D. J. W. Geldard, *Adv. Quantum Chem.* **21**, 235 (1990).
144. F. R. Salisbury and R. A. Harris, *Chem. Phys. Lett.* **279**, 247 (1997); F. R. Salisbury and R. A. Harris, *J. Chem. Phys.* **107**, 7350 (1997).
145. L. Olsson and D. Cremer, *J. Phys. Chem.* **100**, 16881 (1996).
146. C. v. Wüllen, *Phys. Chem. Chem. Phys.* **2**, 2137 (2000).
147. D. Sundholm, J. Gauss, and R. Ahlrichs, *Chem. Phys. Lett.* **243**, 264 (1995).
148. J. Gauss and K. Ruud, *Int. J. Quantum Chem. Symp.* **29**, 437 (1995).
149. C. v. Wüllen, *Theor. Chim. Acta* **87**, 89 (1993).
150. S. P. A Sauer, I. Paiderova, and J. Oddershede, *Mol. Phys.* **81**, 87 (1994).
151. J. Gauss, unpublished results.
152. S. Coriani, M. Jaszunski, A. Rizzo, and K. Ruud, *Chem. Phys. Lett.* **287**, 677 (1998).
153. I. J. Solomon, J. N. Keith, A. J. Kacmarek, and J. K. Raney, *J. Am. Chem. Soc.* **90**, 5408 (1968).
154. E. A. Cohen, K. W. H. Hillig, and H. M. Pickett, *J. Mol. Struct.* **352–353**, 273 (1995).
155. P. J. Wilson and D. J. Tozer, *Chem. Phys. Lett.* **337**, 341 (2001).
156. For a detailed discussion, see W. H. Flygare, *Chem. Rev.* **74**, 653 (1974); W. H. Flygare, *Molecular Structure and Dynamics* (Prentice-Hall, Englewood Cliffs, 1978).
157. C. J. Jameson, *Chem. Rev.* **91**, 1375 (1991).
158. J. Gauss, K. Ruud, and T. Helgaker, *J. Chem. Phys.* **105**, 2804 (1996).
159. J. Gauss and D. Sundholm, *Mol. Phys.* **91**, 449 (1997).
160. D. Sundholm and J. Gauss, *Mol. Phys.* **92**, 1007 (1997).
161. M. A. Frerking and W. D. Langer, *J. Chem. Phys.* **74**, 6990 (1981).

162. J. Vaara, J. Lounila, K. Ruud, and T. Helgaker, *J. Chem. Phys.* **109**, 8388 (1998).
163. See, for example, W. Makulski, and K. Jackowski, *Chem. Phys. Lett.* **341**, 369 (2001).
164. S. P. A. Sauer, V. Spirko, and J. Oddershede, *Chem. Phys.* **153**, 189 (1991); S. P. A. Sauer, J. Oddershede, and J. Geertsen, *Mol. Phys.* **76**, 445 (1992); S. P. A. Sauer and J. F. Ogilvie, *J. Phys. Chem.* **98**, 8617 (1994).
165. S. M. Cybulski and D. M. Bishop, *J. Chem. Phys.* **100**, 2019 (1994).
166. A. K. Jameson and C. J. Jameson *Chem. Phys. Lett.* **134**, 461 (1987).
167. A. A. Auer, J. Gauss, and J. F. Stanton, *J. Chem. Phys.*, to be submitted.
168. R. N. Grimes, *J. Am. Chem. Soc.* **88**, 1895 (1966).
169. D. A. Franz and R. N. Grimes, *J. Am. Chem. Soc.* **92**, 1438 (1972).
170. T. Mennekes, P. Paetzold, R. Boese, and D. Bläser, *Angew. Chem. Int. Ed. Engl.* **30**, 173 (1990).
171. L. Ahmed, J. Castilla, and J. A. Morrison, *Inorg. Chem.* **31**, 1858 (1992).
172. See articles in *The Borane, Carborane, Carbocation Continuum*, edited by J. Casanova (John Wiley & Sons, Inc., New York, 1998).
173. H.-U. Siehl, T. Müller, J. Gauss, P. Buzek, and P.v.R. Schleyer, *J. Am. Chem. Soc.* **116**, 6384 (1994).
174. J. Gauss and J. F. Stanton, *J. Mol. Struct. (THEOCHEM)* **398–399**, 73 (1997).
175. H.-U. Siehl, T. Müller, and J. Gauss, unpublished.
176. G. A. Olah, J. S. Staral, G. Asencio, G. Liang, D. A. Forsyth, and G. D. Mateescu, *J. Am. Chem. Soc.* **100**, 6299 (1978).
177. G. A. Olah, R. J. Spear, and D. A. Forsyth, *J. Am. Chem. Soc.* **99**, 2615 (1977).
178. A. I. Biaglow, R. J. Gorte, and D. White, *J. Chem. Soc. Chem. Commun.* 1164 (1993).
179. E. P. Kanter, Z. Vager, G. Both, and D. Zajfman, *J. Chem. Phys.* **85**, 7487 (1986); J. Berkowitz, C. A. Mayhew, and B. Rusic, *J. Chem. Phys.* **88**, 7396 (1988); M. W. Crafton, M.-F. Jagod, B. D. Rehfuss, and T. Oka, *J. Chem. Phys.* **91**, 5138 (1989).
180. See, for example, D. Lenoir and H.-U. Siehl, in *Houben-Weyl Methoden der Organischen Chemie*, Vol. E19c, edited by M. Hanack, (Thieme, Stuttgart, 1990), p. 26.
181. R. Pelliciari, B. Natalini, B. Sadeghpour, M. Marinozzi, J. P. Snyder, B. L. Williamson, J. T. Kuethe, and A. Pawda, *J. Am. Chem. Soc.* **118**, 1 (1996).
182. R. A. King, T. D. Crawford, J. F. Stanton, and H. F. Schaefer, *J. Am. Chem. Soc.* **121**, 10788 (1999).
183. D. A. Kleier, D. A. Dixon, and W. N. Lipscomb, *Theor. Chim. Acta* **40**, 33 (1975); R. C. Haddon and K. Raghavachari, *J. Am. Chem. Soc.* **104**, 3516 (1982); R. C. Haddon and K. Raghavachari, *J. Am. Chem. Soc.* **107**, 289 (1985); N. C. Baird, *J. Am. Chem. Soc.* **94**, 4941 (1972); J. Aihara, *J. Am. Chem. Soc.* **98**, 2750 (1976); R. C. Haddon, *J. Am. Chem. Soc.* **101**, 1722 (1979).
184. H. M. Sulzbach, P. R. Schleyer, H. Jiao, Y. Xie, and H. F. Schaefer, *J. Am. Chem. Soc.* **117**, 1369 (1995).
185. H. M. Sulzbach, H. F. Schaefer, W. Klopper, and H. P. Lüthi, *J. Am. Chem. Soc.* **118**, 3519 (1996).
186. S. Masamune, K. Hojo, G. Bigam, and D. L. Rabenstein, *J. Am. Chem. Soc.* **93**, 4966 (1971).

COMPUTATIONAL CHEMISTRY OF ACIDS

CLARISSA OLIVEIRA SILVA and
MARCO ANTONIO CHAER NASCIMENTO

Departamento de Físico-Química — Instituto de Química
Universidade Federal do Rio de Janeiro
Rio de Janeiro, RJ
BRAZIL 21949-900

CONTENTS

I. Definition
II. Why Study Acids?
III. Gas-Phase Acidity
    A. The Experimental Approach
    B. The Quantum Chemical Approach
    C. Why Are Carboxylic Acids and Phenols More Acidic than the Corresponding Aliphatic Alcohols, in the Gas Phase?
IV. Acidity in Solution
    A. The Continuum-Solvation Models
    B. The Solvated Proton
    C. The Proton Solvation Energy
    D. The Solvation of Anions
    E. pK_a Calculations
        1. The TC1 or Proton Cycle
        2. The TC2 or Hydronium Cycle
    F. Relative pK_a Calculations
V. Molecular Associations
VI. Concluding Remarks
Acknowledgments
References

I. DEFINITION

There are basically three ways of defining "acid", in the specialized literature. The first and oldest one, proposed by Arrhenius in 1887, defines an acid as any compound that loses a proton (H^+) in aqueous solution.

$$AH_{(aq)} + B_{(aq)} \rightleftharpoons A^-_{(aq)} + BH^+_{(aq)} \tag{7.1}$$

Advances in Chemical Physics, Volume 123, Edited by I. Prigogine and Stuart A. Rice.
ISBN 0-471-21453-1 © 2002 John Wiley & Sons, Inc.

From this definition, it is clear that this concept is based on a solute–solvent interaction that dictates how easily a compound can liberate a hydrogen atom without its electron, in aqueous solution. Due to the solvent dependence that is included in the definition, this is not the best starting point to uniquely classify a substance as an acid, according to our purposes.

The second definition proposed by Brønsted [1, 2], considers an acid to be any compound containing a hydrogen atom, since it may be liberated as a proton. Depending on the molecule and its surroundings, the process requires more or less energy and in some cases it may even be spontaneous. Moreover, since this definition does not depend on the medium where the process is taking place, it may be equally applied to processes in the gas phase or in solution. According to this definition, in the reaction

$$AH + B| \rightleftharpoons |A^- + BH^+ \tag{7.2}$$

AH can be identified as an acid and A^- its conjugated base, or else B can be called a base, BH^+ being its conjugated acid. The whole process is illustrated by Eq. (7.2) and defines an acid–base reaction.

The Brønsted definition is very specific, because it depends basically on the atomic constitution of the system, which will be the definition adopted throughout this chapter. Certainly, the effect of the surroundings will have to be considered whenever one needs to quantify how much easier some substances lose their proton in comparison to others.

A third and more general definition was proposed by Lewis, who defined an acid as any substance that is able of accepting at least one pair of electrons. Depending on the nature of the problem being investigated, this can be a very convenient definition, especially when dealing with chemical reactions.

Clearly, the choice of the definition to be used will be dictated by the nature of the problem being investigated. For a discussion of the historical development of acid–base theory, see Bell [3] and Satchell and Satchell [4].

II. WHY STUDY ACIDS?

There are many reasons why one should pay special attention to acids and their properties, as they are involved in a great number of important chemical phenomena. Trying to list them all would be tedious and most certainly the list would be incomplete. Instead, we chose just a few examples of problems, related to several different areas of research, where acids play an important role.

Proton transfer is the way chosen by nature to link the extracellular (EC) and intracellular (IC) media. Information about the EC environment arrives at the cell interior throughout some chemical reactions, mainly proton-transfer reactions. In pharmacokinetics, the pK_a of a certain drug is mandatory information to check its efficiency. Most drugs are weak acids or basis, and are present in solution as both the ionized and nonionized species. Only the nonionized species are able to cross the phospholipidic cellular membrane, because of its lipophilic character. Therefore, the transmembrane distribution of a weak electrolyte is usually determined by its pK_a, which is directly related to the ratio between the nonionized and ionized species, and the pH gradient across the membrane [5]. For example, cocaine is a very powerful drug with a devastating action on the human body. It passes through the blood–brain barrier and binds to receptors in the brain. As already mentioned, the crossing through membranes and the binding processes depend on the charge, and therefore the knowledge of the percentage of cocaine molecules that are in the protonated and nonprotonated forms is very helpful information for making predictions about cocaine action in the human body [6].

Another example related to therapeutics involves hydroxamic acids, which have been widely used as a key functional group of potential therapeutics targeting at zinc-bound matrix metalloproteinases involved in cancers [7]. But the hydroxamic acids are also iron chelators [8], and since iron is the most abundant transition metal ion present in the human body, studies are being conducted in order to find new stable conformers and tautomers of hydroxamic acids that would preferentially bind to zinc.

Due to their ability to act simultaneously as proton donors–acceptors, carboxylic acids can establish very particular interactions that can be extremely important for many biochemical and biological processes. In this respect, theoretical calculations can be useful in modeling such processes. For example, the interaction between acetate ions and water molecules can be used to model hydrogen bonds in serine-protease enzymes [9]. Also, benzoic acid can be used as a model to study the parallel stacking arrangements of the aromatic side chains of proteins [10].

The oxidation of DNA is a known source of genomic instability, possibly caused by the oxidation of thymine and uracil. The oxidation of these compounds would change their pK_a values, which in turn would affect the base pairing and coding. In fact, some thymine derivatives, protected in the positions that undergo oxidation more easily, have demonstrated antitumor and antiviral properties [11]. Thus, studies directed at these processes could be extremely valuable for the understanding and prevention of mutations [12].

Acids also have technological importance. For example, ethyl benzoates and benzoic acids, can be incorporated to a polyester matrix, in order to alter the gas permeability of the material [13]. In some applications, it is important to prevent entry of O_2 into the container while in others it is desired to block the CO_2 escape. The gas permeability can be regulated by such substituents in the polyester matrix.

Protonated zeolites are solid acids of great importance in the oil and petrochemical industries, as they catalyze the cracking, isomerization, and alkylation of hydrocarbons as well as the conversion of alcohols into gasoline. In spite of the enormous technological importance of these reactions, their mechanisms are still not completely elucidated. The understanding of the mechanism of these reactions, at the molecular level, would be extremely important not only to establish better experimental conditions for increasing the yield in some desired product, but also to develop new selective catalysts of industrial interest [14, 15]. Also, the possibility of replacing strong mineral acids, such as H_2SO_4 and HNO_3, often used as catalysts in many industrial processes, would be extremely important from the environmental point of view.

There are many more equally relevant problems involving acids. However, we believe that the few mentioned above will suffice to justify the importance of studying acids. Many of the important chemical phenomena involving acids depend mainly on their ability to transfer protons or, in other words, on their acidity. Thus, among all the properties amenable to theoretical investigation, we will focus on the calculation of acidity. In addition, some interesting problems involving the association of acid molecules will be discussed briefly.

III. GAS-PHASE ACIDITY

A. The Experimental Approach

The gas-phase acidity of a compound AH is defined as the variation of the Gibbs standard free energy ($\Delta G°$) of the reaction

$$AH \to A^- + H^+ \qquad (7.3)$$

which represents the heterolytic cleavage of the AH bond. Conversely, the proton affinity of species A^- can be defined as the $\Delta G°$ for the reverse reaction. However, since in general $S_{AH} \approx S_{A^-}$ and $S_{H^+} \ll S_{AH}, S_{A^-}$, the entropic term is very small compared to the enthalpy of the reaction, even at high temperatures. Thus, to a good approximation one can take $\Delta G° \approx \Delta H°$ for reaction (7.3). Once the experiments are usually performed at constant T and P, $\Delta H = q$ and the acidity of the compound AH can be directly related to the heat of the reaction (7.3). This reaction is endothermic and one practical way to determine the energy needed for the heterolytic cleavage of the A—H bond is through the thermodynamic cycle:

$$AH \rightarrow A + H \quad \text{homolytic cleavage } \Delta H°_{298\,K} = D_0(A-H)$$
$$A + e^- \rightarrow A^- \quad \text{electron affinity (EA) of A, at 298 K}$$
$$H \rightarrow H^+ + e^- \quad \text{ionization potential (IP) of H, at 298 K}$$
$$\overline{AH \rightarrow A^- + H^+} \quad \Delta H°_{298\,K} = \Delta H°_{acid} = D_0(A-H) + IP(H) - EA(A)$$

Scheme 1. Thermodynamic cycle for a practical calculation of gas-phase acidity.

Since the IP of hydrogen is a constant equal to 313.6 kcal/mol [16], the enthalpy change $\Delta H°_{acid}$ depends basically on the difference between the bond dissociation energy $D_0(A-H)$, and the electron affinity EA(A). For many years the experimental way of obtaining $\Delta H°_{acid}$ for a compound was from dissociation energies and electron affinity measurements. For an interesting explanation of details of such procedures, see Bartmess and McIver [17] and Lias and Bartmess [18]. A more recent and comprehensive discussion of the various experimental techniques available for measuring these properties can be found in the NIST webbook [18].

Of course, once one adopts reaction (7.3) as the definition of gas-phase acidity, this quantity becomes temperature dependent. Therefore, the experimental values must always refer to the temperature at which the measurements have been made. Nevertheless, in a very recent study Bartmess et al. [19] concluded that the effect of the temperature on gas-phase acidities is much less important than the uncertainties related to the thermodynamic quantities being measured.

B. The Quantum Chemical Approach

The definition of the gas-phase acidity through reaction (7.3) implies that this quantity is a thermodynamic state function. Thus, one could use quantum chemical approaches to obtain gas-phase acidities from the theoretically computed enthalpies of the species involved. However, two points must be noted before one proceeds: A chemical bond is being broken and an anion is being formed. Thus, one may anticipate the need for a proper treatment of electronic correlation effects and also of basis sets flexible enough to allow the description of these effects and also of the diffuse character of the anionic species, what immediately rules out the semi-empirical approaches. Hence, our discussion will only consider ab initio (Hartree–Fock and post-Hartree–Fock) and DFT (density functional theory) calculations.

The ab initio and DFT calculations furnish values of the internal energies ($\Delta E°_{eq}$) at 0 K for the species involved in Eq. (7.3), at their equilibrium positions. In order to convert ΔE_{eq}° to ΔH_T°, which is experimentally measured, we can use

$$\Delta\Delta H_T^\circ[\Delta H_T^\circ(A^-) - \Delta H_T^\circ(AH)] = \Delta E_{eq}^\circ + \Delta(pV) + \Delta\text{ZPE} \\ + \Delta E_T^{vib} + \Delta E_T^{rot} + \Delta E_T^{trans} \quad (7.4)$$

where $\Delta(pV)$ converts internal energy into enthalpy and ΔZPE is the correction for the zero-point energies. The parameters ΔE_T^{vib}, ΔE_T^{rot}, and ΔE_T^{trans} are corrections to the changes in vibrational, rotational, and translational energies, respectively, when the system is heated from 0 K to temperature T. The latter three correction terms are generally referred to as the thermal corrections because they depend on the temperature. For the low pressures normally used in the measurements of gas-phase acidities, it is reasonable to consider the system as ideal, and therefore

$$\Delta(pV) = \Delta(nRT) = (\Delta n)RT = RT \quad (7.5)$$

The vibrational and rotational components can be calculated from the harmonic oscillator and rigid rotor models, for example, whose expressions can be found in many textbooks of statistical thermodynamics [20]. If a more sophisticated correction is needed, vibrational anharmonic corrections and the hindered rotor are also valid models to be considered. The translational component can be calculated from the respective partition function or approximated, for example, by $3/2RT$, the value found for an ideal monoatomic gas.

A variety of ab initio quantum chemical methodologies has been used to compute the internal energy variation, ΔE_{eq}°, and the zero-point correction energies ΔZPE, needed to calculate gas-phase acidities and proton affinities.

For the nonexpert reader, Table VII.1 lists the most frequently used methods together with the associated acronyms. To understand the nomenclature used for the DFT methods, let us recall that according to Hohenberg–Kohn theorem, the ground-state electronic energy is determined completely by the electron density. In other words, the total energy of the ground state is a functional of the density. This relationship can be written as

$$E[\rho] = T[\rho] + E_{ne}[\rho] + J[\rho] + E_{xc}[\rho] \quad (7.6)$$

$T[\rho]$, $E_{ne}[\rho]$, $J[\rho]$, and $E_{xc}[\rho]$ being, respectively, the kinetic, electron-nuclear attraction, Coulombian electron-repulsion, and the exchange corre-

TABLE VII.1
Most Frequently Used Ab Initio and DFT Methods

Ab Initio Methods	
HF	Hartree–Fock
CI	Configuration interaction
MR-CI	Multireference CI
QCISD	Quadratic CI with singly and double excited configurations
MCSCF	Multiconfigurational self-consistent field
MBPT	Many-body perturbation theory
MPn	Nth-order Møller–Plesset perturbation theory
CC	Coupled cluster

DFT Methods	
BP86	Becke 88 E_x + Perdew 86 E_c
BLYP	Becke 88 E_x + Lee–Yang–Parr E_c
BPW91	Becke 88 E_x + Perdew–Wang 91 E_c
B3P86	Becke 3 parameter hybrid + Perdew 86 E_c
B3PW91	Becke 3 parameter hybrid + Perdew–Wang 91 E_c
B3LYP	Becke 3 parameter hybrid + Lee–Yang–Parr E_c

lation functionals. It is customary to separate $E_{xc}[\rho]$ in two parts

$$E_{xc}[\rho] = E_x[\rho] + E_c[\rho] \qquad (7.7)$$

and the various DFT methods arise from different choices of E_x and E_c. The associated acronyms are directly related to the form chosen for the functionals $E_x[\rho]$ and $E_c[\rho]$ as shown in Table VII.1.

The main problem with the DFT methods is the lack of a recipe to find the appropriate forms of the E_c and E_x functionals. On the other hand, the main problem with the ab initio methods is the computation of the electronic correlation effects neglected in the HF approximation. All the pos-HF methods shown in Table VII.1 exhibit a rather steep increase in computational cost as the size of the basis set increases. In order to obtain a good estimate of the electronic correlation effects without making use of very large basis sets, calculations from different levels of theory can be combined. Examples of such strategy are the complete basis set (CBS) family of methods, developed by Petersson et al. [21] and the Gaussian (Gn) methods, developed by Pople and co-workers [22]. By choosing the appropriate level of theory and size of the basis set to be used at each different step of the calculation, the CBS and Gn methods can produce total energies that differ by <1 kcal/mol from the experiments. For comprehensive discussions about the ab initio and DFT methods, the reader is referred to the specialized literature (McWeeny [23], Szabo and Ostlund [24], Jensen [25], and Parr and Yang [26]).

A final word about the notation used to indicate the level of theory and the basis set employed in the calculation. The level of theory and the basis set are indicated by the respective acronyms separated by a |, for example, HF/6-31G. However, it is common to use different levels of theory and basis sets at different stages of a calculation. For example, the geometry can be optimized at the HF/6-31 level and the correlation effects computed at the MP2/6-31G(d,p) level. In this case, the two strings are put together separated by a double //: MP2/6-31G(d,p)//HF/6-31G.

Gao et al. [27] published an extensive study, at the Hartree–Fock (HF) level, which included geometry optimization with 6-31G(d) basis and correlation treatment using the Møller–Plesset second-order perturbation theory (MP2) in the valence space. Siggel et al. [28a] calculated gas-phase acidities for methane and formic acid at the MP4/6-31 + G(d) level and for several other compounds at lower levels of theory (HF with 3-21 + G and 6-311 + G basis sets). All these calculations provide gas-phase acidity values that systematically differ from the experimental values. Nevertheless, the results show good linear correlation with the experimental data.

Many other calculations of gas-phase acidity using ab initio and DFT methodologies can be found in the literature (Assfeld et al. [29] and Jorgensen et al. [30]). Some recent papers addressed the problem of investigating the effect of substituents on the acidity of certain classes of compounds. For example, Bernasconi and Wenzel [31] examined the gas-phase acidity of $CH_2\!=\!C\!=\!X$ (X = CH_2, NH, O, and S), in order to study how the enhanced unsaturation of $CH_2\!=\!C\!=\!X$ compared to that of $CH_3\!-\!CH\!=\!X$ would affect the acidities. Perez [32] investigated polarizability and inductive effects of substituent groups on the gas-phase acidities of alcohols and silanols. Rustad et al. [33] considered several protonation states of oxyacids of Si, P, V, As, Cr, and S, while Remko [34] studied the gas-phase acidity of carbonic acid (H_2CO_3) and several of its derivatives RCO_2H (R = F, Cl, NH_2, and CH_3). The effect of increasing the chain length for different classes of acids has also been considered. For example, Ligon [35] calculated proton affinities of n-alkylamines, n-alkylthiols, and n-alcohols of increasing size, in order to establish limiting values for longer chains. However, for the present purposes, it would be more appropriate to focus on those papers where systematic studies, involving different classes of acids, basis sets, and methodologies have been conducted.

Very detailed calculations have been performed by Smith and Radom [36] comparing the ability of many different methodologies for predicting

TABLE VII.2
Mean Deviations between Calculated and Experimental Acidity Values[a]

	B-P86	BLYP	B3P86	B3LYP	MP2	MP4	F4	QCISD(T)	G2
6-311+G (d,p)	−3.42	4.25	0.53	−1.05	2.68	4.01	4.44	4.59	
6-311+G (3df,2p)	−2.65	−3.51	1.34	−0.38	−0.72	0.55	0.69		0.53

[a] $\Delta\Delta H_{acid}$, at 298 K, in kcal/mol.

gas-phase acidities. They used second- and fourth-order Møller–Plesset (MP2 and MP4) and fourth-order Feenberg (F4) theories, quadratic configuration interaction (QCISD(T)) and also the G2 theory and some of its variants, G2(MP2,SVP) and G2(MP2). Several density functionals, namely, BP-86, B-LYP, B3-P86, and B3LYP were also considered and the importance of including diffuse functions to properly describe the anionic species was also discussed, although this point had been previously recognized by Chandrasekhar et al. [37], Dunning and Hay [38], and Kollman [39]. A set of 23 compounds, comprising different classes of acids was investigated. The geometries of all species optimized at the MP2/6-31G(d) level were used in all the calculations. The performance of the several methodologies employed in the calculations was compared in terms of the mean deviations (Table VII.2) between the theoretical and experimental results for the whole set of compounds, and for two different basis sets. The results shown in Table VII.2 indicate that the best sets of gas-phase acidity values for the 23 componds studied are those obtained at the G2 and B3LYP levels of calculation, with the larger basis set. The fact that the B3LYP method can

TABLE VII.3
Mean Deviation Error (kcal/mol) between Calculated and Experimental Acidities of Neutral Acids for Various Functionals (X) with Pople (I and II) and Dunning-Type (III and IV) Basis Sets

X =	S-null	S-VWN	S-LYP	B-null	B-VWN	BLYP	B3PW91	G2
I[a]	1.7	2.0	2.5	2.5	1.5	2.1	1.6	
II[b]	2.6	2.4	3.0	2.1	1.6	1.3	1.0	
III[c]	2.3	1.8	2.1	2.4	1.2	1.6	0.9	
IV[d]	2.4	2.1	2.4	2.7	2.0	1.5	0.9	0.8

[a] I: X/6-31 + G(d).
[b] II: X/6-311+ +G(2df,2pd)//X/6-31+G(d).
[c] III: X/aug-cc-pVDZ.
[d] IV: X/aug-cc-pVTZ//X/aug-cc-pVDZ.

produce acidities of comparable accuracy to those of the G2 method is quite auspicious if one considers that the B3LYP calculations are significantly less time consuming.

Another systematic study involving several different density functionals and basis sets was conducted by Merril and Kass [40]. Contrary to the previous study (Smith and Radom [36]), full geometry optimization and vibrational analysis were performed using all the different functionals considered. A set of 35 different compounds was investigated and a summary of their results is shown in Table VII.3. For comparison with non-DFT methods, G2 calculations have also been performed. Irrespective of the basis set used, the best gas-phase acidity values, comparable to the G2 results, were obtained with the hybrid B3PW91 functional.

At this point, it would be instructive to compare the results obtained by Merril and Kass [40] with those of Smith and Radom [36] using the BLYP functional. Although this functional did not furnish the best results, it was the only one used in both studies. From Tables VII.2 and VII.3 one sees that the results by Merril and Kass [40] using that functional, are on average 1–2 kcal/mol more accurate than the ones by Smith and Radom [36]. Since most of the 23 compounds used by Smith and Radom [36] have also been considered by Merril and Kass [40], one may attribute this difference in accuracy to the fact that the latter authors performed the geometry optimizations at the same level of theory used to compute the acidities.

Sauers [41] compared the performance of the B3LYP functional with that of the MP2 method in predicting proton affinities for a series of 28 alkyl carbanions. Although the amount of experimental data for the species studied is limited, the MP2 results showed better agreement with the experiments. Note that the B3LYP results did not show any improvements by increasing the basis set, a result at variance with the one by Smith and Radom [36]. Also note that the largest discrepancies between the MP2 and B3LYP results occurred for the branched and strained alkyl carbanions. For example, the proton affinity of the *tert*-butyl carbanion was found to be equal to 412.1 kcal/mol at the MP2 level, and 406.8 kcal/mol when using the B3LYP functional. The experimental value is 413.1 kcal/mol.

Recently, Burk and Sillar [42] examined the performance of the B3LYP functional, not considered in the work by Merril and Kass [40], for predicting gas-phase acidities. In order to access the performance of that functional, a set of 49 compounds, comprising distinct classes of acids, was investigated using four different basis sets. The list of compounds chosen includes practically all the acids investigated by Merril and Kass [40] using other density functionals. The authors concluded that acidity values with average absolute errors <2.5 kcal/mol could be obtained with that functional and the 6-311 + G(3df,3pd) basis set. This result is very close to the accuracy expected from the more time demanding G2-type calculations. Interesting enough, when this same methodology was applied to compute the acidity of saturated hydrocarbons (Burk and Sillar [42]), absolute errors of the order of 3 kcal/mol were observed for the branched compounds, while for the linear ones the performance was compared to that of the G2 method. On the other hand, the accuracy of the G2 method does not seem to depend on the type of carbon chain.

The gas-phase acidity of strong mineral acids, such as $HClO_4$, CF_3SO_3H, FSO_3H, H_2SO_4, HBF_4, HPO_3, and HNO_3 was recently investigated by Koppel et al. [43] using the G2 and G2(MP2) methodologies and also DFT-B3LYP calculations with 6-311 + G(d,p) basis sets. The acidity values obtained at the G2 and G2(MP2) levels of theory were always better than the B3LYP ones. Nevertheless, the quality of the B3LYP results prompted the authors to use this level of theory to estimate the acidity of other strong acids and also of some superacids, such as HSb_6F_6 and FSO_3SbF_5H, which would be out of reach of either the G2 or the G2(MP2) theory. The DFT results for a series of carborane acids predict that the dodecafluoro substituted carborane acid $CB_{11}F_{12}H$ should exceed the intrinsic acidity of sulfuric acid by almost 90 kcal/mol or by almost 70 powers of 10.

DFT methods are quite attractive as they explicitly include electronic correlation effects at much lower costs than the ab initio methods, thus offering the possibility of carrying out calculations on much larger systems. The HF method scales formally as ON^4 where O is the number of occupied orbitals and N the total number of basis functions needed to describe the system. The MP2 calculations scale formally as ON^5 while higher orders of perturbation and other sophisticated treatments such as QCISD(T) or CCSD(T) scale even less favorably. On the other hand, hybrid functionals such as B3LYP takes into account the correlation effects at approximately the same cost as a HF calculation. The only inconvenience, as far as gas-phase acidity is concerned, resides in the fact that for some classes of compounds the accuracy of the B3LYP functional method seems to depend on the structure of the acid.

One possible way to circumvent the scaling problem of the ab initio methods is to concentrate our efforts in determining just the differential correlation effects, that is, those that are present in the acid but not on its conjugated base, and vice versa. Normally, one computes the total energies and subtract that of the conjugated base from the one of the acid. Since most of the electronic correlation effects are common to both species, a large amount of computer time is wasted as these effects will be canceled out once ΔE_{eq}° is computed. The results in Table VII.4 illustrate the use of a simple strategy (da Silva and Silva [44a], which can be used to compute accurate

TABLE VII.4
Gas-Phase Acidities for Alcohols (in kcal/mol): $ROH \rightleftharpoons RO^- + H^+$

	HF//6-31 + G(d,p)			L-MP2//6-31 + G(d,p)[b]			Experimental
	ΔE°_{eq}	ΔZPE	$\Delta H^\circ_{298 K}$[a]	ΔE°_{eq}	ΔZPE	$\Delta H^\circ_{298 K}$	$\Delta H^\circ_{298 K}$
CH_3	398.29	−9.64	390.10	389.83	−9.94	381.37	379 ± 2[c]; 382.1 ± 0.7[d]
CH_3CH_2	395.80	−9.38	387.89	381.80	−9.75	373.52	376 ± 2[c]; 379.1 ± 0.7[d]
$CH_3CHOHCH_3$	395.10	−9.50	387.07	384.60	−7.01	379.07	374 ± 2[c]
$CH_3CH_2CH_2OH$	396.47	−9.57	388.38	386.64	−9.68	378.43	375 ± 2[c]

[a] The following corrections are common to all compounds, $\Delta(pV) = 0.58$ kcal/mol and $\Delta E^{trans}_{298 K} = 0.88$ kcal/mol.
[b] Localized Møller–Plesset second-order perturbation (L-MP2) theory, algorithm of Pulay and Saebo [45] as found in Murphy et al. [46], which scales as $n^2 N$, where n is the number of occupied orbitals and N the size of the basis set.
[c] Assfeld et al. [29].
[d] Coitiño et al. [47].

gas-phase acidities using ab initio methods but at costs comparable to B3LYP calculations. This strategy makes use of (Pulay and Saebo [45]) localized MP2 theory.

A HF calculation is performed followed by an orbital localization treatment. Electronic correlation effects are then introduced, at the MP2 level of calculation, but just for the bond being broken and its first neighbors bonds. Since only a few bonds are being correlated, larger basis sets can be used, although the results in Table VII.4 indicate that quite good acidity values can be obtained even with more modest basis sets.

C. Why Are Carboxylic Acids and Phenols More Acidic than the Corresponding Aliphatic Alcohols, in the Gas Phase?

Before concluding this section, it would be instructive to examine a problem that motivated a great deal of discussion in the literature. It has to do with the relative gas-phase acidities of alcohols and carboxylic acids, in the gas phase.

Carboxylic acids and phenols are relatively stronger acids than alcohols both in gas phase and in solution. The difference in the gas-phase acidities is usually attributed to a low-energy content of the carboxylate anions, which would be stabilized by the resonance effect, as shown in Figure 7.1

This generally accepted interpretation can be found in any contemporary textbook. However, in a series of papers (Siggel and Thomas [28a–c, 48]) have challenged this view, attributing the large acidity of carboxylic acids to the inductive effect due to the carbonyl group. An intense discussion in the literature was motivated by these papers, with several authors (Exner [49], Deward and Krull [50], Bordwell and Satish [51]) trying to refute Siggel

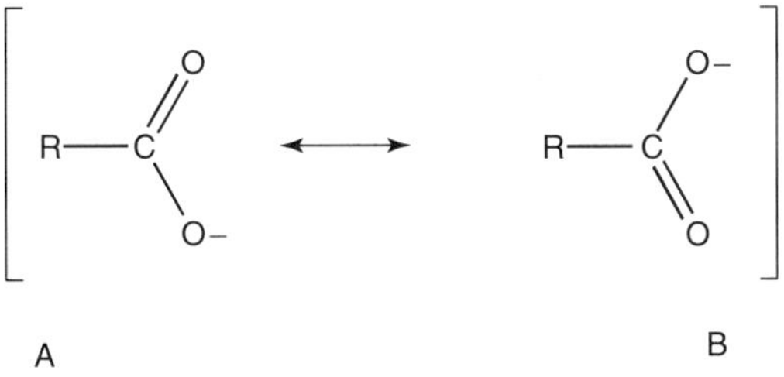

Figure 7.1. Resonance structures for the carboxylate anion.

and Thomas' new interpretation. However, from a theoretical point of view, the contributions of inductive and resonance effects to the acidity of these compounds had not been well quantified. Making use of the generalized multistructural (GMS) wave function, Hollauer and Nascimento [51a] briefly discussed the possibility of quantifying those effects for carboxylic acids, but pointed out that resonance should be the dominant effect. Hiberty and Byrman [52], in a series of calculations aimed not at the calculation of acidities but at the examination of the relative importance of inductive and resonance effects, concluded that they contribute with equal importance to the acidity of carboxylic acids although the calculations would only provide lower bounds for the delocalization effects. da Motta and Nascimento [53] reexamined this problem using GMS wave functions to quantify the contribution of the inductive and resonance effects. With the GMS wave function one can calculate the resonance stabilization of the carboxylate ions by direct computation of the matrix element between the two resonant structures, **A** and **B**, in Figure 7.1. The relative importance of resonance and the inductive effect of the carbonyl group was determined assuming that, in the absence of resonance, the inductive effect of the carbonyl group on the carboxylic acids would be the main responsible for the stronger acidity of these compounds relative to the alcohols. Hence, the difference in acidity between any pair of acid–alcohol can be decomposed into the contributions due to resonance and inductive effects. Table VII.5 shows the relative contributions of resonance and inductive effects for the acidity of formic and acetic acids.

From Table VII.5, one can see that resonance is the dominant effect responsible for the greater acidity exhibited by the carboxylic acids in gas phase, when compared to the acidity of the corresponding alcohols, in agreement with the traditional interpretation. More recently, Burk and Schleyer [54] showed that the application of the Siggel and Thomas approach can lead to contradictory conclusions about whether the difference in acidity between methanol and formic acid is determined by the

TABLE VII.5
Relative Contributions of Resonance and Inductive Effects for the Acidity of Carboxylic Acids (kcal/mol)

	$(\Delta A)_{th}{}^a$	$(\Delta A)_{ind}{}^b$	$(\Delta A)_{res}{}^c$	$(\Delta A)_{exp}{}^d$
Methanol–formic acid	34.59	16.84	17.75	33.90
Methanol–acetic acid	28.83	11.30	17.53	30.67
Ethanol–formic acid	31.13	13.38	17.75	30.67
Ethanol–acetic acid	25.37	7.84	17.53	27.44

$^a(\Delta A)_{th}$: difference in acidity between the two compounds.
$^b(\Delta A)_{ind}$: difference in acidity not considering the resonance effects.
$^c(\Delta A)_{res}$: difference in acidity due to resonance effects, obtained as the difference between the two previous quantities.
$^d(\Delta A)_{exp}$: experimental difference in acidity.

neutral acid or the anion. Hence, their proposed scheme cannot be used to determine the relative importance of resonance and inductive effects to the acidity of a compound.

IV. ACIDITY IN SOLUTION

The gas-phase acidity of a compound AH was defined as the variation in the Gibbs standard free energy of reaction (7.3). The equivalent process in a solvent S may be written as

$$(AH)_S \xrightleftharpoons{K_a} (A^-)_S + (H^+)_S \tag{7.8}$$

and the corresponding $\Delta G° = -2.303RT \log K_a$, where

$$K_a = \frac{[A^-]_S [H^+]_S}{[AH]_S} \tag{7.9}$$

can be used to define the acidity of species AH. One immediate consequence of this definition is that the acidity value becomes solvent dependent. However, the great majority of acid–base processes of chemical and biochemical interest occur in aqueous medium. Therefore most of the computational effort has been directed at determining acidities in aqueous solution.

Defining the pK_a of species AH as

$$pK_a = -\log K_a \tag{7.10}$$

the standard variation of the Gibbs free energy related to process (7.8), taking place in water, at 298.15 K, can be written as

$$\Delta G°(\text{kcal/mol}) = 1.36\,pK_a \qquad (7.11)$$

with $R = 1.98\,\text{cal/mol K}$. Equation (7.11) establishes a linear relationship between the acidity, as defined in terms of $\Delta G°$, and the pK_a value. Thus, one could use either $\Delta G°$ or the pK_a value as a measure of the acidity strength of species AH. However, the pK_a values fall over a narrower range, which makes this property especially convenient for defining an acidity scale. Hence, it became common practice to use the pK_a values as a measure of the acidic strength of an acid, the stronger acids exhibiting the lower values of pK_a.

The calculation of acidities in solution is a considerably more difficult problem than in the gas phase. The first difficulty has to do with the consideration of the solvent (water, in the present case) in the calculations. The other problem is related to the correct description of the proton in water. First, let us consider how to incorporate the solvent molecules in the calculation and after that we will examine the problem of describing the solvated proton.

A. The Continuum-Solvation Models

In recent years, several different approaches have been used to calculate acidity in solution. Explict consideration of the solvent molecules can be made in the context of either the Monte Carlo (MC) statistical mechanics (Jorgensen et al. [55], Jorgensen and Briggs [56]) or the supermolecule (Pullman and Pullman [57], Newton and Ehrenson [58]) approaches, and more recently through a hybrid quantum mechanical/molecular mechanical (QM/MM) method (Gao and Freindorf [59] and references cited therein). Alternatively, the solvent can be represented by a continuum medium, characterized by its bulk dieletric constant, and the acidities calculated using classical electrostatics, the Poisson–Boltzmann (PB) approach (Rashin et al. [60], Lim et al. [61], Yang et al. [62], Antosiewicz et al. [63]) or some self-consistent reaction field model (Rinaldi and Rivail [64], Yomosa [65], Tapia and Goscinski [66], Rinaldi et al. [67], Miertuš et al. [68], Tomasi and Persico [69], Aguilar et al. [70], Wong et al. [71], Wiberg et al. [72], Wiberg et al. [73], Cramer and Truhlar [74], Mikkelsen et al. [75]). Except for the MC calculations, in most cases all of the other approaches start from some ab initio gas-phase description of the solute molecule to which the solvent molecules are incorporated either explicitly or through a continuum model.

Among the several available models, those that represent the solvent molecules through a continuum medium are by far the most extensively used in pK_a and solvation energy calculations. Thus, we will concentrate the discussion on the continuum approaches but since this is not the main objective of this chapter only the basic ideas will be presented. For more comprehensive discussions, the reader is referred to the excellent reviews of Tomasi and Persico [69], Cramer and Thrular [76], and Orozco and Luque [77].

Basically, the continuum solvation models consider the solvent as a uniform polarizable medium with dielectric constant ε, where the solute (M) is immersed inside a cavity. Once placed inside the dielectric, the solute charge distribution polarizes the medium that in turn acts back (reaction field) polarizing the solute molecule. The system is then stabilized by the electrostatic interaction between the polarized solute and the polarized medium. Calling ρ_M the charge distribution of the solute and Φ_σ the solvent reaction potential, the electrostatic solute–solvent interaction can be written as

$$W_{MS} = \int_V \rho_M(r)\Phi_\sigma(r)\,dr^3 \tag{7.12}$$

the integral extending over the whole volume of the system (solute+solvent). In a classical description of the solute, V is equal to the volume inside the solute cavity since ρ_M would be zero outside.

It can be shown (Böttcher [78]) that this electrostatic potential energy is the work necessary to bring the charge ρ_M from infinity into the cavity. However, since from the thermodynamic point of view this quantity is also the maximum work obtainable from the system under isothermal conditions, W_{MS} has the status of a free energy and therefore can be interpreted as the electrostatic component of the solvation free energy (ΔG_{el}). Of course, other terms will contribute to the total free energy of solvation. For example, work must be performed on the system in order to create the solute cavity. On the other hand, dispersion interactions between the solute and the solvent contribute to the stabilization of the system, while repulsion interactions act in an opposite way. Hence, the total solvation free energy can be written as

$$\Delta G^\circ_{solv} = \Delta G^\circ_{el} + \Delta G^\circ_{cav} + \Delta G^\circ_{disp-rep} \tag{7.13}$$

Some models also consider a contribution, ΔG°_{Mm}, due to changes in the vibrational, rotational, and translational motion of solute when transferred from vacuum to the dielectric medium. For ionic solutes, the electrostatic contribution to ΔG_{solv} will be dominant and in many calculations the other components are generally neglected.

The various continuum solvation models may differ in many respects and several classifications have been proposed (Tomasi and Persico [69], Cramer and Truhlar [79]). They may differ on (a) how the size and shape of the cavity is defined; (b) how the nonelectrostatic contributions are computed; (c) how the reaction field is determined; (d) how the solute M is described, classically or quantum mechanically.

The simplest shape for the cavity would be a sphere or possibly an ellipsoid. However, for accurate values of solvation energy more realistic shapes are required. The PCM model (Miertuš et al. [68], Cammi and Tomasi [80]) uses a particular shape of cavity, the solvent excluding (SE) surface (Pomelli and Tomasi [81]), shown in Figure 7.2.

By interlocking spheres centered on each nuclei and with the radius defined in terms of the respective van der Waals radii (Bondi [82]) multiplied by a factor 1.2, the van der Waals surface of the solute is constructed. The SE cavity is then defined as the contact surface of a probe sphere (with radius equal to the molecular radius of the solvent molecule) rolling on the solute van der Waals surface.

The solute charge distribution can be represented by atom centered point charges or as multipole expansions. Of course, if the solute is treated quantum mechanically the charge distribution can be obtained directly from its wave function. Depending on the solvation model, the electrostatic potential derived from the wave function is fitted to atomic charges or multipoles that are then used to construct the solvent reaction field.

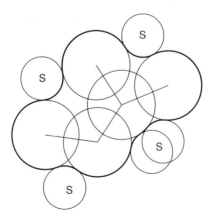

Figure 7.2. Solvent excluding surface (SE).

The simplest reaction field model, known as the Born model, consists of a spherical cavity where only the net charge and the dipole moment of the molecule are considered. When the solute is represented by a set of atomic charges the model is often called Generalized Born (GB) Model. Ellipsoidal cavities can also be employed as in the early Kirkwood–Westheimer model. The main advantage of these simple models is that W_{MS} [Eq. (7.12)] can be computed analytically. Unfortunately, they are of limited accuracy.

The apparent surface charge (ASC) approach appears to be a quite versatile method to calculate the reaction potential $\Phi_\sigma(r)$, using either a quantum or a classical description of the solute molecule. According to classical electrostatics, the reaction potential Φ_σ can be described at any point in space in terms of an apparent charge distribution, σ, spread on the cavity surface. Calling $\sigma(s)$ the apparent charge per unit area, at a point s of the cavity surface Σ, one may write

$$\Phi_\sigma(r) = \int_\Sigma \frac{\sigma(s)}{|\vec{r} - \vec{s}|} d^2s \tag{7.14}$$

One advantage of using the ASC approach is that the Gauss theorem provides an additional condition that must be satisfied by the charge distribution σ, namely,

$$\int_\Sigma \sigma(s) d^2s = -\frac{\varepsilon - 1}{\varepsilon} Q_M \tag{7.15}$$

where Q_M is the total charge of the solute molecule. In order to determine σ, the Σ surface is generally divided into an appropriate number of parts (sometimes called "tesserae") of area Δs_k and containing a charge q_k at its center point s_k:

$$q_k = \Delta s_k \sigma(s_k) \tag{7.16}$$

so that

$$\Phi_\sigma(r) = \sum_k \frac{q_k}{|\vec{r} - \vec{s}|} \tag{7.17}$$

The solvent reaction field can also be determined through the solution of the PB equation:

$$\nabla \cdot \varepsilon(r)\nabla\Phi(r) + 4\pi\{\rho_M(r) + \rho_{ext}(r)\} = 0 \tag{7.18}$$

where $\rho_M(r)$ is the charge distribution of the solute and ρ_{ext} is the charge induced in the solvent, given in the Debye–Huckell/Boltzmann model (for a 1:1 electrolyte) by

$$\rho_{ext}(r) = -\varepsilon(k^2)\sinh[\Phi(k)] \quad (7.19)$$

where k is the Debye–Huckell parameter. Unfortunately, the solution of the PB equation is difficult for systems of interest and some numerical approach, as the finite-difference method must be used (Sharp [83]).

The solution of the PB equation furnishes not only the total electrostatic potential at any point (Φ_i) but also the electrostatic contribution to the solvation free energy given by

$$\Delta G_{el} = \frac{1}{2}\sum_i q_i(\Phi_i^{sol} - \Phi_i^\circ) \quad (7.20)$$

where $\{q_i\}$ represents the point charges used to describe the solute charge distribution. In order to obtain ΔG_{el}, two independent calculations are performed, one with a dielectric constant equal to 1 inside and outside the cavity (Φ°), and the other with a dielectric constant of 1 inside the cavity and ε_s outside the cavity.

In a quantum mechanical treatment of the solute, a perturbational operator, V_σ, representing the solvent reaction field is added to the solute Hamiltonian, H_M°, giving rise to the Schrödinger equation:

$$[H_M^\circ + V_\sigma(\rho_M, \varepsilon)]\Psi_M = E\Psi_M \quad (7.21)$$

The solute–solvent electrostatic interaction contribution to the total energy can be written as

$$W_{MS} = \int_V \Psi^* V_\sigma \Psi \, dV = \int_V \rho_M(r)\Phi_\sigma(r) \, dV \quad (7.22)$$

where the integration is over the whole volume V of the system. The electrostatic contribution to the solvation free energy is then given by

$$\Delta G_{el} = \langle \Psi_M | H_M^\circ + V_\sigma(\Psi_M) | \Psi_M \rangle - \langle \Psi_M^\circ | H^\circ | \Psi_M^\circ \rangle \quad (7.23)$$

Different techniques may be used to solve the quantum mechanical problem, the HF method being one of the most frequently employed. In this case, V_σ is simply added to the Fock operator.

It is clear from Eq. (7.21) that the solute wave function Ψ_M depends on the reaction field, which in turn depends on Ψ_M through the solute charge distribution ρ_M. Hence, Eq. (7.21) has to be solved iteratively.

The Polarizable Continuum Model (PCM), in its original version, uses a quantum description of the solute molecule, an SE molecular cavity, and the ASC approach to determine the reaction field. The quantum mechanical calculation is performed using two nested cycles: in the internal cycle the $q^{(m)}$ charges [Eq. (7.16)] are calculated from the $\rho_M^{(m)}$ distribution while on the external cycle an improved solute charge distribution $\rho_M^{(m+1)}$ is determined.

The Conductor-like Screening Model (COSMO) also uses molecular shaped cavities and represents the electrostatic potential by partial atomic charges (Klamt and Schüürmann [84]). COSMO was initially implemented for semiempirical methods but more recently was also used in conjunction to ab initio methods (Andzelm et al. [85]).

A series of continuum solvation models (SMx, $x = 1-5$) has been developed by Truhlar and co-worker (Cramer and Truhlar [79]), based on the Generalized Born/Surface Area (GB/SA) model (Still et al. [86]). Recall that in the GB approach the molecular shape is taken into account as the solute charge is distributed over a set of atom-centered spheres. For this GB/SA model, the polarization free energy is given by

$$G_{el} = -\frac{1}{2}\left\{1 - \frac{1}{\varepsilon}\right\} \sum_{k,k'} q_k q_{k'} \gamma_{kk'} \qquad (7.24)$$

where q_k is the net atomic charge on center k, and $\gamma_{kk'}$ is a Coulomb integral. The several SMx models differ basically on the functional form used for $\gamma_{kk'}$. Initially developed in conjuction with AM1 and PM3 semiempirical methods, the recently developed SM5.42R solvation model was also parameterized at the HF, BPW91, and B3LYP levels (Li et al. [82]).

As mentioned before, for ionic solutes the electrostatic contribution to the free energy of solvation is the dominant term. However, the other terms may be relevant especially for neutral solutes. The dispersion and repulsion terms in principle can be obtained self-consistently together with Ψ_M (Amovilli and Mennucci [88]), but semiempirical expressions are often used (Floris et al. [89]). In this case, the corresponding energy terms may be taken as being proportional to the total surface area of the cavity or can be parameterized by assigning one parameter (η_i) to each specific atom, and optimizing them by fitting them to the experimental solvation energies:

$$\Delta G_{cav} + \Delta G_{disp} + \Delta G_{rep} = \sum_i^{atoms} \eta_i s_i \qquad (7.25)$$

B. The Solvated Proton

The proton has no electronic structure. Devoided of an electronic shell, a proton can approach far closer to a neighboring molecule than can any other ion or atom. Thus, the proton experiences few steric restrictions in chemical reactions. While other atoms have Angstrom dimensions ($\sim 10^{-10}$ m), the proton has Fermi dimensions ($\sim 10^{-15}$ m). These particular structural characteristics confer to the proton abnormal properties in solution.

The energy needed to remove an electron from a hydrogen atom is much larger than the corresponding energy for the formation of a univalent ion of any other element, as can be seen from Table VII.6 (Bockris and Reddy [90]).

The data in Table VI.6 reflect the strong attraction the proton has for electrons. Thus, protons tend to form covalent bonds by attracting electron pairs. They also tend to form hydrogen bonds, which as a first approximation, may be considered an electrostatic interaction between a proton and an unshared pair of electrons. All this suggests that it is highly unlikely that free protons can be found in solution. They must exist as some kind of molecular aggregate.

Due to the experimental difficulties to obtain structural information about the liquids, the form whereby the proton occurs, in aqueous solution, is one of the most ancient and still open questions in chemistry, and a challenge for theoreticians.

Gas-phase studies of proton solvation indicate the presence of several ionic species, H_3O^+ being the most abundant (Bockris and Reddy [90]). Also, X-ray analysis of acid hydrates, such as $HClO_4 \cdot H_2O$, shows evidence that protons may associate with water molecules giving rise to H_3O^+ and ClO_4^- ions (Bockris and Reddy [90]). Ammonium perchlorate, NH_4ClO_4,

TABLE VII.6
Ionization Energies of the First Group Elements

Z	Element	Ionization Energy (kcal/mol)
1	H	313
3	Li	124
11	Na	118
19	K	100
37	Rb	96
55	Cs	90

a substance isomorphous to $HClO_4 \cdot H_2O$, is an ionic crystal consisting of an assembly of NH_4^+ and ClO_4^- ions. Since an NH_4^+ ion is called an ammonium ion, it would be appropriate to call the H_3O^+ a hydronium ion. However, all this evidence for the presence of H_3O^+ ions in the solid and gas phases does not prove that they exist in solution.

Structural models for the hydrated proton first emerged (Huggins [91]) from attempts at explaining the anomalously high mobility of protons in liquid water. Wicke et al. [92] and Eigen [93] proposed the formation of an $H_9O_4^+$ complex in which an H_3O^+ core is strongly hydrogen bonded to three H_2O molecules, while Zundel [94, 95] supported the idea of an $H_5O_2^+$ complex in which the proton is shared between two H_2O molecules.

In order to explain the observed acidity of aqueous sulfuric acid solutions, Robertson and Dunford [96] proposed a model that considered the presence of several protonated species. They concluded that many species like H_3O^+, $H_5O_2^+$, $H_7O_3^+$, $H_9O_4^+$, can account for the observed acidity. The largest hydrate postulated to exist is $H_{21}O_{10}^+$.

More recently Marx et al. [97] addressed the problem of the proton mobility in water through ab initio path integral simulations. They concluded that both $H_5O_2^+$ and $H_9O_4^+$ are only important as limiting structures and that numerous unclassifiable situations exist in between. Hence, they suggested that the solvated proton should be best visualized as a "fluxional" complex.

According to the majority of the theoretical and experimental studies available in the literature, the simpler reasonable description of a proton in solution is the hydronium (H_3O^+) ion. The shape and size of the H_3O^+ ion is fairly well established from nuclear magnetic resonance (NMR) data on solid acid hydrates (Bockris and Reddy [90]). The H_3O^+ ion exhibits a rather flat trigonal pyramidal structure with the hydrogen atoms at the corners of the pyramid and the oxygen at the center (Fig. 7.3). But no reliable experimental geometric information is available for the hydronium ion in solution.

Figure 7.3 is also the picture for the H_3O^+ ion in water that arises from theoretical calculations by using the PCM (Miertuš et al. [68] and Cammi

Figure 7.3 Structure of the hydronium (H_3O^+) ion.

TABLE VII.7
Geometrical Parameters for Hydronium Ion in Gas Phase and Aqueous Solution at the HF//6-31 + G(d,p) Level

Distance (Å)		Plane Angle (degrees)	
Gas Phase			
O-H1	0.961	H1-O-H2	114.62
O-H2	0.961	H1-O-H3	114.69
O-H3	0.961	H2-O-H3	114.68
Aqueous Solution			
O-H1	0.990	H1-O-H2	110.66
O-H2	0.990	H1-O-H3	109.54
O-H3	0.985	H2-O-H3	110.39

and Tomasi [80]) to represent the solvent. The result of such calculations are given in Table VII.7.

While the hydronium ion has a C_{3v} symmetry in the gas phase, in aqueous solution small but nonnegligible geometric modifications are observed. The molecule is less flat, and the chemical bonds are slightly more elongated. At the level of theory employed, one of the bonds is slightly distinct from the others, reacting differently to the presence of the solvent. Certainly, larger clusters can be treated theoretically but the question is how reliable they are as physical descriptions for the proton in solution. Several important chemical and biochemical processes involve protons in aqueous medium and, from the theoretical point of view, many different methodologies are available to include the solvent molecules in the calculations. Thus, another aspect to be considered is how large should the $H^+(H_2O)_n$ cluster be in order not only to provide a reliable description of the process one is trying to calculate, but also to conform with the basic assumptions of the model being used to represent the solvent. We will return to this point when discussing pK_a calculations in aqueous solutions.

C. The Proton Solvation Energy

From the definition of acidity in solution as presented in Eq. (7.4), it is clear that in order to compute $\Delta G°$ one must know the proton solvation energy. However, from the previous discussion on the structural models for the hydrated proton one may anticipate some difficulties. For example, how many water molecules should be considered in the calculation of the proton solvation energy? In other words: How large should one take the $H^+(H_2O)_n$ cluster? One reasonable approach should be to examine the convergence of the free energy change of the reaction

$$H^+_{(g)} + [H_{2n}O_n]_{(g)} \rightarrow [H^+(H_{2n}O_n)]_{(g)}$$

with increasing n. Of course, as the value of n increases high accurate ab initio calculations turn out to be more intense and may become prohibitive even before convergence is achieved (Tawa et al. [98]). Another possibility would be to represent the proton hydration process as

$$H^+_{(g)} + [H_{2n}O_n]_{(aq)} \rightarrow [H^+(H_{2n}O_n)]_{(aq)}$$

and again compute the convergence of the hydration free energy by increasing n. In the reaction above (aq) represents the species $H^+(H_2O)_n$ embedded in a continuum medium, under the assumption that explicit water molecules are not needed to represent the bulk of the solvent. If this is the case, the convergence should be achieved for a value not too large for n. On the other hand, the diffusion simulations of Marx et al. [97] showed that there are no fixed $H^+(H_2O)_n$ structures in liquid water during a time interval Δt large enough to be characterized.

In spite of all these difficulties, several calculations and experimental determinations of the proton enthalpy of solvation have been reported. Some of the results obtained are summarized in Table VII.8.

The sources of data in Table VII.8 are quite diverse. The value proposed by Reiss and Heller [101] resulted from electrochemical redox potential measurements involving the absolute potential of the standard hydrogen electrode, a type of measurement that requires some assumptions, since just

TABLE VII.8
Proton Solvation Enthalpy Values (kcal/mol)

Reference	$\Delta H^\circ_{solv}(H^+)$
Aue et al. [99]	−269.44 (E)[a]
Marcus [100]	−261.22 (E)
Reiss and Heller [101]	−259.5 (E)
Rashin and Namboodiri [102]	−259.75 – 264.05 (T)
Coe [103]	−275.72 (E)
Bockris and Reddy [90]	−265.30 (E)
Tawa et al. [98]	−267.28 (T)
Tissandier et al. [104]	−274.88 (E)
Mejías and Lago [105]	−274.86 (T)

[a](E) and (T) indicate theoretical or experimental data.

relative quantities are indeed accessible. Tissandier et al. [104], from a set of cluster-ion solvation data obtained the most accepted and used $\Delta H^{\circ}_{solv}(H^+)$ value found in the literature. Recently Mejías and Lago [105], using DFT and continuum solvation models, studied a series of clusters of different sizes, arriving at a value for $\Delta H^{\circ}_{solv}(H^+)$ very close to that of Tissandier et al. [104].

From Table VII.8, three ranges of $\Delta H^{\circ}_{sol}(H^+)$ values, ~ 260, 268, and 275 kcal/mol, can be distinguished. The difference of 15 kcal/mol between the two extreme values is quite large and somehow reflects the difficulty in determining this property. However, the excellent agreement between the Tissandier et al. [104] value, obtained from cluster-ion experiments, and the theoretical prediction of Mejías and Lago [105] could be an indication that we are finally converging to a reliable value of the proton enthalpy of solvation. This agreement is particularly significant because cluster-ion experiments (contrary to the traditional experimental methods) allow the investigation of isolated single ions as found in the theoretical approaches. It is important to mention that the Mejías and Lago [105] best model of a hydrated proton consisted of a H_3O^+ ion surrounded by an explicit hydration shell made up 12 water molecules.

Much less information is available for the proton free energy of solvation, most certainly because of problems related to the determination of the entropic contribution $T\Delta S^{\circ}_{solv}$. From a theoretical point of view, this contribution is much more sensitive than the enthalpic term for the choice of a model for the solvated proton and also to the level of theory employed in the calculations. Table VII.9 summarizes the results for the proton solvation free energy reported in the literature.

The range of values is even wider than the one observed for the enthalpy of solvation. In this case, the theoretical results of Mejías and Lago [105] differ appreciably from the experimental data of Tissandier et al. [104]. The apparent good agreement between the theoretical results of Tawa et al. [98]

TABLE VII.9
Proton Solvation Free Energy Values (kcal/mol)

Reference	$\Delta G^{\circ}_{solv}(H^+)$
Lim et al. [61]	-259.50 (E)[a]
Marcus [100]	-252.36 (E)
Tawa et al. [98]	-262.23 (T)[a]
Tissandier et al. [104]	-263.98 (E)
Mejías and Lago [105]	~ -246.00 (T)

[a](E) and (T) indicate theoretical or experimental data.

TABLE VII.10
Free Energy Values for H_3O^+ Solvation [$\Delta G°_{solv}(H_3O^+)$] (in kcal/mol) Following the Ben-Naim [106] and Ben-Naim and Marcus [108] Definition for a Solvation Process

Different Continuum Solvation Models	HF[a]	MP2[a]
Jaguar solvation module[b]	−98.88	−99.94
PCM[c]	−108.30	
IPCM[d]	−99.53	−98.41
SCIPCM[e]	−106.83	−105.60
Barone et al. [111][f]	−105.34	
Experimental[g]	−102	

[a] HF//6-31 + G(d,p) and MP2/6-31 + G(d,p)//HF/6-31 + G(d,p).
[b] Jaguar 3.a, Schrödinger Inc., Portland, OR, 1998; Tannor et al. [109], Marten et al. [110].
[c] Cammi and Tomasi [80], adopting UATM cavity model (Barone et al. [111]).
[d] Foresman et al. [112].
[e] To the best of our knowledge, this model has not been discussed in the literature.
[f] HF//6-31 + G(d).
[g] Pearson [107].

and the experiments of Tissandier et al. [104] can be attributed to a cancellation of errors in the calculations, since their enthalpy of solvation differ by ∼7 kcal/mol. The accurate calculation of the entropic contribution is still a challenging problem and for the time being the Tissandier et al. [104] value should be taken as the most accurate for the proton solvation energy.

In some circumstances, it might be convenient to take the hydronium ion as the basic unit for a proton in solution. In those cases, what is needed is the hydronium solvation free energy, $\Delta G°_{solv}(H_3O^+)$. Table VII.10 shows the results of theoretical calculations (da Silva [44b]), unless otherwise specified) obtained with different continuum solvation models, according to the Ben-Naim [106] definition of the solvation process. Note that the only experimental value available (Pearson [107]) was obtained assuming $\Delta G°_{solv}(H^+) = -259.5$ kcal/mol, which differs considerable from the more recent experiments of Tissandier et al. [104]. However, if one adopts Tissandier's result for the proton solvation energy, Pearson's scheme furnishes $\Delta G°_{solv}(H_3O^+) = -106.16$ kcal/mol, a value in close agreement with some of the theoretical results shown in Table VII.10.

D. The Solvation of Anions

Before proceeding to the pK_a calculations, it would be instructive to briefly examine the solvation energy of anions and neutral species. Since $\Delta G^\circ_{solv}(H^+)$ is a constant in a given solvent, one should expect that differences in acidities in polar solvents, brought about by structural variations of the species AH, are largely due to changes in the solvation energy of the conjugated bases A^-.

To illustrate this point, let us analyze the theoretical results of ΔG_{solv} in water for some carboxylic acids and their conjugated bases, shown in Table VII.11 (da Silva [44b]).

As expected, the solvation energies of the anions are larger (~ 10 times) than the respective neutral species in a polar solvent. Since the structural changes in the successive members of the homologous series are relatively small, the ΔG°_{solv} of the neutral species do not differ appreciably. On the other hand, the solvation free energy of formate and acetate ions differ by 1.83 kcal/mol, implying a difference of ~ 1.3 pK_a units for the respective neutral acids. Experimentally, the pK_a values for these acids differ by 1.01 pK_a units. For the acetate–propionate pair, the difference in the solvation free energy is much smaller (~ 0.22 kcal/mol) and therefore one should expect a small difference in the pK_a values of the respective acids. This is indeed the case, as their experimental pK_a values differ by 0.11 pK_a units.

More recently Jaworski [114] examined the relative acidities for a series of substituted benzoic acids, in a number of solvents, and concluded that

TABLE VII.11
Solvation Energy Values in Aqueous Solution, Calculated at HF//6-31 + G(d,p) Level, Using the UATM (Barone et al. [111]) Cavity Model[a]

Compound	ΔG_{solv}(kcal/mol)	Experimental
HCOOH	−8.73	−6.58[b]; −5.5[c]
HCOO$^-$	−77.89	
CH$_3$COOH	−8.15	−6.69[b]
CH$_3$COO$^-$	−79.72	−82.2[d]; −77[e]
CH$_3$CH$_2$COOH	−8.49	−6.86[b]
CH$_3$CH$_2$COO$^-$	−79.94	

[a] da Silva [44b].
[b] Cabani et al. [113].
[c] Ben-Naim and Marcus [108].
[d] Lim et al. [61].
[e] Pearson [107].

differences in pK_a values are also dominated by the anion solvations.

The same methodologies used to compute ΔG_{solv} for cations and neutral species can be used as well for the anions. However, care must be exercised when dealing with continumm models. Because of the diffuse character of the anionic species, the tail of the radial part of their wave functions may be long enough to escape from the cavity boundary. Thus, at the end of the self-consistent procedure, which furnishes the solute wave function in the presence of the solvent, the summation of the apparent charges obtained (the charges that represent the solvent polarization) may not be exactly equal to the number of electrons of the solute. This problem, known as tail error, is common to all solutes in continumm solvent descriptions, but is especially important in the case of anions. For continuum models of the apparent surface charge (ASC) type, it is easier to correct such an error. In particular, for the PCM model, Cossi et al. [115] developed mathematical corrections for taking this problem into account.

Systematic studies of anions solvation are scarce except for those directly related to pK_a calculations. Nevertheless, a few examples of ab initio studies on the solvation energy of anions can be found in the literature. Barone et al. [111] examined the solvation energy of a few ions in the process of establishing the best parameters to build molecular cavities in continuum solvation models. Topol et al. [116] studied the influence of including explicit water molecules in the solute cavity, for the solvation energy of the anionic form of halogens. Converged values of solvation energy were obtained with four molecules of water inside the cavity. More recently, Toth et al. [117] examined the influence of diffuse functions on the solvation energy of carboxylic acids and their conjugated bases. By using the CPCM model (Barone and Cossi [118]), they observed that the addition of a diffuse function to the 6-31G(d) basis set causes an average lowering of 0.91 kcal/mol in the solvation energy of the acids and of 0.64 kcal/mol for the respective anions. If a geometry reoptimization is performed with the extended basis set (6-31 + (d)), an additional average lowering of 0.12 kcal/mol is obtained for the ΔG_{solv} of the acids and of 0.44 kcal/mol for the respective anions. The most complete study of anions solvation energy was presented by Li et al. [87] in a paper describing several parameterizations of the SM5.42R solvation model. The parameters furnished solvation energies for 49 ionic solutes in water, with mean unsigned errors (mue) of 3.5–3.8 kcal/mol for the HF, BPW91, and B3LYP parameterizations, and with mue of 4.0–4.1 kcal/mol for the semiempirical AM1 and PM3 parameterizations.

E. pK_a Calculations

In order to establish an acidity scale based on the pK_a values, one needs to determine $\Delta G°$ for the process indicated in Eq. (7.8), which was used to define the acidity of species AH in a solvent S. For the process taking place in aqueous solution:

$$\Delta G°_{aq} = G°(A^-)_{aq} + G°(H^+)_{aq} - G°(AH)_{aq} \quad (7.26)$$

However, absolute values of the Gibbs free energy

$$[G°(X), X = A^-, H^+, AH]$$

can only be computed or experimentally measured if some arbitrary reference state is considered. In general, the thermodynamic reference state is based on hypothetical solutions of species AH, A^-, and H^+, with unit activity, at 298 K and 1 atm of pressure. On the other hand, this is not the most convenient reference state for theoretical calculations. For example, in the continuum models (Tomasi and Persico [69]) the reference state for a solute M corresponds to the nuclei and electrons at infinite separation in the presence of a nonpolarizable dielectric (Böttcher [78]).

To circumvent this problem one can make use of the fact that $\Delta G°$ is a state function, and imagine some other process with the same initial and final states. Alternatively, one can construct thermodynamical cycles from which the desired information may be extracted. Let us examine two of the most used cycles.

1. The TC1 or Proton Cycle

The TC1 or proton cycle is shown in Figure 7.4.

According to this cycle, the acidity of species AH in water can be written as

$$\Delta G°_{aq} = -\Delta G_{solv}(AH) + \Delta G°_g + \Delta G_{solv}(A^-) + \Delta G_{solv}(H^+) \quad (7.27)$$

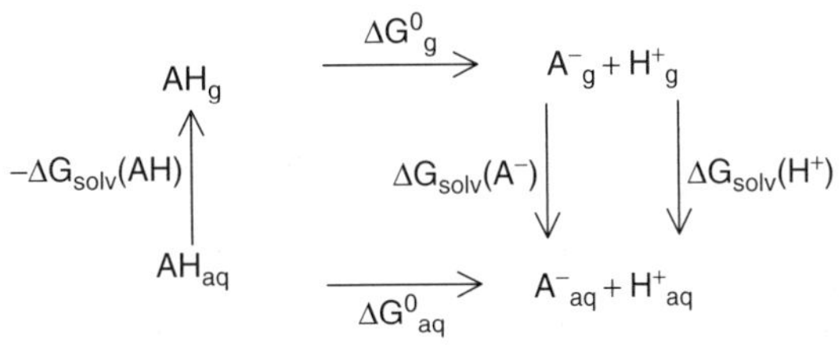

Figure 7.4. Thermodynamic cycle (TC1).

Equation (7.27) expresses the acidity in solution in terms of solvation energies and the gas-phase acidity of species AH. These quantities can be computed using the methodologies described in the previous sections.

At this point, even the nonexpert reader realizes that calculation of acidities in solution are much more difficult than in the gas phase. Not surprisingly, systematic studies such as those reported in Section III are practically nonexistent with the TC1 cycle. Except for the work by Richardson et al. [119], who considered a set not too large but diversified of compounds, most of the other papers report pK_a calculations for small sets of compounds structurally very similar.

Topol et al. [120] investigated some imidazole derivatives at different levels of theory. The results in Table VII.12 show that pK_a values differing from experiments by <0.3 units on average can be obtained with the G2 family of methods and also at the DFT/B3LYP level. The authors used a particular self-consistent reaction field cycle to obtain the solvation energies, and $\Delta G°(H^+) = -262.5$ kcal/mol.

TABLE VII.12
Absolute pK_a Values for the Deprotonation Reaction IMH$^+$—R → IM—R + H$^+$ Calculated at Different Levels

	R			
Method	H[a]	NH$_2$[a]	CH$_3$[a]	Cl[a]
HF/6-31G(d)[b]	10.29(+3.28)	15.48(+7.02)	11.55(+3.61)	5.10(+1.55)
B3LYP/6-31+G (d,p)[b]	5.37(−1.58)	9.64(+1.18)	6.78(−1.16)	1.23(−2.32)
G2MP2[b]	5.14(−1.87)	8.85(+0.39)	6.12(−1.82)	1.43(−2.12)
HF/6-31G(d)[c]	11.79(+4.78)	16.39(+7.93)	12.70(+4.76)	7.16(+3.61)
B3LYP/6-31+G(d,p)[c]	6.88(−0.13)	10.56(+2.10)	7.93(−0.01)	3.28(−0.29)
G2MP2[c]	6.64(−0.37)	9.76(+1.30)	7.26(−0.68)	3.49(−0.06)
G2[c]	6.54(−0.47)			
B3LYP/6-31+G (d,p)[d]	7.48(+0.47)			
G2MP2[d]	7.42(+0.23)			
G2[d]	7.14(+0.13)			
Lim et al. [61]	15.30(+8.29)			
Chen et al. [121]	7.60(+0.59)			
Experiment[e]	6.95 ⇔ 7.07	8.40 ⇔ 8.52	7.88 ⇔ 8.00	3.55

[a]The numbers in parentheses are the differences between the calculated and experimental pK_a values.
[b]Atomic radii from Rashin et al. [122]; the electrostatic component of the solvation energies are calculated at B3LYP/6-311+G(d,p) level.
[c]Atomic radii adopted by the authors (2.23 Å for aromatic carbon).
[d]Experimental solvation free energies (Bartmess et al. [123]) were used.
[e]Gabellin [124].

Shapley et al. [125] computed the pK_a of benzoic and o-, m-, and p-hydroxy benzoic acids in order to understand the higher ability of the o-hydroxy benzoate to complex to the cationic surfactant headgroups in micelles and adsorbed films. Although the pK_a values did not show good agreement with the experimental data, the calculations provided a clear explanation for the unusual behavior of the o-hydroxy benzoate.

Recently Topol et al. [126] calculated pK_a values for a set of weak organic acids in water, using a combination of ab initio, DFT, and continuum solvation methods. The compounds chosen were hydrocarbons whose pK_a values range from 14 to 45 pK_a units. Theoretical calculations are particular valuable in this range of pK_a values because water, a protic solvent ($pK_a = 15.7$), will interfere in the experimental measurements of acids with pK_a values > 15.7. The calculations have been performed at the G2-MP2 and DFT/B3LYP levels of theory and for most of the studied compounds, the results obtained with the two methods are in close agreement. Interesting enough, the authors found that the inclusion of explicit water molecules in the solute cavity of compounds with pK_a values > 40 units leads to innapropriate description of the corresponding anion solvation.

Jang et al. [12] calculated the pK_a values of uracil and some of its five-substituted derivatives, using the DFT/B3LYP method in combination with the PB continuum solvation model (Fig. 7.5).

For this kind of system, the interpretation of the experimental data is complicated by the fact that ionization can occur at either the N1 or N3 sites. By choosing $\Delta G°_{solv}(H^+)$, so as to minimize the root-mean-square deviation between the calculated and experimental pK_a values, the authors were able to show how the substituents change the preference for ionization at N1 and N3. For each one of the derivatives, either the calculated pK_a at N1 or at N3 was found to differ at most by 0.6 pK_a units from the experimental measurements.

Liptak and Shields[127], in a very recent publication, presented extensive pK_a calculations for some carboxylic acids. Using CBS and Gn methods

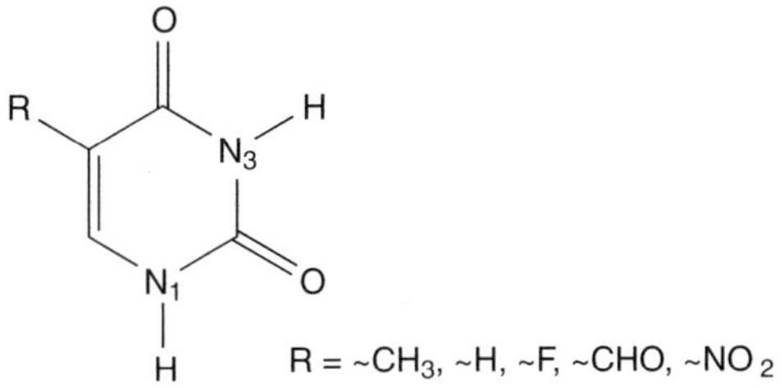

R = ~CH₃, ~H, ~F, ~CHO, ~NO₂

Figure 7.5. Five-substituted uracils.

combined with CPCM continuum solvation method, the authors obtained pK_a values accurate to less than one-half of a pK_a unit. These are undoubtedly the best theoretical pK_a results so far reported for the acids considered. However, it is worth mentioning that while highly accurate methods have been used for the gas-phase calculations, the solvation energies were computed at the HF level.

In another interesting application of the TC1 cycle, Li et al. [128] examined the pK_a of proteins. This paper represents the first attempt at predicting proteins pK_a values using an ab initio QM/MM description of the protein in combination with a polarizable continuum model for the solvent. Briefly, the ionizable residue of the protein is treated quantomechanically while the rest of the structure is represented by an MM force field. This QM/MM representation of the proteins is combined with a linearized Poisson–Boltmann equation (LPBE) description of bulk solvation. By using this procedure, the authors predicted pK_a of Glu 43 (4.4 units) and Lys 55 (11.3 units) in Turkey Ovomucoid third domain that are in very good agreement with the experimental values of 4.8 and 11.1 units, respectively.

2. The TC2 or Hydronium Cycle

In order to introduce the TC2 cycle, let us consider the following process:

$$AH_{aq} + H_2O_{aq} \xrightleftharpoons{K'_{eq}} A^-_{aq} + H_3O^+_{aq} \tag{7.28}$$

with

$$K'_{eq} = \frac{[A^-_{aq}][H_3O^+_{aq}]}{[AH_{aq}][H_2O_{aq}]} \tag{7.29}$$

Recall that the process that defines the gas-phase acidity of species AH is

$$AH_g \xrightleftharpoons{K^g_{eq}} A^-_g + H^+_g$$

with $\Delta G^\circ_g = 2.303RT(pK_a)_g$.

By assuming in Eq. (7.29) that all of the H_3O^+ species come from H^+ solvated by one molecule of water ($[H_3O^+_{aq}] = [H^+_{aq}]$), the constants K'_{eq} and K^g_{eq} can be related

$$K'_{eq} = \frac{K^g_{eq}}{[H_2O_{aq}]} \tag{7.30}$$

Thus, the relationship between $\Delta G°$ for reaction (7.28) and the pK_a in aqueous solution becomes

$$\Delta G°_{aq} = -2.303RT \log\left(\frac{K^g_{eq}}{[H_2O_{aq}]}\right) \quad (7.31)$$

or

$$\Delta G°_{aq}(\text{kcal/mol}) = 1.36\, pK_a + 2.36 \quad (7.32)$$

using R and T equal 1.98 cal/mol K and 298.15 K, respectively.

The TC2 cycle is shown in Figure 7.6 from which one can write

$$\Delta G°_{aq} = -\Delta G_{solv}(AH) - \Delta G_{solv}(H_2O) + \Delta G°_g + \Delta G_{solv}(A^-) + \Delta G_{solv}(H_3O^+) \quad (7.33)$$

where each term can be calculated as previously discussed.

Schüürmann [129] was the first one to use this approach, but without any explicit reference to the thermodynamic cycle. The author examined the pK_a of 16 carboxylic acids and 15 chlorinated phenols, using the semiempirical AM1 method to describe the solutes and different continuum models to represent the solvent. Additional ab initio calculations were performed for the carboxylic acids. His main conclusion was that "current semiempirical continuum solvation models predict solution-phase acidity in qualitative agreement with experiment, but are not sufficiently accurate to yield absolute pK_a values". In another paper, Schüürmann [130] extended his investigation to a set of 52 organic compounds with pK_a values ranging from −11 to 25, still using semiempirical AM1 models. Once more the results were very disappointing. Linear correlations with the experiment were also attempted. From Table VII.13, it is seen that much better linear

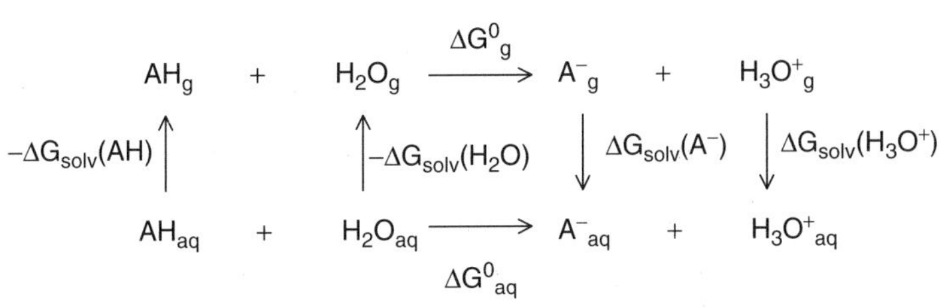

$$\begin{array}{ccccccc}
 & & & \Delta G°_g & & & \\
AH_g & + & H_2O_g & \longrightarrow & A^-_g & + & H_3O^+_g \\
\uparrow -\Delta G_{solv}(AH) & & \uparrow -\Delta G_{solv}(H_2O) & & \downarrow \Delta G_{solv}(A^-) & & \downarrow \Delta G_{solv}(H_3O^+) \\
AH_{aq} & + & H_2O_{aq} & \longrightarrow & A^-_{aq} & + & H_3O^+_{aq} \\
 & & & \Delta G°_{aq} & & &
\end{array}$$

Figure 7.6. Thermodynamic cycle (TC2).

TABLE VII.13
Linear Regression Analysis for Solution Phase pK_a

	Method[a]	r^2	SE[b]
All 52 compounds	SM2	0.69	4.04
Subset of 32 compounds[c]	COSMO	0.93	2.00
Subset of 20 compounds[c]	MST	0.92	1.92

[a] Method refers to the solvation model used, AM1-SM2 (Cramer and Truhlar [131]), AM1-COSMO (Klamt and Schüürmann [84]) or AM1-MST (Miertuš et al. [68], Luque et al. [132]).
[b] Standard error.
[c] The original set of 52 compounds was divided in 2 subsets of 32 and 20 compounds each.

correlations are obtained when the original set of compounds is divided in two smaller subsets, but the standard errors are still too high. However, it should be noted that the best correlation within each set of compounds was reached using different solvation models.

In a subsequent paper, Schüürmann et al. [129] studied a set of 16 compounds, using HF and MP2 descriptions, and the PCM (Miertuš et al. [68]) for calculating the solvation energies. The authors concluded that pK_a calculations are "mainly governed by the level of theory of the underlying gas-phase calculations". Additionally, they concluded that HF and MP2 levels are not sufficient for prediction of absolute pK_a values.

da Silva et al. [44b, 133] used the TC2 approach to investigate the pK_a of 15 compounds. The authors introduced a small modification in the step of the TC corresponding to the process of water solvation in water. In this case, adopting the thermodynamic definition of solvation energy would be physically unsound because there is no way of distinguishing the dissolved water molecule from all the others. Thus, it should be more reasonable to identify the reverse process, that is,

$$H_2O_{aq} \to H_2O_g$$

with the vaporization of water and use $\Delta G°_{vap}(H_2O)$ instead of $\Delta G°_{solv}(H_2O)$. The calculations have been performed at the HF//6-31 + G(d,p) level in combination with the PCM continuun solvent model. This level of calculation was the same used in the PCM model (Barone et al. [111]) to parameterize the cavities such as to obtain solvation energies in close agreement with the experiments. Consequently, in order to keep the same

COMPUTATIONAL CHEMISTRY OF ACIDS 457

TABLE VII.14
pK_a Results Compared to Experimental and Other Theoretical Calculations

Compound	Schüürmann et al. [134][a]	Richardson et al. [119]	da Silva et al. [133]	Experimental
CH_3OH		22.5	20.14	15.5^b; 16^c
CH_3CH_2OH			18.84	15.9^b; 18^c
$CH_3CH_2CH_2OH$			19.37	16.1^b; 18^c
$CH_3CHOHCH_3$			20.56	17.1^b; 18^c
CH_3SH		9.3	5.81	10.33^b
CH_3CH_2SH			6.15	10.61^b
H_2O		16.4	16.77	15.74^b
HCOOH	6.87	1.5	2.92	3.75^d
CH_3COOH	8.17	4.9	4.32	4.76^d
CH_3CH_2COOH	9.21		4.76	4.87^d
$CH_3(CH_2)_2COOH$			4.59	$\approx 4.82^d$
$(CH_3)_3CCOOH$	10.15		5.42	5.05^d
FCH_2COOH	5.25		1.14	2.66^d
$ClCH_2COOH$	5.18	0.8	1.57	2.86^d
$BrCH_2COOH$	5.37		0.63	2.86^d

[a] Values obtained as shown in da Silva et al. [135].
[b] Stewart [136].
[c] McEwen [137].
[d] March [138].

description in both phases, the calculations in the gas phase were also performed at the HF level. Geometry relaxation was also considered, when bringing any solute molecule from the gas to liquid phase.

While the results for the carboxylic acids are in good agreement with the experiment (Table VII.14), the pK_a values for the haloacetic acids and thiols are not so well described. Some discrepancies are also observed for the alcohols. Nevertheless, considering the experimental difficulties for measuring pK_a values of aqueous solutions in a range of values close or above the water pK_a (15.7), the theoretical results should be more reliable for these compounds.

In spite of the fact that correlation effects were neglected in both phases, the TC2 cycle does provide reliable pK_a values for compounds for which differential correlation effects along the cycle are minimal. This happens when the TC proton-transfer steps become isodesmic reactions, as in the case of alcohols, carboxylic acids, and other compounds for which an O—H bond is broken and another O—H bond is being formed (H_3O^+). The reasons for the discrepancies in the pK_a values of thiols and haloacetic acids are also discussed in the original paper.

F. Relative pK_a Calculations

One source of uncertainty in pK_a calculations is undoubtedly the model to be used for the proton in solution. In order to avoid this problem, one may choose to compute relative pK_a values. The basic idea is to consider the thermodynamic cycles for two compounds, AH and BH, and subtract the respective ΔG_{aq}°, to obtain

$$\Delta(\Delta G_{aq}^\circ) = \Delta G_{aq}^\circ(\text{BH}) - \Delta G_{aq}^\circ(\text{AH}) \tag{7.34}$$

The quantity $\Delta(\Delta G_{aq}^\circ)$ can also be related to the difference in the pK_a values of acids AH and BH.

$$\Delta \text{p}K_a = \text{p}K_a(\text{BH}) - \text{p}K_a(\text{AH}) = \Delta(\Delta G_{aq}^\circ)/2.303RT \tag{7.35}$$

It is clear that once the expressions for $\Delta G_{aq}^\circ(\text{AH})$ and $\Delta G_{aq}^\circ(\text{BH})$ derived from either the TC1 or TC2 cycle are inserted into Eq. (7.34), the contribution of the proton solvation energy, $\Delta G_{aq}^\circ(\text{H}^+)$, cancels out.

To illustrate this approach, let us consider a recent paper by Toth et al. [117]. The authors used CBS methods (Petersson et al. [21]) and the G2 family of methods to compute gas-phase energy differences between six different carboxylic acids and their respective anions. Two different continuum solvation methods, SM5.42R (Li et al. [139]) and CPCM (Barone and Cossi [118]) were used to calculate the differences in solvation free energies for the acids and their anions. By using this data, relative pK_a values were determined for each acid using one of the acids as reference. They found that the pK_a value of an unknown from a known molecule can be predicted with an average error of 0.4 pK_a units. It must be noted, however, that this level of accuracy was reached using different levels of theory to describe the processes occurring in the gas phase and in solution.

Relative pK_a calculations are also useful to predict the acidity of electronically excited species, taking the ground-state value as reference. Gao et al. [59] used the TC1 cycle to predict the pK_a of the first singlet excited state of phenol (PhOH*). The calculations were performed using a hybrid quantum mechanical and molecular mechanical method (QM/MM) where the solute is treated quantum mechanically while the solvent molecules are represented by an MM force field. They obtained a ΔpK_a (PhOH $\rightarrow$ PhOH*) of -8.6 ± 0.1 pK_a units, implying a pK_a value of 1.4 for PhOH*.

V. MOLECULAR ASSOCIATIONS

Besides the acidity, much additional important information concerning the structural properties and reactivity of acids can be obtained from theoretical calculations. Depending on the nature of the problem being investigated, theoretical calculations can provide reliable results of geometries, multipole moments, vibrational frequencies, chemical NMR shifts, just to mention a few properties. The degree of accuracy expected is higher for calculations in the gas phase and the reader can certainly appreciate the reason for that. The subject is vast and we will only briefly mention some interesting questions related to molecular associations.

Molecular interactions involving carboxylic acids are very common in molecular recognition. The ability of carboxylic acids to behave as a hydrogen-bond donor–acceptor is responsible for many interesting chemical (Pistolis et al. [140]), Wash et al. [141]) and biochemical (Beveridge and Heywood [142]) properties of these compounds. For example, there is increasing evidence supporting the fact that neutral forms of aspartic and glutamic acids exist inside proteins and play an important role in the mechanism of the action of proteins. These findings stimulated new studies on the molecular association of acids, in particular on the dimerization of carboxylic acids. In the gas-phase, carboxylic acids generally occur as dimers but in solution this situation can change due to competitive solute–solvent interactions.

Colominas et al. [143] published a very detailed study on the dimerization of formic and acetic acids in the gas phase and in aqueous and chloroform solutions. By using quantum mechanical self-consistent reaction field and Monte Carlo calculations, they showed that the dimerization is favored in the gas phase and in a chloroform solution (1 M) basically due to double hydrogen-bonded interaction between the monomers. In aqueous solution, the dimerization does not seem to occur due to the competitive solute–solvent intermolecular interactions. The computed dimerization energies are shown in Table VII.15

The dimerization in the gas phase and in a chloroform solution is slightly favored for CH_3COOH, due to the inductive effect of the methyl

TABLE VII.15
Gibbs Free Energy for the Dimerization of Carboxylic Acids

	$\Delta G_{dim, 298 K}$ (kcal/mol)	
	HCOOH	CH_3COOH
Gas phase	−2.5	−3.2
CH_3Cl (1 M)	−2.3	−3.1
H_2O (1 M)	+5.3	+4.3

TABLE VII.16
Enthalpy of Dimerization for Phosphinic Acids (kcal/mol)

	ΔH_{dim}	Experimental
PA	−23.2	
DMPA	−23.2	−23.9 ± 6[a]

[a] Denisov and Tokhadze [148].

groups, which stabilizes the double hydrogen-bond complex between the monomers.

Siasios and Tiekink [144] and González et al. [145] studied the structure and binding energy of dimers of phosphinic acid (R_2POOH) (PA) and its dimethyl derivative (DMPA) at the MP2 and B3LYP levels of theory. Similar to their carboxylic analogues, both phosphinic acids form cyclic dimers in the gas phase. However, the hydrogen bonds in PA and DMPA dimers are stronger than those of their carboxylic analogues. The estimated dimerization enthalpies, shown in Table VII.16, are almost twice those measured for formic and acetic acids (≈ -14 kcal/mol) (Clague and Bernstein [146], Mathews and Sheets [147]).

Carboxylic acids may also form an orientationally disordered crystalline (ODIC) or plastic phase, with cubic or hexagonal symmetry, usually exhibited by molecular solids of spherical or globular molecules. As pivalic acid [$(CH_3)_3CCOOH$] is one representative of these plastic crystal-forming molecules, its chlorinated derivatives were used to investigate how the substitution of hydrogen atoms by the more electronegative chlorine would affect the hydrogen bonding and consequently the formation of the ODIC phase (Šablinskas et al. [149]). Curiously, while the heats of formation of the cyclic dimers of pivalic and 3-chloropivalic acids in the gas phase were practically the same, the chloro acid does not form the plastic phase.

Aminova et al. [150] investigated the structure of some hydrogen-bonded complexes [$(HCOOH)_n$, $n = 1-4$] for formic acid and for the dimers of acrylic acids using MP2 and B3LYP calculations. For all the polymers of formic acid (dimer, trimer, and tetramer) the cyclic forms were found to be more stable than the chain structures. The hydrogen bond is strongest for the dimers and gets weaker as the number of associated molecules increase. A similar result was found for the dimer of acrylic acid, the cyclic structure being the more stable.

Ionic assemblies of acetic acid and water were investigated by Meot-Ner(Mautner) et al. [151]. Several interesting results emerged from their calculations. For the $CH_3COO^- \cdots (CH_3COOH)(H_2O)_2$ clusters, they found

(a) a variety of geometries very close in energy; and (b) that directly bonded and solvent-bridged isomers have comparable stabilities and that the bridged H_2O molecule furnishes a low-energy pathway for proton transfer through a H_3O^+ bridge intermediate. These results have some important biological implications as discussed by the authors.

Complexes involving benzoic acid (BA), benzene, and water molecules have received much attention in the last years (Sagarik and Rode [152] and references cited therein). Benzoic acid and benzene are used as model compounds in the investigation of the $\pi-\pi$ interaction, supposedly responsible for the parallel stacking arrangements found in aromatic side chains of proteins frequently observed in protein crystal structures. Interation energies and structures of benzoic acid–water (BA–H_2O) 1:1, 1:2, 2:1, 2:2 complexes were investigated using intermolecular potentials derived from the test-particle model (T-model), developed by the authors. In such a model, the BA geometry is optimized at the MP2/6-31G(d,p) level of theory, and the optimum geometry is used to construct the T-model. Based on the T-model potentials, aqueous solutions of BA and BA_2 were investigated through molecular dynamic simulations. The authors found that from the energetic viewpoint the degrees of molecular association in the BA–H_2O–benzene system decrease in the sequence:

$(BA)_2$ > BA–H_2O 1:1 complex > $(H_2O)_2$ > benzene–BA 1:1 complex
> benzene–H_2O 1:1 complex > $(benzene)_2$

VI. CONCLUDING REMARKS

Concerning gas phase acidity calculations, from the results of the systematic studies discussed in Section III, one may conclude that highly accurate values of gas-phase acidities (within ≤ 1 kcal/mol from experiments) can be obtained either by the G2 method or DFT calculations using the B3LYP functional and triple-zeta quality basis sets, including polarization and diffuse functions. However, it is important to keep in mind that this level of accuracy is only reached if the geometry optimizations are conducted at the same level of theory used to compute the acidity. Good estimates can still be obtained at lower levels of theory using either the MP2 method or the B3LYP functional with more modest basis sets. On the other hand, while the accuracy of the G2 method seems to be quite independent of the structure of the acid, the B3LYP functional is less accurate when the proton being abstracted is bound either to a tertiary or to a strained center, as revealed by the calculations of Burk and Sillar [42] and Sauers [41]. From the studies on strong mineral acids and superacids one can also conclude that the DFT/B3LYP method overestimates the acidity of superacidic and strongly acidic compounds ($\Delta G_{ac} \leqslant 340$ kcal/mol) and underestimates the acidity of the weakest acids ($\Delta G_{ac} \geqslant 380$ kcal/mol). In the range of moderately acidic compounds (340 kcal/mol $\leqslant \Delta G_{ac} \leqslant 380$ kcal/mol) the method yields rather good or satisfactory agreement with the available experimental results.

Alternatively, good quality ab initio gas-phase acidities can also be obtained, at costs comparable to B3LYP calculations, by using a localized MP2 description to compute just the differential correlation effects associated to the breaking of an A—H bond.

While highly accurate values of gas-phase acidities can be obtained, the same is not yet true for pK_a calculations in solution.

For aqueous solutions, the choice of a model to describe the solvated protons was always a delicate problem for the computation of absolute pK_a values. On the other hand, the lack of accurate data certainly prevented a more intense use of experimental values of the proton enthalpy and entropy of solvation in theoretical calculations of pK_a values. Fortunately, this situation has changed in the last 2 years. The excellent agreement between a high level calculation (Mejías and Lago [105]) and cluster-ions experiments (Tissandier et al. [104]) points to a value of -274.9 kcal/mol for the proton enthalpy of solvation. Such good agreement has not yet been found for the proton solvation energy, but the Tissandier et al. [104] experimental value ($\Delta G_{solv}(H^+) = -263.98$ kcal/mol) seems accurate enough to be used in pK_a calculations.

It is true that for aqueous solutions many absolute pK_a calculations in good agreement with the experiments have been reported (see preceeding sections). Nevertheless, except for one case, this agreement was achieved through unbalanced calculations. Highly sophisticated levels of theory were employed for the ionization process in the gas phase while more modest levels were used in the solution phase. By choosing levels of theory that minimize the errors in each phase, better values of ΔG_{sol} (and consequently of pK_a) can be obtained. It is most important to devise strategies capable of furnishing very accurate pK_a values in solution. On the other hand, from the theoretical point of view, it would be highly desirable to achieve good accuracy using the same level of description for both phases.

Concerning the best way of treating the solvent molecules, there is still plenty to do before this question can be answered. The supermolecule approach is quite appropriate for describing specific or short-range solvation effects. On the other hand, long-range effects require a large number of solvent molecules and the use of sophisticated ab initio methods to properly describe the solvent–solvent and solute–solvent interactions becomes prohibitive. Hybrid methods (QM/MM), where part of the solvent structure is represented by a MM force field, have not been extensively tested. In particular, the fact that the effect of the solvent is not determined by a single configuration must be taken into account when using this approach.

From the several continuum solvation methods available in the literature, the PCM model and its derivatives seem to be most used for pK_a calculations. In fact, the best pK_a results reported so far have been obtained with one of the PCM-based methods. Nevertheless, the fact that these calculations differ in many respects precludes any comparative evaluation of the different solvation models employed. More systematic studies are needed, taking into consideration the level of theory (method and basis sets) employed, the various continuum solvation models, and different classes of compounds. The introduction of solvent molecules into the solute cavity, in order to better represent the short-range solute–solvent interactions, must also be carefully examined. There is no apparent relationship between the structure of the solute and the number of intracavity solvent molecules that best reproduces the solvation energy. Hence, this best number is generally established on a trial-and-error basis. Also, the use of a such hybrid (discrete + continuum) description of the solvent molecules seems to be inconsistent with the fact that for some of the solvation models employed, the effects of the first solvation shell have been already incorporated when parameterizing the cavity.

Finally, accurate relative pK_a values can be obtained by combining the results of high-level calculations in the gas phase with CPCM continuum calculations of solvation energies.

Acknowledgments

The authors acknowledge CNPq, FAPERJ, and FINEP for financial support. We also acknowledge Dr. J. H. Jensen for making his results available prior to publication.

References

1. J. N. Brønsted, *Rec. Trav. Chim.* **42**, 718 (1923).
2. J. N. Brønsted, *Chem. Ber.* **61**, 2049 (1928).
3. R. P. Bell, in *The Proton in Chemistry*, 2nd ed. (Cornell University Press, Ithaca, NY, 1973), p. 4.
4. D. P. N. Satchell and R. S. Satchell, *J. Chem. Soc. Q. Rev.* **25**, 171 (1971).
5. A. G. Guilman, T. W. Rall, A. S. Nies, and P. Taylor, edited by, *The Pharmacological Basis of Therapeutics*, 8th ed. (Pergamon Press, New York, 1990).
6. E. C. Sherer, G. Yang, G. M. Turner, G. C. Shields, and D. W. J. Landry, *J. Phys. Chem. A* **101**, 8526 (1997).
7. J. El Yazal and Y.-P. Pang, *J. Phys. Chem. A*, **103**, 8346 (1999).
8. M. J. Miller, *Chem. Rev.* **89**, 1563 (1989).
9. M. Meot-Ner(Mautner), *J. Am. Chem. Soc.* **110**, 3075 (1988).
10. M. M. Flocco and S. L. Mowbray, *J. Mol. Biol.* **235**, 709 (1994).

11. S. M. Morris, *Mutat. Res.* **297**, 39 (1993).
12. Y. H. Jang, L. C. Sowers, T. Çagin, and W. A. Goddard, III, *J. Phys. Chem. A* **105**, 274 (2001).
13. M. R. Nelson and R. F. Borkman, *J. Mol. Struc. (THEOCHEM)* **432**, 247 (1998).
14. C. R. A. Catlow, edited by, *Modelling of Structure and Reactivity of Zeolites* (Academic Press, New York, 1992).
15. M. A. C. Nascimento, edited by, *Theoretical Aspects of Heterogeneous Catalysis* (Kluwer Academic Publishers, 2001).
16. D. R. Stull and H. Prophet, edited by, *JANAF Thermochemical Tables*, Natl. Stand. Ref. Data Ser., Natl. Bur. Stand., NSRDS-NBS 37, U.S. Govt. Print. Off. Washington, D.C., 1971.
17. J. E. Bartmess and R. T. McIver, Jr., "The Gas Phase Acidity Scale", in *Gas Phase Ion Chemistry*, Vol. 2, edited by M. T. Bowers (Academic Press, New York, 1979), Chap. 11.
18. S. G. Lias and J. E. Barmess (1998), NIST webbook (http://webbook.nist.gov)
19. J. E. Bartmess, J. L. Pittman, J. A. Aeschleman, and C. A. Deakyne, *Int. J. Mass Spectrom.* **196**, 215 (2000).
20. D. A. McQuarrie, *Statistical Thermodynamics* (Harper & Row, New York, 1976).
21. G. A. Petersson, D. K. Malick, W. G. Wilson, J. W. Ochterski, J. A. Montgomery, and M. J. Frisch, *J. Chem. Phys.* **109**, 10570 (1998).
22. L. A. Curtiss, P. C. Redfern, K. Ragavachari, V. Rasslov, and J. A. Pople, *J. Chem. Phys.* **110**, 4703 (1999).
23. R. McWeeny, *Methods of Molecular Quantum Mechanics* (Academic Press, New York, 1989).
24. A. Szabo and N. S. Ostlund, *Modern Quantum Chemistry* (McGraw-Hill, 1982).
25. F. Jensen, in *Introduction to Computational Chemistry* (John Wiley & Sons, Inc., New York, 1999).
26. R. G. Parr and W. Yang, in *Density Functional Theory* (Oxford University Press, Oxford, 1989).
27. J. Gao, D. S. Garner, and W. L. Jorgensen, *J. Am. Chem. Soc.* **108**, 4784 (1988).
28. (a) M. R. F. Siggel, A. Streitwieser, and D. T. Thomas, *J. Am. Chem. Soc.* **110**, 8022 (1988); (b) 91 (1988); (c) M. R. F. Siggel, D. T. Thomas, and A. Streitwieser, *J. Mol. Struc. (THEOCHEM)* **165**, 309 (1988).
29. X. Assfeld, M. F. Ruiz-Lopes, J. Gonzalez, R. Lopes, and T. L. Sordo JA, *J. Comp. Chem.* **5**, 479 (1994).
30. W. L. Jorgensen, J. F. Blake, D. Lim, and D. L. Severanece, *J. Chem. Soc. Faraday Trans.* **90**, 1727 (1994).
31. C. F. Bernasconi and P. J. Wenzel, *J. Am. Chem. Soc.* **123**, 7146 (2001).
32. P. Perez, *J. Phys. Chem. A* **105**, 6182 (2001).
33. J. R. Rustad, D. A. Dixon, J. D. Kubicki, and A. R. Felmy, *J. Phys. Chem. A* **104**, 4051 (2000).
34. M. Remko, *J. Mol. Struct. (THEOCHEM)* **492**, 203 (1999).
35. A. P. Ligon, *J. Phys. Chem. A* **104**, 8739 (2000).
36. B. J. Smith and L. Radom, *Chem. Phys. Lett.* **245**, 123 (1995).
37. J. Chandrasekhar, J. C. Andrade, and P. V. R. Schleyer, *J. Am. Chem. Soc.* **103**, 5609 (1981).

38. T. H. Dunning and P. J. Hay, *Modern Theoretical Chemistry*, Vol. 3, edited by H. Schaefer, III (Plenum, New York, 1977).
39. H. Kollman, *J. Am. Chem. Soc.* **100**, 2665 (1978).
40. G. N. Merrill and S. R. Kass, *J. Phys. Chem.* **100**, 17465 (1996).
41. R. R. Sauers, *Tetrahedron* **55**, 10013 (1999).
42. P. Burk and K. Sillar, *J. Mol. Struct. (THEOCHEM)* **535**, 49 (2001).
43. I. A. Koppel, P. Burk, I. Koppel, I. Leito, T. Sonoda, and M. Mishima, *J. Am. Chem. Soc.* **122**, 5114 (2000).
44. (a) C. O. da Silva, E. C. Silva, and M. A. C. Nascimento, *Int. J. Quant. Chem.* **74**, 417 (1999); (b) C. O. da Silva, Ph.D. Thesis, Universidade Federal do Rio de Janeiro (1999).
45. P. Pulay and S. Saebo, *J. Chem. Phys.* **81**, 1901 (1984).
46. R. B. Murphy, M. D. Beachy, R. A. Friesner, and M. N. Ringnalda, *J. Chem. Phys.* **103**, 1481 (1995).
47. E. L. Coitiño, R. Cammi, and J. Tomasi, *J. Comp. Chem.* **16**, 20 (1995).
48. M. R. F. Siggel and D. T. Thomas, *J. Am. Chem. Soc.* **108**, 4360 (1986).
49. O. Exner, *J. Org. Chem.* **53**, 1810 (1988).
50. M. J. S. Deward and K. L. Krull, *J. Chem. Soc. Chem. Commun.* 344 (1990).
51. F. G. Bordwell and A. V. Satish, *J. Am. Chem. Soc.* **116**, 8885 (1994).
51a. Hollauer and M. A. C. Nascimento, *J. Chem. Phys.* **99**, 1207 (1993).
52. P. C. Hiberty and C. P. Byrman, *J. Am. Chem. Soc.* **117**, 9875 (1995).
53. J. D. da Motta and M. A. C. Nascimento, *J. Phys. Chem.* **100**, 15105 (1996).
54. P. Burk and P. V. Schleyar, *J. Mol. Struct. (THEOCHEM)* **505**, 161 (2000).
55. W. L. Jorgensen, J. M. Briggs, and J. Gao, *J. Am. Chem. Soc.* **109**, 6857 (1987).
56. W. L. Jorgensen and J. M. Briggs, *J. Am. Chem. Soc.* **111**, 4190.
57. A. Pullman and B. Pullman, *Q. Rev. Biophys.* **7**, 505 (1975).
58. M. D. Newton and S. Ehrenson, *J. Am. Chem. Soc.* **93**, 4971 (1971).
59. J. Gao, N. Li, and M. Freindorf, *J. Am. Chem. Soc.* **118**, 4912 (1996).
60. A. A. Rashin, J. R. Rabinowitz, and J. R. Banfelder, *J. Am. Chem. Soc.* **112**, 4133 (1990).
61. C. Lim, D. Bashford, D., and M. Karplus, *J. Phys. Chem.* **95**, 5610 (1991).
62. A. S. Yang, M. R. Gunner, R. Sampogna, K. Sharp, and B. Honig, *Proteins: Struct. Func. Gen.* **15**, 252 (1993).
63. J. Antosiewica, J. A. McCammon, and M. K. Gilson, *J. Mol. Biol.* **238**, 415 (1994).
64. D. Rinaldi and J. L. Rivail, *Theoret. Chim. Acta* **32**, 57 (1973).
65. S. Yomosa, *J. Phys. Soc. Jpn.* **35**, 1738 (1973).
66. O. Tapia and O. Goscinski, *Mol. Phys.* **29**, 1653 (1975).
67. D. Rinaldi, M. F. Ruiz-Lopez, and J. L. Rivali, *J. Chem. Phys.* **78**, 834 (1983).
68. S. Miertuš, E. Scrocco, and J. Tomasi, *Chem. Phys.* **55**, 117 (1981).
69. J. Tomasi and M. Persico, *Chem. Rev.* **94**, 2027 (1994).
70. M. A. Aguilar, F. Olivares del Vale, and J. Tomasi, *J. Chem. Phys.* **98**, 7375 (1993).
71. M. W. Wong, M. J. Frisch, and K. B. Wiberg, *J. Am. Chem. Soc.* **113**, 4776 (1991).
72. K. B. Wiberg, P. R. Rablen, D. J. Rush, and T. A. Keith, *J. Am. Chem. Soc.* **117**, 4261 (1995).
73. K. B. Wiberg, H. Castejon, and T. A. Keith, *J. Comp. Chem.* **17**, 185 (1996).

74. C. J. Cramer and D. G. Truhlar, in *Reviews in Computational Chemistry*, Vol. VI, edited by K. B. Lipkowitz and D. B. Boyd (VHC Publishers, 1995).
75. K. V. Mikkelsen, H. Ågren, H. J. A. Jensen, and T. Helgaker, *J. Chem. Phys.* **89**, 3086 (1988).
76. C. J. Cramer and D. G. Truhlar, *Chem. Rev.* **99**, 2161 (1999).
77. M. Orozco and F. J. Luque, *Chem. Rev.* **100**, 4187 (2000).
78. C. J. F. Böttcher, *Theory of Electron Polarization*, 2nd ed., Vol. 1, edited by O. C. van Belle, P. Bordewijk, and A. Rip) (Elsevier, Amsterdam, 1973).
79. C. J. Cramer and D. G. Truhlar, in *Solvent Effects and Chemical Reactivity*, edited by O. Tapia and J. Bertrán (Kluwer, Dordrecht, 1996).
80. R. Cammi and J. Tomasi, *J. Comp. Chem.* **16**, 1449 (1995).
81. C. S. Pomelli and J. Tomasi, *Theor. Chem. Acc.* **99**, 34 (1998).
82. A. Bondi, *J. Phys. Chem.* **68**, 441 (1964).
83. K. Sharp, *J. Comp. Chem.* **12**, 454 (1991).
84. A. Klamt and G. Schüürmann, *J. Chem. Soc. Perkin Trans.* 2, 799 (1993).
85. J. Andzelm, C. Kölmel, and A. Klamt, *J. Chem. Phys.* **103**, 9312 (1995).
86. W. C. Still, A. Tempczyrk, R. C. Hawley, and T. Henricksons, *J. Am. Chem. Soc.* **112**, 6127 (1990).
87. J. Li, T. Zhu, G. D. Hawkins, D. A. Liotard, C. J. Cramer, and D. G. Truhlar, *Theor. Chem. Acc.* **103**, 9 (1999).
88. C. Amovilli and B. Mennucci, *J. Phys. Chem. B* **101**, 1051 (1997).
89. F. M. Floris, J. Tomasi, and J. L. Pascual-Ahuir, *J. Comp. Chem.* **12**, 784 (1991).
90. J. O'M. Bockris and A. K. M. Reddy, *Modern Electrochemistry 1* (Plenum/Rosetta, 1998), Chap. 5.
91. M. L. Huggins, *J. Phys. Chem.* **40**, 723 (1936).
92. E. Wicke, M. Eigen, and Th. Ackermann, *Z. Phys. Chem. (N.F.)* **1**, 340 (1954).
93. M. Eigen, *Angew. Chem. Int. Ed. Engl.* **3**, 1 (1964).
94. G. Zundel and H. Metzger, *Z. Physik, Chem. (N.F.)* **58**, 225 (1968).
95. G. Zundel, in *The Hydrogen Bond — Recent Developments in Theory and Experiments II. Structure and Spectroscopy*, edited by P. Schuster, G. Zundel, and C. Sandorfy (North-Holland, Amsterdam, 1976).
96. E. B. Robertson and H. B. Dunford, *J. Am. Chem. Soc.* **87**, 5080 (1964).
97. D. Marx, M. E. Tuckerman, J. Hutter, and M. Parrinelo, *Nature (London)* **397**, 601 (1999).
98. G. J. Tawa, I. A. Topol, S. K. Burt, R. A. Caldwell, and A. A. Rashin, *J. Chem. Phys.* **109**, 4852 (1998).
99. D. J. Aue, H. M. Webb, and M. T. Bowyers, *J. Am. Chem. Soc.* **98**, 318 (1976).
100. Y. Marcus, in *Ion Solvation* (John Wiley, New York, 1985).
101. H. Reiss and A. Heller, *J. Phys. Chem.* **89**, 4207 (1985).
102. A. A. Rashin and K. Namboodiri, *J. Phys. Chem.* **91**, 6003 (1987).
103. J. V. Coe, *Chem. Phys. Lett.* **229**, 161 (1994).
104. M. D. Tissandier, K. A. Cowen, W. Y. Feng, E. Gundlach, M. H. Cohen, A. D. Earhart, J. V. Coe, and T. R. Tuttle, Jr., *J. Phys. Chem. A* **102**, 7787 (1998).
105. J. A. Mejias and S. Lago, *J. Chem. Phys.* **113**, 7306 (2000).

106. A. Ben-Naim, *Solvation Thermodynamics* (Plenum, New York, 1987).
107. R. G. Pearson, *J. Am. Chem. Soc.* **108**, 6109 (1986).
108. A. Ben-Naim and Y. Marcus, *J. Chem. Phys.* **81**, 2016 (1984).
109. D. J. Tannor, B. Marten, R. Murphy, R. A. Friesner, D. Sitkoff, A. Nicholls, M. Ringnalda, W. A. Goddard, III, and B. Honig, *J. Am. Chem. Soc.* **116**, 11875 (1994).
110. B. Marten, K. Kim, C. Cortis, R. A. Friesner, R. B. Murphy, M. N. Ringnalda, D. Sitkoff, and B. Honig, *J. Phys. Chem.* **100**, 11775 (1996).
111. V. Barone, M. Cossi, and J. Tomasi, *J. Chem. Phys.* **107**, 3210 (1997).
112. B. Foresman, T. A. Keith, K. B. Wiberg, J. Snoonian, and M. J. Frisch, *J. Phys. Chem.* **100**, 16098 (1996).
113. S. Cabani, P. Giannni, V. Mollica, and L. Lepori, *J. Sol. Chem.* **10**, 563 (1981).
114. J. S. Jaworski, *J. Chem. Soc. Perkins Trans. II* **5**, 1029 (2000).
115. M. Cossi, B. Mennucci, J. Pitarch, and J. Tomasi, *J. Comp. Chem.* **19**, 833 (1998).
116. I. A. Topol, G. J. Tawa, S. K. Burt, and A. A. Rashin, *J. Phys. Chem.* **111**, 10998 (1999).
117. A. N. Toth, M. D. Liptak, D. L. Phillips, and G. C. Shields, *J. Chem. Phys.* **114**, 4595 (2001).
118. V. Barone and M. Cossi, *J. Phys. Chem. A* **102**, 1995 (1998).
119. W. H. Richardson, C. Peng, D. Bashford, L. Noodleman, and D. A. Case, *Int. J. Quant. Chem.* **61**, 209 (1997).
120. I. A. Topol, G. J. Tawa, S. K. Burt, and A. A. Rashin, *J. Phys. Chem. A* **101**, 10075 (1997).
121. J. L. Chen, L. Noodleman, D. A. Case, and D. Bashford, *J. Phys. Chem.* **98**, 11059 (1994).
122. A. A. Rashin, L. Yang, and I. A. Topol, *A Biophys. Chem.* **51**, 359 (1994).
123. J. E. Bartness, J. A. Scott, and R. T. McIver, *J. Am. Chem. Soc.* **101**, 6046 (1979).
124. C. R. Ganellin, *Molecular and Quantum Pharmacology*, Proceedings of the 7th Jerusalem Symposium on Quantum Chemistry and Biochemistry, edited by E. D. Bergmann and B. Pullman (D. Reidel Publishing Co., Dordrecht/Boston, 1974), p. 43.
125. W. A. Shapley, G. B. Bacskay, and G. G. Warr (1998), *J. Phys. Chem. B* **102**, 1938.
126. I. A. Topol, G. J. Tawa, R. A. Caldwell, M. A. Eissenstat, and S. K. Burt, *J. Phys. Chem. A* **104**, 9619 (2000).
127. M. D. Liptak and G. C. Shields, *J. Am. Chem. Soc.* **123**, 7314 (2001).
128. H. Li, A. W. Hains, J. E. Everts, A. D. Robertson, and J. H. Jensen, *J. Phys. Chem. B* (in preparation) (2001).
129. G. Schüürmann, *Quant. Struct.-Act. Relat.* **15**, 121 (1996).
130. G. Schüürmann, *Quantitative Structure-Activity Relationships in Environmental Sciences* **VII**, edited by F. Chen and G. Schüürmann (SETAC Press, Pensacola FL, 1997), Chap. 16.
131. C. J. Cramer and D. G. Truhlar, *Science* **256**, 213 (1992).
132. F. J. Luque, M. Bachs, and M. Orozco, *J. Comp. Chem.* **15**, 847 (1994).
133. C. O. da Silva, E. C. Silva, and M. A. C. Nascimento, *J. Phys. Chem. A* **104**, 2402 (2000).
134. G. Schüürmann, M. Cossi, V. Barone, and J. Tomasi, *J. Phys. Chem. A* **102**, 6706 (1998).
135. C. O. da Silva, E. C. Silva, and M. A. C. Nascimento, *J. Phys. Chem. A.* **103**, 11194 (1999).
136. R. Stewart, in *The Proton: Applications to Organic Chemistry*, edited by H. H. Wasserman, Vol. 46, of Organic Chemistry A Series of Monographs (Academic Press, New York, 1985).

137. W. K. McEwen, *J. Am. Chem. Soc.* **58**, 1124 (1936).
138. J. March, Advances Organic Chemistry—Reaction, Mechanisms and Structure, 4 ed. (Wiley-Interscience, New York, 1992).
139. J. Li, C. J. Cramer, and D. G. Truhlar, *Chem. Phys. Lett.* **288**, 293 (1998).
140. G. Pistolis, C. M. Paleos, and A. Malliaris, *J. Phys. Chem.* **99**, 8896 (1995).
141. P. L. Wash, E. Maverik, J. Chiefari, and D. A. Lightner, *J. Am. Chem. Soc.* **119**, 3802 (1997).
142. A. J. Beveridge and G. C. Heywood, *Biochemistry* **32**, 3325 (1993).
143. C. Colominas, J. Teixido, J. Cemeli, F. J. Luque, and M. Orozco, *J. Phys. Chem. B* **102**, 2269 (1998).
144. G. Siasios and E. R. T. Tiekink, *Z. Kristallogr.* **209**, 547 (1994).
145. L. González, O. Mó, M. Yáñez, and J. Elguero, *J. Chem. Phys.* **109**, 2685 (1998).
146. A. D. H. Clague and H. J. Bernstein, *Spectrochim. Acta* **25**, 593 (1969).
147. D. M. Mathews and R. W. Sheets, *J. Chem. Soc. Chem. Commun.* 2203 (1969).
148. G. S. Denisov and K. G. Tokhadze, *Dokl. Phys. Chem.* **337**, 117 (1994).
149. V. Šablinskas, B. Mikulskiene, and L. Kimtys. *J. Mol. Struct.* **482–483**, 263 (1999).
150. R. M. Aminova, G. A. Schamov, and A. V. Aganov, *J. Mol. Struct. (THEOCHEM)* **498**, 233 (2000).
151. M. Meot-Ner(Mautner), D. E. Elmore, and S. Scheiner, *J. Am. Chem. Soc.* **121**, 7625 (1999).
152. K. Sagarik and B. M. Rode, *Chem. Phys.* **260**, 159 (2000).

COOPERATIVE EFFECTS IN HYDROGEN BONDING

ALFRED KARPFEN

Institute for Theoretical Chemistry and Structural Biology, University of Vienna, Vienna, Austria

CONTENTS

I. Introduction
II. Cooperativity
    A. The Definition
    B. The Origin
III. Examples
    A. Perfectly Linear Chains and Rings
    B. Zigzag Chains and Rings
        1. $(HF)_n$
        2. $(HCl)_n$ and $(HBr)_n$
    C. Substituted Zigzag Chains and Rings
        1. $(CH_3O)_n$
        2. $(H_2O)_n$
    D. Miscellaneous
IV. Conclusions
References

I. INTRODUCTION

Hydrogen bonding is probably the most important type of anisotropic and directional intermolecular interaction. Hydrogen bonds are encountered in nearly all organic and in many inorganic crystals. The presence of hydrogen bonds is decisive for the physicochemical properties of almost all polar, associating liquids, notably that of water. The role of hydrogen bonds is central for the structure and the chemical dynamics of biopolymers such as proteins, nucleic acids, cellulose, and many others. Of particular significance for the physical chemistry of gases and liquids, is the occurrence of hydrogen bonds in dimers and small clusters. The understanding of the properties of

---

Advances in Chemical Physics, Volume 123, Edited by I. Prigogine and Stuart A. Rice.
ISBN 0-471-21453-1 © 2002 John Wiley & Sons, Inc.

isolated hydrogen bonds in vapor-phase dimers, in inert matrices or in solution, the change and the gradual modification of these properties upon going via clusters or oligomers of increasing size toward solvated clusters, liquids, molecular crystals, or even ionic clusters or ionic crystals with charged hydrogen bonds is a goal that is pursued with still increasing intensity both from the experimental and from the theoretical side. *Property* in this context means each and every structural, spectroscopic, thermodynamic, and also conceptual aspect of the hydrogen bond. An amazing number of books and quite extended collections of review articles [1–19] have been devoted to thorough and comprehensive treatments of the various subfields of this topic.

The characteristic properties of a single hydrogen-bond A—H---B formed between a proton-donor A—H and a proton-acceptor B are often [1, 20] described by a set of structurally and spectroscopically observable features. These features emphasize the physically measurable modifications of the properties of the A—H bond in the hydrogen-bonded complex relative to that in the unperturbed A—H monomer. The formation of a hydrogen-bond A—H---B is generally accompanied by the following trends:

1. The intermolecular distance R(A---B) is shorter than the sum of the van der Waals radii of A and B, despite the presence of the H atom.
2. The intramolecular distance r(A—H) is longer in the complex than in the monomer.
3. The fundamental vibrational frequency v(A—H) is shifted to lower wavenumbers in the complex than in the monomer.
4. The infrared (IR) intensity of v(A—H) is larger in the complex than in the monomer.
5. Proton nuclear magnetic resonance (^{1}H NMR) chemical shifts in the complex are significantly shifted downfield as compared to the monomer.

The extent to which these and related trends (as, e.g., an increase in the polarity of the A—H bond, an increase of the dipole moment of the complex larger than resulting from the assumption of vectorial addition of monomer dipole moments, and others) are actually observable is usually considered as an indication of the strength of the hydrogen bond. In the case of very weak hydrogen bonds, for example, of the type C—H---O, C—H---N, hydrogen bonds to rare gas atoms, X—H---Rg, hydrogen bonds to π bonds, or related examples, criterion (1) must be replaced by the less stringent requirement that R(H---B) is shorter than the sum of the van der Waals radii of H and B. Evidently, a strict definition for the *onset* of hydrogen bonding is not possible. In the case of very strong hydrogen bonds encountered in gas-phase ions like FHF^- or $H_5O_2^+$, the distinction between *intramolecular* properties of the A—H moiety and *intermolecular* properties of the H---B unit tends to vanish or becomes less useful or even meaningless. These *strong* hydrogen bonds can equally well be considered as conventional chemical bonds. Again, a sharp separation between *intermolecular interaction* via hydrogen bonding and the *chemical bond* is impossible and the categorization is necessarily arbitrary.

In many cases, for isolated, single hydrogen bonds formed between neutral molecules either in the vapor phase or in cryogenic matrices, the above-mentioned effects are significantly less pronounced than in the liquid or solid state, in specific, often cyclic dimers, in neutral clusters and oligomers, or in charged species. Hydrogen bonding very often shows the features of distinctly *nonadditive* or *cooperative* behavior. Whenever cooperative effects pertinent to the just mentioned properties of the A—H bond and the H---B distance are important, there is also a general strengthening of the intermolecular interaction energy, accompanied by significant increases but also decreases, that is, splittings of the vibrational frequencies originating from the various intermolecular modes.

The aim of this chapter is to survey recent quantum chemical calculations that attempt to track and to monitor the gradual differences between isolated hydrogen bonds, hydrogen bonds in clusters of increasing size, and hydrogen bonds in solids, thereby describing some of the aspects of nonadditive or cooperative behavior in hydrogen-bonded systems. Particular emphasis is on a comparison to the available experimental data and not on a detailed evaluation of different methodical approaches. In each case discussed, the best state-of-the-art calculations are selected for the confrontation with experiment. The discussion will also be limited to the above list of features connected directly to the A—H bond, that is, a full description of the vibrational spectroscopy of the hydrogen-bonded clusters, in particular, a detailed discussion of the intermolecular modes will not be given here. Similarly, all aspects connected with proton transfer, single or concerted, dynamical interchange between structural isomers, tunneling dynamics, and so on, are omitted. Because of the enormous wealth of literature that has been accumulated in this field, this chapter will further be restricted to the conceptually more simple examples of hydrogen bonds in *linear chains* and in *rings*. The properties of these species can frequently be compared to or extrapolated to the properties of infinitely extended one-dimensional (1D) hydrogen-bonded polymers encountered as quasi-1D chains embedded in anisotropic molecular crystals. More complicated structural forms, such as two-dimensional (2D) or three-dimensional (3D) hydrogen-bonded networks, and the large variety of hydrogen-bond patterns in biopolymers are not treated in this chapter. In particular, the case of larger H_2O clusters and the transition to liquid water and the various ice forms are excluded. Progress reports in the very important research field of structure, dynamics, and vibrational spectroscopy of larger water clusters may be found in a number of recent review articles [21–25].

Molecular crystals in which hydrogen-bonded chains form the constituing elements are quite frequently observed in nature. In these molecular crystals, the neighboring chains are separated by the usual van der Waals distances, resulting in very anisotropic crystal structures. Perfectly linear chains built from linear molecules are observed in the crystal structures of hydrogen cyanide [26] and cyanoacetylene [27]. Zigzag chains are formed in the solid state of hydrogen halides, HF [28, 29], HCl [30], and HBr [31]. Most of the primary alcohols, such as methanol [32] and ethanol [33], the two simplest carbonic acids, (formic acid [34–36] and acetic acid [37–40]), imidazole [41], a variety of pyrazoles [42], and many others also crystallize in the form of hydrogen-bonded catemers. More complicated chain-like motifs with cyclic dimer hydrogen bonds as links between monomers may, for example, be found among the dicarboxylic acids [43], pyrazole-4-carboxylic acids [44], 2-aminopyrimidines [45], 2-pyridone derivatives [46, 47], and other organic cocrystals [48, 49].

Strictly taken, a prerequisite for the discussion of cooperativity or nonadditivity requires the definition of the *additive* or *noncooperative* case [50]. Generally, in the field of intermolecular interaction, the additive model is a model based on the concept of *pairwise additive* interactions. For atomic clusters per definition, but also for molecular clusters, the use of pairwise additive interactions is almost always used in combination with the assumption of structurally frozen interaction partners. Even in cases of much stronger intermolecular interactions the concept of pair potentials modified to that of *effective pair potentials* is often used. Most of the molecular dynamics calculations of liquids and molecular solids take advantage of this concept.

The assumption of a strict, vapor-phase derived pair potential appears acceptable only in those cases where a weak intermolecular interaction does not cause appreciable structural relaxations in the monomers. In the case of hydrogen-bonded systems, the use of the frozen monomer assumption precludes, however, almost always the investigation of all the observable structural and spectroscopic features of the A—H moiety. Therefore, the reference system for the discussion of cooperative, nonadditive effects is exclusively the structurally fully optimized hydrogen-bonded dimer with a single isolated hydrogen bond and with all the properties derivable from the global $3N$-6 dimensional potential energy surface of the dimer.

II. COOPERATIVITY

A. The Definition

There are several ways to *measure* cooperativity. One very often applied strategy is to use only an energetic criterion. The total energy of the system is then expressed as a many-body or cluster expansion

$$E(n) = \sum_{i=1}^{n} E(i) + \sum_{i<j}^{n} E_{ij} + \sum_{i<j<k}^{n} E_{ijk} + \cdots \qquad (8.1)$$

with $E(n)$ representing the total energy of the cluster consisting of n interacting subsystems. The energies of the subsystems are $E(i)$, the pair interaction energies are E_{ij}, the three-body contributions are E_{ijk}, and so on. This kind of expansion is particularly suitable for cases where the frozen molecule approximation can meaningfully be applied, that is, atomic clusters and molecular aggregates built from very weakly interacting molecules. In cases where the intermolecular interaction results in substantial intramolecular structure relaxations, this expansion soon becomes impractical for larger clusters because an increasingly cumbersome number of reference systems has to be calculated. Moreover, for the case of the chain-like structures and rings considered in this chapter, many-body energies beyond the three-body term are anyway exceedingly small.

A somewhat incomplete and less ambitious analysis of the hydrogen-bond energy increase in clusters of identical molecules consists of defining a stabilization energy per hydrogen bond in the following two alternative ways [51, 52]:

$$\Delta E_a(n) = E(n) - E(n-1) - E(1) \qquad (8.2)$$

$$\Delta E_b(n) = (E(n) - n*E(1))/m \qquad (8.3)$$

In the case of a series of cyclic clusters $m = n$, for a series of chain-like clusters $m = n - 1$. As will be demonstrated, these two quantities are well suited for extrapolations toward $n \to \infty$ for the case of extended polymers. In the case of linear, chain-like clusters and for large n, both $\Delta E_a(n)$ and $\Delta E_b(n)$ evidently converge toward the same limiting value. For short-chained linear clusters, $\Delta E_a(n)$ may be interpreted as the hydrogen-bond energy gained by the insertion of a molecule in the center of an already preformed chain, whereas $\Delta E_b(n)$ is the traditional average hydrogen-bond energy. The quantities $\Delta\Delta E_a(n)$ and $\Delta\Delta E_b(n)$ defined as

$$\Delta\Delta E_a(n) = \Delta E_a(n) - \Delta E(2) \qquad (8.4)$$

$$\Delta\Delta E_b(n) = \Delta E_b(n) - \Delta E(2) \qquad (8.5)$$

are easily calculated and are more practical measures for the energy increase per hydrogen bond in chain-like clusters. Clearly, all many-body effects and also the two-body interactions between more distant molecules are included in this definition. The same kind of analysis can also be performed using zero-point energy (ZPE) corrected stabilization energies, enthalpies, or free energies.

The quantitative experimental determination of interaction energies in hydrogen-bonded clusters is still a very difficult problem. The quantum chemically calculated energies, therefore, provide useful and invaluable information. However, more interesting and more informative than the mere energies and their analysis are the detailed structural and vibrational spectroscopic trends observed upon increasing the cluster size, either determined by experiment or calculated. Cooperative or nonadditive effects can then even be more loosely described as the deviations of the various structural and spectroscopic quantities calculated for the clusters from those calculated for the dimer. An eventual separation of additive and nonadditive contributions to the various property changes again could not be free of arbitrary model assumptions. A large part of this chapter will be devoted to a description of these deviations for a few selected cases.

B. The Origin

The well-established perturbation theory of intermolecular interaction [53–59] can be applied to hydrogen-bonded systems in combination with the frozen molecule approximation, when the interaction is either sufficiently weak [60–62], or when the interaction is treated at a more qualitative level. When the interaction becomes larger, structural relaxations become sizable. Then the more usual approach to treat the hydrogen-bonded complex or cluster as a supermolecule becomes more practical and also more appropriate. However, also in this case, the detailed analysis of the interaction energy is often done with the aid of different variants of energy partitioning techniques [63, 64] which closely follow the lines of intermolecular perturbation theory.

Hydrogen-bonded chains or rings differ from the case of nonpolar clusters exemplified, for example, by rare gas trimers, $(Rg)_3$. Among the dominating, low-order perturbation theory terms, *electrostatic, exchange, polarization* or *induction, dispersion,* and *charge transfer*, the most important nonadditive contribution by far arises from the polarization or induction energy in the case of hydrogen-bonded clusters, that is, the permanent dipole–induced dipole interaction. This is particularly so in case of chain like clusters and in rings, because the intermolecular distances between nonneighboring molecules are large. The dominant deviation from an additive mechanism in a linear or ring-like trimer is, therefore, the interaction of the third molecule with the, as a consequence of the polarization or induction interaction, modified electronic distribution in the already preformed hydrogen-bonded dimer. The consequences of this nonadditive behavior for the different spectroscopically observable quantities going from the dimer to the anisotropic molecular crystal will be the topic of Section III, in which a few selected cases of homomolecular hydrogen-bonded oligomers $(H-X)_n$ are investigated in detail.

III. EXAMPLES

A. Perfectly Linear Chains and Rings

Particularly instructive cases for a demonstration of the gradual modification of the molecular properties characteristic of hydrogen-bond formation are oligomers built from strictly linear molecules with the added feature that the global minimum of the dimer is also perfectly linear. The unique advantage of these molecular series is that the properties of molecules embedded in the linear as well as in the cyclic oligomers converge for large n smoothly toward the same bulk limit. Moreover, all cyclic (C_{nh}) and linear ($C_{\infty v}$) oligomers are true minima on the energy surfaces. Examples are hydrogen cyanide (H—C≡N), and cyanoacetylene (H—C≡C—C≡N). These two cases will be dealt with in some detail, because much experimental data has already been collected for these molecular clusters and allows for systematic comparisons with some recent quantum chemical investigations.

Both vapor-phase dimers, $(HCN)_2$ and $(HC_3N)_2$, have linear structures. In the case of $(HCN)_2$, microwave studies [65-69] and rotationally resolved IR spectra of the C—H stretching region [70-74] have convincingly proven that $(HCN)_2$ is indeed linear. The linearity of $(HC_3N)_2$ was demonstrated with the aid of high-resolution IR spectroscopy, again in the C—H stretching region [75].

For both molecules, the structure of the corresponding molecular crystal is also known. Hydrogen cyanide [26] and cyanoacetylene [27] have extremely anisotropic crystal structures. They crystallize in perfectly linear, hydrogen-bonded chains. In the solid state, hydrogen cyanide has two modifications. The high-temperature, tetragonal form has space group C_{4v}^9-$I4mm$ with two molecules in the unit cell. The low-temperature, orthorhombic form has space group C_{2v}^{20}-Imm, again with two molecules in the unit cell. In both modifications, the neighboring chains are oriented parallel to each other with the hydrogen bonds pointing in the same direction. Crystalline cyanoacetylene is monoclinic with the space group $P2_1/m$ and two molecules per elementary cell. Each chain is surrounded by six neighboring chains, two in parallel and four in antiparallel orientation. Since the X-ray investigations cannot precisely locate the H atoms, the structural analyses have been performed under the assumption of frozen monomers. The same also applies, in general, to the interpretation of the vapor-phase dimer structures. This approximation is almost always used when interpreting the rotational spectroscopic data on hydrogen-bonded dimers. The only measured *structural* sign of cooperativity is thus the difference in the experimentally determined C—H---N distances between the vapor-phase dimer and the molecular crystal. The observed reductions in the intermolecular C—H---N distances upon going from the dimer to the crystal amount to ~0.1 and 0.05 Å for HCN and HC_3N, respectively.

For hydrogen-bonded, high-symmetry species larger than the dimer, there are mainly two structural possibilities: Either linear chains with $C_{\infty v}$ symmetry or cyclic clusters with C_{nh} symmetry are formed. The former have a very large dipole moment and should, therefore, be observable with the aid of microwave rotational spectroscopy. The latter have no dipole moment and can only be detected and analyzed with high-resolution vibrational spectroscopic techniques or eventually by electron diffraction. However, electron diffraction data are not yet available for these species. The structures of these oligomers are sketched in Figure 8.1. Clearly, for larger oligomers other cluster configurations may also play a role. In particular, equilibrium structures that are not only stabilized by hydrogen bonds, but additionally by dispersion-dominated interactions between the C≡N or C≡C—C≡N π-systems of the neighboring chains, are also bound to occur, when approaching the bulk structure of the molecular crystal.

The trimers of hydrogen cyanide and of cyanoacetylene are of particular interest. The existence of a linear trimer $(HCN)_3$ has been revealed by microwave spectroscopy [76,77], by IR spectroscopy [78,79], and by Fourier transform infrared (FTIR), photoacoustic Raman (PARS), and coherent anti-Stokes Raman (CARS) spectroscopy [80,81]. Additionally, a cyclic trimer of hydrogen cyanide has also been detected by high-resolution IR spectroscopy [78]. In the case of cyanoacetylene, there is experimental evidence that the very same types of trimer structures have also been found to occur in the vapor phase [82]. Rotational resolution could, however, only be achieved for the linear trimer. It is a quite rare phenomenon among hydrogen-bonded trimers of small molecules that two isomers, cyclic and linear, are minima on the energy surface. The experimental vapor-phase IR data in the C—H stretching region [78,82] of tetramers to hexamers have been interpreted as originating predominantly from cyclic clusters in the case of $(HCN)_n$ and $(HC_3N)_n$. That the energetic balance between cyclic and linear clusters is very subtle has been demonstrated by very recent invest-

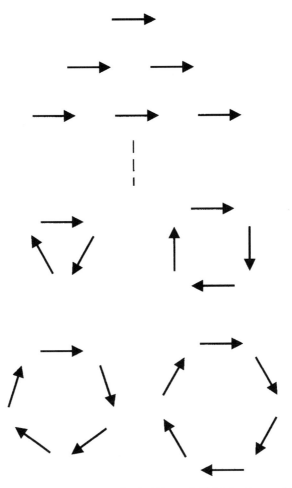

Figure 8.1. Sketch of linear and cyclic (HCN)$_n$ and (HC$_3$N)$_n$ clusters. Monomers are represented by arrows.

igations of HCN [83, 84] and HC$_3$N [85] clusters in superfluid helium droplets. The solvation by helium atoms and the modified conditions for cluster growth result in the exclusive formation of linear clusters. In addition to the vapor-phase work, the vibrational spectra of hydrogen-bonded clusters of HCN were also investigated by low-temperature matrix-spectroscopic studies [86–93]. In addition, the vibrational spectra of solid HCN [94, 95] and solid HC$_3$N [96–98] are also available.

Parallel to these experimental efforts, numerous quantum mechanical investigations have been carried out on trimeric and larger clusters of HCN

[52, 79, 99–109]. Even infinitely extended chains of HCN have been treated at the ab initio level [106, 110–113]. In contrast, only a few systematic investigations of the corresponding HC_3N clusters [107, 114, 115] have been performed so far. In the following, the general picture emerging from these theoretical investigations for the series of linear and cyclic clusters of HCN and HC_3N will be discussed and compared to experimental data.

A qualitative illustration of the convergence of the interaction energy or hydrogen-bond energy, $\Delta E_a(n)$ and $\Delta E_b(n)$, in the series of linear and cyclic HCN and HC_3N oligomers is given in Figure 8.2. The energies have been taken from [107] and originate from calculations performed at the B3LYP/6-31G(d,p) level. Corresponding self-consistent field (SCF) and Møller–Plesset second-order perturbation theory (MP2) data may be found therein. Figure 8.2. shows immediately that all four series, $\Delta E_a(n)$ and $\Delta E_b(n)$, each

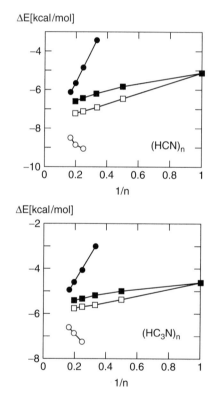

Figure 8.2. Plots of $\Delta E_a(n)$ and $\Delta E_b(n)$ versus $1/n$ for linear and cyclic HCN and HC_3N clusters as obtained at the B3LYP/6-31G(d,p) level [107]. Open symbols: $\Delta E_a(n)$; filled symbols: $\Delta E_b(n)$; circles: cyclic structures; squares: linear structures.

for cyclic and linear oligomers, tend toward the same limit. Not unexpectedly, the behavior with increasing n is qualitatively very similar for both molecular series. It also shows that the stabilization energy per hydrogen bond is more negative for $(HCN)_n$ than for $(HC_3N)_n$ for all n and in particular for $n = 2$. A compilation of the total stabilization energies, $\Delta E_{total}(n)$, defined as

$$\Delta E_{total}(n) = E(n) - nE(1) \quad (8.6)$$

is reported in Table VIII.1. The data illustrate that with increasing n the cyclic clusters become more stable than the linear clusters. In both cases, for $(HCN)_n$ and $(HC_3N)_n$, the turning point occurs between $n = 3$ and $n = 4$. For $n = 3$, the linear trimers are still more stable, from $n = 4$ on, the cyclic configurations have lower energies. The same qualitative picture is actually already obtained at the SCF and also at the MP2 level [107]. The finding that linear and cyclic trimers have very similar stabilization energies is in agreement with the experimental observation that both species have been detected for $(HCN)_3$ [76–78] and $(HC_3N)_3$ [82]. Roughly extrapolated $\Delta E_a(n)$ values for $n \to \infty$ are about -7.5 and -6 kcal/mol for $(HCN)_\infty$ and $(HC_3N)_\infty$, respectively, at the B3LYP/6-31G(d,p) level, which corresponds to increases of ~ 42 and 30% relative to the dimer stabilization energy. From earlier model considerations on a linear chain of Stockmeyer molecules (dipole–dipole plus Lennard–Jones interaction) [116, 117], a purely additive $\Delta E_a(n)$ would be expected at $\sim 28\%$ below $\Delta E(2)$. Therefore, the effects of nonadditivity on the stabilization or hydrogen-bond energy can be justifiably characterized as very weak for both types of chains. From this energetic point of view, HCN chains and particularly HC_3N chains behave essentially like chains of dipole molecules.

TABLE VIII.1
Total Stabilization Energies of Linear and Cyclic $(HCN)_n$ and $(HC_3N)_n$ Clusters as Obtained at the B3LYP/6-31G(d,p) Level[a]

n	$(HCN)_n$ Linear	Cyclic	$(HC_3N)_n$ Linear	Cyclic
2	−5.2		−4.6	
3	−11.6	−10.3	−10.0	−8.9
4	−18.5	−19.4	−15.6	−16.2
5	−25.6	−28.2	−21.3	−23.0
6	−32.8	−36.6	−27.0	−29.6

[a]Data taken from [107]. All Values in kilocalories per mole (kcal/mol).

We now turn to the equilibrium structures of these clusters. We discuss the linear clusters first. In both molecules, the only internal coordinate appreciably modified upon formation of the hydrogen bond is the C—H distance. The C≡N, C≡C, and C—C distances are modified only to a very small extent. Therefore, we restrict the discussion to the intramolecular C—H distances and the intermolecular H---N distances. The trends in the structural relaxation of all C—H and H---N distances taking place upon increasing the chain length up to hexamers are displayed in Figure 8.3. The graphs for the r(C—H) distances nicely confirm the expected picture. With increasing length, we should observe convergence to two different kinds of edge (surface) effects, one to the non-hydrogen-bonded, *free* edge, and one

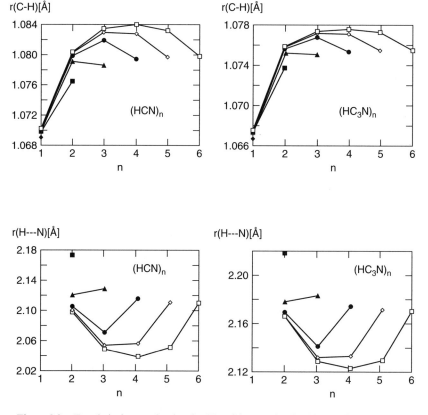

Figure 8.3. Trends in intramolecular C—H and intermolecular H---N distances in linear $(HCN)_n$ and $(HC_3N)_n$ clusters as obtained at the B3LYP/6-31G(d,p) level [107]. Distances belonging to the same oligomer are marked by identical symbols and connected by lines.

to the hydrogen-bonded edge. In addition, the properties of the central molecules must convergence to the *bulk* properties. For both oligomer series, convergence of the *free* end is reached quickly, or in other words, the difference in r(C—H) between the isolated molecule and the free end of a long chain is only on the order of 0.001 Å. The parameter r(C—H) at the hydrogen-bonded edge is approximately converged for hexamers. The increase of r(C—H) amounts to ~ 0.011 Å for (HCN)$_n$ and to ~ 0.009 Å for (HC$_3$N)$_n$. The *bulk* structure is not yet reached in the center of linear hexamers. Complementary results are obtained for the intermolecular distance r(H---N). At both edges, r(H---N) converges approximately toward the same limit, about half-way from the dimer to the bulk intermolecular distance.

In the case of the cyclic oligomers with C_{nh} symmetry, there is only one r(C—H) and one r(H---N) for a given n. The evolution of these two structural parameters with increasing n can be compared to the evolution of the central r(C—H) and r(H---N) distances of the linear clusters. This comparison is shown in Figure 8.4. Although the reliable extrapolation to $n \to \infty$ would require additional calculations on still larger clusters, it is evident that linear and cyclic cluster geometries tend toward the same limit. Both the calculated reduction of the intermolecular H---N distance and the widening of the intramolecular C—H distance taking place upon going from the dimer to the polymer are larger for (HCN)$_n$ than for (HC$_3$N)$_n$. In the smaller cyclic oligomers of both series, the intramolecular bond angles deviate only by <3° from linearity, and this deviation tends to become progressively smaller for the larger rings.

The shifts of vibrational frequencies induced by hydrogen bonding in dimers, oligomers, and solids are as important as the effects on the equilibrium structures. In particular, IR spectroscopy is probably the most important experimental technique for detecting hydrogen bonds and, within certain limits, for measuring the strength of hydrogen bonding.

Tables VIII.2 and VIII.3 contain compilations of calculated vibrational frequency shifts of the C—H stretching vibrations in linear and cyclic (HCN)$_n$ and (HC$_3$N)$_n$ oligomers together with the calculated IR intensity enhancement factors relative to the corresponding monomer as obtained at the B3LYP/6-31G(d,p) and MP2/6-31G(d,p) levels, respectively [107, 114]. Both sets of data qualitatively display the same characteristic features. In the case of the cyclic clusters up to hexamers, only one (a doubly degenerate) C—H stretching vibrational mode is IR active. In the linear oligomers, all C—H stretching modes are in principle IR active. They differ, however, strongly in their intensities. The calculated IR intensity of the most intense, linear HCN hexamer C—H stretching mode, for example, is increased by a factor of 32 relative to the monomer at the MP2/6-31G(d,p) level. The MP2

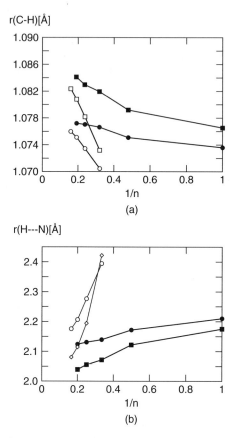

Figure 8.4. Plots of $r(C\!-\!H)$ (a) and $r(H\!\cdot\!\cdot\!\cdot N)$ (b) versus $1/n$ for linear and cyclic clusters of HCN and HC_3N. Linear clusters: filled symbols; cyclic clusters: open symbols; squares: $(HCN)_n$; circles: $(HC_3N)_n$.

calculated shifts are more accurate, B3LYP overshoots considerably. In the dimer, the corresponding intensity increase amounts to a factor of 6.8. This most intense oligomer mode is always the one in which all C—H groups move in phase. It is also always the lowest frequency C—H stretching mode, and is the one that for large n, finally converges toward the optically active $k = 0$ phonon in the infinitely extended $(HCN)_\infty$ and $(HC_3N)_\infty$ polymer. In case of the cyclic oligomers, the lowest frequency mode is also the one (nondegenerate) in which all C—H stretches move in-phase, and the series of these lowest frequencies must converge toward the same limit. However, for symmetry reasons, this mode is not IR active. Evidently, the doubly degenerate IR active mode of the cyclic oligomers must also converge

TABLE VIII.2
Calculated Harmonic C—H Stretching Vibrational Frequency Shifts and IR Intensity Enhancement Factors in Linear and Cyclic (HCN)$_n$ and (HC$_3$N)$_n$ Clusters as Obtained at the B3LYP/6-31G(d,p) Level[a]

	(HCN)$_n$		(HC$_3$N)$_n$	
n	Linear	Cyclic	Linear	Cyclic
1[b]	3476 [57]		3482 [81]	
2	−4 (1.1)		−4 (1.0)	
	−101 (6.8)		−94 (7.2)	
3	−6 (1.2)	−43 (5.2)	−5 (1.1)	−41 (4.6)
	−122 (0.0)	−49 (0.0)	−108 (0.3)	−44 (0.0)
	−140 (17.5)		−115 (17.1)	
4	−7 (1.2)	−104 (0.0)	−5 (1.1)	−83 (0.0)
	−137 (6.9)	−111 (15.9)	−115 (6.1)	−85 (13.0)
	−140 (0.4)	−125 (0.0)	−117 (4.4)	−89 (0.0)
	−177 (23.0)		−134 (18.2)	
5	−7 (1.2)	−140 (0.0)	−6 (1.1)	−106 (0.0)
	−142 (3.4)	−153 (30.8)	−116 (5.7)	−110 (26.4)
	−145 (6.5)	−169 (0.0)	−118 (6.3)	−115 (0.0)
	−175 (0.0)		−135 (0.0)	
	−198 (34.2)		−142 (28.4)	
6	−7 (1.2)	−158 (0.0)	−6 (1.1)	−118 (0.0)
	−144 (4.8)	−164 (0.0)	−117 (6.0)	−119 (0.0)
	−146 (5.4)	−180 (47.2)	−119 (6.3)	−124 (38.3)
	−180 (4.7)	−196 (0.0)	−137 (4.1)	−129 (0.0)
	−190 (0.0)		−140 (0.0)	
	−214 (43.7)		−148 (35.9)	

[a]Frequency shifts are in reciprocal centimeters (cm^{-1}), enhancement factors are in parentheses. Data taken from [107].
[b]Monomer frequencies (cm^{-1}) and IR intensities (km/mol) are in brackets.

toward this limit. A plot of the B3LYP and MP2 calculated lowest C—H stretching frequencies of linear and cyclic (HCN)$_n$ and (HC$_3$N)$_n$ oligomers versus $1/n$ is depicted in Figure 8.5. This plot demonstrates that indeed for both series of molecular clusters the lowest modes of linear and cyclic clusters converge toward the same limiting frequency for a given calculation method. From a comparison of the experimental C—H stretching vibrational frequencies of monomers and of the molecular crystals, monomer to crystal shifts of -182 cm^{-1} for hydrogen cyanide and of -132 cm^{-1} for cyanoacetylene can be deduced. The corresponding MP2/6-31G(d,p) calculated and extrapolated shift for (HCN)$_\infty$ of -212 cm^{-1} somewhat overshoots [107]. By comparing the experimental dimer and crystal frequencies, the shifts amount only to -112 and -66 cm^{-1} for HCN and HC$_3$N, respectively.

TABLE VIII.3

Calculated Harmonic C—H Stretching Vibrational Frequency Shifts and IR Intensity Enhancement Factors in Linear and Cyclic $(HCN)_n$ and $(HC_3N)_n$ Clusters as Obtained at the MP2/6-31G(d,p) Level[a]

n	(HCN)$_n$ Linear	(HCN)$_n$ Cyclic	(HC$_3$N)$_n$ Linear	(HC$_3$N)$_n$ Cyclic
1[b]	3529 [66]		3531 [84]	
2	−1 (1.1) [−4][c]		−4 (1.1) [−3][c]	
	−74 (5.2) [−70][c]		−72 (6.0) [−66][c]	
3	−2 (1.2) [−5][c]	−29 (4.6) [−38][c]	−5 (1.1) [−3][c]	−33 (4.4) [−33][c]
	−90 (0.0)	−33 (0.0)	−82 (0.3)	−35 (0.0)
	−103 (13.0) [−99][c]		−86 (13.7) [−80][c]	
4	−3 (1.2)	−68 (0.0)	−6 (1.1)	−62 (0.0)
	−100 (4.5)	−73 (12.5)	−86 (4.6)	−64 (11.1)
	−103 (0.8)	−84 (0.0)	−88 (3.3)	−67 (0.0)
	−129 (16.9)		−99 (13.9)	
5	−3 (1.2)	−94 (0.0)		
	−102 (2.9)	−104 (22.9)		
	−105 (4.3)	−116 (0.0)		
	−126 (0.0)			
	−143 (24.6)			
6	−4 (1.2)	−107 (0.0)		
	−103 (3.5)	−111 (0.0)		
	−106 (3.8)	−123 (34.1)		
	−128 (2.9)	−134 (0.0)		
	−136 (0.0)			
	−153 (31.6)			

[a] Frequency shifts are in reciprocal centimeters (cm^{-1}), enhancement factors are in parentheses. Data taken from [107 and 114].
[b] Monomer frequencies (cm^{-1}) and IR intensities (km/mol) are in brackets.
[c] Shifts of experimental anharmonic C—H stretching fundamentals as taken from [71, 72, 82].

Infrared frequencies of the C—H stretching bands have been observed for a wide variety of $(HCN)_n$ and $(HC_3N)_n$ clusters. Particularly impressive are the series of *free* C—H stretches and the series of the most intense (the lowest for each chain length) hydrogen-bonded C—H stretches measured in superfluid helium for linear $(HCN)_n$ [84] and $(HC_3N)_n$ [85] chains. In both cases, the experimentally observed frequency shifts relative to the unperturbed monomer compare favorably to MP2 calculated frequency shifts [107, 114].

Other properties showing indications of cooperativity in hydrogen cyanide oligomers were also investigated from the theoretical side. The trends in

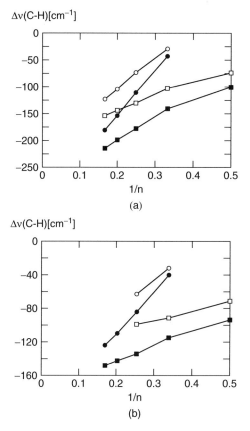

Figure 8.5. Plots of calculated C—H stretching vibrational frequency shifts versus $1/n$ for linear and cyclic clusters of HCN (*a*) and HC_3N (*b*). Linear clusters: squares; cyclic clusters: circles; open symbols: MP2/6-31G(d,p); filled symbols B3LYP/6-31G(d,p).

the nuclear quadrupole coupling constants [105] and in the chemical shifts [52] have been calculated. The ^{14}N nuclear quadrupole coupling constants turn out be most sensitive to hydrogen bonding. Calculated proton NMR (^1H NMR) chemical shieldings, $\sigma(^1H)$, indicate systematic, but small changes from the monomer to the polymer on the order of 2 ppm. Experimental NMR chemical shifts are not yet available for HCN or HC_3N clusters.

Quite apart from the linear and cyclic oligomers just discussed, other cluster geometries have also been found to be of importance in the case of small HC_3N oligomers. From the theoretical side, it could be shown that the antiparallel stacked dimer of HC_3N with C_{2h} symmetry is also a minimum, not too far away in energy from the global, linear, hydrogen-bonded dimer, whereas the analogous antiparallel stacked dimer of HCN is a first-order saddle point [114, 118]. Interestingly, and as a side remark, the next member of this series, the dimer of cyanodiacetylene $(HC_5N)_2$, is predicted to have an antiparallel stacked dimer structure as the global minimum that is more stable than the linear, hydrogen-bonded dimer by ~2 kcal/mol [118]. Moreover, in the case of HC_3N, a C_{2h}-symmetric tetramer built from two antiparallel stacked, linear dimers has a stabilization energy comparable to that of linear and cyclic tetramers [114, 115]. The calculations [114] also yielded evidence that the C—H stretching frequency of this particular tetramer was actually observed in the experiment [82].

By concluding this section on linear molecules forming preferentially linear hydrogen bonds, we can summarize the available experimental and theoretical data as follows: The signature of cooperativity is clearly present in these two series. This observation is valid for the hydrogen-bond energies, for the structural parameters characterizing the C—H---N hydrogen bonds, for the C—H vibrational frequencies, and for their IR intensities. The cooperative effects are stronger in the HCN oligomers than in the HC_3N oligomers. The simple explanation is that in spite of the much larger longitudinal size, a larger dipole moment, and a larger polarizability of the HC_3N molecule [114, 118], the dimer binding energies of linear $(HCN)_2$ and of linear $(HC_3N)_2$ are very close. Cooperative effects are then considerably weaker in the HC_3N polymers because of the much larger distance of nonneighboring molecules. Compared to the systems treated in Section III.B, the cooperative effects in both series may be characterized as weak.

B. Zigzag Chains and Rings

In general, there is a tendency that the optimal, mutual orientation of the interacting monomers, as it is observed in the equilibrium structure of the vapor-phase dimer, is retained in the equilibrium structures of larger clusters as well, whenever possible. In case of the C_{nh} symmetric, cyclic $(HCN)_n$ and $(HC_3N)_n$ clusters just discussed, this optimal (linear) dimer orientation is approached only asymptotically. The energetically unfavorable distortion from the optimal dimer orientation in small rings is, however, counteracted by the energetic gain of an additional hydrogen bond.

When the optimal orientation of the dimer is not linear, two additional features may, eventually, complicate the straightforward extrapolation of the trends from the dimer to the polymer: (a) The series of cyclic structures will not converge smoothly toward the extended polymer, but there is often an *optimal* ring size in which the relative orientation of the hydrogen-bonded

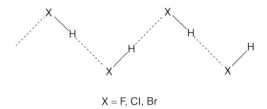

X = F, Cl, Br

Figure 8.6. Sketch of the zigzag chain structure in the molecular crystals of hydrogen halides.

molecules is close to that in the dimer. Beyond that ring size, the stabilization energy per hydrogen bond may decrease again, and the optimal cluster configurations are then no longer planar, or not even ring-like. (b) Although the series of chain-like clusters would conceptually converge toward the structure of the extended chains in the crystals, the short chains need not necessarily be minima on the energy surface.

The most prominent examples for molecules forming zigzag chains in the crystal are the hydrogen halides HF, HCl, and HBr. The chain structure encountered in the molecular crystals is sketched in Figure 8.6. In each case, the chain structures are reminiscent of the structure of the vapor-phase dimer. In all three molecular crystals, the hydrogen bonds are essentially linear. The XXX angle in the crystal is close to the XXH angle in the vapor-phase dimer, indicating that the optimal orientation in the dimer is preserved in the crystal. In case of solid HF and in the $(HF)_2$ vapor-phase dimer, the FFF and the FFH angles are close to 120°, whereas in the case of HCl and HBr, the corresponding angles are close to 90°.

1. $(HF)n$

Among the hydrogen halides, the hydrogen fluoride, $(HF)_n$, clusters are by far the most widely studied species. The hydrogen fluoride dimer, $(HF)_2$, is one of the best and most intensively investigated hydrogen-bonded system. The literature on the vapor-phase dimer is enormous and cannot be discussed here. The experimental and theoretical investigations on the dimer $(HF)_2$ have been dealt with in an exhaustive review by Truhlar [119] covering the available literature up to 1990, in the comprehensive book by Scheiner [12], and in a more recent review article [120]. Therefore, only a few of the most recent experimental and theoretical papers are mentioned [121–126].

The trimer of hydrogen fluoride, $(HF)_3$, does not have an open-chain structure. All experimental evidence points to a C_{3h} symmetric cyclic cluster

[127–132]. Similarly, all the available theoretical investigations confirm that (HF)$_3$ has indeed C_{3h} symmetry [133–158]. It has already been very early recognized [159–161] that the larger clusters with $n = 4$–6 also have cyclic structures. Since then, this finding has been corroborated by more detailed spectroscopic investigations for ring sizes from 4 up to 8 [127, 130, 147, 162–171]. In agreement with experimental evidence, the theoretical calculations on HF clusters also show that the ring-like (HF)$_n$ clusters are minima on the energy surfaces [52, 141, 145, 149–157, 172–180]. More detailed surveys of experimental and theoretical results on HF clusters may be found in recent reviews [12, 120, 181]. The vibrational spectroscopic data as obtained from low-temperature matrix investigations was interpreted in terms of cyclic *and* chain-like clusters [182–186].

The structure [27, 28], NMR spectra [187], and vibrational spectra [188–198] of solid HF have been investigated. Adjusted valence force fields and phonon dispersion curves and the corresponding phonon density of states have also been discussed [197, 199]. Quite a number of ab initio studies have already treated the infinitely extended zigzag chain of hydrogen fluoride molecules [200–206]. In contrast to the previously discussed case of HCN and HC$_3$N, the zigzag open-chain HF oligomers are not minima on the energy surface. The planar zigzag open-chain HF oligomer structures beyond the dimer are all calculated to be first-order saddle points [52, 138, 146, 207]. Normal mode analysis reveals, however, that the single imaginary frequency results from a torsional or out-of-plane motion of the hydrogen atom at the non-hydrogen-bonded edge of the chain. This particular saddle point feature may thus be treated as an edge effect, not related to the convergence behavior toward the bulk properties of a molecule embedded in the center of the chain.

Selected sets of more recent, high-level calculated stabilization energies of hydrogen fluoride clusters are compiled in Table VIII.4. A plot of $\Delta E_a(n)$ and $\Delta E_b(n)$ versus $1/n$ for rings and chains is shown in Figure 8.7. As with the previously discussed cases of HCN and HC$_3$N clusters, the convergence of the chain-like clusters with increasing n is perfectly smooth. Rough extrapolation leads to a stabilization energy around -8.6 kcal/mol for an HF molecule embedded in an infinitely extended chain. Compared to the dimer stabilization energy of -4.9 kcal/mol, as obtained at the same level of approximation, this corresponds to an increase of $\sim 75\%$, a much larger gain than in the previously discussed cases. The parameters $\Delta E_a(n)$ and $\Delta E_b(n)$ for the hexamer ring, (HF)$_6$, amount to -9.1 and -8.1 kcal/mol, respectively. Similarly large enhancements have also been obtained from different DFT calculations on the same set of chain- and ring-like clusters [52] and on periodic polymers [205].

TABLE VIII.4
Calculated Total Stabilization Energies of
Chain-like and Cyclic (HF)$_n$ Oligomers[a]

References		[52]	[149, 181]	[152]
	n			
Chains	2	−4.9	−4.6	−4.6
	3	−11.3		
	4	−18.5		
	5	−26.1		
	6	−34.0		
Rings	2	−3.8	−3.6	
	3	−15.5	−15.1	−15.3
	4	−28.4	−28.0	−27.7
	5	−39.2	−38.5	−37.8
	6	−48.4	−47.6	

[a]All values are in kilocalories per mole (kcal/mol).

Parallel to these quite significant energetic changes, the modifications of the intramolecular r(HF) distances and of the intermolecular r(FF) distances are also much larger than in the HCN and HC$_3$N clusters. This finding is demonstrated in Tables VIII.5, VIII.6 and in Figure 8.8. The calculated widening of the intramolecular r(HF) distance from the monomer to the cyclic hexamer amounts to 0.032 Å, the corresponding reduction of the intermolecular r(FF) distances from the dimer to the hexamer is calculated

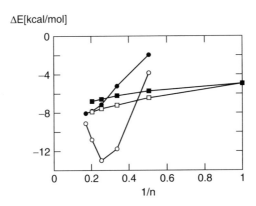

Figure 8.7. Plots of $\Delta E_a(n)$ and $\Delta E_b(n)$ versus $1/n$ for zigzag chains and for cyclic hydrogen fluoride clusters as obtained at the MP2/6-311++G(2d,p) level [52]. Open symbols $\Delta E_a(n)$; filled symbols $\Delta E_b(n)$; circles: cyclic structures; squares: chains.

TABLE VIII.5
Calculated Intramolecular $r(HF)$ Distances in the Center of Chain-Like and in Ring-Like $(HF)_n$ Oligomers[a]

References		[52]	[149, 181]
	n		
Chains	1	0.922	0.917
	2	0.928	0.923
	3	0.934	
	4	0.938	
	5	0.941	
	6	0.944	
Rings	2	0.926	
	3	0.938	0.933
	4	0.949	0.944
	5	0.955	0.948
	6	0.956	0.949

[a] All values are in angstroms (Å).

as 0.33 Å. This finding is in excellent agreement with the experimentally determined structural parameters of the vapor-phase dimer [161] and hexamer [160] and with the structure of the molecular crystal [28, 29] from which reductions for $r(FF)$ of 0.29 and 0.33 Å from the dimer to the hexamer and to the crystal can be inferred.

TABLE VIII.6
Calculated Intramolecular $r(FF)$ Distances in the Center of Chain-Like and in Ring-Like $(HF)_n$ Oligomers[a]

References		[52]	[149, 181]
	n		
Chains	2	2.770	2.735
	3	2.686	
	4	2.621	
	5	2.587	
	6	2.555	
Rings	2	2.716	2.71
	3	2.634	2.59
	4	2.537	2.51
	5	2.496	2.48
	6	2.484	2.47

[a] All values are in angstroms (Å).

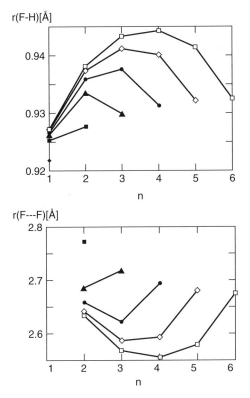

Figure 8.8. Trends in intramolecular F—H and intermolecular F---F distances in zigzag chain-like (HF)$_n$ clusters as obtained at the MP2/6-311G++(2d,p) level [52]. Distances belonging to the same oligomer are marked by identical symbols and connected by lines.

The mutual orientation of the monomers in the cyclic hexamer and also in the pentamer is very close to that in the dimer (110–120°) and in the crystal (116°). The calculated structural properties of pentamer and hexamer are, therefore, very similar to each other and are expected to be very close to the results of calculations on periodic HF chains. The DFT [B3LYP/6-311++G(d,p)] calculations on (HF)$_\infty$ [205] resulted in an r(FF) distance of 2.483 Å, which is in excellent agreement with the X-ray and neutron diffraction results [28, 29]. The calculations on the shorter zigzag chains show that the zigzag hexamer is still way off from convergence toward bulk properties. Lower level ab initio SCF calculations on chains up to (HF)$_{19}$ [146, 207] demonstrated this slow convergence in greater detail.

In view of the considerably larger hydrogen-bond energy gains and the stronger structural relaxations in HF clusters, more prominent shifts of the

TABLE VIII.7
Calculated Harmonic F—H Stretching Vibrational Frequency Shifts and IR Intensity Enhancement Factors in Zigzag Open-chain and in Cyclic $(HF)_n$ Clusters as Obtained at the MP2/6-311++G(2d,p) Level.[a]

n	Chains	Rings
1[b]	4134 [137]	
2	−41 (1.0) [−31][c]	−53 (2.4)
	−112 (3.5) [−93][c]	−73 (0)
3	−56 (1.2)	−253 (9.8) [−249][c]
	−147 (2.1)	−358 (0)
	−232 (7.1)	
4	−62 (1.3)	−431 (0)
	−178 (3.2)	−504 (26.4) [−516][c]
	−247 (1.4)	−687 (0)
	−339 (12.3)	
5	−67 (1.3)	−534 (0)
	−196 (3.4)	−655 (44.7) [−661][c]
	−277 (3.8)	−858 (0)
	−336 (0.7)	
	−434 (18.3)	
6	−69 (1.3)	−533 (0)
	−205 (3.5)	−573 (0)
	−294 (3.7)	−709 (60.4) [−716][c]
	−357 (4.6)	−893 (0)
	−409 (0.1)	
	−516 (24.5)	

[a] Frequency shifts are in reciprocal centimeters (cm^{-1}), enhancement factors are in parentheses. Data taken from [52].
[b] Monomer frequency (cm^{-1}) and IR intensity (km/mol) are in brackets.
[c] Experimentally observed shifts of anharmonic fundamental frequencies as taken from [149].

HF stretching vibrations must be expected than for the C—H stretching frequency shifts in the two previously discussed cases. This finding is indeed the case. The MP2/6-311++G(2d,p) calculated [52] HF stretching frequencies and the corresponding IR intensities of HF clusters are collected in Table VIII.7. These data must be confronted with the analogous sets of data for HCN and HC_3N clusters of Table VIII.3. The comparison reveals a number of features: (a) The shift of the intramolecular X—H stretching frequency of the dimers originating from the X—H group involved in the hydrogen bond is comparable in all three cases. The experimentally observed shift in $(HF)_2$ is $-93 \, cm^{-1}$, which is only slightly larger than the

-70 and $-66\,\text{cm}^{-1}$ shifts in $(HCN)_2$ and $(HC_3N)_2$, respectively. The shifts in the cyclic trimers are already significantly different. In the case of cyclic $(HCN)_3$ and $(HC_3N)_3$ the experimental and calculated shifts are smaller than those in the linear dimers. The opposite is true for $(HF)_3$. The calculated and experimental red-shifts of the IR active H—F stretching mode and, even more so, the calculated shift of the IR inactive, in-phase stretching mode are larger by factors of ~ 2.5 and 3.5 compared to the observed shift in the HF dimer. This trend is continued up to the hexamer for which the experimentally observed shift amounts to $-716\,\text{cm}^{-1}$. The calculated shift of the in-phase stretching mode of $-893\,\text{cm}^{-1}$ is even larger. The cyclic hexamer and the heptamer with an experimental shift of the IR active mode of $-746\,\text{cm}^{-1}$ are structurally already quite close to the infinitely extended HF chains in the molecular crystals. The experimentally observed shift of the in-phase $k = 0$ fundamental of solid HF relative to the monomer ranges from -894 to $-934\,\text{cm}^{-1}$ depending on whether IR [190] or Raman [191–194] frequencies are chosen for that comparison. This observation contrasts with the modest monomer-to-crystal shifts of -182 and $-132\,\text{cm}^{-1}$ for HCN and HC_3N. The IR intensity enhancements are also much larger in the case of HF oligomers. Recent density functional calculations on the vibrational spectra of periodic HF chains also successfully reproduce the experimental shifts [205].

In addition, ^1H NMR chemical shifts, $\sigma(^1H)$, of chain-like [52] and cyclic [52, 149] HF clusters have been evaluated. From the monomer to the cyclic hexamer a reduction of ~ 7–8 ppm has been calculated, again considerably more than the 2 ppm calculated for cyclic $(HCN)_6$ [52].

If we summarize the results on $(HF)_n$ clusters, we observe substantial signs of cooperativity. The hydrogen-bond energy in the polymer is $\sim 75\%$ larger than that in the dimer, the intermolecular distance is reduced by $\sim 0.3\,\text{Å}$ upon going from the dimer to the molecular crystal, there is almost a factor of 10 between the H—F stretching vibrational frequency shift in the dimer and in the molecular crystal, there is a large IR intensity increase of the in-phase HF stretching mode, and there is a strong reduction of the ^1H NMR chemical shift. Hydrogen fluoride is perhaps the molecular crystal with the largest detectable effects of cooperativity.

2. *$(HCl)_n$ and $(HBr)_n$*

Compared to the case of $(HF)_n$ clusters, much less is known about clusters of hydrogen chloride molecules, $(HCl)_n$, and even less about clusters of hydrogen bromide molecules, $(HBr)_n$. The dimer $(HCl)_2$ has, however, been well characterized by a variety of vibrational spectroscopic investigations [208–227] in the gas phase, by extended quantum mechanical investigations of large parts of the energy surface [228–234], and by various fits to these energy surfaces [235–238] followed by investigations of different aspects of vibrational dynamics [125, 222, 235–240]. The structure of the dimer (DBr)$_2$ was recently determined by microwave spectroscopy [241]. A few ab initio studies dealing with the equilibrium structure, vibrational spectroscopy, and the characterization of stationary points of (HBr)$_2$ have also been performed [242–245].

In the case of the trimer, (HCl)$_3$, vibrational spectroscopic investigations in solid Ne and Ar matrices [246–252] and in liquid rare gases [253] have been reported. To date, two gas-phase studies are available [254, 255]. Several quantum chemical calculations [139, 142, 243, 256–259] on the structure, harmonic vibrational frequencies, and other properties of the HCl trimer have also appeared. The first high-resolution IR study on the tetramer, (HCl)$_4$, has been communicated very recently [255]. Infrared spectra of (HCl)$_4$ in solid and liquid noble gases [247, 248, 253] have also been analyzed. Only a few theoretical investigations dealt with (HCl)$_4$ [243, 258, 259], even larger HCl clusters [258, 259], or with an infinitely extended chain [260]. The cyclic configurations of (HBr)$_3$ and (HBr)$_4$ have been studied theoretically [243]. Infrared investigations on HBr clusters trapped in cryogenic matrices have been investigated [247, 248]. Gas-phase HBr clusters have been generated, size-selected, and subjected to ultraviolet (UV) photodissociation [261–265]. To date, there is no structural or high-resolution spectroscopic information available for HBr vapor-phase clusters.

In addition to the investigations on small clusters, rich material on the structure and vibrational spectra of solid HCl and HBr have been available for some time. Both have several different phases in the solid state, the low-temperature phases have, however, orthorhombic structures, isomorphous to the structure of solid HF with the space group C_{2v}^{12} [30, 31, 266–268]. Infrared and Raman spectra of the orthorhombic phases of HCl and HBr have been amply discussed [269–275].

The main differences between HF clusters on the one hand and HCl and HBr clusters on the other can already be guessed at the stage of the dimer and from a knowledge of some of the monomers electric properties. One ingredient is that the dipole moment of the HF molecule is much larger than the dipole moments of HCl or HBr and that the order of the quadrupole moments is just reversed. The second ingredient is the much smaller polarizability of HF as compared to that of HCl and HBr (see, e.g., [117]). These monomer properties in combination with the considerably smaller intermolecular distance in (HF)$_2$ has the consequence that the interaction energy in (HF)$_2$ is essentially acceptably well determined at the SCF level, that is, by electrostatic and polarization effects, whereas the dispersion energy contribution is a comparatively small attractive correction of the order of ~10%. On the contrary, the dimers $(HCl)_2$ and $(HBr)_2$ cannot be reasonably described at the SCF level at all. The dispersion energy contribution amounts to about two-thirds of the total interaction energy. As usual, the relative orientation of the molecules is, however, well described by electrostatics alone, pointing to a dominating quadrupole–quadrupole interaction in the case of $(HCl)_2$ and $(HBr)_2$. The overall interaction energy is largest in $(HF)_2$ and smallest in $(HBr)_2$. From different detailed analyses of the nonadditivities in $(HF)_3$ and $(HCl)_3$ [243, 256, 257, 276] it turns out that the nonadditivity is largely determined by the polarization nonadditivity that is already obtained at the SCF level and that nonaddivities of the dispersion energy are not too important. Therefore, the nonadditivities are expected to be much larger in HF clusters than in HCl or HBr clusters. Figure 8.9 shows a comparison of calculated $\Delta E_a(n)$ and $\Delta E_b(n)$ values for cyclic $(HCl)_3$ and $(HCl)_4$ [243] with $\Delta H_a(n)$ and $\Delta H_b(n)$ values for cyclic $(HF)_n$ and $(HCl)_n$, $n = 3$–6 [259]. From Figure 8.9 it is immediately obvious that the role of cooperative contributions to the interaction energy is indeed less important for $(HCl)_n$ clusters. The other signatures of cooperativity are, therefore, also expected to be considerably weaker in $(HCl)_n$ and $(HBr)_n$.

In Table VIII.8, the computed $r(Cl\text{---}H)$ and $r(Cl\text{---}Cl)$ distances as obtained from the quantum chemical calculations on $(HCl)_n$, $n = 3$–6,

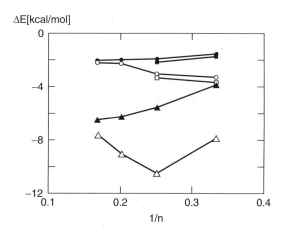

Figure 8.9. Plots of $\Delta E_a(n)$ and $\Delta E_b(n)$ values of cyclic $(HCl)_3$ and $(HCl)_4$ [243] and of $\Delta H_a(n)$ and $\Delta H_b(n)$ values for cyclic $(HF)_n$ and $(HCl)_n$, $n = 3$–6 [259] versus $1/n$. Open symbols: $\Delta E_a(n)$ or $\Delta H_a(n)$; filled symbols: $\Delta E_b(n)$ or $\Delta H_b(n)$; triangles: $(HF)_n$; circles and squares: $(HCl)_n$.

TABLE VIII.8
Calculated Intramolecular r(HCl) and Intermolecular r(ClCl)
Distances in (HCl)$_n$ Oligomers[a]

References	[243]	[259]	Experimental	References
n				
r(HCl)				
1	1.274	1.273	1.275	[277]
2	1.278	1.275		
3	1.282	1.278		
4	1.285	1.278		
5		1.280		
6		1.280		
r(ClCl)				
2	3.832	3.937	3.80	[208][b]
3	3.736	3.855	3.693	[254][b]
4	3.683	3.810		
5		3.791		
6		3.771		
Solid			3.688	[30]

[a] All values are in angstroms (Å).
[b] Center-of-mass distances.

[243, 259] are confronted with the available experimental data. The overall elongation of r(HCl) from the monomer to the tetramer is ~ 0.01 Å similar to the HCN case and much smaller than for HF clusters. The corresponding trend is observed in the smaller reduction of the experimental intermolecular distance from the dimer to the solid, which turns out to be ~ 0.16 Å. The experimental and calculated [243] H—Cl stretching frequency shifts are compiled in Table VIII.9 and show the very same trends. Nonadditivity effects in HCl are much weaker than in HF and are only slightly larger and comparable in magnitude to the case of HCN.

The corresponding effects in structural relaxations and vibrational frequency shifts are calculated to be of comparable magnitude in (HBr)$_n$ complexes [243] as well. From the (HBr)$_2$ dimer to the molecular HBr crystal, the experimentally determined intermolecular Br—Br distance contracts from 4.13 Å [241] to 3.93 Å [31], or by 0.20 Å. The calculated reductions of r(Br—Br) from the dimer to the trimer and tetramer amount to 0.09 and 0.12 Å, respectively [243]. The vibrational frequency shift from the monomer to the in-phase H—Br stretching frequency of solid HBr amounts to -154 cm^{-1} [273]. The calculated shifts to the dimer and trimer are -88, and -91 and -117 cm^{-1}, respectively, and that to the tetramer amount to -96, -128, and -158 cm^{-1} [243].

TABLE VIII.9
Calculated and Experimental Cl—H Stretching
Vibrational Frequency Shifts in $(HCl)_n$ Clusters[a]

n	Calculated [243]	Experimental[b]	References
2	−53	−29	[208]
	−17	−6	[208]
3	−86	−116	[254]
	−116		
4	−111		
	−127	−150	[255]
	−164		
Solid		−184	[273][c]
		−229	[273][d]

[a] Frequency shifts are in reciprocal centimeters in (cm^{-1}).
[b] Experimentally observed shifts of anharmonic fundamental frequencies.
[c] Out-of-phase stretch.
[d] In-phase stretch.

By concluding the section on HCl and HBr clusters, we observe the expected weaker features of cooperativity compared to the case of HF clusters. The reason for this behavior may be found in the comparatively small role of the electrostatic and polarization interaction to the stabilization energy in the dimers $(HCl)_2$ and $(HBr)_2$ and the much more important role of dispersion interaction. The nonadditivities, largely determined by polarization effects, are, therefore, smaller as well.

C. Substituted Zigzag Chains and Rings

1. $(CH_3OH)_n$

Similar to the crystal structures of hydrogen halides, many of the primary alcohols crystallize in anisotropic lattices with zigzag hydrogen-bonded chains as the dominating structural element. This structure is sketched in Figure 8.10. The simplest representative is solid methanol [32] with the space group D_{2h}^{17} − Cmcm and four molecules in the unit cell. The O—H---O hydrogen-bond configurations are again nearly linear. An intermolecular r(O---O) distance of 2.66 Å has been determined by X-ray investigations. From microwave studies on the methanol dimer [278], an H---O distance of 1.96 Å has been deduced that yields an approximate dimer r(O---O) of 2.91 Å and a reduction of 0.25 Å from the dimer to the solid. Vibrational

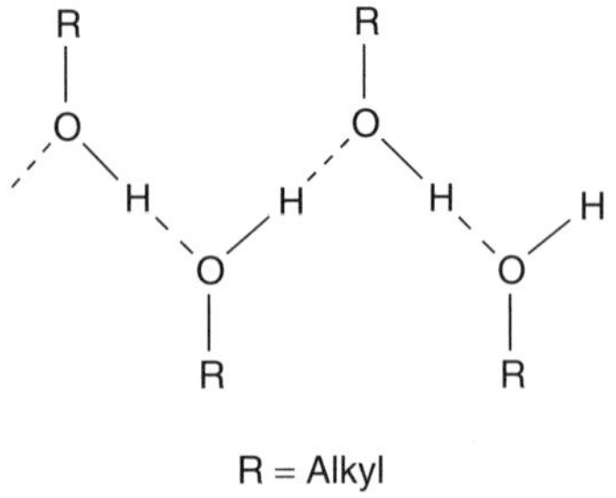

Figure 8.10. Sketch of the zigzag chain structure in solid methanol and other primary alcohols.

spectroscopic investigations of solid methanol [279, 280] led to an observed $k = 0$ in-phase O—H stretching frequency of 3187 cm^{-1}, and thus a shift of about -500 cm^{-1} from the methanol monomer O—H stretch of 3682 cm^{-1} [281]. Within these limits the r(O---O) distances and O—H stretching frequencies of methanol clusters are bound to occur.

In recent reviews [23, 282], the available experimental and theoretical results on methanol clusters are extensively discussed. Apart from the open-chain structure of the methanol dimer, the higher oligomers up to $n = 7$ or 8 are all cyclic, as was the case for the hydrogen halides. However, already from the trimer on, the rings are nonplanar. The experimentally observed highest intensity IR O—H stretching peaks are shifted for $(CH_3OH)_n$, $n = 2$-6 [283–287], by -112, -212, -402, -442, and -452 cm^{-1}, respectively. Quantum chemically computed shifts [286–295] are in good agreement with experimental data. Also, the structural predictions born out by these calculations, in particular, the contraction of r(O---O) and the widening of r(O—H), are in the expected range.

Overall, judging from the calculated and experimentally observed contractions of the intermolecular distance and the O—H stretching vibrational frequency shifts, the cooperative features of $(CH_3OH)_n$ clusters are distinctly smaller than in those in $(HF)_n$, but larger than those in the other hydrogen halide clusters.

2. $(H_2O)_n$

An extended discussion of the large number of experimental data, ab initio calculations and dynamical calculations on $(H_2O)_n$ clusters goes far beyond the scope of this chapter. The recent review articles on $(H_2O)_n$ clusters [21–25] cover most of the aspects that were treated in the previous sections for the other hydrogen-bonded clusters and also discuss the properties of larger clusters with 3D hydrogen-bonded networks. Therefore, only two very recent experimental investigations on the ring-like water hexamer in liquid helium [296] and in solid parahydrogen [297] and a few of the most recent experimental and quantum mechanical studies on $(H_2O)_n$, $n = 2-6$, are mentioned [298–311]. Ignoring the fact that water molecules have the ability to form 3D clusters, it would not be very surprising if the ring-like structures up to $(H_2O)_6$ would have properties similar to those of the corresponding methanol clusters, which is indeed the case. From the open-chain water dimer with an experimentally determined $r(O\text{---}O)$ of 2.952 Å [312] to a $r(O\text{---}O)$ distance of ~ 2.74 Å in ice I [313], the contraction is similar and actually a bit smaller than that in methanol. Earlier small basis set SCF calculations on infinitely extended $(CH_3OH)_\infty$ and $(H_2O)_\infty$ chains [314] and more advanced MP2 and MP4 calculations on $(H_2O)_\infty$ chains [315] were already successful in reproducing these trends. The experimental vibrational frequency shift from the monomer to ice I amounts to about $-600\,\text{cm}^{-1}$ [313], again considerably smaller than in the $(HF)_n$ case, but of similar magnitude to $(CH_3OH)_n$.

D. Miscellaneous

More complicated catemer patterns are frequently encountered in organic crystals. Particularly interesting examples are the first two carbonic acids, formic acid and acetic acid. Both, as with most other carbonic acids, form doubly hydrogen-bonded, cyclic dimers in the vapor phase, but chains with one hydrogen bond linking the monomers in the molecular crystal [34–40]. The structures are sketched in Figure 8.11. Propionic acid [316] and others

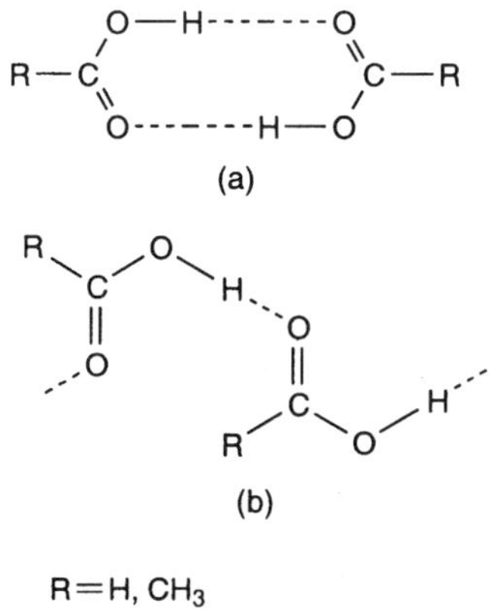

R = H, CH₃

Figure 8.11. Sketch of the (*a*) cyclic dimer and (*b*) chain structure of formic and acetic acid.

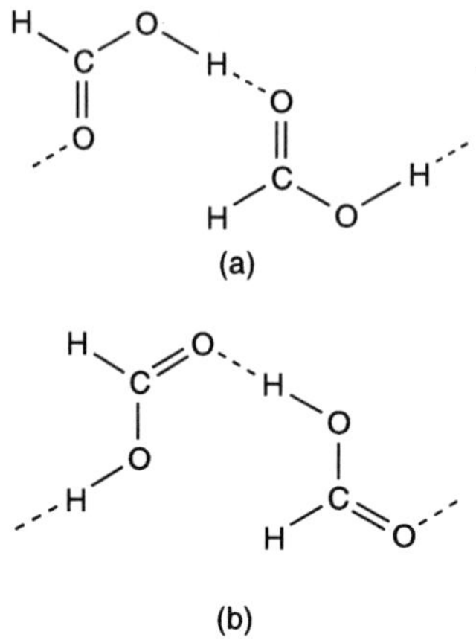

Figure 8.12. Catemeric motifs of formic acid built from (a) syn and (b) anti formic acid monomers

with still larger alkyl or aromatic residues, however, retain the cyclic dimer configuration in the solid state or form more complicated hydrogen-bonded networks [43]. A $r(O\text{---}O)$ distance of 2.703 Å in the vapor-phase cyclic dimer of formic acid has been determined by electron diffraction [317, 318]. In the solid-state catemers, the $R(O\text{---}O)$ distance is 2.63–2.65 Å [35, 36], which is distinctly shorter. These trends could also already be reproduced by early, rather small basis set SCF calculations on the infinitely extended formic chain [319]. Another interesting aspect on the crystal structure of formic acid is the possibility that a hypothetical, second catemer motif built from anti formic acid monomers (see Fig. 8.12) could be close in energy to the experimentally determined structure built from syn formic acid monomers, despite the large energy difference between syn and anti monomers of formic acid. Indications for this behavior have been found in connection with a phase transition in solid formic acid [320] and by a rather small calculated energy difference between these two structure types [319]. Very recently, it was found that this second catemeric motif is actually realized in a cocrystal of formic acid with hydrogen fluoride [321].

Cooperative effects have also theoretically been investigated on infinite chains [322] and finite oligomers [323, 324] of formamide, on oligomers of N-methylacetamide [325, 326], on infinite chains [327] and oligomers of imidazole [328, 329], and on cyclic clusters of pyrazole [330] and tetrazole [331].

IV. CONCLUSIONS

The aim of this chapter was to give a survey and a concise description of cooperative effects in hydrogen bonding as they are encountered in a small selected segment of conceivable systems: dimers, chain- and ring-like oligomers or clusters, and regular periodic 1D-chains as structural elements of molecular crystals, in each case built from one type of molecule. As an additional restriction, only very small molecules were chosen as monomers. Within this class of systems cooperative effects occur on a wide scale of different strengths. The two limiting cases, at least when considering the few systems discussed here, are the clusters of cyanoacetylene and the clusters of hydrogen fluoride. Cyanoacetylene clusters and the cyanoacetylene molecular solid behave nearly additively. The structural and vibrational spectroscopic changes that are observed when comparing dimer and molecular crystals are very small indeed, with a reduction of the intermolecular distance of 0.05 Å, a red-shift of the C—H stretching frequency of -66cm^{-1}, and a calculated increase of the hydrogen-bond energy of 30%. In the case of $(HF)_n$, the corresponding effects are much larger: The reduction of the intermolecular distance amounts to 0.33 Å, the red-shift is larger than -800cm^{-1}, and the increase of the hydrogen-bond energy is on the order of 75%. All other systems discussed fall in between these limits.

Considerably stronger cooperative effects than those discussed here would be observable in mixed hydrogen-bonded clusters built from heterodimers or in charged hydrogen-bonded clusters. For many of these clusters, gradual transitions to the very strong hydrogen bonds in ionic crystals are to be expected or even to crystals where a complete proton transfer has taken place, as, for example, in ammonium chloride. As an instructive example, in mixed crystals of HF and H_2O with different stoichiometry, ionic structures reminiscent of small charged hydrogen-bonded systems are encountered [332, 333]. Other situations, where such gradual transition to the very short hydrogen bonds may be studied systematically, are investigations of protonated defects in hydrogen-bonded chains [146, 207, 334, 335] or networks. Still other objects that lend themselves to a systematic study of shorter hydrogen bonds are the first-order saddle point configurations with symmetric hydrogen bonds arising on the pathway of concerted proton transfer between equi-energetic isomers in the few-membered hydrogen-bonded rings discussed in this chapter. In the case of $(HF)_n$ and $(H_2O)_n$ rings, these transition states with symmetric hydrogen bonds have already been looked at in detail [136, 141, 144, 146, 149, 150, 152, 158]. With increasing ring size the energy difference to the ring minima decreases systematically and drastically. Molecular crystals under high external hydrostatic pressure display phase transitions to structures with symmetric hydrogen bonds, as has been demonstrated for hydrogen fluoride [194] and ice [336]. The systematic description of these phenomena with high-level quantum mechanical approaches will be certainly further attacked in future investigations.

References

1. *Hydrogen Bonding*, edited by D. Hadzi and H. W. Thompson (Pergamon Press, London, 1959).
2. G. W. Pimentel and A. C. McClellan, *The Hydrogen Bond* (Freeman, San Francisco, 1960).
3. W. C. Hamilton and J. A. Ibers, *Hydrogen Bonding in Solids* (Benjamin, New York, 1968).
4. S. N. Vinogradov and R. H. Linell, *Hydrogen Bonding* (Van Nostrand-Reinhold, New York, 1971).
5. M. D. Joesten and L. J. Schaad, *Hydrogen Bonding* (Dekker, New York, 1974).
6. *The Hydrogen Bond-Recent Developments in Theory and Experiment*, Vols. 1–3, edited by P. Schuster, G. Zundel, and C. Sandorfy (North-Holland, Amsterdam, 1976).
7. P. Schuster, *Top. Curr. Chem.* **120** (1984).
8. P. Hobza and R. Zahradnik, "Intermolecular Complexes, The Role of van der Waals Systems in Physical Chemistry and in the Biodisciplines" in *Studies in Physical and Theoretical Chemistry 52* (Elsevier, Amsterdam, 1988).
9. J. Michl and R. Zahradnik, the entire volume *Chem. Rev.* **88**, 6 (1988).
10. G. A. Jeffrey and W. Saenger, *Hydrogen Bonding in Biological Structures* (Springer-Verlag, Berlin, 1991).
11. A. W. Castleman, Jr., and P. Hobza, the entire volume *Chem. Rev.* **94**, 7 (1994).
12. S. Scheiner, *Hydrogen Bonding — A Theoretical Perspective* (Oxford University Press, New York, 1997).
13. G. A. Jeffrey, *An Introduction to Hydrogen Bonding* (Oxford University Press, New York, 1997).
14. *Molecular Interactions — From van der Waals to Strongly Bound Molecular Complexes*, edited by S. Scheiner (John Wiley & Sons, Ltd., Chichester, 1997).
15. *Theoretical Treatment of Hydrogen Bonding*, edited by D. Hadzi (John Wiley & Sons, Inc., New York, 1997).
16. G. R. Desiraju and T. Steiner, *The Weak Hydrogen Bond* (Oxford University Press, Oxford, 1999).
17. *Hydrogen Bond Research*, edited by P. Schuster and W. Mikenda, the entire volume *Mh. Chem.* **130**, 8 (1999)
18. *van der Waals Molecules III*, edited by B. Brutschy and P. Hobza, the entire volume *Chem. Rev.* **100**, 11 (2000).
19. *Recent Theoretical and Experimental Advances in Hydrogen Bonded Clusters*, edited by S. S. Xantheas NATO-ASI Ser. C561 (Kluwer, Dordrecht, 2000).
20. P. Schuster, in *Intermolecular Interactions: From Diatomics to Biopolymers*, edited by B. Pullman, (John Wiley & Sons, Ltd., Chichester, 1978), p. 363.
21. S. S. Xantheas and Th. H. Dunning, Jr., in *Advances in Molecular Vibrations and Collision Dynamics*, edited by J. M. Bowman and Z. Bacic, (JAI Press Inc, Stamford, 1998), p. 281.

22. D. J. Wales, in *Advances in Molecular Vibrations and Collision Dynamics*, edited by J. M. Bowman and Z. Bacic, (JAI Press Inc, Stamford, 1998), p. 365.
23. U. Buck and F. Huisken, *Chem. Rev.* **100**, 3863 (2000).
24. J. M. Ugalde, I. Alkorta, and J. Elguero, *Angew. Chem. Int. Ed. Engl.* **39**, 717 (2000).
25. R. Ludwig, *Angew. Chem. Int. Ed. Engl.* **40**, 1809 (2001).
26. W. J. Dulmage and W. N. Lipscomb, *Acta Crystallogr.* **4**, 330 (1951).
27. V. F. Shallcross and G. B. Carpenter, *Acta Crystallogr.* **11**, 490 (1958).
28. M. Atoji and W. N. Lipscomb, *Acta Crystallogr.* **7**, 173 (1954).
29. M. W. Johnson, E. Sandor, and E. Arzi, *Acta Crystallogr.* **B31**, 1998 (1975).
30. E. Sandor and R. F. C. Farrow, *Nature (London)*, **213**, 171 (1967).
31. V. A. Simon, *J. Appl. Crystallogr.* **4**, 138 (1971).
32. K. J. Tauer and W. N. Lipscomb, *Acta Crystallogr.* **5**, 606 (1952).
33. P.-G. Jönsson, *Acta Crystallogr.* **B32**, 232 (1976).
34. F. Holtzberg, B. Post, and I. Fankuchen, *Acta Crystallogr.* **6**, 127 (1953).
35. I. Nahringbauer, *Acta Crystallogr.* **B34**, 315 (1978).
36. A. Albinati, K. D. Rouse, and M. W. Thomas, *Acta Crystallogr.* **B34**, 2188 (1978).
37. R. E. Jones and D. H. Templeton, *Acta Crystallogr.* **11**, 484 (1958).
38. I. Nahringbauer, *Acta Chem. Scand.* **24**, 453 (1970).
39. P.-G. Jönsson, *Acta Crystallogr.* **B27**, 893 (1971).
40. A. Albinati, K. D. Rouse, and M. W. Thomas, *Acta Crystallogr.* **B34**, 2184 (1978).
41. S. Martinez-Carrera, *Acta Crystallogr.* **20**, 783 (1966).
42. C. Foces-Foces, I. Alkorta, and J. Elguero, *Acta Crystallogr.* **B56**, 1018 (2000).
43. L. Leiserowitz, *Acta Crystallogr.* **B32**, 775 (1976).
44. C. Foces-Foces, A. Echevarria, N. Jagerovic, I. Alkorta, J. Elguero, U. Langer, O. Klein, M. Minguet-Bonhevi, and H.-H. Limbach, *J. Am. Chem. Soc.* **123**, 7898, 2001.
45. M. C. Etter, *Acc. Chem. Res.* **23**, 120 (1990).
46. Y. Ducharme and J. D. Wuest, *J. Org. Chem.* **53**, 5789 (1988).
47. C. B. Aakeröy and K. R. Seddon, *Chem. Soc. Rev.* 400 (1993).
48. J. C. McDonald and G. M. Whitesides, *Chem. Rev.* **94**, 2383, (1994).
49. R. E. Melendez, and A. D. Hamilton, *Top. Curr. Chem.* **198**, 97 (1998).
50. P. Schuster, A. Karpfen, and A. Beyer, in *Molecular Interactions*, Vol. 1, edited by H. Ratajczak and W. J. Orville-Thomas, (John Wiley & Sons, Ltd., Chichester, 1980), p. 117.
51. A. Karpfen, J. Ladik, P. Russegger, P. Schuster, and S. Suhai, *Theor. Chim. Acta (Berlin)*, **34**, 115 (1974).
52. A. Karpfen, in *Molecular Interactions-From van der Waals to Strongly Bound Molecular Complexes*, edited by S. Scheiner, (John Wiley & Sons, Ltd., Chichester, 1997), p. 265.
53. J. O. Hirschfelder, C. F. Curtiss, and R. B. Bird, *Molecular Theory of Gases and Liquids* (John Wiley & Sons Inc., New York, 1964).
54. J. O Hirschfelder, *Adv. Chem. Phys.*, **12** (1967)
55. H. Margenau and N. R. Kestner, *Intermolecular Forces* (Pergamon Press, London, 1971).
56. A. D. Buckingham, in *Intermolecular Interactions: From Diatomics to Biopolymers*, edited by B. Pullman, (John Wiley & Sons Ltd., Chichester, 1978), p. 1.
57. P. Claverie, in *Intermolecular Interactions: From Diatomics to Biopolymers*, edited by B. Pullman, (John Wiley & Sons Ltd., Chichester, 1978), p. 71.
58. G. C. Maitland, M. Rigby, E. B. Smith, and W. A. Wakeham, *Intermolecular Forces. Their Origin and Determination* (Clarendon Press, Oxford, 1981).
59. A. J. Stone, *The Theory of Intermolecular Forces* (Clarendon Press, Oxford, 1961).
60. A. D. Buckingham, P. W. Fowler, and J. M. Hutson, *Chem. Rev.* **88**, 963 (1988).
61. G. Chalasinski and M. M. Szczesniak, *Chem. Rev.* **94**, 1723 (1994).
62. G. Chalasinski and M. M. Szczesniak, *Chem. Rev.* **100**, 4227 (2000).

63. K. Morokuma, *J. Chem. Phys.* **55**, 1236 (1971).
64. M. M. Szczesniak and G. Chalasinski, in *Molecular Interactions-From van der Waals to Strongly Bound Molecular Complexes*, edited by S. Scheiner, (John Wiley & Sons, Ltd., Chichester, 1997), p. 45.
65. A. C. Legon, D. J. Millen, and P. J. Mjöberg, *Chem. Phys. Lett.* **47**, 589 (1977).
66. L. W. Buxton, E. J. Campbell, and W. H. Flygare, *Chem. Phys.* **56**, 399 (1981).
67. R. D. Brown, P. D. Godfrey, and D. A. Winkler, *J. Mol. Spectrosc.* **89**, 352 (1981).
68. A. J. Fillery-Travis, A. C. Legon, L. C. Willoughby, and A. D. Buckingham, *Chem. Phys. Lett.* **102**, 126 (1983).
69. K. Georgiou, A. C. Legon, D. J. Millen, and P. J. Mjöberg, *Proc. R. Soc. (London) Sci. A* **399**, 377 (1985).
70. B. A. Wofford, J. W. Bevan, W. B. Olson, and W. J. Lafferty, *J. Chem. Phys.* **85**, 105 (1986).
71. K. W. Jucks and R. E. Miller, *Chem. Phys. Lett.* **147**, 137 (1988).
72. K. W. Jucks and R. E. Miller, *J. Chem. Phys.* **88**, 6059 (1988).
73. H. Meyer, E. R. Th. Kerstel, D. Zhuang, and G. Scoles, *J. Chem. Phys.* **90**, 4623 (1989).
74. E. R. Th. Kerstel, K. K. Lehmann, J. E. Gambogi, X. Yang, and G. Scoles, *J. Chem. Phys.* **99**, 8559 (1993).
75. E. R. Th. Kerstel, G. Scoles, and X. Yang, *J. Chem. Phys.* **99**, 876 (1993).
76. R. S. Ruoff, T. Emilsson, C. Chuang, T. D. Klots, and H. S. Gutowsky, *Chem. Phys. Lett.* **138**, 553 (1988).
77. R. S. Ruoff, T. Emilsson, T. D. Klots, C. Chuang, and H. S. Gutowsky, *J. Chem. Phys.* **89**, 138 (1988).
78. K. W. Jucks and R. E. Miller, *J. Chem. Phys.* **88**, 2196 (1988).
79. D. S. Anex, E. R. Davidson, C. Douketis, and G. E. Ewing, *J. Phys. Chem.* **92**, 2913 (1988).
80. M. Maroncelli, G. A. Hopkins, J. W. Nibler, and Th. R. Dyke, *J. Chem. Phys.* **83**, 2129 (1985).
81. G. A. Hopkins, M. Maroncelli, J. W. Nibler, and Th. R. Dyke, *Chem. Phys. Lett.* **114**, 97 (1985).
82. X. Yang, E. R. Th. Kerstel, G. Scoles, R. J. Bemish, and R. E. Miller, *J. Chem. Phys.* **103**, 8828 (1995).
83. K. Nauta and R. E. Miller, *J. Chem. Phys.* **111**, 3426 (1999).
84. K. Nauta and R. E. Miller, *Science*, **283**, 1895 (1999).
85. K. Nauta, D. Moore, and R. E. Miller, *Faraday Discuss.* **113**, 261 (1999).
86. Ch. M. King and E. R. Nixon, *J. Chem. Phys.* **48**, 1685 (1968).
87. J. Pacansky, *J. Phys. Chem.* **81**, 2240 (1971).
88. B. Walsh, A. J. Barnes, S. Suzuki, and W. J. Orville-Thomas, *J. Mol. Spectrosc.* **72**, 44 (1978).
89. E. Knözinger, H. Kollhoff, and W. Langel, *J. Chem. Phys.* **85**, 4881 (1986).
90. O. Schrems, M. Huth, H. Kollhoff, R. Wittenbeck, and E. Knözinger, *Ber. Bunsen-Ges. Phys. Chem.* **91**, 1261 (1987).
91. W. Langel, H. Kollhoff, and E. Knözinger, *J. Chem. Phys.* **90**, 3430 (1989).
92. P. Beichert, D. Pfeiler, and E. Knözinger, *Ber. Bunsen-Ges. Phys. Chem.* **99**, 1469 (1995).
93. K. Satoshi, M. Takayanagi, and M. Nakata, *J. Mol. Struct.* **413**, 365 (1997).
94. M. Pezolet and R. Savoie, *Can. J. Chem.* **47**, 3041 (1969).
95. H. B. Friedrich and P. F. Krause, *J. Chem. Phys.* **59**, 4942 (1973).
96. M. Uyemura and S. Maeda, *Bull. Chem. Soc. Jpn.* **47**, 2930 (1974).
97. C. Nolin, J. Weber, and R. Savoie, *J. Raman Spectrosc.* **5**, 21 (1976).
98. K. Aoki, Y. Kakudate, M. Yoshida, S. Usuha, and S. Fujiwara, *J. Chem. Phys.* **91**, 2814 (1989).
99. M. Kofranek, A. Karpfen, and H. Lischka, *Chem. Phys.* **113**, 53 (1987).

100. M. Kofranek, H. Lischka, and A. Karpfen, *Mol. Phys.* **61**, 1519 (1987).
101. W. B. de Almeida, J. S. Craw, and A. Hinchliffe, *J. Mol. Struct.* **184**, 381 (1989).
102. W. B. de Almeida, *Can. J. Chem.* **69**, 2044 (1991).
103. I. J. Kurnig, H. Lischka, and A. Karpfen, *J. Chem. Phys.* **92**, 2469 (1990).
104. B. F. King, F. Weinhold, *J. Chem. Phys.* **103**, 333 (1995).
105. B. F. King, T. C. Farrar, and F. Weinhold, *J. Chem. Phys.* **103**, 333 (1995).
106. S. Suhai, *Intern. J. Quantum Chem.* **52**, 395 (1994).
107. A. Karpfen, *J. Phys. Chem.* **100**, 13474 (1996).
108. C. E. Dykstra, *J. Mol. Struct. (THEOCHEM.)* **362**, 1 (1996).
109. E. M. Cabaleiro-Lago, and M. A. Rios, *J. Chem. Phys.* **108**, 3598 (1998).
110. M. Kertész, J. Koller, and A. Azman, *Chem. Phys. Lett.* **41**, 576 (1975)
111. A. Karpfen, *Chem. Phys.* **79**, 211 (1983).
112. M. Springborg, *Ber. Bunsen-Ges. Phys. Chem.* **95**, 1238 (1991).
113. M. Springborg, *Chem. Phys.* **195**, 143 (1995).
114. A. Karpfen, *J. Phys. Chem. A* **102**, 9286 (1998).
115. E. M. Cabaleiro-Lago and M. A. Rios, *J. Chem. Phys.* **108**, 8398 (1998).
116. A. Beyer, A. Karpfen, and P. Schuster, *Top. Curr. Chem.* **120** (1984).
117. P. Schuster, in *Encyclopedia of Physical Science and Technology*, Vol. 6 (Academic Press, New York, 1987).
118. A. Karpfen, *Mh. Chem.* **130**, 1017 (1999).
119. D. G. Truhlar, in *Dynamics of Polyatomic Van der Waals Complexes*, edited by N. Halberstadt and K. C. Janda, NATO-ASI Series **227** (Plenum, New York, 1990), p.159.
120. M. Quack and M. A. Suhm, in *Advances in Molecular Vibrations and Collision Dynamics*, edited by J. M. Bowman and Z. Bacic, (JAI Press Inc, Stamford, 1998), p. 205.
121. K. Nauta and R. E. Miller, *J. Chem. Phys.* **113**, 10158 (2000).
122. W. Klopper, M. Quack, and M. A. Suhm, *J. Chem. Phys.* **108**, 10096 (1998).
123. W. Klopper and H. P. Lüthi, *Mol. Phys.* **96**, 559 (1999).
124. G. S. Tschumper, M. D. Kelty, and H. F. Schaefer, III, *Mol. Phys.* **96**, 493 (1999).
125. Z. Bacic and Y. Qiu, in *Advances in Molecular Vibrations and Collision Dynamics*, edited by J. M. Bowman and Z. Bacic, (JAI Press Inc, Stamford, 1998), p 182.
126. Y. Volobuev, W. C. Necoechea, and D. G. Truhlar, *Chem. Phys. Lett.* **330**, 471 (2000).
127. J. M. Lisy, A. Tramer, M. F. Vernon, and Y. T. Lee, *J. Chem. Phys.* **75**, 4733 (1981).
128. D. W. Michael and J. M. Lisy, *J. Chem. Phys.* **85**, 2528 (1986).
129. K. D. Kolenbrander, C. E. Dykstra, and J. M. Lisy, *J. Chem. Phys.* **88**, 5995 (1988).
130. H. Sun, R. O. Watts, and U. Buck, *J. Chem. Phys.* **96**, 1810 (1992).
131. M. A. Suhm, J. T. Farreell, jr, S. H. Ashworth, and D. J. Nesbitt, *J. Chem. Phys.* **98**, 5985 (1993).
132. M. A. Suhm and D. J. Nesbitt, *Chem. Soc. Rev.* **24**, 45 (1995).
133. A. Karpfen, A. Beyer, and P. Schuster, *Intern. J. Quantum Chem.* **19**, 1113 (1981).
134. A. Karpfen, A. Beyer, and P. Schuster, *Chem. Phys. Lett.* **102**, 289 (1983).
135. J. F. Gaw, Y. Yamaguchi, M. A. Vincent, and H. F. Schaefer, III, *J. Am. Chem. Soc.* **106**, 3133 (1984).
136. D. Heidrich, H.-J. Köhler, and D. Volkmann, *Intern. J. Quantum Chem.* **27**, 781 (1985).
137. S.-Y. Liu, D. W. Michael, C. E. Dykstra, and J. M. Lisy, *J. Chem. Phys.* **84**, 5032 (1986).
138. G. E. Scuseria and H. F. Schaefer, III, *Chem. Phys.* **107**, 33 (1986).
139. Z. Latajka and S. Scheiner, *Chem. Phys.* **122**, 413 (1988).
140. G. Chalasinski, S. M. Cybulski, M. M. Szczesniak, and S. Scheiner, *J. Chem. Phys.* **91**, 7048 (1989).
141. A. Karpfen, *Intern. J. Quantum Chem. (Quantum Chem. Symp.)* **24**, 129 (1990).
142. J. E. Del Bene and I. Shavitt, *J. Mol. Struct. (THEOCHEM)*, **234**, 499 (1991).

143. A. Komornicki, D. A. Dixon, and P. R. Taylor, *J. Chem. Phys.* **96**, 2920 (1992).
144. D. Heidrich, N. J. R. Van Eikema Hommes, and P. v. R. Schleyer, *J. Comp. Chem.* **14**, 1149 (1993).
145. M. Quack, J. Stohner, and M. A. Suhm, *J. Mol. Struct.* **294**, 33 (1993).
146. A. Karpfen and O. Yanovitskii, *J. Mol. Struct. (THEOCHEM)*, **314**, 211 (1994).
147. D. Luckhaus, M. Quack, U. Schmitt, and M. A. Suhm, *Ber. Bunsen-Ges. Phys. Chem.* **99**, 457 (1995).
148. G. S. Tschumper, Y. Yamaguchi, and H. F. Schaefer, III, *J. Chem. Phys.* **106**, 9627 (1997).
149. C. Maerker, P. v. R. Schleyer, K. R. Liedl, T.-K. Ha, M. Quack, M. A. Suhm, *J. Comp. Chem.* **18**, 1695 (1997).
150. K. R. Liedl, S. Sekusak, R. T. Kroemer, and B. M. Rode, *J. Phys. Chem. A* **101**, 4707 (1997).
151. K. R. Liedl, *J. Chem. Phys.* **108**, 3199 (1997).
152. W. Klopper, M. Quack, and M. A. Suhm, *Mol. Phys.* **94**, 105 (1998).
153. M. P. Hodges, A. J. Stone, and E. Caballeiro-Lago, *J. Phys. Chem. A* **102**, 2455 (1998).
154. T. Loerting, K. R. Liedl, and B. M. Rode, *J. Am. Chem. Soc.* **120**, 404 (1998).
155. K. R. Liedl and R. T. Kroemer, *J. Phys. Chem. A* **102**, 1832 (1998).
156. N. D. Sokolov and V. A. Savel'ev, *Pol. J. Chem.* **72**, 377 (1998).
157. B. L. Grigorenko, A. A. Moskovsky, and A. V. Nemukhin, *J. Chem. Phys.* **111**, 4442 (1997).
158. T. Loerting and K. R. Liedl, *J. Phys. Chem. A* **103**, 9022 (1999).
159. D. F. Smith, *J. Chem. Phys.* **28**, 1040 (1958).
160. J. Janzen and L. S. Bartell, *J. Chem. Phys.* **50**, 3611 (1969).
161. T. R. Dyke, B. J. Howard, and W. Klemperer, *J. Chem. Phys.* **56**, 2442 (1972).
162. J. J. Hinchen and R. H. Hobbs, *J. Opt. Soc. Am.* **69**, 1546 (1979).
163. R. L. Redington, *J. Chem. Phys.* **75**, 4417 (1981).
164. M. F. Vernon, J. M. Lisy, D. J. Krajnovich, A. Tramer, H.-S. Kwok, Y. R. Shen, and Y. T. Lee, *Faraday Discuss.* **73**, 387 (1982).
165. R. L. Redington, *J. Phys. Chem.* **86**, 552 (1982).
166. K. v. Puttkamer and M. Quack, *Chem. Phys.* **139**, 31 (1989).
167. M. A. Suhm, *Ber. Bunsen-Ges. Phys. Chem.* **99**, 1159 (1995).
168. F. Huisken, M. Kaloudis, A. Kulcke, and D. Voelkel, *Infrared Phys. Technol.* **36**, 171 (1995).
169. F. Huisken, M. Kaloudis, A. Kulcke, C. Laush, and J. M. Lisy, *J. Chem. Phys.* **103**, 5366 (1995).
170. L. Oudejans and R. E. Miller, *J. Chem. Phys.* **113**, 971 (2000).
171. T. A. Blake, S. W. Sharpe, and S. S. Xantheas, *J. Chem. Phys.* **113**, 707 (2000).
172. J. E. Del Bene and J. A. Pople, *J. Chem. Phys.* **55**, 2296 (1971).
173. C. E. Dykstra, *Chem. Phys. Lett.* **141**, 159 (1987).
174. C. Zhang, D. L. Freeman, and J. D. Doll, *J. Chem. Phys.* **91**, 2489 (1989).
175. C. E. Dykstra, *J. Phys. Chem.* **94**, 180 (1990).
176. M. Quack, U. Schmidt, and M. A. Suhm, *J. Mol. Struct.* **294**, 33 (1993).
177. K. R. Liedl, R. T. Kroemer, and B. M. Rode, *Chem. Phys. Lett.* **246**, 455 (1995).
178. F. Huisken, E. G. Tarakanova, A. A. Vigasin, and G. V. Yukhnevich, *Chem. Phys. Lett.* **245**, 319 (1995).
179. M. Ovchinnikov and V. A. Apkarian, *J. Chem. Phys.* **110**, 9842 (1999).
180. L. Rincon, R. Almeida, D. Garcia-Aldea, and H. Diez y Riega, *J. Chem. Phys.* **114**, 5552 (2001).
181. M. Quack and M. A. Suhm, in *Conceptual Perspectives in Quantum Chemistry*, Vol. 3, edited by J.-L. Calais and E. S. Kryachko, (Kluwer, Dordrecht, 1997), p. 415,

182. L. Andrews and G. L. Johnson, *Chem. Phys. Lett.* **96**, 133 (1983).
183. L. Andrews and G. L. Johnson, *J. Phys. Chem.* **88**, 425 (1984).
184. L. Andrews, *J. Phys. Chem.* **88**, 2940 (1984).
185. L. Andrews, V. E. Bondybey, and J. H. English, *J. Chem. Phys.* **81**, 3452 (1984).
186. L. Andrews, S. R. Davis, and R. D. Hunt, *Mol. Phys.* **77**, 993 (1992).
187. S. P. Habuda and Yu. V. Gagarisnky, *Acta Crystallogr. B* **27**, 1677 (1971).
188. P. A. Giguère and N. Zengin, *Can. J. Chem.* **36**, 1013 (1958).
189. M. L. N. Sastri and D. F. Hornig, *J. Chem. Phys.* **39**, 3497 (1963).
190. J. S. Kittelberger and D. F. Hornig, *J. Chem. Phys.* **46**, 3099 (1967).
191. A. Anderson, B. H. Torrie, and W. S. Tse, *Chem. Phys. Lett.* **70**, 300 (1980).
192. S. A. Lee, D. A. Pinnick, S. M. Lindsay, and R. C. Hanson, *Phys. Rev. B* **34**, 2799 (1986).
193. R. W. Jansen, R. Bertoncini, D. A. Pinnick, A. I. Katz, R. C. Hanson, O. F. Sankey, and M. O'Keeffe, *Phys. Rev. B* **35**, 9830 (1987).
194. D. A. Pinnick, A. I. Katz, and R. C. Hanson, *Phys. Rev. B* **39**, 8677 (1989).
195. H. Boutin and G. J. Safford, in *Inelastic Scattering of Neutrons in Solids and Liquids*, Vol. 2 (IAEA, Vienna, 1965), p. 393.
196. H. Boutin, G. J. Safford, and V. Brajovic, *J. Chem. Phys.* **39**, 3135 (1963).
197. A. Axmann, W. Biem, P. Borsch, F. Hoszfeld, and H. Stiller, *Faraday Discuss.* **7**, 69 (1969).
198. A. Anderson, B. H. Torrie, and W. S. Tse, *J. Raman Spectrosc.* **10**, 148 (1981).
199. R. Tubino and G. Zerbi, *J. Chem. Phys.* **51**, 4509 (1969).
200. M. Kertész, M. Koller, and A. Azman, *Chem. Phys. Lett.* **36**, 576 (1975).
201. A. Beyer and A. Karpfen, *Chem. Phys.* **64**, 343 (1982).
202. M. Springborg, *Phys. Rev. Lett.* **59**, 2287 (1987).
203. M. Springborg, *Phys. Rev. B* **38**, 1483 (1988).
204. S. Berski and Z. Latajka, *J. Mol. Struct. (THEOCHEM)*, **389**, 147 (1997).
205. S. Hirata and S. Iwata, *J. Phys. Chem. A* **102**, 8426 (1998).
206. D. Jacquemin, J.-M. André, and B. Champagne, *J. Chem. Phys.* **111**, 5324 (1999).
207. A. Karpfen and O. Yanovitskii, *J. Mol. Struct. (THEOCHEM)*, **307**, 81 (1994).
208. N. Ohashi and A. S. Pine, *J. Chem. Phys.* **81**, 73 (1984).
209. A. S. Pine and B. J. Howard, *J. Chem. Phys.* **84**, 590 (1986).
210. A. Furlan, S. Wulfert, and S. Leutwyler, *Chem. Phys. Lett.* **153**, 291 (1988).
211. M. D. Schuder, C, M. Lovejoy, D. D. Nelson, Jr., and D. J. Nesbitt, *J. Chem. Phys.* **91**, 4418 (1989).
212. N. Moazzen-Ahmadi, A. R. W. McKellar and J. W. C. Johns, *Chem. Phys. Lett.* **151**, 318 (1988).
213. G. A. Blake, K. L. Busarow, R. C. Cohen, K. B. Laughlin, Y. T. Lee, and R. J. Saykally, *J. Chem. Phys.* **89**, 6577 (1988).
214. N. Moazzen-Ahmadi, A. R. W. McKellar, and J. W. C. Johns, *J. Mol. Spectrosc.* **138**, 282 (1989).
215. G. A. Blake and R. E. Bumgarner, *J. Chem. Phys.* **91**, 7300 (1989).
216. M. D. Schuder, D. D. Nelson, Jr., R. Lascola, and D. J. Nesbitt, *J. Chem. Phys.* **99**, 4346 (1993).
217. M. D. Schuder, D. D. Nelson, Jr., and D. J. Nesbitt, *J. Chem. Phys.* **99**, 5045 (1993).
218. J. Serafin, H. Ni, and J. J. Valentini, *J. Chem. Phys.* **100**, 2385 (1994).
219. M. D. Schuder and D. J. Nesbitt, *J. Chem. Phys.* **100**, 7250 (1994).
220. R. F. Meads, A. L. McIntosh, J. I. Arnó, Ch. L. Hartz, R. R. Lucchese, and J. W. Bevan, *J. Chem. Phys.* **101**, 4593 (1994).
221. C. L. Hartz, B. A. Wofford, A. L. McIntosh, R. F. Meads, R. R. Lucchese, and J. W. Bevan, *Ber. Bunsenges. Phys. Chem.* **99**, 447 (1995).
222. M. J. Elrod and R. J. Saykally, *J. Chem. Phys.* **103**, 933 (1995).

223. K. Imura, T. Kasai, H. Ohoyama, H. Takahashi, and R. Naaman, *Chem. Phys. Lett.* **259**, 356 (1996).
224. K. Liu, M. Dulligan, I. Bezel, A. Kolessov, and C. Wittig, *J. Chem. Phys.* **108**, 9614 (1998).
225. R. Naaman and Z. Vager, *J. Chem. Phys.* **110**, 359 (1999).
226. H. Ni, J. M. Serafin, and J. J. Valentini, *J. Chem. Phys.* **110**, 3055 (2000).
227. K. Imura, H. Ohoyama, R. Naaman, D.-C. Che, M. Hashinokuchi, and T. Kasai, *J. Mol. Struct.* **552**, 137 (2000).
228. C. Votava and R. Ahlrichs, in *Intermolecular Forces*, Proceedings of the 14th Jerusalem Symposium, edited by B. Pullman, (Reidel, Dordrecht, 1981).
229. C. Votava, R. Ahlrichs, and A. Geiger, *J. Chem. Phys.* **78**, 6841 (1983).
230. A. Karpfen, P. R. Bunker, and P. Jensen, *Chem. Phys.* **149**, 299 (1991).
231. P. C. Gomez and P. R. Bunker, *J. Mol. Spectrosc.* **168**, 507 (1994).
232. F.-M. Tao and W. Klemperer, *J. Chem. Phys.* **103**, 950 (1995).
233. A. W. Meredith, *Chem. Phys.* **220**, 63 (1997).
234. J. M. Hermida-Ramón, O. Engkvist, and G. Karlström, *J. Comp. Chem.* **19**, 1816 (1998).
235. P. W. Jensen, M. D. Marshall, P. R. Bunker, and A. Karpfen, *Chem. Phys. Lett.* **180**, 594 (1991).
236. P. Jensen, P. R. Bunker, V. C. Epa, and A. Karpfen, *J. Mol. Spectrosc.* **151**, 384 (1992).
237. M. J. Elrod and R. J. Saykally, *J. Chem. Phys.* **103**, 921 (1995).
238. Y. Qiu, J. Z. H. Zhang, and Z. Bacic, *J. Chem. Phys.* **108**, 4804 (1998).
239. Y. Qiu, and J. Z. Bacic, *J. Chem. Phys.* **106**, 2158 (1997).
240. X. Sun and W. H. Miller, *J. Chem. Phys.* **108**, 8870 (1998).
241. W. Chen, A. R. H. Walker, S. E. Novick, and F.-M. Tao, *J. Chem. Phys.* **106**, 6240 (1997).
242. Y. Hannachi and B. Silvi, *J. Mol. Struct. (THEOCHEM)*, **200**, 483 (1989).
243. Z. Latajka and S. Scheiner, *Chem. Phys.* **216**, 37 (1997).
244. A. Rauk and D. A. Armstrong, *J. Phys. Chem. A* **104**, 7651 (2000).
245. K. N. Rankin and R. J. Boyd, *J. Comp. Chem.* **22**, 1590 (2001).
246. A. J. Barnes, H. E. Hallam, and G. F. Scrimshaw, *Trans Faraday. Soc.* **65**, 3150 (1969).
247. D. Maillard, A. Schriver, J. P. Perchard, and C. Girardet, *J. Chem. Phys.* **71**, 505 (1979).
248. D. Maillard, A. Schriver, J. P. Perchard, and C. Girardet, *J. Chem. Phys.* **71**, 517 (1979).
249. J. Obriot, F. Fondere, and Ph. Marteau, *J. Chem. Phys.* **85**, 4925 (1997).
250. L. Andrews and R. B. Bohn, *J. Chem. Phys.* **90**, 5205 (1989).
251. R. B. Bohn, R. D. Hunt, and L. Andrews, *J. Phys. Chem.* **93**, 3979 (1990).
252. A. Engdahl and B. Nelander, *J. Phys. Chem.* **94**, 8777 (1990).
253. B. J. van der Veken and F. R. de Munck, *J. Chem. Phys.* **97**, 3060 (1992).
254. J. Han, Z. Wang, A. L. McIntosh, R. R. Lucchese, and J. W. Bevan, *J. Chem. Phys.* **100**, 7101 (1994).
255. M. Fárnik, S. Davis, and D. J. Nesbitt, *Faraday Discuss.* **118**, 63 (2001).
256. G. Chalasinski, S. M. Cybulski, M. M. Szczesniak, and S. Scheiner, *J. Chem. Phys.* **91**, 7048 (1989).
257. M. M. Szczesniak and G. Chalasinski, *J. Mol. Struct. (THEOCHEM)*, **261**, 37 (1992).
258. W. D. Chandler, K. E. Johnson, B. D. Fahlman, and J. L. E. Campbell, *Inorg. Chem.* **36**, 776 (1997).
259. K. N. Rankin, W. D. Chandler, and K. E. Johnson, *Can. J. Chem.* **77**, 1599 (1999).
260. S. Berski and Z. Latajka, *J. Mol. Struct.* **450**, 259 (1998).
261. R. Baumfalk, U. Buck, C. Frischkorn, S. R. Gandhi, and C. Lauenstein, *Ber. Bunsenges. Phys. Chem.* **101**, 606 (1997).
262. R. Baumfalk, U. Buck, C. Frischkorn, S. R. Gandhi, and C. Lauenstein, *Chem. Phys. Lett.* **269**, 321 (1997).

263. R. Baumfalk, U. Buck, C. Frischkorn, N. H. Nahler, and L. Hüwel, *J. Chem. Phys.* **111**, 2595 (1999).
264. R. Baumfalk, N. H. Nahler, and U. Buck, *Faraday Disc.* **118**, 247 (2001).
265. R. Baumfalk, N. H. Nahler, and U. Buck, *Phys. Chem. Chem. Phys.* **3**, 2372 (2001).
266. E. Sandor and R. F. C. Farrow, *Faraday, Discuss.* **48**, 78 (1969).
267. E. Sandor and M. W. Johnson, *Nature (London)*, **217**, 541 (1968).
268. E. Sandor and M. W. Johnson, *Nature (London)*, **233**, 730 (1969).
269. D. F. Hornig and W. E. Osberg, *J. Chem. Phys.* **23**, 662 (1955).
270. R. E. Carlson and H. B. Friedrich, *J. Chem. Phys.* **54**, 2794 (1971).
271. J. E. Vesel and B. H. Torrie, *Can. J. Phys.* **55**, 592 (1977).
272. J. E. Vesel and B. H. Torrie, *Can. J. Phys.* **55**, 975 (1977).
273. A. Anderson, B. H. Torrie, and W. S. Tse, *J. Raman Spectrosc.* **10**, 148 (1981).
274. J. Obriot, F. Fondère, Ph. Marteau, and M. Allavena, *J. Chem. Phys.* **79**, 33 (1971).
275. J. F. Higgs, W. Y. Zeng, and A. Anderson, *Phys. Stat. Sol (B)* **133**, 475 (1986).
276. A. Beyer, A. Karpfen and P. Schuster, *Chem. Phys. Lett.* **67**, 369 (1979).
277. K.-P. Huber and G. Herzberg, *Molecular Spectra and Molecular Structure IV. Constants of Diatomic Molecules* (Van Nostrand, New York, 1979).
278. F. J. Lovas and H. Hartwig, *J. Mol. Spectrosc.* **185**, 98 (1997).
279. M. Falk and E. Whalley, *J. Chem. Phys.* **34**, 1554 (1961).
280. M. Falk and E. Whalley, *J. Chem. Phys.* **34**, 1569 (1961).
281. R. G. Inskeep, J. M. Kelliker, P. E. McMahon, and B. G. Somers, *J. Chem. Phys.* **28**, 1033 (1958).
282. U. Buck, in *Advances in Molecular Vibrations and Collision Dynamics*, edited by J. M. Bowman and Z. Bacic, (JAI Press Inc, Stamford, 1998), p 127.
283. Huisken, A. Kulcke, C. Laush, and J. M. Lisy, *J. Chem. Phys.* **95**, 3924 (1996).
284. F. Huisken, M. Kaloudis, M. Koch, and O. Werhahn, *J. Chem. Phys.* **105**, 8965 (1996).
285. U. Buck and I. Ettischer, *J. Chem. Phys.* **108**, 33 (1998).
286. R. A. Provencal, J. B. Paul, K. Roth, C. Chapo, R. N. Casaes, R. J. Saykally, G. S. Tschumper, and H. F. Schaefer, III., *J. Chem. Phys.* **110**, 4258 (1999).
287. T. Häber, U. Schmitt, and M. Suhm, *Phys. Chem. Chem. Phys.* **1**, 5573 (1999).
288. O. Mó, M. Yánez, and J. Elguero, *J. Mol. Struct. (THEOCHEM)*, **314**, 73 (1994).
289. D. Peeters and G. Leroy, *J. Mol. Struct. (THEOCHEM)*, **314**, 39 (1994).
290. O. Mó, M. Yánez, and J. Elguero, *J. Mol. Struct.* **107**, 3592 (1997).
291. F. C. Hagemeister, C. J. Gruenloh, and T. J. Zwier, *J. Phys. Chem. A* **102**, 82 (1998).
292. U. Buck, J. G. Siebers, and R. J. Wheatley, *J. Chem. Phys.* **108**, 20 (1998).
293. R. D. Parra and X. C. Zeng, *J. Chem. Phys.* **110**, 6329 (1999).
294. G. S. Tschumper, J. M. Gonzales, and H. F. Schaefer, III., *J. Chem. Phys.* **111**, 3027 (1999).
295. M. V. Vener and J. Sauer, *J. Chem. Phys.* **114**, 2623 (2001).
296. K. Nauta and R. E. Miller, *Science*, **287**, 293 (2000).
297. M E. Fajardo and S. Tam, *J. Chem. Phys.* **115**, 6807 (2001).
298. S. S. Xantheas, *Chem. Phys.* **258**, 225 (2000).
299. E. M. Mas, R. Bukowski, K. Szalewicz, G. C. Groenenboom, P. E. S. Wormer, and A. van der Avoird, *J. Chem. Phys.* **113**, 6687 (2000).
300. G. C. Groenenboom, P. E. S. Wormer, A. van der Avoird, E. M. Mas, R. Bukowski, and K. Szalewicz, *J. Chem. Phys.* **113**, 6702 (2000).
301. L. B. Braly, J. D. Cruzan, K. Liu, R. S. Fellers, and R. J. Saykally, *J. Chem. Phys.* **112**, 10293 (2000).
302. L. B. Braly, K. Liu, M. G. Brown, F. N. Keutsch, R. S. Fellers, and R. J. Saykally, *J. Chem. Phys.* **112**, 10314 (2000).

303. W. Klopper, J. G. C. M. van Duijneveldt-van de Rijdt, and F. B. van Duijneveldt, *Phys. Chem. Chem. Phys.* **2**, 2227 (2000).
304. P. E. S. Wormer and A. van der Avoird, *Chem. Rev.* **100**, 4109 (2000).
305. M. J. Smit, G. C. Groenenboom, P. E. S. Wormer, A. van der Avoird, R. Bukowski, and K. Szalewicz, *J. Phys. Chem. A* **105**, 6212 (2001).
306. L. A. Montero, J. Molina, and J. Fabian, *Intern. J. Quantum Chem.* **79**, 8 (2000).
307. C. Capelli, B. Mennucci, C. O. da Silva, and J. Tomasi, *J. Chem. Phys.* **112**, 5382 (2000).
308. H. M. Lee, S. B. Suh, J. Y. Lee, P. Tarakeshwar, and K. S. Kim, *J. Chem. Phys.* **112**, 9759 (2000).
309. C. Kozmutza, E. S. Kryachko, and E. Tfirst, *J. Mol. Struct. (THEOCHEM)*, **501**, 435 (2000).
310. D. M. Upadhyay, M. K. Shukla, and P. C. Mishra, *Intern. J. Quantum Chem.* **81**, 90 (2001).
311. F. N. Keutsch, R. S. Fellers, M. G. Brown, M. R. Viant, P. B. Petersen, and R. J. Saykally, *J. Am. Chem. Soc.* **123**, 5938 (12001).
312. R. S. Fellers, C. Leforestier, L. B. Braly, M. G. Brown, and R. J. Saykally, *Science*, **284**, 945 (1999).
313. D. Eisenberg and W. Kauzman, *The Structure and Properties of Water* (Oxford University Press, Oxford, 1969).
314. A. Karpfen and P. Schuster, *Can. J. Chem.* **63**, 809 (1985).
315. S. Suhai, *J. Chem. Phys.* **101**, 9766 (1994).
316. F. J. Strieter, D. H. Templeton, R. F. Scheuerman, and R. L. Sass, *Acta Crystallogr.* **15**, 1233 (1962).
317. A. Almenningen, O. Bastiansen, and T. Motzfeld, *Acta Chem. Scand.* **23**, 2848 (1969).
318. A. Almenningen, O. Bastiansen, and T. Motzfeld, *Acta Chem. Scand.* **24**, 747 (1970).
319. A. Karpfen, *Chem. Phys.* **88**, 415 (1984).
320. H. R. Zelsmann, F. Bellon, Y. Maréchal and B. Bullemer, *Chem. Phys. Lett.* **6**, 513 (1970).
321. D. Wiechert, D. Mootz, and T. Dahlems, *J. Am. Chem. Soc.* **119**, 12665 (1997).
322. S. Suhai, *J. Chem. Phys.* **103**, 7030 (1995).
323. N. Kobko, L. Paraskevas, E. del Rio, and J. J. Dannenberg, *J. Am. Chem. Soc.* **123**, 4348 (2001).
324. R. Ludwig, F. Weinhold, and T. C. Farrar, *J. Chem. Phys.* **102**, 9746 (1995).
325. R. Ludwig, *J. Mol. Liq.* **84**, 65 (2000).
326. M. Akiyama and H. Torii, *Spetrochim. Acta A* **56**, 137 (1999).
327. J.-L. Brédas, M. P. Poskin, J. Delhalle, J.-M. André, and H. Chojnacki, *J. Phys. Chem.* **88**, 5882 (1984).
328. T. Ueda, S. Nagamoto, H. Masui, N. Nakamura, and S. Hayashi, *Z. Naturforsch. A* **54**, 437 (1999).
329. N. Nakamura, H. Masui, and T. Ueda, *Z. Naturforsch. A* **55**, 315 (2000).
330. J. L. G. de Paz, J. Elguero, C. Foces-Foces, A. J. Lamas-Saiz, F. Aguillar-Parillas, O. Klein, and H.-H. Limbach, *J. Chem. Soc. Perkin Trans.* **2**, 109 (1997).
331. M. A. Garcia, C. López, O. Peters, R. M. Claramunt, O. Klein, D. Schagen, H.-H. Limbach, C. Foces-Foces, and J. Elguero, *Magn. Res. Chem.* **38**, 604 (2000).
332. D. Mootz, U. Ohms, and W. Poll, *Z. Anorg. Allg. Chem.* **479**, 75 (1981).
333. D. Mootz and W. Poll, *Z. Anorg. Allg. Chem.* **484**, 158 (1982).
334. S. Scheiner and J. Nagle, *J. Phys. Chem.* **87**, 4267 (1983).
335. J. Nagle, in *Proton Transfer in Hydrogen-Bonded Systems*, edited by T. Bountis, NATO-ASI Series **B291** (Plenum, New York, 1992) p. 17.
336. K. R. Hirsch and W. B. Holzapfel, *Phys. Lett. A* **101**, 142 (1984).

SOLVENT EFFECTS IN NONADIABATIC ELECTRON-TRANSFER REACTIONS: THEORETICAL ASPECTS

A. V. BARZYKIN, P. A. FRANTSUZOV, K. SEKI, and M. TACHIYA

National Institute of Advanced Industrial Science and Technology (AIST) Tsukuba, Ibaraki 305-8565 Japan

CONTENTS

I. Introduction
II. Golden Rule Limit
    A. Rate Constant
        1. General
        2. Spectral Density
        3. Reorganization Energy and Free Energy Change
        4. Electronic Coupling
    B. Beyond Standard Recipes
        1. Nonlinear Response
        2. Non-Condon Effects
        3. Nonexponential Kinetics
III. Beyond the Golden Rule Limit
    A. Padé Approximation
        1. General
        2. Classical Nonadiabatic Limit
    B. Markovian Bath Model and Its Generalizations
        1. Substitution Approximation and Time-Dependent Diffusion Coefficient
        2. Statistics of Crossing Points
        3. Effective Sink Approximation
    C. Variational Transition State Theory
        1. Solvent Control Regime
        2. Intermediate Regime
        3. Another Example of Apparent Adiabaticity
    D. Nonadiabatic Transitions in the Strong Coupling Limit
        1. Landau–Zener Approach
        2. Breakdown of the Landau–Zener Theory in the Presence of Relaxation
        3. Back to the Golden Rule
        4. Non-Marcus Free Energy Gap Dependence

E. Numerical Methods
IV. Effect of Diffusion on ET Observables
   A. Bulk versus Geminate Reaction
      1. Bulk Reaction
      2. Geminate Recombination after Photoionization
   B. Nonhomogeneous Effects
      1. Hydrodynamic Hindrance and Liquid Structure
      2. Microheterogeneous Environments
      3. Stochastic Gating and Anisotropic Reactivity
V. Concluding Remarks
Acknowledgments
References

I. INTRODUCTION

Electron-transfer (ET) reactions in solution exhibit rich dynamical behaviors, on timescales from femtoseconds to far longer than microseconds, as a result of coupling to multiple degrees of freedom of the environment as well as to the internal vibrational modes. The fundamentals of the theory were established in the 1950s [1–7] and the central role played by the solvent in determining the ET rate was then recognized. The theories of Marcus [5] and Hush [6] were the first to incorporate solvent nonequilibrium polarization correctly in a classical adiabatic formulation. Marcus used a linear response dielectric continuum approach to show that the reaction energetics can be described in terms of a single quantity, the classical solvent reorganization energy E_r, and the reaction activation energy is given by $E_a = (\Delta G + E_r)^2/(4E_r)$, where ΔG is the standard free energy change of the reaction. This novel Gaussian free energy dependence led to a universal classification of the ET reactions to normal $(-\Delta G < E_r)$, activationless $(-\Delta G = E_r)$, and inverted $(-\Delta G > E_r)$ regions (see Fig. 9.1), providing a cornerstone for the analysis of the ET kinetics. These basic ideas were confirmed later in a series of elegant experiments [8–13], although certain problems appeared in the interpretation of the results of the early stationary studies of bimolecular ET in solutions [14, 15], where diffusion can come into full control of the reaction and thus mask the "true" rate constant [16–19].

The first quantum-mechanical consideration of ET is due to Levich and Dogonadze [7]. According to their theory, the ET system consists of two electronic states, that is, electron donor and acceptor, and the two states are coupled by the electron exchange matrix element, V, determined in the simplest case by the overlap between the electronic wave functions localized on different redox sites. Electron transfer occurs by quantum mechanical tunneling but this tunneling requires suitable bath fluctuations that bring reactant and product energy levels into resonance. In other words, ET has

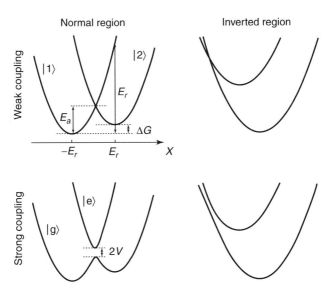

Figure 9.1. Effective energy profiles along the reaction coordinate X for the initial and final diabatic states, $|1\rangle$ and $|2\rangle$, in the case of weak electronic coupling (top), indicating the reorganization energy (E_r), the activation energy (E_a), and the free energy change of reaction (ΔG). In the case of strong electronic coupling (bottom), relevant ET states are adiabatic, $|g\rangle$ and $|e\rangle$, emerging as a result of quantum mixing of diabatic states. The parameter V denotes the coupling strength. Marcus's normal (left) and inverted (right) regimes correspond to $-\Delta G < E_r$ and $-\Delta G > E_r$, respectively.

a dual quantum classical nature: Before quantum mechanics can come into play, classical thermal activation is necessary. The theory naturally clarified the question of strong and weak coupling, corresponding to adiabatic and nonadiabatic limits of ET.

An important achievement of the early theories was the derivation of the exact quantum mechanical expression for the ET rate in the Fermi Golden Rule limit in the linear response regime by Kubo and Toyozawa [4b], Levich and co-workers [20a] and by Ovchinnikov and Ovchinnikova [21], in terms of the dielectric spectral density of the solvent and intramolecular vibrational modes of donor and acceptor complexes. The solvent model was improved to take into account time and space correlation of the polarization fluctuations [20, 21]. The importance of high-frequency intramolecular vibrations was fully recognized by Dogonadze and Kuznetsov [22], Efrima and Bixon [23], and by Jortner and co-workers [24, 25] and Ulstrup [26]. It was shown that the main role of quantum modes is to effectively reduce the activation energy and thus to increase the reaction rate in the inverted region. The Golden Rule approach was extensively used [27–67] and further extended to analyze nonlocal [26–29] and nonlinear [52–58] dielectric response in realistic systems [32–37, 41–43], as well as nonexponential kinetic behaviors [45, 47] resulting from fluctuations of the matrix element [59–61] nonequilibrium initial conditions [62–65], and/or interaction with external fields [66, 67].

The Golden Rule theory applies provided that the electron tunneling rate associated with the coupling V is much smaller than any other relaxation rate in the system, that is, the solvent is equilibrated during the transition. For the opposite adiabatic limit in which the electronic coupling is large (see Fig. 9.1), the reaction in the normal region is controlled by the motion on the lower electronic state due to slow solvent fluctuations and is well described by the activation rate theory [68–70]. These two limits are commonly thought to be well understood while the "crossover region", where the electronic matrix element changes from the weak to the strong coupling, still presents a considerable challenge [71]. In both nonadiabatic and adiabatic theories, there is an assumed separation of timescales between the motions of the electron and the solvent, while in the crossover region it is necessarily violated. Moreover, in virtually all condensed-phase ET reactions the system–bath interaction is characterized by a broad frequency spectrum or, in other words, a broad distribution of timescales that can compete with electronic transitions.

Early attempts [7, 72] to bridge between nonadiabatic and adiabatic ET reaction theories were based on the semiclassical Landau–Zener formula [73] for the transition probability. An alternative semiclassical approach was suggested independently by Zusman [74] and Burshtein and coworkers [75], but it was Zusman who explicitly defined the reaction coordinate relevant to ET. The model assumes a Markovian stochastic motion along this coordinate in a potential corresponding to the quantum state. Quantum transitions between the reactant and the product states take place at the crossing point of the diabatic potential surfaces. The model demonstrates natural behavior of the reaction rate as a function of the electronic coupling. In the weak coupling limit, the reaction rate is proportional to V^2, while in the large coupling limit the adiabatic Kramers result is reproduced with the rate inversely proportional to the longitudinal polarization relaxation time τ_L of the solvent.

There has been a large number of theoretical studies since the inspiring work of Zusman devoted to dynamical solvent effects or, more generally, to effects associated with the competition between electronic transitions at strong coupling and various relaxation modes in the system [76–144]. Ovchinnikova [79] introduced an effective sink approximation to incorporate fast classical vibrational modes into the stochastic model of ET. She assumed that timescales of vibrations and polarization relaxation are well separated so that for each value of the slow polarization coordinate there is a certain Marcus-type transition probability, instead of a δ-function. Helman [80] considered the effect of quantum vibrational modes. Zusman generalized this model further by including a full quantum mechanical golden rule expression for the high-frequency part of the system relaxation spectrum as a sink term for the diffusion equation along the slow relaxation coordinate [81a]. Initial experimental studies [145–148] seemed to verify the predicted strong dependence of the electron-transfer rate on τ_L. However, there are still relatively few systems in which solvent control has been unambiguously demonstrated [149, 150]. In recent systematic experimental studies [151–160], ultrafast components of the ET kinetics have been observed, mainly for low-barrier reactions and in the inverted region. It was shown that in some cases the ET rate can be two or even more orders of magnitude higher than τ_L^{-1}, being almost completely decoupled from solvent polarization relaxation [151, 156], although interpretation of those ultrafast components as ET is not always clear. Moreover, the observed kinetics is usually strongly nonexponential [153], with the concurrent appearance of vibrational coherence effects [153a,b, 154].

Experimental findings stimulated new theoretical developments aimed at understanding the role of ultrafast solvation modes as well as intramolecular vibrational modes in ET. In their influential paper, Sumi and Marcus [89] demonstrated that an irreversible stochastic model of ET with a wide sink due to a classical vibrational mode shows no solvent-control plateau of the rate constant as a function of V but a monotonic increase. The model predicts rich nonexponential dynamic behaviors [89–93, 108, 161], often observed in experiments [151c], but still grossly underestimating the ET rates in the inverted region [127, 156]. Subsequently, Jortner and Bixon pointed out that the quantum high-frequency modes can effectively open up new reaction channels with lower activation energy and thus significantly speed up the transfer in the inverted region [96, 97]. Walker et al. [156b] combined the ideas of Sumi and Marcus and those of Jortner and Bixon into a hybrid model including a slow solvent polarization mode, a low-frequency classical vibrational mode, and a high-frequency quantum vibrational mode, and observed reasonable agreement of the theory with experiment. Recent developments, originating from the work of Hynes [84], also incorporate the non-Markovian nature of the solvation dynamics [103–107].

If the electronic coupling is further increased beyond the range of applicability of the local Golden Rule, the reaction in the normal region becomes adiabatic. The situation in the inverted region is radically different. Here the transfer is nonadiabatic for any value of electronic coupling and the reaction can be limited by the rate of transitions between the adiabatic states for large V and not by the transport to the transition region (see Fig. 9.1). Semiclassical consideration [162, 163] based on the Landau–Zener model predicts an exponential decrease of the rate for large V. However, the Landau–Zener formula can be inapplicable in the presence of dissipation, as was shown by Shushin [164] and others [165–168]. The character of transitions is governed by the details of the relaxation spectrum.

A large number of excellent reviews have appeared, with extensive reference lists and detailed discussions on various aspects of ET reactions [26, 169–179]. Therefore, our contribution has no intent of being comprehensive. Our main emphasis is on homogeneous solutions. Biological applications are mentioned briefly, while a rapidly developing area of interfacial electrochemical ET is not discussed at all. After summarizing the Golden Rule results, we will focus on the effects of solvent dynamics. In particular, we will analyze the limitations of the existing approaches, which are often overlooked, and discuss possible ways to clear these limitations. Solvent effects on the ET rate can be static (via the reorganization energy), dynamic (via coupling of tunneling with solvation modes), or they can be indirect, that is, the solvent can assist in the diffusive (also conformational) delivery of the donor and acceptor molecules to the reaction zone. We will discuss the role of translational diffusion in some detail, both in homogeneous and microheterogeneous systems.

II. GOLDEN RULE LIMIT

A. Rate Constant

1. General

Electron transfer is usually treated as a transition between two electronic states, $|1\rangle$ and $|2\rangle$, in the diabatic representation. The nondiagonal matrix element V, which couples the two states, is assumed to be sufficiently small. The standard Golden Rule prescription for the mean rate constant k_{12} of nonadiabatic transitions can be written as [4, 21]

$$k_{12} = \frac{2}{\hbar^2} \mathrm{Re} \int_0^\infty dt \left\langle \exp\left(\frac{i}{\hbar} t H_1\right) V^* \exp\left(-\frac{i}{\hbar} t H_2\right) V \right\rangle_{H_1} \quad (9.1)$$

where H_1 and H_2 are the diabatic Hamiltonians and the averaging is over the initial equilibrium state. It is usually assumed that V is independent of time and the nuclear configuration (Condon approximation), which is what we also assume for now. The reaction coordinate can be defined as the energy difference between the two states, which is mainly due to the electrostatic interaction between the reactants and the solvent,

$$X = H_2 - H_1 + \Delta G \tag{9.2}$$

where ΔG is the standard free energy change (the driving force) of reaction, included into the definition of the reaction coordinate for symmetry. The time evolution occurs in the initial state,

$$X(t) = \exp\left(\frac{i}{\hbar}tH_1\right) X \exp\left(-\frac{i}{\hbar}tH_1\right)$$

From the known identity [3],

$$\exp\left(\frac{i}{\hbar}tH_1\right)\exp\left(-\frac{i}{\hbar}tH_2\right) = \exp_+\left[-\int_0^t dt' \frac{i}{\hbar}(H_2(t') - H_1(t'))\right]$$

where the subscript $+$ specifies the time ordering, and using the second-order cumulant expansion [180, 181] we can write k_{12} in terms of the bath autocorrelation function, $C(t) = \langle\langle X(t)X(0)\rangle\rangle$, which is expressed in terms of the spectral density $J(\omega)$ as follows:

$$C(t) = \frac{\hbar}{\pi}\int_0^\infty d\omega\, J(\omega)[\coth[\hbar\omega/(2k_BT)]\cos\omega t - i\sin\omega t] \tag{9.3}$$

Note that $C(t)$ is essentially the solvation time correlation function. Both $C(t)$ and $J(\omega)$ are molecular properties and can be related to the dielectric properties of the solvent, in the long wavelength approximation (see [39] for a comprehensive discussion). In terms of the dielectric permittivity $\varepsilon(\omega)$ [7, 21],

$$J_{out}(\omega) = \frac{1}{4\pi}\frac{\mathrm{Im}\,\varepsilon(\omega)}{|\varepsilon(\omega)|^2}\int d\mathbf{r}[\Delta\mathbf{D}(\mathbf{r})]^2 \tag{9.4}$$

where $\Delta\mathbf{D}(\mathbf{r})$ is the difference in the electric field at point $\mathbf{r}$ created by the reactants in the final and the initial state in vacuum. The subscript out emphasizes that we are dealing with the so-called outer-sphere electron transfer [5, 7, 21]. One can consider a nonhomogeneous environment with

$\varepsilon(\mathbf{r}, \omega)$, if necessary [20, 21, 26–28, 39]. Finally, for the rate

$$k_{12} = k_Q(V, \Delta G, J(\omega))$$
$$\equiv \frac{2V^2}{\hbar^2} \operatorname{Re} \int_0^\infty dt \exp\left[-\frac{i}{\hbar}(\Delta G + E_r)t + \left(\frac{i}{\hbar}\right)^2 \int_0^t d\tau \int_0^\tau d\tau' C(\tau') \right]$$
$$\equiv \int_0^\infty dt\, K_{12}(t) \qquad (9.5)$$

where E_r is the reorganization energy characterizing the system–bath interaction strength,

$$E_r = \frac{1}{\pi} \int_0^\infty \frac{d\omega}{\omega} J(\omega) \qquad (9.6)$$

and the rate kernel is given by

$$K_{12}(t) = \frac{2V^2}{\hbar^2} \exp[-Q_2(t)] \cos[\Delta G t/\hbar + Q_1(t)] \qquad (9.7)$$

with the following definitions:

$$Q_1(t) = \frac{1}{\hbar\pi} \int_0^\infty \frac{d\omega}{\omega^2} J(\omega) \sin \omega t \qquad (9.8)$$

$$Q_2(t) = \frac{1}{\hbar\pi} \int_0^\infty \frac{d\omega}{\omega^2} J(\omega) \coth[\hbar\omega/(2k_B T)][1 - \cos \omega t] \qquad (9.9)$$

The Golden Rule formula (9.5) for the mean rate constant assumes the linear response regime of solvent polarization and is completely equivalent in this sense to the result predicted by the spin-boson model, where a two-state electronic system is coupled to a thermal bath of harmonic oscillators with the spectral density of relaxation $J(\omega)$ [38, 71]. One should keep in mind that the actual coordinates of the solvent are not necessarily harmonic, but if the collective solvent polarization follows the linear response, the system can be effectively represented by a set of harmonic oscillators with the spectral density derived from the linear response function [39, 182]. Another important point we would like to mention is that the Golden Rule expression is in fact equivalent [183] to the so-called noninteracting blip approximation [71] often used in the context of the spin-boson model. The perturbation theory can be readily applied to harmonic systems with bilinear coupling [4, 184]. Berne and co-workers [185] have recently extended these ideas in their studies of nonradiative relaxation processes [186].

In the classical high-temperature limit, one recovers Marcus's activation energy and the Levich–Dogonadze prefactor in the celebrated formula,

$$k_{12} = k_M(V, \Delta G, E_r) \equiv \frac{2\pi V^2}{\hbar}(4\pi E_r k_B T)^{-1/2} \exp\left[-\frac{(\Delta G + E_r)^2}{4E_r k_B T}\right] \quad (9.10)$$

In deriving Eq. (9.10), it is also important that $E_r \gg \hbar\Omega$, which justifies the use of the steepest descent method in evaluating the time integral [21]. Here, Ω represents the average frequency of harmonic modes contributing to the reaction coordinate [see Eq. (9.63)]. The effect of quantum tunneling can be important for certain solvents, such as water, even at room temperature, particularly in the inverted region, as shown in Fig. 9.2. Many useful limiting cases of the general Golden Rule expression have been considered in the literature, particularly in the low-temperature regime [4, 21, 178, 187, 188]. It was found that the energy gap law (the dependence of $\ln k_{12}$ on ΔG) is not always parabolic. For example, for large $|\Delta G| \gg \hbar\Omega \gtrsim E_r$ in the weak coupling limit and at low temperatures, $\ln k_{12}$ in the inverted region is approximately a linear function of $|\Delta G|/(\hbar\Omega)$ [187].

2. Spectral Density

The recipe of Section II.A.1 solves the problem of nonadiabatic electron transfer, that is, if we somehow obtain the relaxation spectrum $J(\omega)$ and the coupling V, we can immediately calculate the transfer rate for a given ΔG. The underlying assumptions are linear response, constant and small V, fast relaxation, and exponential kinetics. These assumptions are well justified for many electron-transfer systems, although one must admit that very often they are not. Later, we will discuss possible ways to go beyond these approximations. Now, we consider viable means of obtaining the explicit form of the spectral density. One way is to relate $J_{\text{out}}(\omega)$ to the experimentally measured dielectric spectrum. This route was taken by Bogdanchikov et al. [32] and by Song and Marcus [37] using available $\varepsilon(\omega)$ data for water [189, 190] in the wide frequency range. The linear response assumption with the harmonic bath model seems to be quite accurate for ET in water [33–36, 41], primarily due to the long-range nature of the electrostatic interaction. A similar approach, albeit in the context of solvation dynamics, was also discussed by Bagchi and co-workers [191]. Time-dependent Stokes shift and fluorescence spectra calculations from solvent dielectric dispersion data have been performed recently by the Marcus group for water,

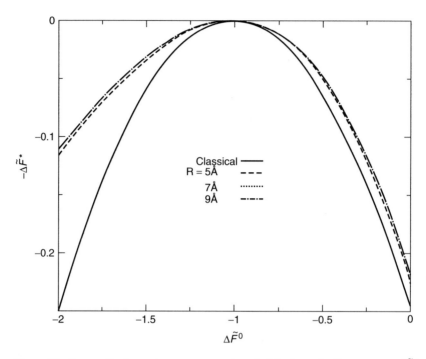

Figure 9.2. Renormalized quantum energy gap law, that is, the activation energy $-\Delta \tilde{F}^*$ versus the free energy change of reaction $\Delta \tilde{F}^0$ ($-E_a/E_r$ vs. $\Delta G/E_r$ in our notation), for H_2O with immersed donor and acceptor molecules of radii 3.5 Å (point charges in spherical cavities) at different separation, as compared against the classical Marcus law. The parameter $\Delta \tilde{F}^*$ was evaluated from the Golden Rule formula by the method of steepest descents. The corresponding simulated relaxation spectrum is shown in Figure 9.3. (Reproduced from [41c] with permission. Copyright (1997) by the American Institute of Physics.)

methanol, and acetonitrile [38] showing an encouraging agreement with experiment [192].

Two points should be mentioned here. First, the effect of solutes on the solvent dielectric response can be important in solvents with nonlocal dielectric properties. In principle, this problem can be handled by measuring the spectrum of the whole system, the solvent plus the solutes. Theoretically, the spatial dependence of the dielectric response function, $\varepsilon(\mathbf{r}, \omega)$, which includes the molecular nature of the solvent, is often treated by using the dynamical mean spherical approximation [28, 36a, 147a, 193–195]. A more advanced approach is based on a molecular hydrodynamic theory [104, 191, 196, 197]. These theoretical developments have provided much physical insight into solvation dynamics. However, reasonable agreement between the experimentally measured Stokes shift and emission line shape can be obtained using the measured $\varepsilon(\omega)$ in pure solvents without explicitly considering the spatial dependence of dielectric relaxation, that is, $\varepsilon(\mathbf{r}, \omega)$ [38].

Second, the reaction can be coupled to the intramolecular vibrational modes. Theoretically, these modes can be readily included as an additive contribution to the overall spectral density,

$$J_{\text{in}}(\omega) = \pi \sum_n \lambda_n \omega_n \delta(\omega - \omega_n) \quad (9.11)$$

where λ_n is the reorganization energy of the nth mode of frequency ω_n, so that

$$J(\omega) = J_{\text{out}}(\omega) + J_{\text{in}}(\omega) \quad (9.12)$$

It is probably possible to determine experimentally all the relevant frequencies but it may be hard to judge whether they are coupled to the reaction coordinate and which values one should assign to the corresponding reorganization energies. Moreover, these modes will most likely be "dressed" due to interactions with the solvent. It should also be noted that the term "outer-sphere" ET usually refers to ET assisted by the medium outside the sphere of strong (nonlinear) ion–dipole interaction. Therefore, J_{in} should generally contain a nonlinear contribution due to librations in the inner sphere (see below).

An alternative way to obtain the spectral density is by numerical simulation. It is possible, at least in principle, to include the intramolecular modes in this case, although it is rarely done [198]. A standard approach [33–36, 41] utilizes molecular dynamics (MD) trajectories to compute the classical real time correlation function of the reaction coordinate from which the spectral density is calculated by the cosine transformation [classical limit of Eq. (9.3)]. The correspondence between the quantum and the classical densities of states via $J(\omega)$ is a key for the evaluation of the quantum rate constant, that is, one can use the quantum expression for k_{12} with the classically computed $J(\omega)$. This is true only for a purely harmonic system [199]. Real solvent modes are anharmonic, although the response may well be linear. The spectral density of the harmonic system is temperature independent. For real nonlinear systems, $J(\omega)$ can strongly depend on temperature [200]. Thus, in a classical simulation one cannot assess equilibrium quantum populations correctly, which may result in serious errors in the computed high-frequency part of the spectrum. Song and Marcus [37] compared the results of several simulations for water available at that time in the literature [34, 201] with experimental data [190]. The comparison was not in favor of those simulations. In particular, they failed to predict

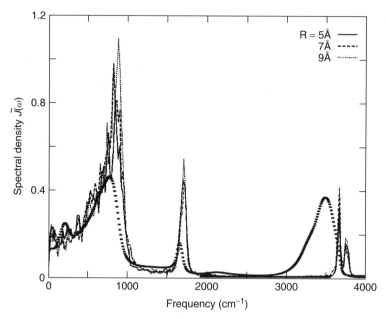

Figure 9.3. The simulated normalized spectral density function, $\tilde{J}(\omega) = J(\omega)/\int_0^\infty d\omega\, \omega^{-1} J(\omega)$, for water with immersed donor and acceptor molecules of radii 3.5 Å at different separation (from [41c] with permission. Copyright (1997) by the American Institute of Physics), as compared against the experimental results (circles, data taken from [202a]). The corresponding time correlation function is shown in Figure 9.4.

any high-frequency modes, which was largely due to the rigid model of the solvent molecules employed. Recent MD simulations of Ando [41c] using a flexible model of water show qualitative agreement with experiment [202a] in the whole frequency range, as illustrated in Figure 9.3. Not only the low-frequency broad band due to solvent translational, rotational, and librational motions is predicted but also the high-frequency peaks due to intramolecular bending and stretching modes of water. The reason for the observed quantitative inconsistencies can be understood by looking at the corresponding time correlation function in Figure 9.4. High-frequency components are very weak and thus hard to resolve. On the other hand, the low-frequency band is noisy because of the increasing simulation uncertainties at long times. Figure 9.2 shows the energy gap law with $J(\omega)$ obtained from simulations. Intramolecular vibrations of water, responsible for the high-frequency modes in the spectrum, are seen to play a significant role in the quantum aspects of ET, despite the fact that their contributions in the static reorganization energy are small.

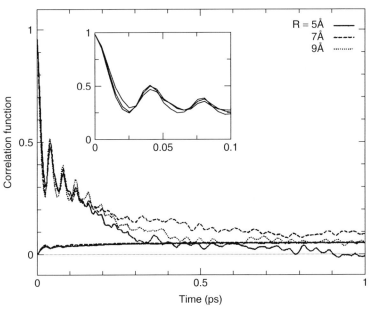

Figure 9.4. The time correlation function of the solvent coordinate for water with immersed donor and acceptor molecules of radii 3.5 Å at different separation. The insert magnifies the short-time behavior. One can distinguish an initial rapid Gaussian stage due to inertial motions, the subsequent oscillatory behavior due to the libration of water, and the final exponential diffusive tail. Very weak high-frequency components come from intramolecular vibrations. The lower curves starting from the origin are the estimated simulation uncertainties. (Reproduced from [41c] with permission. Copyright (1997) by the American Institute of Physics.)

In theoretical studies, one usually deals with two simple models for the solvent relaxation, namely, the Debye model with the Lorentzian form of the frequency dependence, and the Ohmic model with an exponential cut-off [71, 85, 188, 203]. The Debye model can work well at low frequencies (long times) but it predicts nonanalytic behavior of the time correlation function at time zero. Exponential cut-off function takes care of this problem. Generalized sub- and super-Ohmic models are sometimes considered, characterized by a power dependence on ω (the dependence is linear for the usual Ohmic model) and the same exponential cut-off [203]. All these models admit analytical solutions for the ET rate in the Golden Rule limit [46, 48]. One sometimes includes discrete modes or shifted Debye modes to mimic certain properties of the real spectrum [188]. In going beyond the Golden Rule limit, simplified models are considered, such as a frequency-independent (strict Ohmic) bath [71, 85, 203], or a sluggish (adiabatic) bath [204], or a bath in which both very fast and very sluggish modes coexist [118, 119].

Experimentalists are often eager to use the classical ET rate expression for analyzing their data, which is not only because it is simple but mainly because it works. When it comes to the effects of quantum tunneling, a simple modified expression due to Jortner and co-workers [24, 25] incorporating a single vibrational mode of frequency ω_{vib}, has proven to be useful,

$$k_{12} = \sum_n P(E_n) k_M(V, \Delta G + E_n, E_r) \tag{9.13}$$

where $P(E_n)$ is the weight factor describing the probability that the energy exchange $E_n = n\hbar\omega_{\text{vib}}$ between the tunneling electron and the vibrational mode will occur,

$$P(E_n) = I_n(\bar{n}/\sinh \tilde{\omega}) \exp(n\tilde{\omega} - \bar{n} \coth \tilde{\omega}) \tag{9.14}$$

where $\tilde{\omega} = \hbar\omega_{\text{vib}}/(2k_B T)$, $\bar{n} = \lambda_{\text{vib}}/(\hbar\omega_{\text{vib}})$, $I_n(z)$ is the modified Bessel function, λ_{vib} denotes the vibrational reorganization energy, and n spans all integer numbers. In the quantum limit (high-frequency mode, $\tilde{\omega} \gg 1$), negative values of n can be neglected (no thermal excitation, $n \geqslant 0$) and the weight factor takes the Poisson form, $P(E_n) = (\bar{n}^n/n!) \exp(-\bar{n})$. Figure 9.5 shows a beautiful experimental plot of the energy gap dependence for the intramolecular ET rates obtained by Closs and Miller [8c] for a series of eight compounds in methyltetrahydrofuran. Equation (9.13) in the quantum limit for the intramolecular vibrational mode is seen to describe the data well. Note that Eq. (9.13) has a general form that can incorporate any number of intramolecular modes via the modified energy spectrum, E_n, and the corresponding weight factor, $P(E_n)$. The full quantum rate expression k_Q can be used instead of k_M.

3. Reorganization Energy and Free Energy Change

Reorganization energy is one of the central quantities in the ET theory [5, 6]. It arises from changes in the equilibrium geometries of the redox species (inner-sphere reorganization energy) and changes in solvent dielectric displacement (outer-sphere reorganization energy) when an electron is transferred from donor to acceptor. Here, we will briefly discuss the ways of calculating the solvent part, E_r, keeping in mind that the values inferred from experimental data may include contributions from both solvent and solutes [156, 205, 206].

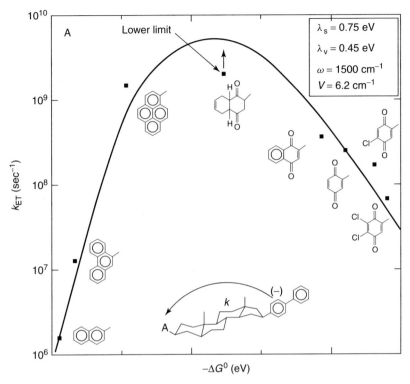

Figure 9.5. Logarithmic plot of the experimental rate constants obtained at room temperature in methyltetrahydrofurane versus free energy changes of the ET reactions indicated in the figure. Structures are the various acceptors. The solid line is computed from Eq. (9.13) in the quantum limit ($\hbar\omega_{vib} \gg k_B T$) with the parameters listed in the figure. The parameters λ_s, λ_v, and ω correspond to E_r, λ_{vib}, and ω_{vib} in our notation. (Reproduced from [8c] with permission. Copyright (1988) by the American Association for the Advancement of Science.)

According to Eqs. (9.4) and (9.6),

$$E_r = \frac{c_p}{8\pi} \int d\mathbf{r} [\Delta \mathbf{D}(\mathbf{r})]^2 \tag{9.15}$$

where the integration is performed over the continuum space filled with the solvent, c_p is the Pekar factor,

$$c_p = \frac{1}{\varepsilon_\infty} - \frac{1}{\varepsilon_s} = \frac{2}{\pi} \int_0^\infty \frac{d\omega}{\omega} \frac{\mathrm{Im}\,\varepsilon(\omega)}{|\varepsilon(\omega)|^2} \tag{9.16}$$

ε_∞ and ε_s are the optical and the static dielectric constants, respectively. Note that the experimental dielectric spectrum of water displayed in Figure 9.3 reproduces the corresponding Pekar factor, $c_p = 0.543$ with excellent accuracy. In order to proceed with spatial integration, we require the specification of the cavity in the dielectric medium, which contains the solute species. Various simplified shapes have been considered [207, 208]. Marcus [5] provided a simple expression for E_r under the assumption of a spatially local response (within a linear dielectric continuum model for a solvent without spatial dispersion). The solutes were modeled either as charged metallic spheres or as spherical vacuum cavities with appropriate point charges in their centers. His final result,

$$E_r = \frac{1}{2} c_p e^2 \left(\frac{1}{r_D} + \frac{1}{r_A} - \frac{2}{r} \right) \qquad (9.17)$$

still proves itself very useful for qualitative analysis. Here r_D and r_A are the radii of donor and acceptor and r is the center-to-center distance between them. The functional dependence of the solvent reorganization energy on the Pekar factor c_p and on the distance r has often been confirmed [209].

Rigorously, Eq. (9.17) is valid within the dielectric continuum model only when the two ions are far apart. If the ions are close to each other, the polarization of the medium, and thus the reorganization energy, change due to the finite size and image effects. There have been several attempts to improve Eq. (9.17) [210, 211], but the exact analytic calculation (of the solvation energy, neglecting solute polarizability) has been performed only recently [212] by solving the corresponding boundary value problem for the electrostatic potential in bispherical coordinates. The conclusion is that for the vacuum cavity model Eq. (9.17) provides an excellent estimate. In fact, previously developed corrected expressions were shown to work even worse. For the conductor model, deviations from Eq. (9.17) are not negligible but only at very small ion separations. A similar calculation was carried out by Song et al. [213] who evaluated the time dependence of the solvation energy. They chose to calculate the susceptibility tensor, $\chi(\mathbf{r}, \mathbf{r}')$, rather than to solve the boundary value problem for the potential. Both studies emphasize the importance of the image effects that may lead to multiple relaxation times of the solvation energy of a pair of ions even if the surrounding medium is descibed by a simple Debye-type single-exponential relaxation model.

With the image effects due to dielectric boundary conditions properly taken into account, the reorganization energy can be viewed as the difference between the equilibrium solvation free energy of the difference electron density, $\Delta\rho$, in a medium with only an optical dielectric response (ε_∞) and that with the full response (ε_s) [214–217]. This approach allows for a more general formulation than the dielectric continuum model [39]. Inclusion of proper boundary conditions leads to an ε-dependence of a **D** field for an equilibrium system. These effects were shown to be of major significance in interfacial ET processes [214a, 218–221].

In situations of nonlinear coupling of the charge density to the medium, nonparabolic free energy profiles might be expected and E_r is thus not uniquely defined [52, 54, 214a, 222, 223]. Computer simulations have shown examples of complex systems with inherent nonlinear microscopic interactions but with the free energy profiles conforming remarkably closely to the parabolic form [33, 34, 41, 224, 225]. In general, a mean E_r can be defined as one-half of the Stokes shift [214a], that is, in terms of the two vertical gaps at the respective profile minima.

In addition to orientational polarization response, E_r may have translational contributions arising from solvent density fluctuations [226]. As it was shown by Matyushov et al. [226, 227], molecular translation of the solvent permanent dipoles is the principal source of temperature dependence for both the solvent reorganization energy and the solvation energy. In fact, the standard dielectric continuum model does not predict the proper temperature dependence of E_r in highly polar solvents: It predicts an increase in contrast to the experimentally observed decrease in E_r with temperature. A molecular model of a polarizable, dipolar hard-sphere solvent with molecular translations remedies this deficiency of the continuum picture and predicts correct temperature dependence of E_r, in excellent agreement with experiment [227a].

Solvents with vanishing molecular dipole moments but finite higher order multipoles, such as benzene, toluene, or dioxane, can exhibit much higher polarity, as reflected by its influence on the ET energetics, than predicted by the local dielectric theory [228]. Full spatially dispersive solvent response formulation is required in this case [27–29, 104, 229–233]. There are different approaches to the problem of spatial dispersion. The original formulation by Kornyshev and co-workers [27c, 28] introduces the frequency-dependent screening effect on the basis of heuristic arguments. More recent approaches are based on the density-function theory [104, 197].

There have been many studies of the effect of added electrolytes on ET rates [171, 234, 235]. The main effect of ionic atmosphere is electrostatic screening, which is usually accounted for in terms of Debye–Hückel theory (mean-field, low-concentration approximation). At sufficiently low ionic strength, the corresponding component of the activation energy is simply proportional to the ratio of the Onsager radius (also referred to as the

Bjerrum length) to the Debye radius (inversely proportional to the square root of the ionic strength). Ionic strength dependence of the ET rates was used to estimate the charge of metalloproteins [171]. Another (weaker) effect of electrolytes is to change the reorganization energy, which can be of some importance if one of the reactants is neutral [236]. Some of the studies have established that ion pairing can strongly influence the observed ET rates, particularly for intramolecular reactions [235]. Marcus has recently considered several scenarios for the effect of ion pairing arising from the competition between ion-pair dissociation, fluctuational ionic motion, and actual ET [237].

Numerically, it is now a common practice to calculate E_r within the dielectric continuum formulation but employing cavities of realistic molecular shape determined by the van der Waals surface of the solute. The method is based upon finite-difference solution of the Poisson–Boltzmann equation for the electrostatic potential with the appropriate boundary conditions [214, 238, 239]. An important outcome of such studies is that even in complex systems there exists a strong linear correlation between the calculated outer-sphere reorganization energy and the inverse donor–acceptor distance, as anticipated by the Marcus formulation (see Fig. 9.6). More

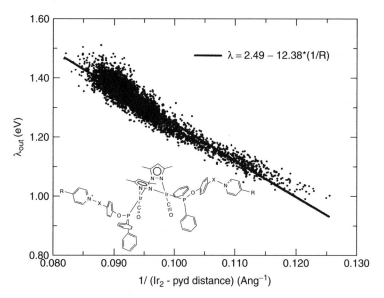

Figure 9.6. Linear correlation between the outer-sphere reorganization energy, calculated by finite-difference solution of the Poisson–Boltzmann equation, and the inverse of the donor–acceptor distance in PC1-Am (X = CH_2, R = Am) complex in acetonitrile. (Reproduced from [239] with permission. Copyright (1997) by the American Chemical Society.)

details of the medium reorganization can be obtained from molecular-level models [33–36, 41, 223, 227, 240]. The full nonlocal response arises naturally in this approach. We can see its manifestation in Fig. 9.3 where the normalized spectral density turns out to be sensitive to the donor–acceptor separation. It was shown that continuum models can provide quantitatively reliable results, even in low polarity solvents, but one may require distinct effective solute radii for the optical and static components of the solvent response [206].

The free energy change of reaction ΔG is calculated from the energy balance equation [14, 52, 211, 241],

$$\Delta G = \Delta G_\infty + \zeta \frac{e^2}{\varepsilon_s r} \tag{9.18}$$

where $\Delta G_\infty = E_{ox} - E_{red} - \Delta E$ is the free energy change at infinite separation of reactants, E_{ox} and E_{red} are the oxidation–reduction potentials in a given solvent, ΔE is the excitation energy, and ζ denotes the change in the product of donor–acceptor charges (in units of e). The distance dependence of ΔG is determined by the Coulomb interaction. In polar solvents, this dependence can be neglected. Weller [242] proposed a practical scheme of calculating ΔG in a solvent with known dielectric properties when redox potentials are known for a different solvent. All energetic calculations are very similar to those for the reorganization energy.

External electric field $\mathbf{E}_{ex}$ changes the energetics of the system [243]. It can also modify the wave functions of donor and acceptor and thus change the electronic coupling. This effect is difficult to access quantitatively, however. The reorganization energy is not affected by $\mathbf{E}_{ex}$ in the linear response regime but ΔG certainly is. If the position of the acceptor with respect to the donor is in the direction of the field, the free energy change should increase, while it decreases if the orientation of the pair is reverse. From the dielectric continuum model, one obtains [244],

$$\Delta G(\mathbf{E}_{ex}) = \Delta G(0) + e\mathbf{E}_{ex} \cdot \mathbf{r} + \frac{r_D^3 + r_A^3}{r^3} \frac{\varepsilon_s - 1}{2\varepsilon_s + 1} e\mathbf{E}_{ex} \cdot \mathbf{r} \tag{9.19}$$

where $\Delta G(0)$ is the free energy change in the absence of the field. The last term in Eq. (9.19) represents the interaction energy between the solutes and the solvent polarization induced by $\mathbf{E}_{ex}$. It decreases rapidly with distance and can usually be neglected (for reactants of comparable size, its maximum effect at contact is $\sim 10\%$ of the second term). The correction factor due to the "dielectric image" effects and thermally averaged ion-dipole forces was

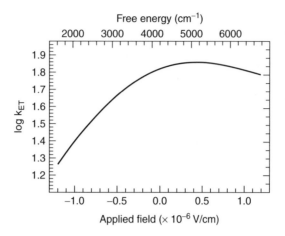

Figure 9.7. The rate constant k_{ET} as a function of the applied electric field for the initial stage of charge recombination between the oxidized special pair donor and the reduced ubiquinone acceptor in bacterial photosynthetic reaction centers of *Rb. sphaeroides* at 80 K. (Reproduced from [243a] with permission. Copyright (1990) by the American Chemical Society.)

also investigated for the interaction of ion pairs and shown to be at most 10% of the main Coulomb energy [245]. Linear dependence of ΔG on $\mathbf{E}_{ex}$ gives a convenient opportunity to monitor the energy gap law for the ET rate constant by varying the external electric field. Figure 9.7 demonstrates experimental realization of this idea for charge recombination in bacterial reaction center.

4. Electronic Coupling

So far, we have considered the nuclear contribution to the ET rate, which results in the activation factor describing the frequency at which donor and acceptor states come into resonance in the course of thermal fluctuations. The electronic factor V describes the probability of transfer in the transition state [170, 171, 217, 246, 247]. The requirement of the resonance is a consequence of the conservation of energy in the tunneling process (vibronic effects allow certain deviations from strict electronic energy matching). In short distance reactions, electronic orbitals of donor and acceptor directly overlap, and the electronic coupling is determined by a direct exchange contribution. In long distance reactions, the intervening medium (either bridge or solvent) can mediate the tunneling, providing sequentially overlapping virtual orbitals for superexchange coupling [248]. This mediating role is crucial in long-range ET in proteins [249] (molecular wires [250], etc.), where an electron can tunnel over distances up to 20–30 Å, which greatly exceeds the range of the direct overlap of the atomic orbitals of the redox centers. The process can be sequential for resonance coupling [251–253], that is, when the intermediate state with the electron on the bridge is energetically favorable.

The transfer matrix element V may be evaluated directly in terms of the charge-localized wave functions [246, 247],

$$V = (H_{12} - E_0 S_{12})/(1 - S_{12}^2) \qquad (9.20)$$

where $H_{12} = \langle 1|H|2 \rangle$ with H being the total Hamiltonian of the system at fixed nuclear coordinates, $S_{12} = \langle 1|2 \rangle$ is the electronic overlap integral, and $E_0 = \langle 1|H|1 \rangle = \langle 2|H|2 \rangle$ is the tunneling energy. Such a procedure has been used successfully in calculations of relatively small ET systems [217, 246, 247, 254], using variationally determined Hartree–Fock self-consistent wave functions. For complex systems with competing superexchange pathways, such direct calculations are not yet feasible and one has to use approximations. An extended Hückel method [255] is often applied and the coupling is defined using the Green's function approach [256–258]. The simplest expression of superexchange coupling is the McConnell product of the form [248, 256],

$$V = V_{D,1} \frac{\Pi_{j=1}^{N-1} V_{j,j+1}}{\Pi_{j=1}^{N} \Delta E_j} V_{N,A} \qquad (9.21)$$

derived in the perturbative limit. Here $V_{j,j+1}$ are the couplings between directly overlapping atomic orbitals of the neighboring atoms along the tunneling path, N is the number of bridge atoms, and ΔE_j is the energy difference between the D–A level E_0 and the energy of the promoting orbital j. This fundamental idea has been extended in various ways to a remarkable degree of sophistication [256–270]. Modern theories are able to account for the structural and dynamical features of the intervening medium at the atomic level. There are several basic strategies. A direct generalization of McConnell's idea within the Green's function formulation is realized in the pathway model of Beratan et al. [261]. The model assumes that a few pathways or "tubes" of orbitals control the electronic coupling between donor and acceptor, and as such seeks these dominant paths by approximating the matrix element by a product of "decay" factors assigned to each type of chemical contact: covalent bonds, hydrogen bonds, and through space (solvent). Siddarth and Marcus [270] developed an artificial intelligence algorithm to search for a better pathway. Mukamel and co-workers [88, 252] introduced density-matrix Liouville-space pathways. Wolynes and co-workers [95, 271] related V to an imaginary-time Feynman path integral with ensuing definition of the tunneling paths.

A very efficient way to handle complex many-electron systems was introduced by Stuchebrukhov [265], who reduced the usual volume integral over the whole space in Eq. (9.20) to a surface integral. As a result, the transfer matrix element is evaluated as the total flux of interatomic stationary currents through the dividing surface anywhere in the central region of the barrier between donor and acceptor. Provided the donor and acceptor diabatic states, $|1\rangle$ and $|2\rangle$, are calculated using standard methods of quantum chemistry, the tunneling currents in the system can be expressed via expansion coefficients of molecular orbitals of $|1\rangle$ and $|2\rangle$ states, and matrix elements of the Hamiltonian of the system in a suitable atomic basis set. This procedure is nonperturbative and turns out to be much more efficient than direct diagonalization of the same size of a problem. The strategy of the matrix element calculation due to Stuchebrukhov et al. involves initial use of the perturbation theory [272, 273] (or its extended version [274], which accounts for possible resonances with the bridge states but still avoids diagonalization of the full Hamiltonian matrix) to find relevant atom sequences on which the tunneling pathways are localized (protein pruning [275]). Further atomic details are examined on the pruned molecules, which are much smaller in size, with the method of interatomic tunneling currents.

Electronic coupling usually decays exponentially with donor–acceptor distance r,

$$V = V_0 \exp[-(\beta/2)r] \qquad (9.22)$$

This dependence follows immediately from simple theories, such as the Gamow WKB model [276] or the Hopfield [30] model, which are based on tunneling through an electronically homogeneous medium represented by an effective one-dimensional (1D) rectangular barrier, as well as from the electronically inhomogeneous superexchange model of McConnell [248] and its generalizations [277], where the distance is measured in terms of the number of covalent bonds in the molecular bridge. More recent approaches that take explicit account of the atomic structure of the intervening medium also predict the exponential distance dependence with characteristic average decay parameter [256, 257, 263, 267]. Experiments confirm this prediction even for very complex systems, such as the photosynthetic reaction center or ET in proteins [249]. Only in situations where essentially resonant transfers occur [278], exponential dependence is not expected to be exhibited [250]. Figure 9.8 illustrates the results of calculation of the electronic matrix element for ET between Zn and Zn^+ in water [279] using the

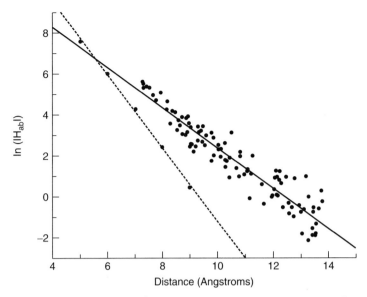

Figure 9.8. Log plot of H_{ab} (in cm^{-1}, V in our notation) versus donor-acceptor distance for Zn_2^+ with (solid line, circles, $\beta = 1.96$ Å^{-1}) and without (dashed line, squares, $\beta = 3.57$ Å^{-1}) water present. Scattered data points correspond to different solvent configurations probed by molecular dynamics. (Reproduced from [279] with permission. Copyright (1999) by the American Chemical Society.)

generalized Mulliken–Hush approach [254]. Besides exponential distance dependence, it is seen that solvent can significantly increase the electronic coupling relative to vacuum tunneling [280, 281].

B. Beyond Standard Recipes

1. Nonlinear Response

Nonlinear response and molecularity of real solvents [52–58] can be accounted for through computer simulations [33–36, 41, 42, 225]. Analytical studies are rare [282, 283]. We have already made a few comments in this respect. Now, we summarize some basic ideas of how to handle the nonlinearity. The situation is quite clear in the classical limit. Assuming that H_1 and H_2 commute, one can rewrite the integrand in Eq. (9.1) as [204] $\langle \exp[i(H_1 - H_2)t/\hbar] \rangle_{H_1}$ and finally obtain for the rate [52c],

$$k_{12} = \frac{2\pi V^2}{\hbar} \langle \delta(H_1 - H_2) \rangle_{H_1} = \frac{2\pi V^2}{\hbar} \phi_{eq}(X^\ddagger) \qquad (9.23)$$

where $\phi_{eq}(X)$ is the probability distribution of the reaction coordinate and $X^{\ddagger}$ corresponds to the solvent cofigurations where the total energies of the states $|1\rangle$ and $|2\rangle$ are the same, so that ET can occur. The parameter $X^{\ddagger}$ is obtained through the energy balance relation in terms of the free energy change of reaction and the appropriate solvation energy [52]. In practice, one calculates the distribution of the potential difference between donor and acceptor sites by using Monte Carlo or molecular dynamics simulations [33–36, 41, 42, 225]. A simple algorithm has been proposed recently by Zhou and Szabo [54], which is essentially the method of stationary phase in the classical analogue of the evaluation of the imaginary-time path integral representation of the rate constant [95].

Matyushov and Voth [283] have recently developed a simple analytical three-parameter model of ET for a solute linearly coupled to a classical, collective solvent mode that force constants differ in the initial and final electronic states. Although a general quantum mechanical solution of this so-called Q model was given by Kubo and Toyozawa [4b], it did not allow a closed-form analytical representation for the diabatic free energies of ET along the reaction coordinate [284]. The exact analytical solution of Matyushov and Voth for the ET free energy surfaces shows a limited band for the energy gap fluctuations, in contrast to the Marcus parabolic model where the fluctuations are unrestricted. The three parameters of the Q model, namely, the equilibrium energy gap, the reorganization energy (defined as the second cumulant of the reaction coordinate; the reorganization energies for the two states are different but interrelated), and the asymmetry parameter of the force constant, can be connected to spectroscopic observables, that is, the Stokes shift and the first two spectral moments. The Q-model provides a simple analytical framework to map physical phenomena conflicting with the Marcus two-parameter model, including nonlinear solvation in ET.

Nonlinear response regime presents a much more difficult task in the presence of quantum modes. Various semiclassical approximations have been developed to estimate the quantum effects using classical computer simulations. We have already mentioned the second-order cumulant expansion method or, equivalently, mapping onto the spin-boson model. Another approximation is the "semiclassical trajectory" approach [33] based on a time-dependent quantum mechanical treatment using the classical trajectory of the energy gap. The frozen Gaussian approximation [42a, 285] assumes that the wave function can be expressed by a product of Gaussians and the center of the Gaussians moves along a classical trajectory with a constant width. To include the quantum effect on the average for the initial values, the frozen Gaussian approximation is combined with the local harmonic approximation [42b]. The validity conditions of all these approximation schemes have been established by Yoshimori [56] recently using the second-order logarithmic $\hbar$ expansion of the rate constant. It turns out that in many practically relevant situations all these approximations give quantitatively reasonable predictions. Recent large-scale computer simulation strategies combine the classical and quantum path integral molecular dynamics methods. [198].

2. Non-Condon Effects

The formalism above has rested on the assumption that the electronic and nuclear factors are separable and that the tunneling matrix element is independent of the nuclear configuration and dynamics (Condon approximation). The validity of this assumption in the context of ET theory has been questioned by many authors [7, 59–63, 66, 67, 261, 269, 286–289]. The Condon approximation is justifiable for tight redox couples. It is not, however, a priori adequate for solutes with weakly localized electronic clouds [289a] or generally for long-range ET [61, 258]. The crucial feature here is that the electronic wave functions are extended into the interreactant space where they are certainly exposed to the environmental polarization fluctuations [289b]. Structural fluctuations play important role in protein functionality [290]. They open pathways for molecular motions and reactive interactions that are not available in rigid proteins (the so-called gating [291–293]). There are two effects of the fluctuations in the intervening medium on the electronic coupling, static and dynamic. The static effect results in different values of V for different configurations in the transition state, and an appropriate averaging of V^2 has to be performed in the rate expression to account for such inhomogeneity [261, 287, 294]. The dynamic effect manifests itself in inelastic tunneling due to electronic–vibrational interaction [61, 258, 289], and can be phenomenologically described via a time-dependent, randomly fluctuating electronic coupling matrix element [59–61, 66, 67, 121, 269].

Recently Daizadeh et al. [61a] examined the tunneling matrix element for different configurations of the nuclear coordinates along the molecular dynamics trajectory of Ru modified azurin. As it can be seen from Figure 9.9, thermal fluctuations of the protein structure are sufficient to cause significant fluctuations of the electronic coupling. These fluctuations are in general very fast, ~ 10 fs, giving rise to a large energy quantum of $\sim 10^3$, cm^{-1}. The short-time fluctuations are of the order of the matrix element itself, which indicates a strong coupling of the tunneling electron to the fast nuclear modes of the protein, such as CN, CC, CO, and CH stretch vibrations.

A simple model [61], which accounts for the breakdown of the Condon approximation in long-range ET, considers two independent contributions

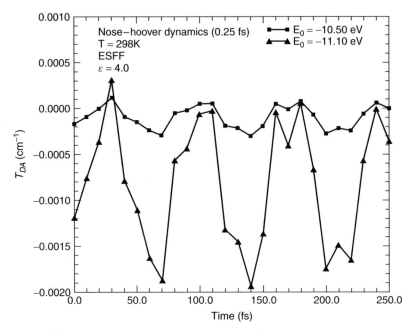

Figure 9.9. Fluctuations of the matrix element T_{DA} (V in our notation) in short-term dynamics of the Phe 110 (C_α—N) bond in Ru-modified azurin system for two different tunneling energies, $E_0 = -11.1$ eV (close to a gap edge) and $E_0 = -10.5$ eV (close to the middle of the gap). (Reproduced from [61] with permission. Copyright (1997) by the National Academy of Sciences, U.S.A.)

to the system Hamiltonian. The first contribution is usual and comes from the solvent polarization modes and vibrational modes of donor and acceptor complexes that shift their equilibrium position upon the change of electronic state. The second contribution describes the motion of the bridge atoms through generalized modes with coordinates $\mathbf{x} = x_1, \ldots, x_L$, which do not change their equilibrium position upon ET. The parameter L is proportional to the tunneling distance. The electronic coupling term is assumed to be a function of $\mathbf{x}$. In the absence of this coordinate dependence, the bridge modes would be uncoupled and irrelevant to the rate constant. All the modes are assumed to be harmonic. Independence of the bridge oscillators of the other modes in the system leads to their factorized contribution to the rate constant. By starting from Eq. (9.1) one then arrives at Eq. (9.13) with $\langle V^2 \rangle^{1/2}$ instead of V, that is, tunneling becomes inelastic due to non-Condon effects. If one further assumes, in accord with McConnell's model, Eqs. (9.21) and (9.22), that $V(\mathbf{x}) = V_{eq} \exp(-\alpha \sum_{j=1}^{L} x_j)$ and that all $L = N + 1$ frequencies of the bridge are the same and equal to ω_b, then $E_n = n\hbar\omega_b$ and Eq. (9.14) for the weight factor is reproduced with $\tilde{\omega} = \hbar\omega_b/(2k_B T)$ and $\bar{n} = \hbar\alpha^2 L/(2m\omega_b)$. Here, α is a constant characterizing the strength of coupling to the bridge modes (assumed to be the same for all modes), V_{eq} is the tunneling matrix element corresponding to the equilibrium configuration of the bridge, and m is the oscillator mass. Note that it follows from Eq. (9.14) that the probability of elastic tunneling ($n = 0$) quickly decreases as the length of the tunneling path L increases.

Analogy with the usual quantum effects can be noticed. In both cases, part of the electronic energy is transferred to quantum vibrations upon a tunneling transition, modifying the classical energy gap law (see Fig. 9.2). The only principal difference is that inelastic tunneling is distance dependent. It is expected that at large donor–acceptor distances distortions of the energy gap law should come mainly from inelastic tunneling since the number of bridge modes that can participate in energy exchange is much larger than the number of high-frequency quantum modes of donor and acceptor complexes. In the quantum limit, $\hbar\omega_n \gg k_B T$, the weight factor becomes Poissonian with $n \geq 0$. For multiphonon processes ($\bar{n} > 1$), it can be further approximated by a Gaussian that results in a modified Marcus form for the activation energy,

$$E_a = (\Delta G + E_r + \bar{n}\hbar\omega_b)^2/(4E'_r) \qquad (9.24)$$

where $E'_r = E_r + \bar{n}\hbar\omega_b\tilde{\omega}$. Here, $\bar{n}$ corresponds to the average number of quanta dissipated in the bridge. We have considered only the simplest situation where the bridge modes that govern fluctuations of the matrix element are independent from the solvent polarization modes, which is generally not the case. Self-consistent variational calculation of both the electron tunneling factor and the polarization configuration in the transition state is a possible approach [289b]. As a result, peculiar dependence of the matrix element on distance and on the reaction free energy is observed. Electronic–vibrational interaction is expected to facilitate ET by spatial extension of the electronic wave functions and also by lowering the nuclear activation energy factor. However, note that so far it is not clear how efficient inelastic tunneling (i.e., how large α) can be.

3. Nonexponential Kinetics

The definition (9.1) of the ET rate constant assumes that the kinetics is exponential. In general, of course, it is not, particularly at low temperatures. Possible sources of nonexponential behavior are complex spectral density and, as a result, complex relaxation dynamics of the participating bath modes [45, 47], fluctuating tunneling matrix element [59–61], time-dependent external field modulating the energy gap [66, 67], and nonequilibrium initial condition [62–65, 137]. Instead of the mean rate constants, one has to work with the time-dependent rate kernels, $K_{12}(t, t')$ and $K_{21}(t, t')$, describing the evolution of the state populations, $P_1(t)$ and $P_2(t)$, via the nonequilibrium generalized master equations [45, 47, 62, 66, 67, 88, 99, 144, 180, 184, 295–297],

$$\dot{P}_1(t) = -\dot{P}_2(t) = -\int_0^t dt' K_{12}(t, t') P_1(t') + \int_0^t dt' K_{21}(t, t') P_2(t') \quad (9.25)$$

where

$$K_{12}(t, t') = \frac{2}{\hbar^2} \operatorname{Re} \operatorname{Tr} \left\{ \exp\left[\frac{i}{\hbar} \tau H_1\right] V^* \exp\left[-\frac{i}{\hbar} \tau H_2\right] V \rho_1^0(t) \right\} \quad (9.26)$$

and similarly for $K_{21}(t, t')$ with $1 \leftrightarrow 2$ in Eq. (9.26). Here,

$$\tau = t - t' \quad \text{and} \quad \rho_1^0(t) = \exp\left(-\frac{i}{\hbar} t H_1\right) \rho_1^0 \exp\left(\frac{i}{\hbar} t H_1\right)$$

is the density matrix associated with the evolution of the initial density matrix ρ_1^0 according to H_1. After nuclear coordinate equilibration occurs within the initially populated state, the rate kernels become functions of the time difference τ only. For constant electronic coupling V, Eq. (9.26) reduces to Eq. (9.7) for K_{12}. Similarly for K_{21}, just the sign of $Q_1(t)$ is changed. In this quasiequilibrium situation, the integrals in Eq. (9.25) become convolution products and the populations can be calculated via Laplace transforms. The long-time populations achieve full Boltzmann equilibrium,

$$P_2(\infty)/P_1(\infty) = \int_0^\infty d\tau K_{12}(\tau) / \int_0^\infty d\tau K_{21}(\tau) \equiv k_{12}/k_{21}$$

Note that master equations for dissipative systems have been derived and used in a variety of contexts, even beyond the weak coupling limit.

If the curve crossing process is highly exothermic, the backward rate is negligible and Eq. (9.25) simplifies into

$$\dot{P}_1(t) = -\int_0^t dt' K_{12}(t, t') P_1(t') \simeq -k_{12}(t) P_1(t) \quad (9.27)$$

where

$$k_{12}(t) = \int_0^t dt' K_{12}(t, t') \simeq \int_0^\infty dt' K_{12}(t, t') \quad (9.28)$$

The reduction to local (in time) kinetics implies that $P_1(t)$ does not change much on the timescale of decay of the memory kernel. This condition is often fulfilled in the limit of small coupling V. Equation (9.28) is equivalent to the nonequilibrium Golden Rule formula, as proposed by Coalson et al. [62]. Cho and Silbey [63] emphasized an important role of optical preparation of nonequilibrium donor population in photoinduced ET. A three-state system has to be considered, with an additional ground state optically coupled to an excited donor state [see Fig. 9.10(a)]. The coordinate associated with the optical transition is not necessarily correlated with the coordinate describing the ET event. Unless one ignores the effect of the initial relaxation of the photoexcited state on the ET process, one has to deal with a truly multidimensional solvation coordinate system in this case. Only if the excitation pulse is short enough to ignore any ET during it, the nonequilibrium Golden Rule formula of Coalson et al. [62] is reproduced.

In the classical limit, one can invoke the stationary phase approximation and obtain the following nonequilibrium generalization of the Marcus expression [63]:

$$k_{12}(t) = k_M(V, \Delta G + \Delta X(t), E_r) \tag{9.29}$$

where $\Delta X(t)$ is the time-dependent energy gap shift,

$$\Delta X(t) = \frac{1}{\pi} \int_0^\infty \frac{d\omega}{\omega} [J_{Ag}(\omega) - J_{DA}(\omega) - J_{Dg}(\omega)] \cos \omega t \tag{9.30}$$

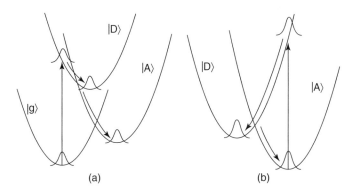

Figure 9.10. Diabatic free energy curves illustrating (a) photoinduced ET reaction and (b) back ET reaction in a 1D solvation coordinate system. A resonant optical pulse brings a stationary nuclear wave packet from the ground potential surface to the donor surface, where it relaxes toward equilibrium with concomitant ET to the acceptor state.

Here, E_r and $J_{DA}(\omega)$ are the usual reorganization energy and the corresponding spectral density associated with the equilibrium ET process. They are related by Eq. (9.6). The other spectral densities necessarily appear due to involvement of the donor ground state $|g\rangle$. In particular, $J_{Dg}(\omega)$ is responsible for the energy broadening effect of the optical spectra. Time zero corresponds to the moment right after the excitation δ pulse. At long times, $\Delta X(\infty) = 0$, and the usual Marcus expression is recovered. In the case of a back ET reaction, in which the electron is initially localized on the acceptor site prior to photoexcitation [see Fig. 9.10(b)], Eq. (9.29) also works [64] but with $\Delta X(t)$ defined by

$$\Delta X(t) = -\frac{2}{\pi} \int_0^\infty \frac{d\omega}{\omega} J(\omega) \cos \omega t \qquad (9.31)$$

Figure 9.11 illustrates time evolution of the nonequilibrium rate constant $k_{12}(t)$ and the electronic population in the donor state $P_1(t)$ in the course of the back-transfer reaction in a model ET complex (rigid collinear triatomic molecule with equivalent donor and acceptor sites separated by a neutral spacer) in a polar solvent. The long-time value of $k_{12}(t)$ is seen to be much smaller than the maximum values it achieves during the relaxation process. Hence, it is the evolution of $k_{12}(t)$ that dominates the electronic state population evolution at short times. After this rapid nonequilibrium stage of nuclear relaxation is over, with its characteristic sequence of plateaus and dips, ET proceeds further very slowly in its usual way with a small (activated) equilibrium rate constant.

III. BEYOND THE GOLDEN RULE LIMIT

The Golden Rule rests on the assumption of weak electronic coupling. Weak, first of all, means nonadiabatic, that is, the following condition for the Massey parameter $\bar{\xi}$ should be fulfilled [298]:

$$\bar{\xi} = \frac{V^2}{\hbar \langle \dot{X}^2 \rangle^{1/2}} \equiv \left(\frac{V}{E_{na}}\right)^2 \ll 1 \qquad (9.32)$$

where $\langle \dot{X}^2 \rangle$ is the mean squared velocity, $X(t)$ is the reaction coordinate defined in Section II.A, and

$$E_{na} = \hbar^{1/2} \langle \dot{X}^2 \rangle^{1/4} \qquad (9.33)$$

defines the size of the nonadiabatic transition region [167]. The character-

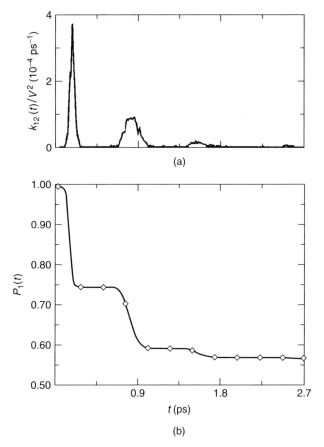

Figure 9.11. Time evolution of (a) nonequilibrium rate constant $k_{12}(t)/V^2$ and (b) electronic population in the donor state $P_1(t)$ for $V = 120 \, \text{cm}^{-1}$. The parameter $k_{12}(t)$ was obtained from the molecular dynamics simulation data for the model back ET reaction in a rigid collinear triatomic molecule with equivalent donor and acceptor sites separated by a neutral spacer in a polar solvent. The parameter $P_1(t)$ is a result of solution of the master equation. (Reproduced from [62c] with permission. Copyright (1996) by the American Institute of Physics.)

istic transition time τ_{na} is related to E_{na} via the uncertainty principle, that is, $\tau_{na} = \hbar^{1/2} \langle \dot{X}^2 \rangle^{-1/4}$.

The nonadiabaticity condition itself may not be sufficient to justify the use of the Golden Rule Eq. (9.1), where $k_{12} \sim V^2$, because the timescale of the bath relaxation is another important factor to be taken into account. If the bath relaxation is very fast, the survival time of the reactive state created by a thermal fluctuation at the curve crossing region is very short, the second-order electronic transition becomes the "bottleneck" of the reaction, and the rate is indeed proportional to V^2 and independent of the bath dynamics. On the other hand, if the solvent relaxation is very slow, multiple crossings of the nonadiabatic transition region, associated with the higher order terms in V, become possible. Ultimately, the reaction can become independent of the electronic coupling, that is, solvent controlled, while still being nonadiabatic.

A. Padé Approximation

1. General

Formally, the theory can be extended beyond the Golden Rule limit by considering higher order terms in the perturbation expansion of the rate in V. Starting with the Liouville equation for the density matrix, one can derive reduced equations of motion for the state populations (generalized master equations), formally exact for arbitrary coupling [88, 99, 295, 296],

$$s\hat{P}_1(s) - P_1(0) = -s\hat{P}_2(s) + P_2(0) = \hat{K}_{21}(s)\hat{P}_2(s) - \hat{K}_{12}(s)\hat{P}_1(s) \quad (9.34)$$

where

$$\hat{f}(s) = \int_0^\infty dt\, e^{-st} f(t) \quad (9.35)$$

denotes the Laplace transform. Note that Eq. (9.34) is just a Laplace transformed Eq. (9.25) but with generalized nonperturbative rate kernels. Laplace factorization of the kernels is a consequence of the assumed equilibrium initial condition (stationary process). Although nonequilibrium effects distroy this simple factorization (see Section II.B.3), this is of no essential consequence for the present discussion.

According to Eq. (9.34), the kernels $K_{12}(t)$ and $K_{21}(t)$ (or their Laplace images) completely define the ET kinetics. It is common practice to use mean rate constants, $\hat{K}_{12}(0)$ and $\hat{K}_{21}(0)$, as simpler characteristics of the process. Note that this simplification makes sense only if the kinetics are exponential (or nearly exponential). We will discuss this point in more detail below for the Markovian bath model, where exact expressions are available for the rate kernels.

One can obtain the following formally exact perturbative expansion of the rate kernel:

$$\hat{K}_{12}(s) = \sum_{n=1}^{\infty} \hat{K}_{12}^{(2n)}(s) \quad (9.36)$$

where $\hat{K}_{12}^{(2n)}(s) \sim V^{2n}$ and the odd terms in V vanish. The parameter $K_{12}^{(2)}(t)$ is the familiar Golden Rule rate kernel. Evaluation of the higher order terms is a matter of increasing complexity. Formally, one can express $K_{12}^{(2n)}(t)$ in terms of the function $Q(t) = Q_2(t) + iQ_1(t)$ for the spin-boson Hamiltonian [71, 85, 203] [see Eqs. (9.8) and (9.9) for the definition of $Q_1(t)$ and $Q_2(t)$]. However, this expression involves multiple irreducible time integrations and usually one only goes as far as $K_{12}^{(4)}(t)$ [124a], unless certain simplifying assumptions are made. Resummation of the perturbatively expanded rate kernel with only a finite number of terms known can be achieved by constructing the corresponding Padé approximant [88, 139, 140]. The simplest possible resummation scheme is based on the first two terms in the expansion and leads to the [1, 0]-Padé approximant,

$$\hat{K}_{12}(s) = \frac{\hat{K}_{12}^{(2)}(s)}{1 - \hat{K}_{12}^{(4)}(s)/\hat{K}_{12}^{(2)}(s)} \quad (9.37)$$

This result can be derived using the diagram technique [124a–c]. As shown by Cho and Silbey [139], the [1, 0]-Padé approximant arises naturally as the simplest (first-order) approximation for the rate kernel satisfying the Schwinger's stationary variational principle. Equation (9.37) is exact in the weak coupling limit up to the fourth order in V. In the opposite limit of strong coupling, Eq. (9.37) predicts saturation [recall that $\hat{K}_{12}^{2n}(s) \sim V^{2n}$]. However, the value of the rate kernel at saturation ($V \to \infty$) is judged by the behavior of the kernel at small V. Such an extrapolation procedure is not generally expected to give good accuracy (see below).

2. Classical Nonadiabatic Limit

In the classical limit ($\hbar\omega_i \ll k_B T$), the reaction coordinate $X(t)$ in each quantum state can be described as a Gaussian stochastic process [203]. It is Gaussian because of the assumed linear response. As follows from the discussion in Section II.A, if the collective solvent polarization follows the linear response, the ET system can be effectively represented by two sets of harmonic oscillators with the same frequencies but different equilibrium positions corresponding to the initial and final electronic states [26, 203]. The reaction coordinate, defined as the energy difference between the reactant and the product states, is a linear combination of the oscillator coordinates, that is, it is a linear combination of harmonic functions and is, therefore, Gaussian. The mean value is $\langle X_1(t)\rangle = -E_r$ for state 1 and $\langle X_2(t)\rangle = E_r$ for state 2, respectively. We can represent $X_1(t)$ and $X_2(t)$ in terms of a single Gaussian stochastic process $x(t)$ with zero mean as follows:

$$X_1(t) = x(t) - E_r, \quad X_2(t) = x(t) + E_r \quad (9.38)$$

Strictly speaking, $X_1(t)$ and $X_2(t)$ correspond to different realizations of $x(t)$. Any nonstationary effects of this kind are neglected in Eq. (9.38). The correlation function $C(t) = \langle x(t)x(0)\rangle$ is obtained as the classical limit of Eq. (9.3),

$$C(t) = \frac{2k_BT}{\pi}\int_0^\infty \frac{d\omega}{\omega} J(\omega)\cos\omega t \equiv \langle x^2\rangle \Delta(t) \qquad (9.39)$$

where

$$\langle x^2\rangle = C(0) = 2E_r k_B T \qquad (9.40)$$

and $\Delta(t)$ is the normalized correlation function to be used later on. The correlation function and the mean completely characterize the Gaussian stochastic process [297].

It can be shown by considering the stationary coordinate-dependent reaction flux for the spin-boson Hamiltonian derived within the Golden Rule (see Fig. 9.12) [114] that in the classical nonadiabatic limit, where the condition (9.32) is satisfied with

$$\langle \dot{x}^2\rangle = \frac{2k_BT}{\pi}\int_0^\infty d\omega\, \omega J(\omega) \qquad (9.41)$$

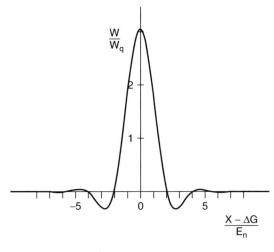

Figure 9.12. Reaction "sink" W (W_{12} in our notation) as a function of the reaction coordinate X in the classical limit. $W_q = V^2/\hbar E_n$. The parameter W is defined in terms of the stationary reaction flux $j(X)$ as follows: $j(X) = \phi_{eq}(X)\,W(X)$, where $\phi_{eq}(X)$ is the equilibrium distribution function in the initial state. The parameter E_n is the size of the nonadiabatic region (E_{na} in our notation). (Reproduced from [114] with permission. Copyright (1999) by the American Institute of Physics.)

the transitions between diabatic states are localized in the vicinity of the crossing point, $X^\ddagger = \Delta G$, within a narrow nonadiabatic zone E_{na}, defined by Eq. (9.33). In other words, it is possible to introduce a delta-functional sink for each passage of the crossing point (local Golden Rule):

$$W_{12}(X) = W_{21}(X) = \kappa \delta(X - \Delta G) \tag{9.42}$$

where $\kappa = 2\pi V^2/\hbar$. Rigorously, localization of the stationary flux also implies that $E_{na} \ll k_B T$ [114], which may not necessarily be the case.

Under the above conditions, we can immediately obtain for the first two terms in the perturbation expansion of the rate kernel,

$$K_{12}^{(2)}(t) = \kappa \phi_{eq}(x_1^\ddagger)\delta(t) \equiv k_M(V, \Delta G, E_r)\,\delta(t) \tag{9.43}$$

$$K_{12}^{(4)}(t) = -\kappa^2 \phi_{eq}(x_1^\ddagger)[g(x_1^\ddagger, x_1^\ddagger, t) + g(x_2^\ddagger, x_2^\ddagger, t)] \tag{9.44}$$

where $x_1^\ddagger = \Delta G + E_r$, $x_2^\ddagger = \Delta G - E_r$,

$$\phi_{eq}(x) = \frac{1}{\sqrt{2\pi\langle x^2\rangle}}\exp\left(-\frac{x^2}{2\langle x^2\rangle}\right) \equiv \Phi(x, \langle x^2\rangle) \tag{9.45}$$

is the equilibrium Gaussian distribution (we will use this notation $\Phi(x, \langle x^2\rangle)$ for the Gaussian function frequently hereafter),

$$g(x, x_0, t) = G(x, x_0, t) - \phi_{eq}(x) \tag{9.46}$$

is the reduced distribution function defined in terms of the two-point conditional probability distribution function $G(x, x_0, t)$ of the Gaussian coordinate $x(t)$,

$$G(x, x_0, t) = \Phi\{x - x_0\Delta(t), \langle x^2\rangle[1 - \Delta^2(t)]\} \tag{9.47}$$

Here, $x_0 = x(0)$.

Clearly, the nth term in the perturbation expansion of the rate kernel $K_{12}^{(2n)}(t)$, associated with n crossings of the nonadiabatic transition region, can be expressed in terms of the Gaussian n-point probability distribution function $G_n(x_1, t_1, \ldots, x_n, t_n)$, which is determined solely by the correlation function $C(t)$ of the process (the mean is zero) [297]. Note that $\phi_{eq}(x) = G_1(x, t)$ and $G(x, x_0, t) = G_2(x_0, 0, x, t)/G_1(x_0, 0)$. Thus, the perturbation expansion can be completely defined and one can construct $[n, n-1]$-Padé approximants if there is no other strategy for resummation at hand. The $[1, 0]$-Padé approximant of Eq. (9.37) along with Eqs. (9.43)–(9.47) constitute the simplest approximation for the rate kernel, sometimes referred to as the substitution approximation (see below). Qualitatively, it solves the problem of nonadiabatic ET kinetics in the classical limit for strong coupling. However, as we shall see below, this solution is exact only for Debye solvents.

B. Markovian Bath Model and Its Generalizations

The long-time behavior of the correlation function is exponential [197],

$$\Delta(t) = \exp(-t/\tau_L) \tag{9.48}$$

and characterized by the longitudinal polarization relaxation time τ_L, defined by

$$\tau_L = \frac{1}{2E_r} \lim_{\omega \to 0} \frac{J(\omega)}{\omega} \tag{9.49}$$

In terms of the usual (constant electric field) dielectric relaxation time τ_D, we have $\tau_L = \tau_D \varepsilon_\infty / \varepsilon_s$. Let us assume that the whole relaxation kinetics is exponential (the Debye model). Gaussian process with exponential correlation is Markovian [297]. In this case, the n-point distribution function G_n can be factorized into a product of two-point distribution functions G_2, and the corresponding convolution-type perturbation series for the rate kernel can be summed up exactly [88] leading to Eq. (9.37), that is,

$$\hat{K}_{12}^{-1}(s) = k_M^{-1}(V, \Delta G, E_r) + \hat{k}_d^{-1}(s) \tag{9.50}$$

where $k_M(V, \Delta G, E_r)$ is the usual Marcus rate constant and

$$\hat{k}_d^{-1}(s) = \phi_{eq}^{-1}(x_1^\ddagger)[\hat{g}(x_1^\ddagger, x_1^\ddagger, s) + \hat{g}(x_2^\ddagger, x_2^\ddagger, s)] \tag{9.51}$$

Equation (9.50) can also be derived (as it originally was) using a seemingly different approach. In the case of Markovian bath relaxation (and only in this case), the partial probability distributions $\rho_1(X, t)$ and $\rho_2(X, t)$ of the two ET states satisfy the following coupled diffusion equations [74, 75b]:

$$\frac{\partial}{\partial t}\rho_1(X, t) = \mathcal{L}_1 \rho_1(X, t) - W_{12}(X)\rho_1(X, t) + W_{21}(X)\rho_2(X, t)$$

$$\frac{\partial}{\partial t}\rho_2(X, t) = \mathcal{L}_2 \rho_2(X, t) - W_{21}(X)\rho_2(X, t) + W_{12}(X)\rho_1(X, t) \tag{9.52}$$

where the rates $W_{12}(X)$ and $W_{21}(X)$ are defined by Eq. (9.42) and the diffusion operators are given by

$$\mathscr{L}_{1,2} = D \frac{\partial}{\partial X}\left(\frac{\partial}{\partial X} + \frac{X \pm E_r}{2E_r k_B T}\right) \qquad (9.53)$$

Here, $D = \langle x^2 \rangle/\tau_L$ is the diffusion coefficient. The state populations are defined as

$$P_n(t) = \int dX \rho_n(X, t) \qquad n = 1, 2 \qquad (9.54)$$

and satisfy Eq. (9.34), as usual. Equation (9.50) for the rate kernel can be obtained by using the Dyson-type identity relating the Green's function in the presence of reaction to the Green's function of free diffusion. This approach has recently been extended to three-level systems and successfully applied to intramolecular light-induced transitions in hydrogen-bonded species [299].

According to Eq. (9.34), the rate kernels completely define the ET kinetics. The mean rate constants $\hat{K}_{12}(0)$ and $\hat{K}_{21}(0)$ can be obtained analytically. We have for the reduced Green's function [74, 75, 87],

$$\hat{g}(x, x, 0)\tau_L^{-1} = \begin{cases} \phi_{eq}(0) \ln 2 & \text{for } x = 0 \\ 1/|x| & \text{for } |x|/\langle x^2 \rangle^{1/2} \gg 1 \end{cases} \qquad (9.55)$$

while for arbitrary x it is expressed in terms of the hypergeometric function $_2F_2$ [113]. Note that the dependence of $\hat{g}(x, x, 0)$ on x (i.e., the activation energy) is only weak, which means that generally the contributions of both forward and backward reactions are important in Eq. (9.51). If, however, the forward reaction is activationless ($-\Delta G = E_r$), we have $\hat{g}(x_2^\ddagger, x_2^\ddagger, 0)/\hat{g}(x_1^\ddagger, x_1^\ddagger, 0) = (\pi k_B T/E_r)^{1/2}/\ln 2 \ll 1$, that is, for polar solvents ($E_r \gg k_B T$) the contribution of the back reaction can be safely neglected. Then $\hat{k}_d(0) \simeq 1/(\tau_L \ln 2)$. In the case of activationless irreversible reaction with a "black" sink ($\kappa \to \infty$), the survival probability, $P_1(t) = \int dX \rho_1(X, t)$, has a particularly simple form [75a],

$$P_1(t) = (2/\pi) \arcsin[\exp(-t/\tau)] \qquad (9.56)$$

The kinetics is nonexponential but has a long-time exponential tail characterized by

$$k_{12} = -\lim_{t \to \infty} [\dot{P}_1(t)/P_1(t)] = 1/\tau_L \qquad (9.57)$$

Note that the long-time rate and the mean rate differ by a factor of $\ln 2 \approx 0.7$. This discrepancy indicates that in general the mean reaction rate can be a poor characteristic of nonexponential kinetics. For activated reactions, the kinetics is practically exponential and the mean reaction rate coincides with the long-time rate. Note also that for a reversible reaction, the decay is usually normalized with respect to the stationary population, $P_1(\infty) = [1 + \exp(\Delta G)]^{-1}$.

Let us summarize the results in the original notation. The ET rate constant for activated reactions is approximately given by

$$k_{12} = \frac{\kappa^*}{\sqrt{4\pi E_r k_B T}} \exp\left[-\frac{(\Delta G + E_r)^2}{4 E_r k_B T}\right] \tag{9.58}$$

where

$$\frac{1}{\kappa^*} = \frac{\hbar}{2\pi V^2} + \tau_L \left(\frac{1}{|\Delta G + E_r|} + \frac{1}{|\Delta G - E_r|}\right) \tag{9.59}$$

while for activationless reactions,

$$\frac{1}{k_{12}} = \frac{\hbar}{2\pi V^2}\sqrt{4\pi E_r k_B T} + \tau_L \tag{9.60}$$

The Marcus activation energy stays intact, but the frequency prefactor is significantly modified.

The usual classical Golden Rule term dominates for weak coupling. Since $\hat{k}_d(s)$ is independent of V, it dominates when the coupling becomes sufficiently strong. The parameter $\hat{k}_d(s)$ describes diffusive passage to the transition state on the timescale of τ_L. Therefore, the ET rate itself, having started with a V^2 dependence, finally saturates as a function of V at a certain plateau value determined by τ_L^{-1} (besides the usual activation factor), as shown in Figure 9.13. This is the so-called dynamical solvent control regime. Crossover from the Golden Rule to the solvent control is determined by the Zusman parameter,

$$\zeta \equiv k_M/\hat{k}_d(0) = \frac{2\pi V^2}{\hbar}\tau_L\left(\frac{1}{|\Delta G + E_r|} + \frac{1}{|\Delta G - E_r|}\right) \tag{9.61}$$

where the second equality is written for activated reactions [for activationless reactions, obvious modification follows from Eq. (9.60)]. The condition $\zeta \gg 1$ corresponds to the solvent control regime. This is often confused with

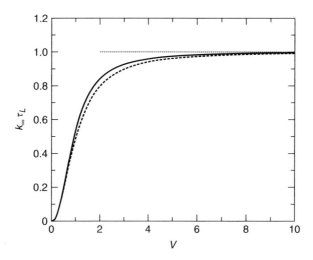

Figure 9.13. Long-time rate constant k_{12} as a function of electronic coupling V for an irreversible activationless ET reaction within Zusman's model (solid). The parameter k_{12} is normalized to its value at large coupling, that is, τ_L^{-1}. The parameter V is scaled so that the values of the dimensionless parameter $(k_M \tau_L)^{1/2}$ proportional to V are shown. Dashed curve is a simple approximation, $k_{12}^{-1} = k_M^{-1} + \tau_L$.

the adiabaticity condition. It should be stressed that there is, so far, no real adiabaticity. As we have seen, the formulation of Zusman's model is essentially nonadiabatic (local Golden Rule), and the seemingly adiabatic result obtained is a consequence of multiple crossings of the nonadiabatic transition region [75–78]. True adiabatic limit is reached when the condition for the Massey parameter in Eq. (9.32) is reversed. For example, the crossover to solvent control regime for symmetric reactions ($\Delta G = 0$) with a typical reorganization energy of $E_r = 20k_B T$ and a typical polarization relaxation time of $\tau_L = 1$ ps occurs at couplings $V \simeq 0.2k_B T$, as estimated from the condition $\zeta \simeq 1$. Longer relaxation times (such as for 1-decanol [194], where $\tau_L \simeq 100$ ps) bring the crossover value of coupling even lower. At the same time, the transitions are nonadiabatic up to $V \simeq k_B T$ (a typical frequency of $\Omega = 10^{13}$ s^{-1} was used for estimation of the mean-squared velocity in Eq. (9.32), see Section III.B.1 for the definition of Ω).

In the normal region, the crossover between the nonadiabatic dynamic solvent control regime and the adiabatic regime is smooth, because in both cases the reaction rate is determined by the diffusive "delivery of the reactive wave packet" to the transition state. However, the situation is radically different in the inverted region, where the reaction is always nonadiabatic (even when the term crossing is adiabatic), and for very high coupling it is limited by the rate of transitions between adiabatic states and not by the transport to the transition region. We will discuss this point below in more detail.

1. Substitution Approximation and Time-Dependent Diffusion Coefficient

The above theory is based on the Debye model of solvent dielectric relaxation. In real systems, the exponential stage of the decay of the correlation function is achieved only at long times. [197, 300]. As one can see in Figure 9.4, the amplitude of this long-time tail can be rather small, while the major part of the decay is inertial. Thus, at short times, as follows from Eq. (9.39),

$$\Delta(t) \simeq \exp[-\Omega^2 t^2/2] \tag{9.62}$$

where

$$\Omega^2 = \frac{\langle \dot{x}^2 \rangle}{\langle x^2 \rangle} \equiv \frac{\int_0^\infty d\omega \omega J(\omega)}{\int_0^\infty d\omega \omega^{-1} J(\omega)} \tag{9.63}$$

is the characteristic frequency of harmonic modes contributing to the reaction coordinate. What is the relevant timescale for the ET process, τ_L or Ω^{-1}? How can we define the rate kernel in terms of nonexponential correlation function with a broad range of timescales? These questions call for generalized non-Markovian formulation of the stochastic ET model.

Perhaps, it was Hynes who initiated two of the most popular so far semiclassical non-Markovian approximations [84]. The first approximation was inspired by the success of the [1, 0]-Padé approximant, which turns out to be exact in the Markovian limit. This approximation is sometimes referred to as the "substitution" approximation, because effectively one substitutes non-Markovian two-point distribution function (9.46)–(9.47) into the Markovian expressions (9.50)–(9.51) for the rate kernel. The substitution approximation was shown to work rather well for the case of biexponential relaxation with similar decay times [102]. However, as Bicout and Szabo [142] recently demonstrated, it considerably overestimates the reaction rate when the two relaxation timescales become largely different (see Fig. 9.14). They also showed that for a non-Markovian process with a multiexponential correlation function, which can be mapped onto a multi-dimensional Markovian process [301], the substitution approximation is equivalent to the well-known Wilemski–Fixman closure approximation [302–304]. A more serious problem arises when we try to deal with the

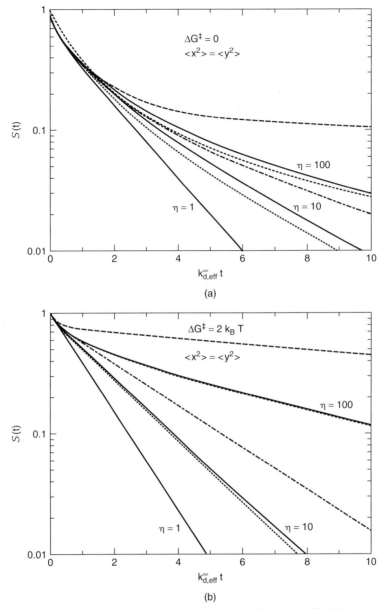

Figure 9.14. Survival probability $S(t)$ of an irreversible ET reaction [$P_1(t)$ in our notation] in the solvent-controlled regime ($\kappa \to \infty$) for biexponential solvent relaxation model with the normalized correlation function $\Delta(t) = \frac{1}{2}e^{-t/\tau_1} + \frac{1}{2}e^{-t/\tau_2}$. The parameter $\Delta G^{\ddagger}$ is the activation energy (E_a in our notation), $\eta = \tau_1/\tau_2$, and $k_{d,\text{eff}}^{\infty}$ is the mean rate constant for ET in an effective Debye solvent with $\tau_L = \mathring{\Delta}^{-1}(0) = 2\tau_1\tau_2/(\tau_1 + \tau_2)$. Solid lines correspond to Brownian dynamics results (exact), the dash–dotted ($\eta = 10$) and long-dashed ($\eta = 100$) lines to the substitution approximation, and the dashed lines to the effective sink approximation. For $E_a = 0$ and $\eta = 1$ the decay is given by Eq. (9.56). (Reproduced from [142] with permission. Copyright (1998) by the American Institute of Physics.)

initial Gaussian stage of relaxation, typical of any realistic system. If we use Eq. (9.62) in Eq. (9.44) and make appropriate expansions, we will see that $K_{12}^{(4)}(t)$ is proportional to $1/t$ at short times and thus its Laplace image in Eq. (9.37) diverges! This is a consequence of the delta-functional sink localization. If the sink was blurred for some reason (e.g., due to the presence of faster modes), the divergence would disappear. Nevertheless, it is clear that the Padé (substitution) approximation should be used with extreme care.

Another approximation is obtained by observing that the two-point conditional probability distribution function $G(x, x_0, t)$ of a Gaussian process exactly satisfies the following diffusion equation [297, 301]:

$$\frac{\partial}{\partial t} G(x, x_0, t) = D(t) \frac{\partial}{\partial x} \left(\frac{\partial}{\partial x} + \frac{x}{\langle x^2 \rangle} \right) G(x, x_0, t) \qquad (9.64)$$

where the time-dependent diffusion coefficient is given by

$$D(t) = -\langle x^2 \rangle [\dot{\Delta}(t)/\Delta(t)] \qquad (9.65)$$

The idea is to substitute this $D(t)$ into the rate equations (9.52). This approach may be useful for the description of the short-time behavior [84, 103–107]. However, it can be shown that it predicts the stationary value for the rate constant totally determined by the long-time value of $D(t)$, which is $\langle x^2 \rangle / \tau_L$. In other words, it simply reproduces Zusman's result in this limit, which is certainly not the case (see Fig. 9.14, e.g.).

There is a very powerful semiclassical path integral approach due to Wang and Wolynes [305] developed for the cases of moderate dependence of the reaction rate on the environmental coordinates, such as "geometrical bottlenecks". Unfortunately, this semiclassical scheme does not work for a δ-functional sink (and it was not intended to).

2. Statistics of Crossing Points

Makarov and Topaler [120] compared the simulated trajectories of a Gaussian stochastic process with exponential and Gaussian correlations and realized that reactive events (i.e., occurrences of the favorable configurations that allow the reaction to proceed) take place in a completely different fashion in these two cases, as shown in Figure 9.15. They tend to occur in clusters separated by wide gaps for exponential correlation. In contrast, Gaussian correlation leads to "repulsive" statistics of the reactive events in a sense that the probability for two crossings to happen during time t goes to zero as $t \to 0$. The formalism for the analysis of a system of random points

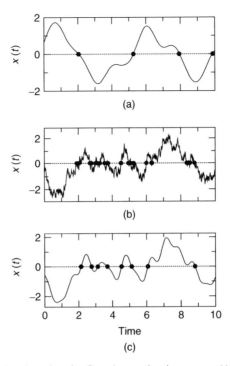

Figure 9.15. Typical trajectories of a Gaussian stochastic process $x(t)$ with zero mean and Gaussian (a) or exponential (b) correlation function. Circles are crossing points of $x^{\ddagger} = 0$. Trajectories were generated by regular sampling in the frequency domain. (c) corresponds to the Debye relaxation spectrum with a cutoff frequency. Reorganization energy of the discarded part of the spectrum is 7% of the total. The sampling pattern was the same as in (b).

was developed by Stratonovich [306] in terms of a sequence of distribution functions $f_n(t_1, \ldots, t_2)$, which define the probability densities of having at least one crossing point at each of the time moments t_j. If one interprets crossings in a classical sense, then

$$f_n(t_1, \ldots, t_2) = \int_{-\infty}^{\infty} dv_1 \cdots \int_{-\infty}^{\infty} dv_n |v_1| \cdots |v_n| G_n(x^{\ddagger}, v_1, t_1, \ldots, x^{\ddagger}, v_n, t_n)$$

(9.66)

where $v_j = \dot{x}_j$, $x^{\ddagger}$ is the crossing point, and G_n is the usual n-point Gaussian coordinate-velocity conditional probability distribution function determined solely by $C(t)$ [297]. For example, we obtain [307] for $f_2(t)$ in the short-time limit, $t = |t_2 - t_1| \to 0$, for $x^\ddagger = 0$,

$$f_2(t) \simeq t \frac{\Delta^{(4)}(0) - \ddot{\Delta}^2(0)}{8\pi\sqrt{|\ddot{\Delta}(0)|}} \tag{9.67}$$

where $\Delta^{(4)}(0)$ stands for the fourth-order derivative. Equation (9.67) proves that indeed the statistics is repulsive for Gaussian and attractive for exponential correlation (time derivatives diverge in the latter case). As we can see, the statistics of crossing points is very sensitive to the short-time behavior of $C(t)$ or, equivalently, the high-frequency part of the relaxation spectrum. Nonanalyticity of the exponential correlation function at time zero corresponds to unrealistically slow decay of the spectral density in the limit of $\omega \to \infty$, that is, $J(\omega) \sim \omega^{-1}$. If we set a certain cutoff frequency ω_c in the Debye spectrum, trajectories will become smooth and crossing points will no longer be clustered, as shown in Figure 9.15. Note that ω_c can be set rather large so that the reorganization energy of the discarded high-frequency part of the spectrum would be negligible ($\sim k_B T$), while the effect on the statistics of crossing points would be significant.

In order to obtain a tractable analytical solution for the reaction rate in the case of Gaussian relaxation, Makarov and Topaler invoked an effective velocity approximation and an approximation of nonapproaching points due to Stratonovich. While the latter is a consequence of Eq. (9.67) and equivalent to the cluster expansion, the former assumes that the transition probability is constant. It is evaluated so as to give a correct value of the short-time rate. Actually, the problem can be handled even without this approximation. As a result, one obtains for the long-time rate constant of an irreversible reaction,

$$k_{12} = -T_c^{-1} \ln(1 - k_0 T_c) \tag{9.68}$$

where $T_c = \int_{-\infty}^{\infty} dt R(t)$,

$$R(t) = 1 - f_2(t)/f_1^2$$

is the correlation coefficient, and $k_0 = -\dot{P}_1(0)$ is the short-time rate constant given by [308],

$$k_0 = \int_{-\infty}^{\infty} dv |v| p_{LZ}(v) \, \Phi(x^\ddagger, \langle x^2 \rangle) \, \Phi(v, \langle v^2 \rangle) = \langle p_{LZ} \rangle_v \Phi\langle x^\ddagger, \langle x^2 \rangle\rangle \tag{9.69}$$

Here, $\langle v^2 \rangle = |\ddot{C}(0)|$, Φ is the Gaussian function defined by Eq. (9.45), and

$$p_{LZ}(v) = 1 - \exp(-\kappa/|v|) \tag{9.70}$$

is the Landau–Zener transition probability at the crossing point.

This simple theory reproduces numerical results for the Gaussian-correlated process very well [120]. One of the reasons is that the decay is practically exponential. Note that $T_c \simeq 1.34\Omega^{-1}$ and the maximum value of k_0 for $x^{\ddagger} = 0$ and $\kappa \to \infty$ is $k_0 \simeq \Omega/\pi$. Therefore, we can expand the logarithm in Eq. (9.68) and obtain $k_{12} \approx k_0$. Actually, k_{12} is somewhat larger than k_0, reflecting the repulsive statistics of crossings. Importantly, the transition to the solvent-controlled regime for the Gaussian-correlated process is determined by k_0, that is, by the Landau–Zener mechanism. Note that the Landau–Zener theory may not be valid in the presence of dissipation, and this can be of particular importance in the activationless and inverted regimes, as will be shown below. Instead of Stratonovich's approximation of nonapproaching points we could make use of the ubiquitous Padé approximation and obtain $k_{12} \simeq k_0/(1 - k_0 T_c/2)$, which coincides with Eq. (9.68) up to the second order in $k_0 T_c$. Interestingly, direct substitution of the exponential correlation function formally leads us back to Zusman's result.

3. Effective Sink Approximation

Let us assume that the reaction coordinate can be represented as a superposition of two independent stochastic processes,

$$x(t) = x_s(t) + x_f(t) \tag{9.71}$$

with well-separated timescales. The slow mode is assumed to be Markovian, with exponential correlation determined by the relaxation time τ_L and the amplitude $\langle x_s^2 \rangle = 2E_r^s k_B T$, where E_r^s is the corresponding reorganization energy. Total reorganization energy E_r is given by the sum of E_r^s and E_r^f, where E_r^f is the reorganization energy of the fast mode. The fast mode is assumed to be equilibrated for each value of the slow coordinate so that we can use the transition state theory for x_f. In the original formulation by Ovchinnikova [79] and by Sumi and Marcus [89] the fast mode was ascribed to intramolecular vibrational relaxation, but it can be related to the solvent polarization, too. The dynamics of this mode is not important (so far).

Electronic populations $\rho_1(x_s, t)$ and $\rho_2(x_s, t)$ of the reactant and the product states, respectively, averaged over the fast coordinate satisfy diffusion equations (9.52) along the slow coordinate (x_s instead of X) with effective potentials,

$$u_1(x_s) = \frac{(x_s + E_r^s)^2}{4E_r^s} \qquad u_2(x_s) = \frac{(x_s - E_r^s)^2}{4E_r^s} + \Delta G \tag{9.72}$$

The diffusion equations are coupled by the sink terms,

$$W_{12,21}(x) = \kappa \, \Phi(\Delta G \pm E_r^f - x_s, \langle x_f^2 \rangle) \qquad (9.73)$$

where $\langle x_f^2 \rangle = 2E_R^f k_B T$ and Φ is the Gaussian function, as before.

The transition probabilities (9.73) obey the principle of detailed balance in the local form,

$$\frac{W_{12}(x_s)}{W_{21}(x_s)} = \exp\left[\frac{u_1(x_s) - u_2(x_s)}{k_B T}\right] = \exp\left(\frac{-\Delta G + x_s}{k_B T}\right) \qquad (9.74)$$

As a result, forward reaction dominates for $x_s > \Delta G$, that is, to the right of the crossing point, while backward reaction dominates for $x_s < \Delta G$, that is, to the left of the crossing point. Note that this is not a real but an effective crossing point. The free energy surfaces are, in fact, paraboloids in coordinates x_s and x_f crossing along the line $x_s + x_f = \Delta G$. As the transitions along the fast coordinate are "vertical" (i.e., the coordinate is always equilibrated), only a projection of the whole picture onto the slow coordinate is relevant. In Figure 9.16, the reactant state term corresponds to the equilibrium value of $x_f = -E_r^f$, while the product state term corresponds to $x_f = E_r^f$. They cross at the point of equal free energy, but the actual transition between the states at this point will also involve activation along the fast coordinate. Figure 9.16 shows that there is a principal difference between the normal and the inverted regime of term crossing. In the normal regime, both sinks are located higher along the corresponding effective potentials than the crossing point, where their strengths are equal. In the 2D representation, the maxima of the sinks are at the intersection of the equilibrated free energy profiles with the transition state line, $x_s + x_f = \Delta G$. As a result, even though the local rate of forward ET increases with coupling promoting reaction without activation, backward ET rate also increases in such a way that full activation up to the crossing point is necessary to complete the transition. In the strong coupling limit, the reaction rate is controlled by diffusion along an effective adiabatic potential [81c],

$$\frac{u_{12}(x_s)}{k_B T} = -\ln\left[\exp\left(-\frac{u_1(x_s)}{k_B T}\right) + \exp\left(-\frac{u_2(x_s)}{k_B T}\right)\right] \qquad (9.75)$$

and does not depend on the reorganization energy of the fast mode. In contrast, in the inverted regime the forward-reaction sink is located below while the backward-reaction sink above the term crossing point, and the overall ET process becomes effectively irreversible with the rate increasing as a function of both V and E_r^f.

SOLVENT EFFECTS IN NONADIABATIC ELECTRON-TRANSFER REACTIONS 557

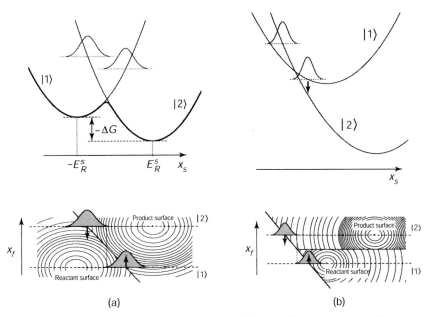

Figure 9.16. Effective energy profiles along the reaction coordinate x_s corresponding to the slow solvent polarization mode (top) and two-dimensional (x_s, x_f) contour plots (bottom) in the Marcus normal (a) and inverted (b) regions. The energy profile of the initial state $|1\rangle$ corresponds to the equilibrium value of the fast coordinate $x_f = -E_r^f$, while the final $|2\rangle$ state profile corresponds to $x_f = E_r^f$ (shown by dashed lines in the contour plot). Sinks illustrate the transition probabilities due to the presence of the fast mode and are shown at their proper position and width for $E_r^f = 0.5 E_r^s$. Their maxima are at the intersection of the corresponding effective potential profiles (dashed) with the transition state line (bold slanted), $x_s + x_f = \Delta G$. The crossing point of the effective potential profiles corresponds to the dotted line in the contour plot. In the limit of strong coupling the transition between the two states in the normal regime is controlled by diffusion along an effective adiabatic potential $u_{12}(x_s)$ (bold). Arrow in (b) indicates that in the inverted regime ET becomes irreversible. (Reproduced from [309] with permission. Copyright (2001) by the American Institute of Physics.)

The above conclusions are illustrated in Figures 9.17 and 9.18 for the long-time rate constant of forward reaction k_{12} obtained from the numerically evaluated [309] lowest mixed eigenvalue $\lambda = k_{12} + k_{21}$ of the operator matrix in the right-hand side of Eq. (9.52) using the principle of detailed balance,

$$k_{12}/k_{21} = \exp(-\Delta G/k_B T) \qquad (9.76)$$

Numerically, one can take advantage of the matrix method of solution of the reaction-diffusion equation using eigenfunctions of the diffusion operator as a basis [101d, 103, 138, 143, 161, 301]. This method is computationally

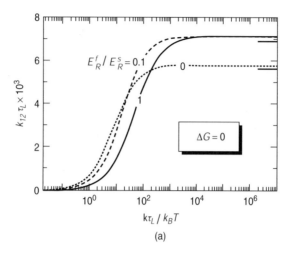

Figure 9.17. Long-time rate constant of forward reaction as a function of coupling in a stochastic Markovian model of reversible ET assisted by a fast vibrational mode for $\Delta G = 0$, $E_r^s = 18 k_B T$, and several values of E_r^f/E_r^s (numbers attached to the curves). Bold lines illustrate the large-coupling limit estimated from Eq. (9.77). Note that here we have E_r^s fixed. (Reproduced from [309].)

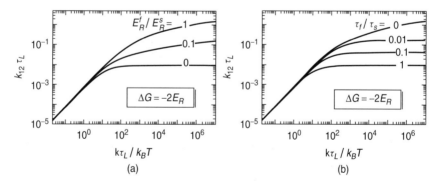

Figure 9.18. (a) Long-time rate constant of forward reaction as a function of coupling in a stochastic Markovian model of reversible ET assisted by a fast vibrational mode for $\Delta G = -2E_r$, $E_r = 18 k_B T$, and several values of E_r^f/E_r^s (numbers attached to the curves). (b) The results for a 2D Markovian model with biexponential relaxation for $E_r^f = E_r^s = 9 k_B T$, $\Delta G = -2E_r$, and several values of τ_f/τ_s (numbers attached to the curves). The parameters τ_f and τ_s denote the relaxation times of the fast and the slow mode, respectively. Note that here we have full $E_r = E_r^s + E_r^f$ fixed. (Reproduced from [309].)

superior over the usual discretization procedure, particularly when the process is irreversible.

In the normal region, ET is controlled by diffusive motion along the slow polarization coordinate for strong coupling ($\kappa \to \infty$) and is indeed independent of the reorganization energy of the fast mode. The only difference is seen when the fast mode is absent and the effective adiabatic potential is cusped, resulting in a bit higher activation energy and, therefore, a bit smaller rate constant. Approximate Kramers-type expression for the rate,

$$k_{12}^d = \frac{1}{\tau_L}\sqrt{\frac{E_r^s k_B T}{\pi}} \frac{\exp[u_{12}(-E_r^s)/k_B T]}{\int_{-E_r^s}^{E_r^s} dx \exp[u_{12}(x)/k_B T]} \tag{9.77}$$

works very well if the activation energy is high enough. For intermediate coupling, one can construct a standard interpolation for k_{12} using Eq. (9.77) for $\kappa \to \infty$ and the Marcus expression for $\kappa \to 0$ [81c, 87],

$$\frac{1}{k_{12}} = \frac{1}{k_M(V, \Delta G, E_r)} + \frac{1}{k_{12}^d} \tag{9.78}$$

Note that although similar in appearance, this approximation is different from (and better than) the [1,0]-Padé approximation defined early in this section. It bridges between two known limits, while [1,0]-Padé extrapolates.

Equation (9.78) can be used to define the Zusman parameter ζ as a ratio of k_M and k_{12}^d, similar to Eq. (9.61). However, in contrast to Eq. (9.61), since k_{12}^d now depends only on the reorganization energy of the slow mode while k_M is a function of the total reorganization energy, ζ will contain a certain activation factor. For example, for symmetric reactions ($\Delta G = 0$) we obtain,

$$\zeta \simeq \frac{2\pi V^2}{\hbar} \tau_L \frac{\exp(-E_r^f/4k_B T)}{\sqrt{E_r E_r^s}} \tag{9.79}$$

Let us emphasize that the condition $\zeta \gg 1$ for the solvent control regime is different from the adiabaticity condition, $\bar{\xi} \equiv (V^2/\hbar\Omega)(2E_r k_B T)^{-1/2} \gg 1$ [reversed Eq. (9.32)]. The latter is determined by the mean-squared velocity of the full reaction coordinate, characterized by Ω, while the former involves only the timescale of the slow component, characterized by τ_L. It is clear from Eq. (9.79) that although τ_L is typically $\gg \Omega^{-1}$, the increasing contribution of fast modes may lead to that nonadiabatic solvent control regime will gradually disappear.

It is important to get a feeling of what the values of the parameter $\kappa \tau_L/k_B T$ can be for actual experimental systems. Clearly, nonadiabatic frequency prefactor $v = \kappa/\sqrt{4\pi E_r k_B T}$ in Eq. (9.10) should not exceed its adiabatic value, which is $\sim 10^{13} \text{s}^{-1}$. Therefore, for $E_r \sim 20 k_B T$, $\kappa/k_B T$ should not exceed $\sim 10^{14} \text{s}^{-1}$. Longitudinal polarization relaxation time τ_L ranges from 1 to 100 ps for typical solvents [145–151, 156–158, 194] (e.g., 0.2–0.5 ps for acetonitrile, 5–7 ps for benzonitrile, ~ 100 ps for 1-decanol). In special cases, such as for triacetin at low temperatures, τ_L can be as large as several microseconds [156]. Thus, the parameter $\kappa \tau_L/k_B T$ can reach 10^2–10^4 for usual solvents and extend to even several orders of magnitude higher values for very viscous solvents, which explains the choice of the range of $\kappa \tau_L/k_B T$ in Figures 9.17 and 9.18.

Clearly, the effect of back reaction is crucial in the normal region [309]. Even if $-\Delta G \gg k_B T$ and the fast mode is present, local detailed balance ensures that the ET rate is controlled by the slow solvent mode when coupling is sufficiently large. An elegant demonstration of the dynamic solvent effect has been reported by Grampp et al. [150] for the electron self-exchange reaction, as shown in Figure 9.19. The parameter ΔG is strictly 0 in this case and one does not have to worry that it will change from solvent to solvent. No correlation is observed in the sense of the classical Marcus theory, where $\ln k_{12}$ should depend linearly on the Pekar factor, but instead good correlation with respect to τ_L.

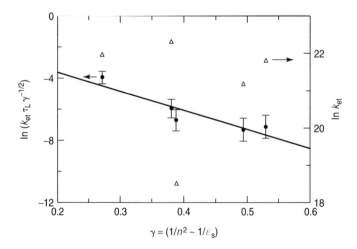

Figure 9.19. Solvent dependence of $\ln k_{et}$ and $\ln(k_{et}\tau_L\gamma^{-1/2})$ versus the Pekar factor $\gamma = \varepsilon_\infty^{-1} - \varepsilon_s^{-1}$ (c_p in our notation). The parameter k_{et} (k_{12} in our notation) is the electron self-exchange rate constant between 2,3-dicyano-5,6-dichloro-p-benzoquinone and its radical anion measured by means of electron paramagnetic resonance (EPR) line broadening effects at $T = 293$ K and corrected for translational diffusion. The data points correspond to $CHCl_3$, CH_2Cl_2, benzonitrile, acetone, and acetonitrile (left to right). (Reproduced from [150] with permission. Copyright (1999) by the Royal Society of Chemistry.)

The situation is different in the inverted regime, where back reaction has no effect on the ET rate, except when the fast mode is absolutely absent. In their influential paper, Sumi and Marcus [89] demonstrated and many others confirmed [90–93, 161] that an irreversible stochastic model of ET with a wide sink (classical vibrational mode) predicts nonexponential kinetic behavior for large coupling and a considerable increase of the long-time rate constant over the usual solvent-control plateau, indicating the breakdown of the Padé approximation, as shown in Figures 9.18 and 9.20. However, this increase is still rather weak, asymptotically logarithmic [161], as can be understood by converting the reaction-diffusion equation into a Schrödinger-type equation with a self-adjoint operator and invoking a local harmonic approximation after locating the minimum of the effective potential [310]. When the coupling is very large, most part of the population decays statically before diffusion can take effect. The resulting strong nonexponentiality of the decay leads to the fractional power law dependence of the mean reaction rate $\hat{K}(0)$ on the sink strength κ [89–91, 93]. Different effect of the fast mode in the normal and the inverted regimes leads to asymmetry of the free energy gap law for strong coupling, as illustrated in Figure 9.21. Quantum modes will further enhance this asymmetry.

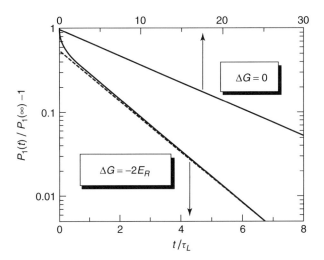

Figure 9.20. Population of the reactant state as a function of time normalized with respect to its stationary value in the normal ($\Delta G = 0$) and the inverted ($\Delta G = -2E_r$) regimes calculated on the basis of a stochastic Markovian model of reversible ET assisted by a fast vibrational mode with $E_r^f = E_r^s = 9k_BT$, and $\kappa\tau_L/k_BT = 10^5$. In the normal regime, the reaction is always activated and well described by a single exponential. Dashed line corresponds to the long-time limit in the inverted regime, described by k_{12}. (Reproduced from [309] with permission. Copyright (2001) by the American Institute of Physics.)

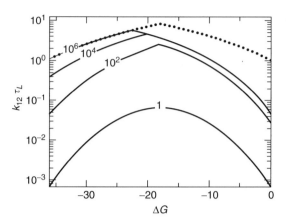

Figure 9.21. Long-time rate constant k_{12} as a function of ΔG (in units of $k_B T$) for $E_r^f = E_r^s = 9k_B T$, and several values of $\kappa \tau_L / k_B T$ (numbers attached to the curves). Circles corresponds to an irreversible reaction with $\kappa \tau_L / k_B T = 10^6$. The cusp for strong coupling values is not a numerical error but the result of crossing of two lowest eigenvalue branches. (Reproduced from [309] with permission. Copyright (2001) by the American Institute of Physics.)

Yoshihara and co-workers [151c] reported good agreement of their experimental results on ET between oxazine and an electron-donating solvent aniline with theoretical simulations based on the Sumi–Marcus model. However, for systems like betaine-30, the Sumi–Marcus model was found to grossly underestimate the observed rates exceeding τ_L^{-1} by several orders of magnitude [151–160]. Clearly, this is where the quantum modes come into play, as pointed out by Jortner and co-workers [96, 97]. Walker et al. [156b] proposed a "hybrid" model including a slow solvent polarization mode, a classical vibrational mode, and a quantum vibrational mode. In such a way, they were able to predict rates close to the experimental values for betaine in a wide range of solvent environments. Figure 9.22 shows the comparison of the theoretical models discussed with experiment [156]. This kind of activity aimed at improving the Sumi–Marcus model is very similar in a way to the quantum modifications of the Marcus transition state formula, only now the sink function is modified. Since we are still within the local Golden Rule limit, we can write a general expression encompassing all these theories by simply representing the sink by the full quantum mechanical Golden Rule rate constant of Eq. (9.5) [81a],

$$W_{12}(x_s) = k_Q(V, \Delta G + E_r^s - x_s, J_f(\omega)) \qquad (9.80)$$

where $J_f(\omega)$ corresponds to the part of the relaxation spectrum regarded as fast.

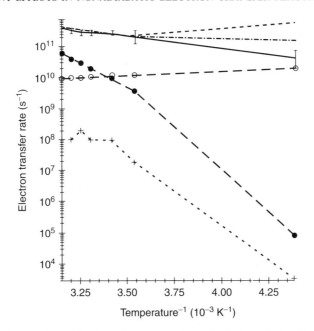

Figure 9.22. Arrhenius plot of experimental data [156b] (dash–dotted) and several theoretical predictions for the ET reaction rate in betaine-30 in glycerol triacetine: Sumi–Marcus [89] (+), Jortner–Bixon [96a] (●), Bixon–Jortner [96b] (○), Walker et al. [156b] (dashed), and Fuchs–Schriber [127] (solid and error bars). The latter is based on the Markovian discretized reduced density matrix approach. (Reproduced from [127] with permission. Copyright (1996) by the American Institute of Physics.)

One should keep in mind that the rate constant increases monotonically with coupling only if the relaxation time of the fast mode is infinitely small. If it is finite, as it should be, the rate versus coupling dependence starts as the Golden Rule prescribes, then continues according to the Sumi–Marcus theory, but finally saturates at a certain value determined by the relaxation time of the fast mode. This effect is illustrated in Figure 9.18(b) for the case where both fast and slow modes relax exponentially. In this case, the problem is reduced to solution of a system of 2D diffusion equations along parabolic potential surfaces corresponding to two coordinates x_s and x_f,

$$\frac{\partial}{\partial t}\rho_1(x_s, x_f, t) = [\mathscr{L}_1^s + \mathscr{L}_1^f - w_{12}(x_s, x_f)]\rho_1(x_s, x_f, t) + w_{21}(x_s, x_f)\rho_2(x_s, x_f, t) \tag{9.81}$$

and similarly for $\rho_2(x_s, x_f, t)$, coupled by a δ-functional sink term,

$$w_{12}(x_s, x_f) = w_{21}(x_s, x_f) = \kappa\delta(x_s + x_f - \Delta G) \tag{9.82}$$

Here, the diffusion operators are defined as before but now supplied with appropriate indexes s or f for all the parameters involved. Standard finite-difference scheme was used to evaluate the lowest eigenvalue for this problem. In the normal regime, the reaction rate is always saturated for high coupling values. When $\tau_f = \tau_s$, the activation energy is determined by the total reorganization energy E_r. When $\tau_f/\tau_s \to 0$, it is determined by the reorganization energy E_r^s of the slow component, which is smaller, and so the corresponding limiting rate constant k_{12}^d is larger. By varying the ratio of τ_f/τ_s from 0 to 1, the rate constant is changed monotonically between the two limits. We will return to this dependence later in a different context but note for now that the transition state expressions (9.73) for the sink terms are justified when τ_f/τ_s is not just less but $\ll 1$.

A considerable improvement of the effective sink approximation can be achieved if the sink terms for the slow mode are determined from solution of the coupled reaction–diffusion equations for the fast mode [81d, 142]. Assume that we know how to calculate the decay when the slow mode is frozen, that is, we know the rate kernels $K_{12,21}^{\text{eff}}(t, x^{\ddagger} - x_s)$ [e.g., Eq. (9.50) for exponential relaxation]. Then, instead of Eq. (9.52) we write [307],

$$\frac{\partial}{\partial t}\rho_1(x_s, t) = \mathscr{L}_1\rho_1(x_s, t) - \int_0^t dt' K_{12}^{\text{eff}}(t - t', x^{\ddagger} - x_s)\rho_1(x_s, t')$$
$$+ \int_0^t dt' K_{21}^{\text{eff}}(t - t', x^{\ddagger} - x_s)\rho_2(x_s, t') \qquad (9.83)$$

and similarly for $\rho_2(x_s, t)$. If the timescales of the fast and slow components are very different, one can use the long-time rate constant $k_{12}(x_s)$ instead of the rate kernel.

Bicout and Szabo [142] found that the effective sink approximation works very well for the case of biexponential relaxation. They neglected the memory effects, however, and used the mean rate constant for the fast stage instead of $k_{12}(x_s)$, which resulted in some discrepancies for the activationless reaction (see Fig. 9.14), where the decay is nonexponential even when the correlation function is exponential. Similar applications of the effective sink approximation can be found in the literature [311–315].

Consideration of the solvent relaxation with a fast inertial component and a slow exponential component led to the conclusion that the reaction kinetics is mainly governed by the fast Gaussian-correlated component, while being only weakly sensitive to the presence of the slowly relaxing component in the activationless regime (see Fig. 9.23), in accord with experiment [156]. This conclusion is supported by the observation of the crossing statistics in the mixed relaxation regime: A typical "waiting time"

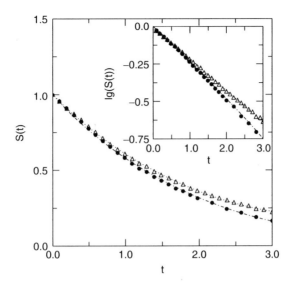

Figure 9.23. Survival probability $S(t)$ of an irreversible activationless ET reaction [$P_1(t)$ in our notation] in the solvent-controlled regime ($\kappa \to \infty$) obtained by Monte Carlo sampling for the Gaussian-correlated process (●) and the two-component process with $C(t)/C(0) = 0.8 \exp[-(t/\tau_g)^2] + 0.2\exp[-t/(10\tau_g)]$ (△) and $C(0) = 16$ in both cases. Time is in units of τ_g. Inset: the same data replotted on a semilog scale. (Reproduced from [120b] with permission. Copyright (1995) by Elsevier Science.)

for a single crossing, which should be the dominant timescale of the reaction, turns out to be of the order of Ω^{-1}, not τ_L [120]. One should admit that in Figure 9.23 the comparison is not entirely fair. First, the Gaussian component has an 80% amplitude, and second, the decay is shown only down to 0.2 for the survival probability, so that nothing can be said from this plot about the long-time rate constant. Anyway, it is clear that ultrafast solvent modes can strongly enhance ET, particularly if the initial condition is nonequilibrium [84, 103–107, 173]. Besides nonequilibrium solvent effects, a variety of nonequilibrium intramolecular vibrational effects have been observed for ultrafast ET reactions, such as vibrational cooling of the hot ET products and coherent oscillations [153a,b, 154, 173, 316].

C. Variational Transition State Theory

1. Solvent Control Regime

Analysis of the Markovian model of ET with a fast mode has shown that in the normal region the rate constant saturates for high coupling and the saturation value is determined solely by the dynamics and the reorganization energy of the slow solvent mode. However, by analyzing the model with biexponential relaxation it has also become clear that the saturation value of the rate constant changes monotonically when the ratio of the relaxation times is varied. Thus, separation of the relaxation dynamics into fast and slow modes is not that trivial. A general procedure that can handle this problem to a good degree of accuracy is offered by the variational transition state theory (VTST) developed in the context of ET by Pollak and co-workers [136, 317]. In order to understand the main idea of the VTST, we will have to change our way of thinking.

In the usual formulation [79, 81a, 89], the slow mode is diffusive with diverging average velocity. However, due to the presence of the strong sink there are virtually no recrossings of the transition point and once the system has overcome the barrier, it may be considered to be in the product state. Therefore, instead of diffusive we may search for a collective quasiballistic (ballistic within the transition region) slow mode by discarding a high-frequency part of the relaxation spectrum. The simplest way to do so is just to introduce a certain cutoff frequency λ, and the slow mode will correspond to the part of the spectrum $J(\omega)$ with $\omega < \lambda$, that is,

$$J_s(\omega) = J(\omega)\theta(\lambda - \omega) \tag{9.84}$$

where $\theta(x)$ denotes the step function. This sepration of the relaxation spectrum is quite natural within the oscillator model of ET. As we recall from Section II.A, if the collective solvent polarization follows the linear response, the reaction coordinate can be effectively represented by a set of harmonic oscillators. Equation (9.84) just defines the slow coordinate in terms of low-frequency oscillators and discards the high-frequency ones. Average velocity no longer diverges, it decreases with decreasing λ.

Now let us write the usual transition state expression for the rate constant,

$$k_{12}^d(\lambda) = \frac{\Omega_s}{2\pi} \exp(-E_a^s/k_B T) \tag{9.85}$$

where

$$E_a^s = E_a^s(\lambda) = \frac{(\Delta G + E_r^s)^2}{4E_r^s} - k_B T \ln 2 \tag{9.86}$$

and E_r^s and Ω_s are defined in a usual way in terms of the spectral density $J_s(\omega)$, see Eqs. (9.6), (9.63). The second term in the right-hand side of Eq.

(9.86) arises because we have to use the effective potential of Eq. (9.75), which predicts lower activation energy than the cusped potential [81c], and $k_B T \ln 2$ is the difference between them at the crossing point. One should note that the maximum of the effective potential corresponds strictly to the crossing point only in the symmetric case ($\Delta G = 0$), but the error of the above estimation of the activation energy for $\Delta G \neq 0$ is negligible.

Both Ω_s and E_r^s decrease with decreasing λ, although in a different way. We have already seen in Section III.B.3 that discarding a negligibly small (in terms of the reorganization energy) high-frequency part of the Debye spectrum at $\lambda > \tau_L^{-1}$ significantly reduces the average velocity, and hence alters the trajectories (see Fig. 9.15). The residual mode indeed appears to be quasiballistic in the vicinity of the crossing point. Decrease in E_r^s leads to a decrease in the activation energy, and thus we have two opposing tendencies in the TST expression (9.85): a decreasing frequency factor and an increasing activation factor. Since TST is known to give an upper bound for the rate constant, this leaves us with the only reasonable choice for the cutoff frequency λ—the one that minimizes $k_{12}^d(\lambda)$. As we shall see below, this "naive" approach works surprisingly well.

Note that quasiballistic is an effective more than a rigorous definition. Consider the mean free path, $l = \langle x_s^2 \rangle \tau_L^{-1} \langle \dot{x}_s^2 \rangle^{-1/2}$, defined as the product of the mean squared velocity $\langle \dot{x}_s^2 \rangle^{1/2}$ and the velocity relaxation time $\tau_v = \Omega_s^{-2} \tau_L^{-1}$. It will certainly increase with respect to its zero value for pure diffusion after the high-frequency part of the spectrum is cut off, and the trajectories will become smooth, but one cannot say that l will be much larger than the transition region. For example, by applying the above minimization procedure to the Debye spectrum, the slow coordinate with the mean free path of about $k_B T$ is obtained. Recrossings do not occur along the slow coordinate not because it is really ballistic but because of the effective sinks formed due to the fast coordinate. For each value of the slow coordinate, reactant and product states are equilibrated along the fast coordinate, as in Figure 9.16, and once the crossing point of the effective slow-coordinate potential profiles is passed, this equilibrium shifts toward the products. After the transition to the product state has taken place along the fast coordinate, the corresponding potential will guide the system one-way downhill along the slow coordinate toward the final equilibrium.

A more elaborate approach, which involves global functional minimization of the rate constant, is offered by VTST [136, 317]. For outer-sphere ET with a harmonic bath, the reaction coordinate is represented as a linear combination of bath coordinates. The coefficients are optimized variationally to minimize the one-way reactive flux at the crossing point of two diabatic surfaces. A clear explanation of this optimization procedure has been presented recently by Benjamin and Pollak [136c]. The basic idea of the method is the same as discussed above—to cut off an irrelevant high-frequency part of the relaxation spectrum and VTST proves that this should optimally be a soft rather than an abrupt cutoff. It can be shown that the VTST optimization procedure is equivalent to the following transformation of the spectrum [309]:

$$J_s(\omega) = \left(\frac{\Omega_s^2 + \lambda^2}{\omega^2 + \lambda^2}\right)^2 J(\omega) \tag{9.87}$$

where λ now is the soft cutoff parameter, which minimizes the TST rate constant of Eq. (9.85). It is practically convenient to rewrite all the definitions in terms of the Laplace transform of the full correlation function,

$$\hat{C}(s) = \int_0^\infty dt \exp(-st) C(t) = \frac{2k_B T}{\pi} \int_0^\infty \frac{d\omega}{\omega} \frac{s J(\omega)}{\omega^2 + s^2} \tag{9.88}$$

We obtain,

$$\Omega_s^2 = \lambda^2 \frac{\hat{C}(\lambda) + \lambda \hat{C}'(\lambda)}{\hat{C}(\lambda) - \lambda \hat{C}'(\lambda)} \tag{9.89}$$

$$E_r^s = \frac{1}{k_B T} \frac{\lambda \hat{C}(\lambda)}{\hat{C}(\lambda) - \lambda \hat{C}'(\lambda)} \tag{9.90}$$

where prime denotes the derivative. The VTST procedure gives a nonlinear equation for λ in terms of $\hat{C}(\lambda)$ [136]. An alternative, albeit completely equivalent, way is just to find the minimum of k_{12}^d as a function of λ for a given spectrum (3.56) [309],

$$k_{12}^{\text{VTST}} = \min_\lambda k_{12}^d(\lambda) \tag{9.91}$$

Let us consider biexponential relaxation as an example,

$$C(t) = 2k_B T \sum_{i=1}^2 E_{ri} \exp(-t/\tau_i) \tag{9.92}$$

Note that we intentionally used different subscripts than before (1 and 2 instead of s and f) because now separation into slow and fast coordinates is different. Figure 9.24 compares numerically exact rate constant k_{12}^d as a function of $\eta = \tau_2/\tau_1$ against VTST results and our simple spectrum separation procedure. The agreement is quite reasonable, which makes us believe

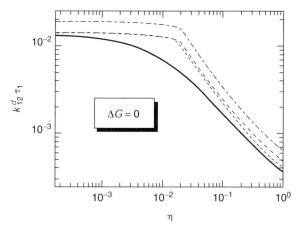

Figure 9.24. Long-time rate constant of forward symmetric ET reaction ($\Delta G = 0$) in the strong coupling limit for the case of biexponential relaxation with $E_{r1} = E_{r2} = 15k_B T$ as a function of the time ratio $\eta = \tau_2/\tau_1$ (τ_1 is fixed). Solid line corresponds to the numerical solution of the 2D reaction–diffusion equation (9.81), dashed line represents VTST with the activation energy given by Eq. (9.86), dotted line is VTST with the activation energy given by Eq. (9.93), and dash–dotted line is the result of the direct spectrum separation TST approach. (Reproduced from [309] with permission. Copyright (2001) by the American Institute of Physics.)

that such an approach should work well in the normal region for any realistic relaxation dynamics, provided that the activation energy is sufficiently high. Starobinets et al. [136d] already reported good agreement of VTST derived in the framework of the generalized Langevin equation approach with numerical results for the case of Ohmic dissipation. The correlation function $C(t)$ is then given by the difference of two exponentials. Since we used a different approach in this work, based on the displaced harmonic oscillator model of ET [136c], it is instructive to compare predictions of the two VTST procedures. It can be done in the common strong damping limit (Debye relaxation). Our predictions for the rate constant turn out to be somewhat highers because in the adiabatic problem the effective potential of mean force along the slow coordinate is different [313, 317c]. This results in a somewhat higher activation energy,

$$E_a^s = \frac{(\Delta G + E_r^s)^2}{4E_r^s} - k_B T \ln\left[1 + \mathrm{erf}\left(\frac{1}{2}\sqrt{E_r^f/k_B T}\right)\right] \qquad (9.93)$$

as compared to Eq. (9.86). In this way, both VTST models predict the same value for the rate constant in the Debye limit, exceeding Kramers' exact result by some 8% [317a]. The above correction to the activation energy is important only if the reorganization energy of the fast mode is small. However, by considering the overall accuracy of VTST in Figure 9.24, we can say that this correction is not significant. Note that the predictions of the direct spectrum separation procedure lie not too far from VTST. As another example of a nonoptimized but simple quasiballistic mode, one can use the expression for the renomalized barrier frequency obtained by Calef and Wolynes [83] by approximating the cusped double-well potential by the parabolic barrier.

When V is further increased beyond the range of applicability of the local Golden Rule, one should use VTST in the adiabatic basis [136]. The literature on adiabatic reactions is vast [68–70] and we are not in a position to go into any detail here. There is one thing we need to mention, though. As we stated several times in the text, Zusman's model shows natural behavior of the rate constant as a function of coupling: from the Golden Rule limit to the adiabatic limit. However, as follows from molecular dynamics simulation studies of adiabatic ET by Hynes et al. [318], Zusman's theory fails significantly even in the overdamped solvent regime, where it is supposed to work. Instead, an excellent agreement was found with the Grote–Hynes theory that assumes the reaction is governed by the solvent dynamics in the barrier top region rather than within the well [319], and a conclusion was drawn that solvent dynamical effect on the ET rate is mainly related to the short-time rather than the long-time frictional dynamics. In our opinion, the comparison with Zusman's theory in [318] was not quite fair, because some average solvation time was used instead of τ_L and full activation energy instead of the activation energy for the slow mode. Before the slow mode is correctly identified, one cannot really say that Zusman's theory will seriously fail. Since temporal separation of the solvent relaxation dynamics into fast and slow modes is generally not well defined, one should expect certain deviations. How big—remains to be tested. A consistent way of spectral separation of the slow mode is offered by VTST. It would be interesting to compare the VTST predictions with MD results. The MD transmission coefficient (the ratio of the actual rate constant to the usual TST rate constant) in model systems of [318b] was found to vary from 0.72 to 0.98 and VTST should give something <1 (due to minimization) but at the same time it is an estimate from above. Thus, it is expected to work well. A study by Starobinets et al. [136d] for the case of Ohmic dissipation using Langevin dynamics simulation has shown that VTST is consistently preferable to the Grote–Hynes theory.

2. Intermediate Regime

A reasonable interpolation between the Golden Rule limit and the solvent control limit can be achieved by the interpolation formula similar to Eq.

(9.78), that is,

$$k_{12}^{-1} = k_M^{-1} + (k_{12}^{VTST})^{-1} \qquad (9.94)$$

On the other hand, Rips and Pollak noticed [136a] that since the VTST optimization procedure effectively minimizes the number of recrossings and therefore ensures the largest mean-free path in the crossing point region for the optimized quasiballistic coordinate, this mean-free path may well be larger than the characteristic Landau–Zener length. Then it should be possible to apply the conventional Landau–Zener theory for evaluation of the transition rate, that is, the rate constant is obtained by averaging the TST flux multiplied by the nonadiabatic transmission coefficient over the Maxwell velocity distribution for the quasiballistic mode (here we somewhat simplified the original formulation by Rips and Pollak):

$$k_{12} = \int_0^\infty dv_s v_s p_H(v_s)\, \Phi(v_s, \langle v_s^2 \rangle) \frac{\exp(-E_a^s/k_B T)}{\sqrt{4\pi E_r^s k_B T}} \qquad (9.95)$$

where Φ denotes the Gaussian function and

$$p_H(v) = 2p_{LZ}[1 + p_{LZ}]^{-1} \qquad (9.96)$$

is the Holstein nonadiabatic transmission coefficient that accounts for multiple recrossings during one passage of the crossing point [72]. Compare this result with Eq. (9.69). Rips [136b] discussed relative merits of the diabatic versus adiabatic VTST optimization procedures and concluded that in the small coupling limit the diabatic procedure should be used.

In the limit of large κ, Eq. (9.95) reproduces the TST expression (9.85) for k_{12}^d in the solvent control regime (the integral over velocities is evaluated as $\sqrt{\langle v_s^2 \rangle/2\pi}$, or $(\Omega_s/2\pi)\sqrt{4\pi E_r^s k_B T}$, if we recall the definitions of $\Omega_s \equiv \langle v_s^2 \rangle^{1/2}/\langle x_s^2 \rangle^{1/2}$ and $\langle x_s^2 \rangle \equiv 2E_r^s k_B T$). On the other hand, in the limit of small κ one obtains,

$$k_{12} \simeq k_M(V, \Delta G, E_r^s) \qquad (9.97)$$

which can strongly differ from the correct Golden Rule result (9.10), that is, $k_M(V, \Delta G, E_r)$. Only under the condition of $E_r^f \lesssim k_B T$ will Eq. (9.97) be consistent with Eq. (9.10). Equation (9.95) will then qualitatively coincide with the simple interpolation formula (9.94) for all parameter values, which is exactly what happens for the oscillator model with Ohmic friction, considered by Rips and Pollak, where $E_r^f \sim k_B T$ even in the overdamped limit. In this case, $\langle v_s^2 \rangle \simeq x^\dagger/\tau_L$, and the crossover from the Golden Rule to the solvent control regime is determined by the original Zusman parameter ζ of Eq. (9.61). Let us emphasize again that although it is very tempting to regard the solvent control regime as being adiabatic, particularly within the VTST formulation, this is only an apparent adiabaticity. The genuine adiabaticity condition [reversed Eq. (9.32)] may not be satisfied.

3. Another Example of Apparent Adiabaticity

Before closing this section, it is instructive to consider a simple example where an apparently adiabatic transition rate between diabatic states is formed as a result of multiple crossings of the transition region. Suppose that in the vicinity of the crossing point the reaction coordinate can be represented as a sum of a slow undamped (truly ballistic) coordinate, $x_b(t) = v_b t$, with a velocity v_b, and a strongly damped rapidly fluctuating stochastic coordinate, $x_f(t)$, of large amplitude (larger than the size of the nonadiabatic transition region). Kayanuma [165] showed by analytically evaluating and summing up all terms of the semiclassical perturbation expansion with respect to κ for Gaussian Markovian fluctuations, that in the strong damping limit the transition rate between diabatic states for one passage of the slow coordinate has the following form:

$$p_K(v_b) = \frac{1}{2}[1 - \exp(-2\kappa/|v_b|)] \tag{9.98}$$

Ao and Rammer [166] obtained the same result (and more) on the basis of a fully quantum mechanical treatment. Frauenfelder and Wolynes [78] derived it from simple physical arguments. Equation (9.98) predicts a quasiadiabatic result, $p_K = \frac{1}{2}$, for $\kappa/|v_b| \gg 1$ and the Golden Rule result, $p_K = \kappa/|v_b|$, in the opposite limit, which is qualitatively similar to the Landau–Zener behavior of the transition probability but the implications are different. Equation (9.98) is the result of multiple nonadiabatic crossings of the delta sink although it does not depend on details of the stochastic process $x_f(t)$. This can be understood from the following consideration. For each moment of time, the fast coordinate has a Gaussian distribution, $\rho(x_f, t) = \Phi(x_f - x_b, \langle x_f^2 \rangle)$. When the slow coordinate approaches the transition region, the fast coordinate crosses it very frequently and thus forms an effective sink for the slow coordinate,

$$W(x_b) = \kappa \Phi(x_b, \langle x_f^2 \rangle) \tag{9.99}$$

both for forward and backward transitions. The populations of diabatic states, $P_{1,2}(t)$, obey the equation,

$$\dot{P}_1 = -\dot{P}_2 = -W(v_b t)(P_1 - P_2) \tag{9.100}$$

whose solution leads to Eq. (9.98). The limit of one-half is reached for $\kappa/|v_b| \gg 1$ because for slow passage and/or strong coupling in the presence of rapid fluctuations of large amplitude, the system looses memory of which diabatic state it has come from and becomes equally distributed between the two states after passing the crossing point. Averaging Eq. (9.98) over the velocity distribution will give the reaction rate. Qualitatively, this is the same idea as used by Rips and Pollak. Let us emphasize again that it is applicable only for small reorganization energy of the fast mode.

D. Nonadiabatic Transitions in the Strong Coupling Limit

1. Landau–Zener Approach

Padé formula (9.37) demonstrates the transiton from the Golden Rule regime to the regime of solvent control, where the ET rate is independent of V and limited by the motion along the reaction coordinate to the transition region. The transport can be governed by the coordinate diffusion or the energy diffusion, depending on the relationship between the effective frequency of solvent fluctuations, Ω, and the velocity relaxation time, τ_v. The solvent-controlled rate is formally identical to the adiabatic rate, although it may be a result of multiple nonadiabatic crossings [75–78]. In the normal region, further increase of V beyond the solvent-controlled regime leads to a genuine adiabatic situation. Nonadiabatic transitions take place on the way, and their multiple nature [170] is often described by Holstein's expression for the transmission coefficient (9.96). This finding is not entirely correct, since the Landau–Zener formula itself does not work for high coupling in the presence of dissipation [164–167]. However, this effect is negligible as long as the overall ET rate is of interest, since it is already controlled by transport to the transition state.

The situation is radically different in the inverted region, as well as in certain cases of nonequilibrium back transfer (see below), which are always nonadiabatic whatever the coupling strength is. For large V, the ET rate is no longer controlled by transport to the transition region but rather by nonadiabatic transitions between adiabatic states (see Fig. 9.1). Therefore, one should expect a decrease of the ET rate with increasing V to follow the solvent-controlled plateau. Usually, the Landau–Zener formula is used for the description of nonadiabatic transitions in the classical limit [162, 163].

One obtains,

$$k_{na} = \langle p_{LZ}^{ad}(v) \rangle_v \Phi(x^\ddagger, \langle x^2 \rangle)$$

$$\simeq \left(\frac{2}{3\pi}\right)^{1/2} \Omega \, (2\pi\bar{\xi})^{1/3} \exp\left[-\frac{3}{2}(2\pi\bar{\xi})^{2/3}\right] \exp\left(-\frac{E_a}{k_B T}\right) \quad (9.101)$$

which is nothing but the thermally averaged transition probability between adiabatic states, $p_{LZ}^{ad}(v) = 1 - p_{LZ}(v) = \exp(-\kappa/|v|)$, multiplied by the corresponding activation factor, $\Phi(x^\ddagger, \langle x^2 \rangle) = (4\pi k_B T E_r)^{-1/2} \exp(-E_a/k_B T)$. Compare this result with the nonadiabatic transition rate between diabatic states, given by Eq. (9.69), where the transition probability is $p_{LZ}(v)$. Approximation in the right-hand side of Eq. (9.101) is obtained in the limit of large V using the steepest descent method [162]. Here $\bar{\xi} = V^2/\hbar \langle v^2 \rangle^{1/2}$ is the average Massey parameter, as in Eq. (9.32). A reasonable interpolation between the weak and strong coupling limits is achieved by writing the transition probability as $p_{LZ}(v)[1 - p_{LZ}(v)]$, which accounts for double crossing of the transition region [73].

2. *Breakdown of the Landau–Zener Theory in the Presence of Relaxation*

The Landau–Zener model assumes the ballistic motion along the reaction coordinate with constant velocity in the vicinity of the crossing point [73]. The applicability condition of this approach is

$$E_{MFP} \gg E_{tr} \quad (9.102)$$

which means that the mean free path of the classical motion, E_{MFP}, should be much larger than the region of transitions, E_{tr}. It is usually assumed that transitions occur within the region of strong interaction of size V [85]. It can be shown, however, that the transition region is $> V$ in both adiabatic and nonadiabatic conditions [167]. If the Massey parameter $\xi = V^2/(\hbar v)$ is small, the transition probability is formed within the nonadiabatic region, $E_{tr} = \sqrt{\hbar v}$. In the adiabatic case, $\xi \gg 1$, the population of the target adiabatic state goes through a pronounced maximum in time before the final exponentially small Landau–Zener value $p_{LZ}^{ad}(v)$ is reached on distances as large as

$$E_{tr} = V\xi^{-1/3} \exp(\pi\xi/3) \quad (9.103)$$

Both situations are illustrated in Figure 9.25. It can be shown that perturbation of the ballistic motion by additive noise can cause dramatic changes in the transition probability [164–167]. So the applicability of the

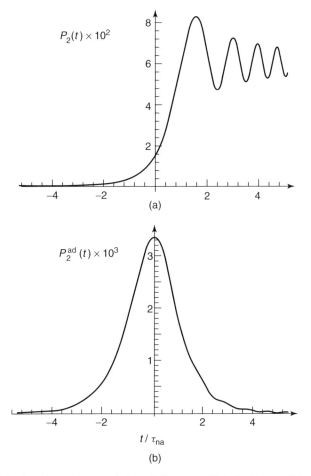

Figure 9.25. Time-dependent population of the target diabatic (*a*) or adiabatic (*b*) state obtained from exact solution of the Zener problem for two values of the Massey parameter: $\xi = 0.01$ [(*a*) nonadiabatic crossing] and $\xi = 2.25$ [(*b*) adiabatic crossing]. Time is in units of $\tau_{na} = \sqrt{\hbar/v}$.

Landau–Zener formula in the presence of relaxation is very much limited. The most significant effect on the ET rate is expected in the inverted region, where the exponential decrease with V predicted by Eq. (9.101) should become slower.

Relaxation effects in level crossing can be modeled using a stochastic approach, where an appropriate noise term is introduced into the Zener Hamiltonian [165]. A more rigorous treatment is possible in the limit of very large coupling, where a new small parameter of nonadiabatic interaction emerges providing the opportunity to take advantage of the Golden Rule [130, 164, 166–168].

3. Back to the Golden Rule

For large coupling ($\bar{\xi} \gg 1$), ET has to be considered in the adiabatic representation (see Fig. 9.1). Instead of H_1, H_2, and V we introduce the Hamiltonians of the ground and excited adiabatic states [298],

$$H_{g,e} = \frac{1}{2}(H_1 + H_2) \mp \frac{1}{2}\hbar\Omega_{ad}(X) \qquad (9.104)$$

and the nonadiabatic interaction operator,

$$V_{na} = \frac{i\hbar}{2}\left[\dot{X}, \frac{V}{\hbar^2\Omega_{ad}^2}\right]_+ \qquad (9.105)$$

where we returned to the original coordinate X and defined the gap between the adiabatic states as follows:

$$\hbar\Omega_{ad}(X) = [(X - \Delta G)^2 + 4V^2]^{1/2} \qquad (9.106)$$

The reaction rate can be obtained using the perturbation theory in V_{na}, that is,

$$k_{na} = k_{eg} = \frac{2}{\hbar^2}\mathrm{Re}\int_0^\infty dt \left\langle \exp\left(\frac{i}{\hbar}tH_e\right) V_{na}^* \exp\left(-\frac{i}{\hbar}tH_g\right) V_{na} \right\rangle_{H_e} \qquad (9.107)$$

Dykhne used this approach to calculate the semiclassical nonadiabatic transition probability for the ballistic case in the strong interaction limit [320]. His final result,

$$P_{ad} = \frac{\pi^2}{9}\exp(-\kappa/|v|) \qquad (9.108)$$

essentially coincides with the Landau–Zener solution except for some 10% difference in the numerical prefactor. The origin of this discrepancy was shown to be in the absence of a small parameter for the perturbation theory [321].

The perturbation theory approach is well justified in the presence of friction. Shushin [164a] used this approach in the reaction rate calculation within the semiclassical approximation, where the reaction coordinate can be considered as a stochastic process. The corresponding Golden Rule formula,

$$k_{eg} = \frac{2}{\hbar^2} \int_0^\infty dt \left\langle V_{na}(t) \exp\left[-2i \int_0^t dt' \Omega_{ad}(t')\right] V_{na}(0) \right\rangle_{H_e} \quad (9.109)$$

can be handled analytically in the two limiting cases. In the case of low friction (ballistic regime), Eq. (9.109) reduces to Eq. (9.101), that is, thermally averaged Landau–Zener result. In the opposite limit of high friction (diffusive regime), where the mean free path along X is $\ll V$ and the velocity relaxation is very fast, the reaction coordinate can be considered as static during formation of the rate and the following approximate formula can be obtained

$$k_{eg} = \int dX \rho_e(X) W_{eg}(X) \quad (9.110)$$

where $\rho_e(X)$ is the coordinate distribution for the excited adiabatic state and $W_{eg}(X)$ is the coordinate-dependent sink,

$$W_{eg}(X) = \frac{V^2}{\hbar^4 \Omega_{ad}^4(X)} \tilde{R}[\Omega_{ad}(X)] \quad (9.111)$$

with $\tilde{R}(\omega)$ denoting the Fourier transform of the velocity correlation function, $R(t) = \langle \dot{X}(t)\dot{X}(0) \rangle$, that is,

$$\tilde{R}(\omega) = \int_{-\infty}^\infty dt R(t) e^{i\omega t} \quad (9.112)$$

The correlation function can be expressed in terms of the relaxation spectrum in a usual way, via $R(t) = -\ddot{C}(t)$, with $C(t)$ given by Eq. (9.3). It is straightforward to derive a quantum extension of Eq. (9.110) [168].

In order to interpolate between the two limiting cases, Shushin used an approximation,

$$\exp\left[-2i \int_0^t dt' \Omega_{ad}(t')\right] \approx \exp(-2iVt/\hbar) \quad (9.113)$$

Detailed analysis conducted within this approximation shows that Eq. (9.110) works well in a much wider range than a strictly diffusional regime, while the range of validity of the averaged Landau–Zener formula (9.101) is very narrow [167b]. Nonadiabatic rate decays much slower with V in the presence of relaxation than is predicted by the Landau–Zener approach. For example, in the case of exponential velocity relaxation, $k_{eg} \sim \bar{\xi}^{-3/2}$. A reasonable interpolation between the limits of strong and weak relaxation can be obtained by simply summing up the two rates [167b]. The effect of transport to the transition state can be included in a usual way by adding the corresponding inverse rate constants.

The above ideas are illustrated in Figure 9.26, where the reaction rate prefactor, k_{12}/k_{TST}, often referred to as the transmission factor, is plotted as a function of coupling V (in terms of the average Massey parameter $\bar{\xi}$) for different values of the diffusion factor, $h = k_d/k_{TST}$. Here $k_d = \hat{k}_d(0)$ of Eq. (9.51), assuming a simplified form of Eq. (9.55) for high activation energies, and $k_{TST} = (\Omega/2\pi)\exp(-E_a/k_BT)$, as usual. The following interpolation expression for the rate constant in the whole range of V was used

$$\frac{1}{k_{12}} = \frac{1}{k_M} + \frac{1}{k_d} + \frac{1}{k_{LZ} + k_R} \tag{9.114}$$

where k_M is the Marcus rate for small V and the last term corresponds to

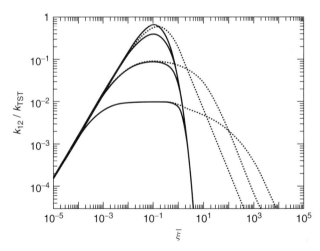

Figure 9.26. Transmission coefficient k_{12}/k_{TST} in the inverted region as a function of the Massey parameter, $\bar{\xi} = V^2/\hbar \langle v^2 \rangle^{1/2}$, for different values of the diffusion factor, $h = k_d/k_{TST} = \infty$ (effectively, meaning that $k_{12} = k_{na}$), 1, 0.1, and 0.01 (top to bottom). Solid lines correspond to the Landau–Zener approach and basically reproduce a plot from [163b]. Dotted lines illustrate the effect of relaxation estimated using the adiabatic Golden Rule formula for exponential velocity correlation function, as explained in the text.

k_{eg} given by a sum of the thermally averaged Landau–Zener transition probability k_{LZ} of Eq. (9.101) and the relaxation counterpart in the diffusive regime, k_R of Eq. (9.110), for exponential velocity correlation function, $R(t) = \langle v^2 \rangle \exp(-\gamma t)$, with $\gamma = \tau_v^{-1}$. Georgievskii et al. [163b] discussed this kind of interpolation for the total ET rate by neglecting the effect of relaxation on nonadiabatic transitions in the strong-coupling limit. They used the following formula: $k_{12}^{-1} = k_d^{-1} + k_{na}^{-1}$, where k_{na} was defined as a thermally averaged $p_{LZ}(v)[1 - p_{LZ}(v)]$ factor, which accounts for double crossing of the transition region. This is not much different from our Eq. (9.114) with k_R set to zero, in a sense that all physically relevant limits are represented in exactly the same fashion. In order to calculate k_R for exponential velocity relaxation we need to make one more approximation, that is, to assume that $\rho_e(X)$ is constant in the transtion region (or, in other words, the transition region is small). Note that this assumption is a consequence of our choice of a model relaxation spectrum [164a, 167]. It would not be needed if a real spectrum was taken [168]. Thus we obtain,

$$\frac{k_R}{k_{TST}} = (2\bar{\xi})^{-1} \left(\frac{\pi}{2\alpha}\right)^{3/2} (\alpha + 2[(1 + \alpha)^{-1/2} - 1]) \qquad (9.115)$$

where

$$\alpha = \frac{\hbar^2 \gamma^2}{4V^2} = \frac{1}{\bar{\xi} \hbar^2} \frac{\hbar \Omega}{k_B T} \frac{\pi E_a}{4\sqrt{2 E_r k_B T}} \qquad (9.116)$$

By considering typical values of parameters in ET, we chose $\alpha = 0.1/(\bar{\xi} h^2)$ for our illustration. Equation (9.115) predicts $k_R \sim \bar{\xi}^{-3/2}$ in the limit of $\xi \to \infty$. A good example of this asymptotic behavior is seen in Figure 9.26. We can also see that the relaxation mechanism always dominates over the Landau–Zener mechanism of nonadiabatic transitions, except for the nearly ballistic situation with almost no friction ($\gamma \to 0$ or $h \to \infty$). Therefore, in the strong damping regime the solvent-control plateau is somewhat extended. Remembering that as far as the chosen value of k_d is concerned, the above illustration can only be considered as qualitative. While in the normal region k_d is determined by the slow relaxation mode and this slow mode can be reasonably defined using VTST, one cannot use the same arguments in the inverted region for an arbitrary relaxation spectrum. Perhaps, a correct way to approach this problem would be to consider the diffusive motion along the adiabatic potential of the excited state in the presence of the coordinate-dependent sink defined by Eq. (9.111).

4. Non-Marcus Free Energy Gap Dependence

Let us demonstrate how the above theory works in practice and consider the case of photoinduced back ET as an example. Here a nonequilibrium population of the excited state is created by an optical pulse, which then gradually relaxes and transfers back to the ground state (see Fig. 9.27). In their experimental study of the recombination kinetics in contact ion pairs produced by excitation of the charge-transfer complexes, Asahi and Mataga [322a] observed no evidence of the normal region. Instead, the kinetic rate was found to depend monotonically (nearly exponentially) on the free energy gap in a large interval from -0.5 to -3 eV. Similar dependences were observed later in other systems [322b]. An explanation of this phenomenon was suggested by Tachiya and Murata [115] and subsequently elaborated upon by Gayathri and Bagchi [105b, 106]. It is based on the nonequilibrium Zusman model. However, in order to fit the data, a very high value of the electronic coupling had to be taken, $V = 0.1–0.3$ eV, which is far beyond the applicability condition of the local Golden Rule, which is at the heart of Zusman's model. Besides, exact solution of this nonequilib-

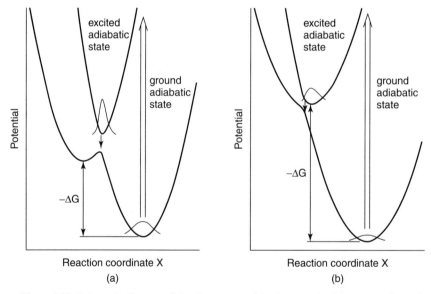

Figure 9.27. Schematic diagram of the charge recombination reaction in a contact ion pair (back ET) in the normal (*a*) and inverted (*b*) regions. Laser pulse (indicated by a block arrow) induces an initial nonequilibrium population in the excited adiabatic state, which quickly relaxes to its equilibrium distribution and then decays via stationary nonadiabatic transitions to the ground adiabatic state. (Reproduced from [168] with permission. Copyright (2000) by the American Institute of Physics.)

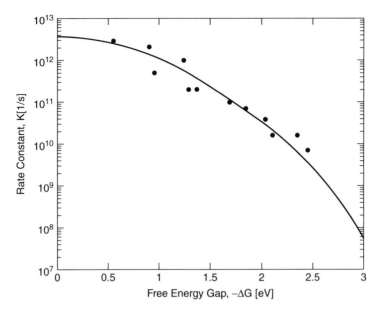

Figure 9.28. Free energy gap dependence for the rate constant of the charge recombination reaction in contact ion pairs. Circles represent experimental results of Asahi and Mataga from [322a], solid line is a result of theoretical fitting with $E_r = 1.6\,\text{eV}$ and $V = 0.195\,\text{eV}$ (details in the text). (Reproduced from [168] with permission. Copyright (2000) by the American Institute of Physics.)

rium model predicts strongly nonexponential kinetics of the ion pair population [105b, 106, 161], in contradiction with experiments where simple exponential decays were observed [322].

An alternative explanation has been suggested recently [168]. Since the electronic coupling is large, as follows from the above consideration as well as from independent measurements in similar systems [323], relevant ET states are adiabatic. An initial nonequilibrium population in the excited adiabatic state induced by the laser pulse quickly relaxes to its equilibrium distribution, as shown in Figure 9.27, and what one observes in experiment is the kinetics of stationary nonadiabatic transitions to the ground adiabatic state. In the normal region, the excited-state depopulation is activationless, while in the inverted region an activation is necessary. This difference explains qualitatively the results of Asahi and Mataga [322a]. Quantitative fit of their free energy gap dependence is shown in Figure 9.28 using the quantum mechanical extension of the Golden Rule expression (9.110). Since no information on the contribution of the intramolecular modes is available, ET was assumed to be governed only by the outer-sphere reorganization.

Experimentally determined IR spectrum of the solvent (acetonitrile) was used [202b]. As a result, quite reasonable values of $E_r = 1.6$ eV and $V = 0.195$ eV were obtained in the fit.

E. Numerical Methods

While simple approximate considerations of ET are useful in many cases, it is important to have methods for the rigorous solution of the general quantum mechanical problem. Exact methods without any approximation can be realized at present only for small molecular systems involving a few atoms, and the reason is that the finite basis used in quantum mechanical calculations grows exponentially as the number of degrees of freedom increases. Therefore, certain approximation is always necessary to handle complex systems with many degrees of freedom. One has to balance between the accuracy and the computational cost of the calculation. Considerable progress in this direction has been made over the last few years [116–118, 123–134].

A very powerful and rigorous way to simulate a quantum system is the quantum Monte Carlo (QMC) algorithm. It is based on the path integral representation of quantum mechanics and prescribes to perform relevant summations over stochastically chosen system trajectories. The main difficulty of this algorithm is its slow convergence because of the highly oscillating terms in the sum (dynamical sign problem). This problem can be avoided in several different ways. One possibility is to use imaginary time. The idea of applying the imaginary time QMC algorithm to nonadiabatic transitions was suggested by Wolynes [95b]. He also introduced a semiclassical approximation for this approach (nonadiabatic instanton) [95a]. The idea was realized by Cao et al. [123a–c] for the calculation of ET rates. Real time QMC methods have also been developed. Topaler and Makri [126] used real time QMC for the bath of harmonic oscillators, where the summation over the oscillator trajectories can be performed analytically. They obtained predictions for the ET reaction rate consistent with Zusman's model (see Fig. 9.29). Stockburger and Mak [124d] suggested introducing an auxiliary Gaussian stochastic process after a Hubbard–Stratonovich transformation of the path functional, which makes it more "local". Realizations of the Gaussian stochastic process can be readily obtained by taking advantage of its spectral resolution via Fourier transformation (see Fig. 9.15 for sample trajectories). This method is most powerful for the Ohmic bath [124g], where it needs no other path summation. Mak and Egger [124f,h] also recently suggested a very promising procedure to resolve the dynamical sign problem—the so-called multilevel blocking approach. For the description of the short time evolution ($tV < 10$) it is possible to use direct summation of all the paths [132]. Figure 9.30 illustrates the dynamics of

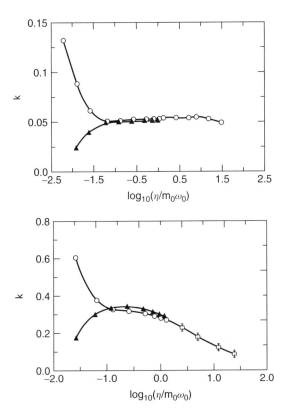

Figure 9.29. Quantum transmission coefficient κ (k_{12}/k_{TST} in our notation) as a function of the dimensionless damping strength $\eta/m_0\omega_0$ ($\Omega\tau_L$ in our notation) obtained via real time path integral QMC calculations for nonadiabatic transitions between harmonic diabatic states with frequency $\Omega = 500\,\text{cm}^{-1}$ and electronic coupling $V = 0.1\hbar\Omega$ in the presence of Ohmic dissipation with exponential cutoff at Ω. Circles and triangles correspond to $-\Delta G = 0$ and $0.5\hbar\Omega$, respectively, $k_B T = 0.64\hbar\Omega$ (*a*) and $0.32\hbar\Omega$ (*b*). Quantum modes manifest themselves at low damping. In the intermediate damping regime at high temperature one observes the Golden Rule plateau. It becomes narrower and raises up due to quantum effects when the temperature is lowered. Large friction leads to solvent control. (Reproduced from [126] with permission. Copyright (1996) by the American Chemical Society.)

back ET reaction for large electronic coupling, where both the exact enumeration method and an approximate transfer matrix path integral approach of Makarov and Makri [125] are compared. As a matter of fact, this work of Evans et al. [132] is one of the rare simulations in the inverted region. Note that coherent oscillations observed in Figure 9.30 are not directly related to those observed in experiments on ET in photoexcited mixed-valence compounds [173, 316]. They are quantum beats resulting

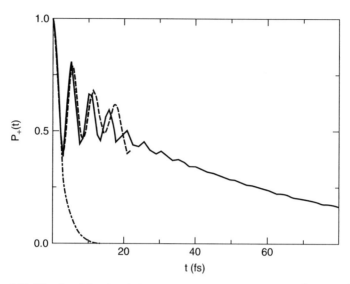

Figure 9.30. Kinetics of the photoinduced back ET reaction for $(NH_3)_5Fe^{II}(CN)(Ru)^{III}(CN)_5$ in water at 300 K with $V = 2500\,cm^{-1}$, $-\Delta G = 3900\,cm^{-1}$, and $E_r = 3800\,cm^{-1}$. Ohmic spectral density with exponential cutoff at $220\,cm^{-1}$ is assumed for the solvent. Exact enumeration method (dashed) and the transfer matrix path integral approach (solid) are compared with the Golden Rule prediction (dash–dotted). (Reproduced from [132] with permission. Copyright (1998) by the American Institute of Physics.)

from diabatic initial condition. Since electronic coupling is present in the system irrespective of photoexcitation, perhaps adiabatic initial condition should be chosen.

Simulations of quantum systems are always computationally demanding, which explains the appearance of numerous semiclassical methods for the description of the processes involving nonadiabatic transitions [101, 111, 129, 134, 143, 324]. The applicability conditions of these methods should be checked in each particular case [114, 325]. For example, the Markovian kinetic equations for a quantum two-state system coupled to a classical bath widely used to describe ET beyond the strict classical limit [101, 141, 143, 326], were shown to face certain difficulties for $E_r > k_B T$ (negative reaction rates and detailed balance violation) [114]. Semiclassical initial value representation combined with the classical analogue model for the discrete quantum states suggested by Wang et al. [134] looks most promising. In particular, they reproduced the solvent-controlled regime for the spin-boson Hamiltonian with Debye relaxation spectrum. The most recent development in this direction is the self-consistent hybrid method [327]. The overall system is first partitioned into a "core" and a "reservoir". The former is treated via an accurate quantum mechanical method, and the latter — via a more approximate method, such as semiclassical initial value representation. The number of "core" degrees of freedom, as well as other variational parameters, is systematically increased to achieve numerical convergence for the overall quantum dynamics.

Due to intrinsic computational complexity of quantum and even semiclassical calculations only the simplest models of bath relaxation are usually employed, namely, the Debye and Ohmic models, as discussed in Section II.A.2. It should be emphasized that these models fail to reproduce Marcus's energy gap dependence in the inverted region. For example, the Debye model predicts in the Golden Rule limit that $k_{12} \sim (\Delta G + E_r)^{-3}$ for $\Delta G \ll -E_r$ [21]. The Ohmic model gives minus fifth power. This illustrates the importance of the cutoff function for the spectral density.

IV. EFFECT OF DIFFUSION ON ET OBSERVABLES

A. Bulk versus Geminate Reaction

A major prediction of the classical Marcus theory is the free energy gap law. It was checked experimentally quite a number of times and appears to be valid for the rate constants spanning over as much as 13 orders of magnitude [249a]. Intramolecular ET and intermolecular ET in solids, where the distance between donor and acceptor is fixed, show well-developed normal and inverted regions [8, 9]. Possible asymmetry of the gap law is usually explainable in terms of an effective high-frequency quantum mode. In liquid solutions, the situation is far less optimistic. Even presently, no clear manifestation of the inverted region has been reported for the charge separation reaction (forward ET) [15, 19], while the normal region is not always observed in geminate ion recombination (back ET) [10]. It was realized very early that all these complications arise due to the distance dependence of the ET rate and the masking role of diffusion [14, 31, 169, 328, 329]. What is actually observed in experiment is not the ET rate but the overall bimolecular reaction rate that becomes diffusion-limited in the region of favorable ΔG values for the bulk charge separation process [14, 17]. Any quantitative characteristics of geminate recombination are even more indirectly related to the ET rate [330, 331]. It was shown that there can be a significant diffusional distortion of what is an effective measure of the back ET efficiency and is expected to follow the free energy gap law [18b]. A series of recent elegant experiments on bimolecular charge shift reactions have demonstrated that by carefully selecting a system one can elevate the diffusion-controlled limit high enough (and/or reduce the ET rate low enough) to observe an undistorted Marcus parabola [13] We will discuss these strategies below.

Our main focus here is on two typical situations arising in experimental studies of intermolecular ET in solutions. One is the bulk bimolecular reaction, where many particles are involved, and the other is the geminate reaction in an isolated donor–acceptor pair. For definiteness, we consider photoinduced ionization followed by geminate recombination that proceeds according to the scheme:

$$D^* + A \to [D^+ \cdots A^-] \tag{9.117}$$

$$[D \cdots A] \leftarrow [D^+ \cdots A^-] \to D^+ + A^- \tag{9.118}$$

The first bulk stage is usually monitored by time-resolved or steady-state donor fluorescence and can be considered independently of the second stage. On the contrary, geminate recombination should be considered together with the precursor stage that generates the initial condition for the back ET. Having escaped geminate recombination (and effectively having lost memory of where they were created), the charges may still recombine at later times with each other or with other charges in solution. This bimolecular recombination stage is neglected here for simplicity, although it can be readily included [332]. Theoretical approaches to the kinetics of multistage diffusion-influenced reactions, ET in particular, have recently been comprehensively reviewed by Burshtein [333].

1. Bulk Reaction

The simplest theory of diffusion-assisted ET assumes that the reaction occurs only when donor and acceptor make contact (Collins–Kimball or the "gray sphere" model) [334–336]. Some experiments were analyzed on the basis of such theory [16, 337]. However, according to the Marcus expression, the ET rate can exhibit a peculiar dependence on the interparticle distance [17, 329, 338] as a result of the interplay between different dependencies of V, E_r, and ΔG, as shown in Figure 9.31. While in the normal region conventional exchange-type exponential dependence is at least qualitatively valid, $k_M(r)$ is generally nonmonotonic and acquires a bell-shaped form with increasing exothermicity.

The theory of diffusion-assisted noncontact reactions was developed [339] and applied to ET [31b] long ago. If the acceptor concentration c is sufficiently small, the normalized population of excited donors $P(t)$ obeys the non-Markovian kinetic equation [339],

$$\dot{P}(t) = -[ck_I(t) + \tau_D^{-1}]P(t) \tag{9.119}$$

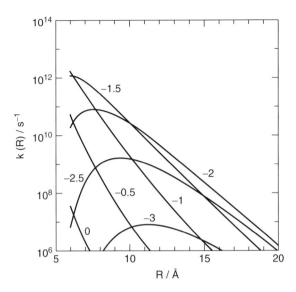

Figure 9.31. First-order rate constant $k(R) \equiv k_M(V, \Delta G, E_r)$ in acetonitrile as a function of the donor–acceptor distance R for different values of the free energy change ΔG (the values attached to the curves, in units of eV). Distance dependence of ΔG is neglected. The reorganization energy is calculated from Eq. (9.17) with $r_D = r_A = 3$ Å, $\varepsilon_\infty = 1.76$, and $\varepsilon_s = 35.92$; the electronic coupling is assumed to be given by Eq. (9.22) with $V_0 = 100 \text{ cm}^{-1}$ and $\beta = 1 \text{ Å}^{-1}$. (Reproduced from [17] with permission. Copyright (1992) by the American Chemical Society.)

where τ_D is the excitation lifetime,

$$k_I(t) = \int_{R_c}^{\infty} dr 4\pi r^2 \rho(r, t) k_f(r) \qquad (9.120)$$

is the time-dependent ionization rate constant, $R_c = r_D + r_A$ is the contact distance, $k_f(r)$ is the Marcus rate constant for the forward ET at a given distance r, and $\rho(r, t)$ is the pair distribution function satisfying the reaction-diffusion equation,

$$\frac{\partial}{\partial t} \rho(r, t) = [\mathscr{L}_f - k_f(r)] \rho(r, t) \qquad (9.121)$$

$\mathscr{L}_f$ is the usual radial diffusion operator,

$$\mathscr{L}_f = \frac{D_f}{r^2} \frac{\partial}{\partial r} r^2 \exp[-U_f(r)/k_B T] \frac{\partial}{\partial r} \exp[U_f(r)/k_B T] \qquad (9.122)$$

$D_f = D$ is the mutual diffusion coefficient, and $U_f(r)$ is the interaction potential assumed to be zero for the process (9.117). Equation (9.121) should be solved with the uniform initial condition, $\rho(r, 0) = 1$, and the reflecting boundary condition at encounter. An equivalent formulation defines $P(t)$ in terms of the pair survival probability [335, 340], a function of time and initial separation, which also satisfies the reaction-diffusion equation but with the adjoint operator $\mathscr{L}_f^\dagger$ ($\mathscr{L}^\dagger = \mathscr{L}$ in the absence of interaction).

There have been numerous attempts to handle this kinetic scheme analytically or by simple numerical means by assuming various model functions for $k_f(r)$, such as a step [338c, 341, 342], an exponential [328, 339d, 343], or a bell-shaped function [338c]. Several approximations have been designed for an arbitrary functional dependence of $k_f(r)$ [302, 344]. The most popular so far is the Wilemski–Fixman closure approximation [302], which is essentially the [1, 0]-Padé approximation discussed in detail in the preceding section, albeit in a little different context. Detailed analysis shows [161, 303, 304] that although the closure approximation is exact for contact reactions (reducing to the Collins–Kimball model [334] in this limit) and works well for short-range reactions, it progressively deviates from exact results as the sink becomes wider and diffusion becomes slower. Therefore, in order to be able to extract reliable ET parameters involved in the Marcus expression from experimental kinetic data, one has to solve the whole problem numerically. It is not too difficult a task and this approach has become a common practice nowadays [18b, 19, 159, 330b, 345, 346]. We would like to advertise in this respect a very efficient program implemented by Krissinel and Agmon [347].

Qualitatively, what happens is that the reaction starts with the kinetic rate,

$$k_0 \equiv k_I(0) = \int_{R_c}^{\infty} dr\, 4\pi r^2 k_f(r) \qquad (9.123)$$

the one that enters the Collins–Kimball boundary condition in the contact model. A growing hole is "burnt" in $\rho(r, t)$ around contact, which spreads out statically [348, 349], that is, with no diffusion basically occurring, until its radius reaches the maximum value of R_Q at time R_Q^2/D. This is an effective radius for quasistationary ionization, where reaction and diffusion balance each other, with the rate constant given by

$$k_\infty = 4\pi R_Q D \qquad (9.124)$$

R_Q is not only larger than R_c but also grows monotonically with decreasing D (logarithmically [338c, 339d]), in contrast to the contact model, where $R_Q = k_0 R_c/(k_0 + 4\pi D R_c) \leqslant R_c$, and to the closure approximation that predicts saturation of R_Q for small diffusion coefficients. The binary encounter theory remains valid under the condition that R_Q is much less than the average distance between acceptors, that is, $cR_Q^3 \ll 1$.

When diffusion is fast enough to restore the equilibrium distribution, the reaction is kinetically controlled and proceeds with k_0 at all times. The parameter k_0 follows the free energy gap law of $k_f(r) \equiv k_M(V, \Delta G, E_r)$, at least qualitatively [18]. When diffusion is slow, the reaction eventually becomes diffusion controlled with $k_\infty \ll k_0$. The parameter k_∞ only weakly depends on ΔG in this limit. Therefore, for reactions with high ET rates in the region of favorable ΔG values, the top of the free energy gap parabola for the observed stationary reaction rate constant is cut by a diffusion-controlled plateau [31b]. As far as the free energy gap law is concerned, detailed model calculations are not of great significance. While R_Q may exceed R_c by a few angstroms, it is still of the same order. Thus, on the usual logarithmic scale of the rate constant the diffusion-controlled plateau is well approximated by Smoluchowski's $k_d = 4\pi D R_c$. Figure 9.32 compares

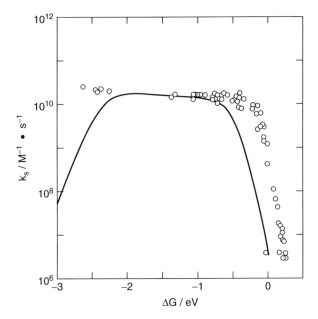

Figure 9.32. Second-order diffusion-mediated ET rate constant k_s (k_∞ in our notation) in acetonitrile as a function of the free energy change ΔG. Solid line is calculated using the classical Marcus expression for the first-order ET rate constant combined with the closure approximation to obtain the second-order rate constant for the following typical set of parameter values: $r_D = r_A = 3\,\text{Å}$, $\varepsilon_\infty = 1.76$, $\varepsilon_s = 35.92$, $V_0 = 100\,\text{cm}^{-1}$, $\beta = 1\,\text{Å}^{-1}$, and $D = 3 \times 10^{-5}\,\text{cm}^2\,\text{s}^{-1}$. Circles show experimental results of Rehm and Weller [14]. (Reproduced from [17] with permission. Copyright (1992) by the American Chemical Society.)

theoretical predictions for a typical set of ET parameter values with the experimental results of Rehm and Weller [14]. Neither this or other [234, 350] stationary experiments nor the time-resolved measurements of Nishikawa et al. [16] under similar conditions show any evidence of the inverted region even for very large negative values of ΔG, while the diffusion model always predicts a drop after the plateau, even in the presence of quantum modes [338d]. This inconsistency still has no clear explanation. It was suggested [17] that a possible reason for deviations in the normal regime is the exciplex formation [351]. Note that in principle one can easily fit the Rehm–Weller data from this side without any additional reaction channels, but either the reorganization energy should be set smaller or the frequency prefactor should be set very large meaning that ET is actually adiabatic. Note that asymmetric free energy gap law is predicted for strong coupling even in the absence of quantum modes (see Figure 9.21). The asymmetry can also be due to the effect of the solute polarizability varying in the course of electronic transition [227b]. We are not in a position here to argue which explanation is correct.

In order to observe the free energy gap law for bimolecular reactions in solution, one should be in the kinetically controlled regime [13], which can be achieved by elevating the diffusion-controlled plateau and/or by lowering the ET rate. This idea was most nicely realized by Guldi and Asmus [13d] in their recent study of ET from the ground-state fullerenes to radical cations of various arenes that showed a beautiful Marcus parabola (see Figure 9.33). The size of fullerenes is much larger than the size of arenes. If one takes into account the Stokes–Einstein relationship for the diffusion coefficient, it is clear that the diffusion-controlled rate is proportional to $(r_D + r_A)(1/r_D + 1/r_A)$, which increases with increasing difference between r_D and r_A. On the other hand, electronic coupling may become smaller for large fullerenes, thereby lowering the ET rate. For C_{76} and naphthalene, for example, a value of $k_d = 3.0 \times 10^{10} M^{-1} s^{-1}$ is obtained, which is definitely higher than the observed rate constants identifying the latter as a true measure of k_0.

2. Geminate Recombination after Photoionization

Spatial distribution of ions created by photoionization can be calculated from [330, 345b,c]

$$\sigma(r, \infty) = ck_f(r) \int_0^\infty dt \rho(r, t) P(t) \qquad (9.125)$$

assuming that there is no recombination. The parameter $\sigma(r, \infty)$ reproduces

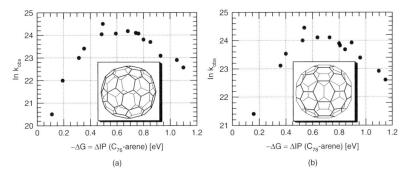

Figure 9.33. Plot of $\ln k_{obs}$ for ET from (a) C_{76} and (b) C_{78} to (arene)$^{\bullet+}$ in dichloromethane at room temperature as a function of the free energy change ΔG. The parameter k_{obs} is the observed stationary bimolecular rate constant. (Reproduced from [13d] with permission. Copyright (1997) by the American Chemical Society.)

the shape of $k_f(r)$ in the kinetically controlled limit, while the diffusion-controlled ionization results in a bell-shaped ion distribution centered at $R_Q > R_c$ (after the excitations close to contact are statically quenched, further diffusion-controlled ionization proceeds only at the border R_Q of the reaction zone). Effective ET distance increases with decreasing ΔG, as shown in Figure 9.34. Long-range ET is suggested to occur even in nonpolar solvents [345d], where the quenching mechanism is known to be different from polar solvents in a way that a contact exciplex state is formed [352].

Photoionization generates the initial condition for geminate recombination. Now the difference of a few angstroms between R_Q and R_c can be of crucial importance, increasing the survival chances of the newborn ions attracted by the Coulomb potential. The distribution of ions provided by Eq. (9.125) can be used as the initial condition only if ionization is so fast that it is completed before the recombination actually starts. In general, one should consider the backward and forward ET simultaneously, by including a nonlocal source term into the evolution equation for the ion pair probability distribution $\sigma(r, t)$, as shown by Burshtein et al. [330] and by Dorfman and Fayer [331]. As a result, one obtains for the evolution of the total population $I(t)$ of ion pairs,

$$I(t) = c \int_{R_c}^{\infty} dr\, 4\pi r^2 k_f(r) \int_0^t dt'\, S(r, t - t')\rho(r, t')P(t') \quad (9.126)$$

where $S(r, t)$ is the survival probability and represents the fraction of ions that survive at time t having initial separation r. It satisfies an equation

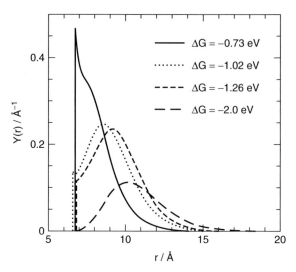

Figure 9.34. Distribution of the electron-transfer distance $Y(r) = 4\pi r^2 \sigma(r, \infty)$ for different values of the free energy change ΔG in ethyleneglycol solution with the acceptor concentration $c = 0.3$ M. (Reproduced from [345b] with permission. Copyright (1996) by the American Chemical Society.)

similar to Eq. (9.121) [335, 340]

$$\frac{\partial}{\partial t} S(r, t) = [\mathscr{L}_b^\dagger - k_b(r)] S(r, t) \qquad (9.127)$$

where

$$\mathscr{L}_b^\dagger = \frac{D_b}{r^2} e^{-r_o/r} \frac{\partial}{\partial r} r^2 e^{r_o/r} \frac{\partial}{\partial r} \qquad (9.128)$$

is the adjoint diffusion operator, $r_o = e^2/(\varepsilon_s k_B T)$ is the Onsager radius, and the subscript b indicates that the corresponding parameters are for backward ET reaction.

Initially, $I(t)$ increases due to ionization but begins to decrease as soon as recombination starts to prevail, and eventually comes down to a constant value that determines the photoseparation quantum yield, $I(\infty) = \phi$. It is ϕ that is usually measured experimentally. Since ϕ depends on the parameters of both forward and backward ET reactions in a rather complicated way, one generally cannot expect to observe the usual Marcus-type free energy gap law by plotting ϕ (or any other characteristic of geminate recombination) as a function of ΔG_b. But at least qualitative resemblance is expected. Experimental data of Niwa et al. [350a] on geminate recombination in solution indeed reproduce nonmonotonic behavior inherent to the Marcus mechanism of ET (see Fig. 9.35). The minimum of the free ion yield in the observed dependence corresponds to the maximum of the recombination rate (activationless ET). Comparison versus theory [353] can be regarded as reasonable, because variation of ΔG in experiment was achieved by changing the reactants, that is, by changing all the molecular parameters, while the theoretical curves were obtained with only one set of parameter values.

Experimentalists often analyze their data within a simple model [10–12, 234, 350, 354], which implies that ions are created at contact and also recombine at contact with the first-order rate constant k_{-e} (essentially the Marcus ET rate constant at contact) or separate (escape the reaction zone and never come back) with the first-order rate constant k_{sep} provided by Eigen [355], $k_{\text{sep}} = (3D_b r_o / R_c^3)[e^{r_o/R_c} - 1]^{-1}$. The model is known as exponential because it predicts an exponential decrease of the ion population with time. In fact, the kinetics is never exponential in the presence of diffusion, even in the contact approximation [340, 356–358]. Therefore, one

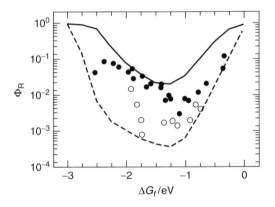

Figure 9.35. Free ion yield per one photogenerated geminate ion pair, $\Phi_R = \phi / \int_{R_c}^{\infty} dr\, 4\pi r^2 \sigma(r, \infty)$, as a function of the free energy change ΔG_f of forward ET in acetonitrile (● experiment, —— theory) and dichlormethane (○ experiment, - - - theory). Theoretical curves were calculated in [353] using Eq. (9.13) for the ET rate, Eq. (9.17) for the reorganization energy, and Eq. (9.22) for the electronic coupling, with the following parameters: $r_D = r_A = 3.4\,\text{Å}$, $V_f = 40\,\text{cm}^{-1}$, $V_b = 120\,\text{cm}^{-1}$, $\beta_f = 1.1\,\text{Å}^{-1}$ and $\beta_b = 1\,\text{Å}^{-1}$, $\lambda_{\text{vib}} = 0.3\,\text{eV}$, and $\omega_{\text{vib}}/(2\pi) = 1500\,\text{cm}^{-1}$. Dielectric constants for acetonitrile are $\varepsilon_s = 37.7$ and $\varepsilon_\infty = 1.8$, for dichlormethane $\varepsilon_s = 8.93$ and $\varepsilon_\infty = 2.02$. The corresponding viscosities are 0.34 and 0.41 cP, respectively (used to calculate the diffusion coefficients). In addition, it was assumed that $\Delta G_f + \Delta G_b = -3\,\text{eV}$ for all systems. Experimental data were taken from [350a].

actually cannot consider charge separation as a first-order rate process. Nevertheless, the model defines the quantum yield of separation as $\phi = (1 + k_{-e}/k_{\text{sep}})^{-1}$, and this simple prescription is used to evaluate k_{-e} from the experimentally measured ϕ and thus study its free energy gap dependence. Basically, the same expression can be derived rigorously on the basis of the contact approximation [333, 340, 356, 357] and it can be shown that k_{-e} so defined (in terms of ϕ) should qualitatively follow the Marcus free energy gap law [18], as it was indeed observed in a number of experiments [12, 350]. This does not mean, however, that the exponential model itself is justified. The fact that forward ET usually displays a diffusion-controlled plateau while back ET does not, simply implies that the backward ET rate is much slower than the forward ET rate for the systems chosen.

B. Nonhomogeneous Effects

1. Hydrodynamic Hindrance and Liquid Structure

It is well known [359] that the local density of particles in a liquid is highly structured and far from being a homogeneous continuum. As a result, the distribution of acceptors around a donor is not uniform. Rather it exhibits an oscillatory behavior as a function of separation and is peaked near the contact radius. Since the first-order ET rate is extremely sensitive to distance, the liquid structure factor should be generally taken into account when formulating the diffusion-influenced ET theory. Coupling of translational and reactive dynamics and the effect of liquid structure on the transient kinetics have received due attention recently, both for contact [360] as well as for long-range reactions [346]. The structure factor was shown to be of particular importance at short distances and early times, immediately after the reaction initiation, and its general effect is to enhance the observed ET rate. The structure factor is included into theory via a potential of mean force entering the diffusion operator. Alternatively, the structural enhancement of ET can be accounted for through the short distance dependence of $\varepsilon(r)$ in the Coulomb potential [361]. Although nowadays one often tends to solve such problems numerically, approximate analytical methods are also available originating from the scattering theory. In particular, the kinetics of diffusion-influenced reactions in the presence of an arbitrarily shaped attractive short-range potential have been analyzed in great detail [362]. It was shown that strong attractive interaction results in a quasistationary state within a well, the so-called cage effect. At short times, this state evolves exponentially but a crossover to a $t^{-3/2}$ long-time tail is observed, typical of a free diffusion process.

At long times, when the initial static stage of the kinetics is over, the reactants diffuse to a closer proximity and start to feel the granularity of the solvent that hinders their motion. This hydrodynamic hindrance [363] leads to a space-dependent diffusion on a radial distance scale where the ET rate is most affected. Consequently, ET slows down. This factor is usually included into theory of diffusion-assisted reactions via a simple analytical form due to Northrup and Hynes [363e], $D(r)/D = 1 - \frac{1}{2}\exp(1 - r/R_c)$. Proper account of both the hydrodynamic hindrance and the liquid structure factor proves to be determinant for comparative analysis with experiment [346]. Note that in general the diffusion coefficient is also time-dependent [197b, 359, 364]. All these effects and many other aspects of the modern theory of diffusion-mediated reactions have been discussed in great detail in the literature [336, 365].

2. Microheterogeneous Environments

ET reactions in organized and constrained media have attracted growing research interest aimed at understanding peculiar kinetic behaviors influenced by the systems topology and ultimately finding ways to control the charge separation yield by suitably manipulating the microscopic environment [366]. Examples of microheterogeneous media in which ET reactions have been investigated include molecular crystals, liquid crystals, micelles, and related assemblies, Langmuir–Blodgett films, monolayers, polymers, clays, silica gels and other porous glasses, zeolites, semiconductor nanomaterials, and natural systems such as DNA. This area is too large to go into any detail here. We would only like to mention a few important points inherent in theoretical treatment of ET reactions in microheterogeneous media, and consider micelles as a typical example [367]. Micelles are self-assembled structures of amphiphilic molecules (surfactant or lipid, composed of a polar head and a hydrophobic tail) in solvents, usually water. The simplest possible structures are spherical aggregates of molecular dimensions (a few nm in diameter). Micelles have long been of interest to the ET community as convenient model systems with a certain range of practical applications [220, 368, 369].

First, heterogeneous geometry can have an impact on the solvent reorganization energy and the free energy change of reaction [219, 220]. Recently, photoinduced ET between N,N-dimethylaniline (DMA) and octadecylrhodamine B (ODRB) has been studied in three alkyltrimethylammonium bromide micelles: dodecyl- (DTAB), tetradecyl- (TTAB), and hexadecyltrimethylammonium bromide (CTAB), which only differ in the length of their surfactant chains [220a]. The DMA and ODRB molecules are localized at the micelle surface. Despite the similarities of the micelles, the experiments show that the ET dynamics vary with micelle size. As the micelle becomes larger, ET is faster for the same DMA surface packing density. This result was attributed to differences in the solvent reorganization energy and described within the dielectric spherical shell model of a micelle [220]. Qualitatively, as the radius of the micelle is increased, the amount of polar solvent in the vicinity of the reactants is decreased and thus the reorganization energy is also decreased leading to an increase in the ET rate. Also, the headgroup regions of the three micelles have different static dielectric constants. It is argued that the differences in dielectric constants of the headgroup shells arise from different extents of water penetration (more water in smaller micelles with higher curvature). The analysis shows that ET is very sensitive to the molecular level details of the solvent. It should always be kept in mind when considering reactions in microheterogeneous structures.

Second, since micelles have molecular dimensions, the number of reactants they comprise is usually small. Therefore, one has to deal with a discrete statistical distribution of reactants among the micelles instead of conventional concentrations. The overall kinetics in the ensemble of micelles is obtained by averaging the microscopic intramicellar kinetics with a given number of reactants over the occupancy distribution [367]. In the low occupancy limit this distribution is Poissonian. The dynamic nature of amphiphilic aggregates means that the number of reactants in a given micelle fluctuate with time. These fluctuations are slow (microseconds), so that the reaction inside the micelle can normally be treated as kinetically independent.

Third, diffusion space of reactants in a micelle is restricted, which means that intramicellar bimolecular diffusion-mediated reactions follow pseudo-first-order kinetics [367]. Detailed formulation is based on the reaction–diffusion equation for the pair survival probability with an appropriate restricted diffusion operator [370]. It can be integrated numerically [368], but a more convenient and computationally superior way is to expand the solution into a discrete set of eigenfunctions of the diffusion operator, as we did in Section III.B. The eigenvalues grow very rapidly in finite-volume systems, ensuring fast convergence of the resulting truncated matrix solution [371]. Note that since the boundary condition for the survival probability does not include the interaction potential, we can use the eigenfunctions of the diffusion operator without interaction as a basis for our expansion. The latter are usually well known. For example, in the practically relevant case of spherical surface diffusion the eigenfunctions can be expressed in terms of Legendre functions. It should be emphasized that virtually any reaction in any geometry, even if it involves multiple stages and reverse channels, can be formulated in terms of matrices, provided the system is confined. We only need to somehow calculate the corresponding eigenmodes, which offers a possibility to analyze systems close to real, taking into account diffusion, the arbitrary distance dependence of the reaction rate, interaction potential, and all that by rather simple numerical means. More details can be found elsewhere [367].

3. Stochastic Gating and Anisotropic Reactivity

For ET reactions in liquids, it is not uncommon that stereoselectivity [372], mutual orientations [373], conformational fluctuations [121, 293, 374], or spin multiplicity restrictions [361, 375] set a kinetic limit to the overall system reactivity. In complex systems, such as proteins, diffusive motion of reactants and thus the kinetics of diffusion-mediated ET reactions can be essentially controlled by conformational dynamics of the embedding medium. Perhaps the simplest way to handle such effects is to model them as a stochastic gate by introducing the corresponding effective gating coordinate [311, 312]. The reactivity depends on the state of the gate. Gating can be modeled as having a discrete number of states [291, 377, 378] or a continuous number of states [305, 376, 377e, 378b–d, 379], characterized by Markovian [291, 376–379] or non-Markovian [305, 377c,e, 378e] relaxation dynamics, it can modulate the first-order ET rate [121, 293] or the second-order diffusion-mediated rate [291, 305, 376–379], there can be donor gating and acceptor gating [377b, 378g–j]. In addition to all these complexities, there is translational–rotational motion of the molecules and anisotropic reactivity [291b, 377a,d, 378b].

Let us consider the simplest case of a two-state stochastic gating [291b, 378j]. It is usually assumed that ET and the reactive–nonreactive configurational turnover take place sequentially. This turnover allows us to write the gated first-order ET rate as a product of the gate interconversion rate and the first-order ET rate. The reaction dynamics is governed by a standard reaction–diffusion equation with the gated first-order ET rate as a sink and a sum of the diffision operator and the two-state interconversion operator describing nonreactive system evolution. The presence of the gate will evidently slow down the reaction. However, as the rate of interconversions between the gate states becomes faster, it no longer limits the reactivity. Originally, it was assumed that the reaction kinetics will follow the same law irrespective of the physical location of the gate, on the donor or on the acceptor. This finding is indeed true for an isolated reacting pair. Recently, however, it has been shown [377b] that the location of the gate has an important bearing on the kinetics of a many-particle system. Namely, the decay laws were found to follow different sets of equations. The principal cause of this difference is that for the donor-gated situation, the reactivity of all the acceptors is simultaneously interrupted or else facilitated, while for the acceptor-gated case the survival probability of the donor is a product of independent pair probabilities corresponding to isolated gating dynamics.

As follows from the general Jensen's inequality, the acceptor-gated kinetics proceeds faster than the donor-gated kinetics for an initial equilibrium distribution of the two gate states [378g]. Such an asymmetry in the ET kinetics diminishes for faster gate interconversion rates and for low acceptor concentrations.

Whenever one deals with diffusion-assisted reactions of complex molecules, one is faced with the problem of anisotropic reactivity. In proteins, the active site normally occupies only a small part of the protein surface (the so-called surface-active site) or the entire active site may be buried in a concave region termed molecular crevice or pocket (the so-called cavity-active site). The usual way to deal with anisotropic reactivity theoretically is to assume a localized reactive patch on the otherwise inert surface of the macromolecule and to formulate the diffusion-reaction problem with a mixed boundary condition [291b, 377a,d, 380]. An important outcome of theoretical analysis is that the reaction rate of a small isotropically reactive particle with an anisotropically reactive macromolecule is proportional to the radius r_a of the active site but not to its square, as one would expect from simple geometrical arguments. Diffusion-controlled reaction of two macromolecules with reactive patches of radii r_a occurs with the rate constant $\sim r_a^3$. Since the ET rate depends sharply on the distance between the active sites (via the transfer integral), the model of flat surface patches may not be adequate. A method has been developed allowing for rigorous analysis of the problem for strongly anisotropic reactivity and interaction [381]. It is based on the approximation of curvilinear coordinates, describing relative position and orientation of the reacting molecules, by the simpler Cartesian coordinates in the vicinity of reactive relative orientations, where the pair distribution function strongly deviates from equilibrium. The method can deal with realistic models of reactivity such as distance-dependent ET. A significant enhancement of the rate constant is predicted for the case of two anisotropic molecules, as compared against the model of patches. By approximating the reactive site as a semisphere of radius r_a, one obtains the rate constant proportional to r_a^2 instead of r_a^3.

V. CONCLUDING REMARKS

Qualitative predictions of the classical Marcus theory of ET have been confirmed for a wide variety of systems. Given this success, theory is now challenged to calculate absolute ET rates on the basis of separately obtainable parameters. Regarding the solvent, the input is the dielectric relaxation spectrum. It can be measured or even calculated independently to a high degree of accuracy. Intramolecular modes participating in the ET event contribute additively to the overall spectral density, although their effect is more difficult to assess. Anyway, we can assume that the spectrum is known. How to use this information depends on the value of the electronic coupling.

Our attention has been focused on nonadiabatic ET reactions. The term "nonadiabatic ET" is very broad. First of all, it includes the usual situation where level crossing is nonadiabatic, which implies that the Massey parameter is small. Here, one can distinguish between the strong nonadiabatic regime (weak coupling), where the reaction is limited by electronic transitions and the usual Golden Rule applies, and the solvent-controlled regime, where the Golden Rule applies only locally and the reaction rate is limited by transport to the transition state. Adiabatic level crossing (large Massey parameter) does not necessarily imply that ET is also adiabatic, which is true only in the normal region, while in the inverted region the reaction is always nonadiabatic (given equilibrium initial condition). When the coupling is high, transitions between adiabatic states become the bottleneck of ET.

In the weak-coupling limit where the reaction is slower than dielectric relaxation, the solvent contibution can be taken into account rigorously by including the whole spectrum into the quantum mechanical Golden Rule formula. The situation becomes much more difficult when the reaction and relaxation occur on comparable timescales. We have analyzed available theoretical approaches and it seems that there is still no adequate recipe that would cover all possible situations. A promising strategy is based on the separation of timescales into fast and slow modes. The slow modes are to be treated as a (non-Markovian) stochastic process, while the fast modes would enter as a (widespread) sink. Although often one can explain experimental findings using this methodology, the separation ansatz is not well defined and can hardly be used for absolute rate predictions. On the other hand, numerical quantum mechanical methods that have this predictive power are still computationally demanding and limited to model spectral densities. Fast progress in this area makes one believe that there will be a breakthrough in the nearest future. In the normal region, a consistent separation ansatz (spectral rather than temporal) is offered by the variational transition state theory. It is expected (not thoroughly checked, though) to approximate the rate constant quite well (from above) for any relaxation spectrum, provided that the activation energy is sufficiently high. Another limit where relatively simple but fully quantum mechanical theory can be used again is for nonadiabatic transitions in the regime of adiabatic level crossing. When the electronic coupling is strong, nonadiabatic interaction between relevant adiabatic states is weak and can be considered as a perturbation.

Chemical kinetics is all about bottlenecks. They determine the reaction mechanism on different timescales. We have seen that ET can be limited by nonadiabatic transitions, solvent dynamics, intramolecular vibrational relaxation, translational diffusion, and conformational fluctuations either of the reactants themselves or of the embedding medium. The interplay between different mechanisms as well as nonequilibrium initial condition may result in rich kinetic behaviors, with strong nonexponentiality and coherence effects observed in recent experiments.

Acknowledgments

We would like to thank B. Bagchi, A. I. Burshtein, J. Jortner, R. A. Marcus, M. D. Newton, I. Rips, and A. I. Shushin for critically reading the manuscript and for their numerous comments and suggestions that helped to improve this chapter. We are also grateful to K. Ando for discussions concerning numerical simulations and for providing us his data files, S. Iwai and S. Murata for sharing with us the details of ET experiment in solutions.

References

1. K. Huang and A. Rhys, *Proc. R. Soc. London Ser. A* **204**, 406 (1950).
2. W. Libby, *J. Phys. Chem.* **56**, 863 (1952).
3. M. Lax, *J. Chem. Phys.* **20**, 1752 (1952).
4. (a) R. Kubo, *Phys. Rev.* **86**, 929 (1952); (b) R. Kubo and Y. Toyozawa, *Prog. Theor. Phys.* **13**, 160 (1955).
5. (a) R. A. Marcus, *J. Chem. Phys.* **24**, 966 (1956); (b) **26**, 867 (1957); (c) **43**, 679 (1965).
6. (a) N. S. Hush, *J. Chem. Phys.* **28**, 962 (1958); (b) *Trans. Faraday Soc.* **57**, 557 (1961); (c) *Progr. Inorg. Chem.* **8**, 391 (1967).
7. (a) V. G. Levich and R. R. Dogonadze, *Dokl. Akad. Nauk SSSR* **124**, 123 (1959); (b) **133**, 158 (1960); (c) V. G. Levich, *Adv. Electrochem. Eng.* **4**, 249 (1965).
8. (a) J. R. Miller, L. T. Calcaterra, and G. L. Closs, *J. Am. Chem. Soc.* **106**, 3047 (1984); (b) G. L. Closs, L. T. Calcaterra, N. J. Green, K. W. Penfield, and J. R. Miller, *J. Phys. Chem.* **90**, 3673 (1986); (c) G. L. Closs and J. R. Miller, *Science* **240**, 440 (1988).
9. (a) K. Kemnitz, N. Nakashima, and K. Yoshihara, *J. Phys. Chem.* **92**, 3915 (1988); (b) J. C. Moser and M. Grätzel, *Chem. Phys.* **176**, 493 (1993); (c) J. M. Rehm, G. M. McLendon, Y. Nagasawa, K. Yoshihara, J. Moser, and M. Grätzel, *J. Phys. Chem.* **100**, 9577 (1996); (d) B. Burfeindt, T. Hannappel, W. Storck, and F. Willig, *J. Phys. Chem.* **100**, 16463 (1996).
10. (a) I. R. Gould, D. Ege, S. L. Mattes, and S. Farid, *J. Am. Chem. Soc.* **109**, 3794 (1987); (b) I. R. Gould, D. Ege, J. E. Moser, and S. Farid, *J. Am. Chem. Soc.* **112**, 4290 (1990); (c) I. R. Gould, R. H. Young, R. E. Moody, and S. Farid, *J. Phys. Chem.* **95**, 2068 (1991); (d) I. R. Gould, and S. Farid, *J. Phys. Chem.* **97**, 3567 (1993); (e) *Acc. Chem. Res.* **29**, 522 (1996).
11. (a) N. Mataga, Y. Kanda, and T. Okada, *J. Phys. Chem.* **90**, 3880 (1986); (b) N. Mataga, T. Asahi, Y. Kanda, T. Okada, and T. Kakitani, *Chem. Phys.* **127**, 249 (1988); (c) T. Asahi and N. Mataga, *J. Phys. Chem.* **93**, 6575 (1989).
12. (a) P. P. Levin, P. F. Pluzhnikov, and V. A. Kuzmin, *Chem. Phys. Lett.* **147**, 283 (1988); (b) *Chem. Phys.* **137**, 331 (1989); (c) G. Grampp and G. Hetz, *Ber. Bunsenges. Phys. Chem.* **96**, 198 (1992).

13. (a) T. M. McCleskey, J. R. Winkler, and H. B. Gray, *J. Am. Chem. Soc.* **114**, 6935 (1992); (b) *Inorg. Chim. Acta* **225**, 319 (1994); (c) C. Turró, J. M. Zaleski, Y. M. Karabastos, and D. G. Nocera, *J. Am. Chem. Soc.* **118**, 6060 (1996); (d) D. M. Guldi and K.-D. Asmus, *J. Am. Chem. Soc.* **119**, 5744 (1997).
14. D. Rehm and A. Weller, *Isr. J. Chem.* **8**, 259 (1970).
15. (a) N. Mataga and H. Miyasaka, *Prog. React. Kinet.* **19**, 317 (1994); (b) T. Kakitani, N. Matsuda, A. Yoshimori, and N. Mataga, *Prog. React. Kinet.* **20**, 347 (1995).
16. S. Nishikawa, T. Asahi, T. Okada, N. Mataga, and T. Kakitani, *Chem. Phys. Lett.* **185**, 237 (1991).
17. M. Tachiya and S. Murata, *J. Phys. Chem.* **96**, 8441 (1992).
18. (a) A. I. Burshtein, *J. Chem. Phys.* **103**, 7927 (1995); (b) A. I. Burshtein and E. Krissinel, *J. Phys. Chem.* **100**, 3005 (1996).
19. N. Mataga and H. Miyasaka, *Adv. Chem. Phys.* **107**, 431 (1999).
20. (a) R. R. Dogonadze, A. M. Kuznetsov, and V. G. Levich, *Dokl. Akad. Nauk SSSR* **188**, 383 (1969); (b) R. R. Dogonadze and A. M. Kuznetsov, *Sov. Elektrochem.* **7**, 735 (1971).
21. (a) A. A. Ovchinnikov and M. Y. Ovchinnikova, *Dokl. Akad. Nauk SSSR* **186**, 76 (1969); (b) *Sov. Phys. JETP* **29**, 688 (1969).
22. R. R. Dogonadze and A. M. Kuznetsov, *Sov. Elektrochem.* **3**, 1189 (1967).
23. (a) S. Efrima and M. Bixon, *Chem. Phys. Lett.* **25**, 34 (1974); (b) *Chem. Phys.* **13**, 447 (1976); (c) *J. Chem. Phys.* **64**, 3639 (1976).
24. N. R. Kestner, J. Logan, and J. Jortner, *J. Phys. Chem.* **78**, 2148 (1974).
25. (a) J. Ulstrup and J. Jortner, *J. Chem. Phys.* **63**, 4358 (1975); (b) J. Jortner, *J. Chem. Phys.* **64**, 4860 (1976).
26. J. Ulstrup, *Charge Transfer Processes in Condensed Media* (Springer, Berlin, 1979).
27. (a) M. A. Vorotyntsev, R. R. Dogonadze, and A. M. Kuznetsov, *Dokl. Akad. Nauk SSSR* **195**, 1135 (1970); (b) R. R. Dogonadze, and A. M. Kuznetsov, *Sov. Elektrochem.* **7**, 735 (1971); (c) R. R. Dogonadze, A. A. Kornyshev, and A. M. Kuznetsov, *Theor. Math. Phys.* **15**, 4047 (1973).
28. (a) A. A. Kornyshev, *Electrochim. Acta* **26**, 1 (1981); (b) A. A. Kornyshev and J. Ulstrup, *Chem. Phys. Lett.* **126**, 74 (1986); (c) A. A. Kornyshev, A. M. Kuznetsov, D. K. Phelps, and M. Weaver, *J. Chem. Phys.* **91**, 7159 (1989).
29. D. V. Matyushov and B. M. Ladanyi, *J. Chem. Phys.* **108**, 6362 (1998).
30. J. J. Hopfield, *Proc. Natl. Acad. Sci. USA* **71**, 3640 (1974).
31. (a) P. Siders and R. A. Marcus, *J. Am. Chem. Soc.* **103**, 741, 748 (1981); (b) R. A. Marcus and P. Siders, *J. Phys. Chem.* **86**, 622 (1982).
32. (a) G. A. Bogdanchikov, A. I. Burshtein, and A. A. Zharikov, *Chem. Phys.* **86**, 9 (1984); (b) **107**, 75 (1986).
33. (a) A. Warshel, *J. Phys. Chem.* **86**, 2218 (1982); (b) A. Warshel and J.-K. Hwang, *J. Chem. Phys.* **84**, 4938 (1986); (c) J.-K. Hwang and A. Warshel, *J. Am. Chem. Soc.* **109**, 715 (1987); (d) A. Warshel, Z. T. Chu, and W. W. Parson, *Science* **246**, 112 (1988); (e) A. Warshel and Z. T. Chu, *J. Chem. Phys.* **93**, 4003 (1990); (f) G. King and A. Warshel, *J. Chem. Phys.* **93**, 8682 (1990); (g) A. Warshel and W. W. Parson, *Annu. Rev. Phys. Chem.* **42**, 287 (1991); (h) C. F. Jen and A. Warshel, *J. Phys. Chem. A* **103**, 11378 (1999).
34. (a) R. A. Kuharski, J. S. Bader, D. Chandler, M. Sprik, M. L. Klein, and R. W. Impey, *J. Chem. Phys.* **89**, 3248 (1988); (b) J. S. Bader, R. A. Kuharski, and D. Chandler, *J. Chem. Phys.* **93**, 230 (1990).

35. (a) M. Marchi and D. Chandler, *J. Chem. Phys.* **95**, 889 (1991); (b) M. Marchi, J. N. Gehlen, D. Chandler, and M. Newton, *J. Am. Chem. Soc.* **115**, 4178 (1993); (c) J. N. Gehlen, M. Marchi, and D. Chandler, *Science* **263**, 499 (1994).
36. (a) M. Maroncelli and G. R. Fleming, *J. Chem. Phys.* **89**, 5044 (1988); (b) M. Maroncelli, *J. Chem. Phys.* **94**, 2084 (1991); (c) M. Cho, G. R. Fleming, S. Saito, I. Ohmine, and R. M. Stratt, *J. Chem. Phys.* **100**, 6672 (1994).
37. X. Song and R. A. Marcus, *J. Chem. Phys.* **99**, 7768 (1993).
38. (a) C.-P. Hsu, X. Song, and R. A. Marcus, *J. Phys. Chem. B* **101**, 2546 (1997); (b) Y. Georgievskii, C.-P. Hsu, and R. A. Marcus, *J. Chem. Phys.* **108**, 7356 (1998); (c) C.-P. Hsu, Y. Georgievskii, and R. A. Marcus, *J. Phys. Chem. A* **102**, 2658 (1998).
39. Y. Georgievskii, C.-P. Hsu, and R. A. Marcus, *J. Chem. Phys.* **110**, 5307 (1999).
40. M. D. Todd, A. Nitzan, M. A. Ratner, and J. T. Hupp, *J. Photochem. Photobiol. A* **82**, 87 (1994).
41. (a) K. Ando and S. Kato, *J. Chem. Phys.* **95**, 5966 (1991); (b) K. Ando, *J. Chem. Phys.* **101**, 2850 (1994); (c) **106**, 116 (1997); (d) **107**, 4585 (1997).
42. (a) E. Neria and A. Nitzan, *J. Chem. Phys.* **99**, 1109 (1993); (b) *Chem. Phys.* **183**, 351 (1994).
43. O. V. Prezhdo and P. J. Rossky, *J. Chem. Phys.* **107**, 5863 (1997).
44. (a) C. Zheng, J. A. McCammon, and P. Wolynes, *Proc. Natl. Acad. Sci. USA* **86**, 6441 (1989); (b) *Chem. Phys.* **158**, 261 (1991).
45. S. Efrima and M. Bixon, *J. Chem. Phys.* **70**, 3531 (1979).
46. (a) J. Tang, *J. Chem. Phys.* **99**, 5828 (1993); (b) J. Tang and S. H. Lin, *J. Chem. Phys.* **107**, 3485 (1997).
47. J. Tang and S. H. Lin, *Chem. Phys. Lett.* **254**, 6 (1996).
48. (a) R. Egger, C. H. Mak, and U. Weiss, *J. Chem. Phys.* **100**, 2651 (1994); (b) G. Lang, E. Paladino, and U. Weiss, *Chem. Phys.* **244**, 111 (1999).
49. M. V. Basilevsky and G. E. Chudinov, *J. Chem. Phys.* **103**, 1470 (1995).
50. A. Lami and F. Santoro, *J. Chem. Phys.* **106**, 94 (1997).
51. T. Cheche, *Chem. Phys.* **229**, 193 (1998).
52. (a) M. Tachiya, *Chem. Phys. Lett.* **159**, 505 (1989); (b) M. Tachiya, *J. Phys. Chem.* **93**, 7050 (1989); (c) **97**, 5911 (1993).
53. (a) J. Tang, *Chem. Phys.* **179**, 105 (1994); (b) **188**, 143 (1994).
54. H.-X. Zhou and A. Szabo, *J. Chem. Phys.* **103**, 3481 (1995).
55. (a) A. Yoshimori and T. Kakitani, *J. Chem. Phys.* **93**, 5140 (1990); (b) A. Yoshimori, *Chem. Phys. Lett.* **225**, 494 (1994).
56. (a) A. Yoshimori, *Chem. Phys. Lett.* **235**, 303 (1994); (b) *J. Electroanal. Chem.* **438**, 21 (1997); (c) *J. Chem. Phys.* **109**, 8790 (1998).
57. T. Ichiye, *J. Chem. Phys.* **104**, 7561 (1996).
58. D. V. Matyushov and B. M. Ladanyi, *J. Chem. Phys.* **107**, 1375 (1997).
59. A. I. Shushin, *J. Chem. Phys.* **97**, 3171 (1992).
60. J. Tang, *J. Chem. Phys.* **98**, 6263 (1993).
61. (a) I. Daizadeh, E. S. Medvedev, and A. A. Stuchebrukhov, *Proc. Natl. Acad. Sci. USA* **94**, 3703 (1997); (b) E. S. Medvedev and A. A. Stuchebrukhov, *J. Chem. Phys.* **107**, 3821 (1997).

62. (a) R. D. Coalson, D. G. Evans, and A. Nitzan, *J. Chem. Phys.* **101**, 436 (1994); (b) D. G. Evans and R. D. Coalson, *J. Chem. Phys.* **102**, 5658 (1995); (c) **104**, 3598 (1996).
63. M. Cho and R. J. Silbey, *J. Chem. Phys.* **103**, 595 (1995).
64. A. I. Ivanov and V. V. Potovoi, *Chem. Phys.* **247**, 245 (1999).
65. R. G. Alden, W. D. Cheng, and S. H. Lin, *Chem. Phys. Lett.* **194**, 318 (1992).
66. (a) I. A. Goychuk, E. G. Petrov, and V. May, *Phys. Rev. E* **51**, 2982 (1995); (b) **56**, 1421 (1997); (c) *Chem. Phys. Lett.* **253**, 428 (1996); (d) *J. Chem. Phys.* **103**, 4937 (1995); (e) **106**, 4522 (1997).
67. (a) Y. Dakhnovskii, *J. Chem. Phys.* **100**, 6492 (1994); (b) D. G. Evans, R. D. Coalson, H. J. Kim, and Y. Dakhnovskii, *Phys. Rev. Lett.* **75**, 3649 (1995); (c) Y. Dakhnovskii and R. D. Coalson, *J. Chem. Phys.* **103**, 2908 (1995); (d) D. G. Evans, R. D. Coalson, and Y. Dakhnovskii, *J. Chem. Phys.* **104**, 2287 (1996); (e) Y. Dakhnovskii, V. Lubchenko, and R. D. Coalson, *Phys. Rev. Lett.* **77**, 2917 (1996); (f) *J. Chem. Phys.* **109**, 691 (1998).
68. H. A. Kramers, *Physica* **7**, 284 (1940).
69. P. Hänggi, P. Talkner, and M. Borkovec, *Rev. Mod. Phys.* **62**, 251 (1990).
70. V. I. Melnikov, *Phys. Rep.* **209**, 1 (1991).
71. (a) S. Chakravarty and A. J. Leggett, *Phys. Rev. Lett.* **52**, 5 (1984); (b) A. J. Leggett, S. Chakravarty, A. T. Dorsey, M. P. A. Fisher, A. Garg, and W. Zwerger, *Rev. Mod. Phys.* **59**, 1 (1987); (c) M. Grifoni and P. Hänggi, *Phys. Rep.* **304**, 229 (1998).
72. T. Holstein, *Ann. Phys. (Leipzig)* **8**, 325; 343 (1959).
73. (a) L. Landau, *Phys. Z. Sow.* **2**, 46 (1932); (b) C. Zener, *Proc. R. Soc. London Ser. A* **137**, 696 (1932).
74. L. D. Zusman, *Chem. Phys.* **49**, 295 (1980).
75. (a) A. I. Burshtein and A. G. Kofman, *Chem. Phys.* **40**, 289 (1979); (b) B. I. Yakobson and A. I. Burshtein, *Chem. Phys.* **49**, 385 (1980).
76. J. E. Straub and B. J. Berne, *J. Chem. Phys.* **87**, 6111 (1987).
77. J. N. Onuchic and P. G. Wolynes, *J. Phys. Chem.* **92**, 6495 (1988).
78. H. Frauenfelder and P. G. Wolynes, *Science* **229**, 337 (1985).
79. M. Ya. Ovchinnikova, *Teor. Eksp. Khim.* **17**, 651 (1981) [*Theor. Exper. Chem.* **17**, 507 (1982)].
80. A. B. Helman, *Chem. Phys.* **65**, 271 (1982).
81. (a) L. D. Zusman, *Chem. Phys. Lett.* **86**, 547 (1982); (b) *Teor. Eksp. Khim.* **19**, 413 (1983); (c) *Chem. Phys.* **80**, 29 (1983); (d) **119**, 51 (1988); (e) *J. Chem. Phys.* **102**, 2580 (1995).
82. (a) L. D. Zusman, *Russ. Chem. Rev. (Engl. Transl.)* **61**, 1 (1992); (b) *Z. Phys. Chem.* **186**, 1 (1994).
83. (a) D. E. Calef and P. G. Wolynes, *J. Phys. Chem.* **87**, 3387 (1983); (b) *J. Phys. Chem.* **78**, 470 (1983).
84. J. T. Hynes, *J. Phys. Chem.* **90**, 3701 (1986).
85. A. Garg, J. N. Onuchic, and V. Ambegaokar, *J. Chem. Phys.* **83**, 4491 (1985).
86. (a) J. N. Onuchic, D. N. Beratan, and J. J. Hopfield, *J. Phys. Chem.* **90**, 3707 (1986); (b) J. N. Onuchic, *J. Chem. Phys.* **86**, 3925 (1986); (c) D. N. Beratan and J. N. Onuchic, *J. Chem. Phys.* **89**, 6195 (1988).
87. (a) I. Rips and J. Jortner, *J. Chem. Phys.* **87**, 2090, 6513 (1987); (b) **88**, 818 (1988).
88. (a) M. Sparpaglione and S. Mukamel, *J. Phys. Chem.* **91**, 3938 (1987); (b) *J. Chem. Phys.* **88**, 3263 (1988).

89. (a) H. Sumi and R. A. Marcus, *J. Chem. Phys.* **84**, 4272 (1986); (b) *J. Phys. Chem.* **84**, 4894 (1986); (c) H. Sumi, *J. Phys. Chem.* **95**, 3334 (1991); (d) *Chem. Phys.* **212**, 9 (1996).
90. W. Nadler and R. A. Marcus, *J. Chem. Phys.* **86**, 3906 (1987).
91. H. Sumi, *Adv. Chem. Phys.* **107**, 601 (1999).
92. (a) O. B. Spirina and A. B. Doktorov, *Chem. Phys.* **203**, 177 (1996); (b) O. B. Jenkins and A. B. Doktorov, *Chem. Phys.* **234**, 121 (1998).
93. A. Okada, *J. Chem. Phys.* **111**, 2665 (1999).
94. K. Ando and H. Sumi, *J. Phys. Chem. B* **102**, 10991 (1998).
95. (a) P. G. Wolynes, *J. Chem. Phys.* **86**, 1957 (1987); (b) **87**, 6559 (1987).
96. (a) J. Jortner and M. Bixon, *J. Chem. Phys.* **88**, 167 (1988); (b) M. Bixon and J. Jortner, *Chem. Phys.* **176**, 521 (1993); (c) M. Bixon, J. Jortner, and J. W. Verhoeven, *J. Am. Chem. Soc.* **116**, 7349 (1994); (d) J. Jortner and M. Bixon, *Ber. Bunsenges. Phys. Chem.* **99**, 296 (1995).
97. M. Bixon and J. Jortner, *Adv. Chem. Phys.* **106**, 35 (1999).
98. (a) H. L. Friedman and M. D. Newton, *Faraday Discuss. Chem. Soc.* **74**, 73 (1982); (b) M. D. Newton and H. L. Friedman, *J. Chem. Phys.* **88**, 4460 (1988).
99. (a) Y. J. Yan and S. Mukamel, *J. Chem. Phys.* **89**, 5160 (1988); (b) *Acc. Chem. Res.* **22**, 301 (1989); (c) Y. Hu and S. Mukamel, *J. Chem. Phys.* **91**, 6973 (1989).
100. A. Okada, V. Chernyak, and S. Mukamel, *Adv. Chem. Phys.* **106**, 515 (1999).
101. (a) R. I. Cukier, *J. Chem. Phys.* **88**, 5594 (1988); (b) M. Morillo and R. I. Cukier, *J. Chem. Phys.* **89**, 6736 (1988); (c) M. Morillo, D.-Y. Yang, and R. I. Cukier, *J. Chem. Phys.* **90**, 5711 (1989); (d) D. Y. Yang and R. I. Cukier, *J. Chem. Phys.* **91**, 281 (1989); (e) R. I. Cukier and D. G. Nocera, *J. Chem. Phys.* **97**, 7371 (1992).
102. T. Fonseca, *J. Chem. Phys.* **91**, 2869 (1989).
103. (a) B. Bagchi, G. R. Fleming, and D. Oxtoby, *J. Chem. Phys.* **78**, 7375 (1983); (b) B. Bagchi, A. Chandra, and G. R. Fleming, *J. Phys. Chem.* **94**, 9 (1990).
104. (a) B. Bagchi, *Annu. Rev. Phys. Chem.* **40**, 115 (1989); (b) B. Bagchi and A. Chandra, *Adv. Chem. Phys.* **80**, 1 (1991).
105. (a) N. Gayathri and B. Bagchi, *J. Phys. Chem.* **100**, 3056 (1996); (b) *J. Phys. Chem. A* **103**, 8496 (1999).
106. B. Bagchi and N. Gayathri, *Adv. Chem. Phys.* **107**, 1 (1999).
107. (a) S. Roy and B. Bagchi, *J. Phys. Chem.* **98**, 9207 (1994); (b) *J. Chem. Phys.* **100**, 8802 (1994); (c) **102**, 6719, 7937 (1995).
108. (a) J. Zhu and J. C. Rasaiah, *J. Chem. Phys.* **95**, 3325 (1991); (b) **96**, 1435 (1992); (c) **98**, 1213 (1993); (d) **101**, 9966 (1994).
109. (a) A. Chandra, *Chem. Phys.* **238**, 285 (1998); (b) *J. Chem. Phys.* **110**, 1569 (1999).
110. (a) Z. Wang, J. Tang, and J. R. Norris, *J. Chem. Phys.* **97**, 7251 (1992); (b) J. Tang, *J. Chem. Phys.* **104**, 9408 (1996).
111. (a) Y. Tanimura and S. Mukamel, *Phys. Rev. B* **47**, 118 (1993); (b) *J. Chem. Phys.* **101**, 3049 (1994).
112. K. L. Sebastian, *Phys. Rev. A* **46**, R1732 (1992).
113. A. I. Burshtein, P. A. Frantsuzov, and A. A. Zharikov, *J. Chem. Phys.* **96**, 4261 (1992).
114. P. A. Frantsuzov, *J. Chem. Phys.* **111**, 2075 (1999).
115. M. Tachiya and S. Murata, *J. Am. Chem. Soc.* **116**, 2434 (1994).

116. (a) J. Jean, R. A. Friesner, and G. R. Fleming, *J. Chem. Phys.* **96**, 5827 (1992); (b) W. T. Pollard, A. K. Felts, and R. A. Friesner, *Adv. Chem. Phys.* **93**, 77 (1996).
117. G. A. Voth, D. Chandler, and W. H. Miller, *J. Chem. Phys.* **91**, 91 (1989).
118. (a) J. N. Gehlen and D. Chandler, *J. Chem. Phys.* **97**, 4958 (1992); (b) J. N. Gehlen, D. Chandler, H. J. Kim, and J. T. Hynes, *J. Phys. Chem.* **96**, 1748 (1992).
119. (a) X. Song and A. A. Stuchebrukhov, *J. Chem. Phys.* **99**, 969 (1993); (b) A. A. Stuchebrukhov and X. Song, *J. Chem. Phys.* **101**, 9354 (1994).
120. (a) D. E. Makarov and M. Topaler, *Phys. Rev. E* **52**, R2125 (1995); **54**, 2174 (1996); (b) *Chem. Phys. Lett.* **245**, 343 (1995).
121. (a) M. Pudlak, *Chem. Phys. Lett.* **221**, 86 (1994); (b) **235**, 126 (1995); (c) *J. Chem. Phys.* **108**, 5621 (1998).
122. (a) M. J. Hornbach and Y. Dakhnovskii, *J. Chem. Phys.* **111**, 5073 (1999); (b) Y. Dakhnovskii, *J. Chem. Phys.* **111**, 5418 (1999).
123. (a) J. Cao, C. Minichino, and G. A. Voth, *J. Chem. Phys.* **103**, 1391 (1995); (b) J. Cao and G. A. Voth, *J. Chem. Phys.* **105**, 6856 (1996); (c) **106**, 1769 (1997); (d) G. A. Voth, *Adv. Chem. Phys.* **93**, 135 (1996); (e) C. D. Schwieters and G. A. Voth, *J. Chem. Phys.* **108**, 1055 (1998).
124. (a) R. Egger and C. H. Mak, *J. Chem. Phys.* **99**, 2541 (1993); (b) C. H. Mak and R. Egger, *Adv. Chem. Phys.* **93**, 39 (1996); (c) J. T. Stockburger and C. H. Mak, *J. Chem. Phys.* **105**, 8126 (1996); (d) J. T. Stockburger and C. H. Mak, *Phys. Rev. Lett.* **80**, 2567 (1998); (e) A. Lucke, C. H. Mak, R. Egger, J. Ankerhold, J. Stockburger, and H. Grabert, *J. Chem. Phys.* **107**, 8397 (1997); (f) C. H. Mak and R. Egger, *J. Chem. Phys.* **110**, 12 (1999); (g) J. T. Stockburger and C. H. Mak, *J. Chem. Phys.* **110**, 4983 (1999); (h) R. Egger, L. Muhlbacher, and C. H. Mak, *Phys. Rev. E* **61**, 5961 (2000).
125. (a) D. E. Makarov and N. Makri, *Chem. Phys. Lett.* **221**, 482 (1994); (b) N. Makri and D. E. Makarov, *J. Chem. Phys.* **102**, 4600, 4611 (1995).
126. M. Topaler and N. Makri, *J. Phys. Chem.* **100**, 4430 (1996).
127. C. Fuchs and M. Schreiber, *J. Chem. Phys.* **105**, 1023 (1996).
128. M. Takasu, *Phys. Rev. E* **52**, 418 (1995).
129. E. R. Bittner and P. J. Rossky, *J. Chem. Phys.* **103**, 8130 (1995).
130. K. Wynne and R. M. Hochstrasser, *Adv. Chem. Phys.* **107**, 263 (1999).
131. K. V. Mikkelsen and M. A. Ratner, *J. Chem. Phys.* **90**, 4237 (1989).
132. D. G. Evans, A. Nitzan, and M. A. Ratner, *J. Chem. Phys.* **108**, 6387 (1998).
133. G. Ashkenazi, R. Kosloff, and M. A. Ratner, *J. Am. Chem. Soc.* **121**, 3386 (1999).
134. H. Wang, X. Song, D. Chandler, and W. H. Miller, *J. Chem. Phys.* **110**, 4828 (1999).
135. Y. Dakhnovskii, V. Lubchenko, and P. Wolynes, *J. Chem. Phys.* **104**, 1875 (1996).
136. (a) I. Rips and E. Pollak, *J. Chem. Phys.* **103**, 7912 (1995); (b) I. Rips, *J. Chem. Phys.* **104**, 9795 (1996); (c) I. Benjamin and E. Pollak, *J. Chem. Phys.* **105**, 9093 (1996); (d) A. Starobinets, I. Rips, and E. Pollak, *J. Chem. Phys.* **104**, 6547 (1996).
137. I. Rips, *Chem. Phys.* **235**, 243 (1998).
138. (a) S.-Y. Sheu and D.-Y. Yang, *Phys. Rev. E* **49**, 4704 (1994); (b) D.-Y. Yang and S.-Y. Sheu, *J. Chem. Phys.* **107**, 9361 (1997).
139. M. Cho and R. J. Silbey, *J. Chem. Phys.* **106**, 2654 (1997).
140. M. Cho and G. R. Fleming, *Adv. Chem. Phys.* **107**, 311 (1999).

141. (a) M. V. Basilevsky and A. I. Voronin, *J. Chem. Soc. Faraday Trans.* **93**, 989 (1997); (b) M. V. Basilevsky, A. V. Soudackov, and A. I. Voronin, *Chem. Phys.* **235**, 281 (1998).

142. D. J. Bicout and A. Szabo, *J. Chem. Phys.* **109**, 2325 (1998).

143. (a) Y. Jung, R. J. Silbey, and J. Cao, *J. Phys. Chem. A* **103**, 9460 (1999); (b) J. Cao and Y. Jung, *J. Chem. Phys.* **112**, 4716 (2000).

144. J. Cao, *J. Chem. Phys.* **112**, 6719 (2000).

145. (a) E. M. Kosower and D. Huppert, *Annu. Rev. Phys. Chem.* **37**, 127 (1986); (b) D. Huppert, V. Ittah, and E. M. Kosower, *Chem. Phys. Lett.* **144**, 15 (1988); (c) D. Huppert, V. Ittah, A. Masad, and E. M. Kosower, *Chem. Phys. Lett.* **150**, 349 (1988).

146. (a) G. E. McManis and M. J. Weaver, *J. Chem. Phys.* **91**, 1720 (1989); (b) M. J. Weaver and G. E. McManis, *Acc. Chem. Res.* **23**, 294 (1990); (c) G. E. McManis, A. Gochev, and M. J. Weaver, *Chem. Phys.* **152**, 107 (1991); (d) M. J. Weaver, *Chem. Rev.* **92**, 463 (1992).

147. (a) M. A. Kahlow, W. Jarzęba, T. J. Kang, and P. F. Barbara, *J. Chem. Phys.* **90**, 151 (1989); (b) T. J. Kang, W. Jarzęba, P. F. Barbara, and T. Fonseca, *Chem. Phys.* **149**, 81 (1990); (c) K. Tominaga, G. C. Walker, T. J. Kang, P. F. Barbara, and T. Fonseca, *J. Phys. Chem.* **95**, 10485 (1991).

148. G. Grampp and W. Jaenicke, *Ber. Bunsenges. Phys. Chem.* **95**, 904 (1991).

149. M. L. Horng, K. Dahl, G. Jones, and M. Maroncelly, *Chem. Phys. Lett.* **315**, 363 (1999).

150. G. Grampp, S. Landgraf, and K. Rasmussen, *J. Chem. Soc., Perkin Trans.* 2, 1897 (1999).

151. (a) T. Kobayashi, Y. Takagi, H. Kandori, K. Kemnitz, and K. Yoshihara, *Chem. Phys. Lett.* **180**, 416 (1991); (b) H. Kandori, K. Kemnitz, and K. Yoshihara, *J. Phys. Chem.* **96**, 8042 (1992); (c) Y. Nagasawa, A. P. Yartsev, K. Tominaga, A. E. Johnson, and K. Yoshihara, *J. Chem. Phys.* **101**, 5717 (1994); (d) K. Yoshihara, K. Tominaga, and Y. Nagasawa, *Bull. Chem. Soc. Jpn.* **68**, 696 (1995); (e) H. Pal, Y. Nagasawa, K. Tominaga, and K. Yoshihara, *J. Phys. Chem.* **100**, 11964 (1996), (f) H. Shirota, H. Pal, K. Tominaga, and K. Yoshihara, *J. Phys. Chem.* **102**, 3089 (1998); (g) H. Pal, H. Shirota, K. Tominaga, and K. Yoshihara, *J. Phys. Chem.* **110**, 11454 (1999).

152. K. Yoshihara, *Adv. Chem. Phys.* **107**, 371 (1999).

153. (a) I. V. Rubtsov and K. Yoshihara, *J. Phys. Chem. A* **101**, 6138 (1997); (b) **103**, 10202 (1999); (c) I. V. Rubtsov, H. Shirota, and K. Yoshihara, *J. Phys. Chem. A* **103**, 1801 (1999).

154. (a) M. Seel, S. Engleitner, and W. Zinth, *Chem. Phys. Lett.* **275**, 363 (1997); (b) S. Engleitner, M. Seel, and W. Zinth, *J. Phys. Chem. A* **103**, 3013 (1999).

155. (a) F. Pöllinger, H. Heitele, M. E. Michel-Beyerle, C. Anders, M. Fulscher, and H. A. Staab, *Chem. Phys. Lett.* **198**, 642 (1992); (b) H. Heitele, F. Pöllinger, T. Häberle, M. E. Michel-Beyerle, and H. A. Staab, *J. Phys. Chem.* **98**, 7402 (1994); (c) T. Häberle, J. Hirsch, F. Pöllinger, H. Heitele, M. E. Michel-Beyerle, C. Anders, A. Döhling, C. Krieger, A. Rückemann, and H. A. Staab, *J. Phys. Chem.* **100**, 18269 (1996); (d) P. Gilch, F. Pöllinger-Dammer, U. E. Steiner, and M. E Michel-Beyerle, *Chem. Phys. Lett.* **275**, 339 (1997); (e) H. Heitele, *Angew. Chem. Int. Ed. Engl.* **32**, 359 (1993).

156. (a) E. Åkesson, G. C. Walker, and P. F. Barbara, *J. Chem. Phys.* **95**, 4188 (1991); (b) G. C. Walker, E. Åkesson, A. E. Johnson, N. E. Levinger, and P. F. Barbara, *J. Phys. Chem.* **96**, 3728 (1992); (c) E. Åkesson, A. E. Johnson, N. E. Levinger, G. C. Walker, T. P. DeBruil, and P. F. Barbara, *J. Chem. Phys.* **96**, 7859 (1992).

157. (a) K. Tominaga, G. C. Walker, W. Jarzęba, and P. F. Barbara, *J. Phys. Chem.* **95**, 10475 (1991); (b) D. A. V. Kliner, K. Tominaga, G. C. Walker, and P. F. Barbara, *J. Am. Chem. Soc.* **114**, 8323 (1992); (c) K. Tominaga, D. A. V. Kliner, A. E. Johnson, N. E. Levinger, and P. F. Barbara, *J. Chem. Phys.* **98**, 1228 (1993).

158. P. J. Reid and P. F. Barbara, *J. Phys. Chem.* **99**, 17311 (1995).
159. (a) S. Iwai, S. Murata, and M. Tachiya, *J. Chem. Phys.* **109**, 5963 (1998); (b) W. Jarzęba, S. Murata, and M. Tachiya, *Chem. Phys. Lett.* **301**, 347 (1999).
160. Q.-H. Xu, G. D. Scholes, M. Yang, and G. R. Fleming, *J. Phys. Chem. A* **103**, 10348 (1999).
161. K. Seki, A. V. Barzykin, and M. Tachiya, *J. Chem. Phys.* **110**, 7639 (1999).
162. H. Sumi, *J. Phys. Soc. Jpn.* **49**, 1701 (1980).
163. (a) A. I. Burshtein and Y. Georgievskii, *J. Chem. Phys.* **100**, 7319 (1994); (b) Y. Georgievskii, A. I. Burshtein, and B. M. Chernobrod, *J. Chem. Phys.* **105**, 3108 (1996).
164. (a) A. I. Shushin, *Chem. Phys.* **60**, 149 (1981); (b) *Chem. Phys. Lett.* **146**, 297 (1988); (c) A. I. Shushin and M. Tachiya, *Chem. Phys.* **235**, 267 (1998).
165. (a) Y. Kayanuma, *J. Phys. Soc. Jpn.* **53**, 108, 118 (1984); (b) Y. Kayanuma and H. Nakayama, *Phys. Rev. B* **57**, 13099 (1998).
166. (a) P. Ao and J. Rammer, *Phys. Rev. Lett.* **62**, 3004 (1989); (b) *Phys. Rev. B* **43**, 5397 (1991).
167. (a) A. A. Zharikov and P. A. Frantsuzov, *Chem. Phys. Lett.* **220**, 319 (1994); (b) P. A. Frantsuzov, S. F. Fischer, and A. A. Zharikov, *Chem. Phys.* **241**, 95 (1999).
168. P. A. Frantsuzov and M. Tachiya, *J. Chem. Phys.* **112**, 4216 (2000).
169. D. Mauzerall and S. G. Ballard, *Annu. Rev. Phys. Chem.* **33**, 377 (1982).
170. M. D. Newton and N. Sutin, *Annu. Rev. Phys. Chem.* **35**, 437 (1984).
171. R. A. Marcus and N. Sutin, *Biochim. Biophys. Acta* **811**, 265 (1985).
172. R. A. Marcus, *Rev. Mod. Phys.* **65**, 599 (1993).
173. (a) P. F. Barbara and W. Jarzęba, *Adv. Photochem.* **15**, 1 (1990); (b) P. F. Barbara, T. J. Meyer, and M. A. Ratner, *J. Phys. Chem.* **100**, 13148 (1996).
174. *Photoinduced Electron Transfer*, edited by M. A. Fox and M. Chanon (Elsevier, Amsterdam, 1988).
175. *Electron Transfer in Inorganic, Organic, and Biological Systems*, edited by J. R. Bolton, N. Mataga, and G. McLendon, Advances in Chemistry Series **228** (ACS, Washington, DC, 1991).
176. *Chem. Rev.* **92**, 365–490, Special Issue, edited by M. A. Fox (1992).
177. *Chem. Phys.* **176**, Special Issue, edited by T. J. Meyer and M. D. Newton (1993).
178. A. M. Kuznetsov and J. Ulstrup, *Electron Transfer in Chemistry and Biology* (John Wiley & Sons, Ltd., Chichester, 1998).
179. *Adv. Chem. Phys.* **106–107**, *Electron Transfer: From Isolated Molecules to Biomolecules*, edited by J. Jortner and M. Bixon (John Wiley & Sons, Inc., New York, 1999).
180. R. Kubo, M. Toda, and N. Hashitsume, *Statistical Physics II* (Springer, Berlin, 1985).
181. D. R. Reichman, F. L. H. Brown, and P. Neu, *Phys. Rev. E* **55**, 2328 (1997).
182. (a) S. Mukamel, *J. Phys. Chem.* **89**, 1077 (1985); (b) G. R. Fleming and M. Ado, *Annu. Rev. Phys. Chem.* **47**, 109 (1996).
183. C. Aslangul, N. Pottier, and D. Saint-James, *J. Phys.* **47**, 1657 (1986).
184. J. L. Skinner and D. Hsu, *J. Phys. Chem.* **90**, 4931 (1986).
185. (a) J. S. Bader and B. J. Berne, *J. Chem. Phys.* **104**, 1293 (1995); (b) S. A. Egorov and B. J. Berne, *J. Chem. Phys.* **107**, 6050 (1997); (c) S. A. Egorov, E. Ribani, and B. J. Berne, *J. Chem. Phys.* **108**, 1407 (1998); (d) E. Ribani, S. A. Egorov, and B. J. Berne, *J. Chem. Phys.* **109**, 6379 (1998); (e) S. A. Egorov, E. Ribani, and B. J. Berne, *J. Chem. Phys.* **110**, 5238 (1999).

186. S. H. Lin, *J. Chem. Phys.* **44**, 3759 (1966).
187. R. Englman and J. Jortner, *Mol. Phys.* **18**, 145 (1970).
188. (a) R. R. Dogonadze, E. M. Itskovich, A. M. Kuznetsov, and M. A. Vorotyntsev, *J. Phys. Chem.* **79**, 2827 (1975); (b) R. R. Dogonadze, A. M. Kuznetsov, M. A. Vorotyntsev, and M. G. Zaqaraia, *Electroanal. Chem.* **75**, 315 (1977).
189. V. M. Zolotarev and A. V. Demin, *Opt. Spectrosc.* **43**, 157 (1977).
190. (a) G. M. Hale and M. R. Querry, *Appl. Opt.* **12**, 555 (1973); (b) M. N. Asfar and J. B. Haster, *Infrared Phys.* **18**, 835 (1978); (c) J. B. Hasted, S. K. Husain, F. A. M. Frescura, and J. R. Birch, *Infrared Phys.* **27**, 11 (1987).
191. (a) S. Roy and B. Bagchi, *J. Chem. Phys.* **99**, 9938 (1993); (b) N. Nandi, S. Roy, and B. Bagchi, *J. Chem. Phys.* **102**, 1390 (1995); (c) R. Biswas, N. Nandi, and B. Bagchi, *J. Phys. Chem.* **101**, 2968 (1997).
192. R. Jimenez, G. R. Fleming, P. V. Kumar, and M. Maroncelli, *Nature (London)* **369**, 471 (1994).
193. (a) P. G. Wolynes, *J. Chem. Phys.* **86**, 5133 (1987); (b) I. Rips, J. Klafter, and J. Jortner, *J. Chem. Phys.* **88**, 3246 (1988); (c) **89**, 4288 (1988).
194. M. L. Horng, J. A. Gardecki, A. Papazyan, and M. Maroncelli, *J. Phys. Chem.* **99**, 17311 (1995).
195. (a) M. V. Basilevsky and D. F. Parsons, *J. Chem. Phys.* **108**, 9114 (1998); (b) D. F. Parsons, M. V. Vener, and M. V. Basilevsky, *J. Phys. Chem. A* **103**, 1171 (1999).
196. (a) P. A. Bopp, A. A. Kornyshev, and G. Sutmann, *Phys. Rev. Lett.* **76**, 1280 (1996); (b) A. A. Kornyshev and G. Sutmann, *Phys. Rev. Lett.* **79**, 3435 (1997); (c) P. A. Bopp, A. A. Kornyshev, and G. Sutmann, *J. Chem. Phys.* **109**, 1939 (1998).
197. (a) B. D. Fainberg and D. Huppert, *Adv. Chem. Phys.* **107**, 191 (1999); (b) B. Bagchi and R. Biswas, *Adv. Chem. Phys.* **109**, 207 (1999).
198. L. W. Ungar, M. D. Newton, and G. A. Voth, *J. Phys. Chem. B* **103**, 7367 (1999).
199. S. A. Egorov and J. L. Skinner, *Chem. Phys. Lett.* **293**, 469 (1998).
200. (a) Y. Nagasawa, S. A. Passimo, T. Joo, and G. R. Fleming, *J. Chem. Phys.* **106**, 4840 (1997); (b) S. A. Passimo, Y. Nagasawa, and G. R. Fleming, *J. Chem. Phys.* **107**, 6094 (1997).
201. M. Neumann, *J. Chem. Phys.* **85**, 1567 (1986).
202. (a) J. E. Bertie and Z. Lan, *Appl. Spectrosc.* **50**, 1047 (1996); (b) *J. Phys. Chem. B* **101**, 4111 (1997); experimental data can be downloaded from http://www.ualberta.ca/~jbertie/jebhome.htm.
203. U. Weiss, *Quantum Dissipative Systems* (World Scientific, Singapore, 1993).
204. D. Chandler, in *Liquids, Freezing, and Glass Transition*, edited by D. Levesque, J. P. Hansen, and J. Zinn-Justin (Elsevier, Amsterdam, 1991), Part 1, p. 193.
205. J. R. Miller, B. P. Paulson, R. Bal, and G. L. Closs, *J. Phys. Chem.* **99**, 6923 (1995).
206. (a) M. V. Basilevsky, G. E. Chudinov, I. V. Rostov, Y.-P. Liu, and M. D. Newton, *J. Mol. Struct.* **371**, 191 (1996); (b) M. D. Newton, M. V. Basilevsky, and I. V. Rostov, *Chem. Phys.* **232**, 201 (1998).
207. E. D. German and A. M. Kuznetsov, *Electrochim. Acta* **26**, 1595 (1981).
208. (a) B. Brunschwig, S. Ehrenson, and N. Sutin, *J. Phys. Chem.* **90**, 3657 (1986); (b) **91**, 4714 (1987).

209. (a) M. J. Powers and T. J. Meyer, *J. Am. Chem. Soc.* **100**, 4393 (1978); (b) G. E McMannis, A. Gochev, R. M. Nielson, and M. J. Weaver, *J. Phys. Chem.* **93**, 7733 (1983); (c) R. L. Blackbourn and J. T. Hupp, *J. Phys. Chem.* **94**, 1788 (1990).

210. (a) Y. I. Kharkats, *Electrokhimiya* **9**, 881 (1973); (b) **10**, 612 (1974).

211. M. Tachiya, *Chem. Phys. Lett.* **230**, 491 (1994).

212. K. Miyazaki and M. Tachiya, *J. Chem. Phys.* **109**, 7424 (1998).

213. X. Song, D. Chandler, and R. A. Marcus, *J. Phys. Chem.* **100**, 11954 (1996).

214. (a) Y.-P. Liu and M. D. Newton, *J. Phys. Chem.* **98**, 7162 (1994); (b) **99**, 12382 (1995).

215. R. A. Marcus, *J. Phys. Chem.* **98**, 7170 (1994).

216. A. M. Kuznetsov and I. G. Medvedev, *J. Phys. Chem.* **100**, 5721 (1996).

217. M. D. Newton, *Adv. Chem. Phys.* **106**, 303 (1999).

218. Y. I. Kharkats and L. I. Krishtalik, *J. Theor. Biol.* **112**, 221 (1985).

219. (a) Y. I. Kharkats and J. Ulstrup, *Chem. Phys.* **141**, 117 (1990); (b) *Chem. Phys. Lett.* **303**, 320 (1999).

220. (a) K. Weidemaier, H. L. Tavernier, and M. D. Fayer, *J. Phys. Chem. B* **101**, 9352 (1997); (b) H. L. Tavernier, A. V. Barzykin, M. Tachiya, and M. D. Fayer, *J. Phys. Chem. B* **102**, 6078 (1998); (c) A. V. Barzykin and M. Tachiya, *Chem. Phys. Lett.* **285**, 150 (1998).

221. Y. Q. Gao, Y. Georgievskii, and R. A. Marcus, *J. Chem. Phys.* **112**, 3358 (2000).

222. H.-X. Zhou, *J. Chem. Phys.* **105**, 3726 (1996).

223. B.-C. Perng, M. D. Newton, F. O. Raineri, and H. L. Friedman, *J. Chem. Phys.* **104**, 7153, 7177 (1996).

224. R. G. Alden, W. W. Parson, Z. T. Chu, and A. Warshel, *J. Am. Chem. Soc.* **117**, 12284 (1995).

225. (a) M. Hilczer and M. Tachiya, *J. Mol. Liq.* **64**, 113 (1995); (b) T. Kato, M. Hilczer, and M. Tachiya, *Chem. Phys. Lett.* **284**, 350 (1998).

226. (a) D. V. Matyushov, *Chem. Phys.* **174**, 199 (1993); (b) **251**, 1 (1996); (c) D. V. Matyushov and R. Schmid, *J. Phys. Chem.* **98**, 5152 (1994).

227. (a) P. Vath, M. Zimmt, D. V. Matyushov, and G. A. Voth, *J. Phys. Chem. B* **103**, 9130 (1999); (b) D. V. Matyushov and G. A. Voth, *J. Phys. Chem. A* **103**, 10981 (1999).

228. L. Reynolds, J. A. Gardecki, S. J. V. Frankland, M. L. Horng, and M. Maroncelli, *J. Phys. Chem.* **100**, 10337 (1996).

229. T. Fonseca and B. Ladanyi, *J. Chem. Phys.* **93**, 8148 (1990).

230. F. O. Raineri, H. Resat, and H. L. Friedman, *J. Chem. Phys.* **96**, 3068 (1992).

231. D. V. Matyushov, *Mol. Phys.* **84**, 533 (1995).

232. H. J. Kim, *J. Chem. Phys.* **105**, 6818, 6833 (1996).

233. M. V. Basilevsky and D. F. Parsons, *J. Chem. Phys.* **105**, 3734 (1996).

234. C. D. Clark and M. Z Hoffman, *J. Phys. Chem.* **100**, 14688 (1996).

235. (a) A. M. Kuznetsov, D. K. Phelps, and M. Weaver, *Int. J. Chem. Kinet.* **22**, 815 (1990); (b) R. L. Blackbourn, Y. Dong, A. Lyon, and J. T. Hupp, *Inorg. Chem.* **33**, 4446 (1994); (c) P. Piotrowiak and J. R. Miller, *J. Phys. Chem.* **97**, 13052 (1993); (d) P. Piotrowiak, R. Kobetic, T. Schatz, and G. Stratti, *J. Phys. Chem.* **99**, 2250 (1995); (e) B. R. Arnold, D. Noukakis, S. Farid, J. L. Goodman, and I. R. Gould, *J. Am. Chem. Soc.* **117**, 4399 (1995).

236. E. D. German and A. M. Kuznetsov, *Elektrokhimiya* **23**, 1671 (1987).

237. R. A. Marcus, *J. Phys. Chem. B* **102**, 10071 (1998).

238. (a) D. Sitkoff, K. A. Sharp, and B. Honig, *J. Phys. Chem.* **98**, 1978 (1994); (b) D. Sitkoff, A. Nicholls, M. Ringnalda, W. A. Goddard, and B. Honig, *J. Am. Chem. Soc.* **116**, 11875 (1994); (c) B. Marten, K. Kim, C. Cortis, R. A. Friesner, R. B. Murphy, M. N. Ringnalda, D. Sitkoff, and B. Honig, *J. Phys. Chem.* **100**, 11775 (1996).
239. I. V. Kurnikov, L. D. Zusman, M. G. Kurnikova, R. S. Farid, and D. N. Beratan, *J. Am. Chem. Soc.* **119**, 5690 (1997).
240. F. O. Raineri and H. L. Friedman, *Adv. Chem. Phys.* **107**, 81 (1999).
241. J. R. Bolton and M. D. Archer, in [175], p. 7.
242. A. Weller, *Z. Phys. Chem.* **133**, 93 (1982).
243. (a) S. Franzen, R. F. Goldstein, and S. G. Boxer, *J. Phys. Chem.* **94**, 5135 (1990); (b) S. Franzen and S. G. Boxer, *J. Phys. Chem.* **97**, 6304 (1993).
244. K. Seki, S. D. Traytak, and M. Tachiya, The 5th AIST International Symposium on Photoreaction Control and Photofunctional Materials 2002 (Japan), Extended Abstracts, p. 80.
245. (a) S. Levine and H. E. Wrigley, *Discuss. Faraday Soc.* **24**, 43 (1957); (b) R. A. Marcus, *J. Chem. Phys.* **43**, 58 (1965).
246. M. D. Newton, *Chem. Rev.* **91**, 767 (1991).
247. (a) C. X. Liang and M. D. Newton, *J. Phys. Chem.* **96**, 2855 (1992); (b) **97**, 3199 (1993).
248. H. M. McConnell, *J. Chem. Phys.* **35**, 508 (1961).
249. (a) C. C. Moser, J. M. Keske, K. Warncke, R. S. Farid, and P. L. Dutton, *Nature (London)* **355**, 796 (1992); (b) H. B. Gray and J. R. Winkler, *Ann. Rev. Biochem.* **65**, 537 (1996); (c) P. F. Barbara and E. J. C. Olson, *Adv. Chem. Phys.* **107**, 647 (1999).
250. (a) V. Mujica, M. Kemp, A. Roitberg, and M. A. Ratner, *J. Chem. Phys.* **104**, 7296 (1996); (b) M. Kemp, A. Roitberg, V. Mujica, T. Wanta, and M. A. Ratner, *J. Phys. Chem.* **100**, 8349 (1996); (c) V. Mujica, A. Nitzan, Y. Mao, W. Davis, M. Kemp, A. Roitberg, and M. A. Ratner, *Adv. Chem. Phys.* **107**, 403 (1999).
251. S. H. Lin, *J. Chem. Phys.* **90**, 7103 (1989).
252. S. S. Skourtis and S. Mukamel, *Chem. Phys.* **197**, 367 (1995).
253. (a) H. Sumi and T. Kakitani, *Chem. Phys. Lett.* **252**, 85 (1996); (b) A. Kimura and T. Kakitani, *Chem. Phys. Lett.* **298**, 241 (1998).
254. (a) R. J. Cave and M. D. Newton, *Chem. Phys. Lett.* **249**, 15 (1996); (b) *J. Chem. Phys.* **106**, 9213 (1997).
255. (a) S. Larsson, *J. Am. Chem. Soc.* **103**, 4034 (1981); (b) *J. Chem. Soc. Faraday Trans 2* **79**, 1375 (1983).
256. (a) J. N. Onuchic, D. N. Beratan, J. Winkler, and H. B. Gray, *Annu. Rev. Biophys. Biomol. Struct.* **21**, 349 (1992); (b) *Science* **258**, 1740 (1992).
257. S. S. Skourtis and D. N. Beratan, *Adv. Chem. Phys.* **106**, 377 (1999).
258. G. Iversen, Y. I. Kharkats, A. M. Kuznetsov, and J. Ulstrup, *Adv. Chem. Phys.* **106**, 453 (1999).
259. J. Tang, *Chem. Phys. Lett.* **217**, 55 (1994).
260. M. Bixon and J. Jortner, *J. Chem. Phys.* **107**, 5154 (1997).
261. D. N. Beratan, J. N. Onuchic, and J. J. Hopfield, *J. Chem. Phys.* **86**, 4488 (1987).
262. I. V. Kurnikov and D. N. Beratan, *J. Chem. Phys.* **105**, 9561 (1996).
263. J. J. Regan and J. N. Onichic, *Adv. Chem. Phys.* **107**, 497 (1999).
264. (a) A. J. A. Aquino, P. Beroza, J. Reagan, and J. N. Onuchic, *Chem. Phys. Lett.* **275**, 181 (1997); (b) P. C. P. de Andrade and J. Onuchic, *J. Chem. Phys.* **108**, 4292 (1998).

265. (a) A. A. Stuchebrukhov, *J. Chem. Phys.* **104**, 8424 (1996); (b) **105**, 10819 (1996); (c) **107**, 6495 (1997); (d) **108**, 8499, 8510 (1998); (e) I. Daizadeh, J.-X. Guo, and A. A. Stuchebrukhov, *J. Chem. Phys.* **110**, 8865 (1999).

266. O. V. Prezhdo, J. T. Kindt, and J. C. Tully, *J. Chem. Phys.* **111**, 7818 (1999).

267. C.-P. Hsu and R. A. Marcus, *J. Chem. Phys.* **106**, 584 (1997).

268. M. Galperin, D. Segal, and A. Nitzan, *J. Chem. Phys.* **111**, 1569 (1999).

269. (a) E. Gudowska-Nowak, *Chem. Phys.* **212**, 115 (1996); (b) E. Gudowska-Nowak, G. Papp, and J. Brickmann, *Chem. Phys.* **232**, 247 (1998).

270. (a) P. Siddarth and R. A. Marcus, *J. Phys. Chem.* **94**, 2985 (1990); (b) **94**, 8430 (1990); (c) **96**, 3213 (1992); (d) **97**, 2400 (1993); (e) **97**, 6111 (1993).

271. A. Kuki and P. G. Wolynes, *Science* **236**, 1647 (1987).

272. S. Priyadarshy, S. S. Skourtis, D. N. Beratan, and S. M. Risser, *J. Chem. Phys.* **104**, 9473 (1996).

273. (a) P. Siddarth and R. A. Marcus, *J. Phys. Chem.* **97**, 13078 (1993); (b) A. A. Stuchebrukhov, *Chem. Phys. Lett.* **225**, 55 (1994); (c) A. A. Stuchebrukhov and R. A. Marcus, *J. Phys. Chem.* **99**, 7581 (1995).

274. D. J. Katz and A. A. Stuchebrukhov, *J. Chem. Phys.* **109**, 4960 (1998).

275. J. N. Gehlen, I. Daizadeh, A. A. Stuchebrukhov, and R. A. Marcus, *Inorg. Chim. Acta* **243**, 271 (1996).

276. G. Gamow, *Z. Phys.* **51**, 204 (1961).

277. (a) A. S. Davydov, *Phys. Stat. Solid. B* **90**, 457 (1978); (b) A. S. Davydov and Y. B. Gaididei, *Phys. Stat. Solid. B* **132**, 189 (1985); (c) E. G. Petrov, *Int. J. Quant. Chem.* **16**, 133 (1978).

278. T. Kakitani and N. Mataga, *J. Phys. Chem.* **89**, 4762 (1985).

279. N. E. Miller, M. C. Wander, and R. J. Cave, *J. Phys. Chem. A* **103**, 1084 (1999).

280. (a) J. R. Miller, K. W. Hartman, and S. Abrash, *J. Am. Chem. Soc.* **104**, 4296 (1982); (b) I. R. Gould, R. H. Young, L. J. Mueller, A. C. Albrecht, and S. Farid, *J. Am. Chem. Soc.* **116**, 3147, 8188 (1994); (c) K. Kumar, Z. Lin, D. H. Waldeck, and M. B. Zimmt, *J. Am. Chem. Soc.* **118**, 243 (1996).

281. (a) S. Larsson, *J. Phys. Chem.* **88**, 1321 (1984); (b) M. D. Newton, *J. Phys. Chem.* **92**, 3049 (1988); (c) M. D. Newton, R. J. Cave, K. Kumar, and M. B. Zimmt, *J. Phys. Chem.* **99**, 17501 (1995); (d) R. J. Cave and T. M. Henderson, *J. Phys. Chem.* **109**, 7414 (1998); (e) K. Kumar, I. V. Kurnikov, D. N. Beratan, D. H. Waldeck, and M. B. Zimmt, *J. Phys. Chem. A* **102**, 5529 (1998); (f) I. Benjamin, D. Evans, and A. Nitzan, *J. Chem. Phys.* **106**, 1291, 6647 (1997).

282. A. I. Burshtein and B. I Yakobson, *Khim. Fiz.* **2**, 479 (1982).

283. D. V. Matyushov and G. A. Voth, *J. Chem. Phys.* **113**, 5413 (2000).

284. (a) R. Islampour and S. H. Lin, *Chem. Phys. Lett.* **179**, 147 (1991); (b) *J. Phys. Chem.* **95**, 10261 (1991); (c) R. Islampour, R. G. Alden, G. Y. C. Wu, and S. H. Lin, *J. Phys. Chem.* **97**, 6793 (1993); (d) A. M. Mebel, M. Hayashi, K. K. Liang, and S. H. Lin, *J. Phys. Chem.* **103**, 10674 (1999).

285. (a) E. J. Heller, *J. Chem. Phys.* **62**, 1544 (1975); (b) **68**, 2066 (1978); (c) **75**, 2923 (1981).

286. M. A. Ratner and A. Madhukar, *Chem. Phys.* **30**, 201 (1978).

287. (a) A. I. Burshtein, G. K. Ivanov, and M. A. Kozhushner, *Chem. Phys. Lett.* **84**, 135 (1981); (b) *Khim. Fiz.* **2**, 195 (1982).

288. (a) S. Franzen, R. F. Goldstein, and W. Bialek, *J. Phys. Chem.* **97**, 3040 (1993); (b) R. F. Goldstein, S. Franzen, and W. Bialek, *J. Phys. Chem.* **97**, 11168 (1993).
289. (a) A. A. Belousov, A. M. Kuznetsov, and J. Ulstrup, *Chem. Phys.* **129**, 311 (1989); (b) A. M. Kuznetsov, M. D. Vigdorovich, and J. Ulstrup, *Chem. Phys.* **176**, 539 (1993).
290. R. Elber and M. Karplus, *J. Am. Chem. Soc.* **112**, 9161 (1990).
291. (a) J. A. McCammon and S. H. Northrup, *Nature (London)* **293**, 316 (1981); (b) A. Szabo, D. Shoup, S. H. Northrup, and J. A. McCammon, *J. Chem. Phys.* **77**, 4484 (1982).
292. J. Feitelson and G. McLendon, *Biochemistry* **30**, 5051 (1991).
293. (a) B. Cartling, *J. Chem. Phys.* **83**, 5231 (1985); (b) **95**, 317 (1991).
294. S. B. Sachs, S. P. Dudek, R. P. Hsung, L. R. Sita, J. F. Smalley, M. D. Newton, S. W. Feldberg, and C. E. D. Childsey, *J. Am. Chem. Soc.* **119**, 10563 (1997).
295. (a) S. Nakajima, *Prog. Theor. Phys.* **20**, 948 (1958); (b) R. Zwanzig, *J. Chem. Phys.* **33**, 1338 (1960).
296. H. Grabert, *Projection operator techniques in nonequilibrium statistical mechanics* (Springer, Berlin, 1982).
297. N. G. van Kampen, *Stochastic Processes in Physics and Chemistry* (Elsevier, Amsterdam, 1992).
298. E. E. Nikitin, *Theory of Elementary Atomic and Molecular Processes in Gases* (Clarendon, Oxford, 1974).
299. (a) A. I. Burshtein, B. M. Chernobrod, and A. Yu. Sivachenko, *J. Chem. Phys.* **108**, 9796 (1998); (b) **110**, 1931 (1999); (c) A. I. Burshtein and A. Yu. Sivachenko, *J. Phys.: Condens. Matter* **12**, 173 (2000); (d) *J. Chem. Phys.* **112**, 4699 (2000).
300. E. A. Carter and J. T. Hynes, *J. Chem. Phys.* **94**, 5961 (1991).
301. H. Risken, *The Fokker-Planck Equation* (Springer, Berlin, 1989).
302. (a) G. Wilemski and M. Fixman, *J. Chem. Phys.* **58**, 4009 (1973); (b) **60**, 866, 878 (1974).
303. (a) M. Doi, *Chem. Phys.* **11**, 107, 115 (1975); (b) S. Sunagawa and M. Doi, *Polym. J.* **7**, 604 (1975); (c) **8**, 239 (1976).
304. G. H. Weiss, *J. Chem. Phys.* **80**, 2880 (1984).
305. (a) J. Wang and P. G. Wolynes, *Chem. Phys. Lett.* **212**, 427 (1993); (b) *Chem. Phys.* **180**, 141 (1994).
306. R. L. Stratonovich, *Topics in the Theory of Random Noise* (Gordon and Breach, New York, 1981).
307. A. V. Barzykin and P. A. Frantsuzov, unpublished results.
308. L. D. Zusman, *Sov. Phys. JETP* **42**, 794 (1975).
309. A. V. Barzykin and P. A. Frantsuzov, *J. Chem. Phys.* **114**, 345 (2001).
310. P. Pechukas and J. Ankerhold, *J. Chem. Phys.* **107**, 2444 (1997).
311. (a) N. Agmon and J. J. Hopfield, *J. Chem. Phys.* **78**, 6947 (1983); (b) N. Agmon and S. Robinovich, *J. Chem. Phys.* **97**, 7270 (1992); (c) N. Agmon and G. M. Sastry, *Chem. Phys.* **212**, 207 (1996).
312. R. Zwanzig, *Acc. Chem. Res.* **23**, 148 (1990).
313. (a) A. M. Berezhkovskii and V. Y. Zitserman, *Chem. Phys. Lett.* **158**, 369 (1989); (b) *Physica A* **166**, 585 (1990); (c) *Chem. Phys.* **157**, 141 (1991).
314. O. B. Spirina and R. I. Cukier, *J. Chem. Phys.* **104**, 538 (1996).
315. R. A. Marcus, *J. Chem. Phys.* **105**, 5446 (1996).

316. (a) H. Lu, J. N. Prieskorn, and J. T. Hupp, *J. Am. Chem. Soc.* **115**, 4927 (1993); (b) H. Lu, V. Petrov, and J. T. Hupp, *Chem. Phys. Lett.* **235**, 521 (1995); (c) P. J. Reid, C. Silva, P. F. Barbara, L. Karki, and J. T. Hupp, *J. Phys. Chem.* **99**, 2609 (1995); (d) L. Karki, H. Lu, and J. T. Hupp, *J. Phys. Chem.* **100**, 15637 (1996).

317. (a) E. Pollak, *J. Chem. Phys.* **93**, 1116 (1990); (b) E. Pollak, S. C. Tucker, and B. J. Berne, *Phys. Rev. Lett.* **65**, 1399 (1990); (c) A. M. Berezhkovskii, E. Pollak, and V. Yu. Zitserman, *J. Chem. Phys.* **97**, 2422 (1992); (d) A. M. Berezhkovskii, A. M. Frishman, and E. Pollak, *J. Chem. Phys.* **101**, 4778 (1994).

318. (a) D. A. Zichi, G. Ciccotti, J. T. Hynes, and M. Ferrario, *J. Phys. Chem.* **93**, 6261 (1989); (b) B. B. Smith, A. Staib, and J. T. Hynes, *Chem. Phys.* **176**, 521 (1993).

319. R. F. Grote and J. T. Hynes, *J. Chem. Phys.* **76**, 2715 (1980).

320. A. M. Dykhne, *Zh. Eksp. Teor. Fiz.* **41**, 1324 (1961) [*Sov. Phys. JETP* **14**, 941 (1962)].

321. J. P. Davis and P. Pechukas, *J. Chem. Phys.* **64**, 3129 (1976).

322. (a) T. Asahi and N. Mataga, *J. Phys. Chem.* **95**, 1956 (1991); (b) H. Segawa, C. Takehara, K. Honda, T. Shimizu, T. Asahi, and N. Mataga, *J. Phys. Chem.* **96**, 503 (1992).

323. (a) I. R. Gould, D. Noukakis, L. Gomez-Jahn, J. L. Goodman, and S. Farid, *J. Am. Chem. Soc.* **115**, 4405 (1993); (b) A. C. Benniston, A. Harriman, D. Philp, and J. F. Stoddart, *J. Am. Chem. Soc.* **115**, 5298 (1993).

324. (a) J. C. Tully, *J. Chem. Phys.* **93**, 1061 (1990); (b) D. F. Coker and L. Xiao, *J. Chem. Phys.* **102**, 496 (1995); (c) D. S. Sholl and J. C. Tully, *J. Chem. Phys.* **109**, 7702 (1998).

325. S. A. Egorov, E. Rabani, and B. J. Berne, *J. Phys. Chem. B* **103**, 10978 (1999).

326. (a) L. Hartmann, I. Goychuk, and P. Hänggi, *J. Chem. Phys.* **113**, 11159 (2000); (b) I. Goychuk, L. Hartmann, and P. Hänggi, *Chem. Phys.* **268**, 151 (2001).

327. (a) H. Wang, M. Thoss, and W. H. Miller, *J. Chem. Phys.* **115**, 2979 (2001); (b) M. Thoss, H. Wang, and W. H. Miller, *J. Chem. Phys.* **115**, 2991 (2001).

328. M. J. Pilling and S. A. Rice, *J. Chem. Soc. Faraday Trans.* 2 **71**, 1563 (1975).

329. B. S. Brunschwig, S. Ehrenson, and N. Sutin, *J. Am. Chem. Soc.* **106**, 6858 (1984).

330. (a) A. I. Burshtein, *Chem. Phys. Lett.* **194**, 247 (1992); (b) A. I. Burshtein, E. Krissinel, and M. S. Mikhelashvili, *J. Phys. Chem.* **98**, 7319 (1994).

331. R. C. Dorfman and M. D. Fayer, *J. Chem. Phys.* **96**, 7410 (1992).

332. A. I. Burshtein and P. A. Frantsuzov, *J. Chem. Phys.* **107**, 2872 (1997).

333. A. I. Burshtein, *Adv. Chem. Phys.* **114**, 419 (2000).

334. (a) F. C. Collins and G. E. Kimball, *J. Colloid Sci.* **4**, 425 (1949); (b) R. M. Noyes, *Prog. React. Kinet.* **1**, 129 (1961).

335. M. Tachiya, *Radiat. Phys. Chem.* **21**, 167 (1983).

336. S. A. Rice, *Diffusion-Limite Reactions*, in *Comprehensive Chemical Kinetics*, Vol. 25, edited by C. H. Bamford, C. F. H. Tipper, and R. G. Compton (Elsevier, Amsterdam, 1985).

337. (a) G. C. Joshi, R. Bhatnagar, S. Doraiswamy, and N. Periasamy, *J. Phys. Chem.* **94**, 2908 (1990); (b) S. A. Angel and K. S. Peters, *J. Phys. Chem.* **95**, 3606 (1991); (c) S. Murata, M. Nishimura, S. Y. Matsuzaki, and M. Tachiya, *Chem. Phys. Lett.* **219**, 200 (1994).

338. (a) A. I. Burshtein, P. A. Frantsuzov, and A. A. Zharikov, *Chem. Phys.* **155**, 91 (1991); (b) A. I. Burshtein and P. A. Frantsuzov, *J. Lumin.* **51**, 215 (1992); (c) P. A. Frantsuzov, N. V. Shokhirev, and A. A. Zharikov, *Chem. Phys. Lett.* **236**, 30 (1995); (d) A. I. Burshtein and P. A. Frantsuzov, *Chem. Phys.* **212**, 137 (1996); (e) *Chem. Phys. Lett.* **263**, 513 (1996).

339. (a) N. N. Tunitskii and Kh. S. Bagdasar'yan, *Opt. Spectrosc.* **15**, 303 (1963); (b) S. F. Kilin, M. S. Mikhelashvili, and I. M. Rozman, *Opt. Spectrosc.* **16**, 576 (1964); (c) I. Z. Steinberg and E. Katchalsky, *J. Chem. Phys.* **48**, 2404 (1968); (d) A. B. Doktorov and A. I. Burshtein, *Sov. Phys. JETP* **41**, 4 (1975).

340. H. Sano and M. Tachiya, *J. Chem. Phys.* **71**, 1276 (1979).

341. A. Szabo, *J. Phys. Chem.* **93**, 6929 (1989).

342. (a) D. D. Eads, B. G. Dismer, and G. R. Fleming, *J. Chem. Phys.* **93**, 1136 (1990); (b) C. F. Shannon and D. D. Eads, *J. Chem. Phys.* **103**, 5208 (1995).

343. (a) L. Song, R. C. Dorfman, S. F. Swallen, and M. D. Fayer, *J. Phys. Chem.* **95**, 3454 (1991); (b) L. Song, S. F. Swallen, R. C. Dorfman, K. Weidemaier, and M. D. Fayer, *J. Phys. Chem.* **97**, 1374 (1993).

344. A. A. Zharikov and N. V. Shokhirev, *Chem. Phys. Lett.* **190**, 423 (1992).

345. (a) S. Murata, S. Y. Matsuzaki, and M. Tachiya, *J. Phys. Chem.* **99**, 5354 (1995); (b) S. Murata and M. Tachiya, *J. Phys. Chem.* **100**, 4064 (1996); (c) *J. Chim. Phys.* **93**, 1577 (1996); (d) L. Burel, M. Mostafavi, S. Murata, and M. Tachiya, *J. Phys. Chem. A* **103**, 5882 (1999).

346. (a) S. F. Swallen, K. Weidemaier, and M. D. Fayer, *J. Phys. Chem.* **104**, 2976 (1996); (b) S. F. Swallen, K. Weidemaier, H. L. Tevernier, and M. D. Fayer, *J. Phys. Chem.* **100**, 8106 (1996).

347. E. B. Krissinel and N. Agmon, *J. Comput. Chem.* **17**, 1085 (1996).

348. M. Inokuti and F. Hirayama, *J. Chem. Phys.* **43**, 1978 (1965).

349. (a) R. P. Domingue and M. D. Fayer, *J. Chem. Phys.* **83**, 2242 (1985); (b) Y. Lin, R. C. Dorfman, and M. D. Fayer, *J. Chem. Phys.* **90**, 159 (1989); (c) R. C. Dorfman, Y. Lin, and M. D. Fayer, *J. Chem. Phys.* **94**, 8007 (1990); (d) R. C. Dorfman, M. Tachiya, and M. D. Fayer, *Chem. Phys. Lett.* **179**, 152 (1991).

350. (a) T. Niwa, K. Kikuchi, N. Matsusita, M. Hayashi, T. Katagiri, Y. Takahashi, and T. Miyashi, *J. Phys. Chem.* **97**, 11960 (1993); (b) T. Niwa Inada, C. Sato Miyazawa, K. Kikuchi, M. Yamauchi, T. Nagata, Y. Takahashi, H. Ikeda, and T. Miyashi, *J. Am. Chem. Soc.* **121**, 7211 (1999).

351. (a) K. Kikuchi, T. Niwa, Y. Takahashi, H. Ikeda, T. Miyashi, and M. Hoshi, *Chem. Phys. Lett.* **173**, 421 (1990); (b) K. Kikuchi, Y. Takahashi, M. Hoshi, T. Niwa, T. Katagiri, and T. Miyashi, *J. Phys. Chem.* **95**, 2378 (1991).

352. (a) J. B. Birks, *Photophysics of aromatic molecules* (Wiley, New York, 1970); (b) N. Mataga, *Pure Appl. Chem.* **56**, 1255 (1984).

353. (a) S. Murata, S. Iwai, and M. Tachiya, Annual Meeting on Photochemistry 1996 (Japan), Book of Abstracts, p. 81; (b) S. Murata and M. Tachiya, unpublished results.

354. Y.-X. Weng, K.-C. Chan, B.-C. Tzeng, and C.-M. Che, *J. Chem. Phys.* **109**, 5948 (1998).

355. M. Eigen, *Z. Phys. Chem.* **NF1**, 176 (1954).

356. K. M. Hong and J. Noolandi, *J. Chem. Phys.* **68**, 5163 (1978).

357. M. Tachiya, Annual Meeting on Photochemistry 1980 (Japan), Book of Abstracts, p. 256.

358. (a) S. D. Traytak, *Chem. Phys.* **140**, 281 (1990); (b) *Chem. Phys. Lett.* **183**, 327 (1991).

359. (a) J. P. Hansen and I. R. McDonald, *Theory of Simple Liquids* (Academic, London, 1976); (b) M. P. Allen, and D. J. Tidesley, *Computer Simulation of Liquids* (Clarendon, Oxford, 1987).

360. (a) H.-X. Zhou and A. Szabo, *J. Chem. Phys.* **95**, 5948 (1991); (b) T. Bandyopadhyay, *J. Chem. Phys.* **102**, 9557 (1995).

361. A. I. Burshtein and A. Yu. Sivachenko, *Chem. Phys.* **235**, 257 (1998).
362. (a) A. I. Shushin, *Chem. Phys. Lett.* **118**, 197 (1985); (b) **170**, 78 (1990).
363. (a) R. Zwanzig, *Adv. Chem. Phys.* **15**, 325 (1969); (b) J. M. Deutch and B. U. Felderhof, *J. Chem. Phys.* **59**, 1669 (1973); (c) P. G. Wolynes and J. M. Deutch, *J. Chem. Phys.* **65**, 450 (1976); (d) P. G. Wolynes and J. A. McCammon, *Macromolecules* **10**, 86 (1977); (e) S. H. Northrup and J. T. Hynes, *J. Chem. Phys.* **71**, 871 (1979).
364. A. Morita and B. Bagchi, *J. Chem. Phys.* **110**, 8643 (1999).
365. E. Kotomin and V. Kuzovkov, *Modern Aspects of Diffusion-Controlled Reactions*, in *Comprehensive Chemical Kinetics*, edited by R. G. Compton and G. Hancock (Elsevier, Amsterdam, 1996), Vol. 34.
366. (a) K. Kalyanasundaram, *Photochemistry in Microheterogeneous Systems* (Academic Press, Orlando, 1987); (b) M. Grätzel, *Heterogeneous Photochemical Electron Transfer* (CRC Press, Boca Raton, 1989); (c) *Photochemistry in Organized and Constrained Media*, edited by V. Ramamurthy (VCH, New York, 1991); (d) M. A. Fox, *Acc. Chem. Res.* **25**, 569 (1992); (e) V. Ramamurthy, R. G. Weiss, and G. S. Hammond, *Adv. Photochem.* **18**, 67 (1993).
367. (a) M. Tachiya, in *Kinetics of Nonhomogeneous Processes*, edited by G. R. Freeman (John Wiley & Sons, Inc., New York, 1987), p. 575; (b) M. Almgren, in *Kinetics and Catalysis in Microheterogeneous Systems*, edited by M. Grätzel and K. Kalyanasundaram (Marcel Dekker, New York, 1991), p. 63; (c) M. Almgren, *Adv. Colloid Interface Sci.* **41**, 9 (1992); (d) M. H. Gehlen and F. C. De Schryver, *Chem. Rev.* **93**, 199 (1993); (e) A. V. Barzykin and M. Tachiya, *Heterogeneous Chem. Rev.* **3**, 105 (1996); (f) A. V. Barzykin, K. Seki, and M. Tachiya, *Adv. Colloid Interface Sci.* **89-90**, 47 (2001).
368. (a) K. Weidemaier and M. D. Fayer, *J. Chem. Phys.* **102**, 3820 (1995); (b) *J. Phys. Chem.* **100**, 3767 (1996).
369. (a) J. E. Hansen, E. Pines, and G. R. Fleming, *J. Phys. Chem.* **96**, 6904 (1992); (b) K. Hamasaki, H. Ikeda, A. Nakamura, A. Ueno, F. Toda, I. Sozuki, and T. Osa, *J. Am. Chem. Soc.* **115**, 5035 (1993); (c) C. Lee, C. Kim, and J. W. Park, *J. Electroanal. Chem.* **374**, 115 (1994); (d) M. G. Kuzmin and I. V. Soboleva, *J. Photochem. Photobiol. A* **87**, 43 (1995); (e) C. H. Evans, S. DeFeyter, L. Viaene, J. van Stam, and F. C. De Schryver, *J. Phys. Chem.* **100**, 2129 (1996); (f) K. Weidemaier, H. L. Tavernier, K. T. Chu, and M. D. Fayer, *Chem. Phys. Lett.* **276**, 309 (1997); (g) J. W. Hackett II and C. Turro, *J. Phys. Chem. A* **102**, 5728 (1998); (h) I. V. Soboleva, J. van Stam, G. B. Dutt, M. G. Kuzmin, and F. C. De Schryver, *Langmuir* **15**, 6201 (1999).
370. H. Sano and M. Tachiya, *J. Chem. Phys.* **75**, 2870 (1981).
371. A. V. Barzykin, K. Seki, and M. Tachiya, *J. Phys. Chem. B* **103**, 6881, 9156 (1999).
372. (a) K. Tsukahara, C. Kimura, J. Kaneko, K. Abe, M. Matsui, and T. Hara, *Inorg. Chem.* **36**, 3520 (1997); (b) R. A. Marusak, T. P. Shields, and A. G. Lappin, in *Inorganic Compounds with Unusual Properties. Electron Transfer in Biology and the Solid State*, edited by M. K. Johnson, R. B. King, D. M. Kurts, Jr., C. Kutal, M. L. Norton, and R. A. Scott, Advances in Chemistry Series, Vol. 226 (American Chemical Society, Washington, DC, 1990), p. 237.
373. (a) R. J. Cave, P. Siders, and R. A. Marcus, *J. Phys. Chem.* **90**, 1436 (1986); (b) J.-P. Dodelet, M. F. Lawrence, M. Ringuet, and R. M. Leblanc, *Photochem. Photobiol.* **33**, 713 (1981).
374. (a) B. M. Hoffman and M. A. Ratner, *J. Am. Chem. Soc.* **109**, 6237 (1987); (b) L. Qin and N. M. Kostic, *Biochemistry* **35**, 3379 (1996); (c) M. M. Ivkovic-Jensen, G. M. Ullmann, S.

Young, Ö. Hansson, M. M. Crnogorac, M. Ejdebäk, and N. M. Kostic, *Biochemistry* **37**, 9557 (1998); (d) M. M. Ivkovic-Jensen and N. M. Kostic, *Biochemistry* **35**, 15095 (1996); (e) R. Bechtold, C. Kuehn, C. Lepre, and S. S. Isied, *Nature (London)* **322**, 286 (1986); (f) J. M. Nocek, N. Liang, S. A. Wallin, A. G. Mauk, and B. M. Hoffman, *J. Am. Chem. Soc.* **112**, 1623 (1990); (g) S. A. Wallin, D. A. Stemp, A. M. Everest, J. M. Nocek, T. L. Netzel, and B. M. Hoffman, *J. Am. Chem. Soc.* **113**, 1842 (1991); (h) M. M. Crnogorac, C. Shen, S. Young, Ö. Hansson, and N. M. Kostic, *Biochemsitry* **35**, 16465 (1996); (i) B. S. Brunschwig and N. Sutin, *J. Am. Chem. Soc.* **111**, 7454 (1989).

375. (a) E. Buhks, M. Bixon, J. Jortner, and G. Navon, *Inorg. Chem.* **18**, 2014 (1979); (b) L.-H. Zang and A. H. Maki, *J. Am. Chem. Soc.* **112**, 4346 (1990); (c) T. Ramasami and J. F. Endicott, *J. Am. Chem. Soc.* **107**, 389 (1985); (d) S. Larsson, K. Ståhl, and M. C. Zerner, *Inorg. Chem.* **25**, 3033 (1986).

376. R. Zwanzig, *J. Chem. Phys.* **97**, 3587 (1992).

377. (a) H.-X. Zhou and A. Szabo, *Biophys. J.* **71**, 2440 (1996); (b) *J. Phys. Chem.* **100**, 2597 (1996); (c) J. L. Spouge, A. Szabo, and G. H. Weiss, *Phys. Rev. E* **54**, 2248 (1996); (d) H.-X. Zhou, *J. Chem. Phys.* **108**, 8146 (1998); (e) D. J. Bicout and A. Szabo, *J. Chem. Phys.* **108**, 5491 (1998).

378. (a) A. I. Burshtein and B. I. Yakobson, *Chem. Phys.* **28**, 415 (1978); (b) S. Lee and M. Karplus, *J. Chem. Phys.* **86**, 1904 (1987); (c) Y. Jung, C. Hyeon, S. Shin, and S. Lee, *J. Chem. Phys.* **107**, 9864 (1997); (d) A. Shushin, *J. Phys. Chem. A* **103**, 1704 (1999); (e) M. O. Cáceres, C. E. Budde, and M. A. Ré, *Phys. Rev. E* **52**, 3462 (1995); (f) W.-S. Sheu, D.-Y. Yang, and S.-Y. Sheu, *J. Chem. Phys.* **106**, 9050 (1997); (g) A. M. Berezhkovskii, D.-Y. Yang, S.-Y. Sheu, and S. H. Lin, *Phys. Rev. E* **54**, 4462 (1996); (h) A. M. Berezhkovskii, D.-Y. Yang, S. H. Lin, Y. A. Makhnovskii, and S.-Y Sheu, *J. Chem. Phys.* **106**, 6985 (1997); (i) Y. A. Makhnovskii, A. M. Berezhkovskii, S.-Y. Sheu, D.-Y. Yang, J. Kuo, and S. H. Lin, *J. Chem. Phys.* **108**, 971 (1998); (j) S.-Y. Sheu and D. Y. Yang, *J. Chem. Phys.* **112**, 408 (2000). (k) T. Bandyopadhyay, K. Seki, and M. Tachiya, *J. Chem. Phys.* **112**, 2849 (2000).

379. (a) N. Eizenberg and J. Klafter, *Chem. Phys. Lett.* **243**, 9 (1995); (b) *J. Phys. Chem.* **104**, 6796 (1996); (c) *Physica A* **249**, 424 (1998); (d) A. M. Berezhkovskii, Yu. A. D'yakov, J. Klafter, and V. Yu. Zitserman, *Chem. Phys. Lett.* **287**, 442 (1998).

380. (a) K. Solc and W. H. Stockmayer, *J. Chem. Phys.* **54**, 2971 (1971); (b) R. Samson and J. M. Deutch, *J. Chem. Phys.* **68**, 285 (1978); (c) S. I. Temkin and B. I. Yakobson, *J. Phys. Chem.* **88**, 2679 (1984); (d) O. G. Berg, *Biophys. J.* **47**, 1 (1985); (e) H.-X. Zhou, *Biophys. J.* **64**, 1711 (1993); (f) S. D. Traytak, *Chem. Phys.* **192**, 1 (1995).

381. (a) A. I. Shushin, *Chem. Phys. Lett.* **130**, 452 (1986); (b) *Chem. Phys.* **120**, 91 (1988); (c) *Mol. Phys.* **64**, 65 (1988); (d) *J. Chem. Phys.* **110**, 12044 (1999).

AUTHOR INDEX

Numbers in parentheses are reference numbers and indicate that the author's work is referred to although his name is not mentioned in the text. Numbers in *italic* show the pages on which the complete references are listed.

Aa, H. J., 357(42), *418*
Aakeröy, C. B., 472(47), *503*
Aartsma, T. J., 201(44), *264*
Abe, K., 597(372), *615*
Abragam, A., 358(73), *419*
Abrash, S., 533(280), 535(280), *611*
Ackermann, Th., 441(92), *466*
Adamowicz, L., 369(90), *419*
Adamson, G. W., 184–185(19), *197*
Ado, M., 518(182), *607*
Aeschleman, J. A., 427(19), *464*
Aganov, A. V., 460(150), *468*
Agmon, N., 200(32), *264*; 564(311), 588(347), 597(311), *612, 614*
Ågren, H., 357(42,65), *418*; 437(75), *466*
Aguilar, M. A., 437(70), *465*
Aguillar-Parillas, F., 500(330), *510*
Ahlrichs, R., 356(16-17,19), 357(45), 371(91), 374(45), 385(127), 386(45,134,136), 393(147), *417–419, 418, 421*; 494(228-229), *508*
Ahmed, L., 407(171), *422*
Aihara, J., 412(183), *422*
Åkesson, E., 515(156), 524(156), 560(156), 562(156), 563(156b), 564(156), *606*
Akiyama, M., 500(326), *510*
Alavi, A., 49(62), *78*
Albinati, A., 472(36,40), 499(36,40), 500(36), *503*
Albrecht, A. C., 533(280), 535(280), *611*
Alden, R. G., 514(65), 527(224), 534(284), 538(65), *603, 609, 611*
Alkorta, I., 472(24,42,44), 498(24), *503*
Allavena, M., 494(274), *509*
Allen, L., 253–254(108), *266*
Allen, M., 10(39), *77*
Allen, M. P., 593–595(359), *614*
Allison, T. C., 117(70), *152*

Almeida, R., 205(81), 250(81), *265*; 488(180), *506*
Almenningen, A., 500(317-318), *510*
Almgren, M., 595–596(367), *615*
Almlöf, J., 385(133), 386(134), *421*
Altenberger, A. R., 279(10), 288(15), 293(10), 300(19), 310(15), 311(15,19), 312(15), 313(21), 324(29), 332(35), *353*
Ambegaokar, V., 514(85), 543(85), 574(85), *603*
Ambrose, W. P., 200(6-7), 201(6-7), 210(7), 223(6-7), 229(92), *263, 266*
Aminova, R. M., 460(150), *468*
Amitrano, C., 80–81(16), *150*
Amos, R. D., 357(62), 360(79), 375(62,79,119), 380(122), 393(62,79), 397–398(79), *418–420*
Amovilli, C., 442(88), *466*
Anders, C., 515(155), 562(155), *606*
Andersen, H. C., 24(47), 42(53), *77*
Andersen, T., 357(42), *418*
Anderson, A., 488(191,198), 493(191), 494(273, 275), 496–497(273), *507, 509*
Anderson, J. B., 4(3), 59(3), *76*
Anderson, P. W., 201(58), 204(58), 242–243(94), *264, 266*
Andersson, K., 372(98), *420*
Ando, K., 514(41, 94), 520(41c), 527(41), 529(41), 533–534(41), *602, 604*
Andrade, J. C., 431(37), *464*
André, J.-M., 488(206), 500(327), *507*
Andrews, L., 488(182-186), 494(250-251), *507–508*
Andzelm, J., 442(85), *466*
Anex, D. S., 476(79), 478(79), *504*
Angel, S. A., 586(337), *613*
Ankerhold, J., 514(124), 561(310), 582(124), *605, 612*

Antosiewica, J., 437(63), *465*
Ao, P., 516(166), 573–574(166), 576(166), *607*
Aoki, K., 477(98), *504*
Apkarian, V. A., 488(179), *506*
Aquino, A. J. A., 513(264), *610*
Archer, M. D., 529(241), *610*
Armstrong, D. A., 494(244), *508*
Arnó, J. I., 493(220), *507*
Arnold, B. R., 527–528(235), *609*
Arzi, E., 472(29), 490–491(29), *503*
Asahi, T., 512(11,16), 580(322), 581(322), 586(16), 590(16), 592(11), *600–601, 613*
Asencio, G., 409(176), *422*
Asfar, M. N., 519(190), *608*
Ashkenazi, G., 514(133), *605*
Ashworth, S. H., 488(131), *505*
Aslangul, C., 518(183), *607*
Asmus, K.-D., 512(13), 585(13), 590(13), *601*
Assfeld, X., 430(29), 434(29), *464*
Åstrand, P.-O., 357(67), *418*
Atoji, M., 472(28), 488(28), 490–491(28), *503*
Aue, D. J., 447(99), *466*
Auer, A. A., 403–405(167), *422*
Auwera, J. V., 184(18), *197*
Axmann, A., 488(197), *507*
Azman, A., 478(110), 488(200), *505, 507*

Baba, T., 357(49), 368(49), *418*
Bach, H., 200–201(17), 223(17), *263*
Bachs, M., 456–457(132), *467*
Bacic, Z., 487(125), 494(125,238-239), *505, 508*
Bacskay, G. B., 453(125), *467*
Bader, J. S., 514(34), 519(34, 185), 527(34), 529(34), 533–534(34), *601, 607*
Bader, R. F. W., 357(36-37), *418*
Baer, T., 159–160(5), *197*
Bagchi, B., 514–515(103-107), 519(191), 520(103-104, 191, 197), 527(104, 197), 546(197), 550(197), 552(103-107), 557(103), 565(103-107), 580–581(105-106), 595(197, 364), *604, 608, 615*
Bagdasar'yan, Kh. S., 586(339), 588(339), *614*
Baird, N. C., 412(183), *422*
Bak, K. L., 357(42,51), 395–397, *418*
Bakken, V., 357(42), *418*
Bal, R., 524(205), *608*
Ball, K. D., 119(75), *152*
Ballard, S. G., 516(169), 585(169), *607*
Bandyopadhyay, T., 597(378), *616*

Banfelder, J. R., 437(60), *465*
Bär, M., 357(45), 374(45), 386(45), *418*
Barbara, P. F., 200(25-26), 202(26), 245(25-26), *264*; 515(147,156-158), 516(173), 520(147a), 524(156), 530(249), 532(249), 560(147,156-157), 562(156-158), 563(156b), 564(156), 565(173, 316), 583(173, 316), *606–607, 610, 613*
Barber, M. N., 274(6), *353*
Barkai, E., 200(16), 201(16,63), 203(79), 204(16, 63), 242–243(16), 247(104), *263, 265–266*
Barnes, A. J., 477(88), 494(246), *504, 508*
Baron, H. P., 371(91), *419*
Barone, V., 448(111), 450(111,118), 456(111), 457(134), 458(118), *467*
Bartell, L. S., 488(160), 490(160), *506*
Bartholomew, J., 334(36), *353*
Bartlett, R. J., 357(41), 369(90), 372(93,95), 374(41), 377(118), 394(92), *418–420*
Bartmess, J. E., 427(17-19), *464*
Bartness, J. E., 452(123), *467*
Barzykin, A. V., 515(161), 527(220), 552–553(307), 557(161, 309), 558(309), 560–562(309), 564(307), 568(309), 581(161), 595(220,367), 596(220,367,371), *607, 609, 612, 615*
Basché, T., 200(7,19), 201(7), 202(69), 210(7), 223(7), 245(19,69), *263, 265*
Bashford, D., 437(61), 447(61), 449(61), 452(61,119,121), 457(119), *465, 467*
Basilevsky, M. V., 514(49,141), 520(195), 524(206), 527(233), 529(206), 584(141), *602, 606, 608–609*
Bastiansen, O., 500(317-318), *510*
Batchelor, G. K., 347(44), *354*
Bauer, M., 200(13), *263*
Baumfalk, R., 494(261-265), *508–509*
Baumgärtner, A., 312(20), *353*
Bawendi, M. G., 200(22-23), 202(43), 209(43), 247(22-23), *263–264*
Beachy, M. D., 434(46), *465*
Bechtold, R., 597(374), *616*
Beck, C., 196(30), *198*
Beck, T. L., 80–81(17), *150*
Beenakker, C. W. J., 347(45), *354*
Beichert, P., 477(92), *504*
Bell, A. T., 8(32), *77*
Bell, R. P., 424(3), *463*
Bellon, F., 500(320), *510*

AUTHOR INDEX

Belousov, A. A., 535(289), *612*
Bemish, R. J., 476(82), 479(82), 484(82), 486(82), *504*
Benito, R. N., 118(73), *152*
Benjamin, I., 514(136), 533(281), 566–571(136), 597(281), *605, 611*
Ben-Naim, A., 448(106,108), 449(108), *467*
Bennett, C. H., 4(3), 59(3), *76*
Benniston, A. C., 581(323), *613*
Beratan, D. N., 514(86), 528(239), 531(256-257,261-262), 532(256-257,272), 533(281), 535(261), 597(281), *603, 610–611*
Berestetskii, V. B., 326(31), *353*
Berezhovskii, A. M., 200(29), *264*; 564(313), 566–567(317), 569(313,317), 597(378-379), 598(378b), *612–613, 616*
Berg, B. A., 44(57), *78*
Berg, O. G., 598(380), *616*
Berkowitz, J., 408(179), *422*
Bernard, J., 200(3,8-10), 201(8-9), 202(8), 203(8-9), 212(87), 223(8-9), 229(8-9), 242–244(9), *263*
Bernasconi, C. F., 430(31), *464*
Bernasconi, M., 49(62), *78*
Berne, B. J., 80(1), *150*; 514(76), 519(185), 549(76), 566–567(317), 569(317), 573(76), 584(325), *603, 607, 613*
Bernstein, H. J., 460(146), *468*
Beroza, P., 513(264), *610*
Berry, R. S., 80(16-19,25-28), 81(16-19, 25-28), 82–83(40-44), 93(40-44), 94(26, 40), 95(40), 99(43), 103(25,28), 105(27,62), 106(64), 107(41), 109(27,41), 110(43-44), 112(40), 119(75,80), 149(44), *150–152*
Berryman, J. G., 317(22), 324(22), 348(22), *353*
Berski, S., 488(204), 494(260), *507–508*
Bertie, J. E., 522(202), 582(202), *608*
Bertoncini, R., 488(193), 493(193), *507*
Bevan, J. W., 475(70), 493(220-221), 494(254), 497(254), *504, 507–508*
Beveridge, A. J., 459(142), *468*
Beyer, A., 472(50), 479(116), 488(133-134, 201), 494(276), *503, 505, 507, 509*
Bezel, I., 493(224), *508*
Bhatnager, R., 586(337), *613*
Biaglow, A. I., 408(178), *422*
Bialek, W., 535(288), *612*
Bicout, D. J., 514(142), 550–551(142),
564(142), 597(377), *606, 616*
Biem, W., 488(197), *507*
Bigam, G., 414–415(186), *422*
Binder, K., 11(41), 53(41), *77*; 312(20), *353*
Binkley, J. S., 378(121), *420*
Birch, J. R., 519(190), *608*
Bird, R. B., 474(53), *503*
Birkhoff, G. D., 83(54), 87–88(54), *151*
Birks, J. B., 591(352), *614*
Bishop, D. M., 356(29), 364(29), 374(29), 402(165), *417, 422*
Biswas, R., 519(191), 520(191,197), 527(197), 546(197), 550(197), 595(197), *608*
Bittner, E. R., 514(129), 584(129), *605*
Bixon, M., 513(23), 514(45,96-97), 515(96-97), 516(179), 531(260), 562(96-97), 563(96b), 597(375), *601–602, 604, 607, 610, 616*
Blackbourn, R. L., 526(209), 527–528(235), *609*
Blake, G. A., 493(213,215), *507*
Blake, J. F., 430(30), *464*
Blake, T. A., 488(171), *506*
Bläser, D., 407(170), *422*
Blatt, R., 202(72), 245(72), *265*
Blümel, R., 174(15), *197*
Blumenfeld, R., 119(76), *152*
Boatz, J. A., 375(110), *420*
Bockris, J. O'M., 443–444(90), *466*
Boens, N., 201(57), *264*
Boese, R., 407(170), *422*
Bogdanchikov, G. A., 514(32), 519(32), *601*
Boger, D. V., 348(48), *354*
Bogoliubov, N. N., 274(3), 286(3), 289(3), 292(3), 326(3), *352*
Bohn, R. B., 494(250-251), *508*
Boiron, A.-M., 200(12), 201(12), 210–211(12), 223(12), 242–243(12), *263*
Bolhuis, P. G., 4(5-7,11,17-19,22), 5(18), 10(6), 26(11), 29(17), 30–31(6), 41–43(5), 44(22), 61(6), 68(17), *76–77*
Bolton, J. R., 516(175), 529(241), *607, 610*
Bondi, A., 439(82), 442(82), *466*
Bondybey, V. E., 488(185), *507*
Bopp, M. A., 201(45), *264*; 520(196), *608*
Bordwell, F. G., 434–435(51), *465*
Borkman, R. F., 425(13), *464*
Borkovec, M., 80(1), *150*; 514(69), 570(69), *603*
Born, R., 356(22-23), *417*

Borondo, F., 118(73), *152*
Borsch, P., 488(197), *507*
Both, G., 408(179), *422*
Böttcher, C. J. F., 438(78), 451(78), *466*
Bouman, T. D., 356(32), 357(32), 364(32), 366(32), 374(103), *417, 420*
Boutin, H., 488(195-196), *507*
Bowyers, M. T., 447(99), *466*
Boxer, S. G., 529(243), *610*
Boyd, R. J., 494(245), *508*
Brajovic, V., 488(196), *507*
Braly, L. B., 499(301-302,312), *509–510*
Bräuchle, C., 200(19), 245(19), *263*
Brédas, J.-L., 500(327), *510*
Brewer, R. G., 208(84), 229(84), *265*
Brickmann, J., 531(269), 535(269), *611*
Briggs, J. M., 437(55-56), *465*
Brij, A., 348(46), *354*
Brønsted, J. N., 424(1-2), *463*
Brouwer, A. C. J., 200(20), 245(20), *263*
Brown, F. L. H., 200–201(15), 204(15), *263*; 517(181), *607*
Brown, M. G., 499(302,311-312), *509–510*
Brown, R., 200(8-9,12), 201(8-9,12), 202(8), 203(8-9), 210-211(12), 223(8-9,12), 229(8-9), 242–243(9,12), 244(9), 245(18), *263*
Brown, R. D., 475(67), *504*
Brown, S. P., 356(24,26), 371(24), *417*
Brunel, C., 201(52), 285(83), *264–265*
Bruns, M., 356(18), 371(18), *417*
Brunschwig, B., 526(208), 585–586(329), 597(374), *608, 613, 616*
Brutschy, B., 470(18), *502*
Buck, U., 472(23), 488(130), 498(23), 494(261-265), 498(282,285,292), *503, 505, 508–509*
Buckingham, A. D., 474(56,60), 475(68), *503–504*
Budde, C. E., 597(378), *616*
Buhks, E., 597(375), *616*
Bühl, M., 356(10,13-14,21), 357(10), 374(14), 405(13), 406(13-15), 407(14-15), *417*
Bukowski, R., 499(299-300,305), *509–510*
Bullemer, B., 500(320), *510*
Bumgarner, R. E., 493(215), *507*
Bunker, P. R., 494(230-231,235-236), *508*
Buratto, S. K., 202(70,75), 245(70), *265*
Burel, L., 588(345d), 590–592(345d), *614*
Burfeindt, B., 512(9), *600*
Bürk, M., 356(20), *417*

Burk, P., 432–433(42), 461(42), *465*; 433(43), *465*; 435(54), *465*
Burshtein, A. I., 247(105), *266*; 512(18), 514(32,75,113), 516(163), 519(32), 533(282), 535(287), 546(75b), 547(75a,113,299), 549(75), 573(75,163), 578–579(163), 585(18b,330), 586(332-333,338-339), 588(18,330b,339), 589(18), 590(330,338d), 591(330), 594(18,330,361), 597(361,378), *601, 603–604, 607, 611–616*
Burt, S. K., 446–447(98), 450(116), 452(120), 453(126), *466–467*
Busarow, K. L., 493(213), *507*
Buxton, L. W., 475(66), *504*
Buzek, P., 356(12), 374(107), 408(107,173), 410(107), *417, 420, 422*

Caballeiro-Lago, E. M., 478(109,115), 486(115), 488(153), *505–506*
Cabani, S., 449(113), *467*
Caceres, M. O., 597(378), *616*
Cagin, T., 425(12), *464*
Calcaterra, L. T., 512(8), 524–525(8), *600*
Caldwell, R. A., 446–447(98), 453(126), *466 467*
Calef, D. E., 514(86), 570(83), *603*
Callen, H. B., 279(12), *353*
Cammi, R., 357(71), *418*; 434(47), 439(80), 445(80), 448(80), *465–466*
Campargue, A., 184(18), *197*
Campbell, E. J., 475(66), *504*
Campbell, J. L. E., 494(258), *508*
Cao, J., 200(33-34), *264*; 514(123,143-144), 538(144), 557(143), 582(123), 584(143), *605–606*
Capelli, C., 499(307), *510*
Car, R., 8(36), 38(36), *77*
Carlson, R. E., 494(270), *509*
Carnahan, N. F., 319(24), *353*
Carpenter, G. B., 472(27), 475(27), 488(27), *503*
Carson, P. J., 202(70,75), 245(70), *265*
Carter, E. A., 550(300), *612*
Cartling, B., 535(293), 597(293), *612*
Cary, J. R., 82–83(45), 141–142(45), *151*
Casaes, R. N., 498(286), *509*
Case, D. A., 452(119,121), 457(119), *467*
Casida, M. E., 357(64), 360(78), 375(78), 392–393(78), 397(78), *419*

Castejon, H., 437(73), *465*
Castilla, J., 407(171), *422*
Castin, Y., 245(100-101), *266*
Castleman, A. W., 470(11), *502*
Catlow, C. R. A., 426(14), *464*
Cave, R. J., 531(254), 533(254,279,281), 597(281,373), *610–611, 615*
Cemeli, J., 459(143), *468*
Ceperley, D. M., 41(52), 43(52), *77*
Cerjan, C. J., 3(1), *76*
Chakravarty, S., 514(71), 518(71), 543(71), *603*
Chalasinski, G., 474(61-62,64), 488(140), 494(256–257), 495(256-257), *503–505, 508*
Champgagne, B., 488(206), *507*
Chan, K.-C., 593(354), *614*
Chandler, D., 4(5-12,15-19,21-22,24-26), 5(18), 10(6), 11(8), 22(10), 23(10,15,25), 24–25(16), 26(11), 29(17), 30–31(6), 38(15,24,26), 41–42(5), 43(5,8,55), 44(15,22), 51(61), 53(61), 61(6,12), 68(17), 73(12), *76–78*; 514(34-35,117-118,134), 519(34-35), 524(118,204), 526(213), 527(34), 529(34-35), 533(34-35,204), 534(34-35), 582(117-118), 584(134), *601–602, 605, 608–609*
Chandler, W. D., 494(258-259), 495–496(259), *508*
Chandra, A., 514(103-104,109), 515(103-104), 520(104), 527(104), 552(103-104), 557(103), 565(103-104), *604*
Chandra, P., 357(64), *418*
Chandrasekhar, J., 431(37), *464*
Chandrasekhar, S., 8(38), 10(40), 39(38), *77*
Chanon, M., 516(174), *607*
Chapo, C., 498(286), *509*
Chatfield, D. C., 81(32), 117(70), *150, 152*
Che, C.-M., 593(354), *614*
Che, D.-C., 493(227), *508*
Cheche, T., 514(51), *602*
Cheeseman, J. R., 357(60), 375(60), 390(60), 403(60), *418*
Chemla, D. S., 200(24,35-36), *264*
Chen, J. L., 452(121), *467*
Chen, W., 494(241), 496(241), *508*
Chen, Y., 201(51), 208(51), 229(51), *264*
Cheng, W. D., 514(65), 538(65), *603*
Chernobrod, B. M., 516(163), 547(299), 573(163), 578–579(163), *607, 612*
Chernyak, V., 200(28), *264*; 514(100), *604*

Chestnut, D. B., 356(3,5), 357(5), *416*
Chiefari, J., 459(141), *468*
Child, M. S., 81(37), *151*
Childsey, C. E. D., 535(294), *612*
Chipman, D. M., 357(69), *418*
Cho, M., 514(36,63,139-140), 519(36), 520(36a), 529(36), 533–534(36), 535(63), 538–539(63), 543(139-140), *602–603, 605*
Chojnacki, H., 500(327), *510*
Christe, K., 375(110), *420*
Christiansen, O., 357(42), *418*
Chu, Z. T., 514(33), 519(33), 527(33,224), 529(33), 533–534(33), *601, 609*
Chuang, C., 476(76-77), 479(76-77), *504*
Chudinov, G. E., 514(49), 524(206), 529(206), *602, 608*
Ciccotti, G., 71(69), *78*; 570(318), *613*
Clague, A. D. H., 460(146), *468*
Claramunt, R. M., 500(331), *510*
Clark, C. D., 527(234), 590(234), 593(234), *609*
Claverie, P., 474(57), *503*
Closs, G. L., 512(8), 524(8,205), 525(8), *600, 608*
Coalson, R. D., 514(62,67), 535(62,67), 537(67), 538(62,67), 539(62), 541(62), *603*
Coe, J. V., 446(103), 447–448(104), 462(104), *466*
Cogdell, R. J., 201(45), *264*
Cohen, E. A., 397(154), *421*
Cohen, M. H., 447(104), 448(104), 462(104), *466*
Cohen, R. C., 493(213), *507*
Cohen-Tannoudji, C., 202(73), 205(73), 224(73), *265*
Coitiño, L., 434(47), *465*
Coker, D. F., 584(324), *613*
Collins, F. C., 586(334), 588(334), *613*
Colmenares, P. J., 205(81), 250(81), *265*
Colominas, C., 459(143), *468*
Colwell, S. M., 357(61), 375(61), 392(61), 398(61), *418*
Comeau, M., 374(104), *420*
Cooperman, B. S., 220(89), *266*
Coriani, S., 397(152), *421*
Cortis, C., 448(110), *467*; 528(238), *610*
Cossi, M., 448(111), 450(111,115,118), 456(111), 457(134), 458(118), *467*
Cotlet, M., 201(46), *264*
Cowen, K. A., 447(104), 448(104), 462(104), *466*

Crafton, M. W., 408(179), *422*
Cramer, C. J., 437(74), 438(76), 439(79), 442(79), 450(87), 456(131), 458(139), *466–468*
Craw, J. S., 478(101), *505*
Crawford, T. D., 412(182), *422*
Cremer, D., 357(68), 372(92), 374(106), 375(116), 376(117), 377(120), 393(116, 144), 394(72), 398(116), 403(116), *418–421*
Crnogorac, M. M., 597(374), *615–616*
Crooks, G. E., 4(26), 38(26), *77*
Cruzan, J. D., 499(301), *509*
Csajka, F. S., 4(8,20-21), 11(8), 43(8), *76*
Cukier, R. I., 514(101), 557(101d), 564(314), 584(101), *604, 612*
Curtiss, C. F., 474(53), *503*
Curtiss, L. A., 429(22), *464*
Cybulski, S. M., 356(29), 364(29), 374(29), 402(165), *417, 422*; 488(140), 494–495(256), *505, 508*

Dahl, K., 515(149), 560(149), *606*
Dahle, P., 357(42), *418*
Dahlems, T., 500(321), *510*
Dahler, J. S., 279(10), 288(15), 293(10), 300(19), 310(15), 311(15,19), 312(15), 313(21), 324(29), 332(35), 347(42), *353–354*
Daizadeh, I., 514(61), 531(265), 532(265,275), 535–537(61), *602, 611*
Dakhovskii, Y., 514(67,122,135), 535(67), 537–538(67), 569(135), *603, 605*
Dalgarno, A. M., 369(87), *419*
Dalibard, J., 245(100-101), *266*
Dalskov, E. K., 357(42), *418*
Da Motta, J. D., 435(53), *465*
Dannenberg, J. J., 500(323), *510*
Da Silva, C. O., 433(44), 456(133), 457(135), *465, 467*; 499(307), *510*
Da Silva, E. C., 433(44), 456(133), 467(135), *465, 467*
Daudey, J.-P., 357(39), *418*
Davidson, E. R., 383(125), *420*; 476(79), 478(79), *504*
Davis, J. P., 576(321), *613*
Davis, M. J., 80–81(20), 116(20,68-69), 118(20,68), *150–151*; 152(2), 170(2), 183(2), 184(17), *197*
Davis, S., 494(255), 497(255), *508*

Davis, S. R., 488(186), *507*
Davis, W., 530(250), 532(250), *610*
Davydov, A. S., 532(277), *611*
Deakyne, C. A., 427(19), *464*
de Almeida, W. B., 478(101-102), *5050*
de Andrade, P. C. P., 513(264), *610*
deBruil, T. P., 515(156), 524(156), 560(156), 562(156), 564(156), *606*
de Dios, A. C., 356–357(9), *417*
deFeyter, S., 595(369), *615*
De Gennes, P. G., 32(49), *77*; 275(9), 281(9), 289(9), 301(9), *353*
de Kruif, C. G., 348(46), *354*
Del Bene, J. E., 488(142,172), 494(142), *505–506*
Dellago, C., 4(5-7,10-15,17-19,24-25), 5(18), 10(6), 19(13), 22(10), 23(10,15,25), 26(11), 29(17), 30–31(6), 38(15,24-25), 41–43(5), 44(15), 61(6,12), 68(17), 73(12), *76–77*
Dellhalle, J., 500(327), *510*
DeLon, N., 80(21), 82(21), *150*
Del Rio, E., 500(323), *510*
Demin, A. V., 519(189), *608*
de Munck, F. R., 494(253), *508*
Denisov, G. S., 460(148), *468*
de Pablo, J. J., 43(54), *77*
de Paz, J. L. G., 500(330), *510*
Deprit, A., 82–83(46), 87(46), 140–141(46), *151*
de Rafael, E., 327(32), *353*
de Schryver, F. C., 595(367,369), 596(367), *615*
desCloizeaux, J., 284(14), 304(14), *353*
de Silva, V., 75(70), *78*
Desiraju, G. R., 470(16), *502*
Deutch, J. M., 594–595(363), 598(380), *615–616*
Deutsch, T., 49(62), *78*
Dewar, R. L., 87(56), 141(56), *151*
Deward, M. J. S., 434(50), *465*
Diez y Riega, H., 488(180), *506*
Ditchfield, R., 357(34), 364(34), *417*
Dixon, D. A., 412(183), *422*; 430(33), *464*; 488(143), *506*
Dodd, M. D., 514(40), *602*
Dodelet, J.-P., 597(373), *615*
Dogonadze, R. R., 512(7), 513(20,22), 514(27), 518(27), 519(188), 527(27), 578(20), 514(7), 517(7), 535(7), *600–601, 608*
Döhling, A., 515(155), 562(155), *606*
Döhmeier, C., 356(16-17), *417*

Doi, M., 550(303), 588(303), *612*
Doktorov, A. B., 514–515(92), 561(92), 586(339), 588(339), *604, 614*
Doll, J. D., 488(174), *506*
Domingue, R. P., 588(349a), *614*
Donati, C., 75(73), *78*
Dong, Y., 527–528(235), *609*
Doniach, S., 8(35), *77*
Doraiswamy, S., 586(337), *613*
Dorfman, R. C., 585(331), 588(343,349), 591(331), *613–614*
Dorsey, A. T., 514(71), 518(71), 543(71), *603*
Dougherty, T. J., 347(40), *354*
Douketis, C., 476(79), 478(79), *504*
Doye, J. P. K., 3(2), *76*
Dragt, A. J., 82–83(47), 87(47), 141(47), *151*
Ducharme, Y., 472(46), *503*
Dudek, S. P., 535(294), *612*
Dulligan, M., 493(224), *508*
Dulmage, W. J., 472(26), 475(26), *503*
Dum, R., 245(98), *266*
Dunford, H. B., 441(96), *466*
Dunning, T. H., 431(38), *465*; 470–472(21), 498(21), *502*
Dupon–Roc, J., 202(73), 205(73), 224(73), *265*
Dupuis, M., 383(124), *420*
Dutt, G. B., 595(369), *615*
Dutton, P. L., 530(249), 532(249), *610*
D'yakov, Yu. A., 597(379), *616*
Dyke, J. M., 4(14), *76*
Dyke, Th. R., 476(80-81), 488(161), 490(161), *504, 506*
Dykhne, A. M., 576(320), *613*
Dykman, M. I., 38(50), *77*
Dykstra, C. E., 478(108), 488(129,137,173,175), *505–506*

Eads, D. D., 588(342), *614*
Earhart, A. D., 447(104), 448(104), 462(104), *466*
Eastman, P., 8(35), *77*
Eberly, J. H., 253–254(108), *266*
Echevarria, A., 472(44), *503*
Edman, L., 200–201(37), *264*
Efrima, S., 513(23), 514(45), 537–538(45), *601–602*
Ege, D., 512(10), 589(10), 593(10), *600*
Egger, R., 514(48,124), 543(124a-b), 582(124), *602, 605*

Egorov, S. A., 519(185), 522(199), 584(325), *607–608, 613*
Ehrenberg, M., 201(48), *264*
Ehrenson, S., 437(58), 458(58), *465*; 526(208), 585–586(329), *608, 613*
Eigen, M., 441(92-93), *466*; 593(355), *614*
Eilers, V. M., 348(49), *354*
Eisenberg, D., 499(313), *510*
Eissenstat, M. A., 453(126), *467*
Eizenberg, N., 597(379), *616*
Ejdebäk, M., 597(374), *615–616*
Elber, R., 8(28-29), 47(28), 75(29,71), *77–78*; 535(290), *612*
Elguero, J., 460(145), *468*; 472(24,42,44), 498(24), 497(288,290), 500(330-331), *503, 509–510*
Elliott, S. D., 385(127), *421*
Elmore, D. E., 460(151), *468*
Elrod, M. J., 493(222), 494(222,237), *507–508*
Elson, E. L., 201(47), *264*
Emilsson, T., 476(76-77), 479(76-77), *504*
Empedocles, S. A., 200(22-23), 201(43), 209(43), 247(22-23), *263–264*
Enderle, T., 200(24,35-36), *264*
Enderlein, J., 229(92), *266*
Endicott, J. F., 597(375), *616*
Enevoldsen, T., 357(42), *418*
Engdahl, A., 494(252), *508*
Engkvist, O., 494(234), *508*
Engleiter, S., 515(154), 562(154), 565(154), *606*
English, J. H., 488(185), *507*
Englman, R., 519(187), *608*
Epa, V. C., 494(236), *508*
Epstein, S. T., 362(82), *419*
Erpenbeck, J. J., 324–325(26), *353*
Estebaranz, J. M., 118(73), *152*
Etter, M. C., 472(45), *503*
Ettischer, I., 498(285), *509*
Evans, C. H., 595(369), *615*
Evans, D. G., 514(62,67,132), 533(281), 535(62,67), 538–539(62,67), 541(62), 582–584(132), 597(281), *603, 605, 611*
Evans, M. G., 80(3), *150*
Everest, A. M., 597(374), *616*
Everts, J. E., 454(128), 456(128), *467*
Ewing, G. E., 476(79), 478(79), *504*
Exner, O., 434(49), *465*
Eyring, H., 80(2), *150*

Ezra, G. S., 80–81(23), 82(48-50), 90(48-50), 92(48-50), 116(68), 118(68), 148(48-50), *150–151*; 156(3), 172(3), 184(17), *197*

Fabian, J., 499(306), *510*
Facelli, J. C., 356–357(7), *417*
Fahlman, B. D., 494(258), *508*
Fainberg, B. D., 520(197), 527(197), 546(197), 550(197), 595(197), *608*
Fair, J. R., 80(22), 82(22), *150*
Fajardo, M. E., 499(297), *509*
Falk, M., 498(279-280), *509*
Fankuchen, I., 472(34), 499(34), *503*
Faratos, S. C., 196(30), *198*
Farid, R. S., 528(239), 530(249), 532(249), *610*
Farid, S., 512(10), 527(235), 528(235), 533(280), 535(280), 581(323), 589(10), 593(10), *600, 609, 611, 613*
Fárnik, M., 494(255), 497(255), *508*
Farrar, T. C., 478(105), 485(105), 500(324), *505, 510*
Farrell, J. T., 488(131), *505*
Farrow, R. F. C., 472(30), 494(30), *503*
Fayer, M. D., 527(220), 585(331), 588(343,346,349), 591(331), 594(346), 595(220,346,368), 596(220,368), *609, 613–615*
Feitelson, J., 535(292), *612*
Feldberg, S. W., 535(294), *612*
Felderhof, B. U., 594–595(363), *615*
Fellers, R. S., 499(301-302,311-312), *509–510*
Felmy, A. R., 430(33), *464*
Felts, A. K., 514(116), 582(116), *605*
Feng, W. Y., 447(104), 448(104), 462(104), *466*
Fernandez, B., 357(42), *418*
Ferrario, M., 570(318), *613*
Fetter, A. L., 335(37), 337(37), *354*
Feyereisen, M. W., 386(134), *421*
Field, R. W., 184(19-20), 185(19), 186(22), *197*
Fillery-Travis, A. J., 475(68), *504*
Finn, J. M., 82–83(47), 87(47), 141(47), *151*
Fischer, A., 356(20), *417*
Fischer, S. F., 516(167), 540(167), 573–574(167), 576(167), 578–579(167), *607*
Fisher, M. P. A., 514(71), 518(71), 543(71), *603*

Fixman, M., 347(43), *354*; 550(302), 588(302), *612*
Flannery, B. P., 95(61), *151*
Fleischer, U., 356(1,6), 357(6), 375(112), 397(112), *416–417, 420*
Fleming, G. R., 514(36,103,116,140), 515(103), 516(160,182), 519(36), 520(36a,192), 522(200), 529(36), 533–534(36), 543(140), 552(103), 557(103), 561–562(160), 565(103), 582(116), 588(342), 595(369), *602, 604–605, 607–608, 614–615*
Fleury, L., 200(8-9,18), 201(8-9,53), 202(8,53), 203(8-9), 212(86), 223(8-9), 229(8-9), 234(53), 242(9), 243(9,53), 244(9), 245(18), *263, 265*
Fliszar, S., 374(104), *420*
Flocco, M. M., 425(10), *463*
Floris, F. M., 442(89), *466*
Flygare, W. H., 399(156), *421*; 475(66), *504*
Foces-Foces, C., 472(42,44), 500(330-331), *503, 510*
Foley, H. M., 369(87), *419*
Fomin, P. I., 334(36), *353*
Fondère, F., 494(249,274), *508–509*
Fonseca, T., 514(102), 515(147), 527(229), 550(102), 560(147), *604–606, 609*
Foresman, B., 448(112), *467*
Forsyth, D. A., 408(177), 409(176-177), *422*
Fowler, P. W, 474(60), *503*
Fox, M., 356(15), 374(15), 406–407(15), *417*
Fox, M. A., 516(174,176), 595(366), *607, 615*
Franke, R., 375(112), 397(112), *420*
Frankland, S. J. V., 527(228), *609*
Frantsuzov, P. A., 514(113-114), 516(167-168), 540(167), 544–545(114), 547(113), 552–553(307), 557–558(309), 560–562(309), 564(307), 568(309), 573–574(167), 576(167-168), 577(168), 578(167), 579(167-168), 580–581(168), 584(114), 586(332,338), 588(338c), 590(338d), *604, 607, 612–613*
Frantz, D. D., 44(56), *78*
Franz, D. A., 406–407(169), *422*
Franzen, S., 529(243), 535(288), *610, 612*
Frauenfelder, H., 514(78), 549(78), 572–573(78), *603*
Freeman, D. L., 44(56), *78*; 488(174), *506*
Freindorf, M., 437(59), *465*
Frenkel, D., 11(44), 26(44), 43(54), 53(44), 66(66), 71(44), *77–78*

Frerking, M. A., 402(161), *421*
Frescura, F. A. M., 519(190), *608*
Fried, L. E., 82(48-50), 90(48-50), 92(48-50), 148(48-50), *151*
Friedman, H. L., 514(98), 527(223,230), 529(223,240), *604, 609–610*
Friedman, R. S., 81(32), 117(70), *150, 152*
Friedrich, H. B., 477(95), 494(270), *504, 509*
Friesner, R. A., 434(46), 448(109-110), *465, 467*; 514(116), 528(238), 582(116), *605, 610*
Frisch, M. J., 357(43,60), 374(43), 375(60), 386(135-136), 390(60), 403(60), *418*; 429(21), 437(71), 448(112), 458(21), *464–465, 467*
Frischkorn, C., 494(261-263), *508–509*
Frishman, A. M., 566–567(317), 569(317), *613*
Fromm, D. P., 200(21), 247(21), *263*
Fu, D., 200(25), 245(25), *264*
Fuchs, C., 514–515(127), 563(127), *605*
Fuchs, D., 356(20), *417*
Fujiwara, S., 477(98), *504*
Fukui, H., 357(40,49), 368(49), *418*
Fukui, K., 117(71), *152*
Fulscher, M., 515(155), 562(155), *606*
Furlan, A., 493(210), *507*
Fuss, M., 374(108), *420*

Gagarisnky, Yu. V., 488(187), *507*
Gaididei, Y. B., 532(277), *611*
Gallagher, A., 200(21), 247(21), *263*
Galperin, M., 531(268), *611*
Gambogi, J. E., 475(74), *504*
Gamow, G., 532(276), *611*
Gandhi, S., 494(261-262), *508*
Ganellin, C. R., 452(124), *467*
Gao, J., 430(27), 437(55,59), *464–465*
Gao, Y.-Q., 527(221), *609*
Garcia, A. E., 119(76-77), *152*
Garcia, M. A., 500(331), *510*
Garcia-Aldea, D., 488(180), *506*
Gardecki, J. A., 520(194), 527(228), 549(194), 560(194), *608–609*
Gardiner, C. W., 245(98), *266*
Garg, A., 514(71,85), 518(71), 543(71,85), 574(85), *603*
Garner, D. S., 430(27), *464*
Garrett, B. C., 80(12-13), 81(32), 105(12), 117(13), *150*
Gasser, U., 66(67), *78*

Gauss, J., 356(4,14,16,18,24), 357(4,41,47-48,52-57,66), 360(75), 367(57,84-85), 369(75), 371(18,24), 372(93), 373(84), 374(14,41,47-48,50,52-57,105-108), 375(50,52-53,55,57,66,109), 376(47-48,117), 377(120), 379(47), 384(47-48), 385(128-130), 386(128-129), 387(129), 388(130), 393(55,57,147-148), 394(52,57), 396(55,66,151), 397(66), 398(55,66), 399(66,158-160), 401(66), 402(66,158-160), 403(47,52,159,167), 404–405(167), 406(14), 407(14,129), 408(105,107,109,128-130, 173-175), 409(105), 410(107,109,130,175), 411(109), 413(175), *416–422*
Gaw, J. F., 380(122), *420*; 488(135), *505*
Gayathri, N., 514–515(105-106), 552 (105-106), 565(105-106), 580–581 (105-106), *604*
Geertsen, J., 360(77), 374(77), 402(77,164), *419, 422*
Gehlen, J. N., 514(35,118), 519(35), 524(118), 529(35), 532(275), 533–534(35), 582(118), 595–596(367), *602, 605, 611, 615*
Geiger, A., 494(229), *508*
Geissler, P. L., 4(10,12,14-16,18-19,24) 5(18), 22(10), 23(10,15), 24–25(16), 38(15,24), 44(15), 61(12), 73(12), *76–77*
Geldard, D. J. W., 390–392(143), *421*
Gelfand, I. M., 340(38), *354*
Gensch, T., 201(46), *264*
Georgievskii, Y., 514(38-39), 516(163), 517(39), 518(38-39), 520(38), 527(221), 573(163), 578–579(163), *602, 607, 609*
Georgiou, K., 475(69), *504*
German, E. D., 526(207), 528(236), *608–609*
Gerratt, J., 378(121), *420*
Geva, E., 200–201(14), 204(14), 242–243(14), *263*
Geyer, C. J., 44(58), *78*
Gianni, P., 449(113), *467*
Giessner-Prettre, C., 357(39), *418*
Giguère, P. A., 488(188), *507*
Gilb, S., 356(20), *417*
Gilch, P., 515(155), 562(155), *606*
Gillilan, R. E., 8(34), *77*; 80–81(23), *150*; 156(3), 172(3), *197*
Gilson, M. K., 437(63), *465*
Girardet, C., 497(247-248), *508*
Gismer, B. G., 588(342), *614*
Glass, J., 200(36), *264*

Glotzer, S. C., 75(73), *78*
Gochev, A., 515(146), 526(209), 560(146), *606, 609*
Goddard, W. A., 425(12), 448(109), *464, 467;* 528(238), *610*
Godfrey, P. D., 475(67), *504*
Goedecker, S., 49(62), *78*
Goldstein, H., 22(46), *77*
Goldstein, R. F., 529(243), 535(288), *610, 612*
Gomez, P. C., 494(231), *508*
Gomez-Jahn, L., 581(323), *613*
Gonzalez, J., 430(29), 434(29), *464*
Gonzalez, J. M., 497(294), *509*
Gonzalez, L., 460(145), *468*
Goodman, J. L., 527–528(235), 581(323), *609, 613*
Goodwin, P. M., 229(92), *266*
Gorte, R. J., 408(178), *422*
Goscinski, O., 437(66), *465*
Gould, I. R., 512(10), 527–528(235), 533(280), 535(280), 581(323), 589(10), 593(10), *600, 609, 611, 613*
Goychuk, I. A., 514(66), 535(66), 537–538(66), 584(326), *603, 613*
Grabert, H., 514(124), 538(296), 542(296), 582(124), *605, 612*
Grampp, G., 512(12), 515(148,150), 560(148,150), 593–594(12), *600, 606*
Gratton, E., 201(51), 208(51), 229(51), *264*
Grätzel, M., 512(9), 595(366), *600, 615*
Gray, H. B., 512(13), 530(249), 531(256), 532(249,256), 585(13), 590(13), *601, 610*
Gray, S. K., 80(20,24), 81(20,24), 116(20), 118(20,24), *150*; 152(2), 170(2), 183(2), *197*
Greatrex, T., 356(15), 374(15), 406–407(15), *417*
Grebenshchikov, Y., 196(30), *198*
Grebogi, C., 174(15), *197*
Green, J. T., 347(44), *354*
Green, N. J., 512(8), 524–525(8), *600*
Grifoni, M., 514(71), 518(71), 543(71), *603*
Grigorenko, B. L., 488(157), *506*
Grimes, R. N., 404(168), 406–407(169), *422*
Groenen, E. J. J., 200(20), 245(20), *263*
Groenenboom, G. C., 499(299-300, 305), *509–510*
Gronbech-Jensen, N., 8(35), *77*
Grosberg, A. Y., 67(68), *78*; 330(34), *353*
Grote, R. F., 570(319), *613*
Gruenloh, C. J., 497(291), *509*

Grynberg, G., 202(73), 205(73), 224(73), *265*
Gudowska-Nowak, E., 531(269), 535(269), *611*
Guilman, A. G., 425(5), *463*
Guldi, D. M., 512(13), 585(13), 590(13), *601*
Gundlach, E., 447(104), 448(104), 462(104), *466*
Gunner, M. R., 437(62), *465*
Guo, J.-X., 531–532(265), *610*
Gustavson, F., 83(55), 86–88(55), 148(55), *151*
Gusynin, V. P., 334(36), *353*
Gutowsky, H. S., 476(76-77), 479(76-77), *504*

Ha, T., 200(24,35-36), *264*
Ha, T.-K., 488–490(149), 492–493(149), 501(149), *506*
Haase, F., 386(136), *421*
Häber, T., 497(287), *509*
Häberle, T., 515(155), 562(155), *606*
Habuda, S. P., 488(187), *507*
Hada, M., 357(63), *418*
Haddon, R. C., 412(183), *422*
Hadzi, D., 470(1,15), *502*
Hagemeister, F. C., 497(291), *509*
Hains, A. W., 454(128), 456(128), *467*
Hale, G. M., 519(190), *608*
Hallam, H. E., 494(246), *508*
Halperin, B. I., 242–243(94), *266*
Hamann, H. F., 200(21), 247(21), *263*
Hamasaki, K., 595(369), *615*
Hameka, H. F., 356(28), 357(33), 364(33), *417*
Hamilton, A. D., 472(49), *503*
Hamilton, W. C., 470(3), *502*
Hammond, G. S., 595(366), *615*
Han, J., 494(254), 497(254), *508*
Handy, N. C., 357(61-62), 360(79), 369(89), 375(61-62,79), 380(122), 392(61), 393(62, 79), 397(79), 398(61,79), *418–420*
Hänggi, P., 514(69), 570(69), 584(326), *603, 613*
Hannachi, Y., 494(242), *508*
Hannappel, T., 512(9), *600*
Hansen, A. E., 356(32), 357(32), 364(32), 366(32), 374(103), *417, 420*
Hansen, J. E., 595(369), *615*
Hansen, J. P., 593–595(359), *614*
Hansmann, U. H. E., 44(60), *78*
Hanson, R. C., 188(192-194), 493(192-194), 502(194), *507*
Hansson, Ö., 597(374), *616*

Hara, T., 597(372), *615*
Harpool, D., 357(39), *418*
Harriman, A., 581(323), *613*
Harris, R. A., 392(144), *421*
Harrowell, P., 75(72), *78*
Hartman, K. W., 533(280), 535(280), *611*
Hartmann, L., 584(326), *613*
Hartwig, H., 497(278), *509*
Hartz, C. L., 493(220-221), *507*
Hase, W. L., 159–160(5), *197*
Häser, M., 357(45), 371(91), 374(45), 385(129), 386(45,129,134), 387(129), 407–408(129), *418–419, 421*
Hashinokuchi, M., 493(227), *508*
Hashitsume, N., 201(61), 250(61), *265*; 517(180), 538(180), *607*
Haster, J. B., 519(190), *608*
Hättig, C., 362(81), *419*
Häussermann, U., 356(18), 371(18), *417*
Hawkins, G. D., 450(87), *466*
Hawley, R. C., 442(86), *466*
Hay, P. J., 431(38), *465*
Hayashi, M., 534(284), 590(350), 593–594(350), *611, 614*
Head-Gordon, M., 372(96), 385(127), 386(135–136), *419, 421*
Hecht, B., 201–202(53), 234(53), 243(53), *264*
Heiberg, H., 357(42), *418*
Heidrich, D., 488(136,144), 501(136,144), *505–506*
Heitele, H., 515(155), 562(155), *606*
Helgaker, T., 356(11), 357(11,42,46,51,62,70), 369(88), 375(62), 393(62), 395–397(51), 399(158), 402(158,162), *417–419, 421*; 437(75), *466*
Heller, A., 446(101), *466*
Heller, E. J., 534(285), *611*
Helman, A. B., 514(80), *603*
Henderson, T. M., 533(281), 597(281), *611*
Henkelman, G., 8(30), 47(30), *77*
Hennrich, F. H., 356(20), *417*
Henricksons, T., 442(86), *466*
Herman, M., 184(18,21), 187(21), *197*
Hermida-Ramón, J. M., 494(234), *508*
Hernandez, R., 91(60), 115(60), *151*
Herrman, A., 201(46), *264*
Herzberg, G., 494(277), *509*
Hess, B., 357(64), *418*
Hettema, H., 357(42), *418*
Hetz, G., 512(12), 593–594(12), *600*

Heuer, A., 243(96), *266*
Heywood, G. C., 459(142), *468*
Higgs, J. F., 494(275), *509*
Hilczer, M., 527(225), 533–534(225), *609*
Hillig, K. W. H., 397(154), *421*
Hinchen, J. J., 488(162), *506*
Hinchliffe, A., 478(101), *505*
Hinde, R. J., 80(25-26), 81(25-26), 94(26), 103(25-26), 105(26), *150*
Hinton, J. F., 356(8), 357(8,35,39), 364(35), 367(83), 370–371(35), 382(35), 415(35), *417–419*
Hirata, S., 488(205), 491(205), 493(205), *507*
Hirayama, F., 588(348), *614*
Hirsch, J., 515(155), 562(155), *606*
Hirsch, K. R., 502(336), *510*
Hirschfelder, J. O., 474(53-54), *503*
Hobbs, R. H., 488(162), *506*
Hobza, P., 470(8,11,18), *502*
Hochstrasser, R. M., 201(45), 220(89), *264, 266*; 514(130), 576(130), *605*
Hodges, M. P., 488(153), *506*
Hoffman, B. M., 597(374), *615–616*
Hoffman, M. Z., 527(234), 590(234), 593(234), *609*
Hofkens, J., 201(46), *264*
Hofmann, M., 356(14), 374(14), 406–407(14), *417*
Hojo, K., 414–415(186), *422*
Hollauer, A., 434(51a), *465*
Holstein, T., 514(72), 571(72), *603*
Holtzberg, F., 472(34), 499(34), *503*
Holzapfel, W. B., 502(336), *510*
Honda, K., 580–581(322), *613*
Hong, K. M., 593(356), *614*
Honig, B., 437(62), 448(109-110), *465, 467*; 528(238), *610*
Hoover, W. G., 8(37), 29(48), 42(37), *77*; 348(51), *354*
Hopfield, J. J., 514(30,86), 531(261), 532(30), 535(261), 564(311), 597(311), *601, 603, 610, 612*
Hopkins, G. A., 476(80-81), *504*
Hori, K., 82–83(51-52), 87(51-52), 136(51), 140–141(51-52), *151*
Horn, H., 357(45), 371(91), 374(45), 386(45), *418–419*
Hornbach, M. J., 514(122), *605*
Horng, M. L., 515(149), 520(194), 527(228), 549(194), 560(149,194), *606, 608–609*

Hornig, D. F., 488(189-190), 493(190), 494(269), *507, 509*
Hoshi, M., 590(351), *614*
Hoszfeld, F., 488(197), *507*
Howard, B. J., 488(161,209), 490(161), *506–507*
Hsu, C.-P., 514(38-39), 517(39), 518(38-39), 520(38), 531–532(267), *602, 611*
Hsu, D., 519(184), 538(184), *607*
Hsung, R. P., 535(294), *612*
Hu, D., 200(25-26), 202(26), 245(25-26), *264*
Hu, Y., 514(99), 538(99), 542(99), *604*
Huang, K., 273(2), *352*; 512(1), *600*
Huber, K.-P., 494(277), *509*
Huggins, M. L., 441(91), *466*
Huisken, F., 472(23), 488(168-169,178), 498(23,283-284), *503, 506, 509*
Hummer, G., 119(76-77), *152*
Huniar, U., 385(127), *421*
Hunt, R. D., 488(186), 494(251), *507–508*
Hupp, J. T., 514(40), 526(209), 527–528(235), 565(316), 583(316), *602, 609, 613*
Huppert, D., 515(145), 520(197), 527(197), 546(197), 550(197), 560(145), 595(197), *606, 608*
Husain, S. K., 519(190), *608*
Hush, N. S., 512(6), 524(6), *600*
Hutchinson, J. S., 80(22), 82(22), *150*
Huth, M., 477(90), *504*
Hutson, J. M., 474(60), *503*
Hutter, J., 4(15,24), 23(15), 38(15,24), 44(15), 49(62), *76–78*; 441(97), 446(97), *466*
Hüwel, L., 494(263), *509*
Hwang, J.-K., 514(33), 519(33), 527(33), 529(33), 533–534(33), *601*
Hyeon, C., 597(378), *616*
Hynes, J. T., 80(4), 105(4,63), *150–151*; 514(84,118), 515(84), 524(118), 550(84,300), 552(84), 565(84), 570(318-319), 582(118), 594–595(363), *603, 605, 612–613, 615*

Ibers, J. A., 470(3), *502*
Ichiye, T., 514(57), 533(57), *602*
Ikeda, H., 590(350-351), 593–594(350), 595(369), *614–615*
Ikeda, N., 184–185(19), *197*
Imamoglu, A., 202(70), 245(70), *265*
Impey, R. W., 514(34), 519(34), 527(34), 529(34), 533–534(34), *601*
Imura, K., 493(223,227), *508*
Inokuti, M., 588(348), *614*
Inskeep, R. G., 498(281), *509*
Isied, S. S., 597(374), *616*
Islampour, R., 534(284), *611*
Itskovich, E. M., 519(188), *608*
Ittah, V., 515(145), 560(145), *606*
Itzykson, C., 326(30), *353*
Ivanov, A. I., 514(64), 535(287), 538(64), 540(64), *603, 611*
Iversen, G., 531(258), 535(258), *610*
Ivkovic-Jensen, M. M., 597(374), *615–616*
Iwai, S., 515(159), 562(159), 588(159), 593(353), *607, 614*
Iwata, S., 488(205), 491(205), 493(205), *507*

Jackowski, K., 402(163), *422*
Jacobson, M. P., 184(20), 186(22), *197*
Jacquemin, D., 488(206), *507*
Jaenicke, W., 515(148), 560(148), *606*
Jaffe, C., 119(74), *152*
Jagerovic, N., 472(44), *503*
Jagod, M.-F., 408(179), *422*
Jameson, A. K., 403(166), 405(166), *422*
Jameson, C. J., 399(157), 403(166), 405(166), *421–422*
Jang, Y. H., 425(12), *464*
Jannik, G., 284(14), 304(14), *353*
Jansen, M., 356(18), 371(18), *417*
Jansen, R. W., 488(193), 493(193), *507*
Janzen, J., 488(160), 490(160), *506*
Jarzeba, W., 515(147,157,159), 516(173), 520(147a), 560(147,157), 562(157,159), 565(173), 583(173), 588(159), *606–607*
Jaszunski, M., 356–357(11), 397(152), *417, 421*
Jaworski, J. S., 449(114), *467*
Jean, J. M., 166(12), *197*; 514(116), 582(116), *605*
Jeffrey, G. A., 470(10,13), *502*
Jelezko, F., 205(82), *265*
Jen, C. F., 514(33), 519(33), 527(33), 529(33), 533–534(33), *601*
Jena, P., 80–81(27), 109(27), *150*
Jenkins, O. B., 514–515(92), 561(92), *604*
Jensen, F., 430(25), *464*
Jensen, H. J. A., 357(42,51), *418*; 437(75), *466*
Jensen, J. H., 454(128), 456(128), *467*

Jensen, P., 494(230, 235–236), *508*
Jia, Y. W., 201(45), 220(89), *264, 266*
Jiao, H., 356(21), 412(184), 415(184), *417, 422*
Jimenez, R., 520(192), *608*
Joesten, M. D., 470(5), *502*
Johannesson, G., 8(30), 47(30), *77*
Johns, J. W. C., 493(212,214), *507*
Johnson, A. E., 515(151,156-157), 560(151,156-157), 562(151,156-157), 563(156b), 597(151), *606*
Johnson, G. L., 488(182-183), *507*
Johnson, K. E., 494(258-259), 495–496(259), *508*
Johnson, M. W., 472(29), 490–491(29), 494(267-268), *503, 509*
Jokisaari, J., 357(65), *418*
Jonas, D. M., 184–185(19), *197*
Jones, D. A. R., 348(48), *354*
Jones, G., 515(149), 560(149), *606*
Jones, R. E., 472(37), 499(37), *503*
Jonsson, D., 357(42), *418*
Jonsson, H., 8(30), 47(30), *77*
Jönsson, P.-G., 472(33,39), 499(39), *503*
Joo, T., 522(200), *608*
Jørgensen, P., 357(42,46,51,70), 360(76), 369(88), 374(76), 395–397(51), *418–419*
Jorgensen, W. L., 430(27,30), 437(55-56), *464–465*
Jortner, J., 513(24-25), 514(87,96-97), 515(96-97), 516(179), 519(187), 520(193), 524(24-25), 531(260), 559(87), 562(96-97), 563(96b), 597(375), *601, 603–604, 607–608, 610, 616*
Joshi, G. C., 586(337), *613*
Jucks, K. W., 475(71-72), 476(78), 479(78), 484(71-72), *504*
Jung, Y., 203(79), 247(104), *265–266*; 514(143), 557(143), 584(143), 597(378), *606, 616*

K

Kacmarek, A. J., 397(153), *421*
Kadanoff, L. P., 4(4), *76*; 275(7-8), *353*
Kador, L., 200(2,13), *263*
Kahlow, M. A., 515(147), 520(147a), 560(147), *606*
Kakitani, T., 512(11,15-16), 514(55-56), 531(253), 532(278), 533(55-56), 535(56), 585(15), 586(16), 590(16), 592(11), *600–602, 610–611*
Kakudate, Y., 477(98), *504*
Kaloudis, M., 488(168-169), 498(283,284), *506, 509*
Kalyanasundaram, K., 595(366), *615*
Kan, I., 174(15), *197*
Kanda, Y., 512(11), 592(11), *600*
Kandori, H., 515(151), 560(151), 561(151), 597(151), *606*
Kaneko, J., 597(372), *615*
Kang, T. J., 515(147), 520(147a), 560(147), *606*
Kanter, E. P., 408(179), *422*
Kappes, M. M., 356(19-20), *417*
Karabastos, Y. M., 512(13), 585(13), 590(13), *601*
Karki, L., 565(316), 583(316), *613*
Karlström, G., 494(234), *508*
Karpfen A., 472(50), 473(51-52), 478(52,99-100,103,107,111,114), 479(107,116), 480(107), 481(107,114), 483(107), 484(107,114), 485(52), 486(114,116,118), 488(52,133-134,141,146,201,207), 491(146,207), 493(52), 494(230,235-236, 276), 499(314), 500(319), 501(141,146), *503–510*
Karplus, M., 8(28), 47(28), 47(61), *77–78*; 437(61), 447(61), 449(61), 452(61), *465*; 535(290), 597(378), *612, 616*
Kasai, T., 493(223, 227), *508*
Kass, S. R., 432(40), *465*
Kassel, L. S., 80(5), 112(5), *150*
Katagiri, T., 590(350-351), 593–594(350), *614*
Katayanagi, H., 165(11), *197*
Katchalsky, E., 586(339), 588(339), *614*
Kato, S., 514(41), 520(41c), 527(41), 529(41), 533–534(41), *602*
Kato, T., 527(225), 533–534(225), *609*
Katz, A. I., 488(193-194), 493(193-194), 502(194), *507*
Katz, D. J., 532(274), *611*
Kaupp, M., 357(64), *418*
Kauzman, W., 499(313), *510*
Kayanuma, Y., 516(165), 572–575(165), *607*
Keck, J. C., 80(6), 105(6), *150*
Keith, T. A., 357(36-37,60), 375(60), 390(60), 397(153), 403(60), *418, 421*; 437(72-73), 448(112), *465, 467*
Keller, H. M., 196(30), *198*
Keller, K. A., 229(92), *266*
Kelliker, J. M., 498(281), *509*

Kellman, M. E., 187(24), *197*
Kelty, M. D., 487(124), *505*
Kemnitz, K., 512(9), 515(151), 560(151), 561(151), 597(151), *600, 606*
Kemp, M., 530(250), 532(250), *610*
Kerstel, E. R. Th., 475(73-75), 476(82), 479(82), 484(82), 486(82), *504*
Kertész, M., 478(110), 488(200), *505, 507*
Keshavamurthy, S., 113(65), *151*
Keske, J. M., 530(249), 532(249), *610*
Kestner, N. R., 474(55), *503*; 513(24), 524(24), *601*
Ketelaars, M., 201(44), *264*
Keutsch, F. N., 499(302,311), *509–510*
Kharkats, Y. I., 526(210), 527(218-219), 531(258), 535(258), 595(219), *609–610*
Khokhlov, A. R., 330(34), *353*
Kikuchi, K., 590(350-351), 593–594(350), *614*
Kilin, S. F., 586(339), 588(339), *614*
Kim, C., 595(369), *615*
Kim, H. J., 514(67,118), 524(118), 527(232), 535(67), 537–538(67), 582(118), *603, 605, 609*
Kim, K., 448(110), *467*; 528(238), *610*
Kim, K. S., 499(308), *510*
Kim, S. K., 80(30), 115(30), *150*; 161(9-10), *197*; 202(78), 245(78), *265*
Kimball, G. E., 586(334), 588(334), *613*
Kimtys, L., 460(149), *468*
Kimura, A., 531(253), *610*
Kimura, C., 597(372), *615*
Kindt, J. T., 531(266), *611*
King, B. F., 478(104-105), 485(105), *505*
King, Ch. M., 477(86), *504*
King, G., 514(33), 519(33), 527(33), 529(33), 533–534(33), *601*
King, H. F., 383(124), *420*
King, R. A., 412(182), *422*
Kirpekar, S., 357(42), *418*
Kittelberger, J. S., 488(190), 493(190), *507*
Klafter, J., 201(40-41), *264*; 520(193), 597(379), *608, 616*
Klamt, A., 442(84-85), 457(84), *466*
Klärner, F.-G., 356(26), *417*
Kleier, D. A., 412(183), *422*
Klein, M. L., 43(55), *78*; 514(34), 519(34), 527(34), 529(34), 533–534(34), *601*
Klein, O., 472(44), 500(330-331), *503, 510*
Klemp, C., 356(18), 371(18), *417*

Klemperer, W., 488(161), 490(161), 494(232), *506, 508*
Kliner, D. A. V., 515(157), 560(157), 562(157), *606*
Klippenstein, S. J., 80(13), 117(13), *150*
Klopper, W., 373(101), 412(185), *420, 422*; 487(122-123), 488–489(152), 499(303), 501(152), *505–506, 510*
Klosek, M. M., 67(68), *78*
Klots, T. D., 476(76-77), 479(76-77), *504*
Knight, P. L., 202(78), 245(78,99), *265–266*
Knowles, P., 385–386(132), *421*
Knözinger, E., 477(89-92), *504*
Kob, W., 75(73), *78*
Kobayashi, K., 356(20), *417*
Kobayashi, R., 357(42,51), 395–397(51), *418*
Kobayashi, T., 515(151), 560(151), 561(151), 597(151), *606*
Kobetic, R., 527–528(235), *609*
Kobko, N., 500(323), *510*
Koch, H., 357(42), 362(82), *418–419*
Koch, M., 498(284), *509*
Kofman, A. G., 514(75), 547(75a), 549(75), 573(75), *603*
Kofranek, M., 478(99-100), *504–505*
Kogut, J., 334(36), *353*
Köhler, H.-J., 488(136), 501(136), *505*
Kohler, J., 201(44), *264*
Kohn, F., 201(46), *264*
Kolenbrander, K. D., 488(129), *505*
Kolessov, A., 493(224), *508*
Koller, J., 478(110), 488(200), *505, 507*
Kollhoff, H., 477(89-91), *504*
Kollman, H., 431(39), *465*
Kollwitz, M., 385(128-129), 386(128-129), 387(129), 407(129), 408(128-129), *421*
Kölmel, C., 357(45), 374(45), 386(45), *418*; 442(85), *466*
Komasa, J., 360(77), 374(77), 402(77), *419*
Komatsuzaki, T., 82(38-44), 83(38-44), 90(38-39), 92(38), 93(38-44), 94–95(40), 99(43-44), 102(38), 105(62), 106(64), 107(41), 109(41), 110(43-44), 112(40), 115(38-39), 116(40), 119(78,80), 149(44), *151–152*; 161(8), 196(29), *197–198*
Komornicki, A., 488(143), *506*
Komuro, M., 174(16), *197*
Koppel, D. E., 201(49), *264*
Koppel, I., 433(43), *465*
Koppel, I. A., 433(43), *465*

Korneyshev, A. A., 514(27-28), 518(27-28), 520(28,196), 527(27-28), *601, 608*
Kosloff, R., 514(133), *605*
Kosower, E. M., 515(145), 560(145), *606*
Kostic, N. M., 597(374), *615–616*
Kotomin, E., 595(365), *615*
Kozankiewicz, B., 200(10), *263*
Kozhushner, M. A., 535(287), *611*
Koziol, F., 356(26), *417*
Kozmutza, C., 499(309), *510*
Krajnovich, D. J., 488(164), *506*
Kraka, E., 357(68), 377(120), *418, 420*
Kramers, H. A., 80(7), 105(7), 119(7), *150*; 514(68), 570(68), *603*
Krause, P. F., 477(95), *504*
Kremer, K., 312(20), *353*
Krieger, C., 515(155), 562(155), *606*
Krieger, I. M., 347(41), *354*
Krishnan, R., 378(121), *420*
Krishtalik, L. I., 527(218), *609*
Krissinel, E., 585(330), 588(330b,347), 590–591(330), *613–614*
Kroemer, R. T., 488(150,155,177), 501(150), *506*
Krull, K. L., 434(50), *465*
Krumhansl, J. A., 119(76), *152*
Kryachko, E. S., 499(309), *510*
Kubicki, J. D., 430(33), *464*
Kubo, R., 210(59-61), 203(60), 204(59-60), 212(60), 232(60), 250(61), *265*; 512(3), 513(3b), 516(3), 517(180), 519(3), 534(3), 538(180), *600, 607*
Kuehn, C., 597(374), *616*
Kuethe, J. T., 408(181), *422*
Kuharski, R. A., 514(34), 519(34), 527(34), 529(34), 533–534(34), *601*
Kuki, A., 532(271), *611*
Kulcke, A., 488(168-169), *506*
Kumar, K., 533(280-281), 535(280), 597(281), *611*
Kumar, P. V., 520(192), *608*
Kummer, S., 200(19), 245(19), *263*
Kuno, M., 200(21), 247(21), *263*
Kuo, J., 597(378), *616*
Kurnig, I. J., 478(103), *505*
Kurnikov, I. V., 528(239), 531(262), 533(281), 597(281), *610–611*
Kurnikova, M. G., 528(239), *610*
Kutzelnigg, W., 356(1,6,30-31), 357(6), 362(80), 364(30-31), 375(112-114),

395(113), 397(112), *416–417, 419–420*
Kuzmin, M. G., 595(369), *615*
Kuzmin, V. A., 512(12), 593–594(12), *600*
Kuznetsov, A. M., 512(20,22), 514(27-28), 516(178), 518(27-28), 519(178,188), 520(28), 526(207), 527(27-28,216,235), 528(235-236), 531(258), 535(258,289), 537(289b), 578(20), *601, 607–610, 612*
Kuzokov, V., 595(365), *615*
Kwok, H.-S., 488(164), *506*

L
Ladanyi, B. M., 514(29,58), 527(29,229), 533(58), *601–602, 609*
Ladik, J., 473(51), *503*
Lafferty, W. J., 475(70), *504*
Lago, S., 447(105), 462(105), *466*
Lai, H., 165(11), *197*
Lai, Y. C., 174(15), *197*
Laidig, W. D., 369(90), *419*
Lamas-Saiz, A. J., 500(330), *510*
Lami, A., 514(50), *602*
Lan, Z., 522(202), 582(202), *608*
Landau, D. P., 11(41), 53(41), *77*
Landau, L. D., 327–328(33), *353*; 358(72), 362(72), *419*; 514(73), 574(73), *603*
Landgraf, S., 515(150), 560(150), *606*
Landry, D. W. J., 425(6), *463*
Lang, G., 514(48), *602*
Langel, W., 477(89-91), *504*
Langer, U., 472(44), *503*
Langer, W. D., 402(161), *421*
Langford, J. C., 75(70), *78*
Lappin, A. G., 597(372), *615*
Laria, D., 4(25), 23(25), *77*
Larsson, S., 531(255), 533(281), 597(281,375), *610–611, 616*
Lascola, R., 493(216), *507*
Laskar, J., 118(72), *152*
Lasker, L., 156(1), 184(1), 193(1), *197*
Laso, M., 43(54), *77*
Latajka, Z., 488(139,204), 494(139,243,260), 495–496(243,260), 497(243), *505, 507–508*
Lauderdale, W. J., 357(41), 374(41), *418*
Lauenstein, C., 494(261-262), *508*
Laughlin, K. B., 493(213), *507*
Laush, C., 488(169), 498(283), *506, 509*
Lawrence, M. F., 597(373), *615*
Lax, M., 512–513(3), *600*
Lazaridis, T., 47(61), *78*

Lazzeretti, P., 357(37), *418*
Leary, B., 348(48), *354*
Leatherdale, C. A., 200(23), 247(23), *263*
le Bellac, M., 289(17), *353*
Leblanc, R. M., 597(373), *615*
Lee, A. M., 357(61), 375(61), 392(61), 398(61), *418*
Lee, C., 595(369), *615*
Lee, E. P. F., 4(14), *76*
Lee, H. M., 499(308), *510*
Lee, J. Y., 499(308), *510*
Lee, S., 597(378), *616*
Lee, S. A., 188(192), 493(192), *507*
Lee, T. J., 372(93), *419*
Lee, Y. T., 488(127,164), 493(213), *505–507*
Leforestier, C., 499(312), *510*
Leggett, A. J., 514(71), 518(71), 543(71), *603*
Legon, A. C., 475(65,68-69), *504*
Lehmann, K. K., 475(74), *504*
Leiserowitz, L., 472(43), 500(43), *503*
Leitner, D. M., 80–81(17), *150*
Leito, I., 433(43), *465*
Lenoir, D., 408(180), *422*
Lepori, L., 449(113), *467*
Lepre, C., 597(374), *616*
Leroy, G., 497(289), *509*
Lester, W. A., 373(102), *420*
Leutwyler, S., 493(210), *507*
Levich, V. G., 512(7), 514(7), 517(7), 535(7), *600*
Levin, P. P., 512(12), 593–594(12), *600*
Levine, S., 530(245), *610*
Levinger, N. E., 515(156-157), 524(156), 560(156-157), 562(156-157), 563(156b), 564(156), *606*
Li, H., 454(128), 456(128), *467*
Li, J., 450(87), 458(139), *466, 468*
Li, L. Q., 201(45), 229(89), *264, 266*
Li, N., 437(59), *465*
Liang, C. X., 530–531(247), *610*
Liang, G., 409(176), *422*
Liang, K. K., 534(284), *611*
Liang, N., 597(374), *616*
Lias, S. G., 427(18), *464*
Libby, W., 512(2), *600*
Lichtenberg, A. J., 82–83(53), 87(53), 116(53), 141(53), *151*
Liedl, K. R., 488(149-151,154-155,158,177), 489–490(149), 492–493(149), 501(149-150,158), *506*

Lievin, J., 184(18), *197*
Lifshitz, E. M., 326(31), *353*; 358(72), 362(72), *419*
Lightner, D. A., 459(141), *468*
Ligon, A. P., 430(35), *464*
Lim, C., 437(61), 447(61), 449(61), 452(61), *465*
Lim, D., 430(30), *464*
Limbach, H.-H., 472(44), 500(330-331), *503, 510*
Lin, S. H., 514(46-47, 65), 519(186), 531(251), 534(47,284), 537(47), 538(47,65), 597(378), 598(378g), *602–603, 608, 610–611, 616*
Lin, Y., 588(349b-c), *614*
Lin, Z., 533(280), 535(280), *611*
Lindsay, S. M., 188(192), 493(192), *507*
Linell, R. H., 470(4), *502*
Liotard, D. A., 450(87), *466*
Lipscomb, W. N., 378(121), 412(183), *420, 422*; 472(26,28,32), 475(26), 488(28), 490–491(28), *503*
Liptak, M. D., 450(117), 453(127), 458(117), *467*
Lischka, H., 478(99-100,103), *504–505*
Lisy, J. M., 488(127-129,137,164,169), 498(283), *505–506, 509*
Liu, K., 493(224), *508*; 499(301–302), *509*
Liu, S.-Y., 488(137), *505*
Liu, Y., 200(11), *263*
Liu, Y.-P., 524(206), 527–528(214), 529(206), *608–609*
Loerting, T., 488(154,158), 501(158), *506*
Logan, J., 513(24), 524(24), *601*
London, F., 357(38), *418*
Lopes, R., 430(29), 434(29), *464*
López, C., 500(331), *510*
Loring, R. F., 232–233(93), *266*
Losada, J. C., 118(73), *152*
Loudon, R., 202(74), *265*
Lounila, J., 402(162), *422*
Lounis, B., 200(4,12), 201(12,52,54), 205(82-83), 210–211(12), 223(12), 242–243(12), *263–265*
Lovas, F. J., 497(278), *509*
Lovejoy, C. M., 493(211), *507*
Lovejoy, E. R., 80(30-31), 115(30), *150*; 161(9-10), *197*
Lu, H., 565(316), 583(316), *613*
Lu, H. P., 200(30-31), 209(30), *264*

Lubchenko, V., 243(97), 266; 514(67,135), 535(67), 537–538(67), 569(135), 603, 605
Lucchese, R. R., 493(220-221), 494(254), 297(254), 507–508
Luchinsky, D. G., 38(50), 77
Lucke, A., 514(124), 582(124), 605
Luckhaus, D., 488(147), 506
Ludwig, R., 472(25), 498(25), 500(324-324), 503, 510
Lum, K., 11(42), 77
Lundberg, J. K., 184–185(19), 197
Luque, F. J., 438(77), 456–457(132), 459(143), 466–468
Luthey-Schulten, Z., 195(27), 197
Lüthi, H. P., 412(185), 422; 487(123), 505
Luzar, A., 11(42), 77
Lynch, G. C., 117(70), 152
Lyon, A., 527–528(235), 609

Maali, A., 200(4), 263
Madhukar, A., 535(286), 611
Maeda, S., 477(96), 504
Maerker, C., 488–490(149), 492–493(149), 501(149), 506
Magde, D., 201(47), 264
Maier, R. S., 38(51), 77
Maillard, D., 497(247-248), 508
Maitland, G. C., 474(58), 503
Mak, C. H., 514(48,124), 543(124a-c), 582(124), 602, 605
Makarov, D. E., 245(102-103), 266; 514(120, 125), 552(120), 555(120), 565(120), 582–583(125), 605
Makhnovskii, Y. A., 597(378), 616
Maki, A. H., 597(375), 616
Makri, N., 514(125-126), 582–583(125-126), 605
Makulski, W., 402(163), 422
Malagoli, M., 357(37), 418
Malick, D. K., 429(21), 458(21), 464
Malkin, V. G., 357(64), 360(78), 375(78,115), 390(115), 392–393(78), 397(78), 418–420
Malkina, O. L., 357(64), 360(78), 375(78,115), 390(115), 392–393(78), 397(78), 418–420
Malliaris, A., 459(140), 468
Mandel, L., 202(68,71), 207–208(68), 215(68), 219(68), 229(68), 245(71), 265
Mao, Y., 530(250), 532(250), 610
March, J., 457(138), 468

Marchi, M., 514(35), 519(35), 529(35), 533–534(35), 602
Marcus, R. A., 56(65), 78; 80(8), 81(33), 91(8), 112(8), 115(33), 150; 512(5), 514(31,37-39,89-90), 515(89-90), 516(171-172), 517(5,39), 518(38-39), 519(37), 520(38), 524(5), 526(5,213,215), 527(221), 528(171, 237), 530(171), 531(267,270), 532(267,273,275), 555(89), 561(89-90), 564(315), 566(89), 585(31), 589(31), 597(373), 600–602, 604, 609, 611–612, 615
Marcus, Y., 447(100), 448–449(108), 466–467
Maréchal, Y., 500(320), 510
Margenau, H., 474(55), 503
Marinari, E., 44(59), 78
Marinozzi, M., 408(181), 422
Maris, H. J., 275(8), 353
Maroncelli, M., 476(80-81), 504; 514(36), 515(149), 519(36), 520(36a,192,194), 527(228), 529(36), 533–534(36), 549(194), 560(149,194), 602, 606, 608–609
Marshall, M. D., 494(235), 508
Marteau, Ph., 494(249,274), 508–509
Marten, B., 448(109-110), 467; 528(238), 610
Martens, C. C., 116(68), 118(68), 151; 184(17), 197
Marti, J., 4(20-21), 76
Martin, J. C., 229(92), 266
Martinez-Carrera, S., 472(41), 503
Marusak, R. A., 597(372), 615
Marx, D., 49(62), 78; 441(97), 446(97), 466
Mas, E. M., 499(299-300), 509
Masad, A., 515(145), 560(145), 606
Masamune, S., 414–415(186), 422
Mason, M. D., 202(70), 245(70), 265
Masui, H., 500(328-329), 510
Mataga, N., 512(11,15-16,19), 516(175), 532(278), 580–581(322), 585(15,19), 586(16), 588(19), 590(16), 592(11), 600–601, 607, 611, 613
Mateescu, G. D., 409(176), 422
Mathews, D. M., 460(147), 468
Matkowsky, B. J., 67(68), 78
Matsuda, H., 357(49), 368(49), 418
Matsuda, N., 512(15), 585(15), 601
Matsui, M., 597(372), 615
Matsunaga, Y., 106(64), 119(78), 150–152
Matsusita, N., 590(350), 593–594(350), 614

Matsuzaki, S. Y., 586(337), 588(345), 590–592(345), *613–614*
Mattes, S. L., 512(10), 589(10), 593(10), *600*
Mattuck, R. D., 289(16), 336–337(16), *353*
Matyushov, D. V., 514(29,58), 527(29,226-227, 231), 529(227), 533(58,283), 534(283), 590(227b), *601–602, 609, 611*
Mauk, A. G., 597(374), *616*
Maus, M., 201(46), *264*
Mauzerall, D., 516(169), 585(169), *607*
Maverik, E., 459(141), *468*
May, V., 514(66), 535(66), 537–538(66), *603*
Mayhew, C. A., 408(179), *422*
McCammon, J. A., 437(63), *465*; 514(44), 535(291), 297(291), 594–595(363), 599(291), *602, 612, 615*
McClellan, A. C., 470(2), *502*
McCleskey, T. M., 512(13), 585(13), 590(13), *601*
McClintock, P. V. E., 38(50), *77*
McConnell, H. M., 530–532(248), *610*
McCoy, B. M., 292(18), 304(18), *353*
McDonald, J. C., 472(48), *503*; 593–595(359), *614*
McEwen, W. K., 457(137), *468*
McIntosh, A. L., 493(220-221), 494(254), 497(254), *507–508*
McIver, J. W. Jr., 91(57), *151*
McIver, R. T., 427(17), 452(123), *464, 467*
McKellar, A. R. W., 493(212,214), *507*
McLendon, G. M., 512(9), 516(175), 535(292), *600, 607, 612*
McMahon, P. E., 498(281), *509*
McManis, G. E., 515(146), 526(209), 560(146), *606, 609*
McMurchie, L., 383(125), *420*
McQuarrie, D. A., 281(13), *353*; 428(20), *464*
McWeeny, R., 378(121), *420*; 430(23), *464*
Meads, R. F., 493(220-221), *507*
Mebel, A. M., 534(284), *611*
Medvedev, E. S., 514(61), 535–537(61), *602*
Medvedev, I. G., 527(216), *609*
Meeker, S. P., 348(47), *354*
Mejias, J. A., 447(105), 462(105), *466*
Melendez, R. E., 472(49), *503*
Meller, J., 8(29), 75(29), *77*
Melnikov, V. I., 514(70), 570(70), *603*
Mennekes, T., 407(170), *422*
Mennucci, B., 357(71), *418*; 442(88), 450(115), *466–467*; 499(307), *510*

Meot-Ner(Mautner), M., 425(9), 460(151), *463, 468*
Meredith, A. W., 494(233), *508*
Merrill, G. N., 432(40), *465*
Metiu, H., 202(75), 245(102-103), *265–266*
Metropolis, A. W., 11(45), *77*
Metropolis, N., 11(45), *77*
Metzgfer, H., 441(94), *466*
Meyer, H., 475(73), *504*
Meyer, T. J., 516(173,177), 526(209), 565(173), 583(173), *607, 609*
Michael, D. W., 488(128,137), *505*
Michel, R. H., 356(20), *417*
Michel-Beyerle, M. E., 515(155), 562(155), *606*
Michl, J., 470(9), *502*
Michler, P., 202(70), 245(70), *265*
Mielke, S. L., 117(70), *152*
Miertus, S., 437(68), 439(68), 444(68), 456(68), *465*
Mihalic, Z., 374(107), 408(107), 410(107), *420*
Mikenda, W., 470(17), *502*
Mikhelashvili, M. S., 585(330), 586(339), 588(330b, 339), 590–591(330), *613–614*
Mikkelsen, K. V., 357(42,70), *418*; 437(75), *466*; 514(131), *605*
Mikulskiene, B., 460(149), *468*
Milburn, G. J., 209(85), *265*
Millen, D. J., 475(65,69), *504*
Miller, J. R., 512(8), 524(8,205), 525(8), 527–528(235), 533(280), 535(280), *600, 608–609, 611*
Miller, M. J., 425(8), *463*
Miller, N. E., 532–533(279), *611*
Miller, R. E., 475(71-72), 476(78,82-85), 479(78,82), 484(71-72, 82, 84-85), 486(82), 487(121), 488(170), 499(296), *504–506, 509*
Miller, W. H., 3(1), *76*; 80(9-10), 91(10,60), 113(65), 115(60), *150–151*; 494(240), *508*; 514(117,134), 582(117), 584(134,327), *605, 613*
Mills, I. M., 378(121), *420*
Mil'nikov, G. V., 113(66), *151*
Minguet-Bonhevi, M., 472(44), *503*
Minichino, C., 514(123), 582(123), *605*
Miransky, V. A., 334(36), *353*
Mishima, M., 433(43), *465*
Mishra, P. C., 499(310), *510*
Miura, K., 357(40,49), 368(49), *418*

AUTHOR INDEX

Miyasaka, H., 512(15,19), 585(15,19), 588(19), *601*
Miyashi, T., 590(350-351), 593–594(350), *614*
Miyazaki, K., 526(212), *609*
Mjöberg, P. J., 475(65,69), *504*
Mnatsakammjan, M. A., 342(39), *354*
Mó, O., 460(145), *468*; 497(288,290), *509*
Moazzen-Ahmadi, N., 493(212,214), *507*
Moerner, W. E., 200(1-2,5-7), 201(6-7,54), 202(54,69), 210(7), 223(6-7), *263–265*
Molina, J., 499(306), *510*
Mollica, V., 449(113), *467*
Møllmer, K., 245(100-101), *266*
Molski, A., 201(56), *264*
Montero, L. A., 499(306), *510*
Montgomery, J. A., 429(21), 458(21), *464*
Moody, R. E., 512(10), 589(10), 593(10), *600*
Mooij, G. C. A. M., 43(54), *77*
Moore, C. B., 80(30-31), 115(30), *150*; 161(9-10), *197*
Moore, D., 476(85), 484(85), *504*
Mootz, D., 500(321), 501(332-333), *510*
Morita, A., 595(364), *615*
Morokuma, K., 196(30), *198*; 474(63), *504*
Morrillo, M., 514(101), 584(101), *604*
Morris, S., 425(11), *464*
Morrison, J. A., 407(171), *422*
Moser, C. C., 530(249), 532(249), *610*
Moser, J. C., 512(9-10), 589(10), 593(10), *600*
Moskovsky, A. A., 488(157), *506*
Mostafavi, M., 588(345d), 590–592(345d), *614*
Motzfeld, T., 500(317-318), *510*
Mourik, T. V., 375(112), 397(112), *420*
Mowbray, S. L., 425(10), *463*
Mueller, L. J., 533(280), 535(280), *611*
Muhlbacher, L., 514(124), 582(124), *605*
Mujica, V., 530(250), 532(250), *610*
Mukamel, S., 200(28), 201–202(62), 205(62), 217(62), 232(62,93), 233(93), *264–266*; 514(99-100,111), 518(182), 531(252), 538(99), 542(99), 584(111), *604, 607, 610*
Mullen, K., 201(46), *264*
Müller, J. D., 201(51), 208(51), 229(51), *264*
Müller, T., 408(173,175), 410(175), 413(175), *422*
Murata, S., 512(17), 514(115), 515(159), 562(159), 580(115), 585(17), 586(17,337), 587(17), 588(159,345), 589(17), 590(17,345), 591–592(345), 593(353), *601, 604, 607, 613–614*
Murphy, R. B., 434(46), 448(109-110), *465, 467*; 528(238), *610*

Naaman, R., 493(223,225,227), *508*
Nagamoto, S., 500(328), *510*
Nagaoka, M., 82–83(38-39), 90(38-39), 92(38), 93(38-39), 102(38), 115(38-39), *151*; 161(8), *197*
Nagasawa, Y., 512(9), 515(151), 522(200), 560(151), 562(151), 597(151), *600, 606, 608*
Nagase, S., 356(20), *417*
Nagata, T., 590(350), 593–594(350), *614*
Nagle, J., 501(334-335), *510*
Nahler, N. H., 494(263-265), *509*
Nahringbauer, I., 472(35,38), 499(35,38), 500(35), *503*
Nakajima, S., 538(295), 542(295), *612*
Nakamura, A., 595(369), *615*
Nakamura, H., 113(67), 116(67), *151*
Nakamura, N., 500(328-329), *510*
Nakashima, N., 512(9), *600*
Nakata, M., 477(93), *504*
Nakatsuji, H., 357(63), *418*
Nakatsuka, H., 200(11), *263*
Nakayama, H., 516(165), 572–575(165), *607*
Namboodiri, K., 446(102), *466*
Nandi, N., 519–520(191), *608*
Nascimento, M. A. C., 426(15), 433(44), 434(51a), 435(53), 448–449(44), 456(133), 457(44,135), *464–465, 467*
Natalini, B., 408(181), *422*
Naumov, A. V., 200(13), *263*
Nauta, K., 476(83-85), 484(84-85), 487(121), 499(296), *504–505, 509*
Navon, G., 597(375), *616*
Nayak, S. K., 80–81(27), 109(27), *150*
Necoechea, W. C., 487(126), *505*
Nelander, B., 494(252), *508*
Nelson, D. D., 493(211,216-217), *507*
Nelson, M. R., 425(13), *464*
Nemukhin, A. V., 488(157), *506*
Neria, E., 514(42), 533–534(42), *602*
Nesbitt, D. J., 200(21), 247(21), *263*; 488(131-132), 493(211,216-217,219), 494(255), 497(255), *505, 507–508*
Netzel, T. L., 597(374), *616*
Neu, P., 517(181), *607*

Neuhaus, T., 44(57), 78
Neuhauser, R. G., 200(22-23), 247(22-23), 263
Neuhauser, W., 202(72), 245(72), 265
Neumann, M., 522(201), 608
Newton, M. D., 437(58), 458(58), 465;
 514(35,98), 516(170,177), 519(35),
 522(198), 524(206), 527(214,217,223),
 528(214), 529(35,206,223),
 530(170,217,246-247), 531(217,246-
 247,254), 533(35,254,281), 534(35),
 535(198, 294), 573(170), 597(281), 602,
 604, 607-612
Ni, H., 493(218,226), 507-508
Nibler, J. W., 476(80-81), 504
Nicholls, A., 448(109), 467; 528(238), 610
Nielson, R. M., 526(209), 609
Nies, A. S., 425(5), 463
Nikitin, E. E., 540(298), 576(298), 612
Ninham, B. W., 274(6), 353
Nishikawa, S., 512(16), 586(16), 590(16), 601
Nishimura, M., 586(337), 613
Nitzan, A., 514(40,42,62,132), 530(250),
 531(268), 532(250), 533(42,281), 534(42),
 535(62), 538-539(62), 541(62),
 582-584(132), 597(281), 602-603, 605,
 610-611
Niwa, T., 590(350-351), 593-594(350), 614
Nixon, E. R., 477(86), 504
Nocek, J. M., 597(374), 616
Nocera, D. G., 512(13), 514(101e), 584(101e),
 585(13), 590(13), 601, 604
Nolin, C., 477(97), 504
Nonn, T., 201(55), 264
Noodleman, L., 452(119,121), 457(119), 467
Noolandi, J., 593(356), 614
Norman, P., 357(42), 418
Norris, D. J., 201(43), 209(43), 264
Norris, J. R., 514(110), 604
Northrup, S. H., 535(291), 594-595(363),
 597(291), 599(291), 612, 615
Nosaka, T., 357(40), 418
Noukakis, D., 527-528(235), 581(323), 609,
 613
Novick, S. E., 494(241), 496(241), 508
Novikov, E., 201(57), 264
Nursey, P. N., 66(67), 78

Obata, S., 383(126), 421
O'Brien, J. P., 186-187(22), 197
Obriot, J., 494(249, 274), 508-509
Ochsenfeld, C., 356(24-26), 371(24-25), 417
Ochterski, J. W., 429(21), 458(21), 464
Oddershede, J., 360(76-77), 374(76-77),
 395(150), 402(77,164), 419, 421-422
Odiot, S., 374(104), 420
Ogilvie, J. F., 402(164), 422
Ogletree, D. F., 200(24), 264
Ohashi, N., 493(208), 497(208), 507
Ohmine, I., 514(36), 519(36), 520(36a),
 529(36), 533-534(36), 602
Ohms, U., 501(332), 510
Ohoyama, H., 493(223,227), 508
Oka, T., 408(179), 422
Okada, A., 514(93,100), 515(93), 516(93), 604
Okada, T., 512(11), 592(11), 600
O'Keeffe, M., 488(193), 493(193), 507
Olah, G. A., 408(177), 409(176-177), 422
Olender, R., 8(29), 75(29), 77
Olivares del Vale, F., 437(70), 465
Olsen, J., 357(42), 418
Olson, E. J., 530(249), 532(249), 610
Olson, W. B., 475(70), 504
Olsson, L., 357(68), 375(116), 393(116,145),
 398(116), 403(116), 418, 420-421
Onuchic, J. N., 195(27-28), 197-198; 514(77,
 85-86), 531(256,261,263-264),
 532(256,263), 535(261), 543(85), 549(77),
 573(77), 574(85), 603, 610
Orlik, P., 191(26), 197
Orozco, M., 438(77), 456-457(132), 459(143),
 466-468
Orrit, M., 200(1,3-5,8-10,12,18), 201(8-
 9,12,52), 202(8,69), 203(8-9), 205(82-83),
 210-211(12), 212(87), 223(8-9,12),
 229(8-9), 242-243(9,12), 244(9), 245(18),
 263-265
Orville-Thomas, W. J., 477(88), 504
Osa, T., 595(369), 615
Osad'ko, I. S., 201(42), 202(76-77), 264-265
Osberg, W. E., 494(269), 509
Ostlund, N. S., 430(24), 464
Otto, A. H., 374(106), 420
Oudejeans, L., 488(170), 506
Ovchinnikov, A. A., 513(21), 516-519(21),
 585(21), 611
Ovchinnikov, M., 488(179), 506
Ovchinnikova, M. Y., 513(21), 514(79),
 516-519(21), 555(79), 566(79), 585(21),
 601, 603

Oxtoby, D., 514–515(103), 552(103), 557(103), 565(103), *604*

Pacansky, J., 477(87), *504*
Packer, M. J., 357(42), *418*
Paetzold, P., 407(170), *422*
Page, M., 91(57), *151*
Paiderova, I., 360(77), 374(77), 395(150), 402(77), *419, 421*
Pal, H., 515(151), 560(151), 562(151), 597(151), *606*
Paladino, E., 514(48), *602*
Paleos, C. M., 459(140), *468*
Pande, V., 67(68), *78*
Pang, Y.-P., 425(7), *463*
Papazyan, A., 520(194), 549(194), 560(194), *608*
Papp, G., 531(269), 535(269), *611*
Paraskevas, L., 500(323), *510*
Parisi, G., 44(59), *78*
Park, J. W., 595(369), *615*
Parkins, A. S., 245(98), *266*
Parr, R. G., 372(100), *420*; 430(26), *464*
Parra, R. D., 497(293), *509*
Parrinello, M., 4(15,24), 8(31,36), 23(15), 38(15,24,36), 44(15), 47(31), 49(62), *76–78*; 441(97), 446(97), *466*
Parson, W. W., 514(33), 519(33), 527(33,224), 529(33), 533–534(33), *601, 609*
Parsons, D. F., 520(195), 527(233), *608–609*
Partt, L. R., 8(27), 67(27), *77*
Pascual-Ahuir, J. L., 442(89), *466*
Passerone, D., 8(31), 47(31), *77*
Passimo, S. A., 522(200), *608*
Patzelt, F., 386(134), *421*
Paul, J. B., 498(286), *509*
Paulson, B. P., 524(205), *608*
Pawda, A., 408(181), *422*
Paz, J. L., 205(81), 250(81), *265*
Pearson, R. G., 448–449(107), *467*
Pechukas, P., 81(37), *151*; 561(310), 576(321), *612–613*
Pedersen, T. B., 362(81), *419*
Peeters, D., 497(289), *509*
Pelliciari, R., 408(181), *422*
Penfield, K. W., 512(8), 524–525(8), *600*
Peng, C., 452(119), 457(119), *467*
Perchard, J. P., 497(247-248), *508*
Perera, D. N., 75(72), *78*

Perez, P., 430(32), *464*
Periaswamy, N., 586(337), *613*
Perng, B.-C., 527(223), 529(223), *609*
Persico, M., 437–439(69), 457(69), *465*
Peters, K. S., 586(337), *613*
Peters, O., 500(331), *510*
Petersen, P. B., 499(311), *510*
Peterson, J. M., 347(43), *354*
Petersson, G. A., 429(21), 458(21), *464*
Petrov, E. G., 514(66), 532(277), 535(66), 537–538(66), *603, 611*
Petrov, V., 565(316), 583(316), *613*
Pezolet, M., 477(94), *504*
Pfeiler, D., 477(92), *504*
Phelps, D. K., 514(28), 518(28), 520(28), 527(28,235), 528(235), *601, 609*
Philip, D., 581(323), *613*
Phillips, D. L., 450(117), 458(117), *467*
Phillips, W. A., 242–243(95), *266*
Pickett, H. M., 397(154), *421*
Pilling, M. J., 585(328), 588(328), *613*
Pimentel, G. W., 470(2), *502*
Pine, A. S., 493(208-209), 497(208), *507*
Pines, E., 595(369), *615*
Pinnick, D. A., 188(192-194), 493(192-194), 502(194), *507*
Piotrowiak, P., 527–528(235), *609*
Pistolis, G., 459(140), *468*
Pitaevskii, L. P., 326(31), *353*
Pitarch, J., 450(115), *467*
Pittman, J. L., 427(19), *464*
Pitzer, R. M., 378(121), *420*
Plakhotnik, T., 201(55,64-66), 202(66), 204(65-66), *264–265*
Plenio, M. B., 245(99), *266*
Plimpton, S. J., 75(73), *78*
Plotkin, S. S., 119(79), *152*
Pluzhnikov, P. F., 512(12), 593–594(12), *600*
Polanyi, M., 80(3), *150*
Poll, W., 501(332-333), *510*
Pollak, E., 81(37), *151*; 514(136), 566–571(136,317), 569(317), *605, 613*
Pollard, W. T., 514(116), 582(116), *605*
Pöllinger, F., 515(155), 562(155), *606*
Pöllinger-Dammer, F., 515(155), 562(155), *606*
Pomelli, C. S., 439(81), *466*
Pomeranchuk, I., 327–328(33), *353*
Poole, P. H., 75(73), *78*
Poon, W. C. K., 348(47), *354*

Pople, J. A., 372(96), 378(121), 386(135-136), 419–420; 429(22), 464; 488(172), 506
Posch, H. A., 29(48), 77
Poskin, M. P., 500(327), 510
Post, B., 472(34), 499(34), 503
Potovoi, V. V., 514(64), 538(64), 540(64), 603
Pottier, N., 518(183), 607
Powers, M. J., 526(209), 609
Press, W. H., 95(61), 151
Prezhdo, O. V., 514(43), 531(266), 602, 611
Price, D. R., 375(111), 414(111), 420
Prieskorn, J. N., 565(316), 583(316), 613
Priyadarshy, S., 532(272), 611
Prophet, H., 427(16), 464
Provencal, R. A., 498(286), 509
Pudlak, M., 514(121), 535(121), 597(121), 605
Pulay, P., 356(8), 357(8,35,44,59), 364(35), 367(83,86), 369(86), 370–371(35), 375(59), 382(35,123), 385–386(131), 388(137), 390(59), 415(35), 417–421; 434(45), 448–449(45), 457(45), 465
Pullman, A., 357(39), 418; 437(57), 465
Pullman, B., 437(57), 465
Purvis, G. D., 372(95), 419
Pusey, P. N., 348(47), 354
Puttkamer, K. V., 488(166), 506

Qian, H., 201(50), 264
Qian, J., 166(12), 197
Qin, L., 597(374), 615
Qiu, Y., 487(125), 494(125,238-239), 505, 508
Quack, M., 487(120,122), 488(145,147,149,152,166,176,181), 489(149,152,181), 490(149,181), 492–493(149), 501(149,152), 505–506
Querry, M. R., 519(190), 608

Rabani, E., 584(325), 613
Rabenstein, D. L., 414–415(186), 422
Rabinowitz, J. R., 437(60), 465
Rablen, P. R., 437(72), 465
Radom, L., 430(36), 464
Ragavachari, K., 372(96), 419; 429(22), 464
Raghavachari, K., 412(183), 422
Raineri, F. O., 527(223,230), 529(223,240), 609–610
Rall, T. W., 425(5), 463
Ramamurthy, V., 595(366), 615

Ramasami, T., 597(375), 616
Rammer, J., 516(166), 573–574(166), 576(166), 607
Ramsay, N. F., 356(27), 360(27), 417
Raney, J. K., 397(153), 421
Rankin, K. N., 494(245,259), 495–496(259), 508
Rasaiah, J. C., 514–515(108), 604
Rashin, A. A., 437(60), 446(98,102), 447(98), 450(116), 452(120,122), 465–467
Rasmussen, K., 515(150), 560(150), 606
Rasolt, M., 390–392(143), 421
Rasslov, V., 429(22), 464
Ratner, M. A., 514(40,131-133), 516(173), 530(250), 532(250), 535(286), 565(173), 582(132), 583(132,173), 584(132), 597(374), 602, 605, 607, 610–611, 615
Rauhut, G., 357(59), 375(59), 390(59), 418
Rauk, A., 494(244), 508
Ré, M. A., 597(378), 616
Reddy, A. K. M., 443–444(90), 466
Redfern, P. C., 429(22), 464
Redington, R. L., 488(163,165), 506
Ree, F. H., 348(51), 354
Regan, J. J., 531(263-264), 610
Rehfuss, B. D., 408(179), 422
Rehm, J. M., 512(9,14), 529(14), 589–590(14), 600–601
Reichel, F., 357(68), 374(106), 418, 420
Reichman, D. R., 517(181), 607
Reid, P. J., 515(158), 560(158), 562(158), 565(316), 583(316), 607, 613
Reilly, P. D., 201(38-39), 203(38-39), 249(39), 264
Reiss, H., 446(101), 466
Remko, M., 430(34), 464
Renn, A., 200–201(17), 223(17), 263
Resat, H., 527(230), 609
Reynolds, L., 527(228), 609
Rhys, A., 512(1), 600
Ribani, E., 519(185), 607
Ribas Prado, F., 357(39), 418
Rice, J. E., 377(119), 380(122), 420
Rice, O. K., 80(11,24,29), 81(24,29), 106(64), 112(11), 118(24,29), 119(80), 150–152
Rice, S. A., 585(328), 586(336), 588(328), 595(336), 613
Richard, S., 356(19), 417
Richard-Schneider, S., 356(20), 417
Richardson, W. H., 452(119), 457(119), 467

Rigby, M., 474(58), *503*
Rigler, R., 201(48), *264*
Rinaldi, D., 437(64,67), *465*
Rincon, L., 488(180), *506*
Ringnalda, M. N., 434(46), 448(109-110), *465, 467*; 528(238), *610*
Ringuet, M., 597(373), *615*
Rios, M. A., 478(109,115), 486(115), *505*
Rips, I., 514(87,136-137), 520(193), 559(87), 566–571(136), *603, 605, 608*
Risken, H., 550–551(301), 557(301), *612*
Risser, S. M., 532(272), *611*
Rivail, J. L., 437(64,67), *465*
Rizzo, A., 397(152), *421*
Robertson, A. D., 454(128), 456(128), *467*
Robertson, E. B., 441(96), *466*
Robertson, H. S., 349(53), *354*
Robinovich, S., 564(311), 597(311), *612*
Robl, C., 356(17), *417*
Rode, B. M., 461(152), *468*; 488(150,154,177), 501(150), *506*
Rodriguez, J., 4(25), 23(25), *77*
Roitberg, A., 530(250), 532(250), *610*
Roos, B. O., 372(97-98), *420*
Rose, J., 119(75), *152*
Rosenbluth, A. W., 43(54), *77*
Rosenbluth, M. N., 11(45), 43(54), *77*
Rosner, J., 327(32), *353*
Rossky, P. J., 514(43,129), 584(129), *602, 605*
Rostov, I. V., 524(206), 529(206), *608*
Rotenberg, A., 325(27), *353*
Roth, K., 498(286), *509*
Rouse, K. D., 472(36,40), 499(36,40), 500(36), *503*
Roweis, S. T., 75(70), *78*
Roy, S., 514–515(107), 519–520(191), 552(107), 565(107), *604, 608*
Rozman, I. M., 586(339), 588(339), *614*
Rubtsov, I. V., 515(153), 562(153), 565(153b), *606*
Rückemann, A., 515(155), 562(155), *606*
Ruiz-Lopes, F., 430(29), 434(29), *464*
Ruiz-Lopez, M. F., 437(67), *465*
Ruiz-Montero, M. J., 66(66), *78*
Ruoff, R. S., 476(76-77), 479(76-77), *504*
Rush, D. J., 437(72), *465*
Rusic, B., 408(179), *422*
Russegger, P., 473(51), *503*
Russell, E. L., 11(43), *77*
Rustad, J. R., 430(33), *464*

Ruud, K., 356(11), 357(11,42,51,65,67,70), 393(148), 395–396(51), 397(51,152), 399(158), 402(158,162), *417–418, 421–422*
Rys, J., 383(124), *420*

Šablinskas, V., 460(149), *468*
Sachs, S. B., 535(294), *612*
Sadeghpour, B., 408(181), *422*
Saebo, S., 385(131), 386(131,135), 388(137), *421*; 434(45), 448–449(45), 457(45), *465*
Saenger, W., 470(10), *502*
Safford, G. J., 488(195-196), *507*
Sagarik, K., 461(152), *468*
Saika, A., 383(126), *421*
Saint-James, D., 518(183), *607*
Saito, S., 514(36), 519(36), 520(36a), 529(36), 533–534(36), *602*
Salahub, D. R., 357(64), 360(78), 375(78,115), 390(115), 392–393(78), 397(78), *419–420*
Saleh, B., 202(67), 209(67), *265*
Salisbury, F. R., 392(144), *421*
Sampogna, R., 437(62), *465*
Samson, R., 598(380), *616*
Sandor, E., 472(29-30), 490–491(29), 494(30,266-268), *503, 509*
Sandorfy, C., 470(6), *502*
Sankey, O. F., 488(193), 493(193), *507*
Sano, H., 588(340), 592–594(340), 596(37), *614–615*
Santoro, F., 514(50), *602*
Sass, R. L., 499(316), *510*
Sastri, M. L. N., 488(189), *507*
Sastry, G. M., 564(311), 597(311), *612*
Satchell, D. P. N., 424(4), *463*
Satchell, R. S., 424(4), *463*
Satish, A. V., 434–435(51), *465*
Sato Miyazawa, C., 590(350), 593–594(350), *614*
Satoshi, K., 477(93), *504*
Saue, T., 357(42), *418*
Sauer, J., 497(295), *509*
Sauer, S. P. A., 395(150), 402(164), *421–422*
Sauers, R. R., 432(41), 461(41), *465*
Saul, L. K., 75(70), *78*
Saunders, M., 356(21), *417*
Savel'ev, V. A., 488(156), *506*
Savoie, R., 477(94,97), *504*
Saykally, R. J., 493(213,222), 494(222,237), 498(286), 499(301-302,310-311), *507–510*

Schaad, L. H., 470(5), *502*
Schaefer, H., 369(89), 412(182,184-185), 415(184), *419, 422*
Schaefer, H. F. III, 487(124), 488(135,138,148), 497(294), 498(286), *505–506, 509*
Schäfer, A., 357(66), 375(66), 397–399(55), 401–402(55), *418*
Schagen, D., 500(331), *510*
Schaller, T., 356(26), *417*
Schamov, G. A., 460(150), *468*
Schatz, T., 527–528(235), *609*
Scheiner, S., 460(151), *468*; 470(12,14), 487(12), 488(12,139-140), 494(139,243,256), 495(243,256), 497(243), 501(334), *502, 505, 508, 510*
Schenter, G. K., 200(31), *264*
Schenzle, A., 208(84), 229(84), *265*
Scheuerman, R. F., 499(316), *510*
Schimmelpfennig, B., 357(64), *418*
Schindler, M., 356(1,31), *416–417*
Schinke, R., 196(30), *198*
Schlegel, H. B., 378(121), *420*
Schleyer, P. V. R., 356(12-15,21), 374(14-15,104-107), 405(13), 406(14-15), 407(14), 408(105,107,173), 409(105), 410(107), 412(184), 415(184), *417, 420, 422*; 431(37), *464*; 435(54), *465* ; 488(144,149), 489(149), 492–493(149), 501(144,149), *506*
Schmid, R., 527(226), *609*
Schmidt, J., 200(20), 201(44), 245(20), *263–264*
Schmidt-Rohr, K., 359(74), *419*
Schmitt, U., 56(64), *78*; 488(147,176), 497(287), *506, 509*
Schneider, U., 356(16-17,19), *417*
Schnell, I., 356(24), 371(24), *417*
Schnöckel, H., 356(16-18), 371(18), *417*
Schofield, A., 66(67), *78*
Scholes, G. D., 515(160), 561–562(160), *607*
Schreckenbach, G., 357(58), 375(58), 390(58), *418*
Schreiber, M., 514–515(127), 563(127), *605*
Schrems, O., 477(90), *504*
Schriver, A., 497(247-248), *508*
Schroeder, M., 268(1), *352*
Schubert, M., 202(72), 245(72), *265*
Schuder, M. D., 493(211,216-217,219), *507*
Schultz, S. L., 166(12), *197*

Schulz, M., 200(28), *264*
Schuss, Z., 67(68), *78*
Schuster, P., 470(6-7,17,20), 472(50), 473(51), 479(116-117), 488(133-134), 494(117,276), 499(314), *502–503, 505, 509–510*
Schütz, M., 385–386(132), 388(138,140-142), *421*
Schüürmann, G., 442(84), 455(129-130), 457(84,134), *466–467*
Schwenke, D. W., 81(32), 117(70), *150, 152*
Schwieters, C. D., 514(123), 582(123), *605*
Scoles, G., 475(73-75), 479(82), 484(82), 486(82), *504*
Scott, J. A., 452(123), *467*
Scrimshaw, G. F., 494(246), *508*
Scroocco, E., 437(68), 439(68), 444(68), 456(68), *465*
Scuseria, G. E., 372(93), *419*; 488(138), *505*
Sebastian, K. L., 514(112), *604*
Seddon, K. R., 472(47), *503*
Seel, M., 515(154), 562(154), 565(154), *606*
Seelbach, U., 356(26), *417*
Segal, D., 531(268), *611*
Segawa, H., 580–581(322), *613*
Segre, P. N., 348(47), *354*
Segura, J.-M., 201–202(53), 234(53), 243(53), *264*
Seideman, T., 80(10), 91(10), *150*
Seki, K., 515(161), 529(244), 557(161), 581(161), 595(367), 596(367,371), 597(378), *607, 610, 615–616*
Sekusak, S., 488(150), 501(150), *506*
Selvin, P. R., 200(24,35), *264*
Semin, D. J., 229(92), *266*
Serafin, J., 493(218,226), *507–508*
Severanece, D. L., 430(30), *464*
Sevic, E. M., 8(32), *77*
Shakhnovich, E. I., 11(43), 67(68), *77–78*
Shallcross, V. F., 472(27), 475(27), *503*
Shannon, C. F., 588(342), *614*
Shapley, W. A., 453(125), *467*
Sharp, K., 437(62), 441(83), *465–466*; 528(238), *610*
Sharpe, S. W., 488(171), *506*
Shavitt, I., 372(94), *419*
Sheehy, J. A., 375(110), *420*
Sheets, R. W., 460(147), *468*
Shen, C., 597(374), *616*
Shen, Y. R., 488(164), *506*
Shenker, S. H., 334(36), *353*

Shepard, R., 372(97), *420*
Sherer, E. C., 425(6), *463*
Sheu, S.-Y., 514(138), 557(138), 597(378), 598(378g), *605, 616*
Sheu, W.-S., 597(378), *616*
Shibata, T., 165(11), *197*
Shida, N., 171(14), *197*
Shields, G. C., 425(6), 450(117), 453(127), 458(117), *463, 467*
Shields, T. P., 597(372), *615*
Shigmetsu, J., 334(36), *353*
Shilov, G. E., 340(38), *354*
Shimada, J., 11(43), *77*
Shimizu, K., 200(22-23), 247(22-23), *263*; 580–581(322), *613*
Shin, S., 597(378), *616*
Shinomoto, S., 324(25), *353*
Shirkov, D. V., 274(3-5), 286(3-4), 289(3-5), 292(3-5), 326(3), 341(4), *352–353*
Shirota, H., 515(151,153), 560(151), 562(151,153), 565(153b), 597(151), *606*
Shokhirev, N. V., 586(338), 588(338c,344), *613–614*
Sholl, D. S., 584(324), *613*
Shore, B. W., 205(80), 247(80,105), 250(80), *265–266*
Short, R., 202(71), 245(71), *265*
Shoup, D., 535(291), 597(291), 599(291), *612*
Shukla, M. K., 499(310), *510*
Shushin, A. I., 514(59), 516(164), 535(59), 537(59), 573–574(164), 576(164), 579(164), 594(362), 597(378), 598(381), *602, 607, 615–616*
Siasios, G., 460(144), *468*
Siddarth, P., 531(270), 532(273), *611*
Siders, P., 514(31), 585(31), 589(31), 597(373), *601, 615*
Sieber, S., 356(12), 374(105-106), 408–409(105), *417, 420*
Siebers, J. G., 497(292), *509*
Siehl, H.-U., 374(108), 408(173,175,180), 410(175), 413(175), *420, 422*
Siemers, I., 202(72), 245(72), *265*
Siepmann, J. L., 43(54), *77*; 288(15), 310–312(15), 313(21), *353*
Siggel, M. R. F., 430(28), 434(28,48), *464–465*
Silbey, R. J., 186–187(22), *197*; 200(15-16), 201(15-16,63), 203(79), 204(15-16,63), 220(90), 242(16), 243(16,96), 247(104), *263, 265–266*; 514(63,143), 535(63), 538–539(63), 557(143), 584(143), *603, 606*

Sillar, K., 432–433(42), 461(42), *465*
Silva, C., 565(316), 583(316), *613*
Silvi, B., 494(242), *508*
Simandiras, E. D., 380(122), *420*
Simon, V. A., 472(31), 494(31), 496–497(31), *503*
Sinclair, D. K., 334(36), *353*
Sita, L. R., 535(294), *612*
Sitkoff, D., 448(109-110), *467*; 528(238), *610*
Sivachenko, A. Yu., 547(299), 594(361), 597(361), *612, 615*
Skinner, J. L., 200(14,38-39), 201(14), 203(38-39), 204(14), 242–243(14), 249(39), *263–264*; 519(184), 522(199), 538(184), *607–608*
Skourtis, S. S., 531(252,257), 532(272), *610–611*
Sloan, J., 334(36), *353*
Smalley, J. F., 535(294), *612*
Smelyanskiy, V. N., 38(50), *77*
Smit, B., 4(13,23), 11(44), 19(13), 26(44), 43(54), 53(44), 71(44), *76–77*
Smit, M. J., 499(305), *510*
Smith, B., 570(318), *613*
Smith, B. J., 430(36), *464*
Smith, D. F., 488(159), *506*
Smith, E. B., 474(58), *503*
Snoonian, J., 448(112), *467*
Snyder, J. P., 408(181), *422*
So, P. T. C., 201(51), 208(51), 229(51), *264*
Soboleva, I. V., 595(369), *615*
Socci, N. D., 195(27-28), *197–198*
Sokolov, N. D., 488(156), *506*
Solc, K., 598(380), *616*
Solomon, I. J., 397(153), *421*
Somers, B. G., 498(281), *509*
Song, L., 588(343), *614*
Song, X., 514(37-38,119,134), 518(38), 519(37), 520(38), 524(119), 526(212), 584(134), *602, 605, 609*
Sonoda, T., 433(43), *465*
Sordo, T. L., 430(29), 434(29), *464*
Soudakov, V. A., 514(141), 584(141), *606*
Sozuki, I., 595(369), *615*
Sparpaglione, M., 514(88), 531(88), 538(88), 542–543(88), 546(88), *603*
Spear, R. J., 408–409(177), *422*

Spiess, H. W., 356(22-24,26), 359(74), 374(24), 417, 419
Spirina, O. B., 514–515(92), 561(92), 564(314), 604, 612
Spirko, V., 402(164), 422
Spouge, J. L., 597(377), 616
Sprik, M., 43(55), 71(69), 78; 514(34), 519(34), 527(34), 529(34), 533–534(34), 601
Springborg, M., 478(112-113), 488(202-203), 505, 507
Squitieri, E., 205(81), 250(81), 265
Staab, H. A., 515(155), 562(155), 606
Ståhl, K., 597(375), 616
Staib, A., 570(318), 613
Stanton, J. F., 357(41,52-56), 367(84-85), 372(93), 373(84), 374(41,52-56), 375(109, 111), 393(55), 394(52), 396(55), 398(55), 403(52,167), 404–405(167), 408(375), 308(174), 410(37), 411(37,111), 412(182), 418–420, 422
Staral, J. S., 409(176), 422
Starling, K., 319(24), 353
Starobinets, A., 514(136), 566–571(136), 605
Stein, D. L., 38(51), 77
Steinberg, I. Z., 586(339), 588(339), 614
Steiner, T., 470(16), 502
Steiner, U. E., 515(155), 562(155), 606
Stemp, D. A., 597(374), 616
Sternheimer, R. M., 369(87), 419
Stevens, R. M., 378(121), 420
Steward, R., 457(136), 467
Stewart, A. L., 369(87), 419
Stienko, Yu. A., 334(36), 353
Still, W. C., 442(86), 466
Stiller, H., 488(197), 507
Stockburger, J. T., 514(124), 543(124c), 582(124), 605
Stockmayer, W. H., 598(380), 616
Stoddart, J. F., 581(323), 613
Stohner, J., 488(145), 506
Stone, A. J., 474(59), 488(153), 503, 506
Stone, M., 334(36), 353
Storck, W., 512(9), 600
Stößer, G., 356(18), 371(18), 417
Stratonovich, R. L., 552–553(306), 612
Stratt, R. M., 514(36), 519(36), 520(36a), 529(36), 533–534(36), 602
Stratti, G., 527–528(235), 609
Straub, J. E., 80(1), 150; 514(76), 549(76), 573(76), 603

Streiter, F. J., 499(316), 510
Streitwieser, A., 430(28), 343(28), 464
Strouse, G. F., 202(70), 245(70), 265
Stuchebrukhov, A. A., 514(61,119), 524(119), 531(265), 532(265,273-275), 535–537(61), 602, 605, 611
Stull, D. R., 427(16), 464
Suarez, A., 220(90), 266
Suh, S. B., 499(308), 510
Suhai, S., 473(51), 478(106), 499(315), 500(322), 503, 505, 510
Suhm, M. A., 487(120,122), 488(131-132,145,147,149,152,167,176,181), 489(149,152,181), 490(149,181), 492–493(149), 497(287), 501(149,152), 505–506, 509
Sulzbach, H. M., 412(184-185), 415(184), 422
Sumi, H., 514(89,91,94), 515(89,91), 516(162), 531(253), 555(89), 561(89,91), 566(89), 573–574(162), 588(172), 604, 607, 610
Sun, H., 488(130), 505
Sun, X., 494(240), 508
Sunada, T., 160(6), 197
Sundholm, D., 357(66), 393(147), 399(159-160), 402(159-160), 403(159), 418, 421
Suter, U. W., 43(54), 77
Sutin, N., 516(170-171), 526(206), 527–528(171), 530(170-171), 573(170), 585–586(329), 597(374), 607–608, 613, 616
Sutmann, G., 520(196), 608
Suzuki, S., 477(88), 504
Suzuki, T., 165(11), 197
Swager, T. M., 200(25), 245(25), 264
Swallen, S. F., 588(343,346), 594–595(346), 614
Sytnik, A., 220(89), 266
Szabo, A., 200(29), 264; 430(24), 464; 514(54, 142), 527(54), 533–534(54), 535(291), 550–551(142), 564(142), 588(341), 594(360), 597(291,377), 598(377a), 599(291), 602, 605, 612, 614–616
Szalay, P. G., 369(88), 372(99), 419–420
Szalewicz, K., 499(299-300,305), 509–510
Szczesniak, M. M., 474(61-62,64), 488(140), 494–495(256-257), 503–505, 508

Tachiya, M., 512(17), 514(52,115), 515(159,161), 516(164,168), 526(211-212), 527(52,220,225), 529(52,211,244), 533–

534(52,225), 557(161), 562(159),
573–574(164), 576(164,168), 577(168),
579(164,168), 580(115,168), 581(161),
585(17), 586(17,335,337c,357), 587(17),
588(159,335,340,345,349), 589(17),
590(17,345), 591(345), 592(335,340,345),
593(340,353), 594(340,357), 595(220,367),
596(220,367,370-371), 597(378), *601–602,
604, 607, 609–610, 613–616*
Takada, S., 113(67), 116(67), *151*
Takagi, Y., 515(151), 560(151), 561(151),
597(151), *606*
Takahashi, H., 493(223), *508*
Takahashi, Y., 590(350-351), 593–594(350),
614
Takano, H., 201(41), *264*
Takanyanagi, M., 477(93), *504*
Takashima, M., 357(63), *418*
Takasu, M., 514(128), *605*
Takehara, C., 580–581(322), *613*
Talkner, P., 514(69), 570(69), *603*
Talon, H., 201(18), 202(69), 212(82),
245(18,69), *263, 265*
Tam, S., 499(297), *509*
Tamarat, P., 200(4,12), 201(12,52), 205(83),
210–211(12), 223(12), 242–243(12),
263–265
Tanaka, T., 67(68), *78*
Tang, J., 514(46-47,53,60,110), 531(259),
533(53), 534(47), 535(60), 537(47,60),
538(47), *602, 604, 610*
Tani, T., 200(11), *263*
Tanimura, Y., 201(41), *264*; 514(111),
584(111), *604*
Tannor, D. J., 448(109), *467*
Tao, F.-M., 494(232,241), 496(241), *508*
Tapia, O., 437(66), *465*
Tarakanova, E. G., 488(178), *506*
Tarakeshwar, P., 499(308), *510*
Tate, T., 160(7), *197*
Tauer, K. J., 472(32), *503*
Tavernier, H. L., 527(220), 588(346), 594(346),
595(220,346), 596(220), *609, 614*
Tawa, G. J., 446–447(98), 450(116), 452(120),
453(126), *466–467*
Taylor, P., 425(5), *463*
Taylor, P. R., 357(42,67), *418*; 488(143), *506*
Teixido, J., 459(143), *468*
Teller, A. H., 11(45), *77*
Teller, E., 11(45), *77*

Temkin, S. I., 598(380), *616*
Tempczyrk, A., 442(86), *466*
Templeton, D. H., 472(37), 499(37), *503*;
499(316), *510*
Temsamani, M. A., 184(21), 187(21), *197*
Tenenbaum, J. B., 75(70), *78*
ten Wolde, P. R., 66(66), *78*
Terao, H., 191(26), *197*
Teukolsky, S. A., 95(61), *151*
Tfirst, E., 499(309), *510*
Theodorou, D. N., 8(32), *77*
Thiel, W., 356(21), *417*
Thomas, D. T., 430(28), 434(28,48), *464–465*
Thomas, M. W., 472(36,40), 499(36,40),
500(36), *503*
Thompson, E. A., 44(58), *78*
Thompson, H. W., 470(1), *502*
Thoss, M., 584(327), *613*
Tidesley, D. J., 593–595(359), *614*
Tiekink, E. R. T., 460(144), *468*
Tildesley, D. J., 10(39), *77*
Tissandier, M. D., 447(104), 448(104),
462(104), *466*
Tisza, L., 279(11), *353*
Toda, F., 595(369), *615*
Toda, M., 81(34-35), 106(64), 119(80),
150–152; 156(4), 176(4), 186–187(23),
190(25), 196(29,31), *197–198*; 201(61),
250(61), *265*; 517(180), 538(180), *607*
Tokhadze, K. G., 460(148), *468*
Tomasi, J., 357(71), *418*; 434(47), 437(68-70),
438(69), 439(68-69,80-81), 442(89),
444(68), 445(80), 448(80,111),
450(111,115), 456(68,111), 457(69,134),
465–467; 499(307), *510*
Tominaga, K., 515(147,151,157),
560(147,151,157), 562(151,157), 597(151),
606
Topaler, M., 514(120,126), 552(120), 555(120),
565(120), 582–583(126), *605*
Topol, I. A., 446–447(98), 450(116), 452(120,
122), 453(126), *466–467*
Torii, H., 500(326), *510*
Torrie, B. H., 488(191,198), 493(191),
494(271-273), 496–497(273), *507, 509*
Toschek, P. E., 202(72), 245(72), *265*
Toth, A. N., 450(117), 458(117), *467*
Toyozawa, Y., 512(3), 513(3b), 516(3), 519(3),
534(3), *600*
Tozer, D. J., 399(155), *421*

Tramer, A., 488(127,164), *505–506*
Traytak, S. D., 529(244), 593(358), 598(380), *610, 614, 616*
Trucks, G. W., 357(60), 372(96), 375(60), 390(60), 403(60), *418–419*
Truhlar, D. G., 80(12-13,15), 81(32), 105(12), 117(13,70), *150, 152*; 437(74), 438(76), 439(79), 442(79), 450(87), 456(131), 458(139), *466–468*; 487(119,126), *505*
Tschumper, G. S., 487(124), 488(148), 497(294), 498(286), *505–506, 509*
Tse, W. S., 488(191,198), 493(191), 494(273), 496–497(273), *507, 509*
Tsuchiya, S., 184–185(19), *197*
Tsukahara, K., 597(372), *615*
Tubino, R., 488(199), *507*
Tucker, S. C., 566–567(317), 569(317), *613*
Tuckerman, M., 49(62), *78*; 441(97), 446(97), *466*
Tully, J. C., 531(266), 584(324), *611, 613*
Tunitskii, N. N., 586(339), 588(339), *614*
Turner, G. M., 425(6), *463*
Turró, C., 512(13), 585(13), 590(13), *601*
Tuttle, T. R., 447(104), 448(104), 462(104), *466*
Tzeng, B.-C., 593(354), *614*

Ueda, T., 500(328-329), *510*
Ueno, A., 595(369), *615*
Ugalde, J. M., 472(24), 498(24), *503*
Ullmann, G. M., 597(374), *615*
Ulstrup, J., 513(25-26), 514(26,28), 516(26,178), 518(26,28), 519(178), 524(24), 527(28,219), 531(258), 535(258,289), 537(289b), 543(26), 595(219), *601, 607, 609–610, 612*
Ungar, L. W., 522(198), 535(198), *608*
Upadhyay, D. M., 499(310), *510*
Usuha, S., 477(98), *504*
Uyemura, M., 477(96), *504*
Uzer, T., 119(74), *152*

Vaara, J., 357(65), 402(162), *418, 422*
Vacha, M., 200(11), *263*
Vager, Z., 408(179), *422*; 493(225), *508*
Vahtras, O., 357(42,65), 386(134), *418, 421*
Vainer, Y. G., 200(13), *263*
Valentini, J. J., 493(218,226), *507–508*
Vancik, H., 374(107), 408(107), 410(107), *420*
Vand, V., 348(50), *354*
vanden Bout, D. A., 200(25-26), 202(26), 245(25-26), *264*
van der Avoird, A., 499(299-300,304-305), *509–510*
van der Veken, B. J., 494(253), *508*
van der Zwan, G., 105(63), *151*
van Duijneveldt, F. B., 499(303), *510*
van Duijneveldt-van de Rijdt, J. G. C. M., 499(303), *510*
Van Eikema Hommes, N. J. R., 488(144), 501(144), *506*
van Kampen, N. G., 538(297), 544–546(297), 551(297), 553(297), *612*
van Lersel, E. M. F., 348(46), *354*
van Oijen, A. M., 201(44), *264*
van Rensburg, J. E. J., 319(23), *353*
van Stam, J., 595(369), *615*
Varandas, A. J. C., 113(66), *151*
Varma, C. M., 242–243(94), *266*
Vath, P., 527(227), 529(227), *609*
Vauthier, E. C., 374(104), *420*
Vekhter, B., 119(75), *152*
Vener, M. V., 497(295), *509*; 520(195), *608*
Verhoeven, J. W., 514–515(96c), 562(96c), *604*
Vernon, M. F., 488(127,164), *505–506*
Vesel, J. E., 494(271-272), *509*
Vetterling, W. T., 95(61), *151*
Viaene, L., 595(369), *615*
Viant, M. R., 499(311), *510*
Vigasin, A. A., 488(178), *506*
Vigdorovich, M. D., 535(289), *612*
Vignale, G., 390–392(143), *421*
Villa, J., 80(15), 105(15), 117(15), *150*
Vincent, M. A., 488(135), *505*
Vinogradov, S. N., 470(4), *502*
Vladimirov, S., 220(89), *266*
Vlugt, T. J. H., 4(13,23), 19(13), *76–77*
Voelkel, D., 488(168), *506*
Volkmann, D., 488(136), 501(136), *505*
Volobuev, Y., 487(126), *505*
Voorhis, T. V., 4(14), *76*
Voronin, A. I., 514(141), 584(141), *606*
Vorotyntsev, M. A., 514(27), 518(27), 519(188), 527(27), *601, 608*
Vosch, T., 201(46), *264*
Votava, C., 494(228-229), *508*
Voter, A. F., 8(33), *77*
Voth, G. A., 56(64), *78*; 514(117,123),

522(198), 527(227), 529(227),
533–534(283), 535(198), 582(117, 123),
590(227b), *605, 608–609, 611*

Wahlgren, U., 357(64), *418*
Wajnryb, E., 347(42), *354*
Wakeham, W. A., 474(58), *503*
Waldeck, D. H., 533(280-281), 535(280), 597(281), *611*
Walecka, J. D., 335(37), 337(37), *354*
Wales, D. J., 3(2), *76*; 80–81(25,28), 103(25,28), *150*; 472(22), 498(22), *503*
Walker, A. R. H., 494(241), 496(241), *508*
Walker, G. C., 515(147,156-157), 524(156), 560(147,156-157), 562(156-157), 563(156b), 564(156), *606*
Wallin, S. A., 597(374), *616*
Walls, D. F., 209(85), *265*
Walser, D., 201(64), *265*
Walsh, B., 477(88), *504*
Wander, M. C., 532–533(279), *611*
Wang, H., 514(134), 552(305), 584(134,327), 597(305), *605, 612–613*
Wang, J., 200(27), *264*
Wang, Z., 494(254), 497(254), *508*; 514(110), *604*
Wanta, T., 530(250), 532(250), *610*
Warncke, K., 530(249), 532(249), *610*
Warr, G. G., 453(125), *467*
Warshel, A., 514(33), 519(33), 527(33,224), 529(33), 533–534(33), *601, 609*
Wash, P. L., 459(141), *468*
Watts, J. D., 357(41), 374(41), *418*
Watts, R. O., 488(130), *505*
Weaver, M. J., 515(146), 526(209), 527–528(235), 560(146), *606, 609*
Webb, H. M., 447(99), *466*
Weber, J., 477(97), *504*
Weeks, E. R., 66(67), *78*
Weidemaier, K., 527(220), 588(343,346), 594(346), 595(220,346,368), 596(220,368), *609, 614–615*
Weigend, F., 386(134), *421*
Weinhold, F., 478(104-105), 485(105), 500(324), *505, 510*
Weis, P., 371(91), *419*
Weiss, G. H., 200(29), 224(91), 226(91), *264, 266*; 550(304), 588(304), 597(377), *612, 616*
Weiss, R. G., 595(366), *615*

Weiss, S., 200(24,35-36), *264*
Weiss, U., 514(48), 522(203), 543(203), *602, 608*
Weitz, D. A., 66(67), *78*
Weller, A., 512(14), 529(14,242), 589–590(14), *601, 610*
Weng, Y.-X., 593(354), *614*
Wenzel, P. J., 430(31), *464*
Werhahn, O., 498(284), *509*
Werner, H.-J., 372(97), 385(130,132), 386(132), 388(130,138-141), 408(130), 410(130), *420–421*
Weston, K. D., 202(75), *265*
Whalley, E., 498(279-280), *509*
Wheatley, R. J., 497(292), *509*
White, D., 408(178), *422*
Whitesides, G. M., 472(48), *503*
Wiberg, K., 437(71-73), 448(112), *465, 467*
Wicke, E., 441(92), *466*
Wiechert, D., 500(321), *510*
Wiesenfeld, L., 119(74), *152*
Wiggins, S., 81(36), 119(74), *151–152*; 171(13), *197*
Wigner, E., 80(14), *150*
Wild, U. P., 200(5,17), 201(17,53), 223(17), 234(53), 243(53), *263–264*
Wilemski, G., 550(302), 588(302), *612*
Williamson, B. L., 408(181), *422*
Willig, F., 512(9), *600*
Willoughby, L. C., 475(68), *504*
Wilson, K. R., 8(34), *77*
Wilson, P. J., 357(62), 360(79), 375(62), 393(62,79), 397–398(79), 399(155), *418–419, 421*
Wilson, W. G., 429(21), 458(21), *464*
Wilson, W. W., 375(110), *420*
Winkler, D. A., 475(67), *504*
Winkler, J. R., 512(13), 530(249), 531(256), 532(249,256), 585(13), 590(13), *601, 610*
Wittenbeck, R., 477(90), *504*
Wittig, C., 493(224), *508*
Wofford, B. A., 475(70), 493(221), *504, 507*
Wolf, E., 202(68), 207–208(68), 215(68), 219(68), 229(68), *265*
Wolfram, S., 248(107), 260(107), *266*
Wolinski, K., 357(35,59), 364(35), 367(83), 370–371(35), 375(59), 382(35), 390(59), 415(35), *417–419*
Wolynes, P. G., 119(79), *152*; 195(27-28), *197–198*; 200(27), 243(97), *264, 266*;

514(44,77-78,83,95,135), 520(193),
532(95,271), 534(95), 549(77-78), 552(305),
569(135), 570(83), 572(78), 573(77-78),
582(95), 594–595(363), 597(305),
602–605, 608, 611–612, 615
Wong, M. W., 437(71), 465
Woo, W. K., 200(22-23), 247(22-23), 263
Wood, W. W., 324–325(26), 353
Woodcock, L. V., 325(28), 348(52), 353–354
Wormer, P. E. S., 499(299-300,304-305),
509–510
Wright, K. R., 80(22), 82(22), 150
Wrigley, H. E., 530(245), 610
Wu, G. Y. C., 534(284), 611
Wu, T. T., 292(18), 304(18), 353
Wuest, J. D., 472(46), 503
Wulfert, S., 493(210), 507
Wüllen, C. V., 356(6,18), 357(6), 371(18),
375(112-114), 393(146), 394(149),
395(113), 397(112), 417, 420–421
Wyld, H. W., 334(36), 353
Wynne, K., 514(130), 576(130), 605

Xantheas, S. S., 470(19,21), 472(21), 488(171),
498(21), 499(298), 502, 506, 509
Xiao, L., 584(324), 613
Xie, X. S., 200(30-31), 209(30), 264; 209(86),
265
Xie, Y., 412(184), 415(184), 422
Xu, Q.-H., 515(160), 561–562(160), 607
Xun, L. Y., 200(30), 209(30), 264

Yakobson, B. I., 514(75), 533(282), 546(75b),
549(75), 573(75), 597(378), 598(380),
603, 611, 616
Yamaguchi, Y., 488(135,148), 505–506
Yamanouchi, K., 184–185(19), 197
Yamashita, K., 196(30), 198
Yamauchi, M., 590(350), 593–594(350), 614
Yamazaki, H., 357(40), 418
Yan, Y. J., 514(99), 538(99), 542(99), 604
Yánez, M., 497(288,290), 509
Yáñez, M., 460(145), 468
Yang, A. S., 437(62), 465
Yang, D.-Y., 514(101,138), 557(101d,138),
584(101), 597(378), 598(378g), 604–605,
616
Yang, G., 425(6), 463
Yang, L., 452(122), 467

Yang, M., 515(160), 561–562(160), 607
Yang, S., 200(34), 264
Yang, W., 372(100), 420; 430(26), 464
Yang, X., 475(74-75), 476(82), 479(82),
484(82), 486(82), 504
Yanoviskii, O., 488(146,207), 491(146,207),
501(146), 506–507
Yartsev, A. P., 515(151), 560(151), 562(151),
597(151), 606
Yazal, J. E., 425(7), 463
Yeager, D. L., 360(76), 374(76), 419
Yershova, L. B., 201(42), 264
Yip, W., 200(25-26), 202(26), 245(25-26), 264
Yomosa, S., 437(65), 465
Yoshida, M., 477(98), 504
Yoshihara, K., 512(9), 515(151-152), 560(151),
561(151), 562(152-153), 565(153a-b),
597(151), 600, 606
Yoshimori, A., 512(15), 514(55-56),
533(55-56), 535(56), 585(15), 601–602
Young, F., 357(39), 418
Young, R. H., 512(10), 533(280), 535(280),
589(10), 593(10), 600, 611
Young, S., 597(374), 615–616
Yu, J., 200(26), 202(26), 245(26), 264
Yukhnevich, G. V., 488(178), 506

Zahradnik, R., 470(8), 502
Zaijfman, D., 408(179), 422
Zaleski, J. M., 512(13), 585(13), 590(13), 601
Zaloj, V., 75(71), 78
Zanasi, R., 357(37), 418
Zang, L.-H., 597(375), 616
Zaqaraia, M. G., 519(188), 608
Zelsmann, H. R., 500(320), 510
Zeng, W. Y., 494(275), 509
Zeng, X. C., 497(293), 509
Zengin, N., 488(188), 507
Zerbi, G., 488(199), 507
Zerner, M. C., 597(375), 616
Zhan, C. G., 357(69), 418
Zhang, C., 488(174), 506
Zhang, H., 484(238), 508
Zhao, M., 80(29), 81(29), 118(29), 150
Zharikov, A. A., 514(32,113), 516(167),
519(32), 540(167), 547(113), 573–574(167),
576(167), 578–579(167), 586(338),
588(338c,344), 601, 604, 607, 613–614
Zheng, C., 514(44), 602

Zhou, H.-X., 514(54), 527(54,222), 533–534(54), 594(360), 597(377), 598(377a,d,380e), *602, 609, 614–616*
Zhu, J., 514–515(108), *604*
Zhu, T., 450(87), *466*
Zhuang, D., 475(73), *504*
Zichi, D. A., 570(318), *613*
Ziegler, T., 357(58), 375(58), 390(58), *418*
Zilker, S., 200(13), *263*
Zimmt, M., 527(227), 529(227), 533(280-281), 535(280), 597(281), *609, 611*
Zinth, W., 515(154), 562(154), 565(154), *606*
Zitserman, V. Y., 564(313), 566–567(317), 569(313,317), 597(379), *612–613, 616*
Zoller, P., 245(98), *266*
Zolotarev, V. M., 519(189), *608*
Zuber, J. B., 326(30), *353*

Zumbusch, A., 200(8-9), 201(8-9), 202(8), 203(8-9), 223(8-9), 229(8-9), 242–244(9), *263*
Zumofen, G., 200(16-17), 201(16-17,40,53,63), 202(53), 204(16,63), 223(17), 234(53), 242(16), 243(16,53), *263–265*
Zundel, G., 441(94-95), *466*; 470(6), *502*
Zusman, L. D., 514(74,81-82), 528(239), 546(74), 554(308), 556(81), 559(81), 562(81a), 564(81), 566(81a), 568(81), *603, 610, 612*
Zwanzig, R., 538(295), 542(295), 564(312), 594–595(363), 597(312,376), *612, 615–616*
Zwerger, W., 514(71), 518(71), 543(71), *603*
Zwier, T. J., 497(291), *509*

SUBJECT INDEX

Ab initio calculations:
  gas-phase acidity:
    limitations of, 462–463
    quantum chemistry, 427–434
  hydrogen bonding cooperativity, substituted zigzag chains and ring structures, water molecules, 498–499
  pK_a values, proton cycle, 453–454
  solution acidity, continuum-solvation models, 437–442
Acceptance probability:
  transition path ensemble:
    shooting moves, 17–19
    trajectory space calculations, 14–15
  transition path sampling, shifting moves, 33–37
Acetic acid, hydrogen bonding cooperativity, 499–500
Acetylene molecules, chemical reaction dynamics, intramolecular vibrational energy redistribution (IVR), 184–190
Acids, computational chemistry:
  DNA oxidation, 425
  future research, 461–463
  gas-phase acidity, 426–436
    carboxylic acids and phenol acidity *vs.* aliphatic alcohol acidity, 434–436
    electron affinity experiments, 426–427
    quantum chemistry techniques, 427–434
  hydroxamic acid therapeutics, 425
  molecular associations, 458–461
  polyester chemistry, 425–426
  proton transfer, 424–425
  solution acidity, 436–458
    anion solvation, 449–450
    continuum-solvation models, 437–442
    pK_a acidity calculations, 450–458
      relativity of, 458
      TC2 or hydronium cycle, 454–457
      TC1 or proton cycle, 451–454
    proton solvation energy, 445–448
    solvated protons, 443–445
  theoretical definitions, 423–424
  zeolite structures, 426

Action-angle coordinate system, chemical reaction dynamics, canonical perturbation theory (CPT), 85–87
Additive formulation:
  hydrogen bonding, cooperativity, 472, 474–475
  renormalization group (RG) theory, self-similarity, 340–348
    hard-sphere suspension viscosity, 345–348
    translational self-similarity, 341–345
Adiabaticity, electron-transfer (ET) reactions, nonadiabatic solvent effects, variational transition state theory (VTST), 572–573
Algebraic quantization:
  calculation techniques, 142–144
  chemical reaction dynamics, canonical perturbation theory (CPT), 91–93
Aliphatic alcohols, gas-phase acidity measurements, 434–436
AM1 methods, pK_a value calculations, hydronium cycle, 455–457
Analytic second derivatives, nuclear magnetic resonance (NMR) chemical shifts, 367–370
Andersen thermostate, transition path ensemble, dynamical algorithm, 42–43
Anion solvation, solution acidity, 449–450
Anisotropic reactivity, electron-transfer (ET) reactions, nonadiabatic solvent effects, nonhomogeneous effects, 597–598
Anomalous dimension, renormalization (RG) theory, self-similarity, quantum electrodynamics, photon propagator, 331–334
Antiparallel stacked dimers, hydrogen bonding cooperativity, perfectly linear chains and rings, 485–486
Anti-Stokes shifts, chemical and chaos reaction dynamics, local equilibrium reactions, 166–168
Apparent surface charge (ASC) approach:
  anion solvation, 450

649

Apparent surface charge (*Continued*)
  solution acidity, continuum-solvation models, 440–442
Aqueous solutions, pK_a calculations, 462–463
Arnold diffusion, chemical reaction and chaos dynamics, 155–158
  resonance hierarchies, 193
Arnold web, chemical reaction dynamics:
  chaos and, basic principles, 157–158
  intramolecular vibrational energy redistribution (IVR), 182–184
    acetylene molecules, 184–190
    resonance hierarchy, 190–193
  transition state theory (TST), hierarchical regularity, 116
Arrhenius law, acid computational chemistry, 423–424
Asymptotic self-similarity:
  generalized propagator renormalization group (GPRG):
    quantum electrodynamics, photon propagator, 330–334
    swelling factor calculation, 307–312
  renormalization group (RG) theory:
    ideal systems, 279–280
    nonideal systems, 280–295
      average squared magnetization, Ising spin open chain, 292–295
      direct mapping solution, functional evolution equations, 286–289
      generalized propagator renormalization group (GPRG) Callan-Symanzik equations, 290–292
      functional evolution equations, 286–289
      Lie equations, 289–290
      linear nonideal polymeric chains, partition function, 314–317
Atomic orbitals (AO), nuclear magnetic resonance (NMR) chemical shifts:
  density functional theory, electron correlation procedures, 385–389
  GAIO-MP2 approach, principles, 376–385
Attrition coefficient, renormalization (RG) theory, self-similarity, linear nonideal polymeric chains, partition function, 313–317

Autonomous cases, Lie transforms, 122–128
Average squared magnetization (ASM), generalized propagator renormalization group (GPRG), 292–295
  swelling factor calculation, 307–312

Back reactions, electron-transfer (ET) reactions, nonadiabatic solvent effects, Markovian bath model, effective sink approximation, 560–565
Bare coupling constant:
  asymptotic self-similarity, nonideal systems, renormalization group (RG) theory, 280–282
  generalized propagator renormalization group (GPRG), swelling factor calculation, 307–312
  renormalization group (RG) theory, self-similarity:
    electron gas screening, quantum field theory, 336–340
    quantum electrodynamics, photon propagator, 329–334
Bare interaction energy, renormalization group (RG) theory, self-similarity, electron gas screening, quantum field theory, 335–340
Basis of attaction, transition path ensemble, path quenching, 73–74
Basis set convergence, nuclear magnetic resonance (NMR) chemical shifts, local gauge-origin methods, 363–367
Bath fluctuations:
  electron-transfer (ET) reactions, nonadiabatic solvent effects:
    Markovian bath model, 546–565
    crossing point statistics, 552–555
    effective sink approximation, 555–565
    historical background, 514–515
    substitution approximation, time-dependent diffusion coefficient, 550–552
  Massey parameter, 541–542
  single molecule spectroscopy (SMS):
    fast modulation regime, 232–237
    intermediate modulation regime, 238
    phase diagram, 238–242
    slow modulation regime, 223–231
    timescale dimensions, 203

Benchmark calculations, nuclear magnetic resonance (NMR) chemical shifts, absolute magnetic shielding constants, 393–403
Benzene, molecular associations, 461
Benzoic acid (BA), molecular associations, 461
Bessel function, electron-transfer (ET) reactions, nonadiabatic solvent effects, spectral density, 524
BH molecule, nuclear magnetic resonance (NMR) chemical shifts, magnetic shielding constants, benchmark calculations, 393–394
Bifurcation patterns, chemical and chaos reaction dynamics, 180–182
  dynamical correlation, 196
Biot-Savart law, nuclear magnetic resonance (NMR) chemical shifts, chemical shielding tensors, 358–359
Birkoff-Gustavson equation, Lie canonical perturbation theory (LCPT) vs., chemical reaction dynamics, 88–90
Bloch equation, single molecule spectroscopy (SMS):
  application, 201–202
  line shape analysis, marginal averaging, 249
  perturbation expansion, 250–255
  Q parameter, three-time correlation function, Rabi frequency (Ω), 216–219
  simulation using, 210–213
  slow modulation regime, 223–231
  stochastic optical equation, 204–207
B3LYP calculations:
  gas-phase acidity:
    limitations of, 461–463
    quantum chemistry, 431–434
  hydrogen bonding cooperativity, perfectly linear chains and rings, 478–486
  molecular associations, 460–461
  pK_a value calculations, proton cycle, 452–454
Boltzmann distribution:
  chemical and chaos reaction dynamics, local equilibrium reactions, 166–168
  transition path ensemble, path quenching, 73–74
Bond configuration integral, renormalization (RG) theory, self-similarity, linear nonideal polymeric chains, 306–317
Bond-vector distribution function, renormalization (RG) theory, self-similarity, ideal self-similar systems, 275–280
Born reaction field model, solution acidity, continuum-solvation models, 440–442
Brønsted acidity definition, acid computational chemistry, 424
Brownian dynamics, transition path ensemble, stochastic dynamics, shooting moves, 30–31
Bulk bimolecular reaction, electron-transfer (ET), nonadiabatic solvent effects, diffusion effects, 585–590

Callan-Symanzik equation:
  positive function renormalization group (PFRG), 301–302
  renormalization group (RG) theory, self-similarity:
    additive formulation, translational self-similarity, 344–345
    generalized propagator renormalization group (GPRG), 291–292
    linear nonideal polymeric chains, partition function, 316–317
    principles of, 273–274
Canonical partition function, renormalization group (RG) theory, self-similarity:
  asymptotic self-similarity, nonideal systems, 281–282
  ideal self-similar systems, 274–280
Canonical perturbation theory (CPT):
  chemical reaction dynamics, 83–93
    algebraic quantization, 91–93
    Lie CPT, 87–90
    regional Hamiltonian, 90–91
    semiclassical theory, 113–114
    "unify" transition state theory vs. Kramers-Grote Hynes theory, 104–106
  Lie transforms, 133–142
Carbonic acids, hydrogen bonding cooperativity, 499–500
Carboxylic acids:
  gas-phase acidity, 434–436
  molecular associations, 459–461

SUBJECT INDEX

Carnahan-Starling formula, renormalization (RG) theory, self-similarity, hard particle compressibility, fluid pressure, 324–326
Carr-Parrinello molecular dynamics (CPMD), transition path ensemble, sampling programs, 49
Cation solvation, acid computational chemistry, 450
CCSD/CCSD(T) technique:
  gas-phase acidity, quantum chemistry, 433–434
  nuclear magnetic resonance (NMR) chemical shifts:
    chemical applications, 404–415
    density functional theory, electron correlation procedures, 385–389
    magnetic shielding constants, benchmark calculations, 394–403
Central limit theorem, single molecule spectroscopy (SMS), classical shot noise, 209–210
Chain structures:
  hydrogen bonding cooperativity:
    perfect linearity, 475–486
    substituted zigzag chains and rings, 497–499
    zigzag chains, 486–497
    hydrogen chloride/hydrogen bromide molecules, 493–497
    hydrogen fluoride, 487–493
  linear, nonideal polymer chains:
    renormalization group (RG) theory, self-similarity, 304–317
    partition function, 312–317
    swelling factor, GPRG calculation, 307–312
  self-similarity, renormalization group (RG) theory, 304–317
    partition function, 312–317
    swelling factor, GPRG calculation, 307–312
Chaos theory, chemical reaction dynamics:
  correlation of reaction processes, 193–196
  global features, 168–182
    crisis, 174–180
    homoclinic intersection, 168–170
    normally hyperbolic invariant manifold, 170–172
    reaction path bifurcation, 180–182
    tangency, 172–174
  intramolecular vibrational-energy redistribution (IVR):
    acetylene, 184–190
    Arnold web, 182–184
    hierarchical structures, 164
    resonance hierarchy, 190–193
  local equilibrium breakdown, reactions in solution, 165–168
  research background, 154–158
  resonant rate structures, 161–163
  statistical reaction theory, radical breakdown, 165
  transition regularity, 93–114
    "extract" invariant stable/unstable manifolds, transition state, 110–112
    many-body phase space reaction bottleneck visualization, 106–110
    "see" trajectory recrossing, configurational dividing surface, phase space, 99–102
    semiclassical theories, 113–114
    transition state actions, invariancy in, 95–99
    "unify" transition state theory/Kramers-Grote-Hynes theory, 102–106
  transition state theory, 158–161
Chapman-Kolmogorov equation, renormalization (RG) theory, self-similarity, ideal self-similar systems, 276–280
Charge-localized wave functions, electron-transfer (ET) reactions, nonadiabatic solvent effects, electronic coupling, 531–533
Chemical reaction dynamics:
  canonical perturbation theory (CPT), 83–93
    algebraic quantization, 91–93
    Lie CPT, 87–90
    regional Hamiltonian, 90–91
  chaos and:
    future correlation of reaction processes, 193–196
    global features, 168–182
      crisis, 174–180
      homoclinic intersection, 168–170
      normally hyperbolic invariant manifold, 170–172
      reaction path bifurcation, 180–182

tangency, 172–174
intramolecular vibrational-energy redistribution (IVR):
  acetylene, 184–190
  Arnold web, 182–184
  hierarchical structures, 164
  resonance hierarchy, 190–193
  local equilibrium breakdown, reactions in solution, 165–168
  research background, 154–158
  resonant rate structures, 161–163
  statistical reaction theory, radical breakdown, 165
  transition regularity, 93–114
    "extract" invariant stable/unstable manifolds, transition state, 110–112
    many-body phase space reaction bottleneck visualization, 106–110
    "see" trajectory recrossing, configurational dividing surface, phase space, 99–102
    semiclassical theories, 113–114
    transition state actions, invariancy in, 95–99
    "unify" transition state theory/ Kramers-Grote-Hynes theory, 102–106
  transition state theory, 158–161
  research background, 80–83
  transition state hierarchical regularity, 114–117
Chemical shielding tensors, nuclear magnetic resonance (NMR) chemical shifts:
  basic principles, 358–359
  research background, 356–357
  second energy derivatives, 359–360
C—H stretching region, hydrogen bonding cooperativity, perfectly linear chains and rings, 476–486
Classical nonadiabatic limit, electron-transfer (ET) reactions, nonadiabatic solvent effects, Padé approximation, 543–546
Classical photon emission, single molecule spectroscopy (SMS):
  Lorentz model, perturbation expansion, 253–255
  shot noise, 207–210
Cluster expansion, hydrogen bonding cooperativity, 473–474
Collins-Kimball model, electron-transfer (ET), nonadiabatic solvent effects, diffusion effects, bulk bimolecular reaction, 586–590
Committor, transition path ensemble:
  computation of, 68–70
  distribution, 71–73
  separatrix and, 67–68
Complete basis set (CBS) methods:
  gas-phase acidity, quantum chemistry, 429–434
  pK_a value calculations, proton cycle, 453–454
Compressibility factor, renormalization (RG) theory, self-similarity:
  asymptotic self-similarity, nonideal systems, 281–282
  fluid pressure self-similarity, 321–326
  hard-particle fluid compressibility, 319–326
Computing rates:
  transition path ensemble, committor computation, 68–70
  transition path sampling, 49–65
    convenient factorization, 55–59
    population fluctuations, 51–52
    reversible work, 52–53
    umbrella sampling, 53–55
Condon approximation, electron-transfer (ET) reactions, nonadiabatic solvent effects, non-Condon effects, 535–537
Conductor-like screening model (COSMO), solution acidity, continuum-solvation models, 442
Configurational bias Monte Carlo, transition path ensemble, 43
Configurational dividing surface, phase space, chemical reaction dynamics, see trajectories, 99–102
Conservation equation, renormalization (RG) theory, self-similarity:
  asymptotic self-similarity, nonideal systems, 282
  ideal self-similar systems, 274–280
Continuum-solvation methods:
  pK_a value calculations, proton cycle, 453–454
  solution acidity, 437–442
Convulsion theorem, single molecule spectroscopy (SMS), two-state jump model, 221–222
Cooperativity, hydrogen bonding:
  definition, 473–474

Cooperative: hydrogen bonding (*Continued*)
  miscellaneous patterns, 499–500
  origins of, 474–475
  perfectly linear chains and rings, 475–486
  research background, 469–472
  substituted zigzag chains and rings, 497–499
  zigzag chains and rings, 486–497
Correlation functions:
  single molecule spectroscopy (SMS), three-time correlation function:
    Q parameter, 213–219
    theoretical background, 202–203
  transition path ensemble, efficiency of shooting moves, 26–29
Coulomb law:
  electron-transfer (ET) reactions, nonadiabatic solvent effects, Fermi Golden Rule, reorganization energy, 530
  gas-phase acidity, quantum chemistry, 428–434
  renormalization group (RG) theory, self-similarity, electron gas screening, 335–340
Coupled-cluster (CC) approach, nuclear magnetic resonance (NMR) chemical shifts:
  analytic second derivatives, 369–370
  chemical applications, 404–415
  magnetic shielding constants, benchmark calculations, 393–403
  research background, 357
Coupled-perturbed Hartree-Fock (CPHF) equations, nuclear magnetic resonance (NMR) chemical shifts:
  analytic second derivatives, 370
  density functional theory (DFT), 391–393
  GAIO-MP2 approach, principles, 379–385
  quantum chemical procedures, 371–393
Coupling function, generalized propagator renormalization group (GPRG), nonideal systems, asymptotic self-similarity:
  Callan-Symanzik equations, 290–292
  direct mapping, evolution equations, 286–289
  functional evolution equations, 286–289
  Lie equations, 289–290

CPU time requirements, nuclear magnetic resonance (NMR) chemical shifts, density functional theory, electron correlation procedures, 385–389
Crisis analysis, chemical reaction and chaos dynamics, 174–180
Crooks-Chandler algorithm, transition path ensemble, stochastic trajectories, random number sequences, 38–40
Crossing points statistics, electron-transfer (ET) reactions, nonadiabatic solvent effects, Markovian bath model, 552–555
  effective sink approximation, 556–565
Crossover region, electron-transfer (ET) reactions, nonadiabatic solvent effects:
  Markovian bath model, 548–565
  theoretical background, 514
Crystalline structures, hydrogen bonding cooperativity, perfectly linear chains and rings, 475–486
CTBP structure, chemical reaction dynamics:
  chaotic transition regularity, 94–95
  invariancy of actions, transition states, 95–99
  reaction bottleneck "visualization," 107–110
  "see" trajectories, configurational dividing surface, phase space, 100–102
Cyanoacetylene, hydrogen bonding cooperativity:
  limits of, 501
  perfectly linear chains and rings, 475–486

Dalton law, renormalization group theory, self-similarity, hard-particle fluid compressibility, 320–326
Debye-Hückel/Boltzmann model, solution acidity, continuum-solvation models, 441–442
Debye-Hückel theory, electron-transfer (ET) reactions, nonadiabatic solvent effects, Fermi Golden Rule, reorganization energy, 527–530
Debye relaxation model, electron-transfer (ET) reactions, nonadiabatic solvent effects:
  Fermi Golden Rule, 526–530
  Markovian bath model, 546–565

SUBJECT INDEX 655

substitution approximation and time-dependent diffusion coefficient, 550–552
numerical calculations, 584–585
variational transition state theory (VTST), solvent control regime, 567–570
Delta functions, transition path sampling, trajectory length, 63–65
Density functional theory (DFT):
  electron-transfer (ET) reactions, nonadiabatic solvent effects, Fermi Golden Rule, reorganization energy, 527–530
  gas-phase acidity:
    limitations of, 461–463
    quantum chemistry, 427–434
  hydrogen bonding cooperativity, zigzag chain/ring structure, hydrogen fluoride, 488–493
  nuclear magnetic resonance (NMR) chemical shifts:
    chemical applications, 405–415
    chemical shielding tensors, second energy derivatives, 360
    electron correlation procedures, 385–389
    future research applications, 416
    principles of, 389–393
    quantum chemical procedures, 372–393
    research background, 357
  pK_a value calculations, proton cycle, 452–454
  proton solvation energy, 447–448
  transition path ensemble, memory requirements, 38
Detailed balance condition, transition path ensemble, trajectory space calculations, 14–15
Deterministic dynamics, transition path ensemble, 8–9
  committor computation, 70
  efficiency of shooting moves, 26–29
  shifting moves, 35–36
  shooting moves, 19–22
Deterministic optical Bloch equation, single molecule spectroscopy (SMS), 205–207
Dielectric continuum model, electron-transfer (ET) reactions, nonadiabatic solvent effects, reorganization energy, 525–530

Dielectric permittivity, electron-transfer (ET) reactions, nonadiabatic solvent effects, Fermi Golden rule, rate constants, 517–519
spectral density, 520–524
Differential equations, positive function renormalization group (PFRG), 301–302
Diffusion effects, electron-transfer (ET), nonadiabatic solvent effects, 585–598
  bulk $vs.$ geminate reaction, 585–594
  bulk reaction, 586–590
  geminate reaction after photoionization, 590–594
  nonhomogeneous effects, 594–598
  hydrodynamic hindrance and liquid structure, 594–595
  microheterogeneous environments, 595–597
  stochastic gating and anisotropic reactivity, 597–598
Dimethyl derivative (DMPA) dimers, molecular associations, 460–461
N,N-Dimethylaniline (DMA), electron-transfer (ET) reactions, nonadiabatic solvent effects, microheterogeneous environments, 595–597
Dirac delta functions:
  chemical reaction dynamics, "unify" transition state theory $vs.$ Kramers-Grote Hynes theory, 102–106
  transition path sampling, trajectory length, 63–65
Direct mapping, evolution equations, generalized propagator renormalization group (GPRG), nonideal systems, asymptotic self-similarity, 286–289
Dissipative systems, chemical reaction and chaos dynamics, crisis analysis, 174–180
DNA oxidation, computational chemistry, 425
Donor-acceptor distance, electron-transfer (ET) reactions, nonadiabatic solvent effects, Fermi Golden Rule, electronic coupling, 532–533
Dunham expansion, chemical reaction dynamics, intramolecular vibrational energy redistribution (IVR), 184–190

Duration of regularity, chemical reaction dynamics, "extract" invariant stable/unstable manifolds, 110–112
Dynamical algorithm, transition path ensemble, 42–43
Dynamical correlation, chemical reaction and chaos dynamics, 193–196
Dynamical path probability, transition path ensemble, 6–7
  stochastic trajectories, random number sequences, 39–40
Dyson equation:
  electron-transfer (ET) reactions, nonadiabatic solvent effects, Markovian bath model, 547–565
  renormalization group (RG) theory, self-similarity, electron gas screening, quantum field theory, 335–340

Effective coupling function:
  generalized propagator renormalization group (GPRG), swelling factor calculation, 308–312
  positive function renormalization group (PFRG), stretched-exponential scaling, 298–304
  renormalization (RG) theory, self-similarity:
    additive formulation, translational self-similarity, 343–345
    asymptotic self-similarity, nonideal systems, 282
    average squared magnetization (ASM), Ising spins, open chain model, 292–295
    hard-sphere suspension viscosity, 345–348
    linear nonideal polymeric chains, partition function, 312–317
    quantum electrodynamics, photon propagator, 331–334
Effective pair potentials, hydrogen bonding, cooperative effects, 472
Electron affinity (EA), gas-phase acidity, 426–427
Electron-correlated calculations, nuclear magnetic resonance (NMR) chemical shifts:
  analytic second derivatives, 367–370
  applications, 370–371
  basic principles, 358–359
  benchmark calculations, absolute nuclear magnetic shielding constants, 393–403
  chemical applications, relative NMR chemical shifts, 403–415
  future research issues, 415–416
  gauge invariance and gauge-origin independence, 361–362
  local gauge-origin methods, 362–367
  molecule Hamiltonian, magnetic field presence, 360–361
  quantum chemical procedures, 371–393
    density functional theory (DFT), 389–393
    GIAO-MP2 approach, 375–385
    larger system applications, 385–389
  research background, 356–357
  shielding tensors as second derivatives, 359–360
Electron gas screening, renormalization group (RG) theory, self-similarity, quantum field theory (QFT), 334–340
Electronic coupling, electron-transfer (ET) reactions, nonadiabatic solvent effects:
  Fermi Golden Rule, 530–533
  historical background, 515–516
Electron-transfer (ET), nonadiabatic solvent effects:
  diffusion effects, 585–598
    bulk vs. geminate reaction, 585–594
    bulk reaction, 586–590
    geminate reaction after photoionization, 590–594
    nonhomogeneous effects, 594–598
    hydrodynamic hindrance and liquid structure, 594–595
    microheterogeneous environments, 595–597
    stochastic gating and anisotropic reactivity, 597–598
  Fermi Golden Rule limit, 516–540
    non-Condon effects, 535–537
    nonexponential kinetics, 537–540
    nonlinear response, 533–535
    rate constant, 516–533
      electronic coupling, 530–533
      reorganization energy and free energy change, 524–530
      spectral density, 519–524
  future research issues, 598–600

SUBJECT INDEX

Markovian bath model, 546–565
  crossing point statistics, 552–555
  effective sink approximation, 555–565
  substitution approximation, time-dependent diffusion coefficient, 550–552
  numerical methods, 582–585
  Padé approximation, 542–546
  classic nonadiabatic limit, 543–546
  research background, 512–516
  strong coupling limit, nonadiabatic transitions, 573–582
  Fermi Golden rule limit, 576–579
  Landfau-Zener approach, 573–574
  non-Marcus free energy gap dependence, 580–582
  relaxation effects, Landau-Zener theory, 574–576
  variational transition state theory, 565–573
  adiabacity, 572–573
  intermediate regime, 570–572
  solvent control regime, 565–570
Electrostatic solute-solvent interaction, solution acidity, continuum-solvation models, 438–442
Enthalpy values:
  phosphinic acid, 460–461
  proton solvation, 446–448
Entropy, transition path sampling, 3–5
Equations of motion:
  Lie transforms, 121–131
  transition path ensemble:
    deterministic dynamics, 8–9
    shifting moves, 35–36
    proofs, 120–121
Ergodicity:
  chemical and chaos reaction dynamics, transition state theory, 159–161
  single molecule spectroscopy (SMS), Q parameter, three-time correlation function, 215–219
Euler equation, renormalization (RG) theory, self-similarity, ideal self-similar systems, 279–280
Evolution equations:
  generalized propagator renormalization group (GPRG), nonideal systems, asymptotic self-similarity:
    Callan-Symanzik equations, 290–292

    functional evolution equations, 286–289
    Lie equations, 289–290
  positive function renormalization group (PFRG), stretched-exponential scaling, 295–304
Evolution operators, Lie transforms, perturbation theory, 132–142
Excess partition function, renormalization (RG) theory, self-similarity:
  linear nonideal polymeric chains, 306–317
  quantum electrodynamics, photon propagator, 327–334
"Excluded-volume" interactions, renormalization group (RG) theory, asymptotic self-similarity, nonideal systems, 281–282
Extracellular (EC) media, acid computational chemistry, 424–425
"Extract" invariant stable/unstable manifolds, chemical reaction dynamics, transition state theory (TST), 110–112

Face-centered cubic lattices, renormalization group (RG) theory, self-similarity, hard-particle fluid compressibility, 317–326
Factorization, transition path sampling, computer rates, reversible work, 55–59
Fast modulation regime:
  single molecule spectroscopy (SMS), 231–237
  Q parameter in, 261–263
  slow modulation regime, phase diagram, 241–242
Fermi Golden Rule, electron-transfer (ET) reactions, nonadiabatic solvent effects, 516–540
  non-Condon effects, 535–537
  nonexponential kinetics, 537–540
  nonlinear response, 533–535
  rate constant, 516–533
    electronic coupling, 530–533
    reorganization energy and free energy change, 524–530
    spectral density, 519–524
  strong coupling limit, 576–579
  theoretical background, 513–514
Fermi-Pasta-Ulam (FPU) model, chemical reaction and chaos dynamics, 155–158

Fermi wave number:
  renormalization group (RG) theory, self-similarity, electron gas screening, quantum field theory, 336–340
  solvated protons, 443–445
Feynman diagrams, renormalization group (RG) theory, self-similarity, electron gas screening, quantum field theory, 336–340
Feynman path integral, electron-transfer (ET) reactions, nonadiabatic solvent effects, Fermi Golden Rule, electronic coupling, 532–533
Finite timescales, chemical reaction and chaos dynamics, crisis analysis, 175–180
First-order perturbation:
  nuclear magnetic resonance (NMR) chemical shifts, density functional theory (DFT), 391–393
  single molecule spectroscopy (SMS), 250–252
"Fixed points," generalized propagator renormalization group (GPRG), nonideal systems, asymptotic self-similarity, direct mapping, evolution equations, 287–289
Fluid pressure, renormalization group theory, self-similarity, hard-particle fluid compressibility, 320–326
Formic acid, hydrogen bonding cooperativity, 499–500
Fourier coefficients, Lie canonical perturbation theory (LCPT), 145–150
Fourier transform (FT):
  electron-transfer (ET) reactions, nonadiabatic solvent effects, strong coupling limit, 577–579
  renormalization (RG) theory, self-similarity:
    electron gas screening, quantum field theory, 335–340
    ideal self-similar systems, 276–280
Fourth-order Feenberg (F4) theories, gas-phase acidity, quantum chemistry, 431–434
Four-wave mixing spectroscopy, fast modulation regime, 232–237
Free energy change, electron-transfer (ET) reactions, nonadiabatic solvent effects:
  Fermi Golden Rule, 524–530
  microheterogeneous environments, 595–597
  non-Marcus free energy gap dependence, 580–582
Free energy surface, transition path ensemble, initial/final region determination, 12–13
Frequency modulation, single molecule spectroscopy (SMS):
  phase diagram, 242
  slow modulation regime, 223–231
Frozen Gaussian approximation, electron-transfer (ET) reactions, nonadiabatic solvent effects, nonlinear response, trajectory approach, 534–535
Frozen molecule approximation, hydrogen bonding, cooperativity origins, 474–475
Full configuration-interaction (FCI) calculations, nuclear magnetic resonance (NMR) chemical shifts, magnetic shielding constants, 393–403
Functional evolution equations, direct mapping, generalized propagator renormalization group (GPRG), nonideal systems, asymptotic self-similarity, 286–289

Gamow WKB model, electron-transfer (ET) reactions, nonadiabatic solvent effects, Fermi Golden Rule, electronic coupling, 532–533
Gas-phase acidity:
  computational chemistry, 426–436
  carboxylic acids and phenol acidity $vs.$ aliphatic alcohol acidity, 434–436
  electron affinity experiments, 426–427
  quantum chemistry techniques, 427–434
  molecular associations, 459–461
  proton solvation, 443–445
Gauge-including atomic orbital (GIAO) approach, nuclear magnetic resonance (NMR) chemical shifts:
  applications, 370–371
  chemical applications, 406–415
  density functional theory, electron correlation procedures, 385–389
  future research applications, 415–416
  GAIO-MP2 approach, principles, 375–385

local gauge-origin methods, 364–367
magnetic shielding constants, benchmark
 calculations, 395–403
quantum chemical procedures, electron
 correlation, 374–375
research background, 357
Gauge invariance, nuclear magnetic
 resonance (NMR) chemical shifts,
 361–362
Gauge-origin independence, nuclear magnetic
 resonance (NMR) chemical shifts,
 361–362
Gauge problem, nuclear magnetic resonance
 (NMR) chemical shifts:
 local gauge-origin methods, 364–367
 research background, 356–357
Gaussian distribution, single molecule
 spectroscopy (SMS):
 classical shot noise, 209–210
 slow modulation regime, 231
Gaussian (Gn) models:
 electron-transfer (ET) reactions,
  nonadiabatic solvent effects:
  Markovian bath model, 546–565
   crossing point statistics, 552–555
   effective sink approximation,
    556–565
   substitution approximation and time-
    dependent diffusion coefficient, 552
  Padé approximation, 543–546
 gas-phase acidity, quantum chemistry,
  429–434
 pK_a value calculations, proton cycle,
  453–454
 solution acidity, continuum-solvation
  models, 440–442
Gell-Mann function, renormalization group
 (RG) theory, self-similarity, 273–274
Gell-Mann-Low function, renormalization
 group (RG) theory, self-similarity, Lie
 equation, 290–291
Geminate donor-acceptor reaction, electron-
 transfer (ET), nonadiabatic solvent
 effects, diffusion effects, 585, 590–594
Generalized Born (GB) model, solution
 acidity, continuum-solvation models,
 440–442
Generalized Born (GB)/surface area (GB/SA)
 model, solution acidity, continuum-
 solvation models, 442

Generalized multistructural (GMS) wave
 function, carboxylic acid gas-phase
 acidity measurements, 435–436
Generalized propagator renormalization
 group (GPRG):
 electron gas screening, quantum field
  theory, 338–340
 hard-particle fluid compressibility, fluid
  pressure self-similarity, 321–326
 linear nonideal polymeric chains:
  self-similarity theory, 306–317
  swelling factor calculation, 307–312
 nonideal systems, asymptotic self-similarity:
  Callan-Symanzik equations, 290–292
  functional evolution equations, 286–289
  Lie equations, 289–290
 quantum electrodynamics, photon
  propagator, 330–334
Generation probability, transition path
 ensemble sampling, shooting moves,
 16–19
Geometric fractals, self-similarity,
 renormalization (RG) theory, 268–274
Gibbs canonical partition function,
 renormalization group (RG) theory,
 self-similarity:
 hard-particle fluid compressibility, 318–326
 linear nonideal polymeric chains, 305–317
Gibbs free standard energy ($\Delta G°$):
 anion solvation, 449–450
 electron-transfer (ET), nonadiabatic
  solvent effects, diffusion effects, bulk
  bimolecular reaction, 589–590
 electron-transfer (ET) reactions,
  nonadiabatic solvent effects, 512–513
 Fermi Golden Rule:
  rate constants, 517–519
  spectral density, 519–524
 reorganization energy, 529–530
 Markovian bath model, effective sink
  approximation, 560–565
 gas-phase acidity, 426–427
 hydrogen bonding cooperativity, zigzag
  chain/ring structure, hydrogen
  fluoride, 488–493
 pK_a calculations, 450–457
 relative calculations, 458
 proton solvation, 445–448
 solution acidity, 436–458
  continuum-solvation models, 438–442

Global properties, chemical reaction and chaos dynamics, 168–182
  crisis, 174–180
  homoclinic intersection, 168–170
  normally hyperbolic invariant manifold, 170–172
  reaction path bifurcation, 180–182
  tangency, 172–174
Green's function, electron-transfer (ET) reactions, nonadiabatic solvent effects:
  Fermi Golden Rule, electronic coupling, 531–533
  Markovian bath model, 547–565

Hard-particle fluid compressibility, renormalization group (RG) theory, self-similarity, 317–326
Hard-sphere suspension viscosity, renormalization (RG) theory, self-similarity, additive formulation, 345–348
Harmonic oscillators, chemical reaction dynamics, intramolecular vibrational energy redistribution (IVR), 183–184
Hartree-Fock (HF) calculations:
  gas-phase acidity, quantum chemistry, 427–434
  nuclear magnetic resonance (NMR) chemical shifts, gauge invariance and gauge-origin independence, 362
  pK_a value calculations, 454
    hydronium cycle, 456–457
  solution acidity, continuum-solvation models, 441–442
Hartree-Fock self-consistent-field (HF-SCF) calculations:
  electron-transfer (ET) reactions, nonadiabatic solvent effects, Fermi Golden Rule, electronic coupling, 531–533
  nuclear magnetic resonance (NMR) chemical shifts:
    chemical applications, 403–415
    GAIO-MP2 approach, principles, 376–385
    local gauge-origin methods, 362–367
    magnetic shielding constants, benchmark calculations, 395–403
    quantum chemical procedures, 371–393

Hasse diagram, chemical reaction dynamics, resonance hierarchies, 191–193
Hausdorf dimension, self-similarity, renormalization (RG) theory, 268–274
Heaviside step function:
  chemical reaction dynamics, "unify" transition state theory vs. Kramers-Grote Hynes theory, 102–106
  transition path sampling:
    computing rates, reversible work, 50–51
    trajectory length, 63–65
Hermite Gaussians, nuclear magnetic resonance (NMR) chemical shifts, GAIO-MP2 approach, principles, 383–385
Heteroclinic tangency, chemical reaction and chaos dynamics:
  research background, 156–158
  simplectic property, 173–174
Hierarchical regularity, chemical and chaos reaction dynamics:
  intramolecular vibrational energy redistribution (IVR), 164
  resonant structures, 190–193
  transition state theory (TST), 114–117
Hohenberg-Kohn theorems:
  gas-phase acidity, quantum chemistry, 428–434
  nuclear magnetic resonance (NMR) chemical shifts, density functional theory (DFT), 390–393
Homoclinic tangency, chemical reaction and chaos dynamics:
  crisis analysis, 175–180
  global features, 168–170
  research background, 156–158
  simplectic property, 173–174
Hydrated protons:
  solvation energy, 445–448
  structural models, 444–445
Hydrodynamic hindrance, electron-transfer (ET) reactions, nonadiabatic solvent effects, 594–595
Hydrogen bonding:
  cooperative effects:
    definition of cooperativity, 473–474
    miscellaneous patterns, 499–500
    origins of, 474–475
    perfectly linear chains and rings, 475–486

research background, 469–472
substituted zigzag chains and rings, 497–499
zigzag chains and rings, 486–497
molecular associations, 460–461
Hydrogen bromide molecules, hydrogen bonding cooperativity, zigzag chain/ring structure, 493–497
Hydrogen chloride molecules, hydrogen bonding cooperativity, zigzag chain/ring structure, 493–497
Hydrogen cyanide, hydrogen bonding cooperativity, perfectly linear chains and rings, 475–486
Hydrogen fluoride, hydrogen bonding cooperativity:
limits of, 501
zigzag chain/ring structure, 487–493
Hydronium ions:
pK_a value calculations, 454–457
proton solvation models, 444–445
solvation free energy, 448
Hydroxamic acids, computational chemistry, 425
Hypergeometric function, electron-transfer (ET) reactions, nonadiabatic solvent effects, Markovian bath model, 547–565
Hypervirial theorem, nuclear magnetic resonance (NMR) chemical shifts, gauge invariance and gauge-origin independence, 362

Ideal gas pressure, renormalization group theory, self-similarity, hard-particle fluid compressibility, 320–326
Ideal systems, self-similarity, renormalization group (RG) theory, 268–274
Imaginary frequency mode, Lie canonical perturbation theory (LCPT), 145–150
Individual gauges for localized molecular orbitals (IGLO), nuclear magnetic resonance (NMR) chemical shifts, 356–357
chemical applications, 406–415
local gauge-origin methods, 364–367
magnetic shielding constants, benchmark calculations, 393–403
quantum chemical procedures, 375

Infinitesimal generator, self-similarity, renormalization (RG) theory, 273–274
Infrared intensity, hydrogen bonding cooperativity, 470
perfectly linear chains and rings, 481–486
substituted zigzag chains and ring structures, methanol clusters, 498
zigzag chain/ring structure, hydrogen fluoride, 491–493
Initial path generation, transition path ensemble, 46–48
Integration algorithms, transition path ensemble, stochastic dynamics, 10
Intermediate modulation regime:
electron-transfer (ET) reactions, nonadiabatic solvent effects, variational transition state theory (VTST), 570–572
single molecule spectroscopy, 237–238
slow modulation regime, phase diagram, 240–242
Intermediate semichaotic region, chemical reaction dynamics, transition state theory (TST), hierarchical regularity, 114–117
Intermolecular interaction:
hydrogen bonding, 470–471
cooperativity origins, 474–475
hydrogen bonding cooperativity, zigzag chain/ring structure, hydrogen fluoride, 489–493
Intracellular (IC) media, acid computational chemistry, 424–425
Intramolecular effects:
electron-transfer (ET) reactions, nonadiabatic solvent effects, theoretical background, 513–514
hydrogen bonding cooperativity, 470–471
zigzag chain/ring structure, hydrogen fluoride, 489–493
Intramolecular vibrational energy redistribution (IVR), chemical reaction and chaos dynamics:
acetylene, 184–190
Arnold web, 182–184
hierarchical structures, 164
potential well dynamical correlation, 193–196
principles of, 154–158
resonance hierarchy, 190–193

Invariant manifolds, chemical reaction dynamics, "extract" invariant stable/unstable manifolds, 110–112
Inverse evolution operator, Lie transforms, perturbation theory, 132–142
Ionic assemblies:
  hydrogen bonding cooperativity, limits of, 501
  molecular associations, 460–461
Ising chain model:
  average squared magnetization (ASM), 292–295
  generalized propagator renormalization group (GPRG):
    nonideal chain, partition function, 312–317
    swelling factor calculation, 310–312
  positive function renormalization group (PFRG), one-dimensional Ising model, partition function, 302–304
Isotropic shielding constants, nuclear magnetic resonance (NMR) chemical shifts, 359
  benchmark calculations, absolute constants, 393–403
Iterative solutions, positive function renormalization group (PFRG), 300–301

Jacobian equations, transition path ensemble, shooting moves, deterministic dynamics, 22

Keck-Anderson techniques, chemical reaction dynamics, chaotic transition regularity, 95
Kellman's constants, chemical reaction dynamics, intramolecular vibrational energy redistribution (IVR), 187–190
Kirkwood-Westheimer model, solution acidity, continuum-solvation models, 440–442
Koch curve, self-similarity, renormalization (RG) theory, 268–274
Kohn-Sham orbitals:
  nuclear magnetic resonance (NMR) chemical shifts, density functional theory (DFT), 390–393
  transition path ensemble, memory requirements, 38

Kolmogorov-Arnold-Moser (KAM) theorem, chemical reaction and chaos dynamics, 155–158
Kolmogorov entropy, chemical reaction dynamics, "unify" transition state theory $vs.$ Kramers-Grote Hynes theory, 104–106
Kramers' coupling limit, electron-transfer (ET) reactions, nonadiabatic solvent effects:
  historical background, 514–515
  Markovian bath model, effective sink approximation, 559–565
Kramers-Grote Hynes theory, chemical reaction dynamics, "unify" transition state theory $vs.$, 102–106
Kramers' low-viscosity theory, chemical reaction dynamics, 119
Kronecker delta, Lie transforms, 122
Kubo-Anderson theory, single molecule spectroscopy (SMS):
  application to, 201–202
  line shape analysis, marginal averaging, 249
  Q parameter in long time limit, 260
  stochastic optical Bloch equation, 204–207
  two-level system (TLS) tunneling, 243–244
Kuhn segment, generalized propagator renormalization group (GPRG):
  linear nonideal polymer chains:
    partition function, 312–317
    swelling factor calculation, 307–312
  nonideal systems, asymptotic self-similarity, direct mapping, evolution equations, 288–289
  quantum electrodynamics, photon propagator, 330–334

Lagrangian equations:
  chemical reaction and chaos dynamics, crisis analysis, 176–180
  nuclear magnetic resonance (NMR) chemical shifts, molecular Hamiltonian, magnetic field, 361
  transition path ensemble, deterministic dynamics, 8–9
Lambda equations, nuclear magnetic resonance (NMR) chemical shifts, analytic second derivatives, 369–370

SUBJECT INDEX 663

Landau ghost pole, renormalization (RG) theory, self-similarity, quantum electrodynamics, photon propagator, 327–334
Landau leading logarithm formula, renormalization (RG) theory, self-similarity, quantum electrodynamics, photon propagator, 327–334
Landau-Zener semiclassical theory, electron-transfer (ET) reactions, nonadiabatic solvent effects:
  historical background, 514–516
  Markovian bath model, crossing point statistics, 554–555
  strong coupling limit, 573–574
    relaxation and breakdown of, 574–576
  variational transition state theory (VTST):
    adiabaticity, 572–573
    intermediate regime, 571–572
Langevin equation:
  chemical reaction dynamics, "unify" transition state theory vs. Kramers-Grote Hynes theory, 105–106
  electron-transfer (ET) reactions, nonadiabatic solvent effects, variational transition state theory (VTST), solvent control regime, 569–570
  transition path ensemble:
    dynamical algorithm, 42–43
    shifting moves, 37
    stochastic dynamics, 9–10
      random number sequences, 39–40
      shooting moves, 31
Laplace transform:
  electron-transfer (ET) reactions, nonadiabatic solvent effects:
    Padé approximation, 542–546
    variational transition state theory (VTST), solvent control regime, 568–570
  single molecule spectroscopy (SMS):
    Q parameter in slow modulation regime, 261
    slow modulation regime, 226–231
    straightforward complex analysis, 256–259
    two-state jump model, 221–222
Legendre functions, electron-transfer (ET) reactions, nonadiabatic solvent effects, microheterogeneous environments, 596–597
Lennard-Jones potentials:
  chemical reaction dynamics, chaotic transition regularity, 93–94
  hydrogen bonding cooperativity, perfectly linear chains and rings, 479–486
  transition path ensemble, parallel tempering, 45–46
Levich-Dogonadze prefactor, electron-transfer (ET) reactions, nonadiabatic solvent effects, Fermi Golden rule, rate constants, 519
Lévy statistics, single molecule spectroscopy (SMS), two-level system (TLS) tunneling, 243–244
Lewis acids, defined, 424
Liapunov analysis, chemical reaction dynamics, canonical perturbation theory (CPT), 86–87
Lie canonical perturbation theory (LCPT):
  chemical reaction dynamics:
    chaotic transition regularity, 94–95
    invariancy of actions, transition states, 96–99
    principles of, 87–90
    research background, 82–83
    "see" trajectories, configurational dividing surface, phase space, 100–102
    semiclassical theory, 113–114
  imaginary frequency mode, 145–150
  Lie transforms, 136–142
Lie derivative generated by W, properties, 122
Lie equation:
  generalized propagator renormalization group (GPRG), electron gas screening, quantum field theory, 338–340
  positive function renormalization group (PFRG), 301–302
  renormalization group (RG) theory, self-similarity, 273–274
    additive formulation, translational self-similarity, 344–345
    generalized propagator renormalization group (GPRG), 289–291
    fluid pressure self-similarity, 322–326
    hard-sphere suspension viscosity, 346–348
    quantum electrodynamics, photon propagator, 331–334

Lie transforms:
  characteristics of, 121–131
    autonomous cases, 122–128
    nonautonomous cases, 128–131
  perturbation theory based on, 131–142
Linear, nonideal polymer chains, renormalization group (RG) theory, self-similarity, 304–317
  partition function, 312–317
  swelling factor, GPRG calculation, 307–312
Linear combination of atomic orbital (LCAO) approximation, nuclear magnetic resonance (NMR) chemical shifts, gauge invariance and gauge-origin independence, 362
Linearized Poisson-Boltzmann equation (LPBE), pK_a value calculations, proton cycle, 454
Line shape theory, single molecule spectroscopy (SMS):
  classical shot noise, 208–210
  marginal averaging technique, 247–249
  Q parameter, three-time correlation function, 216–219
  slow modulation regime, 223–231
  two-state jump model, 219–222
Liouville equation:
  electron-transfer (ET) reactions, nonadiabatic solvent effects:
    Fermi Golden Rule, electronic coupling, 531–533
    Padé approximation, 542–546
  transition path ensemble, shooting moves, deterministic dynamics, 22
Liquid structures, electron-transfer (ET) reactions, nonadiabatic solvent effects, hydrodynamic hindrance, 594–595
Lissajous figures, chemical reaction dynamics, "see" trajectories, configurational dividing surface, phase space, 100–102
Local equilibrium, chemical and chaos reaction dynamics, statistical reaction theory, 165–168
Local gauge-origin methods, nuclear magnetic resonance (NMR) chemical shifts, 362–367
Localized orbital/local origin (LORG) scheme, nuclear magnetic resonance (NMR) chemical shifts:
  local gauge-origin methods, 364–367
  quantum chemical procedures, electron correlation, 374–375
Local Møller-Plesset perturbation (LMP2), nuclear magnetic resonance (NMR) chemical shifts, density functional theory, electron correlation procedures, 388–389
London atomic orbitals (LAOs), nuclear magnetic resonance (NMR) chemical shifts, local gauge-origin methods, 366–367
Long time limit, single molecule spectroscopy (SMS), Q parameter in, 259–260
Long-time rate constants, electron-transfer (ET) reactions, nonadiabatic solvent effects, Markovian bath model, effective sink approximation, 556–565
Lorentzian line shape analysis, single molecule spectroscopy (SMS):
  fast modulation regime, 232–237
  intermediate modulation regime, 238
  marginal averaging, 249
  perturbation expansion, 250–255
  slow modulation regime, 225–231
Lyapunov exponents, chemical reaction and chaos dynamics, crisis analysis, 175–180
Lyapunov instability, transition path ensemble, efficiency of shooting moves, 29

Magnetic field, nuclear magnetic resonance (NMR) chemical shifts, molecular Hamiltonians, 360–361
Magnetic shielding constants, nuclear magnetic resonance (NMR) chemical shifts, benchmark calculations, 393–403
Manifold invariance, chemical reaction and chaos dynamics, 170–172
  crisis analysis, 174–180
Many-body expansion, hydrogen bonding cooperativity, 473–474
Many-body perturbation theory (MBPT/MP), nuclear magnetic resonance (NMR) chemical shifts:
  chemical shielding tensors, 360
  quantum chemical procedures, electron correlation, 374–375

research background, 357
Many-body phase space, chemical reaction dynamics:
  reaction bottleneck "visualization," 106–110
  transition state theory (TST), chaos in, 118–119
Marcus's activation energy:
  electron-transfer (ET), nonadiabatic solvent effects:
    diffusion effects, 585–598
    Fermi Golden Rule:
      nonlinear response, 534–535
      rate constants, 519
      reorganization energy, 528–530
    Markovian bath model, 548–565
      effective sink approximation, 559–565
    non-Condon effects, 537
    nonexponential kinetics, 539–540
    strong coupling limit, 578–579
Marginal averaging technique, line shape theory, single molecule spectroscopy (SMS), 247–249
Markovian bath model, electron-transfer (ET) reactions, nonadiabatic solvent effects, 546–565
  crossing point statistics, 552–555
  effective sink approximation, 555–565
  historical background, 514–515
  substitution approximation, time-dependent diffusion coefficient, 550–552
Markovian processes:
  renormalization group (RG) theory, self-similarity, ideal self-similar systems, 276–280
  single molecule spectroscopy (SMS):
    limits of, 247
    two-state jump model, 220–222
Massey parameter, electron-transfer (ET) reactions, nonadiabatic solvent effects, 540–542
  Markovian bath model, 549–565
  strong coupling limit:
    Golden rule, 578–579
    Landau-Zener theory, 574–576
Maxwellian momentum distribution function, renormalization (RG) theory, self-similarity, ideal self-similar systems, 278–280

McConnell equation, electron-transfer (ET) reactions, nonadiabatic solvent effects:
  Fermi Golden Rule, electronic coupling, 531–533
  non-Condon effects, 536–537
McMurchie-Davidson scheme, nuclear magnetic resonance (NMR) chemical shifts, GAIO-MP2 approach, principles, 383–385
Mean square displacement, single molecule spectroscopy (SMS), slow modulation regime, 225–231
Memory requirements, transition path ensemble, 37–38
Methanols, hydrogen bonding cooperativity, substituted zigzag chains and ring structures, 497–498
Metropolis algorithm, transition path ensemble:
  Monte Carlo simulation, 11
  stochastic dynamics, shooting moves, 31
  stochastic pathway local algorithm, 41–42
  trajectory space calculations, 14–15
Micellar structures, electron-transfer (ET) reactions, nonadiabatic solvent effects, 595–597
Microheterogeneous environments, electron-transfer (ET) reactions, nonadiabatic solvent effects, 595–597
Molecular associations, acids, computational chemistry, 458–461
Molecular dynamics simulations:
  chemical reaction dynamics, research background, 82–83
  transition path ensemble:
    memory requirements, 37–38
    sampling programs, 48–49
Molecular Hamiltonians, nuclear magnetic resonance (NMR) chemical shifts, magnetic fields, 360–361
Molecular orbitals (MO), nuclear magnetic resonance (NMR) chemical shifts:
  density functional theory, electron correlation procedures, 385–389
  GAIO-MP2 approach, principles, 376–385
Molecular translation, electron-transfer (ET) reactions, nonadiabatic solvent effects, Fermi Golden Rule, 527–530

Møller-Plesset perturbation theory (MP2):
  gas-phase acidity:
    limitations of, 461–463
    quantum chemistry, 430–434
  hydrogen bonding cooperativity:
    perfectly linear chains and rings, 478–486
    zigzag chain/ring structure, hydrogen fluoride, 491–493
  molecular associations, 460–461
  nuclear magnetic resonance (NMR) chemical shifts:
    chemical applications, 404–415
    density functional theory, electron correlation procedures, 385–389
    GAIO-MP2 approach, principles, 375–385
    magnetic shielding constants, benchmark calculations, 393–403
  pK_a value calculations, hydronium cycle, 456–457
Momentum rescaling, transition path ensemble, shooting moves, 24–26
Monte Carlo-self-consistent field (MCSCF) calculations, nuclear magnetic resonance (NMR) chemical shifts:
  GAIO-MP2 approach, principles, 384–385
  magnetic shielding constants, benchmark calculations, 393–403
  quantum chemical procedures, 374–375
Monte Carlo simulation:
  electron-transfer (ET) reactions, nonadiabatic solvent effects, quantum Monte Carlo algorithm (QMC), 582–585
  gas-phase acidity, molecular associations, 459–461
  renormalization group (RG) theory, self-similarity, 271–274
    linear nonideal polymeric chains, partition function, 314–317
    quantum electrodynamics, photon propagator, 331–334
  single molecule spectroscopy (SMS), 245–246
  solution acidity, continuum-solvation models, 437–442
  transition path ensemble, 10–11
    configurational bias Monte Carlo, 43
    initial path generation, 47–48

shifting moves, 37
stochastic dynamics, shooting moves, 31
stochastic pathway local algorithm, 41–42
trajectory space calculations, 13–15
transition path sampling, 4–5
Morse oscillators, Lie canonical perturbation theory (LCPT), 145–150
Mulliken-Hush theory, electron-transfer (ET) reactions, nonadiabatic solvent effects, Fermi Golden Rule, electronic coupling, 533
Multidimensional systems, chemical reaction dynamics, transition state theory (TST), 119

Nearest-neighbor bond potential:
  generalized propagator renormalization group (GPRG), swelling factor calculation, 308–312
  linear nonideal polymeric chains, self-similarity, renormalization group (RG) theory, 305–317
Near resonant case, Lie transforms, 137–142
Neighboring barriers, chemical reaction and chaos dynamics, potential well dynamical correlation, 194–196
Newtonian equations of motion, transition path ensemble:
  deterministic dynamics, 8–9
  shooting moves, 23–24
NNO molecule, nuclear magnetic resonance (NMR) chemical shifts, magnetic shielding constants, benchmark calculations, 394–395
Nonadditive behavior, hydrogen bonding, 471–472
Nonadiabatic solvent effects, electron-transfer (ET) reactions:
  diffusion effects, 585–598
    bulk *vs.* geminate reaction, 585–594
    bulk reaction, 586–590
    geminate reaction after photoionization, 590–594
    nonhomogeneous effects, 594–598
    hydrodynamic hindrance and liquid structure, 594–595
    microheterogeneous environments, 595–597

SUBJECT INDEX 667

stochastic gating and anisotropic
  reactivity, 597–598
Fermi Golden Rule limit, 516–540
  non-Condon effects, 535–537
  nonexponential kinetics, 537–540
  nonlinear response, 533–535
  rate constant, 516–533
    electronic coupling, 530–533
    reorganization energy and free energy
      change, 524–530
    spectral density, 519–524
future research issues, 598–600
Markovian bath model, 546–565
  crossing point statistics, 552–555
  effective sink approximation, 555–565
  substitution approximation, time-
    dependent diffusion coefficient, 550–
    552
numerical methods, 582–585
Padé approximation, 542–546
  classic nonadiabatic limit, 543–546
research background, 512–516
strong coupling limit, nonadiabatic
  transitions, 573–582
  Fermi Golden rule limit, 576–579
  Landfau-Zener approach, 573–574
  non-Marcus free energy gap dependence,
    580–582
  relaxation effects, Landau-Zener theory,
    574–576
variational transition state theory, 565–573
  adiabacity, 572–573
  intermediate regime, 570–572
  solvent control regime, 565–570
Nonautonomous cases, Lie transforms, 128–
  131
Non-canonical gauge-origin transformation,
  nuclear magnetic resonance (NMR)
  chemical shifts, local gauge-origin
  methods, 364–367
Non-Condon effects, electron-transfer (ET)
  reactions, nonadiabatic solvent effects,
  535–537
Nonexponential kinetics, electron-transfer
  (ET) reactions, nonadiabatic solvent
  effects, 537–540
Nonhomogeneous diffusion effects, electron-
  transfer (ET) reactions, nonadiabatic
  solvent effects, 594–598
  hydrodynamic hindrance and liquid
    structure, 594–595
  microheterogeneous environments,
    595–597
  stochastic gating and anisotropic reactivity,
    597–598
Nonideal systems, renormalization group
  (RG) theory:
  asymptotic self-similarity, 280–295
    average squared magnetization, Ising
      spin open chain, 292–295
    direct mapping solution, functional
      evolution equations, 286–289
    generalized propagator renormalization
      group (GPRG):
      Callan-Symanzik equations, 290–292
      functional evolution equations,
        286–289
      Lie equations, 289–290
  linear, nonideal polymer chains, 304–317
    partition function, 312–317
    swelling factor, GPRG calculation,
      307–312
  renormalization group theory, self-
    similarity, hard-particle fluid
    compressibility, 320–326
Noninteracting blip approximation, electron-
  transfer (ET) reactions, nonadiabatic
  solvent effects, Fermi Golden rule,
  rate constants, 518–519
Nonlinearity, chemical reaction dynamics,
  intramolecular vibrational energy
  redistribution (IVR), 183–184
Nonlinear response, electron-transfer (ET)
  reactions, nonadiabatic solvent effects,
  Fermi Golden Rule limit, 533–535
Non-(near) resonant case, Lie transforms,
  137–142
Nonuniformity, chemical reaction dynamics,
  research background, 81–83
"No-return" hypothesis, chemical reaction
  dynamics, "unify" transition state
  theory vs. Kramers-Grote Hynes
  theory, 105–106
Normally hyperbolic invariant manifolds,
  chemical reaction and chaos dynamics,
  171–172
  crisis analysis, 176–180
  dynamical correlation, 194–196
Nosé-Hoover thermostat, transition path
  ensemble, dynamical algorithm, 42–43

Nuclear magnetic resonance (NMR) chemical shifts, electron-correlated calculations:
  analytic second derivatives, 367–370
  applications, 370–371
  basic principles, 358–359
  benchmark calculations, absolute nuclear magnetic shielding constants, 393–403
  chemical applications, relative NMR chemical shifts, 403–415
  future research issues, 415–416
  gauge invariance and gauge-origin independence, 361–362
  local gauge-origin methods, 362–367
  molecule Hamiltonian, magnetic field presence, 360–361
  quantum chemical procedures, 371–393
    density functional theory (DFT), 389–393
    GIAO-MP2 approach, 375–385
    larger system applications, 385–389
    research background, 356–357
    shielding tensors as second derivatives, 359–360
Numerical techniques, electron-transfer (ET) reactions, nonadiabatic solvent effects, 582–585

Obara-Saika recursion, nuclear magnetic resonance (NMR) chemical shifts, GAIO-MP2 approach, principles, 383–385
Octadecylrhodamine B (ODRB), electron-transfer (ET) reactions, nonadiabatic solvent effects, microheterogeneous environments, 595–597
Octahedral atomic arrangement, chemical reaction dynamics:
  chaotic transition regularity, 94–95
  invariancy of actions, transition states, 95–99
  reaction bottleneck "visualization," 107–110
Ohmic dissipation, electron-transfer (ET) reactions, nonadiabatic solvent effects:
  numerical calculations, 584–585
  variational transition state theory (VTST), solvent control regime, 569–570
Oligomer clusters, hydrogen bonding cooperativity, perfectly linear chains and rings, 475–486

One-particle density matrix, nuclear magnetic resonance (NMR) chemical shifts, GAIO-MP2 approach, principles, 376–385
Onsager radius, electron-transfer (ET), nonadiabatic solvent effects:
  diffusion effects, geminate recombination, 592–594
  Fermi Golden Rule, reorganization energy, 527–530
Ordered closest packing (OCP), renormalization (RG) theory, self-similarity, hard particle compressibility, fluid pressure, 324–326
Order parameter:
  transition path ensemble:
    initial/final region determination, 12–13
    reaction coordinates, 65–66
  transition path sampling, factorization, 55–59
Orientationally disordered crystalline (ODIC) phase, molecular associations, 460–461
Origin dependence, nuclear magnetic resonance (NMR) chemical shifts, 356–357

Padé approximation, electron-transfer (ET) reactions, nonadiabatic solvent effects, 542–546
  classic nonadiabatic limit, 543–546
  Markovian bath model, effective sink approximation, 561–565
Pairwise additive interactions, hydrogen bonding, cooperative effects, 472
Parabolic barriers, chemical reaction dynamics, chaotic transition regularity, 94–95
Parallel tempering, transition path ensemble, 43–46
Partition function:
  positive function renormalization group (PFRG), one-dimensional Ising model, 302–304
  renormalization (RG) theory, self-similarity:
    ideal self-similar systems, 274–280
    linear nonideal polymeric chains, 305–317

transition path sampling, reversible work, 52–53
Path quenching, transition path ensemble, 73–74
Pauli matrix, single molecule spectroscopy (SMS), stochastic optical Bloch equation, 204–207
"Pearl necklace" model, self-similarity, renormalization group (RG) theory, linear nonideal polymer chains, 304–317
Pekar factor, electron-transfer (ET) reactions, nonadiabatic solvent effects:
 Fermi Golden Rule, reorganization energy and free energy change, 525–530
 Markovian bath model, effective sink approximation, 560–565
Periodic orbit dividing surface (PODS), chemical reaction dynamics:
 reaction bottleneck "visualization," 109–110
 research background, 81–83
Perturbation theory:
 electron-transfer (ET) reactions, nonadiabatic solvent effects, strong coupling limit, Gold Rule provisions, 576–579
 Lie transforms, 131–142
 single molecule spectroscopy (SMS), perturbation expansion, 250–255
Phase diagrams, single molecule spectroscopy (SMS), photon statistics, 238–242
Phase space analysis:
 chemical reaction dynamics:
  reaction bottleneck "visualization," many-body phase space, 106–110
  "see" trajectories, configurational dividing surface, phase space, 99–102
 transition path ensemble:
  shooting moves, 22–24
  deterministic dynamics, 22
Phenolic acidity, gas-phase acidity measurements, 434–436
Phosphinic acid, dimerization enthalpy, 460–461
Photoionization, electron-transfer (ET), nonadiabatic solvent effects, diffusion effects, geminate recombination, 590–594

Photon propagator, quantum electrodynamics, renormalization group (RG) theory, self-similarity, 326–334
pK_a values:
 acid computational chemistry:
  anion solvation, 449–450
  pharmacokinetics, 425–426
 calculation of, 450–457
  hydronium (TC2) cycle, 454–457
  TC1 or proton cycle, 451–454
 limitations of, 462–463
 relative calculations, 458
Poincaré-Von Zeipel technique:
 Lie canonical perturbation theory (LCPT), chemical reaction dynamics, 87–90
 Lie transforms, 141–142
Poisson-Boltzman (PB) approach:
 electron-transfer (ET) reactions, nonadiabatic solvent effects, Fermi Golden Rule, reorganization energy, 528–530
 solution acidity, continuum-solvation models, 437–442
Poisson brackets:
 algebraic quantization, 143–144
 chemical reaction dynamics, algebraic quantization, 93
 Lie canonical perturbation theory (LCPT), chemical reaction dynamics, 87–90
 Lie transforms, 121–131
Poissonian counting statistics:
 electron-transfer (ET) reactions, nonadiabatic solvent effects, non-Condon effects, 537
 single molecule spectroscopy (SMS):
  applications, 202–203
  classical shot noise, 208–210
  deviations from, 244–246
  fast modulation regime, 234–237
  line shape theory, marginal averaging, 247–249
  phase diagram, 238–242
  Q parameter, three-time correlation function, 215–219
  simulation techniques, 211–213
  slow modulation regime, 229–231
  two-level system (TLS) tunneling, 244

Polarizable continuum model (PCM):
  anion solvation, 450
  limits of, 462–463
  pK_a value calculations, 454
    hydronium cycle, 456–457
  solution acidity, continuum-solvation models, 439–442
Polarization function, renormalization group (RG) theory, self-similarity, electron gas screening, quantum field theory, 335–340
Polyester matrix, computational chemistry, 425–426
Polymeric chains:
  hydrogen bonding cooperativity, zigzag chains and rings, 486–497
    hydrogen chloride/hydrogen bromide molecules, 493–497
    hydrogen fluoride, 487–493
  nonideal linearity, self-similarity, renormalization group (RG) theory, 304–317
Population correlation function, transition path sampling, factorization, 58–59
Population fluctuations, transition path sampling, computing rates, reversible work, 51–52
Positive function renormalization group (PFRG):
  linear nonideal polymeric chains:
    partition function, 312–317
    swelling factor calculation, 307–312
  one-dimensional Ising model, partition function, 302–304
  stretched-exponential scaling, 295–304
    differential equations, 301–302
    iterative solution, functional equations, 300–301
    partition function, one-dimensional Ising model, 302–304
Potential barriers, chemical reaction dynamics, 193–196
Potential energy surface:
  chemical reaction dynamics:
    invariancy of actions, transition states, 96–99
    research background, 81–83
    "unify" transition state theory vs. Kramers-Grote Hynes theory, 105–106
  transition path sampling, 3–5
Power law equations:
  generalized propagator renormalization group (GPRG), nonideal systems, asymptotic self-similarity:
    Callan-Symanzik equations, 290–292
    functional evolution equations, 286–289
    Lie equations, 289–290
  self-similarity, renormalization group (RG) theory, 270–274
  ideal self-similar systems, 274–280
Probability density function, single molecule spectroscopy (SMS):
  classical shot noise, 208–210
  Q parameter, three-time correlation function, 213–219
Probability functionals, transition path ensemble, reactive path probability, 7–8
Proportionality values, nuclear magnetic resonance (NMR) chemical shifts, chemical shielding tensors, 358–359
Protein folding, chemical reaction and chaos dynamics, dynamical correlation, 195–196
Protonated water timer, transition path sampling, 5
Proton cycle (TC1), pK_a value calculations, 451–454
Proton nuclear magnetic resonance (H NMR):
  hydrogen bonding, 470
    zigzag chain/ring structure, hydrogen fluoride, 493
  hydrogen bonding cooperativity, perfectly linear chains and rings, 485–486
Proton solvation, solution acidity, 443–445
Proton transfer, acid computational chemistry, 424–425
Pulse shape functions, single molecule spectroscopy (SMS):
  Q parameter, three-time correlation function, 217–219
  two-state jump model, 220–222

Q parameter, single molecule spectroscopy (SMS):
  applications, 202–203
  deviations, 245–246

SUBJECT INDEX

exact solution analysis, limiting cases, 222–242
  fast modulation regime: $v, \Gamma \ll R$, 231–237
  intermediate modulation regime, 237–238
  phase diagram, 238–242
  slow modulation regime: $R \ll v,\Gamma$, 223–231
fast modulation regime, 261–263
long time limit, 259–260
slow modulation regime, 260–261
three-time correlation function, 213–219
two-state jump model, 222
Quadratic configuration interaction (QCISD(T)), gas-phase acidity, quantum chemistry, 431–434
Quantized thresholds, chemical reaction dynamics, research background, 81–83
Quantum chemical procedures:
  gas-phase acidity, 427–434
  hydrogen bonding, cooperative effects, 471–472
  nuclear magnetic resonance (NMR) chemical shifts, electron-correlated calculations, 371–393
    density functional theory (DFT), 389–393
    GIAO-MP2 approach, 375–385
    larger system applications, 385–389
Quantum electrondynamics (QED), renormalization group (RG) theory, self-similarity, photon propagator, 326–334
Quantum field theory (QFT) applications, renormalization group (RG) theory, self-similarity, 326–340
  electron gas screening, 334–340
  photon propagator, quantum electrodynamics, 326–334
Quantum jump theory, single molecule spectroscopy (SMS), 245–246
Quantum mechanical/molecular mechanical (QM/MM) methods:
  gas-phase acidity, molecular associations, 459–461
  hydrogen bonding cooperativity, perfectly linear chains and rings, 477–486
  limits of, 462–463

pK_a value calculations, 454
  relative calculations, 458
  solution acidity, continuum-solvation models, 437–442
Quantum mechanics, electron-transfer (ET) reactions, nonadiabatic solvent effects, 512–513
Quantum Monte Carlo algorithm (QMC), electron-transfer (ET) reactions, nonadiabatic solvent effects, 582–585
Quasiregular region, chemical reaction dynamics, transition state theory (TST), hierarchical regularity, 114–117

Rabi frequency, single molecule spectroscopy (SMS), Q parameter, three-time correlation function, 216–219
Random close packing (RCP), renormalization group (RG) theory, self-similarity, hard particle compressibility, fluid pressure, 324–326
Random number sequences, transition path ensemble, stochastic trajectories, 38–40
Random-phase approximation, renormalization group (RG) theory, self-similarity, electron gas screening, quantum field theory, 336–340
Random walk theory:
  renormalization group (RG) theory, self-similarity:
    ideal self-similar systems, 274–280
    linear nonideal polymeric chains, partition function, 313–317
  single molecule spectroscopy (SMS):
    Q parameter in, 260–261
    slow modulation regime, 223–231
Rate constant:
  electron-transfer (ET) reactions, nonadiabatic solvent effects, variational transition state theory (VTST), solvent control regime, 565–570
  Fermi Golden Rule, electron-transfer (ET) reactions, nonadiabatic solvent effects, 516–533
    electronic coupling, 530–533

Rate constant (*Continued*)
  reorganization energy and free energy change, 524–530
  spectral density, 519–524
Rate quantization, chemical and chaos reaction dynamics, resonant structures, 161–164
Reaction bottleneck "visualization," chemical reaction dynamics, 106–110
Reaction coordinates:
  chemical and chaos reaction dynamics:
    bifurcation in reaction paths, 180–182
    statistical reaction theory, 165–167
    transition state theory, 158–161
  chemical reaction and chaos dynamics, crisis analysis, 176–180
  transition path ensemble:
    initial/final region determination, 12–13
    order parameters, 65–66
Reactive flux theory, transition path sampling, 59–61
  computing rates, reversible work, 50–51
Reactive path probability, transition path ensemble, 7–8
  deterministic dynamics, 9
Recrossing problem, chemical and chaos reaction dynamics, transition state theory, 158–161
Regional Hamiltonian, chemical reaction dynamics, canonical perturbation theory (CPT), 90–91
Rehm-Weller data, electron-transfer (ET), nonadiabatic solvent effects, diffusion effects, bulk bimolecular reaction, 590
Relaxation parameters, electron-transfer (ET) reactions, nonadiabatic solvent effects, strong coupling limit, Landau-Zener theory, 574–576
Renormalization constant, positive function renormalization group (PFRG), stretched-exponential scaling, 299–300
Renormalization group (RG) theory, self-similarity:
  additive formulation, 340–348
    hard-sphere suspension viscosity, 345–348
    translational self-similarity, 341–345
  hard-particle fluid compressibility, 317–326
    fluid pressure self-similarity, 320–326

  ideal systems, theoretical background, 268–280
  limitations and future research, 348–352
  linear, nonideal polymer chains, 304–317
    partition function, 312–317
    swelling factor, GPRG calculation, 307–312
  nonideal systems, asymptotic self-similarity, 280–295
    average squared magnetization, Ising spin open chain, 292–295
    direct mapping solution, functional evolution equations, 286–289
    generalized propagator renormalization group (GPRG):
      Callan-Symanzik equations, 290–292
      functional evolution equations, 282–286
      Lie equations, 289–290
  quantum field theory (QFT) applications, 326–340
    electron gas screening, 334–340
    photon propagator, quantum electrodynamics, 326–334
  stretched-exponential scaling, positive function renormalization group (PFRG), 295–304
    differential equations, 301–302
    iterative solution, functional equations, 300–301
    partition function, one-dimensional Ising model, 302–304
Reorganization energy, electron-transfer (ET) reactions, nonadiabatic solvent effects:
  Fermi Golden Rule, 524–530
  microheterogeneous environments, 595–597
Reptation, transition path sampling, shifting moves, 32–37
Resonance frequency:
  carboxylic acid gas-phase acidity measurements, 435–436
  single molecule spectroscopy (SMS), simulation techniques, 212–213
Resonant Raman spectroscopy, chemical and chaos reaction dynamics, local equilibrium reactions, 166–168
Resonant structures, chemical and chaos reaction dynamics:
  intramolecular vibrational energy

SUBJECT INDEX

redistribution (IVR), hierarchy in, 190–193
rate quantization, 161–164
Reversibility condition, transition path ensemble, shooting moves, deterministic dynamics, 20–22
Reversible work, transition path sampling, 49–65
  computer rates, 52–53
  convenient factorization, 55–59
  population fluctuations, 51–52
  reactive flux theory, 59–61
  umbrella sampling, 53–55
Rice-Ramsperger-Kassel-Marcus (RRKM) theory, chemical reaction dynamics, "extract" invariant stable/unstable manifolds, 112
Ring structures, hydrogen bonding cooperativity:
  perfectly linear chains and rings, 475–486
  properties, 471–472
  substituted zigzag chains and rings, 497–499
  zigzag rings, 486–497
    hydrogen chloride/hydrogen bromide molecules, 493–497
    hydrogen fluoride, 487–493
Rotational energy, gas-phase acidity, quantum chemistry, 428–434
Rovibrational contributions, nuclear magnetic resonance (NMR) chemical shifts, magnetic shielding constants, benchmark calculations, 401–403
Runge-Kutta algorithm, chemical reaction dynamics, chaotic transition regularity, 95
Rys polynomials, nuclear magnetic resonance (NMR) chemical shifts, GIAO-MP2 approach, principles, 383–385

Saddle point analysis:
  chemical reaction and chaos dynamics:
    manifold invariance, 171–172
    transition regularity, 93–114
    transition path ensemble, separatrix and, 67–68
Scaling transformations, renormalization (RG) theory, self-similarity, 273–274
  generalized propagator renormalization group (GPRG), Lie equation, 291
  ideal self-similar systems, 274–280
  linear nonideal polymeric chains, 306–317
  quantum electrodynamics, photon propagator, 328–334
Schrödinger equation, nuclear magnetic resonance (NMR) chemical shifts, gauge invariance and gauge-origin independence, 362
Second energy derivatives, nuclear magnetic resonance (NMR) chemical shifts:
  analytic evaluation, 367–370
  chemical shielding tensors, 359–360
Second-order cumulant approximation, single molecule spectroscopy (SMS), fast modulation regime, 233–237
Second-order perturbation:
  single molecule spectroscopy (SMS), 252–253
Second-order polarization propagator approximation (SOPPA):
  nuclear magnetic resonance (NMR) chemical shifts:
    chemical shielding tensors, 359–360
    magnetic shielding constants, benchmark calculations, 395–403
    quantum chemical procedures, 372–393
    quantum chemical procedures, electron correlation, 374–375
"See" trajectories, chemical reaction dynamics, configurational dividing surface, phase space, 99–102
Self-consistent field (SCF) calculations:
  gas-phase acidity, molecular associations, 459–461
  hydrogen bonding cooperativity:
    carbonic acids, 500
    perfectly linear chains and rings, 478–486
    zigzag chain/ring structure:
      hydrogen chloride/hydrogen bromide molecules, 495–497
      hydrogen fluoride, 490–493
  nuclear magnetic resonance (NMR) chemical shifts, GIAO-MP2 approach, principles, 381–385
  solution acidity, continuum-solvation models, 437–442
Self-similarity hypothesis, renormalization group theory:
  additive formulation, 340–348

Self similarity hypothesis (*Continued*)
  hard-sphere suspension viscosity, 345–348
  translational self-similarity, 341–345
  hard-particle fluid compressibility, 317–326
  fluid pressure self-similarity, 317–326
  ideal systems, theoretical background, 268–280
  limitations and future research, 348–352
  linear, nonideal polymer chains, 304–317
    partition function, 312–317
    swelling factor, GPRG calculation, 307–312
  nonideal systems, asymptotic self-similarity, 280–295
    average squared magnetization, Ising spin open chain, 292–295
    direct mapping solution, functional evolution equations, 286–289
    generalized propagator renormalization group (GPRG):
      Callan-Symanzik equations, 290–292
      functional evolution equations, 282–286
      Lie equations, 289–290
  quantum field theory (QFT) applications, 326–340
    electron gas screening, 334–340
    photon propagator, quantum electrodynamics, 326–334
  stretched-exponential scaling, positive function renormalization group (PFRG), 295–304
    differential equations, 301–302
    iterative solution, functional equations, 300–301
    partition function, one-dimensional Ising model, 302–304
Semiclassical theory:
  chemical reaction dynamics, 113–114
  electron-transfer (ET) reactions, nonadiabatic solvent effects, nonlinear response, trajectory approach, 534–535
Separatrix, transition path ensemble, 67–68
  committor distribution, 71–73
SEP bright states, chemical reaction dynamics, intramolecular vibrational energy redistribution (IVR), 187–190

Shifting moves, transition path sampling, 32–37
  deterministic dynamics, 35–36
  stochastic dynamics, 36–37
Shinomoto formula, renormalization (RG) theory, self-similarity, hard particle compressibility, fluid pressure, 324–326
Shooting moves, transition path ensemble sampling, 15–31
  deterministic dynamics, 19–22
  deterministic efficiency, 26–29
  momentum rescaling, 24–26
  phase-space dynamics, 22–24
  stochastic dynamics, 29–32
Shot noise, single molecule spectroscopy (SMS), stochastic theory, 207–210
Simplectic property, chemical reaction and chaos dynamics, tangency, 173–174
Simulation techniques, single molecule spectroscopy (SMS), stochastic theory, 210–213
Single molecule spectroscopy (SMS), stochastic theory:
  classical shot noise, 207–210
  exact solution analysis, limiting cases, 222–242
    fast modulation regime: $v, \Gamma \ll R$, 231–237
    intermediate modulation regime, 237–238
    phase diagram, 238–242
    slow modulation regime: $R \ll v, \Gamma$, 223–231
  experimental applications, 242–244
  line shape calculation, marginal averaging, 247–249
  optical Bloch equation, 204–207
  perturbation expansion, 250–255
  Q parameter:
    fast modulation regime, 261–263
    long time limit, 259–260
    principles of, 202–203
    slow modulation regime, 260–261
    three-time correlation function, 213–219
  research background, 200–203
  simulation, 210–213
  straightforward complex analysis, 255–259

SUBJECT INDEX

two-state jump model, exact solution, 219–222
validity of, 244–246
Sink approximation, electron-transfer (ET) reactions, nonadiabatic solvent effects:
Markovian bath model, 555–565
strong coupling limit, 577–579
Slow modulation regime, single molecule spectroscopy (SMS):
exact solution analysis, limiting cases, 223–231
strong modulation case, 227
time dependence, 228–231
weak modulation case, 227–228
Q parameter in, 260–261
Smoluchowski's equation, electron-transfer (ET), nonadiabatic solvent effects, diffusion effects, bulk bimolecular reaction, 589–590
SM5.42R model, solution acidity, continuum-solvation models, 442
SOLO scheme, nuclear magnetic resonance (NMR) chemical shifts, quantum chemical procedures, electron correlation, 374–375
Solute-solvent interaction, acid computational chemistry, 424
Solution acidity, computational chemistry, 436–458
anion solvation, 449–450
continuum-solvation models, 437–442
pK_a acidity calculations, 450–458
relativity of, 458
TC2 or hydronium cycle, 454–457
TC1 or proton cycle, 451–454
proton solvation energy, 445–448
solvated protons, 443–445
Solvation time correlation function, electron-transfer (ET) reactions, nonadiabatic solvent effects, Fermi Golden rule, rate constants, 517–519
Solvent control regime, electron-transfer (ET) reactions, nonadiabatic solvent effects, variational transition state theory (VTST), 565–570
Solvent effects:
nonadiabatic electron-transfer (ET) reactions:
diffusion effects, 585–598

bulk vs. geminate reaction, 585–594
bulk reaction, 586–590
geminate reaction after photoionization, 590–594
nonhomogeneous effects, 594–598
hydrodynamic hindrance and liquid structure, 594–595
microheterogeneous environments, 595–597
stochastic gating and anisotropic reactivity, 597–598
Fermi Golden Rule limit, 516–540
non-Condon effects, 535–537
nonexponential kinetics, 537–540
nonlinear response, 533–535
rate constant, 516–533
electronic coupling, 530–533
reorganization energy and free energy change, 524–530
spectral density, 519–524
future research issues, 598–600
Markovian bath model, 546–565
crossing point statistics, 552–555
effective sink approximation, 555–565
substitution approximation, time-dependent diffusion coefficient, 550–552
numerical methods, 582–585
Padé approximation, 542–546
classic nonadiabatic limit, 543–546
research background, 512–516
strong coupling limit, nonadiabatic transitions, 573–582
Fermi Golden rule limit, 576–579
Landfau-Zener approach, 573–574
non-Marcus free energy gap dependence, 580–582
relaxation effects, Landau-Zener theory, 574–576
variational transition state theory, 565–573
adiabacity, 572–573
intermediate regime, 570–572
solvent control regime, 565–570
solution acidity, continuum-solvation models, 437–442
Solvent excluding (SE) surface, solution acidity, continuum-solvation models, 439–442

Spectral density, electron-transfer (ET) reactions, nonadiabatic solvent effects, Fermi Golden rule, rate constants, 519–524
Spectral diffusion, single molecule spectroscopy (SMS):
　applications, 201
　simulation techniques, 210–213
Spin-boson model, electron-transfer (ET) reactions, nonadiabatic solvent effects:
　Fermi Golden rule, rate constants, 518–519
　numerical calculations, 584–585
　Padé approximation, 543–546
Spin-rotation constants, nuclear magnetic resonance (NMR) chemical shifts, magnetic shielding constants, benchmark calculations, 402–403
Stable states, transition path sampling, 2–5
Statistical reaction theory, chemical reaction and chaos dynamics:
　intramolecular vibrational energy redistribution (IVR), 165–167
　potential well dynamical correlation, 194–196
Statistical weight, transition path ensemble, dynamical path probability, 6–7
Stochastic (fully developed chaotic) region:
　chemical reaction dynamics, transition state theory (TST), hierarchical regularity, 116–117
　electron-transfer (ET) reactions, nonadiabatic solvent effects, Padé approximation, 543–546
Stochastic pathway local algorithm, transition path ensemble, 41–42
　future applications, 75–76
Stochastic theory:
　electron-transfer (ET) reactions, nonadiabatic solvent effects:
　　historical background, 515
　　Markovian bath model, effective sink approximation, 555–565
　　nonhomogeneous effects, 597–598
　single molecule spectroscopy (SMS):
　　classical shot noise, 207–210
　　exact solution analysis, limiting cases, 222–242
　　fast modulation regime: $v, \Gamma \ll R$, 231–237

　　intermediate modulation regime, 237–238
　　phase diagram, 238–242
　　slow modulation regime: $R \ll v, \Gamma$, 223–231
　　experimental applications, 242–244
　　line shape calculation, marginal averaging, 247–249
　　optical Bloch equation, 204–207
　　perturbation expansion, 250–255
　　Q parameter:
　　　fast modulation regime, 261–263
　　　long time limit, 259–260
　　　principles of, 202–203
　　　slow modulation regime, 260–261
　　　three-time correlation function, 213–219
　　research background, 200–203
　　simulation, 210–213
　　straightforward complex analysis, 255–259
　　two-state jump model, exact solution, 219–222
　　validity of, 244–246
　transition path ensemble, 9–10
　　committor computation, 70
　　shifting moves, 36–37
　　shooting moves, 29–31
Stochastic trajectories, transition path ensemble, random number sequences, 38–40
Stockmeyer molecules, hydrogen bonding cooperativity, perfectly linear chains and rings, 479–486
Stokes-Einstein relation, electron-transfer (ET), nonadiabatic solvent effects, diffusion effects, bulk bimolecular reaction, 590
Stokes shift calculations, electron-transfer (ET) reactions, nonadiabatic solvent effects, Fermi Golden rule, rate constants, spectral density, 519–524
Straightforward complex analysis, single molecule spectroscopy (SMS):
　exact solution calculations, 255–259
　two-state jump model, 222
Stretched-exponential scaling, positive function renormalization group (PFRG), 295–304

SUBJECT INDEX 677

differential equations, 301–302
iterative solution, functional equations, 300–301
partition function, one-dimensional Ising model, 302–304
Strong coupling limit, electron-transfer (ET) reactions, nonadiabatic solvent effects, 573–582
  Fermi Golden rule limit, 576–579
  Landfau-Zener approach, 573–574
  non-Marcus free energy gap dependence, 580–582
  relaxation effects, Landau-Zener theory, 574–576
Strong modulation:
  fast modulation regime, 235–236
  slow modulation regime:
    phase diagram, 240–242
    single molecule spectroscopy (SMS), 227
Sub-Poissonian nonclassical effects, single molecule spectroscopy (SMS), applications, 202–203
Substitution approximation, electron-transfer (ET) reactions, nonadiabatic solvent effects, Markovian bath model, 550–552
Sumi-Marcus model, electron-transfer (ET) reactions, nonadiabatic solvent effects, Markovian bath model, effective sink approximation, 561–565
Sum-over-states density-functional perturbation theory (SOS-DFPT), nuclear magnetic resonance (NMR) chemical shifts, 393
Superexchange electronic coupling, electron-transfer (ET) reactions, nonadiabatic solvent effects, Fermi Golden Rule, 530–533
Superfluid helium droplets, hydrogen bonding cooperativity, perfectly linear chains and rings, 477–486
Super-Poissonian behavior, single molecule spectroscopy (SMS), applications, 202–203
Swelling factor, renormalization group (RG), self-similarity:
  asymptotic self-similarity, nonideal systems, 281–282

generalized propagator renormalization group (GPRG), 307–312

Tail error, anion solvation, 450
Tangency properties, chemical reaction and chaos dynamics, 156–158
  global features, 172–174
Taylor series:
  Lie transforms, 125–128
  nuclear magnetic resonance (NMR) chemical shifts, molecular Hamiltonian, magnetic field, 361
Tetramethylsilane (TMS), nuclear magnetic resonance (NMR) chemical shifts, 359
Thermodynamic cycle, gas-phase acidity, 426–427
Thermodynamic properties, renormalization group (RG) theory, self-similarity:
  fluid pressure, 321–326
  hard-particle fluid compressibility, 317–326
Three-time correlation function, single molecule spectroscopy (SMS):
  applications, 202–203
  fast modulation regime, 232–237
  Q parameter, 213–219
    fast modulation regime, 261–263
  two-state jump model, 220–222
Threshold values:
  evolution equations, generalized propagator renormalization group (GPRG), nonideal systems, asymptotic self-similarity, 286–289
  generalized propagator renormalization group (GPRG), Lie equation, 291
  self-similarity, renormalization (RG) theory, 271–274
Time correlation function, transition path sampling, factorization, 55–59
Time dependence, slow modulation regime, single molecule spectroscopy (SMS), 228–231
Time-dependent diffusion coefficient, electron-transfer (ET) reactions, nonadiabatic solvent effects, Markovian bath model, 550–552
Time evolution:
  transition path ensemble, shooting moves, deterministic dynamics, 20–22
  transition path sampling, 3–5
    deterministic dynamics, 8–9

Timescale dimensions:
  chemical and chaos reaction dynamics, transition state theory ergodicity, 159–161
  chemical reaction and chaos dynamics: crisis analysis, 175–180
  homoclinic intersections, 168–170
  single molecule spectroscopy (SMS), bath fluctuations, 203
Time slices, transition path ensemble, dynamical path probability, 6–7
Trajectory length, transition path sampling, 61–65
Trajectory space, transition path ensemble:
  sampling, 13–15
  shifting moves, 35–37
Transfer function, renormalization (RG) theory, self-similarity:
  additive formulation, translational self-similarity, 342–345
  fluid pressure self-similarity, 321–326
Transfer matrix element, electron-transfer (ET) reactions, nonadiabatic solvent effects, electronic coupling, 531–533
Transition path ensemble:
  defined, 4–5
  deterministic dynamics, 8–9
  dynamical path probability, 6–7
  initial and final regions, 12–13
  Monte Carlo dynamics, 10–11
  reactive path probability, 7–8
  stochastic dynamics, 9–10
  transition path ensemble, future issues, 75–76
Transition path sampling:
  algorithms:
    configurational bias Monte Carlo, 43
    dynamic algorithm, 42–43
    stochastic pathway algorithm, 41–42
  analytic techniques, 65
    committor distributions, 71–73
    computing committors, 68–71
    path quenching, 73–74
    reaction coordinates and order parameters, 65–66
    separatrix, ensemble state, 67–68
  computing rates, 49–65
    convenient factorization, 55–59
    population fluctuations, 51–52
    reversible work, 52–53
    umbrella sampling, 53–55
  ensemble definition:
    deterministic dynamics, 8–9
    dynamical path probability, 6–7
    initial and final regions, 12–13
    Monte Carlo dynamics, 10–11
    reactive path probability, 7–8
    stochastic dynamics, 9–10
  future research issues, 74–76
  initial path generation, 46–48
  memory requirements, 37–38
  molecular dynamics (MD), 48–49
  parallel tempering, 42–46
  pathway length determination, 61–65
  reactive flux theory, 59–61
  shifting moves, 32–37
    deterministic dynamics, 35–36
    stochastic dynamics, 36–37
  shooting moves, 15–31
    deterministic dynamics, 19–22
    deterministic efficiency, 26–29
    momentum rescaling, 24–26
    phase-space dynamics, 22–24
    stochastic dynamics, 29–32
  stochastic trajectories, random number sequences, 38–40
  theory and methodology, 2–5
Transition state theory (TST). See also Variational transition state theory (VTST)
  chemical reaction dynamics:
    chaos and, 158–161
    "extract" invariant stable/unstable manifolds, 110–112
    hierarchical regularity, 114–117
    invariancy of actions, 95–99
    research background, 80–83
    semiclassical theory, 113–114
    "unify" theory vs. Kramers-Grote-Hynes, 102–106
Translational energy, gas-phase acidity, quantum chemistry, 428–434
Translational self-similarity, renormalization group (RG) theory, additive formulation, 341–345
Transmission coefficient:
  chemical reaction dynamics, "unify" transition state theory vs. Kramers-Grote Hynes theory, 103–106
  transition path sampling, trajectory length, 64–65

trans-Stilben, chemical and chaos reaction dynamics, local equilibrium reactions, 167–168
Tunneling matrix element, electron-transfer (ET) reactions, nonadiabatic solvent effects:
  electronic coupling, 532–533
  non-Condon effects, 535–537
  nonexponential kinetics, 537–540
  spectral density, 522–524
"Twist" structures, nuclear magnetic resonance (NMR) chemical shifts, chemical applications, 410–415
Two-level system (TLS) tunneling, single-molecule spectroscopy, 243–244
Two-state jump model, single molecule spectroscopy (SMS):
  fast modulation regime, 233–237
  model parameters, 219–220
  solution, 220–222
Two-state stochastic gating, electron-transfer (ET) reactions, nonadiabatic solvent effects, nonhomogeneous effects, 597–598

Ultrafast solvation, electron-transfer (ET) reactions, nonadiabatic solvent effects, historical background, 515
Ultraviolet catastrophe, renormalization (RG) theory, self-similarity, quantum electrodynamics, photon propagator, 327–334
Umbrella sampling, transition path sampling, reversible work, 53–55
Uncertainty principle, electron-transfer (ET) reactions, nonadiabatic solvent effects, Massey parameter, 541–542
Uniform random number generator, single molecule spectroscopy (SMS), simulation techniques, 210–213
"Unify" transition state theory, chemical reaction dynamics, *vs.* Kramers-Grote-Hynes, 102–106
Unscaled variables, renormalization (RG) theory, self-similarity, ideal self-similar systems, 275–280

Van der Waals surfaces:
  chemical reaction and chaos dynamics, statistical reaction theory, 165–167

electron-transfer (ET) reactions, nonadiabatic solvent effects, Fermi Golden Rule, reorganization energy, 528–530
  hydrogen bonding, cooperative effects, 472
VanHove function, renormalization (RG) theory, self-similarity, ideal self-similar systems, 278–280
Vapor-phase dimers, hydrogen bonding cooperativity:
  perfectly linear chains and rings, 475–486
  zigzag chain/ring structure, hydrogen fluoride, 490–493
Variational transition state theory (VTST), electron-transfer (ET) reactions, nonadiabatic solvent effects, 565–573
  adiabacity, 572–573
  intermediate regime, 570–572
  solvent control regime, 565–570
  strong coupling limit, 578–579
Vector potential, nuclear magnetic resonance (NMR) chemical shifts, molecular Hamiltonian, magnetic field, 360–361
Vibrational frequency:
  gas-phase acidity, quantum chemistry, 428–434
  hydrogen bonding cooperativity, 470
  perfectly linear chains and rings, 481–486
  substituted zigzag chains and ring structures, methanol clusters, 497–498
  zigzag chain/ring structure, hydrogen chloride/hydrogen bromide molecules, 496–497
Virial coefficients, renormalization (RG) theory, self-similarity:
  hard-particle fluid compressibility, 319–326
  hard-sphere suspension viscosity, 346–348
Viscosity parameters, hard-sphere suspension, renormalization (RG) theory, self-similarity, additive formulation, 345–348

Water molecules:
  hydrogen bonding cooperativity, substituted zigzag chains and ring structures, 498–499

Water molecules (*Continued*)
  molecular associations, 461
Wave function parameters, nuclear magnetic resonance (NMR) chemical shifts, 367–370
Weak-coupling limit, electron-transfer (ET) reactions, nonadiabatic solvent effects, 599–600
Weak modulation:
  fast modulation regime, 236–237
  slow modulation regime:
    phase diagram, 240–242
    single molecule spectroscopy (SMS), 227–228
Wiener-Khintchine theorem, single molecule spectroscopy (SMS), 202–203
  line shape analysis, marginal averaging, 249
  Q parameter, three-time correlation function, 216–219
  two-state jump model, 222
Wilemski-Fixman closure approximation, electron-transfer (ET), nonadiabatic solvent effects, diffusion effects, bulk bimolecular reaction, 588–590

Yukawa shielded-Coulomb contribution, renormalization group (RG) theory, self-similarity, electron gas screening, quantum field theory, 337–340

Zeolites, computational chemistry, 426
Zero-point energies:
  gas-phase acidity, quantum chemistry, 428–434
  hydrogen bonding cooperativity, 474
Zeroth-order transmission coefficient:
  chemical reaction dynamics, "unify" transition state theory *vs.* Kramers-Grote Hynes theory, 103–106
  equations of motion, 120–121
  Lie transforms, 134–143
Zigzag chains and rings, hydrogen bonding cooperativity, 486–497
  hydrogen chloride/hydrogen bromide molecules, 493–497
  hydrogen fluoride, 487–493
  substituted zigzag chains and rings, 497–499
Zusman model, electron-transfer (ET) reactions, nonadiabatic solvent effects:
  Markovian bath model, 548–565
  non-Marcus free energy gap dependence, 580–582
  variational transition state theory (VTST), solvent control regime, 570
Z-vector equations, nuclear magnetic resonance (NMR) chemical shifts:
  analytic second derivatives, 369–370
  GIAO-MP2 approach, principles, 377–385